Errata

Page 9 line 2 of text there is a mysterious period.

Page 24 equation 13

$$\sup_{\vec{\theta}}\left(\langle\vec{\theta},\vec{y}\rangle - g(\vec{\theta})\right) \quad \text{should be} \quad \sup_{\vec{\theta}}\left(\langle\vec{\theta},\vec{a}\rangle - g(\vec{\theta})\right)$$

Page 24 last equation

$$\ldots = \ell(\vec{a}). \quad \text{should be} \quad \ldots = -\ell(\vec{a}).$$

Page 132:

$$S \triangleq \{\vec{r} : \vec{r}(-T) = \vec{q},\ \vec{r}(t) \in G^o \text{ for } t < T,\ \vec{r}(0) \in \partial G,\ T > 0\}, \qquad (6.4)$$

has the wrong range of time. It should be

$$S \triangleq \{\vec{r} : \vec{r}(-T) = \vec{q},\ \vec{r}(t) \in G^o \text{ for } -T < t < 0,\ \vec{r}(0) \in \partial G,\ T > 0\}, \qquad (6.4)$$

Page 133:

In Assumption 6.5, note that G is closed by definition, so that it should begin

Assumption 6.5. There is a point $\vec{q} \in G^o$ and, for every $\varepsilon > 0, \ldots$

Page 264: The equations below the statement of Corollary 11.1 are a bit messed up with respect to "T" and "t". The pair of equations

$$\varphi_t(x, y) \triangleq \inf_{F(x,y,t)} I_0^T(r)$$

(see Definitions 6.14) where

$$F(x, y, t) = \{r : r(0) = x,\ r(t) = y,\ t > 0\}.$$

should be

$$\varphi_t(x, y) \triangleq \inf_{F(x,y,t)} I_0^t(r)$$

for $t > 0$ (see Definitions 6.14) where

$$F(x, y, t) = \{r : r(0) = x,\ r(t) = y\}.$$

Page 517

$$\ell(r(t), r'(t)) - r'(t)\frac{\partial}{\partial r'}\, f(r(t), r'(t)) = \text{constant}. \qquad (C.3)$$

should be

$$f(r(t), r'(t)) - r'(t)\frac{\partial}{\partial r'}\, f(r(t), r'(t)) = \text{constant}. \qquad (C.3)$$

The latest errata sheet may be found at
http://netlib.att.com/netlib/att/math/weiss/LargeDeviations.html.
You can retrieve it by anonymous ftp from
ftp://netlib.att.com/netlib/att/math/weiss/errata.ps.Z

Large Deviations for Performance Analysis

Large Deviations for Performance Analysis
Queues, communications, and computing

STOCHASTIC MODELING SERIES

Adam Shwartz
AT&T Bell Laboratories, USA

Alan Weiss
Israel Institute of Technology, Haifa, Israel

With an appendix by
Robert J. Vanderbei

CHAPMAN & HALL
London • Glasgow • Weinheim • New York • Tokyo • Melbourne • Madras

Published by
Chapman & Hall, 2–6 Boundary Row, London SE1 8HN, UK

Chapman & Hall, 2–6 Boundary Row, London SE1 8HN, UK

Blackie Academic & Professional, Wester Cleddens Road, Bishopbriggs, Glasgow, UK

Chapman & Hall GmbH, Pappelallee 3, 69469 Weinheim, Germany

Chapman & Hall USA, 115 Fifth Avenue, New York, NY 10003, USA

Chapman & Hall Japan, ITP-Japan, Kyowa Building, 3F, 2-2-1 Hirakawacho, Chiyoda-ku, Tokyo 102, Japan

Chapman & Hall Australia, 102 Dodds Street, South Melbourne, Victoria 3205, Australia

Chapman & Hall India, R. Seshadri, 32 Second Main Road, CIT East, Madras 600 035, India

First edition 1995

© 1995 AT&T

ISBN   0 412 06311 5

Library of Congress Cataloging-in-Publication Data

Shwartz, Adam, 1953–
    Large deviations for performance analysis : queues, communication, and computing / Adam Shwartz and Alan Weiss : with an appendix by Robert J. Vanderbei.
        p.   cm.
    Includes bibliographical references and index.
    ISBN 0-412-06311-5
    1. Large deviations.   2. System analysis.   I. Weiss, Alan.   II. Title.
    QA273.67.S48   1994
    003'.8—dc20   94-40138
                                                          CIP

⊗

Printed on acid-free text paper, manufactured in accordance with ANSI/NISO Z39.48-1992 (Permanence of Paper).

Table of contents

vii

Large Deviations for
Performance Analysis

Chapter 0

What this Book Is, and What It Is Not

The field of communication and computer networks is bustling with activity. One of the active areas falls under the rubric "performance." Researchers and development engineers tackle systems that are huge, complex and fast; think of the telephone network in the United States. The resulting models are, for the most part, discrete-event, continuous time stochastic processes, technically known as jump Markov processes. The objective is to analyze the behavior of these systems, with the goal of designing systems that provide better service. "Better" may mean faster, less prone to error and breakdown, more efficient, or improved by many other criteria.

Until quite recently, the tools brought to bear on these problems were appropriate for small, simple systems. Some of these methods take into account only average behavior (or perhaps variances). But this is often not enough, as the performance of many systems is limited by events with a small probability of occurring, but with consequences that are severe. Clearly, new tools are needed. Computer simulation is one relatively new tool. But this method, for all its power, is limited in that it usually does not provide rules of thumb for design, may not give estimates on the sensitivity of results to various parameters, and can be extremely costly in terms of both computer time and programming (especially debugging) time. Analytic methods clearly retain some advantages. This book is about a fairly new analytic method called large deviations.

Large deviations is a mathematical theory that is very active at present. The theory deals with rare events, and is asymptotic in nature; it is thus a natural candidate for analyzing rare events in large systems. The theory of large deviations includes a set of techniques for turning hard probability problems that concern a class of rare events into analytic problems in the calculus of variations. It also provides a nice qualitative theory for understanding rare events. As an asymptotic technique, its effectiveness resides in the relative simplicity with which one may analyze systems whose size may be growing with the asymptotic parameter, or whose "conditioning" may be getting worse. The theory is often useful even when simulation or other numerical techniques become increasingly difficult as the parameter tends to its limit.

However, the theory is noted for being technically (mathematically) very demanding, and solving a problem in the calculus of variations is not typically an engineer's dream. Although the theory is being increasingly used for analyzing rare events in large systems, this is done by a relatively small number of researchers. We believe that the reason for this state of affairs is that the theory is not easily accessible to non-mathematicians, and the final results seem to require an additional translation to engineering lingo. Hence

> Large deviations is useful.
>
> Large deviations is formidably technical.
>
> What's a student to do?

Herein is contained one point of view on what's to do. We develop the theory of large deviations from the beginning (independent, identically distributed (i.i.d.) random variables) through recent results on the theory for processes with boundaries, keeping to a very narrow path: continuous-time, discrete-state processes. By developing only what we need for the applications we present, we try to keep the theory to a manageable level, both in terms of length and in terms of difficulty. We make no particular claim to originality of the theory presented herein, except for the material concerning boundaries, which is the subject of Chapter 8. Most of the trailblazing work of Freidlin and Wentzell [FW], and of Donsker and Varadhan [DV1–DV4] goes further than we do. Also, others have subsequently treated the general theory much more thoroughly; e.g. Ellis [Ell], Wentzell [Wen], Deuschel and Stroock [DeS], Dembo and Zeitouni [DZ], and the recent work of Dupuis and Ellis [DE2]. We have, however, formulated a complete, self-contained set of theorems and proofs for jump Markov processes. Since our scope is limited to a class of relatively simple processes, the theory is much more accessible, and less demanding mathematically. To enhance the pedagogical value of this work, we have attempted to convey as much intuition as we could, and to keep the style friendly. In addition, we present for the first time a complete theory for processes with a flat boundary, and for some processes in a random environment. The level of the book is somewhat uneven, as indicated in the dependence chart Figure 0.1. This is purposeful—we believe that a neophyte would not want to read the difficult chapters, and that an expert doesn't want as much hand holding as a beginner.

We believe that our applications are important enough to require no apologies. As Mark Kac said, "Theorems come and go, but an application is forever." Our applications cover large areas of the theory of communication networks: circuit-switched transmission (Chapter 12), packet transmission (Chapter 13), multiple-access channels (Chapter 14), and the $M/M/1$ queue (Chapter 11). We cover aspects of parallel computation in a much more spotty fashion: basics of job allocation (Chapter 9), rollback-based simulation (Chapter 10), and assorted priority queuing models (Chapters 15 and 16) that may be used in performance models of various computer architectures.

The key word in the phrase "our applications" is "our." We present only our own results concerning the applications. We do not synthesize existing theory except in our narrow fashion for jump processes. We ignore possible improvements in order to remain within the realm of those large deviations bounds that we actually use. For example, Anantharam's beautiful results on the $G/G/1$ queue [An] are certainly relevant to the subjects we address, but his techniques are different. We do not obtain the best results known for jump Markov processes. It is certainly arguable whether this is a wise choice. However, we wanted to present a consistent, fully worked out point of view, avoiding digressions. Furthermore, once a student has learned the limited range of large deviations techniques we present, he or she should find it a much simpler matter to read both more abstract and complete works, and understand more wide-ranging applications. By limiting our range, we are able to give complete proofs for nearly all the results concerning our applications. We were also able to present a "bag of tricks" in the calculus of variations, which allows us to extract concrete information regarding these examples. We try to remedy some of the narrowness of our point of view in the end notes to the chapters and in the appendices.

On a less defensive note, we firmly believe that the large deviations of processes should be taught first for jump Markov processes. Diffusions are complicated objects, and the student does not need the extra burden of a subtle process to hinder the understanding of large deviations. Discrete time presents another unnecessarily difficult process, because the jumps are usually more general than those of the processes we consider. Furthermore, as we believe the book shows, there are many interesting applications of jump Markov processes. After all, we live in continuous time, and the events that occur in digital equipment are discrete.

As mentioned above, our book contains a new exposition of the theory of large deviations for jump Markov processes, but does not contain any new theory except for the results of §7.4 and Chapter 8. The applications contain many new results, though, and new derivations of previously known work. The new results include:

- A large deviations analysis of the $M/M/1$ queue that includes a surprising asymptotic formula for

$$\mathbb{P}\left(x(0) = ny, \ x(nt) = nz\right)$$

  as n gets large, where $x(t)$ is the queue size at time t (§11.7).
- Fully proved large deviations principle for jump Markov processes with a flat boundary (Chapter 8).
- Analysis of a new class of Markov processes, "finite levels," for which both a fluid limit theorem and a large deviations principle are proved (Chapter 8).
- New analysis of an Aloha multiple-access protocol, using finite levels theory, gives the quasi-stability region for instant-feedback, continuous time Aloha (Chapter 14).
- New results for Erlang's model:
  - Transient analysis from any initial load (§12.5).
  - Transient analysis of a finite population model (§12.7.A).
  - Analysis of bulk service (large customers) (§12.7.B).
  - Transient analysis of trunk reservation (§12.8).
- New results for the AMS model:
  - Analysis of bit-dropping models (§13.7).
  - Calculation of buffer asymptotics for the multiple class case (§13.8).

- Analyses of a simple priority queue (§15.1), "serve the longest queue" (§ 15.6), and "join the shortest queue" (§15.10).
- Simple analysis of the Flatto-Hahn-Wright queueing system (Chapter 16).

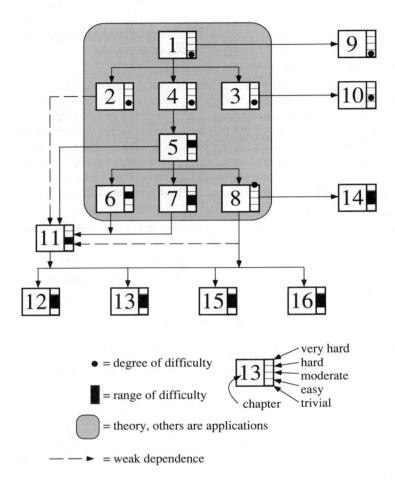

Figure 0.1. Dependence between the chapters.

0.1. What to Do with this Book

This book can be used as a basis for two types of one-semester courses. The first is an introduction to the theory of large deviations, through jump Markov processes. This course should cover most of Chapters 1, 2, 5, 6, Appendix D, and possibly the advanced material in Chapter 8. Such a course would prepare the student to read the more mathematical theory, and to fully appreciate the applications worked out in the rest of the book. It would be wise (in our opinion) to sprinkle such a theory-oriented course with some of the applications.

The second course is application-oriented. Such a course should probably start with Chapter 1 (at least §1.1–1.3), so that some flavor of the theory is provided. The results of §1.4, 2.1, and 2.3, and of Chapters 5–8 can then be stated without proof, with or without intuitive explanations. Some basic tools from the calculus of variations, at least to the extent summarized in Appendix C, should be covered. Then applications can be presented, according to the dependence chart shown in Figure 0.1.

Chapter 3 provides an easy application of the basic theory, and can thus be used to motivate the more general (and more technical) process-level theory. Chapter 4 summarizes some basic results concerning the Poisson process, and more generally jump Markov processes. There is nothing new in that chapter, but it is a strict prerequisite for the rest of the book. Finally, in the appendices we collect, for easy reference, some background material from analysis and probability theory.

In our judgment, the prerequisites for such courses (and for reading the book) are probability and analysis at a level of first-year graduate courses for engineering students, or senior-level courses for students of mathematics. The applications course can be done with much less background, provided the student is willing to believe the material as summarized in the appendices. However, some mathematical maturity (even affinity) is required.

0.2. About the Format of the Book

There are four types of exercises in the book. Some results that are easy to prove, important but not central to our development are presented as exercises. In some cases, extensions are relegated to an exercise when they are deemed not-too-hard but long; this is simply to save space. Examples and special cases are given as exercises, and are intended to help build intuition, or clarify a technical point. These exercises are an integral part of the text, and should at least be read, preferably solved. The last type of exercises are marked JFF ("just for fun"). The end of an Exercise is marked thus:
♠

There are two counters in the book: one for equations, one for all theorems, propositions, lemmas and corollaries, exercises, examples, figures, assumptions, and definitions. Equation numbers are written as (Chapter-Number.Equation-Number), and other numbers as Chapter-Number.Number. References appear in square brackets [ ], and we use either the first two letters of the author's last name, or—in the case of multiple authors—all initials. Conflicts are resolved creatively.

The index identifies definitions by bold page numbers, and includes the frequently used symbols.

We often wish to make a comment, or expand on a particular topic, in such a way that the reader may feel free to skip the comment, but will know that it is there. This is how we do it: in small type, in a narrow paragraph.

0.3. Acknowledgments

This project spanned many more years than we had ever anticipated. In the course of those years we have had help from many, many people. Preeminent among them are Armand Makowski, Debasis Mitra, and S.R.S. Varadhan. Debasis was steadfast in his support: moral, financial, and technical. He believed in us when we weren't sure we believed in ourselves. This project would never have been done without him. And we would never have gotten into the field (it is not certain that there would be much of a field to get into!) without Professor Varadhan. It was a tremendous comfort to know that there was no technical point, however difficult or subtle, that could not be answered almost instantly by a simple visit to NYU. Armand Makowski not only introduced us, and not only is he responsible for stating that the world would benefit from lecture notes on queueing applications of large deviations, but he can also be held accountable for doing something about it. With the support of John Baras, Armand arranged a sabbatical at the Systems Research Center where a first draft of these notes was hammered out by AW, and provided a sabbatical at the Systems Research Center where, somewhat unexpectedly, most of the time of AS was devoted to this project.

There are many more people who have helped over the years. Robert J. Vanderbei was, for a time, a coauthor of the book, and one appendix still bears his sole authorship. Several "field tests" of these ideas were graciously hosted by Armand Makowski at the University of Maryland, College Park, by Elja Arjas and the Finnish Summer School, and by Sid Browne and the Columbia Business School and Department of Mathematics. Within Bell Labs and the Technion, our home institutions, it seems that nearly everyone had something to contribute. Notable among those were co-large deviants Ofer Zeitouni and Amir Dembo. Also, Marty Reiman was a constant source of technical wisdom, moral support (what do you mean you aren't done yet?), and was an invaluable asset in transportation and living accommodations. Howard Trickey was our accessible TEX wizard, and justified his title hands down. Thanks also to Andrew Trevorrow for long-distance TEX help.

There were many students and colleagues who gave suggestions and feedback on everything from typos to approach. They include also those attending several courses given at the Technion, as well as lectures delivered at AT&T. We are particularly grateful for comments from Laurence Baxter, Hong Chen, Bill Cleveland, Ed Coffman, Amir Dembo, Amanda Galtman, Leo Flatto, Ben Fox, Predrag Jelenkovic, Armand Makowski, Colin Mallows, Bill Massey, Jim Mazo, Debasis Mitra, Marty Reiman, Emre Telatar, Stephen R.E. Turner, Yashan Wang, Ward Whitt, Paul Wright, Aaron Wyner, Ofer Zeitouni, and Li Zhang.

The editor-in-chief of this series Laurence Baxter did yeoman's work. Our editor John Kimmel amazed us by answering "yes" to every one of our requests, and promptly, too!

Typists Sue Pope and Lesley Price helped turn scribbled handwriting into beautiful TEX, quickly, accurately, and cheerfully.

This book was produced using TEX, with AS serving as local TEXpert, and was set in Times Roman, with MathTimes and other math fonts from Y&Y. The figures were drawn by AW using Canvas©, and according to the egalitarian tendencies of the authors, was set on Macintosh©, UNIX©, and various PC computers and clones.

I (AS) am delighted for this opportunity to acknowledge Armand Makowski for his role as colleague, collaborator, and catalyst in my professional life and, above all, to express my appreciation for his friendship.

And I (AW) am eternally grateful for my two mentors, Debasis Mitra and Raghu Varadhan. These two fine men have unselfishly nurtured me throughout this and other projects. I hope that in some way they can find some recompense in this volume.

Finally, our families, particularly our wives Shuli Cohen Shwartz and Judy Weiss, deserve thanks for putting up with us during all these years of labor. While we'll never know, it was probably as hard on them as having children; it was certainly longer and with less reward at the end. We promise we'll never do it again.

Chapter 1

Large Deviations of Random Variables

This chapter can be viewed as a guided tour through basic large deviations. Following a heuristic exposition, we derive . large deviations estimates for random variables. We provide proofs when these provide insight, or are typical; otherwise, we provide references. The modern tools and approaches, especially those that have proved useful for the applications, are discussed in Chapter 2 and Appendix D.

The main results, Theorems 1.5, 1.10, and 1.22, as well as the computations of Examples 1.13–1.18 and Exercises 1.6, 1.17–1.25, will be used heavily throughout the book.

1.1. Heuristics and Motivation

Estimates of probabilities of rare events turn out to have an exponential form; i.e., these probabilities decrease exponentially fast as a function of the asymptotic parameter. To motivate the exponential form of the large deviations estimates, consider the following examples. Let $x_1, x_2, \ldots$ be a sequence of independent, identically distributed (i.i.d.) random variables with a common distribution function F and finite mean. Fix a number $a > \mathbb{E}x_1$. Now the probability that $x_1 + \cdots + x_n > na$ is clearly decreasing in n in a long-term sense, since by the (weak) law of large numbers

$$\mathbb{P}\left(\frac{x_1 + \cdots + x_n}{n} \geq a\right) \to 0 \text{ as } n \to \infty.$$

The next question would be: How fast does this probability decrease? Let us perform some quick calculations. First, if for some integer k,

$$x_{jk+1} + \ldots + x_{(j+1)k} \geq ak \quad \text{for all} \quad j = 0, \ldots, (n/k) - 1,$$

then clearly $x_1 + \cdots + x_n \geq na$. Therefore

$$\mathbb{P}\left(\frac{x_1 + \cdots + x_n}{n} \geq a\right)$$

$$\geq \mathbb{P}\left(x_{jk+1} + \ldots + x_{(j+1)k} \geq ak \quad \text{for all} \quad j = 0, \ldots, (n/k) - 1\right)$$

$$= (\mathbb{P}(x_1 + \ldots + x_k \geq ak))^{n/k}$$

by independence. This immediately implies that the rate of convergence is at most exponential. On the other hand, for any positive θ, by Chebycheff's inequality

(Theorem A.113),

$$\mathbb{P}\left(x_1 + \cdots + x_n \geq na\right) = \mathbb{P}\left(e^{\theta(x_1 + \cdots + x_n)} \geq e^{\theta na}\right)$$
$$\leq e^{-\theta na}\,\mathbb{E}e^{\theta(x_1 + \cdots + x_n)}$$
$$= e^{-\theta na}\left(\mathbb{E}e^{\theta x_1}\right)^n$$
$$= \left(e^{-\theta a}\,\mathbb{E}e^{\theta x_1}\right)^n$$

by independence. For the right choice of θ, this exponential expression is decreasing:

Exercise 1.1. Show that if $a > \mathbb{E}x \geq 0$ and if $\mathbb{E}e^{\theta x} < \infty$ for all $|\theta|$ small, then $e^{-\theta a}\mathbb{E}e^{\theta x} < 1$ for some θ. Hint: compute $d\mathbb{E}e^{\theta x}/d\theta$ at $\theta = 0$.   ♠

Thus, probabilities should decay exponentially in n. The questions are: Do the rates in the upper and lower bound agree, and if so, how do we compute the right exponent? In §1.2 we show that they are indeed the same, and give a formula. In §1.3 we compute several examples. Anticipating the shape of things to come, the arguments indicate that

$$\mathbb{P}\left(\sum_1^n x_i \geq na\right) = e^{-n\ell(a) + o(n)}, \tag{1.1}$$

where the function ℓ depends on the distribution F. For the meaning of $o(n)$ see Definition A.14.

Here is another view that some find quite intuitive. If, indeed, $x_1 + \cdots + x_n \geq an$, then probably $x_1 + \cdots + x_n \approx na$ (for an illustration see Exercise 1.2 below). Moreover, it is likely that this happens by nearly-equal splitting, i.e., $x_1 + \ldots + x_{n/2} \approx an/2$ and $x_{(n/2+1)} + \ldots + x_n \approx an/2$, with an error of order $\sqrt{n}$. (This issue, of how improbable things happen, is explained in later chapters.)

Exercise 1.2. Show that in the case of fair coin flips, if x is the number of heads obtained in n flips and $0.8n$ is an integer,

$$\mathbb{E}\left(x - 0.8n \mid x \geq 0.8n\right) \rightarrow \frac{1}{3} \quad \text{as} \quad n \to \infty,$$

and does not grow with n ! Hint: $\binom{n}{0.8n+1} \approx \frac{1}{4} \cdot \binom{n}{0.8n}$.   ♠

Exercise 1.3. Compare the chances of obtaining $(1/2 + \alpha) \cdot n$ heads in n coin flips, with $0 < \alpha < 1/2$ in the following two ways: (i) by getting two series of $n/2$ flips, each with $\alpha \cdot n/2$ heads more than expected. (ii) by obtaining the additional heads in one series of length $n/2$ with the other series being "normal." Hint: use Stirling's formula.   ♠

These heuristics imply that, for large n, we have the rough estimate

$$\mathbb{P}\left(\sum_1^n x_i \geq an\right) \approx \mathbb{P}\left(\sum_1^{n/2} x_i \geq a\frac{n}{2} \, , \, \sum_{n/2+1}^n x_i \geq a\frac{n}{2}\right)$$

$$\approx \left[\mathbb{P}\left(\sum_1^{n/2} x_i \geq a\frac{n}{2}\right)\right]^2 .$$

Similarly, for any k much smaller than n,

$$\mathbb{P}\left(\sum_1^n x_i \geq an\right) \approx \left[\mathbb{P}\left(\sum_1^{n/k} x_i \geq a\frac{n}{k}\right)\right]^k ;$$

If we could choose k to be linear in n, we would see that this probability decreases exponentially fast. However, in general,

$$\mathbb{P}\left(\sum_1^n x_i \geq an\right) \not\approx \left[\mathbb{P}\left(x_1 \geq a\right)\right]^n$$

[for Bernoulli random variables with $a = 1/2$, the left-hand side is $1/2$ while the right is $(1/2)^n$!]. Thus the integer k above cannot quite grow linearly with n. This indicates that indeed (1.1) is to be expected, and that the "error term" $o(n)$ cannot be omitted.

Let us now illustrate some of the ideas from a different angle. To avoid technical difficulties, assume that the distribution function satisfies $F(1) = 0$, $F(2) = 1$. Let $\mu \overset{\triangle}{=} \mathbb{E}x_1$ and $\alpha \overset{\triangle}{=} \mathbb{E}\log x_1$. Then

$$\mathbb{E}\left(x_1 \cdot x_2 \cdots x_n\right) = \mu^n.$$

Let us estimate this expectation in a different way. Write

$$\mathbb{E}\left(x_1 \cdot x_2 \cdots x_n\right) = \mathbb{E}\left(\exp\left(\log x_1 + \ldots + \log x_n\right)\right)$$

$$= \mathbb{E}\left(\exp n\left(\frac{\log x_1 + \ldots + \log x_n}{n}\right)\right). \qquad (1.2)$$

By the strong law of large numbers,

$$\mathbb{P}\left(\frac{\log x_1 + \ldots + \log x_n}{n} \to \alpha\right) = 1$$

so that we expect $\mathbb{E}(x_1 \cdot x_2 \cdots x_n)$ to grow exponentially, roughly as $e^{n\alpha}$. However, by Jensen's inequality (A.11),

$$\mu \overset{\triangle}{=} \mathbb{E}x_1 = \mathbb{E}e^{\log x_1} > e^{\mathbb{E}\log x_1} = e^{\alpha} \, !$$

Clearly, the law of large numbers is not precise enough to estimate this expectation. Indeed, in this case we cannot expect convergence w.p.1 to imply convergence in expectation, since we are taking expectations of something that may grow quickly. Here is a refinement that will consolidate the two calculations. Suppose that, as in (1.1), we have an exponential estimate for the density of the sample averages of the sequence $\log x_1, \log x_2, \ldots$

$$\mathbb{P}\left(a \leq \frac{\log x_1 + \cdots + \log x_n}{n} \leq a + da\right) \approx e^{-n\ell(a)} da$$

for some non-negative function ℓ, and suppose that $\ell(a)/|a| \to \infty$ as $|a| \to \infty$. Then

$$\mathbb{E}(x_1 \cdot x_2 \cdots x_n) = \mathbb{E} \exp\left(n \cdot \frac{\log x_1 + \cdots + \log x_n}{n}\right)$$

$$\approx \int e^{na} e^{-n\ell(a)} \, da = \int e^{n(a-\ell(a))} \, da.$$

Suppose the maximum $m \stackrel{\triangle}{=} \sup_a (a - \ell(a))$ is attained at some point and write

$$\int e^{n(a-\ell(a))} \, da = e^{nm} \int e^{n(a-\ell(a))-m)} \, da.$$

By the assumptions on ℓ, $a - \ell(a)$ diverges to $(-\infty)$ as $|a| \to \infty$, so that it is strictly negative outside a finite interval. Thus, the integrand in the last integral goes to zero (exponentially fast) as $n \to \infty$, except where the maximum is attained, so that

$$\mathbb{E}(x_1 \cdot x_2 \cdots x_n) \leq e^{n(m+\varepsilon)}$$

for all $\varepsilon > 0$ and all n large. By looking at the points where $a - \ell(a) > m - \varepsilon$ we have

$$\mathbb{E}(x_1 \cdot x_2 \cdots x_n) \geq e^{n(m-\varepsilon)}$$

for every positive ε (this idea of estimating the rate of growth of an integral by considering the maximum of the integrand is called Laplace's method). We summarize these two inequalities using the notation

$$\mathbb{E}(x_1 \cdot x_2 \cdots x_n) \asymp \exp\left(n \cdot \sup_a (a - \ell(a))\right), \tag{1.3}$$

where the meaning of $\asymp$ is that the left-hand side grows exponentially fast, with rate $\sup_a (a - \ell(a))$. We will find in §2.2, as part of the derivation of large deviations estimates, that $\sup_a (a - \ell(a)) = \log \mu$, giving the correct exponential growth rate.

But this is just a formal calculation, and you are probably asking yourself now, "How can this be? I know that the mean is μ^n, and I've seen that the strong law of large numbers implies that the mean is almost surely near $e^{\alpha n}$, but how do I reconcile the two?" Let's consider what would happen if you would actually try to run an experiment to estimate $\mathbb{E}(x_1 \cdot x_2 \cdots x_n)$. You would collect n samples of the x_i, and then evaluate the product. You would undoubtedly (law of large numbers) come up with a number in the range of $e^{\alpha n}$. Repeat the experiment, and the results would be similar. However, after a great many experiments, you would come up with an unusually large observation—say something near μ^n. This observation is so large relative to the others that it completely dominates the mean you have been keeping, so that all of a sudden the mean looks like μ^n even though only one observation was of that order. Now what keeps an even more colossal observation from skewing further the observed mean? The answer is that it is too improbable for it to happen often [remember $\ell(a)$ grows quickly with a]. It will happen so rarely, that enough observations have been taken to completely dilute the effect of the "extra large" observation. This is the tradeoff we see between $\ell(a)$ ("the probability") and a ("the size"), and is the reason that $\sup_a (a - \ell(a))$ is the important quantity. It also serves to demonstrate that, sometimes, rare events are the most important to determine what's going on.

Stock and investment models

The last example has more than purely theoretical or pedagogical interest; it has monetary applications. Consider that investments usually pay an amount proportional to the investment. Suppose that an investment is risky; to be precise, an investment of one unit at the beginning of the ith period yields x_i units at the end of the period [which is the beginning of the $(i + 1)^{\text{th}}$ period]. Hence, after n periods, the value of a unit investment made at the beginning of the first period is $\prod_{i=1}^{n} x_i$.

How should we value an investment? This is a complicated question, but we have just seen that the return after a large number of periods is *most likely* to be near $\exp\left(n\mathbb{E} \log x_i\right)$, not near $\mathbb{E} \prod_{i=1}^{n} x_i$. Optimal investment strategies are based on this and related observations. See Kelly [KeJ], Algoet and Cover [AC], and references therein.

Beyond deviations from the mean.

Sanov's Theorem, introduced in §2.4, takes us one step up to "Level 2 Large Deviations." The question we ask there is: What do the random variables $x_1, x_2, \ldots$ look like when they do make a big excursion (such as making $\prod_{i=1}^{n} x_i \geq \mu^n$)? It turns out that *they all look like they are sampled from a "tilted distribution," one for which* $\mathbb{E} \log x = \log \mu$. In other words, the product becomes large because of conspiracies, **not** because of outliers. This conspiracy is a very rare occurrence, but when it occurs, its effect is huge. This is captured by the balance between the size of the effect e^{na}, and the rarity $e^{-n\ell(a)}$. Whereas in §1.1 we ask "How likely is it for the sample mean to deviate from the ensemble mean?," Sanov's Theorem addresses the question "How likely is it for the empirical distribution to deviate from the true distribution?" But let us establish first things first.

1.2. I.I.D. Random Variables

Chernoff's Theorem establishes (1.1) for i.i.d. random variables. The proof consists of an upper bound and a lower bound. The upper bound is just a parameterized version of Chebycheff's inequality (A.9) applied to the function $e^{\theta x}$. The lower bound uses a change of measure argument much as in importance sampling. These ideas generalize to random vectors and to processes, and will be used in all our large deviations proofs.

So, consider a sequence $x_1, x_2, \ldots$ of i.i.d. random variables with common distribution function F, and assume the mean $\mathbb{E}x_1$ exists. Define

$$M(\theta) = \mathbb{E}e^{\theta x_1} \tag{1.4a}$$

$$\ell(a) = -\log\left(\inf_\theta e^{-\theta a} M(\theta)\right) = \sup_\theta (\theta a - \log M(\theta)). \tag{1.4b}$$

$M(\theta)$ is the *moment generating function* of the random variable x_1. The function $\log M(\theta)$ is called the *logarithmic moment generating function* or *cumulant generating function* of x_1. Note that ℓ is non-negative [put $\theta = 0$ in (1.4 b)] and convex (by Theorem A.47, being the supremum of linear, hence convex functions); see Proposition 5.10, §5.2. The transformation applied to $\log M$ in (1.4b) is variously called the convex transform, Fenchel transform, Legendre transform, or Cramér transform.

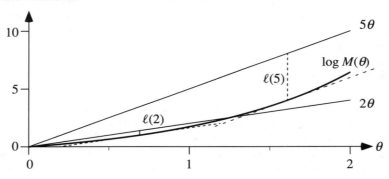

Figure 1.4. The ℓ-function: computing the Legendre transform.

By Exercise A.92, if the supremum in (1.4) is attained at a point θ^* in the interior of the interval where $M(\theta)$ is finite, then $M(\theta)$ is differentiable at θ^*, so that

$$\ell(a) = -\log \mathbb{E}e^{\theta^*(x_1-a)}. \tag{1.5}$$

Theorem 1.5. *Consider the sequence $x_1, x_2, \ldots$ of i.i.d. random variables. For every $a > \mathbb{E}x_1$ and positive integer n,*

$$\mathbb{P}(x_1 + \cdots + x_n \geq na) \leq e^{-n\ell(a)}. \tag{1.6a}$$

Assume that $M(\theta) < \infty$ for θ in some neighborhood of 0 and that (1.5) holds for some θ^* in the interior of that neighborhood. Then for every $\varepsilon > 0$ there exists an integer n_0 such that whenever $n > n_0$,

$$\mathbb{P}(x_1 + \cdots + x_n \geq na) \geq e^{-n(\ell(a)+\varepsilon)}. \tag{1.6b}$$

Equations (1.6a)–(1.6b) imply that

$$\mathbb{P}(x_1 + \cdots + x_n \geq na) = e^{-n\ell(a)+o(n)}. \tag{1.6c}$$

Remark. This result holds, in fact, for i.i.d. random variables without any assumptions. The general case is proved using an approximation argument; see, e.g., Chernoff [Ch], Dembo and Zeitouni [DZ §2.2] and Theorem 1.10 below.

Proof. By virtue of Exercise 1.6 below, it suffices to establish the result when $\mathbb{E}x_1 = 0$. The upper bound is proved using an exponential estimate. First fix $\theta \geq 0$.

$$
\begin{aligned}
\mathbb{P}(x_1 + \cdots + x_n \geq na) &\leq \mathbb{P}\left(e^{\theta(x_1+\cdots+x_n)} \geq e^{\theta na}\right) && e^{\theta x} \text{ is increasing}\\
&\leq e^{-\theta na}\,\mathbb{E}e^{\theta(x_1+\cdots+x_n)} && \text{Chebycheff}\\
&= \left(e^{-\theta a}\,\mathbb{E}e^{\theta x_1}\right)^n && \text{independence.}
\end{aligned}
$$

Equality in the first relation holds whenever $\theta > 0$. Since $\theta \geq 0$ was arbitrary, taking the infimum would yield (1.6a) provided we show that we can ignore negative values of θ in (1.4b).

By Jensen's inequality,

$$M(\theta) = \mathbb{E}e^{\theta x_1} \geq e^{\mathbb{E}\theta x_1} = 1$$

for all θ. Thus, since $a > 0$,

$$e^{-\theta a} M(\theta) \geq 1 \text{ for } \theta \leq 0,$$

with equality for $\theta = 0$. Therefore in computing the middle term of (1.4), we can restrict the range of the infimum to $\theta \geq 0$, i.e.,

$$\inf_{\theta} e^{-\theta a} M(\theta) = \inf_{\theta \geq 0} e^{-\theta a} M(\theta).$$

This completes the proof of the upper bound (1.6a).

The lower bound is established using a change of measure (if you are unfamiliar with the idea of a change of measure, see §A.4). Let F be the distribution of x_1. Then G, defined by

$$G(x) = \left[M(\theta^*)\right]^{-1} \int_{-\infty}^{x} e^{\theta^* y}\, dF(y), \tag{1.7}$$

is a new distribution function (check!) (G is the tilted distribution referred to at the end of §1.1). For any real α, we clearly have

$$
\mathbb{P}(x_1 \geq \alpha) = \int_{-\infty}^{\infty} \mathbf{1}\,[y \geq \alpha]\,dF(y)
$$

$$
= \int_{-\infty}^{\infty} \mathbf{1}\,[y \geq \alpha]\,e^{-\theta^* y} e^{\theta^* y}\,dF(y)
$$

$$
= M(\theta^*) \int_{-\infty}^{\infty} \mathbf{1}\,[y \geq \alpha]\,e^{-\theta^* y}\,dG(y)
$$

by the definition of G. Applying this idea to the left-hand side of (1.6b),

$$
\mathbb{P}(x_1 + \cdots + x_n \geq na) = \int \cdots \int \mathbf{1}[y_1 + \cdots + y_n \geq na]\,dF(y_1)\cdots dF(y_n)
$$

$$
= \int \cdots \int \mathbf{1}[y_1 + \cdots + y_n \geq na]\,e^{-\theta^*(y_1 + \cdots + y_n)}
$$

$$
\times e^{\theta^* y_1}\,dF(y_1)\cdots e^{\theta^* y_n}\,dF(y_n).
$$

Changing to the measure G, we have, for any $\varepsilon' > 0$,

$$
\mathbb{P}(x_1 + \cdots + x_n \geq na) \tag{1.8a}
$$

$$
= \left[M(\theta^*)\right]^n \int \cdots \int \mathbf{1}[y_1 + \cdots + y_n \geq na]\,e^{-\theta^*(y_1 + \cdots + y_n)}
$$

$$
dG(y_1)\cdots dG(y_n)
$$

$$
\geq \left[M(\theta^*)\right]^n \int \cdots \int \mathbf{1}[n(a + \varepsilon') \geq y_1 + \cdots + y_n \geq na]\,e^{-\theta^*(y_1 + \cdots + y_n)}
$$

$$
dG(y_1)\cdots dG(y_n)
$$

$$
\geq \left[M(\theta^*)\right]^n e^{-n\theta^*(a+\varepsilon')} \int \cdots \int \mathbf{1}[n(a + \varepsilon') \geq y_1 + \cdots + y_n \geq na]
$$

$$
dG(y_1)\cdots dG(y_n).
$$

Let $\tilde{x}_1, \ldots, \tilde{x}_n$ be i.i.d. random variables, distributed according to G. Then the probability in the first expression of (1.8a) is bounded below by

$$
\left[M(\theta^*)\right]^n e^{-n\theta^*(a+\varepsilon')}\mathbb{P}\left(n(a + \varepsilon') \geq \tilde{x}_1 + \cdots + \tilde{x}_n \geq na\right). \tag{1.8b}
$$

We now provide a lower bound for the probability on the right of (1.8b). First, since $M(\theta)$ is finite in a neighborhood of θ^*, it is differentiable there by Exercise A.92. Therefore

$$
\frac{d^n}{d\theta^n} M(\theta^*) = \mathbb{E}x_1^n e^{\theta^* x_1} < \infty, \qquad n = 1, 2, \ldots.
$$

and, in particular,

$$
\mathbb{E}\tilde{x}_1^2 \overset{\triangle}{=} \frac{\mathbb{E}x_1^2 e^{\theta^* x_1}}{M(\theta^*)} < \infty.
$$

Since $\inf_\theta \mathbb{E}e^{\theta(x_1-a)} = \mathbb{E}e^{\theta^*(x_1-a)}$ and it is differentiable, the derivative vanishes at θ^*, so that

$$\mathbb{E}(x_1 - a)e^{\theta^*(x_1-a)} = 0, \quad \text{or} \quad \mathbb{E}x_1 e^{\theta^* x_1} = aM(\theta^*).$$

This implies that the change of measure puts the mean of $\tilde{x}$ exactly at a since

$$\mathbb{E}\tilde{x}_1 = \int y\,dG(y) = [M(\theta^*)]^{-1} \int ye^{\theta^* y}dF(y) = a.$$

Consider the sum in (1.8b) of the i.i.d. random variables $\tilde{x}_1, \ldots, \tilde{x}_n$. Since these random variables have mean a and finite variance, the central limit theorem A.112 implies

$$\mathbb{P}\left(n(a + \varepsilon') \geq \tilde{x}_1 + \cdots + \tilde{x}_n \geq na\right) = \mathbb{P}\left(\sqrt{n}\,\varepsilon' \geq \frac{1}{\sqrt{n}}\sum_{i=1}^{n}(\tilde{x}_i - a) \geq 0\right)$$

$$\rightarrow \frac{1}{2} \quad \text{as} \quad n \rightarrow \infty.$$

Let n_0 be such that the probability exceeds $1/4$ whenever $n \geq n_0$ (clearly n_0 depends on ε'). Then for $n \geq n_0$,

$$\mathbb{P}(x_1 + \cdots + x_n \geq na) \geq \frac{1}{4}\left[e^{-\theta^* a}M(\theta^*)\right]^n e^{-n\theta^* \varepsilon'}$$

$$= \frac{1}{4}e^{-n\ell(a)}e^{-n\theta^* \varepsilon'}.$$

Now since $\theta^* > 0$ (why?) we can choose ε' so that $(1/4)e^{-n\theta^* \varepsilon'} \geq e^{-n\theta^* \varepsilon}$ whenever $n \geq n_0$. This proves (1.6b–1.6c). ∎

Exercise 1.6. Let $y_i \overset{\triangle}{=} x_i + \bar{y}$ for each $i \geq 1$, where $\bar{y}$ is a fixed constant. Express the moment generating function M_y and the Cramér transform ℓ_y of its logarithm in terms of M_x and ℓ_x. Write Theorem 1.5 for $y_1, y_2, \ldots$ and conclude that the zero-mean assumption on $x_1, x_2, \ldots$ is without loss of generality. ♠

Exercise 1.7. Assume $\ell(a)$ is continuous, and re-derive the lower bound without invoking the central limit theorem. Hint: use the law of large numbers. ♠

Exercise 1.8. Let $y_1, y_2, \ldots$ be independent (but not necessarily identically distributed!) random variables so that for all i and M, $\mathbb{P}(|y_i| \geq M) \leq e^{-rM}$ for some $r > 0$. Then there exists a continuous function $f(\varepsilon)$, which depends only on r, so that for all $\varepsilon > 0$ we have $f(\varepsilon) > 0$ and

$$\mathbb{P}\left(\left|\sum_{i=1}^{n}(y_i - \mathbb{E}y_i)\right| > n\varepsilon\right) \leq e^{-nf(\varepsilon)}.$$

Hint: compute separately for the case that the sum is larger than $n\varepsilon$ and smaller than $(-n\varepsilon)$, and start with the zero-mean case. Use Chebycheff's inequality as in the proof of Theorem 1.5. Prove $\mathbb{E}e^{|\theta y_i|}$ is finite for all θ small, uniformly in i. Use Exercise A.92 to conclude that the functions $f_i(\varepsilon, \theta) \stackrel{\triangle}{=} e^{-\theta\varepsilon}\mathbb{E}e^{\theta y_i}$ have continuous derivatives (of all orders!) near $(\varepsilon, \theta) = (0, 0)$. Now use a Taylor expansion in the two variables ε, θ to second order and set $\varepsilon = 0$. Obtain a bound of the form $(1 - c\varepsilon^2)^n$ with $c > 0$ that holds for small ε.   ♠

Remark. To compute $\mathbb{P}(x_1 + \cdots + x_n \le na)$ for $a < \mathbb{E}(x_1)$, note from (1.4) that the ℓ-function for the sequence $-x_1, -x_2, \ldots$ is equal to the ℓ-function of $x_1, x_2, \ldots$ with the sign of its argument reversed, so that, for $a < 0$, by Chernoff's Theorem,

$$\mathbb{P}(x_1 + \cdots + x_n \le na) = e^{-n\ell(a)+o(n)}.$$

A more detailed statement of a large deviations theorem in $\mathbb{R}^1$ and under weaker conditions is given in Theorem 1.10 below.

Theorem 1.5 gives us an estimate of the probability that the sample mean lies in the half-line above $a > \mathbb{E}x_1$, and the remark extends this to the half-line below $\mathbb{E}x_1$. This easily extends to more general sets. Define the real-valued function J on sets S in $\mathbb{R}$ by

$$J(S) \stackrel{\triangle}{=} \inf_{a \in S} \ell(a).$$

Corollary 1.9. *Assume $M(\theta) < \infty$ for θ in some neighborhood of zero. Then, for every open set S and positive integer n,*

$$\mathbb{P}\left(\frac{x_1 + \cdots + x_n}{n} \in S\right) = e^{-nJ(S)+o(n)}.$$

Note that such a result is not possible for closed sets: in particular, single points are closed sets, and if x_1 possesses a density, then the probability that the sample mean is in the set is zero.

For a proof of this corollary see Dembo and Zeitouni [DZ]. Here is a heuristic argument (when $\mathbb{E}x_1 = 0$). An application of Jensen's inequality (to the convex function $(-\log\alpha)$: use the definition of ℓ with θ fixed) shows that $\ell(0) \le 0$, and since ℓ is non-negative, $\ell(0) = 0$. Thus the result is just the weak law of large numbers if $(0 =)\mathbb{E}x_1 \in S$. Now ℓ is non-negative and convex, so that it is increasing for $x > 0$ and decreasing for $x < 0$. But then there is a point, say a, in the closure of S so that $\ell(a) = J(S)$. Since S is open, there is an interval in S whose endpoint is a. The argument of the lower bound now applies, since (1.8) uses only a small interval, so that the same lower bound holds for all open sets for which a is a minimum point. For the upper bound, enclose S by the two smallest semi-infinite intervals $(-\infty, a^-]$ and $[a^+, \infty)$ and apply Theorem 1.5.

Actually, this discussion is generic in that lower bounds are usually proved locally, while upper bounds are established by increasing the sets.

The one-dimensional case is unique in that the upper bound holds for open sets. The typical large deviations statement consists of an upper bound for closed sets and a lower bound for open sets. Here is the best result for i.i.d. random variables, stated in generic large deviations form. For a proof, see [DZ §2.2].

Theorem 1.10. *Let* $x_1, x_2, \ldots$ *be i.i.d. random variables. Then the function* ℓ *defined in (1.4) is convex and lower semicontinuous. For any closed set* F,

$$\limsup_{n \to \infty} \frac{1}{n} \log \mathbb{P} \left(\frac{x_1 + \cdots + x_n}{n} \in F \right) \leq - \inf_{a \in F} \ell(a)$$

and for any open set G,

$$\liminf_{n \to \infty} \frac{1}{n} \log \mathbb{P} \left(\frac{x_1 + \cdots + x_n}{n} \in G \right) \geq - \inf_{a \in G} \ell(a).$$

Note that no conditions, not even existence of the mean, are required.

1.3. Examples—I.I.D. Random Variables

In some cases, notably exponential families, the function ℓ of (1.4b) can be calculated explicitly (see, e.g., [MN]). We now present some simple calculations in order to develop a feeling for the scope of the large deviations estimates.

Example 1.11: Normal random variables. Let $x_1, x_2, \ldots$ be standard normal random variables. Then

$$M(\theta) = \frac{1}{\sqrt{2\pi}} \int e^{\theta y} e^{-\frac{1}{2}y^2} dy = e^{\frac{1}{2}\theta^2}$$

by completing the square in the exponent, so that $\ell(a) = \sup_\theta (\theta a - \frac{1}{2}\theta^2) = \frac{1}{2}a^2$. Thus Chernoff's Theorem states that, for any $a > 0$,

$$\mathbb{P}(x_1 + \cdots + x_n \geq na) \approx e^{-n\frac{1}{2}a^2}.$$

In this case, we can also perform a direct calculation: $x_1 + \cdots + x_n$ is a normal random variable distributed as $\sqrt{n}x_1$, so

$$\mathbb{P}(x_1 + \cdots + x_n \geq na) = \mathbb{P}(x_1 \geq \sqrt{n}a)$$
$$= \frac{1}{\sqrt{2\pi}} \int_{\sqrt{n}a}^{\infty} e^{-\frac{1}{2}t^2} dt.$$

Using an estimate of this integral [Mc, p. 5],

$$\frac{1}{y + y^{-1}} e^{-\frac{1}{2}y^2} \leq \int_y^{\infty} e^{-\frac{1}{2}t^2} dt \leq \frac{1}{y} e^{-\frac{1}{2}y^2},$$

we obtain

$$\mathbb{P}(x_1 + \cdots + x_n \geq na) \approx \frac{1}{\sqrt{2\pi na}} e^{-n\frac{1}{2}a^2},$$

which is in agreement with the exponential order of the large deviations estimates. The fact that $1/\sqrt{n}$ appears is also generic, as will be seen in the sequel.

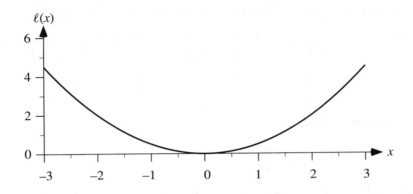

Figure 1.12. The rate function for Standard Normal random variables.

Example 1.13: Poisson random variables. Let x_i be Poisson with mean λ. Then $M(\theta) = e^{\lambda(e^{\theta}-1)}$, and for $a > 0$, $\theta^* = \log(a/\lambda)$. Thus

$$\ell(a) = a\left(\log\left(\frac{a}{\lambda}\right) - 1\right) + \lambda \,,$$

and $\ell(0) = \lambda$, $\ell(a) = \infty$ for $a < 0$, with $|\theta^*| = \infty$ in the last two cases. Thus Chernoff's Theorem implies, for $a > \lambda$,

$$\mathbb{P}(x_1 + \cdots + x_n \geq na) = \left(\frac{a}{\lambda}\right)^{-na} e^{-n(\lambda-a)+o(n)}.$$

Let us compare this with a direct estimate. Since $x_1 + \cdots + x_n$ is a Poisson random variable with mean $n\lambda$,

$$\mathbb{P}(x_1 + \cdots + x_n \geq na) = \sum_{j=na}^{\infty} \frac{(n\lambda)^j}{j!} e^{-n\lambda} \approx \frac{(n\lambda)^{na}}{(na)!} e^{-n\lambda}$$

$$\approx \frac{(n\lambda)^{na}}{\sqrt{2\pi na}(na)^{na}e^{-na}} e^{-n\lambda}$$

$$= \frac{1}{\sqrt{2\pi na}} \left(\frac{a}{\lambda}\right)^{-na} e^{-n(\lambda-a)}$$

using Stirling's formula, and the factor $1/\sqrt{n}$ appears again.

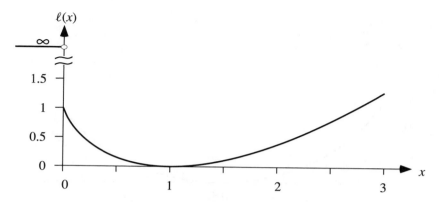

Figure 1.14. The rate function for Standard Poisson random variables.

Example 1.15: Bernoulli random variables. Let x_i take values zero and one with probability 1/2. Then $M(\theta) = \frac{1}{2}\left(1 + e^{\theta}\right)$. When $\frac{1}{2} < a < 1$, straightforward calculus shows that $\theta^* = \log a - \log(1 - a)$, so that in this range

$$\ell(a) = -\log\left(e^{-a\theta^*} M(\theta^*)\right) = a \log a + (1 - a)\log(1 - a) + \log 2. \quad (1.9)$$

Chernoff's Theorem thus implies that, for $1/2 < a < 1$,

$$\mathbb{P}(x_1 + \cdots + x_n \geq na) = e^{-n\ell(a) + o(n)} = 2^{-n} a^{-na}(1 - a)^{-n(1-a)} e^{o(n)}.$$

We can obtain an estimate in a direct way, by approximating the binomial coefficient using Stirling's formula:

$$\mathbb{P}(x_1 + \cdots + x_n \geq na)$$

$$= \sum_{j=na}^{n} \binom{n}{j} 2^{-n} \approx \frac{n!}{(na)!(n - na)!} 2^{-n}$$

$$\approx 2^{-n} \sqrt{2\pi n}(n)^n e^{-n}$$

$$\times \left(\sqrt{2\pi na}(na)^{na} e^{-na} \sqrt{2\pi n(1 - a)}(n(1 - a))^{n(1-a)} e^{-n(1-a)}\right)^{-1}$$

$$= \left(\sqrt{2\pi na(1 - a)}\right)^{-1} 2^{-n} a^{-na}(1 - a)^{-n(1-a)}.$$

The formula for M immediately implies that $\ell(a) = \infty$ whenever $a < 0$ or $a > 1$, and $\ell(0) = \ell(1) = \log 2$, with $|\theta^*| = \infty$ in all these cases. Chernoff's Theorem tells us that $\mathbb{P}(x_1 + \cdots + x_n \geq na) \leq e^{-n\infty} = 0$ whenever $a > 1$. For, $a = 1$ the theorem implies $\mathbb{P}(x_1 + \cdots + x_n \geq n) = \left(\frac{1}{2}\right)^n e^{o(n)}$, which is quite close to the exact probability $\left(\frac{1}{2}\right)^n$.

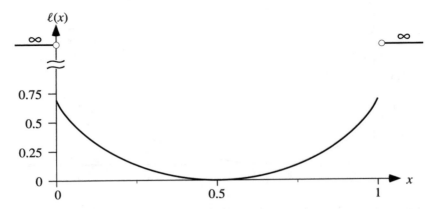

Figure 1.16. The rate function for Bernoulli-1/2 random variables.

Exercise 1.17. For Bernoulli random variables $b_1, b_2, \ldots$ with $\mathbb{P}(b_i = 1) = p$,

$$\ell(a) = a \log \frac{a}{p} + (1 - a) \log \frac{1 - a}{1 - p}. \qquad \spadesuit$$

Example 1.18: Exponential random variables. Let x_i be exponential random variables with mean 1. Then $M(\theta) = 1/(1-\theta)$ for $\theta < 1$ and is infinite otherwise. Therefore $\theta^* = (a - 1)/a$ whenever $a > 1$ and then

$$\ell(a) = a - 1 - \log a.$$

Chernoff's Theorem states that for $a > 1$,

$$\mathbb{P}(x_1 + \cdots + x_n \geq na) = a^n e^{-n(a-1)+o(n)}.$$

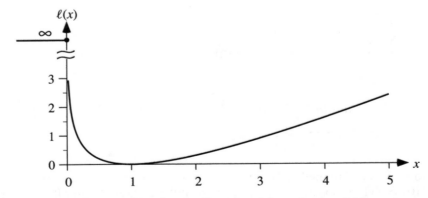

Figure 1.19. The rate function for Exponential random variables, rate one.

Exercise 1.20. For exponential random variables with parameter λ (mean $1/\lambda$),

$$\ell_\lambda(a) = \ell(a\lambda).$$

Interpret this as a time change.   ♠

From the examples, the following should be expected.

Exercise 1.21. Let c be the constant that is the greatest lower bound for a random variable x; that is, $\mathbb{P}(x < c) = 0$ and $\mathbb{P}(x \leq c + \varepsilon) > 0$ for all $\varepsilon > 0$. Then $\ell(a) = \infty$ for $a < c$. Moreover, $\ell(c) < \infty$ if and only if $\mathbb{P}(x = c) > 0$. Hint: take $c = 0$, use dominated convergence, then extend by Exercise 1.6.   ♠

1.4. I.I.D. Random Vectors

Large deviations in $\mathbb{R}^d$ are much more complex than in $\mathbb{R}^1$. The main reason for this is that open and closed sets are more complex. Fortunately, these results are not needed in the development of our theory. Sticking to our principle of proving just what we need, let us state a reasonable large deviations result, and provide rough intuition about a way a proof might go. A modern approach to this problem is discussed in §2.1.

Consider the $\mathbb{R}^d$-valued i.i.d. random vectors $\vec{x}_1, \vec{x}_2, \ldots$ with (vector) mean $\vec{0}$. Now let $\vec{\theta} \in \mathbb{R}^d$ and define (see Example A.8 for the notation)

$$M(\vec{\theta}) = \mathbb{E}e^{\langle \vec{\theta}, \vec{x}_1 \rangle} \tag{1.10a}$$

$$\ell(\vec{a}) = -\log\left(\inf_{\vec{\theta}} e^{-\langle \vec{\theta}, \vec{a} \rangle} M(\vec{\theta})\right) = \sup_{\vec{\theta}}\left(\langle \vec{\theta}, \vec{a} \rangle - \log M(\vec{\theta})\right). \tag{1.10b}$$

Define J as in Corollary 1.9, but for sets S in $\mathbb{R}^d$. That is,

$$J(S) \overset{\triangle}{=} \inf_{\vec{a} \in S} \ell(\vec{a}). \tag{1.11}$$

Theorem 1.22. *Assume $M(\vec{\theta}) < \infty$ for all $\vec{\theta}$. Then, for every closed set C and $\varepsilon > 0$, there exists an integer n_0 such that, whenever $n > n_0$,*

$$\mathbb{P}\left(\frac{\vec{x}_1 + \ldots + \vec{x}_n}{n} \in C\right) \leq e^{-n(J(C) - \varepsilon)} \tag{1.12u}$$

and for every open set G and $\varepsilon > 0$, there exists an integer n_0 such that, whenever $n > n_0$,

$$\mathbb{P}\left(\frac{\vec{x}_1 + \ldots + \vec{x}_n}{n} \in G\right) \geq e^{-n(J(G) + \varepsilon)}. \tag{1.12l}$$

The proof of this theorem in $\mathbb{R}^d$ is much more involved than in $\mathbb{R}^1$. Furthermore, the "weakest assumptions" possible in $\mathbb{R}^d$ are much more restrictive than in $\mathbb{R}^1$. See, for example, §D.1, [DZ] and the remarks in §1.5 below. Although modern proofs rely on the technique of §D.1, we outline the extension of the one-dimensional arguments to $\mathbb{R}^d$.

The lower bound in $\mathbb{R}^1$ was based on estimating the probability of the sample mean being in a small interval around the point where ℓ is smallest. A similar argument works in the d-dimensional case: we need to consider small neighborhoods, or balls, around the minimizing point.

Exercise 1.23. Generalize Exercise 1.7 to $\mathbb{R}^d$. ♠

For the upper bound, consider half-spaces of the form $H_{s,a} \stackrel{\triangle}{=} \{\vec{y} : \langle \vec{s}, \vec{y} \rangle \geq a\}$ for some $\vec{s} \in \mathbb{R}^d$ and $a > 0$. Then $\tilde{x}_i \stackrel{\triangle}{=} \langle \vec{s}, \vec{x}_i \rangle$ are i.i.d. and

$$\frac{\vec{x}_1 + \ldots + \vec{x}_n}{n} \in H_{s,a} \iff \frac{\tilde{x}_1 + \ldots + \tilde{x}_n}{n} \geq a.$$

Exercise 1.24. Assume ℓ is continuous and finite. Prove the upper bound for convex sets in $\mathbb{R}^d$. Extend the proof to finite unions of convex sets. Hint: use Chernoff's Theorem for $\tilde{x}_1, \tilde{x}_2, \ldots$. Note that ℓ is convex as explained below (1.4) so that $C_\varepsilon \stackrel{\triangle}{=} \{\vec{x} : \ell(\vec{x}) \leq J(C) - \varepsilon\}$ is convex, and has empty intersection with C. Therefore there is a half-space containing C that does not intersect C_ε. ♠

The following calculation will be used for our Poisson processes. It follows from (1.10)–(1.12).

Exercise 1.25. Suppose y_{ij} are i.i.d. with $y_{ij} \stackrel{\mathcal{L}}{=} Pois(\lambda_j)$. Define

$$\vec{x}_i \stackrel{\triangle}{=} \sum_{j=1}^{J} y_{ij} \vec{e}_j.$$

Show that $\ell(\vec{a})$ defined in (1.10b) has the form

$$\ell(\vec{a}) \stackrel{\triangle}{=} \sup_{\vec{\theta}} \left(\langle \vec{\theta}, \vec{y} \rangle - g(\vec{\theta}) \right), \tag{1.13}$$

where $g(\vec{\theta}) = \log M(\vec{\theta})$ is given by

$$g(\vec{\theta}) = \sum_{j=1}^{J} \lambda_j \left(e^{\langle \vec{\theta}, \vec{e}_j \rangle} - 1 \right). \tag{1.14}$$

Consequently, Exercise 1.24 implies that

$$\lim_{\varepsilon \downarrow 0} \lim_{n \to \infty} \frac{1}{n} \log \mathbb{P} \left(\left| \frac{\vec{x}_1 + \cdots + \vec{x}_n}{n} - \vec{a} \right| < \varepsilon \right) = \ell(\vec{a}). \quad ♠$$

1.5. End Notes

While the one-dimensional case is simple enough to make the ideas clear, it can (for the same reason) lull the reader into unwarranted carelessness. We conclude this chapter by identifying some potential pitfalls, and then discussing related works and possible extensions.

One-dimensional caveats

The reader should be aware of several delicate points. Some of these are discussed in Chapter 2 and Appendix D.

1. Properties of the rate function.

   a. Convexity. The calculations in Chapter 1 show that rate functions for random vectors are convex. This (unfortunately) does not extend to rate functions for processes, as will be seen in Chapter 5.

   b. Semicontinuity. The calculations in Chapter 1 show that rate functions are lower semicontinuous. Recall that this means that $\lim\inf_{x \to y} \ell(x) \geq \ell(y)$, so that ℓ can only jump down. When we formulate, in §2.1, the "axiomatic" large deviations principle, we shall *require* that rate functions be lower semicontinuous. There are several reasons for this restriction. Under this condition there is a convenient, equivalent formulation of the upper bound (see Lemma 2.11), and it guarantees uniqueness of the rate function (§2.1). This condition also makes the upper bound for closed sets particularly easy to prove. In addition, it implies that if a rate function is strictly positive at every point of a compact set, then the probability of that set decays exponentially fast (lower semicontinuous functions attain their minimum on compact sets).

   c. Compact level sets. The examples in Chapter 1 show that ℓ possesses compact level sets, i.e., the sets

$$\{x : \ell(x) \leq \alpha\}$$

   are closed and bounded for each α, if and only if the probability that the random variable actually takes its smallest possible value is zero: Exercise 1.21. In particular, these sets are closed, which implies lower semicontinuity—see Definition A.28 and Exercise A.29. The compactness condition is necessary in order to establish the important *contraction principle* (§2.3). This is illustrated further in §2.3. In Chapter 7 we provide an example of a non-negative birth-death process with constant drift (-1) for $x > 0$, but with the cost (in terms of the rate function) of going from 0 to ∞ being finite. It is easy to show that the process will explode (transition to infinity) if allowed to run for a long enough (finite) time.

2. Difficulties in higher dimensions. In one dimension, the use of Chernoff's Theorem for semi-infinite sets actually provides enough control to estimate the probability that

$$z_n \overset{\triangle}{=} (x_1 + \cdots + x_n)/n$$

is in a fairly arbitrary set. In higher dimensions, it is more complicated to estimate this probability, because the topology is more complicated. Thus we
need stronger assumptions in $\mathbb{R}^d$ when $d \geq 2$.

3. Difficulties of processes, as opposed to finite dimension. Processes can be
   viewed, if you are so inclined, as random variables with values in some (infinite dimensional) space of functions. The topology that troubled us in $\mathbb{R}^d$ is
   simple compared to the topology in function spaces. In this book we restrict
   our attention to particularly simple processes: jump Markov processes, where
   the topology is well understood. This topology is discussed in §A.1.

Extensions and relations to other methodologies.

Extensions and generalizations of the results of Chapter 1 are discussed in Chapter
2 and Appendix D. Let us now mention briefly some extensions that will not be
touched upon.

The only type of large deviations estimate we obtain in this book is on the order
of

$$\mathbb{P}_n(S) \approx e^{-nI^*(S)}.$$

This is only the first term in an asymptotic expansion, though. Using formal methods such as WKB expansions, one finds [O1] that the series usually continues as
follows:

$$\mathbb{P}_n(S) \approx \frac{1}{n^{d/2}} e^{-nI^*(S)+C+O(1/n)} \, ,$$

where d is the number of dimensions of the process in question. There are a few
cases where the full asymptotic expansion has been worked out, and there are
many more cases where some *formal* terms have been calculated.

a. Formal expansions of singular equations (e.g., WKB methods). Several investigators, notably Knessl, Matkowski, Morrison, Schuss, and Tier [KMS, Mo1,
   Mo2] have calculated quite accurate and explicit asymptotic expressions for
   various large deviations problems using formal expansions. The main criti-
   cism of these techniques (there are several that are employed) is that there is no
   proof of their validity; in contrast, the student will note that in the present book,
   about 50% of the pages are devoted to proving the validity of the methods we
   employ. Martin Day provides rigorous proofs for the validity, in some cases, of
   formulae obtained by formal methods of the WKB type; see e.g., [D2]. Formal
   methods often give more terms in the asymptotic series than the "rough" methods we employ. They do not usually give sample path information, though,
   such as we obtain in Chapter 16.

b. Central limit expansions and moderate deviations. The quantities we estimate
   are generalizations of

$$\mathbb{P}\left(\frac{x_1 + x_2 + \ldots + x_n}{n} \geq a\right)$$

compared to the central limit quantity

$$\mathbb{P}\left(\frac{x_1 + x_2 + \ldots + x_n}{\sqrt{n}} \geq a\right).$$

Clearly there is room to investigate the quantities

$$\mathbb{P}\left(\frac{x_1 + x_2 + \ldots + x_n}{n^{b+1/2}} \geq a\right)$$

for $0 < b < 1/2$. Some of these questions have been approached by Ibragimov and Linnik [IL] and there has been a good deal of activity since then, for both random variables and processes. See, e.g., [DZ, § 3.7].

c. Spectral methods. Many of the applications we analyze may also be examined using spectral methods. For example, the AMS model (Chapter 13) has been investigated by A. Elwalid, D. Mitra, and T. Stern [EMS] among others. Some of the calculations we do are provably equivalent to calculations done on the spectrum; see, e.g., [C1].

d. Calculus of variations methods, optimal control. You can view our approach to probability problems as a method for turning them into problems in the calculus of variations; hence, anything you know about such problems is related to our methods. The type of variational problems arising here also appear in optimal control: of course optimal control and variational problems are themselves inextricably linked, e.g. [Yo, Ce]. In addition, there are several problems in recursive estimation theory, cf. [DK1, DK2], that can be solved via large deviations techniques.

e. Viscosity solutions. Variational problems can often be solved in terms of PDEs (partial differential equations). One of the technical problems that arises is smoothness of solutions. The so-called viscosity solutions turn out to be the correct object (in terms of the degree of smoothness) for many variational problems. Using this concept one can sometimes prove that formal calculations are correct, at least to first order. Barles, Evans, and Souganidis [BES] and Dupuis, Ishii, and Soner [DIS] have used viscosity techniques to prove large deviations principles for certain classes of systems. Reference [DIS] is notable because it proves the principle for the very important case of Jackson networks. In addition, viscosity solutions naturally lead to methods of solving PDEs (and hence variational problems) via successive approximations—a procedure that can facilitate the numerical solution of some large deviations problems.

f. Entropy. Entropy and large deviations are intimately related. We have deliberately avoided this relationship, but others have exploited it to good effect. Ellis [Ell] goes into great detail, proving results on the Ising model among others. All of information theory is based on Chernoff-type estimates; see, for example, Bucklew's book [Bu]. Donsker and Varadhan [DV3] showed how optimal change of measure can be calculated via entropy in a very general Markov process setting. Kullback-Leibler information can be viewed as a large deviations quantity. Again, the reasons we avoid this fruitful subject are lack of time, space, and our choice of applications.

g. Importance sampling. Importance sampling is, in essence, the use of change of measure to improve the accuracy of statistical estimates. It is increasingly important in the simulation of rare events. Our approach to the large deviations lower bound is equivalent to choosing an optimal importance sampling scheme among a class of changed measures. For more details, see [CFM, Bu].

h. Feynman path integrals. The Feynman-Kac formula can be viewed as showing that the exponential martingale we use is indeed a martingale. In other words, our method for proving upper bounds is based on the reasoning behind Feynman path integrals. For a more direct (but, so far as we know, unproven) connection, see Gunther [Gu]. See also Brydges and Maya [BrM].

i. Steepest descent methods. The first large deviations calculations were made by steepest descent methods. It is a natural method, since the transforms (Laplace or Fourier) of sums of independent random variables are simply powers, and steepest descent is then quite accurate for computing the inverse transform.

Chapter 2

General Principles

This chapter contains the definition of the large deviations principle, as well as a few standard introductory results. Varadhan's integral lemma and Sanov's theorem are standard tools of the large deviationist. However, in the remainder of the book we use only the results of §2.1 and, in Chapter 11, of §2.3.

2.1. The Large Deviations Principle

The general theory of large deviations has a beautiful and powerful formulation due to Varadhan, called the "large deviations principle." The axiomatic nature of this formulation makes it very general, with a drawback in that its meaning is not apparent at first glance. With Chernoff's Theorem snugly under our belts, though, we can appreciate the meaning of the "large deviations principle," and then indicate how it may be used in more general frameworks such as Level II large deviations (see §2.4).

The formal statement is given in terms of measures, but do not be alarmed; we shall proceed in steps. First we shall paraphrase the large deviations principle in the language of random variables. Note that we make no substantial assumptions as to the range of these random variables, their distributions, or their dependencies.

Start with a probability space $(\Omega, \mathcal{F})$, and some random vectors $z_1, z_2, \ldots$ with values in some space $\mathcal{X}$. The only requirement on their range we need here is that the definitions below make sense. (They certainly do if $\mathcal{X} = \mathbb{R}^d$, and more general cases include $\mathcal{X} = C[0, T]$, the space of continuous functions. In §2.4 we work with a space of probability measures.) However, to avoid having to keep track of the requirements on $\mathcal{X}$, assume for the rest of this chapter that $\mathcal{X}$ (and $\mathcal{Y}$ below) are complete separable metric spaces (Definition A.18). For a more abstract setting, see [DZ].

Definition 2.1. *A real valued function I on $\mathcal{X}$ is called a "rate function" if it satisfies the following two requirements.*

(i) $I(x) \geq 0$,

(ii) I is lower semicontinuous. That is, if $y_1, y_2, \ldots$ is a sequence so that $y_n \to y$ in $\mathcal{X}$ as $n \to \infty$, then $\liminf_n I(y_n) \geq I(y)$.

(iii) If in addition the set $\{x : I(x) \leq a\}$ is compact for every real a (in short—I has compact level sets), then I is called a "good rate function."

Remarks.

1. We allow $I(x) = +\infty$.

2. Part 2.1 (ii) is equivalent to the requirement that the set $\{x : I(x) \leq a\}$ is closed for every real a (Definition A.28 and Exercise A.29). It means that the values can only jump down, so that I attains its minimum in any compact set. If I is a good rate function, then it attains its minimum in any *closed* set (why?).

3. The terminology "good rate function" is due to Stroock [St]. Note that the distinction between "rate function" and "good rate function" is not standard. In some publications the definition of rate function is taken to imply "goodness."

The function ℓ [defined in (1.4) or (1.10)] is a rate function; it is non-negative [put $\theta = 0$ in (1.4b), (1.10b)] and lower semicontinuous (by a direct application of Fatou's Lemma—Theorem A.93). If $\lim_{|a|\to\infty} \ell(a) = \infty$ then it is a good rate function, for in this case the level sets are bounded, and are closed by lower-semicontinuity.

In Chernoff's Theorem 1.5, 1.10, and in 1.22 we considered the random variables $z_n = n^{-1}(x_1 + \cdots + x_n)$. Keeping this theorem in mind, here is a formal statement of the "large deviations principle."

Definition 2.2. *We say that* $z_1, z_2, \ldots$ *satisfies a large deviations principle with rate function I if, for every closed set $C \subset \mathcal{X}$, we have*

$$\limsup_{n\to\infty} \frac{1}{n} \log \mathbb{P}(z_n \in C) \leq - \inf_{x\in C} I(x) \tag{2.1u}$$

and for every open set $G \subset \mathcal{X}$, we have

$$\liminf_{n\to\infty} \frac{1}{n} \log \mathbb{P}(z_n \in G) \geq - \inf_{x\in G} I(x). \tag{2.1l}$$

Let us see how this relates to Chernoff's Theorem. The bounds (1.6) in Chernoff's Theorem 1.5 (or 1.10) correspond to (2.1), with ℓ playing the role of the I function. It states that (asymptotically) bounds for the probability that z_i is in some set is determined by the minimal value the function I takes on that set. More precisely, the exponential order of decay of this probability is bounded below by the worst point estimate of I on the set. That ℓ is indeed a rate function is also established in Chapter 1.

Thus we can restate Theorems 1.5, 1.10, and 1.22 in terms of Definition 2.2 as follows. If $M(\theta) < \infty$ for θ in some neighborhood of 0, then $z_1, z_2, \ldots$ where $z_n \overset{\triangle}{=} n^{-1}(x_1 + \cdots + x_n)$ satisfies a large deviations principle with rate function ℓ of (1.10).

Here are some immediate consequences of the definitions.

Exercise 2.3. Let I be a rate function satisfying (2.1). Then $\inf_{x\in\mathcal{X}} I(x) = 0$.

Let F be a closed set and define

$$I^*(F) = \inf_{y \in F} I(y) \quad \text{and} \quad \mathcal{O} \overset{\triangle}{=} \{x \in F : I(x) = I^*(F)\}. \qquad (2.2)$$

If I is a good rate function, then $\mathcal{O}$ is not empty, so that the minimum $I^*(F)$ is attained in F. Therefore, if $I(x) > 0$ for all x in F, then $I^*(F) > 0$, so that the probability of this set decays exponentially fast. ♠

Exercise 2.4. There is at most one rate function I satisfying (2.1). That is, the rate function (if it exists) is unique. Hint: assume $I_1(x_0) > I_2(x_0)$ for some point x_0. By lower semicontinuity you can find a closed ball around x_0 and a positive δ so that $I_1(x) > I_2(x_0) - \delta$ for all x in that ball. Now use (2.1). ♠

Why the two separate statements for closed and open sets? Or, why separate upper and lower bounds? It is easy to see that the lower bound cannot hold for a general closed set, since it may be too small. Take a sequence $x_1, x_2, \ldots$ of Bernoulli ($\frac{1}{2}$) random variables (Example 1.15), and let $C \overset{\triangle}{=} \{\sqrt{1/2}\}$. Then $\mathbb{P}(z_n \in C) = \mathbb{P}(x_1 + \cdots + x_n = n/\sqrt{2}) = 0$, so that the left-hand side of (2.11) equals $(-\infty)$, but from (1.9), $\ell\left(\sqrt{1/2}\right) < \infty$! On the other hand, for an open set G,

$$\limsup_{n \to \infty} \frac{1}{n} \log \mathbb{P}(z_n \in G) \leq \limsup_{n \to \infty} \frac{1}{n} \log \mathbb{P}\left(z_n \in \overline{G}\right)$$
$$\leq - \inf_{x \in \overline{G}} I(x).$$

Now, if I is continuous (near the boundary of G), then $\inf_{x \in G} = \inf_{x \in \overline{G}}$, and the upper bound holds for the open set. But in general I may be only lower semicontinuous, and a discontinuity on the boundary of G may invalidate the upper bound. This leads to the following definition.

Definition 2.5. *A set S is called an I-continuity set for a rate function I if*

$$\inf_{x \in S^o} I(x) = \inf_{x \in \overline{S}} I(x).$$

The following is an immediate consequence of Definitions 2.2 and 2.5.

Exercise 2.6. If $z_1, z_2, \ldots$ satisfies a large deviations principle with rate function I and if S is an I-continuity set, then

$$\lim_{n \to \infty} \frac{1}{n} \log \mathbb{P}(z_n \in S) = - \inf_{x \in S} I(x). \qquad ♠$$

In general, the limits may be different, and in general limits may not exist.

Exercise 2.7. Let $x_1, x_2, \ldots$ be i.i.d. Bernoulli ± 1 random variables and define $z_n \overset{\triangle}{=} n^{-1}(x_1 + \cdots + x_n)$. Find a set A that is not an I-continuity set. Find a set

S so that $n^{-1} \log \mathbb{P}(z_n \in S)$ possesses no limit. Hint: for A try an interval, for S a single point.   ♠

The large deviations principle and the variational formulation imply something about the way rare events happen. If the variational problem has a unique optimal point x^*, then a point far from x^* cannot be optimal. Therefore, rare events happen by z_n staying close to x^*. Here is a precise statement.

Lemma 2.8. *Assume I is a good rate function on a metric space $(\mathcal{X}, d)$. Let F be a closed set and define $I^*(F)$ and $\mathcal{O}$ by (2.2). Then for each ε, there is a $\delta > 0$ so that $x \in F$ and $d(x^*, x) \geq \varepsilon$ for all x^* in $\mathcal{O}$ implies $I(x) \geq I^*(F) + \delta$. If $z_1, z_2, \ldots$ satisfies a large deviations principle with rate function I and if F is an I-continuity set, then*

$$\mathbb{P}\left(d(z_n, x^*) \geq \varepsilon \text{ for all } x^* \in \mathcal{O} \ \middle| \ z_n \in F \right) \leq e^{-n\delta + o(n)}.$$

Proof. Assume first that $\mathcal{O}$ contains exactly one point x^*. For a proof by contradiction, assume that there are points x_i in F with $I(x_i) \leq I(x^*) + 2^{-i}$ but $d(x^*, x) \geq \varepsilon$. Since the set $F \cap \{x : I(x) \leq I(x^*) + 1\}$ is compact, we can take a convergent subsequence of $\{x_i\}$ with limit x. Then $x \in F$ since F is closed, $d(x, x^*) \geq \varepsilon$ so that $x \neq x^*$, but, by lower semicontinuity, $I(x) \leq I(x^*)$, contradicting the definition of $\mathcal{O}$.

To establish the second claim note that from Bayes' Rule,

$$\begin{aligned}
\mathbb{P}\left(d(z_n, x^*) \geq \varepsilon \ \middle| \ z_n \in F \right) &\leq \frac{\mathbb{P}(d(z_n, x^*) \geq \varepsilon)}{\mathbb{P}(z_n \in F)} \\
&\leq e^{-n(I(x^*)+\delta)+o(n)} e^{-nI(x^*)+o(n)} \qquad (2.3) \\
&= e^{-n\delta + o(n)}.
\end{aligned}$$

The second inequality in (2.3) comes from the large deviations upper bound applied to the numerator, and the large deviations lower bound and Exercise 2.6 applied to the denominator. The proof for the case when $\mathcal{O}$ is not a single point is relegated to the exercise below.   ■

Exercise 2.9. Extend the proof above to the case that $\mathcal{O}$ contains more than one point.   ♠

We now give a formal statement of the large deviations principle in its standard, abstract form. We are given a sequence of probability spaces $(\mathcal{X}, \mathcal{F}, \{P_n\})$, where the P_n are all measures on $(\mathcal{X}, \mathcal{F})$. In terms of Definition 2.2, P_n is the "distribution" of z_n. We are also given a rate function I on $\mathcal{X}$.

Definition 2.10. *We say $\{P_n\}$ satisfies a large deviations principle with rate function I if, for every closed set $C \subset \mathcal{X}$, we have*

$$\limsup_{n \to \infty} \frac{1}{n} \log P_n(C) \leq - \inf_{x \in C} I(x), \qquad (2.4u)$$

and for every open set $G \subset \mathcal{X}$, we have

$$\liminf_{n\to\infty} \frac{1}{n} \log P_n(G) \geq - \inf_{x\in G} I(x). \qquad (2.4l)$$

This reduces easily to Definition 2.2 if $\mathcal{X} = \mathbb{R}^d$ (cf. Definition A.81) and $P_n(A) \overset{\triangle}{=} \mathbb{P}(z_n \in A)$. The advantage of the not-too-intuitive, abstract formulation will become clear gradually, starting with the next section. We conclude this section with another common formulation of the large deviations upper and lower bounds. Implicit in our previous formulation was the fact that $\mathcal{F}$, the σ-field on the probability space, contains a topology (so that the notion of lower semicontinuity can be defined). This is not always possible (although it is in our case)—and these considerations lead to the more general formulation below. For details and a proof of Lemma 2.11, see [DZ].

Lemma 2.11. *Let I be a rate function: then the upper and lower bounds in (2.1), (2.4) are equivalent to the following. For any (measurable) set S,*

(i) Upper bound: if

$$\overline{S} \cap \{x : I(x) \leq \alpha\} = \emptyset,$$

then

$$\limsup_{n\to\infty} \frac{1}{n} \log \mathbb{P}(z_n \in S) \leq -\alpha.$$

(ii) Lower bound: if $x \in S^o$, then

$$\liminf_{n\to\infty} \frac{1}{n} \log \mathbb{P}(z_n \in S) \geq -I(x).$$

Therefore, (i)–(ii) is equivalent to Definition 2.2.

The sets $\{x : I(x) \leq \alpha\}$ are the *level sets* of I.

2.2. Varadhan's Integral Lemma

The following theorem, due to Varadhan, illustrates the power of the large deviations principle. Indeed, this theorem (really a corollary of Definitions 2.2 and 2.10) is one of the main tools in the theory of large deviations.

Theorem 2.12. *Suppose that $z_1, z_2, \ldots$ satisfy the large deviations principle with rate function I. Then for any bounded continuous function g on $\mathcal{X}$,*

$$\lim_{n\to\infty} \frac{1}{n} \log \mathbb{E}\left(e^{ng(z_n)}\right) = \sup_x (g(x) - I(x)). \qquad (2.5)$$

Proof. We show that the right-hand side of (2.5) is both an upper bound and a lower bound for the left-hand side. Let P_n denote the probability measure on $\mathcal{X}$ associated with z_n (the distribution of z_n); we shall invoke either of Definitions 2.2 or 2.10 as convenient. First, we establish that (2.5) holds with $\leq$ replacing equality. For any fixed $\delta > 0$, we can find closed sets $C_1, \ldots, C_k$ so that

$\sup_{x \in C_i} g(x) - \inf_{x \in C_i} g(x) < \delta$ for $i = 1, \ldots, k$, and such that $\cup_{i=1}^{k} C_i = \mathcal{X}$.
Then

$$\mathbb{E}\left(e^{ng(z_n)}\right) \leq \sum_{i=1}^{k} \int_{C_i} e^{ng(x)} d\mathbb{P}_n(x)$$

$$\leq \sum_{i=1}^{k} \exp\left(n \sup_{x \in C_i} g(x)\right) \mathbb{P}(z_n \in C_i)$$

$$\leq k \max_{1 \leq i \leq k} \exp\left(n \sup_{x \in C_i} g(x)\right) \mathbb{P}(z_n \in C_i)$$

$$= k \max_{1 \leq i \leq k} \exp\left(n\left(\sup_{x \in C_i} g(x) + \frac{1}{n} \log \mathbb{P}(z_n \in C_i)\right)\right).$$

But since $\sup_{x \in C_i} g(x) \leq \inf_{C_i} g(x) + \delta$, we have from Definition 2.2

$$\limsup_{n \to \infty} \frac{1}{n} \log \mathbb{E} e^{ng(z_n)}$$

$$\leq \limsup_{n \to \infty} \left(\frac{\log k}{n} + \max_{1 \leq i \leq k}\left(\sup_{x \in C_i} g(x) + \frac{1}{n} \log \mathbb{P}(z_n \in C_i)\right)\right)$$

$$\leq \max_{1 \leq i \leq k}\left(\sup_{x \in C_i} g(x) - \inf_{x \in C_i} I(x)\right)$$

$$\leq \max_{1 \leq i \leq k}\left(g(x_i) + \sup_{x \in C_i}(-I(x))\right) + \delta$$

for any points x_i in C_i. But then

$$\limsup_{n \to \infty} \frac{1}{n} \log \mathbb{E} e^{ng(z_n)} \leq \max_{1 \leq i \leq k} \sup_{x \in C_i} (g(x) - I(x)) + \delta$$

$$= \sup_{x \in \mathcal{X}} (g(x) - I(x)) + \delta.$$

Since δ was arbitrary, this proves the upper bound.

To obtain the result it remains to establish that (2.5) holds with $\geq$ replacing the equality. This is even easier: let y be a point such that $g(y) - I(y) > \sup(g(x) - I(x)) - \delta$, and let U be an open neighborhood of y. We can take U small enough so that $\sup_{x \in U} |g(x) - g(y)| < \delta$ so, invoking Definition 2.10,

$$\liminf_{n \to \infty} \frac{1}{n} \log \mathbb{E}\left(e^{ng(z_n)}\right) \geq \liminf_{n \to \infty} \frac{1}{n} \log e^{n(g(y)-\delta)} \mathbb{P}_n(U)$$

$$\geq g(y) - \delta - \inf_{x \in U} I(x)$$

$$\geq \sup_{x \in U} g(x) + \sup_{x \in U}\left(-I(x)\right) - 2\delta$$

$$\geq \sup_{x \in U}(g(x) - I(x)) - 2\delta$$

and, since δ is arbitrary, the lower bound follows. ∎

This theorem can be used without modification to examine the second example of §1.1, namely the product (1.2) of random variables. Consider the sequence $x_1, x_2, \ldots$ of i.i.d. random variables from that example, and let

$$z_n = \frac{1}{n} \sum_{i=1}^{n} \log x_i. \tag{2.6}$$

Then $\prod_1^n x_i = e^{nz_n}$, and (since we were assuming that $1 \leq x_i \leq 2$) we have that $z_1, z_2, \ldots$ is bounded. Hence Theorem 2.12 states that $\lim_{n \to \infty} \frac{1}{n} \log \mathbb{E} \prod_1^n x_i = \sup_x (x - I(x))$. Chernoff's Theorem (for the random variables $\log x_i$) tells us that $I(a) = \sup_\theta (\theta a - \log \mathbb{E}(e^{\theta \log x_1}))$ (where the expectation is with respect to the variables x_i). Let us now complete the calculation: from (C.6), $\frac{d}{dx} I(x) = \theta^*(x)$. Therefore $\theta^* = 1$ at any extremal point x^* of $x - I(x)$, so that

$$\sup_x(x - I(x)) = x^* - \left(x^* \cdot 1 - \log \mathbb{E}e^{1 \cdot \log x_1}\right)$$

$$= x^* - (x^* - \log \mathbb{E}x_1)$$

$$= \log \mathbb{E}(x_1).$$

This proves the remarkable fact that

$$\lim_{n \to \infty} \frac{1}{n} \log \mathbb{E} \prod_1^n x_i = \log \mathbb{E}(x).$$

Well, even though we had to go on a long journey to get to this elementary result, we hope you found it worthwhile.

2.3. The Contraction Principle

Suppose $x_1, x_2, \ldots$ is a sequence of random variables with values in a space $\mathcal{X}$. For a continuous function f on $\mathcal{X}$ with values in a space $\mathcal{Y}$, let $y_n = f(x_n)$. We may be interested in a large deviations principle for $y_1, y_2, \ldots$, when such a principle is easier to derive for $x_1, x_2, \ldots$. For example, if $x_1, x_2, \ldots$ is a Markov process, in general $y_1, y_2, \ldots$ may not be. The contraction principle provides a tool to establish the large deviations principle and compute the rate function of $y_1, y_2, \ldots$ from those of $x_1, x_2, \ldots$. Define

$$I'(y) \triangleq \begin{cases} \inf\{I(x) : x \in \mathcal{X}, \ y = f(x)\} & \text{if } y = f(x) \text{ for some } x; \\ \infty & \text{otherwise.} \end{cases} \quad (2.7)$$

In view of our applications, assume that $\mathcal{X}$ and $\mathcal{Y}$ are some metric spaces (although the result holds for Hausdorff topological spaces [DZ §4.2.2]).

Theorem 2.13. *Let f be a continuous function.*

 (i) *If I is a good rate function, then I' is a good rate function.*

 (ii) *If $x_1, x_2, \ldots$ satisfies a large deviations principle with a good rate function I, then $y_1, y_2, \ldots$ satisfies a large deviations principle with the good rate function I'.*

Proof. I' is clearly non-negative. Due to the remark following Definition 2.1, to establish (i) it suffices to show that for any arbitrary α, the set $\{y : I'(y) \le \alpha\}$ is compact. Since I is a good rate function, the set $\{x : I(x) \le \alpha\}$ is compact and since f is continuous, the image of that set,

$$f(\{x : I(x) \le \alpha\}) \triangleq \{f(x) : x \in \mathcal{X}, \ I(x) \le \alpha\}, \quad (2.8)$$

is compact (Theorem A.27). Now

$$\{f(x) : x \in \mathcal{X}, \ I(x) \le \alpha\} = \{y : y = f(x), \ x \in \mathcal{X} \text{ and } I(x) \le \alpha\}.$$

But by Definition A.25, the set $\{x : f(x) = y\}$ is closed since f is continuous. Therefore

$$\{x : I(x) \le \alpha\} \cap \{x : f(x) = y\}$$

is compact for every α and y. Since I is lower semicontinuous, by Theorem A.31 it achieves its infimum on that compact set. Thus

$$\{y : y = f(x) \text{ and } I(x) \le \alpha, \ \text{some } x \in \mathcal{X}\}$$
$$= \{y : \inf\{I(x) : y = f(x), \ x \in \mathcal{X}\} \le \alpha\}$$
$$= \{y : I'(y) \le \alpha\}$$

is compact, and (i) is established.

To establish (ii), let S be a set in $\mathcal{Y}$. By definition,

$$\inf\{I'(y) : y \in S\} = \inf\{I(x) : f(x) \in S\}.$$

Since f is continuous, by Definition A.25 $\{x : f(x) \in S\}$ is open (closed) if S is open (closed), and (ii) now follows from the definition of a large deviations principle.
∎

Note that to apply this result we need a good rate function: for example, take the rate function $I \equiv 0$ and the continuous function $f(x) = e^x$. Then I' is not a rate function.

There are various refinements and extensions to this result: see, e.g., [AR, DZ, FW, Va]. They relax the continuity assumption, and generalize to more abstract spaces.

The contraction principle serves to "bring down" a large deviations principle from a more complicated space to a simpler one. It is interesting that it is also possible to obtain the converse: for example, a large deviations principle for infinite-dimensional spaces can be obtained by establishing the principle for every finite-dimensional projection, together with some "compactness." This is the method of projective limits; see [DZ, DeS].

2.4. Empirical Measures: Sanov's Theorem

Chernoff's Theorem is very useful in telling us how often certain rare events occur. Sanov's Theorem tells us something amazing, and possibly more useful: it tells us **how** these events occur on the occasions when they do. It turns out that with overwhelming probability, rare events happen only one way. This is a sort of strong law of large numbers for rare events. In some situations this is intuitive. Suppose we flip a fair coin 10,000 times and, oddly enough, obtain 8,000 heads. How did this occur? It might have happened by having 2,000 tails in a row, then 8,000 heads; it might have happened by having 4,000 heads, then 2,000 tails, then 4,000 heads—there are many ways it could have happened. However, it should be reasonably clear that the most likely way it happened is for the coin to act as if its probability of coming up heads was .8 all along. In the first 100 tosses we should have seen about 80 heads, and in the next 100 tosses also. This can be made very precise in a number of ways, and Sanov's Theorem is one of the ways. In general, if an odd event happens, then the way it happens is for a "conspiracy" to take place, and for all the observations to behave as if they were samples from a "tilted" distribution.

There is a very close relationship between this theorem and the lower bound of Chernoff's Theorem. In the lower bound, we saw that the probability of a rare event can be estimated by comparing a process with the tilted process. Since estimates obtained this way are tight, we might suspect that something such as Sanov's Theorem should hold.

This section provides a first taste of "level 2 large deviations." Its main purpose is to introduce the concept of large deviations of empirical measures. We shall present some notation, develop a feeling for such results, and even argue that they might be useful. The general formulation of level 2 results is deferred until we study Markov chains in Chapters 7 and 8.

Let $x_1, x_2, \ldots$ be i.i.d. random variables with distribution function F and distribution μ [i.e., $F(y) = \mu((-\infty, y])$]. Let F_n denote the empirical distribution of the sample mean, and μ_n the corresponding empirical measure (Definition A.82).

For each y, the law of large numbers (applied to the i.i.d. random variables $\mathbf{1}[x_i \leq y]$) yields $F_n(y) \to F(y)$ as $n \to \infty$. (In fact, results on the Kolmogorov-Smirnov statistics show [Br §13.6] that $F_n(y) - F(y) = O(1/\sqrt{n})$ for all y.) Now

$$\frac{x_1 + \cdots + x_n}{n} = \int y \, dF_n(y) = \int y \, d\mu_n(y) \, ,$$

i.e., the empirical mean is the mean of the empirical distribution (to see this, multiply both sides by n and use induction on n). Define the set S_a of probability measures ν on $\mathbb{R}$ (distributions) by

$$S_a \overset{\triangle}{=} \left\{ \nu : \int y \, d\nu(y) \geq a \right\}. \tag{2.9}$$

So, we can now make a simple statement look complicated:

$$\mathbb{P}(x_1 + \cdots + x_n \geq na) = \mathbb{P}(\mu_n \in S_a). \tag{2.10}$$

The reason for doing so is that the latter formulation is much more general. For example, assume that we have some measure d of distance between probability measures (let's stay intuitive for a while; it is perhaps too early to consult §A.3). Then we can ask how likely it is that $\mu_n \approx \nu$, or try to estimate $\mathbb{P}(d(\mu_n, \nu) < \varepsilon)$ for some fixed ν. It turns out that the answer can be given "explicitly" in terms of the Radon-Nikodym derivatives $d\nu/d\mu$ (see Theorem A.117):

$$\lim_{\varepsilon \downarrow 0} \lim_{n \to \infty} \frac{1}{n} \log \mathbb{P}(d(\mu_n, \nu) < \varepsilon) = -\int \log\left(\frac{d\nu}{d\mu}(y)\right) d\nu(y). \tag{2.11}$$

Moreover, we will show that for appropriate sets S of measures, we obtain

$$\lim_{n \to \infty} \frac{1}{n} \log \mathbb{P}(\mu_n \in S) = -\inf_{\nu \in S} \int \log\left(\frac{d\nu}{d\mu}(y)\right) d\nu(y). \tag{2.12}$$

If this reminds you of (1.11), it is no accident; as we shall illustrate below, the integral indeed plays the role of ℓ in (1.4). Sanov's Theorem formalizes (2.12). Consider the space $\mathcal{P}$ of distributions on $\mathbb{R}$ (or the associated probability measures; see Theorem A.90). Define

$$I(\nu) \overset{\triangle}{=} \begin{cases} \int \log\left(\frac{d\nu}{d\mu}(y)\right) d\nu(y) & \text{if the integral is well defined;} \\ \infty & \text{otherwise.} \end{cases} \tag{2.13}$$

Note that I depends on μ. In the terminology of information theory, I is usually denoted as $H(\nu \mid \mu)$ and is called the (conditional) entropy of ν relative to μ.

Theorem 2.14 (Sanov) [Sa]. *Consider the sequence* $\mu_1, \mu_2, \ldots$ *. For every closed set* $C \subset \mathcal{P}$ *we have*

$$\limsup_{n \to \infty} \frac{1}{n} \log \mathbb{P}(\mu_n \in C) \leq - \inf_{v \in C} I(v) \qquad (2.14u)$$

and for every open set $G \subset \mathcal{P}$ *we have*

$$\liminf_{n \to \infty} \frac{1}{n} \log \mathbb{P}(\mu_n \in G) \geq - \inf_{v \in G} I(v). \qquad (2.14l)$$

For this to become a precise statement we need to define what we mean by closed or open sets in the space of measures. This will be done in §A.1 and A.3. In this section we shall restrict our attention to multinomial distributions. In this case a simple and intuitive notion of distance of measures is available, so that we can carry the calculations through. Moreover, these distributions (when scaled) are dense, in some sense, in the space of distributions, so that we expect the results and formulae we derive to hold in general. Indeed, this was the way Sanov originally proceeded. We shall proceed to illustrate (2.11), and develop intuition by demonstrating the relation between Sanov's Theorem and the large deviation results in $\mathbb{R}^d$.

So, suppose $x_1, x_2, \ldots$ is a sequence of i.i.d. random variables, with

$$\mathbb{P}(x_i = j) = p_j > 0, \quad \sum_{j=1}^{d} p_j = 1, \qquad 1 \leq j \leq d < \infty.$$

Let $\{\vec{u}_1, \ldots, \vec{u}_d\}$ denote the standard basis for $\mathbb{R}^d$; that is,

$$\vec{u}_j \overset{\triangle}{=} (0, 0, \ldots, 0, 1, 0, \ldots, 0) \in \mathbb{R}^d, \qquad (2.15)$$

where the one is in the jth position. We can now represent our random variables by random vectors in $\mathbb{R}^d$ as follows. Define

$$\vec{y}_i \overset{\triangle}{=} \sum_{j=1}^{d} \vec{u}_j \mathbf{1}[x_i = j] \; ;$$

that is, if $x_i = j$, then $\vec{y}_i$ has all components zero, except for a one in the jth component. Note that $\frac{1}{n}(\vec{y}_1 + \cdots + \vec{y}_n)$ is a representation of the empirical measure of $x_1, x_2, \ldots$ (cf. Definition A.82). Now $\vec{y}_1, \vec{y}_2, \ldots$ is clearly a sequence of i.i.d. random vectors, with $\mathbb{E}\vec{y}_1 = p \overset{\triangle}{=} (p_1, \ldots, p_d)$, so that the strong law of large numbers implies

$$\mathbb{P}\left(\lim_{n \to \infty} \frac{1}{n} \sum_{i=1}^{n} y_i = p \right) = 1.$$

So, the large deviations question would be: given a distribution $\vec{q}$ on $(1, \ldots, d)$, i.e., a vector $(q_1, \ldots, q_d)$ with positive components and with $q_1 + \ldots + q_d = 1$, what is

$$\lim_{\varepsilon \downarrow 0} \lim_{n \to \infty} \frac{1}{n} \log \mathbb{P} \left(\left| \frac{1}{n} \sum_{i=1}^{n} y_i - \vec{q} \right| < \varepsilon \right) ?$$

From Theorem 1.22 for vectors in $\mathbb{R}^d$ we would expect that

$$\mathbb{P} \left(\left| \frac{1}{n} \sum_{i=1}^{n} y_i - \vec{q} \right| < \varepsilon \right) \approx e^{-nJ(q)} \tag{2.16}$$

for all ε small enough, and that, moreover, $J(\vec{q}) > 0$ if and only if $\vec{p} \neq \vec{q}$. Let us show this and identify $J(\vec{q})$ directly; we shall later see that, as mentioned above, Sanov's result (2.12) and Theorem 1.22 are equivalent. It will be convenient to use the definition $|\vec{y}| = \sup_i |y_i|$ for $\vec{y}$ in $\mathbb{R}^d$.

Note that the value of each component of $\frac{1}{n} \sum_{i=1}^{n} y_i$ is a multiple of $1/n$. Assume first that the components q_j are multiples of $1/n$. Then

$$\mathbb{P} \left(\frac{1}{n} \sum_{i=1}^{n} y_i = (q_1, \ldots, q_d) \right) = \frac{n!}{(nq_1)!(nq_2)! \cdots (nq_d)!} p_1^{nq_1} p_2^{nq_2} \cdots p_d^{nq_d},$$

and using Stirling's formula

$$\mathbb{P} \left(\frac{1}{n} \sum_{i=1}^{n} y_i = (q_1, \ldots, q_d) \right) \approx \left(\sqrt{2\pi n} \right)^{(1-d)} \left(\prod_{i=1}^{d} \sqrt{q_i} \right)^{-1}$$

$$\times \left(\frac{ne^{-1}}{(nq_1)^{q_1}(nq_2)^{q_2} \cdots (nq_d)^{q_d} e^{-q_1} e^{-q_2} \cdots e^{-q_d}} p_1^{q_1} p_2^{q_2} \cdots p_d^{q_d} \right)^n.$$

But since $q_1 + \ldots + q_d = 1$, we obtain

$$\mathbb{P} \left(\frac{1}{n} \sum_{i=1}^{n} y_i = q \right) \approx \left(\left(\frac{p_1}{q_1} \right)^{q_1} \left(\frac{p_2}{q_2} \right)^{q_2} \cdots \left(\frac{p_d}{q_d} \right)^{q_d} \right)^n$$

$$= \exp \left(-n \sum_{i=1}^{d} q_i \log \frac{q_i}{p_i} \right).$$

Since $q_1 \log \frac{q_1}{p_1} + \ldots + q_d \log \frac{q_d}{p_d}$ is a continuous function of $\vec{q}$, if $\vec{y}$ is close to $\vec{q}$ and its components are multiples of $1/n$, we can approximate $\mathbb{P} \left(\frac{1}{n} \sum_{i=1}^{n} y_i = \vec{y} \right)$ by $\mathbb{P} \left(\frac{1}{n} \sum_{i=1}^{n} y_i = \vec{q} \right)$. Now the number of points $\vec{y}$ that are possible values of $\frac{1}{n} \sum_{i=1}^{n} y_i$ with $|\vec{y} - \vec{q}| < \varepsilon$ is at most $(2\varepsilon n)^d$ (and at least one, for n large enough). So, summing over the $(2\varepsilon n)^d = \exp(d \log(\varepsilon n) + d \log 2)$ possible events,

$$\mathbb{P} \left(\left| \frac{1}{n} \sum_{i=1}^{n} y_i - \vec{q} \right| < \varepsilon \right) \approx \exp \left(-n \sum_{j=1}^{d} q_j \log \frac{q_j}{p_j} + d \, O(\log \varepsilon n) + n O(\varepsilon) \right).$$

The first error term is due to the number of points in the ε-box, and the second to the approximation of $q_j \log(q_j/p_j)$ by their values at the center of that box. This establishes (2.16) and identifies I for the set under consideration.

Sanov's Theorem seems more powerful than Chernoff's Theorem, since it gives information about the entire empirical distribution, not just its mean. However, let us show that for multinomial random variables they are equivalent. To derive Sanov's Theorem, i.e., the form (2.13) of I, recall that from Theorem 1.22,

$$\frac{1}{n} \log \mathbb{P} \left(\frac{1}{n} \sum_{i=1}^{n} y_i \approx q \right) \approx \sup_{\theta} (\langle \theta, q \rangle - \log M(\theta))$$

where in this case,

$$M(\theta) = \sum_{j=1}^{d} e^{\theta_j} p_j.$$

Differentiating to obtain the maximizer, we obtain $q_j = M(\theta)^{-1} p_j e^{\theta_j}$ so that

$$\theta_j^* = \log M(\theta^*) + \log \frac{q_j}{p_j}.$$

Hence Chernoff's Theorem implies

$$\frac{1}{n} \mathbb{P} \left(\frac{1}{n} \sum_{i=1}^{n} y_i \approx \vec{q} \right) \approx - \left(\sum_{j=1}^{d} q_j \left(\log \frac{q_j}{p_j} + \log M(\theta^*) \right) - \log M(\theta^*) \right)$$

$$= - \sum_{j=1}^{d} q_j \log \frac{q_j}{p_j},$$

which is Sanov's Theorem. This works since the x_is have finite range; otherwise we would have an infinite-dimensional space to consider, and Chernoff's Theorem cannot be invoked.

The derivation of Chernoff's Theorem from Sanov's Theorem is quite general, and applies also for abstract versions of Chernoff's Theorem. With the notation in (2.10) and (2.12) Sanov's Theorem implies that

$$\mathbb{P} \left(\mu_n \in S_a \right) \approx \exp \left(-n \inf_{v \in S_a} \int \log \left(\frac{dv}{d\mu}(x) \right) dv(x) \right),$$

where μ is the distribution of x_i. We will show that the right-hand side equals $\exp(-n\ell(a))$, with ℓ defined in (1.4), which is Chernoff's Theorem. Assume for definiteness that $a > \mathbb{E}x_1$ and write $v(x) \overset{\triangle}{=} \frac{dv}{d\mu}(x)$. Then

$$\inf_{v \in S_a} \int \log \left(\frac{dv}{d\mu}(\vec{y}) \right) dv(\vec{y})$$

$$= \inf \left\{ \int v(x) \log v(x) \, d\mu(x) \; : \; v(\vec{y}) \geq 0, \right.$$

$$\left. \int v(x) \, d\mu(x) = 1, \; \int x v(x) \, d\mu(x) \geq a \right\}. \tag{2.17}$$

The first constraint simply means that v is a probability measure, so that it integrates to one, while the second constraint is that under v, the mean is larger than a. Note that I is non-negative, and equals zero only if $v = \mu$, since by Jensen's inequality and the convexity of the function $x \log x$,

$$\int \log \left(\frac{dv}{d\mu} \right) \frac{dv}{d\mu} \, d\mu \geq \log \left(\int \frac{dv}{d\mu} \, d\mu \right) \cdot \left(\int \frac{dv}{d\mu} \, d\mu \right) = \log 1 \cdot 1 = 0$$

with equality only if $dv/d\mu \equiv 1$. Alternatively, if $d\mu/dv$ exists, then since the function $(-\log \alpha)$ is convex,

$$\int \log \frac{dv}{d\mu} \, dv = \int -\log \frac{d\mu}{dv} \, dv \geq -\log \int \frac{d\mu}{dv} \, dv = 0.$$

Fix any v satisfying the constraints in (2.17), but with $\int x v(x) \, d\mu(x) > a$. Our first step is to show that v cannot be a minimizer in (2.17), and that lower values for I are obtained by modifying v to get equality. This implies that the last inequality in (2.17) can be replaced by an equality. Note that the function $v_1 \equiv 1$ satisfies the first two constraints, but not the third, since by assumption,

$$\int x v_1(x) \, d\mu(x) = \mathbb{E} x_1 < a.$$

Hence, for any $0 < \alpha < 1$, the function $v_\alpha(x) \stackrel{\triangle}{=} \alpha v_1(x) + (1 - \alpha) v(x)$ satisfies the first two constraints. Now the function $x \log x$ is convex in x (for $x \geq 0$), so that $I(\cdot)$ expressed through (2.17) is convex in v. By this convexity,

$$\int \Big(\alpha v_1(x) + (1 - \alpha) v(x) \Big) \log \Big(\alpha v_1(x) + (1 - \alpha) v(x) \Big) \, d\mu(x)$$

$$\leq \alpha \int v_1(x) \log v_1(x) \, d\mu(x) + (1 - \alpha) \int v(x) \log v(x) \, d\mu(x)$$

$$= (1 - \alpha) \int v(x) \log v(x) \, d\mu(x)$$

$$< \int v(x) \log v(x) \, d\mu(x),$$

where the second relation holds since $\log v_1(x) \equiv 0$ and the last since the last expression is positive. Thus $I(v_\alpha) < I(v)$. On the other hand, $\mathbb{E} x_1 < a <$

$\int xv(x)\, d\mu(x)$, so that

$$\int x\Big(\alpha v_1(x) + (1-\alpha)v(x)\Big)\, d\mu(x) = \alpha\mathbb{E}x_1 + (1-\alpha)\int xv(x)\, d\mu(x)$$

$$< \int xv(x)\, d\mu(x).$$

Thus we can choose $\alpha > 0$ so that v_α satisfies the constraints. The conclusion is that by changing v so that the last constraint becomes an equality, we can improve our minimization, and our first step is concluded.

To deal with our constrained minimization problem, introduce the Lagrange multipliers λ_1, λ_2. (For the idea of Lagrange multipliers, see [Ew pp. 111, 116 and 123]; but note that this reference considers only piecewise smooth functions. For an application with proof, see §13.6.) We then look for extremals of

$$\int v(x)\log v(x)\, d\mu(x) + \lambda_1\left(\int v(x)\, d\mu(x) - 1\right) + \lambda_2\left(\int xv(x)\, d\mu(x) - a\right).$$

To that end, fix a function $\delta(x)$, change v to $v + \varepsilon\delta$, and differentiate with respect to ε at $\varepsilon = 0$, to obtain

$$\int \Big(\delta(x)\log v(x) + (1+\lambda_1)\delta(x) + \lambda_2 x\delta(x)\Big)\, d\mu$$

$$= \int \Big(\log v(x) + (1+\lambda_1) + \lambda_2 x\Big)\delta(x)\, d\mu$$

$$= 0$$

over all positive functions δ. But this implies

$$\log v(x) + 1 + \lambda_1 + \lambda_2 x = 0$$

or $v(x) = ce^{\theta x}$ for some c, θ (i.e., the change of measure is exponential!). The constraints imply that

$$c\int e^{\theta x}\, d\mu(x) = c\mathbb{E}e^{\theta x_1} = 1,$$

$$c\int xe^{\theta x}\, d\mu(x) = c\mathbb{E}x_1 e^{\theta x_1} = a$$

$$(2.18)$$

so that for such v we have

$$\int \log\left(\frac{dv}{d\mu}(x)\right)\, dv(x) = \int v(x)\log v(x)\, d\mu(x)$$

$$= c\int e^{\theta x}(\theta x + \log c)\, d\mu$$

$$= a\theta - \log\int e^{\theta x}\, d\mu$$

$$= a\theta - \log M(\theta).$$

We conclude that any ν minimizing $I(\nu)$ over S_a must satisfy $I(\nu) = a\theta - \log M(\theta)$. However, the constraints also determine θ, for writing (2.18) in the notation of the Chernoff bounds, we have

$$cM(\theta) = 1, \qquad c\frac{d}{d\theta}M(\theta) = a.$$

But this means that θ must satisfy

$$\frac{d}{d\theta}(\log M(\theta) - a\theta) = 0.$$

But this θ is exactly the one that appears in the Chernoff estimate

$$\sup_{\theta}(a\theta - \log M(\theta)),$$

and the result is established.   ∎

Chapter 3

Random Walks, Branching Processes

In this chapter we describe several processes, starting with the standard and simple and proceeding to the less so. We introduce large deviations for processes. The results of Chapter 1 are applied to analyze these processes. In particular, a slight modification of the proof of the upper bound in Chernoff's Theorem is used. Applications of the current results are presented in Chapter 10. The results of this chapter are needed only for Chapter 10; however, they serve as a simple introduction to large deviations for processes. The general derivation we provide in Chapter 5 is far more involved.

A random walk, also called a drunkard's walk, denotes the position of a drunk person in a long narrow corridor. Starting from position x_0 he takes a step of size x_t at integer time t. The step sizes are real valued and i.i.d. with distribution function F. Fix some x_0 which is assumed independent of $\{x_1, x_2, \ldots\}$. If $\mathbb{E}x_1$ is finite, then the law of large numbers implies that

$$\frac{x_0 + x_1 + \cdots + x_n}{n} \to \mathbb{E}x_1$$

with probability one as $n \to \infty$, so that on the average our drunkard in fact makes progress, with speed $\mathbb{E}x_1$. Large deviations will help us compute the probability that he arrives at an unlikely position, i.e.,

$$\mathbb{P}\left(x_0 + x_1 + \cdots + x_n \geq na\right)$$

for some $a > \mathbb{E}x_1$. Theorem 1.5 gives estimates of this and related quantities. Unlike previous results, we shall ask questions about the deviations of his whole path from its mean. Specifically, we will compute

$$\mathbb{P}(x_0 + x_1 + \ldots + x_j \geq ja \quad \text{for all} \quad 0 \leq j \leq n).$$

This is a question about the behavior of the *process* $x_1, x_2, \ldots$, not just the sample mean. For this restricted problem, it turns out that Theorem 1.5 together with an interesting observation suffice to prove this "process level" result. In general, we will need to develop more delicate machinery. But first let us address this specific problem.

3.1. The Ballot Theorem

Fix n and define the function $(i)_n$ (i mod n) to have values between one and n:
$$(i)_n = i \bmod n \overset{\triangle}{=} i - n\lfloor (i-1)/n \rfloor.$$

Lemma 3.1. *Suppose the real numbers $r_1, \ldots, r_n$ satisfy $r_1 + \cdots + r_n \geq n\beta$. Let*

$$i^* \overset{\triangle}{=} \arg\inf\{r_1 + \ldots + r_i - i\beta \ : \ 1 \leq i \leq n\}.$$

In other words, i^* is an integer such that the minimum is attained. Then

$$r_{(i^*+1)_n} + \cdots + r_{(i^*+j)_n} \geq j\beta \quad \text{for all} \quad 1 \leq j \leq n.$$

Exercise 3.2. Prove Lemma 3.1. Hint: draw a picture.   ♠

Now let $x_1, x_2, \ldots$ be a sequence of i.i.d. random variables.

Theorem 3.3. *If $a > \mathbb{E}x_1$, then*

$$\mathbb{P}(x_1 + \ldots + x_j \geq ja \quad \text{for all} \quad 1 \leq j \leq n) = e^{-n\ell(a)+o(n)} \tag{3.1}$$

if and only if
$$\mathbb{P}(x_1 + \cdots + x_n \geq na) = e^{-n\ell(a)+o(n)}. \tag{3.2}$$

Proof. Clearly,

$$\mathbb{P}(x_1 + \ldots + x_j \geq ja \quad \text{for all} \quad 1 \leq j \leq n) \leq \mathbb{P}(x_1 + \cdots + x_n \geq na). \tag{3.3}$$

But Lemma 3.1 tells us that if $x_1 + \cdots + x_n \geq na$, then for at least one (random) i^* we have

$$x_{(i^*+1)_n} + \ldots + x_{(i^*+j)_n} \geq ja \quad \text{for all} \quad 1 \leq j \leq n.$$

So

$$\mathbb{P}\left(x_1 + \cdots + x_n \geq na\right)$$
$$= \mathbb{P}\left(x_{(i^*+1)_n} + \ldots + x_{(i^*+j)_n} \geq ja \quad \text{for all} \quad 1 \leq j \leq n\right)$$
$$= \sum_{k=1}^{n} \mathbb{P}\left(x_{(i^*+1)_n} + \ldots + x_{(i^*+j)_n} \geq ja \quad \text{for all} \quad 1 \leq j \leq n, \text{ and } k = i^*\right)$$

$$\tag{3.4}$$

$$\leq \sum_{k=1}^{n} \mathbb{P}\left(x_{(k+1)_n} + \ldots + x_{(k+j)_n} \geq ja \quad \text{for all} \quad 1 \leq j \leq n\right)$$
$$= n\mathbb{P}\left(x_1 + \ldots + x_j \geq ja \quad \text{for all} \quad 1 \leq j \leq n\right)$$

since the sums in the next to last term are identically distributed. Equations (3.3)–(3.4) yield the result. ∎

The condition (3.2) was the subject of §1.2. Theorem 3.3 holds without the assumption $a > \mathbb{E}x_1$ (and in fact without assuming the existence of the mean) if we replace $\ell(a)$ by $J([a, \infty))$; see (1.11), Corollary 1.9 and Theorem 1.10. For a detailed analysis of random walks via large deviations methods, see [DZ §5.1] and references therein.

To analyze the random walk, we need only to incorporate the starting position x_0. Let $x_1, x_2, \ldots$ be i.i.d., and independent of x_0, and write $M_0(\theta) \stackrel{\triangle}{=} \mathbb{E} \exp \theta x_0$.

Theorem 3.4. *Assume $a > \mathbb{E}x_1$. If M_0 is finite for all θ then*

$$\mathbb{P}(x_0 + x_1 + \ldots + x_j \geq ja \quad \text{for all} \quad 0 \leq j \leq n) \leq e^{-n\ell(a)+o(n)}.$$

If $\mathbb{P}(x_0 \geq 0) > 0$, then

$$\mathbb{P}(x_0 + x_1 + \ldots + x_j \geq ja \quad \text{for all} \quad 0 \leq j \leq n) \geq e^{-n\ell(a)+o(n)}.$$

Proof. Following the proof of the upper bound in Theorem (1.5), for $\theta \geq 0$,

$$\mathbb{P}(x_0 + x_1 + \ldots + x_j \geq ja \quad \text{for all} \quad 0 \leq j \leq n)$$
$$\leq \mathbb{P}(x_0 + x_1 + \cdots + x_n \geq na)$$
$$\leq e^{-na\theta} e^{n \log M(\theta)} e^{\log M_0(\theta)}$$
$$\leq e^{-n(\ell(a)-\varepsilon)+\log M_0(\theta_\varepsilon)},$$

where $\theta_\varepsilon \geq 0$ is such that ℓ is within ε of its maximum. This establishes the upper bound. Now if $x_0 < 0$ with probability one, then the left-hand side in Theorem 3.4 is zero, so that the estimate cannot hold. Otherwise, by the independence assumption,

$$\mathbb{P}(x_0 + x_1 + \ldots + x_j \geq ja \quad \text{for all} \quad 0 \leq j \leq n)$$
$$\geq \mathbb{P}(x_0 \geq 0)\mathbb{P}(x_1 + \ldots + x_j \geq ja \quad \text{for all} \quad 1 \leq j \leq n)$$
$$= e^{-n\ell(a)+o(n)}$$

by Theorem 3.3. ∎

For the applications in Chapter 10 we will need to evaluate the probability of staying strictly above the line, i.e.,

$$\mathbb{P}(x_0 + x_1 + \ldots + x_j > ja \text{ for all } 0 \leq j \leq n).$$

This requires only a slight modification. Note that the statement of the upper bound in this case is exactly as in Corollary 1.9, namely we need to replace ℓ with J. However, since $\ell(a) \leq \inf_{x \geq a} \ell(x)$ whenever $a > \mathbb{E}x_1$, the upper bound is valid. Now if $\mathbb{P}(x_0 > 0) = 0$ then $\mathbb{P}(x_0 + x_1 + \ldots + x_j > ja \text{ for all } 0 \leq j \leq n) = 0$. If $\mathbb{P}(x_0 > 0) > 0$, then

$$\mathbb{P}(x_0 + x_1 + \ldots + x_j > ja \text{ for all } 0 \leq j \leq n)$$
$$\geq \mathbb{P}(x_0 > 0 \text{ and } x_1 + \ldots + x_j \geq ja \quad \text{for all} \quad 1 \leq j \leq n)$$
$$\geq \mathbb{P}(x_0 > 0)\mathbb{P}(x_1 + \ldots + x_j \geq ja \quad \text{for all} \quad 1 \leq j \leq n)$$

and the lower bound holds as well. We summarize this as a corollary.

Corollary 3.5. *Assume* $a > \mathbb{E}x_1$. *If* M_0 *is finite for all* θ, *then*

$$\mathbb{P}(x_0 + x_1 + \ldots + x_j > ja \text{ for all } 0 \leq j \leq n) \leq e^{-n\ell(a)+o(n)}.$$

If $\mathbb{P}(x_0 > 0) > 0$, *then*

$$\mathbb{P}(x_0 + x_1 + \ldots + x_j > ja \text{ for all } 0 \leq j \leq n) \geq e^{-n\ell(a)+o(n)}.$$

3.2. Branching Random Walks

A standard branching process, also called a Galton-Watson process, is defined as follows [Big]. Start with a single "parent" node, and to each node assign a random number of "children," with the number drawn independently from a fixed (discrete) distribution $\{p_i, \ i = 0, 1, \ldots\}$. More formally, let $\{y(i, j)\}$ be i.i.d. random variables with distribution $\{p_i\}$, representing the number of children of node j at generation i. Then $k(n)$, the number of siblings at generation n, is given by

$$k(0) = 1$$

$$k(n) = \sum_{i=1}^{k(n-1)} y(n - 1, i). \tag{3.5}$$

Let b denote the mean number of offspring per node, which we assume is finite, i.e., $b \triangleq \sum_{i=1}^{\infty} i p_i$. The following result is standard in the theory of branching processes; see, e.g., the references in [Big].

Theorem 3.6. *(i)* $\mathbb{E}(k(n)) = b^n$. *(ii) If* $b < 1$, *then* $\mathbb{P}(\lim_{n\to\infty} k(n) = 0) = 1$. *(iii) If* $b > 1$ *then* $\mathbb{P}(\lim_{n\to\infty} k(n) = \infty) > 0$.

Exercise 3.7. Prove (i) and (ii) of Theorem 3.6. Hint: (i) is proved by conditioning. For (ii), first prove convergence in probability. Note that if $k(n) \neq 0$, then it is at least one, and use Markov's inequality. Now use the fact that the sets $\{k(n) = 0\}$ are increasing. ♠

Exercise JFF 3.8. Prove the following weak version of Theorem 3.6(iii): If $b > 1$ then $\lim_{n\to\infty} \mathbb{P}(k(n) = j) = 0$ for all $j \neq 0$. Hint: $g(z) \triangleq \sum_{i=0}^{\infty} z^i p_i$ is the generating function of $k(1)$. The generating function of $k(n)$ is the n-fold composition $g^{(n)}(z) \triangleq g(g(g(\cdots(z))))$. Draw $g(z)$ and conclude that $g^{(n)}$ are all convex increasing on $(0, 1)$, and converge to the unique constant z^* in $(0, 1)$ for which $g(z) = z$. ♠

Exercise JFF 3.9. Show that $\lim_{n\to\infty} \mathbb{P}(k(n) = 0) = 1$ if $b = 1$ and $p_1 \neq 1$. Hint: draw g of Exercise 3.8. ♠

Exercise 3.10. Prove that $b^{-n}k(n)$ is a martingale. ♠

Note that Exercise 3.7(ii) is a consequence of Exercise 3.10 when the martingale is uniformly integrable, for then its mean converges.

We can now construct a branching random walk [Big]. Start with a branching process and assign a value $x(n, i)$ to the ith member of the nth generation, as follows. The root, or origin, of the branching process is assigned the value x_0. Each child of a member is displaced from the value of its direct ancestor by a random variable, which is independent of all other random variables and is drawn from a distribution F. Formally, let $\{w(n, i)\}$ be i.i.d. with distribution function F. If the ith member of generation n is a child of the jth member of generation $n - 1$, then

$$x(n, i) = x(n - 1, j) + w(n, i) = x_0 + \sum_{l=1}^{n} w(l, L(l, n, i)), \qquad (3.6)$$

where $L(l, n, i)$ identifies the ancestor, at generation l, of member (n, i). Note that by construction, $\{x(n, 1), x(n, 2), \ldots\}$ are identically distributed (but not independent!). We denote by x_n such a "generic" n^{th} generation random variable, and by w_n a generic step-size with distribution F. Finally, let $Z(n, x)$ denote the number of nth generation members that are positive when $x_0 = x$.

Lemma 3.11.

$$\mathbb{E}Z(n, x) \triangleq \mathbb{E} \sum_{i=1}^{k(n)} \mathbf{1}[x(n, i) \geq 0]$$

$$= \mathbb{E}(k(n))\mathbb{P}(x_n \geq 0).$$

Exercise 3.12. Prove Lemma 3.11. Hint: write the middle term as

$$\sum_{k=1}^{\infty} \sum_{i=1}^{k} \mathbb{E}\mathbf{1}[k(n) = k]\mathbf{1}[x(n, i) \geq 0],$$

condition on the branching process and use independence. ♠

Remark. This lemma makes it easy to calculate $\mathbb{E}Z(n, x)$, by reducing it to two quantities that are easy to calculate.

Theorem 3.6(ii) states that if $b < 1$, the process $Z(n, x)$ will become extinct (i.e. zero for large n). On the other hand, if $b > 1$ and $\mathbb{E}w_n > 0$, then both will contribute to the growth of Z (as n increases). The interesting question here is the balance between an increasing population (since $b > 1$) but decreasing (on the average) size (since the step size satisfies $\mathbb{E}w_n < 0$).

Theorem 3.13 Biggins [Big]. *Consider the case $b > 1$ and $\mathbb{E}w_n < 0$.*

  (i) *For any x,*

$$\lim_{n\to\infty} \frac{1}{n} \log \mathbb{E}Z(n, x) = \log b - \ell(0)$$

   where ℓ is defined in (1.4) using the displacement distribution F.

  (ii) *If $\log b - \ell(0) < 0$, then $\mathbb{P}(\lim_{n\to\infty} Z(n, x) = 0) = 1$.*

  (iii) *If $\log b - \ell(0) > 0$, then $\mathbb{P}(\limsup_{n\to\infty} Z(n, x) = \infty) > 0$.*

Proof. By Lemma 3.11 we have

$$\log \mathbb{E}Z(n, x) = n \log b + \log \mathbb{P}(x_n \geq 0).$$

But by Theorems 1.5 and 3.4, $\mathbb{P}(x_n \geq 0) = e^{-n\ell(0)+o(n)}$, since $0 > \mathbb{E}w_n$, and (i) is established.

If $\log b - \ell(0) < 0$, then (i) together with Markov's inequality imply that $\mathbb{P}(Z(n, x) \geq 1)$ converges to zero geometrically fast. But then the Borel-Cantelli Lemma A.116 implies that $Z(n, x)$ is greater than zero only a finite number of times, establishing (ii).

To prove (iii) let n^* be large enough so that

$$\mathbb{E}Z(n^*, 0) \stackrel{\triangle}{=} b^{n^*} \mathbb{P}(w_1 + \ldots + w_{n^*} \geq 0) > 1.$$

This is possible by Theorem 1.5. Now construct a branching process $K(k)$ as follows. The offspring at generation one are the members of the branching random walk starting at zero which are positive at generation n^*. Thus $K(1) = Z(n^*, 0)$, so that the branching distribution for $K(k)$ is the distribution of $Z(n^*, 0)$. We have arranged for the mean of this branching distribution to be larger than one. Therefore Theorem 3.6(iii) shows that the probability that $K(k) \to \infty$ is strictly positive. Let us construct K so that $K(k) \leq Z(kn^*, 0)$. From $x(kn^*, i)$, ignore not only those who are negative, but also all those who are smaller than at least one of their ancestors at generation ln^*, $l < k$. This guarantees that for all k, $K(k) \leq Z(kn^*, 0)$ and indeed $K(k)$ is a branching process with the specified branching distribution.

Since $Z(n, x) \geq Z(n, 0)$ for all $x \geq 0$, we conclude that the probability that $Z(n, x) \to \infty$ along the sequence kn^* is positive.   ∎

Remark. The result depends only on the mean b, not on the distribution of the number of children. However, the whole step-size distribution of the random walk enters through the computation of $\ell(0)$. The same applies to the rest of the results of this chapter.

The following variation is rather obvious:

Corollary 3.14. *Lemma 3.11 and Theorem 3.13 continue to hold if $Z(n, x)$ is defined as the number of members of the nth generation that are **strictly** positive.*

Branching random walk with a barrier.

A somewhat more interesting variation arises in the rollback application discussed in Chapter 10. A branching random walk with a barrier $z(n, x)$ is obtained from a branching random walk $Z(n, x)$ by deleting all members who have even one ancestor who is not strictly positive. Formally,

$$z(n, x) \stackrel{\triangle}{=} \sum_{i=1}^{k(n)} \mathbf{1} \left[x(j, L(j, n, i)) > 0 \quad \text{for all} \quad 0 \le j \le n \right]$$

where L is defined after (3.6).

Lemma 3.15.

$$\mathbb{E}z(n, x) = \mathbb{E}(k(n))\mathbb{P}(x_0 + w_1 + \ldots + w_j > 0 \quad \textit{for all} \quad 0 \le j \le n).$$

Exercise 3.16. Prove Lemma 3.15. Hint: follow Exercise 3.12; construct the process with a barrier by first constructing a process without a barrier and then deleting members. ♠

Theorem 3.17 Biggins, Lubachevsky, Shwartz and Weiss [BLS]. *Consider the case $b > 1$ and $\mathbb{E}w_n < 0$.*

(i) *For any x,*

$$\lim_{n \to \infty} \frac{1}{n} \log \mathbb{E}z(n, x) = \log b - \ell(0),$$

where ℓ is defined in (1.4) using the displacement distribution F.

(ii) *If $\log b - \ell(0) < 0$ then $\mathbb{P}(\lim_{n \to \infty} z(n, x) = 0) = 1$.*

(iii) *If $\log b - \ell(0) > 0$ then $\mathbb{P}(\limsup_{n \to \infty} z(n, x) = \infty) > 0$.*

The proof is identical to that of Theorem 3.13 and is left to Exercise 3.18. Note that (iii) can be strengthened; see [Big].

Exercise 3.18. Prove Theorem 3.17. Hint: use Corollary 3.5 and modify the proof of Theorem 3.13. ♠

This result is applied to a rollback model in §10.1. To get a feel for the behavior of branching random walks with barrier, refer to the last part (analysis) of §10.1.

Some further results on this model are in [BLS]. They include a proof that (under some conditions on the distribution F) $\sup_{n,i} x(n, i)$ is a finite random variable with exponential tail. It is also possible to obtain, in some cases, explicit expressions for $\sum_{i=1}^{\infty} \mathbb{E}z(n, x)$; see [BLS].

Chapter 4

Poisson and Related Processes

Most of our applications are based on (generalizations of) the Poisson process. There is a very good reason for this: the real world is a continuous-time one, and events in computer and communication systems are generally discrete valued: a message is sent, a task is completed etc. The simplest continuous-time, discrete-state-space stochastic process is the Poisson process. In addition, there are physical reasons to use a modeling approach based on the Poisson process; see, e.g., [GM, Ros1, Ros2, RS, Ta1, Ti].

In order to keep the book self-contained, we collect some basic material in this section. We also present some special tricks that apply to Poisson processes. While the statements of the theorems are precise, the derivation is intentionally heuristic. The purpose of this chapter is to provide sufficient information about these processes so that later chapters can be understood. A precise and more complete treatment is available in Appendix B.

4.1. The One-Dimensional Case

Recall Definition A.133 of point processes and counting processes. A Poisson process is a counting process, and we use the notation $N(t)$ for number of events over the interval $[0, t)$.

Definition 4.1. *A Poisson process $N(t)$ with intensity λ is a counting process with*

(i) $N(0) = 0$,

(ii) $\mathbb{P}(N(t) - N(s) = n) = e^{-\lambda(t-s)} \dfrac{(\lambda(t-s))^n}{n!}$, $n = 0, 1, \ldots$ *for* $t > s > 0$,

(iii) $N(t) - N(s)$ *is independent of* $N(u) - N(v)$ *whenever* $u > v \geq t > s \geq 0$.

We think of the Poisson process as representing *events*, and of $N(t)$ as a count of the number of events (or total number of arrivals) up to time t. There is an obvious one-to-one relation between the counting process and the times between events. The time of the first event τ_1 is the time of the first jump of $N(t)$:

$$\tau_1 \overset{\triangle}{=} \inf\{t > 0 \,:\, N(t) > 0\}. \tag{4.1a}$$

Similarly, the time τ_{n+1} between the nth event and the $(n+1)^{\text{st}}$ event is the time between the nth and the next jump of $N(t)$:

$$\tau_{n+1} \overset{\triangle}{=} \inf\{t > 0 \,:\, N(\tau_1 + \cdots + \tau_n + t) > N(\tau_1 + \cdots + \tau_n)\}. \tag{4.1b}$$

Conversely, if we know only the inter-jump times, we can easily recover $N(t)$: it is just the number of events up to t, so

$$N(t) = \begin{cases} 0 & \text{if } t < \tau_1; \\ \sup\{n \; : \; \tau_1 + \cdots + \tau_n \leq t\} & \text{otherwise.} \end{cases} \tag{4.2}$$

The term "Poisson process" will usually refer to the counting process; however, this will also refer to the associated Point process. There are many characterizations of the Poisson process, each useful in its own domain. Here are two.

Proposition 4.2 [Ros3 Thm. 2.1.2 p. 32, Wo pp. 70–71]. *Let $N(t)$ be a counting process with $N(0) = 0$. The following are equivalent;*

(i) *$N(t)$ is a Poisson process with rate λ,*

(ii) *$N(t)$ has stationary, independent increments, and*

$$\mathbb{P}(\text{there is an event in } (0,t)) = \lambda t + o(t)$$

$$\mathbb{P}(\text{there are two or more events in } (0,t)) = o(t).$$

The properties listed in the next three theorems turn out to be quite useful in both the construction and analysis of systems based on the Poisson process, and we illustrate this below.

Proposition 4.3 [Ros3 pp. 35–36, Wo p. 71 eq. (65)].

(i) *Let $\tau_1, \tau_2, \ldots$ be i.i.d. exponentially distributed random variables with mean λ^{-1}. Then the counting process defined in (4.2) is a Poisson process with rate λ.*

(ii) *Conversely, let N be a Poisson process with rate λ. Then the times between jumps $\tau_1, \tau_2, \ldots$ defined in (4.1) are i.i.d. exponentially distributed random variables with mean λ^{-1}.*

Remark. Exponential random variables are memoryless in the following sense. Let τ be the lifetime of a component, and suppose we observe that the component is still in "good shape." Then for τ exponentially distributed, the distribution of the remaining lifetime is again exponential, with the same parameter. This and Proposition 4.3 imply that the Poisson process is also memoryless in the sense that if we stop the process at any (deterministic) point, what follows is again a Poisson process with the same rate. It is easy to verify that the Poisson process is a Markov process.

Proposition 4.4 [Ros3 Thm. 2.3.1 p. 37, Wo Thm. 3 p. 73]. *Fix a time interval, say $[0, T]$, and condition the Poisson process $N(t)$ to have J jumps in that interval. Then, regardless of the Poisson rate, the event times are distributed as J i.i.d. uniform $[0, T]$ random variables (actually, as their order statistic since you need to rearrange the uniform random variables in increasing order).*

Put another way, this theorem states that the distribution of arrival epochs, conditioned on J arrivals in $[0, T]$, is identical to the distribution of the values of J i.i.d. uniform $[0, T]$ random variables.

Here is a "calculus" for Poisson processes.

Proposition 4.5 [Ros3 Thm. 2.3.2 pp. 38–39, Wo Example 2-5 pp. 74–75].

(i) *Splitting a Poisson process: let N be a Poisson process with rate λ. Every time an event occurs, flip a coin with $\mathbb{P}(head) = p$. If you get a head, throw the event into box 1, and otherwise into box 2. Let $N_i(t)$ be the number of events in box i at time t. Then N_i are independent Poisson processes with rates λp and $\lambda(1 - p)$, respectively.*

(ii) *Merging Poisson processes: let N_i be independent Poisson Processes with rates λ_1 and λ_2, respectively. Then $N = N_1 + N_2$ is a Poisson process with rate $\lambda_1 + \lambda_2$.*

To define N_i more precisely for part (i), let $x_1, x_2, \ldots$ be i.i.d. Bernoulli random variables, independent of the process N, with $\mathbb{P}(x_1 = 1) = p$. Then N_i is defined as the counting process whose jump times are a subset of those of N and that satisfies

$$N_1(\tau_1 + \cdots + \tau_n + \tau_{n+1}) \overset{\triangle}{=} N_1(\tau_1 + \cdots + \tau_n) + x_n$$

$$N_2(\tau_1 + \cdots + \tau_n + \tau_{n+1}) \overset{\triangle}{=} N_2(\tau_1 + \cdots + \tau_n) + (1 - x_n) \, .$$

Obviously, (i) and (ii) extend to any finite number of independent Poisson processes. These statements are both deep and special to the Poisson process.

Indeed, assume that red balls arrive at some assembly line according to a Poisson (λ_1) stream, and green balls arrive at this assembly line according to an independent Poisson (λ_2) stream. A person is placed to sort them into two baskets. A color-blind inspector observes the arrival streams to the baskets. Being mischievous, upon each arrival the sorter just flips a coin, with probability $\lambda_1/(\lambda_1+\lambda_2)$ for heads, and sorts the balls according to the outcome of the coin flips. Then (i)–(ii) imply that *it is impossible* for the inspector to discover the scam. This would not be the case if the streams were not Poisson.

Let us perform some simple calculations with Poisson processes and Poisson random variables. Since by Definition 4.1(ii) the distribution of the number of events in a Poisson process with rate λ for a duration of T is a Poisson random variable with parameter λT, the mean number of events is λT and the variance of the number of events is λT. Now a Poisson random variable with parameter λT is distributed the same as the sum of T i.i.d. Poisson random variables with parameter λ, or the sum of λT i.i.d. Poisson random variables with parameter one. Hence, the central limit theorem shows that a Poisson (λT) random variable is approximately normal, with mean and variance λT.

Let us see an amusing application of this to analysis. How quickly does the series $\sum_{i=0}^{\infty} n^i/i!$ converge to e^n? More precisely, what is the limit

$$\lim_{n \to \infty} e^{-n} \sum_{i=0}^{n} \frac{n^i}{i!} \, ?$$

This can be solved by inspection once we notice that the expression is the probability that a Poisson(n) random variable is less than or equal to n. The central limit theorem shows that this approaches the probability that a normal random variable is less than its mean, which is one half. So the limit equals $\frac{1}{2}$.

4.2. Jump Markov Processes

Let x_t be a stochastic process. Denote by $\mathcal{F}_t \triangleq \sigma\{x_s, \ s \le t\}$ the past information (Definitions A.99, A.123) about the process. By Definition A.134, the process x is called Markov if for all t and all $s > 0$,

$$\mathbb{P}\,(x_{t+s} \in S \mid \mathcal{F}_t) = \mathbb{P}\,(x_{t+s} \in S \mid x_t)$$

for all (Borel) sets S. That the Poisson process is a Markov process is an immediate consequence of the memoryless property; see remark following Proposition 4.3. All the processes of interest to us are Markov, and below we construct several jump Markov processes. The Markov property is essential to our large deviations analysis, as will become clear in Chapter 5.

Since the Poisson process is one-dimensional, it is of limited use for modeling purposes. However, we can easily construct multidimensional processes out of Poisson processes. The construction below will serve as motivation and introduction to more general jump Markov processes.

Let $\mathcal{Z}^d$ be the d-dimensional integer lattice (i.e., the collection of vectors in $\mathbb{R}^d$ with integer components). Let $N_1, \ldots, N_n$ be independent Poisson processes with rates λ_i, and let $\vec{e}_1, \ldots, \vec{e}_n$ be vectors in $\mathcal{Z}^d$. Define

$$\vec{x}_t \triangleq \sum_{i=1}^n N_i(t)\vec{e}_i \ . \tag{4.3}$$

Then $\vec{x}_t$ is a multidimensional jump process, which is easily seen to be Markov. From Proposition 4.5(ii), we know that the total event process $N_1 + \cdots + N_n$ is a Poisson process with rate $\lambda_1 + \cdots + \lambda_n$. Moreover, it is easy to see that, conditioned on a jump occurring, the probability that it occurs in a direction $\vec{e}_i$ is just $\lambda_i(\lambda_1 + \cdots + \lambda_n)^{-1}$. Thus the distribution of this process is characterized by the total event rate and the probability of jump in a certain direction, and is by construction space homogeneous.

The next step is now obvious, and will give us the general discrete space jump Markov process. For each *pair of distinct states* $\vec{x}, \vec{y}$ in $\mathcal{Z}^d$ assign a jump rate $\lambda(\vec{x}, \vec{y})$. If the process is at state $\vec{x}$, the time to leave $\vec{x}$ is defined to be exponential with rate $\sum_{\vec{y} \ne \vec{x}} \lambda(\vec{x}, \vec{y})$, and the probability of going to $\vec{y}$ is defined as

$$\lambda(\vec{x}, \vec{y}) \cdot \left(\sum_{\vec{z} \ne \vec{x}} \lambda(\vec{x}, \vec{z})\right)^{-1} . \tag{4.4}$$

Equivalently, generate a set of exponential random variables with rates $\lambda(\vec{x}, \vec{y})$, and jump to $\vec{y}$ if the corresponding random variable was the smallest in the set. The equivalence of the two constructions follows from Proposition 4.3. These definitions apply to any discrete, countable (or finite) state space $\mathcal{X}$, and we give some simple examples below.

Generators.

Let $\lambda(x, x) \overset{\Delta}{=} -\sum_{y \neq x} \lambda(x, y)$, and denote by L the array $\{\lambda(x, y)\}_{x,y}$. This array is called the *generator* of this process. If the state space $\mathcal{X}$ is finite, then L is just a matrix, and if we represent functions f on $\mathcal{X}$ as column vectors, we have

$$(Lf)(x) = \sum_y \lambda(x, y) f(y) = \sum_y \lambda(x, y)(f(y) - f(x)) \qquad (4.5)$$

where the second equality follows from the definition of $\lambda(x, x)$. Another common (and equivalent) definition for the generator excludes from the second sum the term with $y = x$. The same definition extends to the countable case, so that L is a linear operator on the real-valued functions on $\mathcal{X}$. This generator will prove quite useful, so we make some effort to develop intuition about it. In fact, under rather broad conditions a generator determines a process (probabilistically). The formal definition of a generator is more general; we illustrate below that the definitions agree.

Definition 4.6. *Let $\{x_t, \ t \geq 0\}$ be a Markov process with state space $\mathcal{X}$ and L an operator on real-valued functions f on $\mathcal{X}$. We call L the generator of the process if, whenever the limit on the right exists,*

$$(Lf)(y) = \lim_{t \downarrow 0} \frac{\mathbb{E}\left[f(x_t) - f(y) \mid x_0 = y\right]}{t}. \qquad (4.6)$$

The existence of the limit depends on both the function f and the point y. The set of functions for which this limit exists *for all* y in $\mathcal{X}$ is called the domain of L and is denoted $\mathcal{D}(L)$.

Let us show that these definitions are equivalent for the simple case of a Poisson process with rate λ. In this case $\mathcal{X} = \mathcal{Z}$, and for $z \neq y$

$$\lambda(y, z) = \begin{cases} \lambda & \text{if } z = y + 1 \\ 0 & \text{otherwise.} \end{cases}$$

Our first definition (4.5) thus gives

$$(Lf)(y) = \lambda \left(f(y + 1) - f(y)\right) .$$

Let us compute directly; for bounded f the right-hand side of (4.6) reads

$$
(Lf)(y) = \lim_{t\downarrow 0} \frac{1}{t}\left(\sum_{j=0}^{\infty}(f(y+j)-f(y))\frac{e^{-\lambda t}(\lambda t)^j}{j!}\right)
$$

$$
= \sum_{j=1}^{\infty}(f(y+j)-f(y))\lim_{t\downarrow 0}\frac{1}{t}\frac{e^{-\lambda t}(\lambda t)^j}{j!} \tag{4.7}
$$

$$
= \lambda(f(y+1)-f(y)),
$$

where the interchange between limit and summation is justified by boundedness [Theorem A.91(iii); clearly, weaker tail conditions on f suffice].

We can derive the equivalence of (4.5) and (4.6) for the more general case in about the same way as for the Poisson process. Let us assume that

$$
\lambda \stackrel{\triangle}{=} \sup_x |\lambda(x,x)| < \infty.
$$

Then the time to leave state y is exponential with rate $|\lambda(y,y)|$, and the probability of going from y to z is $\lambda(y,z)|\lambda(y,y)|^{-1}$. Let f be a bounded function. To compute the expectation in (4.6) we separate the first jump from the rest. Since the probability of exactly one jump in $[0,t)$ is $e^{-|\lambda(y,y)|t}(|\lambda(y,y)|t)$,

$$
(Lf)(y) = \lim_{t\downarrow 0}\frac{1}{t}\left(\sum_{z\neq y}\left((f(z)-f(y))e^{-|\lambda(y,y)|t}|\lambda(y,y)|t\frac{\lambda(y,z)}{|\lambda(y,y)|}\right)+e(t)\right)
$$

where $e(t)$ represents the case of two or more transitions in $(0,t)$, so that

$$
|e(t)| < \mathbb{P}(\text{at least two jumps occurred in } (0,t))\cdot 2\sup_z|f(z)|.
$$

Since

$$
\lim_{t\downarrow 0}\frac{1}{t}\mathbb{P}(\text{at least two jumps occurred in } (0,t)) \leq \lim_{t\downarrow 0}\frac{1}{t}\left(1-e^{-\lambda t}(1+\lambda t)\right) = 0,
$$

we have

$$
(Lf)(y) = \sum_{z\neq y}(f(z)-f(y))\lim_{t\downarrow 0}\frac{1}{t}\left(e^{-|\lambda(y,y)|t}|\lambda(y,y)|t\cdot\frac{\lambda(y,z)}{|\lambda(y,y)|}\right)
$$

$$
= \sum_{z\neq y}(f(z)-f(y))\lambda(y,z).
$$

The interchange between limit and summation is justified by bounded convergence, Theorem A.91(iii), since the time t cancels out (except in the exponent), $f\cdot\exp(-|\lambda(y,y)|t)$ is bounded and the measure defined by the rest of the summands is finite (at most λ). Thus we obtain (4.5).

Birth-Death processes.

In most of the applications, the jumps of the process are only to the nearest neighbors. One such jump Markov process is the multidimensional birth-death process. We can loosely think of a d-dimensional birth-death process as representing the number of live creatures of d types. At each event-point, at most one individual of each type may be born or die, while the birth (death) rates depend on the current state. However, we may have a simultaneous birth-death case, where one individual from each of several types is born, and at the same time one individual from each of several other types dies. Other possible phenomena are mutation, coagulation, etc. Moreover, for this process the numbers of individuals of a type is not necessarily positive. Thus we have a finite number of jump directions $\vec{e}_1, \ldots, \vec{e}_m$, where the components of the vector $\vec{e}_i$ include 1 s, (-1) s, and 0 s only, and the rate takes the form

$$\lambda(x, y) = \begin{cases} \lambda_j(x) & \text{if } y = x + \vec{e}_j; \\ 0 & \text{otherwise}. \end{cases} \qquad (4.8)$$

Most of the processes that arise in queueing applications are of this form. As the most basic example, the queue size of an $M/M/1$ queue is a birth-death process where arrivals are described by $e_1 = 1$, with rate $\lambda_1 = \lambda$, and departures by $e_2 = -1$, with rate $\lambda_2(x) = \mu \mathbf{1}[x > 0]$, where μ is the service rate.

A general jump Markov process need not be defined on a discrete state-space. For example, if $\mathcal{X} = \mathbb{R}^d$, we can construct such a process by making it jump from state x after an exponential time with mean $(\lambda(x))^{-1}$ and, when it jumps, have its next position distributed according to some distribution F_x. However, unless the distribution F_x is supported by at most a countable number of points, we lose the interpretation of this process as jumping according to the smallest exponential random variable with rate $\lambda(x, y)$. Definitions (4.5)–(4.6) still make sense, with sums in (4.5) replaced by integrals, and they are equivalent. In fact, it is easy to see that the same calculation establishes the equivalence for the general jump Markov case if the support of F_x is countable; we just have to invoke boundedness to interchange the order of limit and summation.

This trick of switching between interpretations of the process is fairly useful. What we have discovered is that, in some sense, only the first jump counts; after that the process "restarts." Let us specialize to processes of the type described in (4.9) and derive a general tool for approximations from one of the representations. So, consider a process with generator

$$Lf(\vec{x}) = \sum_{i=1}^{k} \lambda_i(\vec{x}) \left(f(\vec{x} + \vec{e}_i) - f(\vec{x}) \right). \qquad (4.9)$$

Assume that the maximal rate of jump out of any state is finite:

$$\bar{\lambda} \overset{\triangle}{=} \sup_{\vec{x}} \sum_{i=1}^{k} \lambda_i(\vec{x}) < \infty.$$

Then by Theorem 4.5 we can construct a process with this generator in the following way. Let $N(t)$ be a Poisson process with rate $\bar{\lambda}$. Given a starting point $\vec{x}$ for the process, whenever the process is at a point $\vec{y}$ the following happens. The process may leave $\vec{y}$ at the next jump of $N(t)$. At that time, an independent $k+1$-sided die is cast. The probability of side $i = 1, \ldots, k$ is $\lambda_i(\vec{y})/\bar{\lambda}$. If side $i = 1, \ldots, k$ appears then the new state is $\vec{y} + \vec{e}_i$. If side $k+1$ appears, the process stays at state $\vec{y}$.

This construction allows us to compare, and therefore bound, one process in terms of another, usually simpler process. The general definition is quite abstract.

Definition 4.7. *Two stochastic processes are called coupled if they are defined on the same probability space.*

Here is a specific example, which will be made more concrete below. This example illustrates the use of coupling of jump Markov processes so that information on one could be used for the other. Let $\vec{u}(t)$ and $\vec{v}(t)$ be two jump processes of the type (4.9), with the same jump directions and with rates $\lambda_i^u(\vec{x})$ and $\lambda_i^v(\vec{x})$, respectively. Let $\bar{\lambda}$ be a finite upper bound for both $\lambda_i^u(\vec{x})$ and $\lambda_i^v(\vec{x})$. We can then construct the processes in the following way.

Example 4.8. Consider a Poisson process $N(t)$ with rate $\bar{\lambda}$, and independent random variables

$$\{d_\alpha(m, \vec{x}), \ \alpha = u, v, \ m = 1, 2, \ldots, \ \vec{x} \in \mathbb{R}^d\}$$

taking values $1, \ldots, k, k+1$, all defined on a probability space $(\Omega, \mathcal{F}, \mathbb{P})$. Assume the $d_\alpha(m, \vec{x})$ are also independent of the process $N(t)$, $t \geq 0$. Represent the processes u and v through $N(t)$ and the $d_\alpha(m, \vec{x})$, where $d_\alpha(m, \vec{x})$ is the outcome of the die corresponding to process α at its mth visit to state $\vec{x}$. Then u and v have the desired generators, are coupled, and moreover they share some (but not all) of the jump times.

Example 4.9. Continuing Example 4.8, let $N(t)$ have rate

$$\lambda \triangleq \sum_{i=1}^{k} \bar{\lambda}_i < \infty \quad \text{where} \quad \bar{\lambda}_i \triangleq \sup_{\vec{x}} \lambda_i^u(\vec{x}).$$

Let $\vec{v}$ be a process with rates $\bar{\lambda}_i$, independent of $\vec{x}$. Construct $\vec{v}$ as before, except now $d(m, \vec{x}) = d(m)$ is thrown at the mth jump of $N(t)$. Assume that, on the same probability space, we also have the independent Bernoulli random variables $b(i, m, \vec{x})$. Note that by definition $\vec{v}$ jumps whenever N does. At each such jump, if $\vec{u}$ is at $\vec{x}$ and if $\vec{v}$ jumps in direction $\vec{e}_i$, then u will either jump in direction $\vec{e}_i$ or stay at $\vec{x}$ according to whether the Bernoulli random variable $b(i, m, \vec{x})$ took the value one or zero. The probability of $b(i, m, \vec{x}) = 1$ is $\lambda_i^u(\vec{x})/\bar{\lambda}_i$. A little reflection shows that the processes have the following property: whenever $\vec{u}$ jumps in direction $\vec{e}_i$, so does $\vec{v}$. Thus at any time t, the total number of jumps of $\vec{v}$ in any given direction in $[0, t)$ dominates that of $\vec{u}$.

Exercise 4.10. Given a process $\vec{u}$ with $\overline{\lambda}_i < \infty$ and with values in $\mathbb{R}^d$, construct a process $\vec{v}$ that dominates it in direction x_1. That is, starting at the same point $\vec{x}$, we have $v_1(t) \geq u_1(t)$ for all $t \geq 0$ and all ω.

It is possible to include an initial distribution in the coupling: there is no need that the two coupled processes start at the same point, or with the same initial distribution. In the case of the example below, such a technique enables one to bound the moments of one process in terms of another process for which explicit formulas are available.

Exercise 4.11. Consider two $M/M/1$ processes $x_1(t)$ and $x_2(t)$ [as defined below Equation (4.8)] with the same parameters, and with $\mu > \lambda$. Construct a coupling so that the following conditions hold. The process $x_1(t)$ is in steady state, that is, its distribution is invariant under time shifts, $x_2(0) = 0$, and $x_1(t) \geq x_2(t)$ for all t and ω. Conclude that $\mathbb{E} f(x_1(t)) \geq \mathbb{E} f(x_2(t))$ for all increasing functions f and all t. ♠

The preceding discussion forms the theoretical foundation for the technique known as uniformization. When simulating or analyzing processes based on Poisson processes, one may assume that the events occur in discrete time, not continuous time, as follows. We consider a discrete-time, discrete-state space Markov chain where, if we are at state y, the probability of a transition $y \to z$ occurring at any integer time is $\lambda(y, z)/\overline{\lambda}$, where

$$\overline{\lambda} \overset{\triangle}{=} \sup_x |\lambda(x, x)|.$$

Thus the transition matrix of this chain is $(\overline{\lambda})^{-1}L + I$. Note that we have introduced "dummy" transitions $y \to y$ which occur with probability $(\overline{\lambda} - \sum_{z \neq y} \lambda(y, z))/\overline{\lambda}$. We relate this to the original system by associating the integer time T with the random time when a Poisson process with rate $\overline{\lambda}$ achieves T events (including "dummy" events!).

The advantages of uniformizing are lower variance and the availability of a host of methods for analyzing discrete time Markov chains. For a precise derivation, see (B.9)—§B.1.

4.3. Martingales and Markov Processes

One of the most useful consequences of the introduction of generators is that they allow us to generate martingales that are related to the jump Markov processes of interest. This will come in very handy in the analysis of large deviations for processes, since tools for the analysis of martingales are well developed.

The exposition in this section is purely heuristic. However, since these results are so useful for the analysis of jump Markov processes, we develop them in detail in §B.4.

Recall the formal definition (4.6) of a generator of a Markov process x_t. Fix some real function f and assume that, at time $t = 0$, the process started from

$x(0) = x$. Now define

$$M_f(t) \triangleq f(x_t) - \int_0^t (Lf)(x_s)\, ds. \qquad (4.10)$$

Then, loosely speaking,

$$\mathbb{E}M_f(t) \triangleq \mathbb{E}f(x_t) - \mathbb{E}\int_0^t \left(Lf\right)(x_s)\, ds$$

$$= \mathbb{E}f(x_t) - \int_0^t \mathbb{E}\left(\lim_{\varepsilon\downarrow 0} \frac{1}{\varepsilon}\mathbb{E}[f(x_{s+\varepsilon}) - f(x_s) \mid x_s]\right) ds$$

$$= \mathbb{E}f(x_t) - \int_0^t \lim_{\varepsilon\downarrow 0} \frac{1}{\varepsilon}\mathbb{E}\mathbb{E}[f(x_{s+\varepsilon}) - f(x_s) \mid x_s]\, ds$$

$$= \mathbb{E}f(x_t) - \lim_{\varepsilon\downarrow 0} \frac{1}{\varepsilon}\left(\int_t^{t+\varepsilon} \mathbb{E}f(x_s)\, ds - \int_0^\varepsilon \mathbb{E}f(x_s)\, ds\right)$$

$$= \mathbb{E}f(x_t) - \mathbb{E}f(x_t) + f(x)$$

$$\triangleq M_f(0),$$

where we have not stated the continuity and other technical conditions needed to justify this sequence of equalities. Now, if we start at time s with initial condition x_s, then this calculation and the Markov nature of the process will yield

$$\mathbb{E}\left(M_f(t) \mid \mathcal{F}_s\right) = M_f(s) \quad a.s.$$

where $\mathcal{F}_s$ is the σ-field of $\{x_u,\ 0 \le u \le s\}$, i.e., it encodes all information about the process x up to time s. But this means that $M_f(t)$ is a martingale. For a precise derivation and sufficient conditions in the jump Markov case, see §B.4. A general derivation can be found in [SV, KS]. The importance of the martingale (4.10) is underlined by the following theorem, which applies to a different but important class of stochastic processes—diffusions. Let b and σ be functions from $\mathbb{R}^d$ to $\mathbb{R}^d$, and let $a(\vec{x}) \triangleq \sigma(\vec{x})\sigma^T(\vec{x})$. We call a uniformly positive definite if for some $\alpha > 0$,

$$\sum_{ij} y_i a_{ij}(\vec{x}) y_j \ge \alpha \sum_i y_i^2 \quad \text{for all} \quad \vec{x} \text{ and } \vec{y}.$$

Define the operator L by

$$(Lf)(\vec{x}) = \sum_{j=1}^d b_j(\vec{x})\frac{\partial f(\vec{x})}{\partial x_j} + \sum_{j,m=1}^d a_{jm}(\vec{x})\frac{\partial^2 f(\vec{x})}{\partial x_j \partial x_m}. \qquad (4.11)$$

Theorem 4.12 [EK, Ku1, SV, KS §5.4]. *Assume b and σ are bounded and Lipschitz continuous, and that a is uniformly positive definite. Let $\vec{x}$ be a Markov process with values in $\mathbb{R}^d$, and with generator L defined in (4.11). Then for every smooth function f (bounded continuous, together with its derivatives up to second order), the process M_f defined in (4.10) is a martingale. Conversely, let L be* the operator (4.11) defined on the smooth functions on $\mathbb{R}^d$. Then there is a continuous (measurable) process $\vec{x}$ so that, for every smooth f, the process M_f is a martingale. This process is Markov and its distribution is unique. Consequently, if for a process $\vec{x}$ the process M_f is a martingale for every smooth f, then $\vec{x}$ is Markov and L is its generator.

We will need to apply Theorem 4.12 only in the degenerate case of Corollary 4.15, and for jump Markov processes with a finite number of jumps. For the latter process, an analogous theorem can be obtained from the development of §B.4, leading to Equation (B.25). In this case the result is

Theorem 4.13. *Theorem 4.12 continues to hold if the generator takes the form (4.5) provided the sums are all finite and $\lambda(\cdot, \cdot)$ is bounded below. The process $\vec{x}$ is then right continuous (rather than continuous). The continuity assumptions on f are not necessary.*

The boundedness assumptions on σ, b (above) and λ (below) can all be discarded. The price to pay is the possibility of explosion: the processes may run off to infinity in finite time. In this case, a "localization argument" is used: the process is considered only up to the time of this explosion, and the "stopped process" satisfies the statements of the theorems.

Since we already know that the generator determines the process (in the sense of distribution), Theorem 4.12 gives us a way to establish the characterization of processes in terms of associated martingales. This is very useful in applications, when we try to obtain approximations or limit theorems. Trying to compute directly the effect of small perturbations in some parameters on the distribution of the process may be quite difficult. However, the effect of such perturbations on the generator of the process can often be written by inspection. Then the limiting process can be found by computing the limiting generator (subject, of course, to technical conditions).

Here is a sample theorem that will allow us to do exactly such an approximation. For details (and proofs) about this powerful approximation technique see Kurtz [Ku1], Ethier and Kurtz [EK], and references therein.

Theorem 4.14. *Let $\vec{x}_n$ be a sequence of Markov processes, and let L be a generator satisfying the conditions of Theorem 4.12 or 4.13. Assume the initial distributions converge to the distribution concentrated, say, at the point $\vec{z}$, in the sense of Definition A.87. Suppose the set $\{\vec{x}_n\}$ is tight in the sense of Definition A.89. If for all m, all $0 \le t_1 \le \ldots \le t_m \le t \le t + s$ and all functions f and h_i which are bounded and continuous, together with their derivatives up to second order,*

$$\lim_{n \to \infty} \mathbb{E}\left[\left(f(\vec{x}_n(t+s)) - f(\vec{x}_n(t)) - \int_t^{t+s} (Lf)(\vec{x}_n(u))du \right) \times \prod_{i=1}^{m} h_i(\vec{x}_n(t_i)) \right]$$

$$= 0, \quad (4.12)$$

then $\vec{x}_n \Rightarrow \vec{x}$, a Markov process with generator L.

The point is that if L in that estimate is replaced with L_n, then by definition the expectation is zero. So, if L_n is close to L in some sense, the estimate is easy to establish. See Exercise 4.16 below.

The scaled process.

Consider, for example, a Poisson process $N(t)$. Let $z_n(t) \stackrel{\triangle}{=} N(nt)/n$ be the process we obtain by making the jumps smaller, but faster. To compute the generator L_n for the new process, apply Definition 4.6. The definition of z_n and the change of variable $nz = y$ give

$$(L_n f)(z) = \lim_{t \downarrow 0} \frac{\mathbb{E}\left[f\left(\frac{1}{n}N(nt)\right) - f\left(\frac{y}{n}\right) \mid \frac{N(0)}{n} = \frac{y}{n} \right]}{t}.$$

Now set $s = nt$ and define the function g through $g(y) = f(y/n)$. Then by (4.5) applied to g,

$$\begin{aligned}
(L_n f)(z) &= n \lim_{s \downarrow 0} \frac{\mathbb{E}\left[g(N(s)) - g(y) \mid N(0) = y \right]}{s} \\
&= n\lambda \left(g(y+1) - g(y) \right) \qquad\qquad (4.13) \\
&= n\lambda \left(f\left(z + \frac{1}{n}\right) - f(z) \right).
\end{aligned}$$

Thus, if there is a limit here, it should satisfy

$$(L_\infty f)(z) = \lambda \frac{d}{dz} f(z).$$

However,

Corollary 4.15. *Under the conditions of Theorem 4.12, first-order generators define deterministic processes.*

Proof. We can easily deduce this from Theorem 4.12 as follows. Suppose a generator L is given by $Lf \stackrel{\triangle}{=} g\frac{d}{dz}f$ for some function g. Let x_t be a process that satisfies

$$\frac{d}{dt}x_t = g(x_t).$$

Then for any smooth f,

$$\begin{aligned}
\int_0^t (Lf)(x_s)\, ds &= \int_0^t g(x_s)\frac{d}{dx} f(x_s)\, ds \\
&= \int_0^t \frac{d}{ds} f(x_s)\, ds \\
&= f(x_t) - f(x_0).
\end{aligned}$$

Thus M_f is a martingale (in fact, it is constant). But this implies that L is indeed the generator of the process x. Since the generator determines the distribution of the process, we conclude that any process whose generator is a first-order differential operator is deterministic (except for possible random initial conditions), and Corollary 4.15 is established. ∎

For the scaled Poisson process $z_n(t)$ we conclude that, under the appropriate technical conditions, the limit $z_\infty(t)$ of the process $N(t)$ scaled by n is a deterministic process. Since here $g \equiv \lambda$ the limit $z_\infty(t)$ must satisfy

$$\dot{z}_\infty(t) = \lambda \quad \text{or} \quad z_\infty(t) = z_\infty(0) + \lambda \cdot t .$$

For *fixed* t, we could obtain this result from the law of large numbers. However, our result is stronger—it deals with the convergence of the process. Therefore, in order to make this convergence result precise, we need to define a space of paths that is appropriate for the paths of the Poisson process, a notion of convergence in that space, and specify in what (probabilistic) sense the sequence converges. Using this technique one typically obtains "convergence in distribution" which, in this context, is called "weak convergence" (Definition A.87).

The process we shall concentrate on throughout the rest of the book is a birth-death process with precisely the same scaling as in (4.13). So, let $\vec{x}(t)$ be a birth-death process with generator as defined in (4.5)–(4.8). The scaled process $\vec{z}_n$ is the jump Markov process such that

$$\vec{z}_n(0) \overset{\mathcal{L}}{=} \frac{1}{n}\vec{x}(0) \quad \text{and} \quad L_n f(\vec{x}) = \sum_{i=1}^{k} n\lambda_i(\vec{x}) \left(f\left(\vec{x} + \frac{\vec{e}_i}{n}\right) - f(\vec{x}) \right). \quad (4.14)$$

Note that a generator defines the *distribution* of the process—not the paths. If the λ_j are constant (or more generally if $\lambda(n\vec{x}) = \lambda(\vec{x})$ for all $n \geq 1$), then we can define $\vec{z}_n$ by scaling time and space (as was done for the Poisson process). But this is not true for the general case; see also Chapter 5.

Exercise 4.16. Let $\vec{z}_n$ is defined through (4.14) and assume $\vec{z}_n(0) = 0$. Assume that $\log \lambda_i$ are bounded and Lipschitz continuous. Show that $\vec{z}_n \Rightarrow \vec{z}_\infty$, where

$$\vec{z}_\infty(0) = 0, \quad \text{and} \quad \frac{d\vec{z}_\infty(t)}{dt} = \sum_{i=1}^{k} \lambda_i(\vec{z}_\infty(t))\vec{e}_i.$$

Hint: use (4.12), write L as $L_n - (L - L_n)$, and bound the "error term," noting that f is smooth and bounded.

Generating martingales.

Using Theorem 4.12 we can generate other martingales. Our first goal is to derive (heuristically) the famous "exponential martingale" (4.16) below. We start with a Markov process x and its generator L. Fix a nice (smooth, bounded) function q and consider the two-dimensional process

$$z(t) \overset{\triangle}{=} \left\{ x_t, \exp\left(-\int_0^t q(x_s)\, ds\right) \right\}.$$

It is easy to see (at least formally) from Definition 4.6 that the generator $\tilde{L}$ for the process z is given by

$$\tilde{L}f(x, y) \stackrel{\triangle}{=} Lf(x, 1) - q(x) \cdot \frac{d}{dy}f(x, y) . \tag{4.15}$$

[Use Taylor expansion near the initial point $(x, 0)$ and expand the exponent to a series.] Consider, for example, a function of the form $f(x) \cdot g(y)$, where $g(y) \equiv y$. Then

$$\tilde{L}\Big(f(x) \cdot g(y)\Big) = Lf(x) - q(x)f(x) .$$

Now let $q \stackrel{\triangle}{=} Lf/f$ (assume f is bounded away from 0). Then $\tilde{L}\Big(f(x) \cdot g(y)\Big) = 0$. Theorem 4.12 and the definition (4.10) now imply that

$$\begin{aligned} M_{f \cdot g}(t) &\stackrel{\triangle}{=} (f \cdot g)(z_t) - \int_0^t \tilde{L}\Big(f \cdot g\Big)(z_t)\, ds \\ &= f(x_t) \exp\left(-\int_0^t \left(\frac{Lf(x_s)}{f(x_s)}\right) ds\right) \end{aligned} \tag{4.16}$$

is a martingale, called an exponential martingale associated with the process x. For a precise derivation of (4.16) see §B.4.

Exercise 4.17. Show that for a birth-death process (4.8), the exponential martingale corresponding to $f(x) = e^{\langle \theta, x \rangle}$ takes the form

$$M_t = \exp\left[\langle x_t, \theta \rangle - \int_0^t \sum_{i=1}^k \lambda_i(x_s)\left(e^{\langle \theta, \vec{e}_i \rangle} - 1\right) ds\right]. \tag{4.17}$$

♠

Change of measure.
In the proof of Chernoff's Theorem we obtained the lower bound on the probability of a rare event by a change of measure. In other words, we constructed a different distribution, or measure, $\tilde{\mathbb{P}}$, and wrote, for a set or event A

$$\mathbb{P}(A) = \int_A d\mathbb{P} = \int_A \left(\frac{d\mathbb{P}}{d\tilde{\mathbb{P}}}\right) d\tilde{\mathbb{P}}.$$

$d\mathbb{P}/d\tilde{\mathbb{P}}$ is the Radon-Nikodym derivative; see §A.4. We would like to use the same idea for Poisson and related processes. Here is a heuristic derivation of a formula for $d\mathbb{P}/d\tilde{\mathbb{P}}$ when $\mathbb{P}$ and $\tilde{\mathbb{P}}$ represent Poisson processes. A rigorous treatment is given in §B.4.

Suppose that we observe a Poisson process with rate λ over a time interval $[0, T]$, and we see n events, at times $t_1 < t_2 < \ldots < t_n$ where $0 < t_1$ and $t_n < T$. From Definition 4.1, the probability that no event would occur in $[0, t_1]$ is $e^{-\lambda t_1}$.

From Proposition 4.2, the probability that an event would occur in $[t_1, t_1 + dt)$ is $\lambda \, dt$. Similarly, the probability of no event in $[t_1, t_2)$ is $e^{-\lambda(t_2 - t_1)}$ and that of a single event in $[t_2, t_2 + dt)$ is $\lambda \, dt$. Similar expressions hold for subsequent intervals. Since the intervals between events are disjoint, the events are independent, and we may compute the likelihood of the path we observed as

$$
\begin{aligned}
d\mathbb{P}(\omega) &= e^{-\lambda t_1} \lambda \, dt \cdot e^{-\lambda(t_2 - t_1)} \lambda \, dt \cdots e^{-\lambda(t_n - t_{n-1})} \lambda \, dt \cdot e^{-\lambda(T - t_n)} \\
&= e^{-\lambda T} \lambda^n \, (dt)^n.
\end{aligned}
$$

Similarly, the likelihood of this path for a Poisson process with rate μ and distribution $\tilde{\mathbb{P}}$ is

$$
d\tilde{\mathbb{P}}(\omega) = e^{-\mu T} \mu^n \, (dt)^n.
$$

Therefore,

$$
\frac{d\mathbb{P}}{d\tilde{\mathbb{P}}}(\omega) = \frac{e^{-\lambda T} \lambda^n \, (dt)^n}{e^{-\mu T} \mu^n \, (dt)^n} = e^{-(\lambda - \mu)T} \left(\frac{\lambda}{\mu} \right)^n ;
$$

which we may also write as

$$
\frac{d\mathbb{P}}{d\tilde{\mathbb{P}}}(\omega) = \exp\left(-\int_0^T (\lambda - \mu) \, ds - \sum_{j=1}^n \log\left(\frac{\mu}{\lambda} \right) \right). \tag{4.18}
$$

Exercise 4.18. There is another way to arrive at (4.18). Divide the interval $[0, T]$ into n equal intervals, and write the Radon-Nikodym derivative for the process sampled at the end of these intervals. ♠

Thus the Radon-Nikodym derivative depends only on the number of jumps in the interval, but not on their times. To get the correct formula we replace n by the (random) number N of jumps on $[0, T]$. Now if we have a pair of Poisson processes $\mathbb{P}$ and $\tilde{\mathbb{P}}$ with non-constant rates, the same reasoning leads to

$$
\frac{d\mathbb{P}}{d\tilde{\mathbb{P}}}(\omega) = \exp\left(-\int_0^T (\lambda(x_s) - \mu(x_s)) \, ds - \sum_{j=1}^N \log\left(\frac{\mu(x_{t_j^-})}{\lambda(x_{t_j^-})} \right) \right).
$$

Here $x_{t_j^-}$ is the state of the process just before the jth jump and N is the (random) number of jumps. Thus when the rates depend on the position, the Radon-Nikodym derivative depends on the number of jumps in the interval, and also on the times of jumps.

Finally, to generalize this to multidimensional jump Markov processes simply note that by uniformization [see the end of §4.2 and Equation (B.9), in §B.1) we can consider the jumps as being driven by a single Poisson process. So if we have processes $\mathbb{P}$ and $\tilde{\mathbb{P}}$ as jump Markov with jump directions $\{\vec{e}_i, \ i = 1, \ldots, k\}$ and corresponding jump rates $\lambda_i(\vec{x})$ and $\mu_i(\vec{x})$, then we should expect

$$\frac{d\mathbb{P}}{d\widetilde{\mathbb{P}}}(\omega) = \exp\left(-\int_0^T \sum_{i=1}^k (\lambda_i(\vec{x}_s) - \mu_i(\vec{x}_s))\,ds - \sum_{j=1}^N \log\left(\frac{\mu_{l(j)}(\vec{x}_{t_j^-})}{\lambda_{l(j)}(\vec{x}_{t_j^-})}\right)\right),$$

(4.19)

where $l(j)$ denotes the directions of the jth jump, i.e., $l(j) = l$ implies $\vec{x}_{t_j} - \vec{x}_{t_j^-} = \vec{e}_l$. This is proved in Theorem B.6, §B.4. It is also established that, as a function of T, $d\mathbb{P}/d\widetilde{\mathbb{P}}$ is a martingale, and in fact every positive mean-one martingale defines a change of measure. These facts hold for much more general processes.

Chapter 5

Large Deviations for Processes

In this chapter we state and prove a large deviations principle for a class of multi-dimensional jump processes. We introduced multidimensional jump Markov processes in Chapter 4; let us briefly recapitulate. We are given a finite set of vectors $\{\vec{e}_1, \ldots, \vec{e}_k\}$ in $\mathbb{R}^d$. We consider processes $\{\vec{x}(t), \ t \geq 0\}$ with generator

$$Lf(\vec{x}) = \sum_{i=1}^{k} \lambda_i(\vec{x})(f(\vec{x} + \vec{e}_i) - f(\vec{x})).$$

Here $\vec{e}_i$ is a jump direction and $\lambda_i(\vec{x})$ the rate of jump in that direction when the position is $\vec{x}$.

Recall that in Chernoff's Theorem applied to a Poisson random variable (Example 1.13), the question of estimating

$$\mathbb{P}(x_1 + \cdots + x_n \geq na), \qquad x_i \overset{\mathcal{L}}{=} \text{Pois}(\lambda)$$

is equivalent to estimating

$$\mathbb{P}(y_n/n \geq a), \qquad y_n \overset{\mathcal{L}}{=} \text{Pois}(n\lambda).$$

As in (4.13), applying this scaling to the process $\vec{x}$ we are led to consider the process $\vec{z}_n(t)$ whose generator is

$$L_n f(\vec{x}) = \sum_{i=1}^{k} n\lambda_i(\vec{x}) \left(f\left(\vec{x} + \frac{\vec{e}_i}{n}\right) - f(\vec{x}) \right). \tag{5.1}$$

In other words, the process $\vec{z}_n(t)$ has jump rates $n\lambda_i$ and jump directions $\vec{e}_i/n$. The process $\vec{z}_n(t)$ is the main object of study in this book. Nearly all the applications have models of the form $\vec{z}_n(t)$.

We shall assume that the logarithms of the rates $\log \lambda_i(\vec{x})$ of (5.1) are bounded and continuous. In particular, this precludes a case where some rates vanish and, in terms of applications, implies that there are no boundaries. However, the results also apply in many cases when there are boundaries; in some cases, use of the contraction principle (§2.3) transforms a problem with boundaries to a problem without boundaries (see, for example, §11.4). Freidlin-Wentzell's theory of Chapter 6 allows us to essentially ignore point discontinuities, i.e., these two chapters put together provide most of the results we need for one-dimensional processes. Furthermore, many interesting questions about higher-dimensional processes with boundaries do not involve the boundaries; in these cases the results of this chapter apply.

There is no theory available, at present, for the case of "general boundaries." Chapter 8 is devoted to the large deviations theory for processes with some specific types of boundaries.

The results we seek concern the question: how likely is it that the process $\vec{z}_n$ stays in a given set of paths? To describe the behavior of the *process* $\vec{z}_n$ in the manner of Chernoff's Theorem or according to the general setup of §2.1, we need to define the applicable space of functions to which the paths of the process belong. We also need the notions of open and closed sets, i.e., a topology. In our case the relevant space is simple: it includes functions with values in $\mathbb{R}^d$ that are piecewise constant and right-continuous (as are the multidimensional jump Markov processes), and their limits, which are continuous functions. The appropriate topology for our applications is the Skorohod topology which, unfortunately, is not simple. In §A.1 we provide first some intuition and then a precise definition for the Skorohod space $D^d[0, T]$, its topology and the metric d_d we shall use. The relation with the more familiar "sup norm," which we will also use, is discussed in the appendix as well.

For a large deviations principle, we also need a rate function I. Heuristically, in order for the process $\vec{z}_n$ to be near some continuous function $\vec{r}$, both processes should start nearly at the same point and also have nearly the same increments, i.e.,

$$\vec{z}_n(0) \approx \vec{r}(0) \quad \text{and} \quad \vec{z}_n(t) - \vec{z}_n(s) \approx \vec{r}(t) - \vec{r}(s) \approx \vec{r}'(s)(t - s)$$

for all $0 \le s \le t \le T$ (at least when $|t - s|$ is small). But locally (i.e., as long as the rates λ_i do not change much), $\vec{z}_n(t) - \vec{z}_n(s)$ is a sum of independent Poisson processes (one for each direction of jump). From Exercise 1.25, if

$$\vec{v}_i \overset{\triangle}{=} \sum_{j=1}^{k} y_{ij} \vec{e}_j$$

where the y_{ij} are i.i.d. and $y_{ij} \overset{\mathcal{L}}{=} \text{Pois}(\lambda_j(\vec{x}))$, then in the notation (1.10) of random variables, the probabilities of rare events are governed by the function

$$\ell(\vec{x}, \vec{y}) \overset{\triangle}{=} \sup_{\vec{\theta}} \left(\langle \vec{\theta}, \vec{y} \rangle - g(\vec{x}, \vec{\theta}) \right), \tag{5.2}$$

where

$$g(\vec{x}, \vec{\theta}) \overset{\triangle}{=} \log M(\vec{\theta}) = \log \mathbb{E} e^{\langle \vec{\theta}, \vec{v}_1 \rangle} \tag{5.3}$$

$$= \sum_{i=1}^{k} \lambda_i(\vec{x}) \left(e^{\langle \vec{\theta}, \vec{e}_i \rangle} - 1 \right). \tag{5.4}$$

If the rates $\lambda_i(\vec{x})$ were constant and $\vec{r}$ linear, we would obtain, over the time interval $[0, T]$, the rate function $T \cdot \ell(\vec{x}, \vec{y})$, where $\vec{y} \overset{\triangle}{=} \vec{r}'(t)$. The local effect of the rates and the generality (non-linearity) of the function $\vec{r}$ are taken into account by the definition of the rate function

$$I_0^T(\vec{r}) = \begin{cases} \displaystyle\int_0^T \ell\left(\vec{r}(s), \vec{r}\,'(s)\right)\, ds & \text{if } \vec{r} \text{ is absolutely continuous,} \\ \infty & \text{otherwise.} \end{cases} \tag{5.5}$$

Thus, ℓ is just the "local" rate function.

The large deviations principle for $\vec{z}_n$ that we prove in this chapter is:

Theorem 5.1. *Assume that for each i, $\log \lambda_i(\vec{x})$ is a bounded and Lipschitz continuous function. Then I_0^T is a good rate function [in $\left(D^d[0, T], d_d\right)$], and*

(i) *For every closed set $F \in D^d[0, T]$ and every $\vec{x}$,*

$$\limsup_{n \to \infty} \frac{1}{n} \log \mathbb{P}_{\vec{x}}\left(\vec{z}_n \in F\right) \leq -\inf\left\{ I_0^T(\vec{r}) : \vec{r} \in F,\ \vec{r}(0) = \vec{x} \right\}.$$

(ii) *For every open set $G \in D^d[0, T]$, uniformly for $\vec{x}$ in compact sets,*

$$\liminf_{n \to \infty} \frac{1}{n} \log \mathbb{P}_{\vec{x}}\left(\vec{z}_n \in G\right) \geq -\inf\left\{ I_0^T(\vec{r}) : \vec{r} \in G,\ \vec{r}(0) = \vec{x} \right\}.$$

The proof of this theorem is in three parts. That I_0^T is a good rate function is established in Proposition 5.49 and Corollary 5.50, §5.2. The lower bound is restated as Theorem 5.51, §5.3, with a proof at the end of that section. The upper bound is restated as Theorem 5.54, §5.4, and proved as a consequence of Theorem 5.64, §5.5.

Remarks.

a. In the notation of §2.1, the space we consider is

$$\left(D^d[0, T] \cap \{\vec{r} : \vec{r}(0) = \vec{x}\}, d_d\right).$$

By Theorem A.9, a set S is open (closed) in this space if and only if it is of the form $S' \cap \{\vec{r} : \vec{r}(0) = \vec{x}\}$, where S' is open (closed) in $\left(D^d[0, T], d_d\right)$. Thus Theorem 5.1 makes a statement that is stronger than a standard large deviations principle: it concerns a family of such principles, with a "parameter" $\vec{x}$.

b. The uniformity in the lower bound means that we can get close to the bound by choosing n large, simultaneously for all $\vec{x}$ in a compact set. Uniformity in the upper bound holds in a weaker sense: the right-hand side, where I is evaluated at paths starting at $\vec{x}$, is a bound not only for the process starting at $\vec{x}$, but is also close to the bound for neighboring starting points. See Theorem 5.64 below, Corollary 5.65, and the remark concerning terminology following the corollary.

c. Although we usually know the starting point for $\vec{z}_n$, the reason we need to state Theorem 5.1 with $\mathbb{P}_{\vec{x}}$ and obtain uniformity in the initial condition $\vec{x}$ is that if the initial condition is not known, it can have a distribution in a small neighborhood without affecting the bounds. This uniformity is also useful in obtaining bounds; see, for example, the derivation below.

d. In (4.13) we defined the processes $\vec{z}_n$ in the one-dimensional, constant coefficient case. For that case, (5.1) was a consequence of the definition $z_n(t) = n^{-1} y(nt)$ where y is a Poisson process. In general, the definition (5.1) is equivalent to a direct, explicit definition $\vec{z}_n(t) = n^{-1} x(nt)$ only in the constant coefficient case. However, we allow the rates to depend on the space variables. Thus definition (5.1) is more general.

e. The derivation applies to the case where the $\vec{e}_i$ are not on a lattice, so it encompasses more than birth-death processes. However, our applications will use only the birth-death case. It is also possible to treat time-inhomogeneous rates with little change, but we will not do so—see [Wen].

To illustrate the statement of this theorem, consider some continuous function (path) $\vec{r}(t)$ and the following sets of paths around it: the "open sausage"

$$G_\varepsilon \overset{\triangle}{=} \left\{ \vec{z} \in D^d[0, T] : \sup_{0 \le t \le T} |\vec{r}(t) - \vec{z}(t)| < \varepsilon \right\}$$

and the "closed sausage"

$$F_\varepsilon \overset{\triangle}{=} \left\{ \vec{z} \in D^d[0, T] : \sup_{0 \le t \le T} |\vec{r}(t) - \vec{z}(t)| \le \varepsilon \right\}.$$

(See Exercise A.63 for information about these sets.) Then the large deviations theorem states that *the probability that the path $\vec{z}_n$ stays close to $\vec{r}$ on the interval $[0, T]$ is roughly*

$$e^{-n I_0^T(\vec{r}) + o(n)} \le \mathbb{P}(G_\varepsilon) \le \mathbb{P}(F_\varepsilon) \le e^{-n I_0^T(\vec{r}) + o(n)},$$

provided $\vec{z}_n(0) = \vec{r}(0) = \vec{x}$ is fixed.

Theorem 5.1 as well as the illustration above deal with sets of paths over a finite time interval. This is an essential restriction: as the exercise below shows, in general even the "law of large numbers" would not work on the infinite-time interval. However, it is possible to "bootstrap" the finite-interval theory: this is the subject of Chapter 6.

Exercise 5.2. Let $x_1, x_2, \dots$ be a sequence of independent Poisson processes with mean λ. Show that

$$\mathbb{P}\left(\sup_{t \ge 0} \left| \frac{x_1(t) + x_2(t) + \dots + x_n(t)}{n} - \lambda \cdot t \right| \ge \varepsilon \right) = 1$$

for every n and every positive ε.   ♠

Here is a heuristic derivation of the statement "the probability that the process $\vec{z}_n$ stays close to a given function $\vec{r}(t)$ is approximately $\exp\left(-n I_0^T(\vec{r})\right)$." Divide the interval $[0, T]$ to J equal subintervals of size Δ. The probability that $\vec{z}_n(t)$ is close to $\vec{r}(t)$ for all $0 \leq t \leq T$ is approximated by the product of the probabilities that it remains close on each interval: it starts near $\vec{r}(j\Delta)$ and goes in the same direction as $\vec{r}$, namely in the direction $\vec{r}'(j\Delta)$ for each j. More precisely, by the Markov property, for any $0 < s < T$,

$$\mathbb{P}_{\vec{r}(0)}\left(\sup_{0 \leq t \leq T} |\vec{z}_n(t) - \vec{r}(t)| < \delta\right)$$

$$= \mathbb{E}_{\vec{r}(0)}\left(\mathbf{1}\left[\sup_{0 \leq t \leq s} |\vec{z}_n(t) - \vec{r}(t)| < \delta\right]\mathbf{1}\left[\sup_{s \leq t \leq T} |\vec{z}_n(t) - \vec{r}(t)| < \delta\right]\right)$$

$$= \mathbb{E}_{\vec{r}(0)}\left(\mathbf{1}\left[\sup_{0 \leq t \leq s} |\vec{z}_n(t) - \vec{r}(t)| < \delta\right]\right.$$

$$\left. \times \, \mathbb{E}_{\vec{z}_n(s)}\left(\mathbf{1}\left[|\vec{z}_n(s) - \vec{r}(s)| < \delta\right]\mathbf{1}\left[\sup_{s \leq t \leq T} |\vec{z}_n(t) - \vec{r}(t)| < \delta\right]\right)\right)$$

$$\leq \mathbb{E}_{\vec{r}(0)}\mathbf{1}\left[\sup_{0 \leq t \leq s} |\vec{z}_n(t) - \vec{r}(t)| < \delta\right]$$

$$\times \sup_{|\vec{x} - \vec{r}(s)| < \delta} \mathbb{E}_{\vec{x}}\mathbf{1}\left[\sup_{0 \leq t \leq T-s} |\vec{z}_n(t) - \vec{r}(t)| < 2\delta\right].$$

Iterating on the J subintervals,

$$\mathbb{P}_{\vec{r}(0)}\left(\sup_{0 \leq t \leq T} |\vec{z}_n(t) - \vec{r}(t)| < \delta\right) \leq \mathbb{P}_{\vec{r}(0)}\left(\sup_{0 \leq t \leq \Delta} |\vec{z}_n(t) - \vec{r}(t)| < \delta\right)$$

$$\times \sup_{|\vec{x} - \vec{r}(\Delta)| < \delta} \mathbb{P}_{\vec{x}}\left(\sup_{\Delta \leq t \leq 2\Delta} |\vec{z}_n(t) - \vec{r}(t)| < 2\delta\right)$$

$$\vdots \qquad\qquad\qquad\qquad\qquad\qquad (5.6)$$

$$\times \sup_{|\vec{x} - \vec{r}((J-1)\Delta)| < (J-1)\delta} \mathbb{P}_{\vec{x}}\left(\sup_{(J-1)\Delta \leq t \leq T} |\vec{z}_n(t) - \vec{r}(t)| < J\delta\right).$$

Since we assume that $\vec{z}_n$ and $\vec{r}$ are close at the beginning of each subinterval, we need to find the probability that a change $\Delta\vec{z}_n(t)$ in $\vec{z}_n(t)$ is about equal to the change $\Delta\vec{r}(t) \approx \vec{r}'(t)\Delta t$ in $\vec{r}(t)$. Now over a small interval of time, the jump rates $\lambda_i(\vec{z}_n(t))$ do not change much. Therefore, Proposition 4.5(ii) (extended to the multidimensional case) indicates that we can represent

$$\Delta \vec{z}_n(t) \stackrel{\Delta}{=} \vec{z}_n(t + \Delta t) - \vec{z}_n(t)$$

$$\stackrel{\mathcal{L}}{=} \frac{1}{n} \sum_{i=1}^{k} \sum_{j=1}^{n\Delta t} y_{ji} \vec{e}_i,$$

where each y_{ji} has a Poisson distribution with rate $\lambda_i(\vec{z}_n(t))$. Then by (5.3) and the large deviations results for random variables,

$$\mathbb{P}\left(|(\Delta \vec{z}_n(t) - \vec{r}'(t)\Delta t| < \delta\right) = \mathbb{P}\left(\left|\frac{\Delta \vec{z}_n(t)}{\Delta t} - \vec{r}'(t)\right| < \frac{\delta}{\Delta t}\right)$$

$$= \mathbb{P}\left(\left|\frac{1}{n\Delta t} \sum_{i=1}^{k} \sum_{j=1}^{n\Delta t} y_{ji} \vec{e}_i - \vec{r}'(t)\right| < \frac{\delta}{\Delta t}\right)$$

$$\approx \exp(-n\Delta t\, \ell(\vec{z}_n(t), \vec{r}'(t)).$$

We must let $\delta \to 0$ before we let $\Delta t \to 0$. This will be apparent during the technical parts of the proofs in this chapter. What makes this argument work is that, to first order, the value of δ is immaterial in the result of Chernoff's Theorem.

Now on the event of interest, $\vec{z}_n(t)$ is close to $\vec{r}(t)$. Since $\lambda_i(\vec{x})$ is smooth in $\vec{x}$ and $\vec{r}$ is continuous,

$$\ell(\vec{z}_n(t), \vec{r}'(t))) \approx \ell(\vec{r}(j\Delta), \vec{r}'(t))) \text{ whenever } t \in [j\Delta, j\Delta + \Delta].$$

Therefore,

$$\mathbb{P}_{\vec{r}(0)}\left(\sup_{0 \le t \le T} |\vec{z}_n(t) - \vec{r}(t)| < \delta\right) \approx \prod_{j=0}^{J-1} \exp\left(-n\ell(\vec{r}(j\Delta), \vec{r}'(j\Delta))\Delta\right)$$

$$\approx \exp\left(-n \int_0^T \ell(\vec{r}(t), \vec{r}'(t))\, dt\right).$$

A similar approximation provides a lower bound of the same form: replace the left-hand side in (5.6) by

$$\mathbb{P}_{\vec{r}(0)}\left(\sup_{0 \le t \le T} |\vec{z}_n(t) - \vec{r}(t)| < J\delta\right)$$

to obtain a lower bound, and proceed along the same lines.

The remainder of this chapter is organized as follows. We start in §5.1 with Kurtz's Theorem, which is a generalization of the strong law of large numbers to the processes $\vec{z}_n$. It concerns a deterministic process denoted $\vec{z}_\infty$ that has the property that $\vec{z}_n(t) \to \vec{z}_\infty(t)$ in a very strong sense. This result is interesting in its own right, and is a key element in the proof of the lower bound. It is used extensively in the proofs and in the applications that follow.

Next, in §5.2, we derive some properties of the rate function. The derivations are complicated enough (and technical enough) to warrant a separate section. If this is your initiation into large deviations for processes, we recommend that you skip this section for now, or at least skip the proofs.

In §5.3 we prove the lower bound for open sets. The proof is very similar to the proof of the lower bound in Chernoff's Theorem: we define a "twisted process" (one with jump rates that are different from the λ_i, but with the same jump directions $\vec{e}_i$). We show that the law of large numbers (Kurtz's Theorem) for the twisted process gives a lower bound. The lower bound is derived in terms of a function J, defined in (5.23), that does not have the form of the rate function I. In Lemma 5.26 of §5.2 we establish the equivalence of I and J.

Finally, in §5.5, we prove the upper bound for closed sets. As in Chernoff's Theorem, the key is an exponential Chebycheff estimate. This estimate is derived here from the exponential martingale, defined in (4.16) and (B.32). Applying (4.17) with the function $e^{\langle \vec{x}, \vec{\theta} \rangle}$ [Exercise 4.17 and Equation (4.17)] we obtain a martingale

$$M_t \overset{\triangle}{=} \exp\left[\langle \vec{z}_n(t), \vec{\theta} \rangle - \int_0^t \sum_{i=1}^k n\lambda_i(\vec{z}_n(s)) \left(e^{\langle \vec{\theta}, \vec{e}_i/n \rangle} - 1 \right) ds \right]$$

for every $\vec{\theta} \in \mathbb{R}^d$. The upper bound is complicated enough that we provide a separate orientation section—§5.4—to help expose the overall plan of the proof.

5.1. Kurtz's Theorem

Kurtz's Theorem is an extension of the law of large numbers. If $x_1, x_2, \ldots$ are i.i.d. random variables with mean μ, then the law of large numbers states that

$$\left| \frac{x_1 + \cdots + x_n}{n} - \mu \right|$$

becomes small as n increases. For random processes, a similar result can be stated: if $x_1(t), x_2(t), \ldots$ are i.i.d. processes with mean $\mu(t)$, then under the appropriate conditions,

$$\sup_{0 \le t \le T} \left| \frac{x_1(t) + \cdots + x_n(t)}{n} - \mu(t) \right|$$

becomes small as n increases. In fact, a slightly more general result is obtained here by analyzing the process $\vec{z}_n$ instead of $n^{-1}(x_1 + \cdots + x_n)$.

To see that $\vec{z}_n$ is indeed a generalization of $n^{-1}(x_1 + \cdots + x_n)$, let $x_1(t), x_2(t), \ldots$ be independent Poisson processes with rate λ. Then, by Proposition 4.5(ii), $x_1(t) + \cdots + x_n(t)$ is a Poisson process with rate $n\lambda$, and so by (4.5) and (5.1),

$$n^{-1}(x_1(t) + \cdots + x_n(t)) = \vec{z}_n(t),$$

where $\vec{e}_1 = 1$ and $\lambda_1 = \lambda$. But we allow $\lambda_1 = \lambda_1(x)$, corresponding to a type of coupling among the processes x_i that physicists call "mean field coupling."

Let us examine the generator of a general $\vec{z}_n$ to see what we might expect as $n \to \infty$. From the definition (5.1),

$$L_n f(\vec{x}) = \sum_{i=1}^{k} n\lambda_i(\vec{x}) \left(f\left(\vec{x} + \frac{\vec{e}_i}{n}\right) - f(\vec{x}) \right)$$

$$= \sum_{i=1}^{k} \lambda_i(\vec{x}) \langle \nabla f(\vec{x}), \vec{e}_i \rangle + O\left(\frac{1}{n}\right)$$

by Taylor's Theorem. We are led to the definition

$$L_\infty f(\vec{x}) = \sum_{i=1}^{k} \lambda_i(\vec{x}) \langle \nabla f(\vec{x}), \vec{e}_i \rangle.$$

Since L_∞ is a first order differential operator, it corresponds to a deterministic process (Corollary 4.15) satisfying

$$\frac{d}{dt}\vec{z}_\infty(t) = \sum_{i=1}^{k} \lambda_i(\vec{z}_\infty(t))\vec{e}_i. \qquad (5.7)$$

The aim of this section is to prove the following result.

Theorem 5.3 Kurtz [Ku2]. *Let* $\lambda_i(\vec{x}) : \mathbb{R}^d \to \mathbb{R}^+$ *be uniformly bounded and Lipschitz continuous, and let* $\vec{z}_\infty$ *be the unique solution of (5.7) with* $\vec{z}_\infty(0) = \vec{x}$. *For each finite* T *there exist a positive constant* C_1 *and a function* C_2 *with*

$$\lim_{\varepsilon \downarrow 0} \frac{C_2(\varepsilon)}{\varepsilon^2} \in (0, \infty) \quad \text{and} \quad \lim_{\varepsilon \uparrow \infty} \frac{C_2(\varepsilon)}{\varepsilon} = \infty$$

such that, for all $n \geq 1$ *and* $\varepsilon > 0$,

$$\mathbb{P}_{\vec{x}} \left(\sup_{0 \leq t \leq T} |\vec{z}_n(t) - \vec{z}_\infty(t)| \geq \varepsilon \right) \leq C_1 e^{-nC_2(\varepsilon)}.$$

Moreover, C_1 *and* C_2 *can be chosen independently of* $\vec{x}$.

Remark. Under the assumptions on λ_i, the solution $\vec{z}_\infty$ exists for all t, is continuously differentiable, and is unique; see Definition A.65 and Theorems A.66 and A.67.

For the Poisson process, we can derive something very close to Kurtz's Theorem by a different approach. Let $\vec{z}_n(t)$ be a Poisson process with rate λ and let $\vec{z}_n(0) = 0$. Then $\vec{z}_n(T)$ is a Poisson random variable with parameter λT, and by Chernoff's Theorem, for $a > \lambda T$ (Example 1.13), $\mathbb{P}(\vec{z}_n(T) \geq a) \leq e^{-n\ell(a)}$, where

$$\ell(a) = a\left(\log\left(\frac{a}{\lambda T}\right) - 1\right) + \lambda T.$$

Now suppose that $\vec{z}_n(T) = b$ for some b. By Proposition 4.4, we know that the nb jumps of $\vec{z}_n$ occur uniformly in $[0, T]$. The Kolmogorov-Smirnov Theorem [Br §13.6] states that for a sequence $U_1, U_2, \ldots$ of i.i.d. uniformly $[0, 1]$ distributed random variables, the empirical distribution

$$F_n(x) \overset{\triangle}{=} \frac{1}{n} \sum_{i=1}^{n} \mathbf{1}\,[U_i \leq x]$$

satisfies

$$\sup_{0 \leq x \leq 1} \sqrt{n}(F_n(x) - x) \overset{d}{\to} \chi,$$

where χ is a known distribution with a Gaussian tail. Representing $\vec{z}_n$ by those nb uniform $[0, 1]$ random variables, a little algebra shows that

$$\vec{z}_n(t) - b\frac{t}{T} = b\left[F_{nb}\left(\frac{t}{T}\right) - \frac{t}{T}\right].$$

Therefore, conditioned on $\vec{z}_n(T) = b$,

$$\sup_{0 \leq t \leq T} \left|\vec{z}_n(t) - b\frac{t}{T}\right| \overset{d}{\approx} \frac{1}{\sqrt{nb}}\, b \cdot \chi.$$

In other words, the deviation between $\vec{z}_n(t)$ and bt/T, given $\vec{z}_n(T) = b$, is about $\sqrt{b/n}$ times a random variable with a Gaussian tail. Hence

$$\mathbb{P}\left(\sup_{0 \leq t \leq T} |\vec{z}_n(t) - \lambda t| > \varepsilon\right)$$

$$\leq \mathbb{P}\left(|\vec{z}_n(T) - \lambda T| > \frac{\varepsilon}{2}\right)$$

$$+ \mathbb{P}\left(\sup_{0 \leq t \leq T}\left|\vec{z}_n(t) - \left(\frac{t}{T}\right)\vec{z}_n(T)\right| \geq \frac{\varepsilon}{2} \,\middle|\, |\vec{z}_n(T) - \lambda T| \leq \frac{\varepsilon}{2}\right)$$

$$\approx \exp\left(-n \min\left\{\ell\left(\lambda T + \frac{\varepsilon}{2}\right), \ell\left(\lambda T - \frac{\varepsilon}{2}\right)\right\}\right) + \mathbb{P}\left(\chi > \frac{\varepsilon}{2}\sqrt{\frac{n}{\lambda T}}\right),$$

where ℓ is the rate function for the random variable $\vec{z}_n(T)$. Since χ has a Gaussian tail, we know that $\mathbb{P}(\chi > a) \leq \exp(-ca^2)$ for some $c > 0$; hence

$$\mathbb{P}\left(\chi > \frac{\varepsilon}{2}\sqrt{\frac{n}{\lambda T}}\right) \leq e^{-nc}.$$

The only reason that the preceding argument is not a proof of exponential (n) decay is that we are interchanging the Kolmogorov-Smirnov limit with the large deviations limit on χ; this interchange should be justified. Instead of doing so, we will prove the general theorem; we hope that our argument has rendered Kurtz's Theorem plausible.

Kurtz's Theorem is the basis of our large deviations results. It has many other uses as well. When analyzing a system that can be modeled by a process such as

$\vec{z}_n$, clearly the first step is to investigate the behavior of $\vec{z}_\infty$. This basic method is increasingly utilized, sometimes under the rubric "fluid limits" or "mean flow" analysis [Cn, CM, Da, DM, Ku1, We2].

As far as large deviations is concerned, after the behavior of $\vec{z}_\infty$ has been obtained, one would like to estimate how often $\vec{z}_n(t)$ behaves differently. This is the subject of the lower and upper bounds given in §5.3, §5.4, and §5.5.

As a final note before we begin the proof, (5.7) can be written as

$$\vec{z}_\infty(t) = \vec{z}_\infty(0) + \int_0^t \sum_{i=1}^k \lambda_i(\vec{z}_\infty(s))\vec{e}_i \, ds. \qquad (5.8)$$

This is the form we will use in the proof.

The proof of Kurtz's Theorem is presented as a series of lemmas. This is done partly to give the proof in bite-size chunks, and partly since we will later need to reference the ideas of these lemmas. Those lemmas in turn rely on two technical results. We will now state the technical results, then provide an outline of the proof of Kurtz's Theorem, and then proceed with the proof.

Lemma 5.4 Gronwall's lemma [Hal]. *Let $u(t)$ be a bounded function on $[0, T]$ satisfying*

$$u(t) \leq \varepsilon + \delta \cdot \int_0^t u(s) \, ds$$

for $0 \leq t \leq T$, where δ and ε are some positive constants. Then $u(t) \leq \varepsilon e^{\delta t}$.

Exercise 5.5. Prove Gronwall's Lemma. Hint: $u(t) \leq \varepsilon + B\delta t$, where B bounds u on $[0, T]$. Now use repeated substitutions. ♠

Lemma 5.6. *Suppose $f : D^d[0, T] \times \mathbb{R} \to \mathbb{R}$ and $G : D^d[0, T] \times \mathbb{R} \times \mathbb{R} \to \mathbb{R}$ are such that*

$$M_t \overset{\triangle}{=} \exp\big(\rho f_t(\vec{z}_n) - G_t(\vec{z}_n, \rho)\big)$$

is a right-continuous, mean one martingale for each $\rho > 0$. Suppose further that $R : \mathbb{R} \times \mathbb{R} \to \mathbb{R}$ is monotone increasing in the first argument and

$$G_t(\vec{x}, \rho) \leq R(t, \rho)$$

for all $\vec{x} \in D^d[0, T]$ and all $\rho > 0$. Then for any $T > 0$ and any a,

$$\mathbb{P}_{\vec{x}}\left(\sup_{0 \leq t \leq T} f_t(\vec{z}_n) \geq a\right) \leq \inf_{\rho > 0} \exp\left(R(T, \rho) - a\rho\right).$$

Proof. Fix $\rho > 0$ and use the assumptions on R and G to obtain

$$\mathbb{P}_{\vec{x}}\left(\sup_{0 \le t \le T} f_t(\vec{z}_n) \ge a\right)$$

$$= \mathbb{P}_{\vec{x}}\left(\sup_{0 \le t \le T} \exp\left(\rho f_t(\vec{z}_n)\right) \ge \exp(\rho a)\right)$$

$$\le \mathbb{P}_{\vec{x}}\left(\sup_{0 \le t \le T} \exp\left(\rho f_t(\vec{z}_n) - G_t(\vec{z}_n, \rho)\right) \ge \exp\left(\rho a - R(T, \rho)\right)\right)$$

$$\le \exp\left(R(T, \rho) - \rho a\right)$$

where the last relation follows from the martingale inequality (A.15). Now minimize over ρ. $\blacksquare$

The idea behind the proof of Kurtz's Theorem is very similar to the idea behind large deviations upper bounds: in each direction $\vec{\theta}$, we estimate how much $\vec{z}_n(t)$ can differ from its mean by using an exponential martingale (the process version of Chebycheff's inequality). Since there are only $2d$ basic directions in $\mathbb{R}^d$, namely, the d coordinate directions and their negatives, if we have an estimate for the (scalar) process in each direction, we can piece together a bound on the process $\vec{z}_n(t)$ being far from its mean $\vec{z}_\infty(t)$. Using the formulation for $\vec{z}_n$ given in Example 4.17 and (5.8) for $\vec{z}_\infty$, we see that for any $\vec{\theta}$,

$$\exp\left(\langle \vec{z}_n(t) - \vec{z}_\infty(t), \vec{\theta}\rangle\right.$$

$$\left. - \int_0^t \sum_{i=1}^k \left[\lambda_i(\vec{z}_n(s))n\left(e^{\langle\vec{\theta},\vec{e}_i/n\rangle} - 1\right) - \lambda_i(\vec{z}_\infty(s))\langle\vec{\theta}, \vec{e}_i\rangle \, ds\right]\right) \quad (5.9)$$

is a mean one martingale. Let $\vec{u}(t) \stackrel{\triangle}{=} \vec{z}_n(t) - \vec{z}_\infty(t)$, and recall Taylor's expansion

$$n\left(e^{\langle\vec{\theta},\vec{e}_i/n\rangle} - 1\right) \approx \langle\vec{\theta}, \vec{e}_i\rangle + O\left(\frac{1}{n}\right).$$

Substituting this into (5.9), we obtain that

$$\exp\left(\langle\vec{\theta}, \vec{u}(t)\rangle - \int_0^T \sum_{i=1}^k \langle\vec{e}_i, \vec{\theta}\rangle \left(\lambda_i(\vec{z}_n(t)) - \lambda_i(\vec{z}_\infty(t))\right) + O\left(\frac{1}{n}\right)\right)$$

is a mean one martingale. Now use the Lipschitz continuity of $\lambda_i(\vec{z})$ to replace the term $\lambda_i(\vec{z}_n(t)) - \lambda_i(\vec{z}_\infty(t))$ with an estimate $K|\vec{u}(t)|$, and we are in a position to use Lemma 5.6, with G being the term $O(\frac{1}{n})$. This will show that an integral expression in $\langle\vec{u}(t), \vec{\theta}\rangle$ is small, and Gronwall's inequality will enable us to conclude that $\langle\vec{u}(t), \vec{\theta}\rangle$ itself is small. Now we use the estimates in each of the $2d$ basic directions to conclude that $|\vec{u}(t)|$ is small.

Lemma 5.7. *Let $\vec{y}$ be a random vector (with values in $\mathbb{R}^d$). Suppose there are numbers a and δ such that, for each $\vec{\theta} \in \mathbb{R}^d$ with $|\vec{\theta}| = 1$,*

$$\mathbb{P}\left(\langle \vec{\theta}, \vec{y} \rangle \geq a\right) \leq \delta.$$

Then

$$\mathbb{P}\left(|\vec{y}| \geq a\sqrt{d}\right) \leq 2d\delta.$$

Proof. For $i = 1, \ldots, d$, let $\vec{\theta}_i \overset{\triangle}{=} \vec{u}_i$, the standard orthonormal basis of $\mathbb{R}^d$, and let $\vec{\theta}_{d+i} \overset{\triangle}{=} -\vec{\theta}_i$, $i = 1, \ldots, d$. Now

$$\left\{\vec{y} : |\vec{y}| \geq a\sqrt{d}\right\} \subset \bigcup_{i=1}^{2d} \left\{\vec{y} : \langle \vec{y}, \vec{\theta}_i \rangle \geq a\right\},$$

so $\mathbb{P}\left(|\vec{y}| \geq a\sqrt{d}\right) \leq 2d\delta$ by a union bound (Lemma A.115). ∎

The following corollary is not used in proving Kurtz's Theorem. It is used in the proof of the upper bound. We give it here since the idea is the same as in the next main lemma. By considering the case of T small, the corollary implies that $\vec{z}_n$ is nearly equicontinuous at $t = 0$.

Corollary 5.8. *If $\lambda_i(\vec{x})$ are all bounded, then there is a function $C_3(a)$ with*

$$\lim_{a \to \infty} C_3(a)/a = \infty$$

such that

$$\mathbb{P}_{\vec{x}}\left(\sup_{0 \leq t \leq T} |\vec{z}_n(t) - \vec{z}_n(0)| \geq a\right) \leq 2d\exp\left(-nTC_3\left(\frac{a}{T}\right)\right).$$

Proof. Since the last inequality holds trivially whenever $C_3(a/T) = 0$, it suffices to prove the corollary for a large. For fixed $\vec{x} \in \mathbb{R}^d$, $\vec{\theta} \in \mathbb{R}^d$ with $|\vec{\theta}| = 1$, take

$$f_t(\vec{y}) = \langle \vec{y}(t) - \vec{x}, \vec{\theta} \rangle$$

$$g(\vec{z}, \vec{\mu}) \overset{\triangle}{=} n\sum_{i=1}^{k} \lambda_i(\vec{z})\left(\exp\langle \vec{\mu}, \vec{e}_i/n \rangle - 1\right)$$

$$G_t(\vec{z}, \rho) = \int_0^t g(\vec{z}(s), \rho\vec{\theta})\,ds.$$

To apply Lemma 5.6, note that by Exercise 4.17, f and G make M_t a right-continuous mean one martingale whenever $\vec{z}_n(0) = \vec{x}$. Defining

$$\bar{\lambda} \overset{\triangle}{=} \{\sup \lambda_i(\vec{z}) : \vec{z} \in \mathbb{R}^d, \ 1 \leq i \leq k\}$$

and $\overline{e} \overset{\triangle}{=} \max_{1 \leq i \leq k} |\vec{e}_i|$, we have

$$G_t(\vec{z}, \rho) \leq tnk\overline{\lambda}e^{\overline{e}\rho/n} \overset{\triangle}{=} R(t, \rho).$$

Hence Lemma 5.6 shows that

$$\mathbb{P}_{\vec{x}}\left(\sup_{0 \leq t \leq T} \langle \vec{z}_n(t) - \vec{z}_n(0), \vec{\theta} \rangle \geq a\right) \leq \inf_{\rho > 0} \exp\left(nT\left[k\overline{\lambda}e^{\overline{e}\rho/n} - \frac{\rho}{n}\frac{a}{T}\right]\right)$$

$$= \inf_{\rho > 0} \exp\left(nT\left[k\overline{\lambda}e^{\overline{e}\rho} - \rho\frac{a}{T}\right]\right).$$

Take

$$\rho = \frac{1}{\overline{e}} \log \frac{a}{Tk\overline{\lambda}\overline{e}},$$

which is positive for all a/T large. Then a little algebra show that if we set

$$C(\varepsilon) \overset{\triangle}{=} \frac{\varepsilon}{\overline{e}}\left(\log \frac{\varepsilon}{k\overline{\lambda}\overline{e}} - 1\right) \tag{5.10}$$

for ε large and $C(\varepsilon) \overset{\triangle}{=} 0$ otherwise, then

$$\mathbb{P}_{\vec{x}}\left(\sup_{0 \leq t \leq T} \langle \vec{z}_n(t) - \vec{z}_n(0), \vec{\theta} \rangle \geq a\right) \leq \exp\left(-nTC\left(\frac{a}{T}\right)\right).$$

Applying Lemma 5.7, the corollary is established with $C_3(\varepsilon) \overset{\triangle}{=} C\left(\varepsilon/\sqrt{d}\right)$. ∎

We now come to the main lemma for Kurtz's Theorem. It is proved in exactly the same way as Corollary 5.8.

Lemma 5.9. *Assume the functions λ_i are all bounded and that $\vec{z}_\infty$ is a right-continuous solution of (5.8) with $\vec{z}_\infty(0) = \vec{x}$. Then for each $\vec{\theta} \in \mathbb{R}^d$ with $|\vec{\theta}| = 1$ and each $T > 0$ there is a function $C_3(\varepsilon) > 0$ so that*

$$\lim_{\varepsilon \downarrow 0} \frac{C_3(\varepsilon)}{\varepsilon^2} \in (0, \infty) \quad and \quad \lim_{\varepsilon \uparrow \infty} \frac{C_3(\varepsilon)}{\varepsilon} = \infty,$$

$$\mathbb{P}_x\left(\sup_{0 \leq t \leq T}\left[\langle \vec{z}_n(t) - \vec{z}_\infty(t), \vec{\theta} \rangle \right.\right.$$

$$\left.\left. - \int_0^t \sum_{i=1}^k (\lambda_i(\vec{z}_n(s)) - \lambda_i(\vec{z}_\infty(s))) \langle \vec{\theta}, \vec{e}_i \rangle \, ds\right] \geq \varepsilon\right) \leq e^{-nC_3(\varepsilon)}.$$

Moreover, C_3 can be chosen independently of $\vec{x}$.

Remark. Since we do not assume that the λ_i are continuous, there is no a priori guarantee that a solution to (5.8) even exists.

Proof. As in Corollary 5.8, for each $\vec{\theta} \in \mathbb{R}^d$ and $\rho > 0$,

$$m_t \overset{\triangle}{=} \exp\left(\langle \vec{\theta}\rho, \vec{z}_n(t) - \vec{z}_n(0) \rangle - \int_0^t g(\vec{z}_n(s), \rho\vec{\theta}) \, ds \right)$$

is a mean one right-continuous martingale. Since $\vec{z}_\infty(0) = \vec{z}_n(0)$, by (5.8),

$$
\begin{aligned}
m_t &= \exp\Bigg(\langle \vec{\theta}\rho, \vec{z}_n(t) - \vec{z}_\infty(t) \rangle \\
&\quad - \int_0^t \sum_{i=1}^k \left[n\lambda_i(\vec{z}_n(s)) \left(\exp\langle \rho\vec{\theta}, \vec{e}_i/n \rangle - 1 \right) - \lambda_i(\vec{z}_\infty(s)) \langle \rho\vec{\theta}, \vec{e}_i \rangle \right] ds \Bigg) \\
&= \exp\Bigg(\langle \vec{\theta}\rho, \vec{z}_n(t) - \vec{z}_\infty(t) \rangle \\
&\quad - \int_0^t \sum_{i=1}^k \left[\lambda_i(\vec{z}_n(s)) - \lambda_i(\vec{z}_\infty(s)) \right] \langle \rho\vec{\theta}, \vec{e}_i \rangle \, ds \\
&\quad - \int_0^t \sum_{i=1}^k \lambda_i(\vec{z}_n(s)) \left[n \left(\exp\langle \rho\vec{\theta}, \vec{e}_i/n \rangle - 1 \right) - \langle \rho\vec{\theta}, \vec{e}_i \rangle \right] ds \Bigg).
\end{aligned}
$$

To apply Lemma 5.6, let f denote the first two terms in the exponent and G the last integral. From Taylor's Theorem, for some s with $|s| \le |y|$,

$$n\left(e^{y/n} - 1 \right) - y = \frac{y^2}{2n} e^{s/n} \le \frac{y^2}{2n} e^{|y|/n}.$$

Using this, we see that

$$G_t(\vec{z}, \rho) \le t\bar{\lambda}k \frac{\rho^2 \bar{e}^2}{2n} e^{\bar{e}\rho/n} \overset{\triangle}{=} R(t, \rho),$$

where $\bar{\lambda}$ and $\bar{e}$ are as in Corollary 5.8, and $|\vec{\theta}| = 1$. Now by Lemma 5.6

$$
\begin{aligned}
\mathbb{P}_{\vec{x}}\Bigg(\sup_{0 \le t \le T} &\bigg[\langle \vec{z}_n(t) - \vec{z}_\infty(t), \vec{\theta} \rangle \\
&- \int_0^t \sum_{i=1}^k (\lambda_i(\vec{z}_n(s)) - \lambda_i(\vec{z}_\infty(s))) \langle \vec{\theta}, \vec{e}_i \rangle \, ds \bigg] \ge a \Bigg) \\
&\le \exp\left(T\bar{\lambda}k \frac{\rho^2 \bar{e}^2}{2n} e^{\bar{e}\rho/n} - a\rho \right) \quad (5.11)
\end{aligned}
$$

for all $\rho > 0$. Rewriting in the form of the claim of this lemma and changing variables from ρ to ρn, let

$$C(a) \overset{\triangle}{=} \sup_{\rho > 0} \left(a\frac{\rho}{n} - T\bar{\lambda}k \frac{\rho^2 \bar{e}^2}{2n^2} e^{\bar{e}\rho/n} \right) \quad (5.12a)$$

$$= \sup_{\rho > 0} \left(\rho \left[a - T \overline{\lambda} k \frac{\rho \overline{e}^2}{2} e^{\overline{e}\rho} \right] \right). \tag{5.12b}$$

Choosing ρ small enough so that $T \overline{\lambda} k \rho \overline{e}^2 e^{\overline{e}\rho} < a$ we conclude that $C(a) > 0$, and satisfies the exponential bound. We will now define the function C_3 for small and for large a, and set $C_3(a) = C(a)$ otherwise. For small a, choose $\rho = a/(\overline{e}^2 T \overline{\lambda} k)$ and substitute in (5.12b). With this choice for C_3, a little algebra shows that

$$\frac{C_3(a)}{a^2/(\overline{e}^2 T \overline{\lambda} k)} \to \frac{1}{2}$$

as $a \downarrow 0$, and, in particular, $C_3(a)$ is positive for all $a \neq 0$ small enough. For large a, choose $\rho = \delta \log a$ where δ is chosen to satisfies $\delta \overline{e} < 1$. Then

$$a - T \overline{\lambda} k \frac{\rho \overline{e}^2}{2} e^{\overline{e}\rho} = a - \frac{1}{2} T \overline{\lambda} k \delta \overline{e}^2 \log a \cdot \left(a^{\delta \overline{e}} \right)$$

which is at least $a/2$ for all large a. With this choice of ρ, for all a large enough

$$C_3(a) \overset{\triangle}{=} \rho \left[a - T \overline{\lambda} k \frac{\rho \overline{e}^2}{2} e^{\overline{e}\rho} \right]$$

$$\geq \frac{\delta}{2} a \log a.$$

Finally, the fact that C_3 does not depend on $\vec{x}$ is evident from the proof. ∎

The preceding lemma and Gronwall's inequality yield Kurtz's Theorem in a manner reminiscent of the basic existence proof for solutions of ODEs. Here are the details.

Proof of Kurtz's Theorem 5.3. By hypothesis, $\lambda_i(\vec{x})$ are Lipschitz continuous, so that there is a constant K such that

$$\sum_{i=1}^{k} |\vec{e}_i| \, |\lambda_i(\vec{z}_n(t)) - \lambda_i(\vec{z}_\infty(t))| \leq K |\vec{z}_n(t) - \vec{z}_\infty(t)|. \tag{5.13}$$

Applying Lemmas 5.7 and 5.9, we obtain

$$\mathbb{P}_{\vec{x}} \left(\sup_{0 \leq t \leq T} \left| \vec{z}_n(t) - \vec{z}_\infty(t) - \int_0^t \sum_{i=1}^{k} (\lambda_i(\vec{z}_n(s)) \right. \right. \tag{5.14}$$

$$\left. \left. - \lambda_i(\vec{z}_\infty(s)) \right) \vec{e}_i \, ds \right| \geq \varepsilon \right) \leq 2d e^{-nC_2'(\varepsilon)}$$

where we obtain C_2' from C_3 as in Corollary 5.8. Now for any a, b, c in $\mathbb{R}^d$,

$$|c| < |b| \quad \text{implies} \quad |a| - |b| \leq |a| - |c| \leq |a - c|.$$

Hence by (5.13), for all t,

$$|\vec{z}_n(t) - \vec{z}_\infty(t)| - \int_0^t K|\vec{z}_n(s) - \vec{z}_\infty(s)|\,ds \tag{5.15}$$

$$\leq |\vec{z}_n(t) - \vec{z}_\infty(t)| - \int_0^t \sum_{i=1}^k |\lambda_i(\vec{z}_n(s)) - \lambda_i(\vec{z}_\infty(s))||\vec{e}_i|\,ds$$

$$\leq |\vec{z}_n(t) - \vec{z}_\infty(t)| - \left| \int_0^t \sum_{i=1}^k (\lambda_i(\vec{z}_n(s)) - \lambda_i(\vec{z}_\infty(s)))\,\vec{e}_i\,ds \right|$$

$$\leq \left| \vec{z}_n(t) - \vec{z}_\infty(t) - \int_0^t \sum_{i=1}^k (\lambda_i(\vec{z}_n(s)) - \lambda_i(\vec{z}_\infty(s)))\,\vec{e}_i\,ds \right|.$$

Combining (5.14) and (5.15) yields

$$\mathbb{P}_{\vec{x}}\left(\sup_{0 \leq t \leq T} |\vec{z}_n(t) - \vec{z}_\infty(t)| - \int_0^t K|\vec{z}_n(s) - \vec{z}_\infty(s)|ds \geq \varepsilon \right) \leq 2de^{-nC_2'(\varepsilon)}.$$

By Gronwall's Lemma 5.4, if $u(t) > \varepsilon e^{\delta t}$ for some t, then necessarily

$$u(t) > \varepsilon + \delta \int_0^t u(s)\,ds$$

for some t. Applying this to the function $|\vec{z}_n - \vec{z}_\infty|$ we conclude that

$$\mathbb{P}_{\vec{x}}\left(\sup_{0 \leq t \leq T} |\vec{z}_n(t) - \vec{z}_\infty(t)| \geq \varepsilon e^{KT} \right)$$

$$\leq \mathbb{P}_{\vec{x}}\left(\sup_{0 \leq t \leq T} |\vec{z}_n(t) - \vec{z}_\infty(t)| - \int_0^t K|\vec{z}_n(s) - \vec{z}_\infty(s)|\,ds \geq \varepsilon \right)$$

$$\leq 2d \cdot e^{-nC_2'(\varepsilon)}$$

and the result follows with $C_2(\varepsilon) \overset{\triangle}{=} C_2'\left(\varepsilon e^{-KT}\right)$. C_1 and C_2 can be chosen independently of $\vec{x}$ since Lemma 5.9 and the derivation do not depend on $\vec{x}$. ∎

5.2. Properties of the Rate Function

In this section we collect a number of properties of the local rate function ℓ, the rate function I, and related quantities. The less technical results are given in Corollary 5.12, Exercises 5.14 and 5.27, and in Lemmas 5.15 and 5.16. In addition to smoothness and approximation results, Proposition 5.49 establishes that I_0^T is a good rate function. It is perhaps best to skip the other parts of this section in first reading, since they provide little intuition; they are rather technical in nature.

Let $\vec{x}$ be a random variable in $\mathbb{R}^d$ with $\mathbb{E}\vec{x} = \vec{m}$ finite, and for $\vec{\theta} \in \mathbb{R}^d$ recall the definitions (1.10) for random variables

$$M(\vec{\theta}) = \mathbb{E}e^{\langle \vec{\theta}, \vec{x} \rangle}$$

$$\ell(\vec{a}) = -\log\left(\inf_{\vec{\theta}} e^{-\langle \vec{\theta}, \vec{a} \rangle} M(\vec{\theta})\right) = \sup_{\vec{\theta}}\left(\langle \vec{\theta}, \vec{a} \rangle - \log M(\vec{\theta})\right).$$

Let us obtain some properties of ℓ.

Proposition 5.10. *Define the function ℓ through (1.10). Then*

(i) ℓ is convex,

(ii) $\ell(\vec{a}) \geq \ell(\vec{m}) = 0$ for all $\vec{a} \in \mathbb{R}^d$,

(iii) ℓ is lower semicontinuous, i.e., if $\vec{a}_n \to \vec{a}$ in $\mathbb{R}^d$ as $n \to \infty$, then

$$\liminf_n \ell(\vec{a}_n) \geq \ell(\vec{a}).$$

Proof. (i) ℓ is the supremum (in $\vec{\theta}$) of the functions $\langle \vec{\theta}, \vec{a} \rangle - \log M(\vec{\theta})$, which are convex in a. Hence Theorem A.47 shows that ℓ is convex.

(ii) By Jensen's inequality, $M(\vec{\theta}) \geq e^{\mathbb{E}\langle \vec{\theta}, x \rangle} = e^{\langle \vec{\theta}, \vec{m} \rangle}$. Thus $\langle \vec{\theta}, \vec{m} \rangle - \log M(\vec{\theta}) \leq 0$ for all $\vec{\theta}$, so that $\ell(\vec{m}) \leq 0$. Now substitute $\vec{\theta} = 0$ in (1.10b) to obtain

$$\ell(\vec{a}) \geq \langle 0, \vec{a} \rangle - \log M(0) = \log 1 = 0.$$

(iii) Clearly, $\langle \vec{\theta}, \vec{a} \rangle - \log M(\vec{\theta})$ is continuous is $\vec{a}$. Therefore ℓ is the supremum of a family of continuous functions. By Exercise A.30 ℓ is lower semicontinuous.∎

Recall that, for $\vec{x}$ and $\vec{y}$ in $\mathbb{R}^d$ and $\vec{r} \in C^d[0, T]$, the rate function for jump Markov processes was defined through (5.2), (5.4), and (5.5);

$$g(\vec{x}, \vec{\theta}) \stackrel{\triangle}{=} \sum_{i=1}^{k} \lambda_i(\vec{x})\left(e^{\langle \vec{\theta}, \vec{e}_i \rangle} - 1\right),$$

$$\ell(\vec{x}, \vec{y}) \stackrel{\triangle}{=} \sup_{\vec{\theta}}\left(\langle \vec{\theta}, \vec{y} \rangle - g(\vec{x}, \vec{\theta})\right),$$

$$I_0^T(\vec{r}) = \begin{cases} \displaystyle\int_0^T \ell\left(\vec{r}(s), \vec{r}'(s)\right) ds & \text{if } \vec{r} \text{ is absolutely continuous,} \\ \infty & \text{otherwise.} \end{cases}$$

If indeed I_0^T is the rate function in the sense of Theorem 5.1, then in particular by Theorem 5.1(ii) we should have $I_0^T(\vec{z}_\infty) = 0$ where $\vec{z}_\infty$ is the most likely path as given in §5.1. In other words, z_∞ solves (5.7).

Exercise 5.11. Show that the definition of ℓ (and therefore of I_0^T) does not depend on the choice of coordinates. That is, make a linear transformation $\vec{u} = L\vec{x}$, where L is a nonsingular matrix. Define ℓ_u for the process in the new coordinates. Show that, if $\vec{u} = L\vec{x}$ and $\vec{v} = L\vec{y}$ then $\ell(\vec{x}, \vec{y}) = \ell_u(\vec{u}, \vec{v})$. ♠

Corollary 5.12. *For each $\vec{x}$, $\ell(\vec{x}, \vec{y})$ is convex and lower semicontinuous, and*

$$\ell(\vec{x}, \vec{y}) \geq \ell(\vec{x}, \vec{m}) = 0, \quad \vec{m} = \vec{m}(\vec{x}) \overset{\triangle}{=} \sum_{i=1}^{k} \lambda_i(\vec{x})\, \vec{e}_i.$$

Proof. Using Exercise 1.25 or the argument of (5.3), for each fixed $\vec{x}$ this is a statement about the rate function ℓ of a multidimensional Poisson random variable. Therefore the corollary follows from Proposition 5.10. ■

Lemma 5.13. *The function $\ell(\vec{x}, \vec{y})$ is strictly convex in $\vec{y}$. Therefore, $\ell(\vec{x}, \vec{y}) = 0$ only at $\vec{y} = \vec{m}$.*

Proof. Recall Definition A.42 and the remark following it. In particular, we need to consider only the case where at least one λ_i is not zero at $\vec{x}$, for otherwise $\ell(\vec{x}, \vec{y}) = \infty$ for all $\vec{y} \neq \vec{0} = \vec{m}$, in which case strict convexity holds trivially. The strict convexity is the subject of Exercise 5.27. The second claim follows from Corollary 5.12, strict convexity, and Theorem A.48(ii). ■

Exercise 5.14. If $\vec{r}$ solves (5.7), then $I_0^T(\vec{r}) = 0$. If (5.7) has a unique solution $\vec{z}_\infty$ and $I_0^T(\vec{r}) = 0$, then $\vec{r}(t) = \vec{z}_\infty(t)$ (for almost all t). Hint: if $\vec{r}$ solves (5.7), then $\vec{r}'(t) = \vec{m}$. For the converse use strict convexity. ♠

Lemma 5.15. *Let k be a convex function. For all $\vec{r}$ absolutely continuous, all $\vec{x}$ and all s,*

$$\int_0^s k(\vec{r}'(t))\, dt \geq s \cdot k\left(\frac{\vec{r}(s) - \vec{r}(0)}{s}\right).$$

If k is strictly convex, then equality holds if and only if $\vec{r}'(t)$ is a constant.

Proof. From Jensen's inequality,

$$s\left(\frac{1}{s}\int_0^s k(\vec{r}'(t))\, dt\right) \geq s \cdot k\left(\frac{1}{s}\int_0^s \vec{r}'(t)\, dt\right)$$

$$= s \cdot k\left(\frac{\vec{r}(s) - \vec{r}(0)}{s}\right)$$

and, if k is strictly convex, equality holds if and only if $\vec{r}'(t)$ is constant. ■

Lemma 5.16. *Assume the $\lambda_i(x)$ are all constant. Then for all $\vec{r}$ absolutely continuous and all T,*

$$I_0^T(\vec{r}) \triangleq \int_0^T \ell(\vec{r}(t), \vec{r}\,'(t))\, dt \geq T \cdot \ell\left(\vec{r}(0), \frac{\Delta \vec{r}}{T}\right),$$

where $\Delta \vec{r} \triangleq \vec{r}(T) - \vec{r}(0)$. Equality holds only if $\vec{r}\,'(t)$ is constant. Consequently, $I_0^T(\vec{r})$ is minimized at a path that is a straight line.

Proof. Since the λ_i are constant, $g(\vec{x}, \vec{\theta})$, and hence $\ell(\vec{x}, \vec{y})$, are independent of $\vec{x}$. By Lemma 5.13 the function $\ell(\vec{x}, \cdot)$ is strictly convex. Now apply Lemma 5.15.■

Lemma 5.16 is more important than might appear. It implies that *locally, the cheapest way to get from point $A = \vec{r}(0)$ to $B = \vec{r}(T)$ in a fixed time T is by a straight line*. Here, "locally" means that λ_i is independent of the position, and "cost" is in terms of the rate function I_0^T. When λ_i can vary, this result may not hold.

Lemma 5.17. *If $\lambda_i(\vec{x})$ are bounded, then there exist constants C_1 and B_1 so that for all $\vec{x}$ and all $|\vec{y}| \geq B_1$,*

$$\ell(\vec{x}, \vec{y}) \geq C_1 |\vec{y}| \log |\vec{y}|.$$

Proof. For each $\vec{y}$, consider $\vec{\theta} = t \cdot \vec{y}/|\vec{y}|$. With $\overline{\lambda} \triangleq \sup_{i,\vec{x}} \lambda_i(\vec{x})$ and $\overline{e} \triangleq \max_i |\vec{e}_i|$, clearly

$$\ell(\vec{x}, \vec{y}) \geq t|\vec{y}| - k\overline{\lambda}e^{\overline{e}t}.$$

Define $t(|\vec{y}|)$ by $t(|\vec{y}|) \triangleq \log(|\vec{y}|)/\overline{e}$. Then

$$\ell(\vec{x}, \vec{y}) \geq (|\vec{y}| \log(|\vec{y}|)/\overline{e}) - |\vec{y}|k\overline{\lambda},$$

which grows like $|\vec{y}| \log |\vec{y}|$ as $|\vec{y}| \to \infty$. ■

Lemma 5.18 (uniform absolute continuity). *Assume the $\lambda_i(\vec{x})$ are bounded, let $I_0^T(\vec{r}) \leq K$, and fix some $\varepsilon > 0$. Then there is a δ, independent of $\vec{r}$, such that for any collection of nonoverlapping intervals in $[0, T]$*

$$\{[t_j, s_j],\ j = 1, \ldots, J\} \quad \text{with} \quad \sum_{j=1}^J s_j - t_j = \delta,$$

with total length δ, we have

$$\sum_{j=1}^J |\vec{r}(t_j) - \vec{r}(s_j)| < \varepsilon.$$

Moreover, we can find a constant B depending only on ε and K so that

$$\int_0^T \mathbf{1}\left[|\vec{r}\,'(t)| \geq B\right] dt \leq \varepsilon.$$

Proof. Define the function $k(t)$ to be equal to one if t is in some interval $[t_j, s_j)$ and zero otherwise. Since r is absolutely continuous, for any $a > 0$,

$$\sum_{j=1}^{J} |\vec{r}(t_j) - \vec{r}(s_j)| \leq \int_0^T |\vec{r}'(u)| k(u) \, du$$

$$\leq \int_0^T a \cdot \mathbf{1} \left[|\vec{r}'(u)| \leq a \right] k(u) \, du$$

$$+ \int_0^T \frac{\ell(\vec{r}(u), \vec{r}'(u))}{\ell(\vec{r}(u), \vec{r}'(u)) / |\vec{r}'(u)|} \mathbf{1} \left[|\vec{r}'(u)| > a \right] k(u) \, du$$

$$\leq a \cdot \delta + \frac{I_0^T(\vec{r})}{f(a)}$$

where, by Lemma 5.17,

$$f(a) \stackrel{\triangle}{=} \inf_{\vec{x}, \vec{y}} \left\{ \frac{\ell(x, y)}{|y|} : |\vec{y}| \geq a \right\} \to \infty \quad \text{as} \quad a \to \infty.$$

The choice $a = 1/\sqrt{\delta}$ establishes the result, since

$$\sqrt{\delta} + \frac{K}{f\left(1/\sqrt{\delta}\right)} \to 0 \quad \text{as} \quad \delta \to 0$$

and δ depends on ε and K, but not directly on $\vec{r}$. To prove the second statement, note that

$$\int_0^T \mathbf{1} \left[|\vec{r}'(t)| \geq B \right] dt \leq \frac{1}{B} \int_0^T |\vec{r}'(t)| \mathbf{1} \left[|\vec{r}'(t)| \geq B \right] dt$$

$$\leq \frac{1}{B} \int_0^T \frac{\ell(\vec{r}(t), \vec{r}'(t))}{\ell(\vec{r}(t), \vec{r}'(t)) / |\vec{r}'(t)|} \mathbf{1} \left[|\vec{r}'(t)| \geq B \right] dt$$

$$\leq \frac{1}{B} \frac{I_0^T(\vec{r})}{f(B)}. \qquad \blacksquare$$

To proceed further we need some geometric definitions and facts.

Definition 5.19. *The positive cone C generated by $\{\vec{e}_i\}$ is*

$$C \stackrel{\triangle}{=} \left\{ \vec{y} : \vec{y} = \sum_{i=1}^{k} a_i \vec{e}_i \text{ for some } \vec{a} \text{ with } a_i \geq 0 \right\}.$$

Lemma 5.20. *For any $B_1 > 0$ there is a B_2 so that any $\vec{y} \in C$ with $|\vec{y}| \leq B_1$ can be represented by a bounded vector $\vec{a}$ with*

$$\vec{y} = \sum_{i=1}^{k} a_i \vec{e}_i, \quad a_i \geq 0 \quad \text{and} \quad |\vec{a}| \leq B_2.$$

In fact, we can choose $B_2 = K B_1$ for some constant K.

Proof. By contradiction. For any $\vec{y}$, choose a representation a that is minimal in the sense that

$$\vec{y} = \sum_{i=1}^{k} a_i' \vec{e}_i, \quad a_i' \geq 0 \quad \text{implies} \quad \max_i a_i' \geq \max_i a_i$$

(a minimal representation exists owing to continuity and compactness). Assume the lemma does not hold. Then there exists a sequence $\vec{y}_j$ with $|\vec{y}_j| \to 0$ so that for the minimal representations,

$$\vec{y}_j = \sum_{i=1}^{k} a_{ij} \vec{e}_i \quad \text{and} \quad \max_i a_{ij} \geq \delta.$$

By taking subsequences if necessary and rescaling, we may assume that the coefficients converge, say $a_{ij} \to a_i$ as $j \to \infty$, and that, in addition, the minimal representations satisfy

$$\max_i a_{ij} = 1, \quad 1 \leq i \leq k, \ j = 1, 2, \ldots.$$

Thus $\max_i a_i = 1$. Taking limits, since $\vec{y}_j \to 0$ we obtain

$$\sum_{i=1}^{k} a_i \vec{e}_i = 0.$$

But since the coefficients are all non-negative, the convergence implies that $a_{ij} \geq a_i/2$ for all i and all j large enough. So

$$\vec{y}_j = \sum_{i=1}^{k} \left(a_{ij} - \frac{a_i}{2} \right) \vec{e}_i.$$

But this implies that $\vec{a}_j$ was not a minimal representation. This contradiction establishes the first claim. The choice $K = B_2/B_1$ then establishes the last claim. ∎

Define

$$\ell(\vec{\theta}, \vec{\lambda}, \vec{y}) \stackrel{\triangle}{=} \langle \vec{\theta}, \vec{y} \rangle - \sum_{i=1}^{k} \lambda_i \left(e^{\langle \vec{\theta}, \vec{e}_i \rangle} - 1 \right) \tag{5.16}$$

and let $\vec{y}$ have the representation $\vec{y} = \sum a_i \vec{e}_i$, with positive a_i.

Lemma 5.21. *Assume the $\log \lambda_i(\vec{x})$ are bounded. Given any c_1, there exists a constant c_2 so that for all $\vec{y} \in C$ with $|\vec{y}| \leq c_1$,*

$$\ell(\vec{\theta}, \vec{\lambda}, \vec{y}) \geq -1 \quad \text{implies} \quad \langle \vec{\theta}, \vec{e}_i \rangle \leq c_2 \quad \text{for all} \quad i.$$

Moreover, if $\vec{\theta}_1, \vec{\theta}_2, \ldots$ is a maximizing sequence, i.e.,

$$\lim_{j \to \infty} \ell(\vec{\theta}_j, \vec{\lambda}(\vec{x}), \vec{y}) = \ell(\vec{x}, \vec{y}),$$

and if for some i we have

$$\liminf_{j \to \infty} \langle \vec{\theta}_j, \vec{e}_i \rangle = -\infty,$$

then, necessarily, $a_i = 0$ in every representation of $\vec{y}$.

Remark. The lemma implies that whenever $\vec{y}$ is in the positive cone generated by those $\{\vec{e}_i\}$ for which $\lambda_i(\vec{x}) > 0$, we can choose a maximizing subsequence of $\{\vec{\theta}_j\}$ so that $a_i \neq 0$ implies that

$$\langle \vec{\theta}_j, \vec{e}_i \rangle \to C_i,$$

with C_i bounded uniformly over compact $\vec{y}$ sets, and moreover $\exp\langle \vec{\theta}_j, \vec{e}_i \rangle$ converges for all i. Note that if $\vec{y}$ is not in the positive cone, then there is a $\vec{z}$ with $\langle \vec{z}, \vec{y} \rangle > 0$ while $\langle \vec{z}, \vec{e}_i \rangle \leq 0$ for all i; e.g., take $\vec{v}$ to be the closest point in the cone to $\vec{y}$, and choose $\vec{z} = \vec{v} - \vec{y}$. Choosing $\vec{\theta} = t\vec{z}$ with $t \to \infty$ shows that, in this case, $\ell(\vec{x}, \vec{y}) = \infty$.

Proof. By elementary calculus, for $\lambda > 0$,

$$\sup_x \left[ax - \lambda \left(e^x - 1 \right) \right] = a \log \frac{a}{\lambda} - a + \lambda.$$

Let $\overline{\lambda} \overset{\Delta}{=} \sup_{i,\vec{x}} \lambda_i(\vec{x})$ and $\underline{\lambda} \overset{\Delta}{=} \inf_{i,\vec{x}} \lambda_i(\vec{x})$. By Lemma 5.20,

$$
\begin{aligned}
\langle \vec{\theta}, \vec{y} \rangle - \sum_{i=1}^{k} \lambda_i(\vec{x}) \left(e^{\langle \vec{\theta}, \vec{e}_i \rangle} - 1 \right) &= \sum_{i=1}^{k} \left(a_i \langle \vec{\theta}, \vec{e}_i \rangle - \lambda_i(\vec{x}) \left(e^{\langle \vec{\theta}, \vec{e}_i \rangle} - 1 \right) \right) \\
&\leq k \left(K c_1 \log \frac{K c_1}{\underline{\lambda}} + \overline{\lambda} \right) \qquad (5.17) \\
&\overset{\Delta}{=} c_3.
\end{aligned}
$$

Now since $\ell(\vec{\theta}, \vec{\lambda}(\vec{x}), \vec{y}) \geq -1$, from the argument of (5.17) we have, for any $1 \leq l \leq k$ for which $\langle \vec{\theta}, \vec{e}_l \rangle > 0$,

$$
\begin{aligned}
c_3 &\geq \ell(\vec{\theta}, \vec{x}, \vec{y}) - \left(a_l \langle \vec{\theta}, \vec{e}_l \rangle - \lambda_l(\vec{x}) \left(e^{\langle \vec{\theta}, \vec{e}_l \rangle} - 1 \right) \right) \\
&\geq -1 + \lambda_l(\vec{x}) \left(e^{\langle \vec{\theta}, \vec{e}_l \rangle} - 1 \right) - a_l \langle \vec{\theta}, \vec{e}_l \rangle \\
&\geq \underline{\lambda} \left(e^{\langle \vec{\theta}, \vec{e}_l \rangle} - 1 \right) - K c_1 \langle \vec{\theta}, \vec{e}_l \rangle - 1.
\end{aligned}
$$

Now choose c_2 large enough so that

$$\underline{\lambda} \left(e^{c_2} - 1 \right) - K c_1 c_2 > c_3 + 1$$

and the first claim follows. Moreover, since $\langle \vec{\theta}_j, \vec{e}_i \rangle$ are bounded above for all i, j, if $a_i > 0$, then

$$\langle \vec{\theta}_j, \vec{e}_i \rangle \to -\infty \quad \text{implies} \quad \ell(\vec{\theta}_j, \vec{\lambda}(\vec{x}), \vec{y}) \to -\infty. \qquad \blacksquare$$

We next establish that in order to obtain a nearly optimal $\vec{\theta}$ for $\ell(\vec{\theta}, \vec{x}, \vec{y})$, it suffices to consider $\vec{\theta}$ in bounded sets. The bound on $|\vec{\theta}|$ is independent of $\vec{x}$ in $\mathbb{R}^d$, and depends only on $|\vec{y}|$. But first, a technical lemma.

Lemma 5.22. *Assume the* $\log \lambda_i(\vec{x})$ *are bounded. Then* $\ell(\vec{x}, \vec{y})$ *is bounded for* $\vec{y}$ *in bounded subsets of* C, *uniformly in* $\vec{x}$. *Moreover, for each* x, $\ell(\vec{x}, \cdot)$ *is continuous as a function from* C *to* $\mathbb{R}$.

Proof. Recall that $\ell(\vec{x}, \vec{y})$ is non-negative. If $\vec{y} \in C$ and $|\vec{y}| \le B$ then, by Lemma 5.20, we can choose a representation $\vec{y} = \sum_{i=1}^{k} a_i \vec{e}_i$ where $|a_i| \le KB$. By Lemma 5.21, $\langle \vec{\theta}_j, \vec{e}_i \rangle \le c_2$ along any maximizing sequence. Therefore

$$\ell(\vec{x}, \vec{y}) = \sup_{\vec{\theta}} \left\{ \sum_{i=1}^{k} \left(a_i \langle \vec{\theta}, \vec{e}_i \rangle - \lambda_i(\vec{x}) \left(e^{\langle \vec{\theta}, \vec{e}_i \rangle} - 1 \right) \right) \right\}$$

$$\le KBkc_2 + k\bar{\lambda}$$

and so ℓ is bounded on bounded subsets of C. The result follows from this, Corollary 5.12, and Theorem A.45. But let us provide an explicit proof (essentially the same proof is used to establish Theorem A.45).

Assume, without loss of generality, that the collection $\{\vec{e}_i\}$ spans $\mathbb{R}^d$, for otherwise the discussion can be reduced to lower dimension. Assume also that $C \ne \mathbb{R}^d$, since otherwise ℓ is obviously continuous (although the argument below covers this case, mutatis mutandis). Now by Corollary 5.12, ℓ is (for each $\vec{x}$) convex and lower semicontinuous in $\vec{y}$. Therefore, by Exercise A.30 it suffices to show that it is also upper semicontinuous at each point $\vec{y} \in C$. So, fix a point $\vec{y} \in C$. Define $\vec{e} \triangleq -\sum_{i=1}^{k} \vec{e}_i$ so that the positive cone of $\{\vec{e}, \vec{e}_i\}$ equals $\mathbb{R}^d$. Let $\vec{e}'$ be the closest vector to $\vec{e}$ such that $\vec{y} + \alpha \vec{e}' \in C$ for some $\alpha > 0$ small enough and set $\vec{e}_0 = \alpha \vec{e}'$. Note that if $\vec{y}$ is in the interior of C, then $\vec{e}' = \vec{e}$, while if $\vec{y} = 0$ then $\alpha = 0$, and if $\vec{y}$ is on a lower-dimensional face of the cone, then the requirements makes $\vec{e}_0$ parallel this face. With this construction, every point $\vec{z}$ in C can be represented as

$$\vec{z} = \vec{y} + \sum_{i=0}^{k} \bar{a}_i \vec{e}_i, \quad a_i \ge 0, \ i = 0, 1, \ldots, k,$$

and $\vec{y} + \vec{e}_i \in C$ for each i. Moreover, by Lemma 5.20 (applied to the new collection $\{\vec{e}_i, i = 0, 1, \ldots, k\}$), we can choose the $\{a_i\}$ so that $|\vec{a}| \le K' |\vec{y} - \vec{z}|$. By convexity of ℓ,

$$\ell(\vec{x}, \vec{z}) \le \left(1 - \sum_{i=0}^{k} a_i \right) \ell(\vec{x}, \vec{y}) + \sum_{i=0}^{k} a_i \ell(\vec{x}, \vec{y} + \vec{e}_i)$$

whenever $\vec{z}$ is close enough to $\vec{y}$ so that the sum of the coefficients is less then one. Now take a sequence $\vec{z}_n \in C$ so that $\vec{z}_n \to \vec{y}$. Then $a_i \to 0$, and so

$$\limsup_{n \to \infty} \ell(\vec{x}, \vec{z}_n) \le \ell(\vec{x}, \vec{y})$$

since ℓ is finite in C. This establishes upper semicontinuity.   ∎

Lemma 5.23. *Assume the* $\log \lambda_i(\vec{x})$ *are bounded. Then for each* $\varepsilon > 0$ *and* c_1 *there exists a bound* B *so that for all* $\vec{y} \in C$ *with* $|\vec{y}| \le c_1$,

$$\sup_{|\vec{\theta}| \le B} \ell(\vec{\theta}, \vec{\lambda}(\vec{x}), \vec{y}) \ge \sup_{\vec{\theta} \in \mathbb{R}^d} \ell(\vec{\theta}, \vec{\lambda}(\vec{x}), \vec{y}) - \varepsilon.$$

Proof. Let ε be given, and fix $\vec{x}$. Recall that C consists of points of the form $\sum_{i=1}^{k} a_i \vec{e}_i$ with $\vec{a}_i \ge 0$. Write $C_1 = C \cap \{\vec{y} : |\vec{y}| \le c_1\}$. We shall find a finite set of points $\{\vec{z}_l\}$, and show that those $\vec{\theta}$ that are nearly optimal at these points will work, in fact, for all points in C_1.

Fix δ and for each $\vec{y} \in C_1$ define

$$\vec{z}(\vec{y}, \delta) \overset{\triangle}{=} \vec{y} + \sum_{i=1}^{k} \delta \vec{e}_i \quad \text{and} \quad B_{\vec{y}, \delta} \overset{\triangle}{=} \left\{ \vec{x} : \vec{x} = \vec{z}(\vec{y}, \delta) - \sum_{i=1}^{k} a_i \vec{e}_i, \ 0 < a_i < 2\delta \right\}.$$

Then, since the $\{\vec{e}_i\}$ span $\mathbb{R}^d$, the set $B_{\vec{y}, \delta}$ is open, and it clearly contains $\vec{y}$. Therefore the collection

$$\left\{ B_{\vec{y}, \delta}, \ \vec{y} \in C \right\}$$

is an open cover of the compact set C_1, for each δ, and so we can extract a finite subcover. Since ℓ is continuous and C_1 is compact, and since each $\vec{z}(\vec{y}, \delta)$ is in C, we can choose δ small enough so that

$$|\ell(\vec{x}, \vec{v}) - \ell(\vec{x}, \vec{z}(\vec{y}, \delta))| \le \varepsilon/4 \quad \text{whenever} \quad \vec{v} \in B_{\vec{y}, \delta}.$$

Choose δ smaller, if necessary, so that it also satisfies $8kc_2\delta \le \varepsilon$. For this value of δ, let $\{B_l\}$ be a finite cover with "centers" $\{\vec{y}_l\}$ and "endpoints" $\{\vec{z}_l\}$.

For each l, let $\vec{\theta}_l$ be such that

$$\ell(\vec{\theta}_l, \vec{\lambda}(\vec{x}), \vec{z}_l) \ge \ell(\vec{x}, \vec{z}_l) - \varepsilon/4$$

and let $B \overset{\triangle}{=} \max_l |\vec{\theta}_l| < \infty$. Then for $\vec{y} \in B_l$,

$$\ell(\vec{x}, \vec{y}) \le \ell(\vec{x}, \vec{z}_l) + \varepsilon/4$$
$$\le \ell(\vec{\theta}_l, \vec{\lambda}(\vec{x}), \vec{z}_l) + \varepsilon/2$$
$$= \ell(\vec{\theta}_l, \vec{\lambda}(\vec{x}), \vec{y}) + \langle \vec{\theta}_l, \vec{z}_l - \vec{y} \rangle + \varepsilon/2.$$

Now, by construction,

$$\vec{z}_l - \vec{y} = \sum_{i=1}^{k} a_i(\vec{y}) \vec{e}_i, \quad \text{where} \quad 0 < a_i(\vec{y}) < 2\delta$$

while the fact that $\vec{\theta}_l$ makes ℓ larger than $(-\varepsilon/4)$ implies, via Lemma 5.21, that $\langle \vec{\theta}_l, \vec{e}_i \rangle \leq c_2$. This and our choice of δ now imply that

$$\ell(\vec{x}, \vec{y}) \leq \ell(\vec{\theta}_l, \vec{\lambda}(\vec{x}), \vec{y}) + 3\varepsilon/4,$$

and the result is established for $\vec{x}$ fixed.

This conclusion certainly holds for all $\vec{x}'$ so that $\vec{\lambda}(\vec{x}') = \vec{\lambda}(\vec{x})$. Since, by assumption, we have the bounds $0 < \underline{\lambda} \leq \lambda_i(\vec{x}) \leq \bar{\lambda} < \infty$, it suffices to establish that B can be chosen independently of $\vec{\lambda}$: in fact, the proof will also establish continuity in $\vec{x}$, uniformly in $\vec{y}$ and $\vec{\theta}$. So, given ε, let B_λ be the bound at $\vec{\lambda}$ corresponding to $\varepsilon/4$. By Lemma 5.21, for any $\vec{\theta}$, $\vec{\lambda}$, $\vec{\lambda}'$, and $\vec{y}$, if $\ell(\vec{\theta}, \vec{\lambda}, \vec{y}) > -1$ then

$$|\ell(\vec{\theta}, \vec{\lambda}, \vec{y}) - \ell(\vec{\theta}, \vec{\lambda}', \vec{y})| \leq \sum_{i=1}^{k} |\lambda_i - \lambda_i'| e^{c_2}. \tag{5.18}$$

Fix $\vec{\lambda}$ and choose δ so that the right-hand side of (5.18) is bounded by $\varepsilon/4$ whenever $|\vec{\lambda}' - \vec{\lambda}| < \delta$. Then for some $\vec{\theta}'$,

$$\begin{aligned}
\sup_{\vec{\theta}} \ell(\vec{\theta}, \vec{\lambda}', \vec{y}) &\leq \ell(\vec{\theta}', \vec{\lambda}', \vec{y}) + \frac{\varepsilon}{4} \\
&\leq \ell(\vec{\theta}', \vec{\lambda}, \vec{y}) + \frac{2\varepsilon}{4} \\
&\leq \sup_{|\vec{\theta}| \leq B_\lambda} \ell(\vec{\theta}, \vec{\lambda}, \vec{y}) + \frac{3\varepsilon}{4} \\
&\leq \sup_{|\vec{\theta}| \leq B_\lambda} \ell(\vec{\theta}, \vec{\lambda}', \vec{y}) + \varepsilon,
\end{aligned} \tag{5.19}$$

where (5.18) was applied twice. But this implies that it suffices to obtain a bound at a finite number of points $\vec{\lambda}$, so take B to be the maximum of B_λ at those points and we are done. $\blacksquare$

Equivalence of upper and lower bounds.

The upper bound for closed sets will be established with the rate function $I_0^T(\vec{r})$, while the lower bound holds with a function $J_0^T(\vec{r})$ defined below. Both are defined by integrals, using the functions ℓ and $\tilde{\ell}$, respectively. In this section we show that $\ell = \tilde{\ell}$, so that $I(r) = J(r)$.

This equivalence has more than theoretical importance: one can use the different forms of the rate function to advantage in calculations. ℓ involves an extremum over $\mathbb{R}^d$, while $\tilde{\ell}$ involves an extremum over $\mathbb{R}^k$. Since usually $d \neq k$, we can choose the smaller space for calculations. Moreover, as seen below, some properties are more easily established in terms of one, but not the other.

Define

$$f(\vec{\mu}, \vec{\lambda}) \overset{\triangle}{=} \sum_{i=1}^{k} \lambda_i - \mu_i + \mu_i \log \frac{\mu_i}{\lambda_i} \qquad (5.20)$$

$$K_y \overset{\triangle}{=} \left\{ \vec{\mu} : \mu_i \geq 0, \ \sum_{i=1}^{k} \mu_i \vec{e}_i = \vec{y} \right\} \qquad (5.21)$$

$$\tilde{\ell}(\vec{x}, \vec{y}) \overset{\triangle}{=} \begin{cases} \inf_{\vec{\mu} \in K_y} f(\vec{\mu}, \vec{\lambda}(\vec{x})) & \text{if } K_y \neq \emptyset; \\ \infty & \text{otherwise .} \end{cases} \qquad (5.22)$$

$$J_0^T(\vec{r}) \overset{\triangle}{=} \begin{cases} \displaystyle\int_0^T \tilde{\ell}(\vec{r}(s), \vec{r}\,'(s)) \, ds & \text{if } \vec{r} \text{ is absolutely continuous;} \\ \infty & \text{otherwise.} \end{cases} \qquad (5.23)$$

Exercise 5.24. For each $\vec{y}$, the function $f(\vec{a}, \vec{\lambda})$ of (5.20) is strictly convex (in $\vec{a}$; Definition A.42) over the convex set K_y. If $K_y \cap \operatorname{dom} f(\cdot, \vec{\lambda}) \neq \emptyset$ (Definition A.37(vi)), then f has a unique finite minimum point $a^* = a^*(\vec{y})$ in K_y. In particular, this is the case if $\vec{y} \in C$ and $\lambda_i > 0$ for all i. Hint: show that it suffices to consider the one-dimensional problem. ♠

Recall the definitions (5.2), (5.5), and (5.16)

$$\ell(\vec{\theta}, \vec{\lambda}, \vec{y}) \overset{\triangle}{=} \langle \vec{\theta}, \vec{y} \rangle - \sum_{i=1}^{K} \lambda_i \left(e^{\langle \vec{\theta}, \vec{e}_i \rangle} - 1 \right)$$

$$\ell(\vec{x}, \vec{y}) \overset{\triangle}{=} \sup_{\vec{\theta} \in \mathbb{R}^d} \ell(\vec{\theta}, \vec{\lambda}(\vec{x}), \vec{y})$$

$$I_0^T(\vec{r}) \overset{\triangle}{=} \begin{cases} \displaystyle\int_0^T \ell\left(\vec{r}(s), \vec{r}\,'(s)\right) \, ds & \text{if } \vec{r} \text{ is absolutely continuous,} \\ \infty & \text{otherwise.} \end{cases}$$

Lemma 5.25. *Let $\vec{y} = \sum_{i=1}^{k} a_i \vec{e}_i$ for some collection of $a_i \geq 0$. Let $\vec{\theta} \in \mathbb{R}^d$ and let λ_i be positive numbers. Then $\ell(\vec{\theta}, \vec{\lambda}, \vec{y}) \leq f(\vec{a}, \vec{\lambda})$.*

Proof. Assume first that $a_i \neq 0$ for all i. Then

$$f(\vec{a}, \vec{\lambda}) - \ell(\vec{\theta}, \vec{\lambda}, \vec{y}) = \sum_{i=1}^{k} \lambda_i - a_i + a_i \log \frac{a_i}{\lambda_i} - a_i \langle \vec{\theta}, \vec{e}_i \rangle + \lambda_i e^{\langle \vec{\theta}, \vec{e}_i \rangle} - \lambda_i$$

$$= \sum_{i=1}^{k} a_i \left[\frac{\lambda_i}{a_i} e^{\langle \vec{\theta}, \vec{e}_i \rangle} - \langle \vec{\theta}, \vec{e}_i \rangle - \log \frac{\lambda_i}{a_i} - 1 \right]$$

$$\overset{\triangle}{=} \sum_{i=1}^{k} a_i U(r_i, x_i)$$

where $r_i \overset{\triangle}{=} \lambda_i/a_i$ and $x_i \overset{\triangle}{=} \langle \vec{\theta}, \vec{e}_i \rangle$, and $U(r, x) \overset{\triangle}{=} re^x - x - \log r - 1$. But $U(r, x) \geq 0$ since, for each x, it is minimized at $r = e^{-x}$ and $U(e^{-x}, x) = 0$. If $a_i = 0$ for some i, then the corresponding term is strictly positive. ∎

Theorem 5.26. *Let $\lambda_i = \lambda_i(\vec{x})$ be positive numbers. Then*

$$\sup_{\vec{\theta} \in \mathbb{R}^d} \ell(\vec{\theta}, \vec{\lambda}, \vec{y}) = \inf \left\{ f(\vec{a}, \vec{\lambda}) : \vec{a} \in \mathbb{R}^d, \ a_i \geq 0, \ \vec{y} = \sum_{i=1}^{k} a_i \vec{e}_i \right\}.$$

Proof. By Lemma 5.22, $\ell(\vec{x}, \vec{y}) < \infty$ if and only if $\vec{y}$ can be written as $\vec{y} = \sum_{i=1}^{k} a_i \vec{e}_i$ for some non-negative constants $\{a_i\}$, while the same is true for $\tilde{\ell}$ by definition. Thus we need to consider only the case where both are finite. By Lemma 5.25, it suffices to produce $\vec{\theta}^*$ and $\vec{a}^*$ with $\ell(\vec{\theta}^*, \vec{\lambda}, \vec{y}) = f(\vec{a}^*, \vec{\lambda})$, and with $\vec{y} = \sum_{i=1}^{k} a_i^* \vec{e}_i$. Let $\vec{\theta}_j$ be a maximizing sequence for $\ell(\vec{\theta}, \vec{x}, \vec{y})$. By Lemma 5.21 and the remark following the lemma, we may assume that, for each i,

$$e^{\langle \vec{\theta}_j, \vec{e}_i \rangle} \to C_i$$

as $j \to \infty$, where the C_i are finite. Therefore

$$\langle \vec{\theta}_j, \vec{y} \rangle \to \ell(\vec{x}, \vec{y}) + \sum_{i=1}^{k} \lambda_i (C_i - 1).$$

Now

$$\nabla_{\vec{\theta}} \ell(\vec{\theta}_j, \vec{\lambda}, \vec{y}) = \vec{y} - \sum_{i=1}^{k} \lambda_i e^{\langle \vec{\theta}_j, \vec{e}_i \rangle} \vec{e}_i.$$

Hence $\nabla_{\vec{\theta}} \ell(\vec{\theta}_j, \vec{\lambda}, \vec{y})$ converges as $j \to \infty$. But since $\vec{\theta}_j$ is a maximizing sequence and $\ell(\vec{x}, \vec{y})$ is finite, the limit must be zero.

Now define

$$a_i^* = \lambda_i C_i, \quad 1 \leq i \leq k.$$

Then $\lim_{j \to \infty} \ell(\vec{\theta}_j, \vec{\lambda}, \vec{y}) = f(\vec{a}^*, \vec{\lambda})$ by direct substitution, and

$$\lim_{j \to \infty} \nabla_{\vec{\theta}} \ell(\vec{\theta}_j, \vec{\lambda}, \vec{y}) = \vec{y} - \lim_{j \to \infty} \sum \lambda_i e^{\langle \vec{\theta}_j, \vec{e}_i \rangle} \vec{e}_i = \vec{0}$$

$$\overset{\triangle}{=} \vec{y} - \sum a_i \vec{e}_i, \tag{5.24}$$

so that $\vec{a}^*$ satisfies the constraint $\vec{y} = \sum a_i^* \vec{e}_i$. ∎

Here are a few consequences.

Exercise 5.27. The function $\ell(\vec{x}, \vec{y})$ is strictly convex in $\vec{y}$ (Definition A.42 and following remark). Hint: the case $\vec{\lambda}(\vec{x}) = \vec{0}$ is immediate from the definitions.

Otherwise, suppose that strict convexity does not hold, so that there are $\vec{y}_1 \neq \vec{y}_2$ and $\gamma \in (0, 1)$ with

$$\ell(\vec{x}, \gamma\vec{y}_1) + \ell(\vec{x}, (1-\gamma)\vec{y}_2) = \gamma\ell(\vec{x}, \vec{y}_1) + (1-\gamma)\ell(\vec{x}, \vec{y}_2).$$

Then

$$\sup_{\vec{\theta}} \left(\langle\vec{\theta}, \gamma\vec{y}_1 + (1-\gamma)\vec{y}_2\rangle - g(\vec{x}, \vec{\theta}) \right) \qquad (5.25)$$

$$= \sup_{\vec{\theta}} \left(\gamma \left(\langle\vec{\theta}, \vec{y}_1\rangle - g(\vec{x}, \vec{\theta}) \right) + (1-\gamma) \left(\langle\vec{\theta}, \vec{y}_2\rangle - g(\vec{x}, \vec{\theta}) \right) \right) \qquad (5.26)$$

$$= \sup_{\vec{\theta}} \left(\gamma \left(\langle\vec{\theta}, \vec{y}_1\rangle - g(\vec{x}, \vec{\theta}) \right) \right) + \sup_{\vec{\theta}} \left(\gamma \left(\langle\vec{\theta}, \vec{y}_2\rangle - g(\vec{x}, \vec{\theta}) \right) \right) \qquad (5.27)$$

Therefore a maximizing sequence for (5.25) must also be maximizing for (5.27). Now use (5.24) to conclude $\vec{y}_1 = \vec{y}_2$.   ♠

Exercise 5.28. The proof above used the specific structure of g. In general, the conclusion will be slightly weaker. Recall Definitions A.37–A.42. Let h be convex and differentiable on $\mathbb{R}^d$. Then h^* is essentially strictly convex. In our case this implies strict convexity in the interior of C. Hint: use $h^{**} = h$ and Theorems A.40–A.43.   ♠

Exercise 5.29. Let $\lambda_i = \lambda_i(\vec{x})$ be positive. Then the numbers C_i of Theorem 5.26 are unique. Moreover, $a_i^* \neq 0$ for all i if and only if there is a unique finite $\vec{\theta}^*$, and $a_i^* = 0$ if and only if $C_i = 0$. Hint: see Exercise 5.24 and Theorem 5.26. Or use Lemma 5.25, let $\vec{\theta}_j$ be a maximizing sequence for ℓ, and let $\vec{a}_j$ be a minimizing sequence for f with $\sum_{i=1}^k a_j^i \vec{e}_i = \vec{y}$, and observe that

$$\lim_{j\to\infty} a_j^i - \lambda_i e^{\langle\vec{\theta}_j, \vec{e}_i\rangle} = 0.$$ ♠

Exercise 5.30. Show that if we adopt the natural convention that, for $b > 0$,

$$\log(b/0) = \infty, \quad \text{and} \quad 0\log(0/0) = 0,$$

then the conclusions of Theorem 5.26 and Exercise 5.29 remain valid even if some of the λ_i are equal to zero. Conclude that both ℓ and $\tilde{\ell}$ are the same as those obtained if we delete all jump directions for which $\lambda_i = 0$.   ♠

Exercise 5.31. If $J_0^T(\vec{r}) < \infty$, then each increment $\Delta\vec{r} = \vec{r}(t) - \vec{r}(s)$, $t > s$, can be represented as

$$\vec{r}(t) - \vec{r}(s) = \Delta\vec{r} = \sum_{i=1}^k a_i \vec{e}_i$$

for some positive a_i. Hint: recall Definition 5.19 and observe that

$$\vec{r}(t) - \vec{r}(s) = \int_s^t \frac{d\vec{r}(u)}{du} \, du \quad \text{and} \quad \int_s^t \mathbf{1}\left[\frac{d\vec{r}(u)}{du} \notin C\right] du = 0$$

since $\vec{r}$ is absolutely continuous and by assumption, respectively.   ♠

Lemma 5.32. *If the* $\log \lambda_i(\vec{x})$ *are bounded then there exist* C_2 *and* B_2 *so that for all* $\vec{y} \in C$ *and all* $\vec{x}$,

$$\ell(\vec{x}, \vec{y}) \leq \begin{cases} C_2 & \text{if } |\vec{y}| \leq B_2, \\ C_2 |\vec{y}| \log |\vec{y}| & \text{if } |\vec{y}| \geq B_2. \end{cases}$$

Proof. Let L be a bound on $|\log \lambda_i|$. By the definition (5.22),

$$\ell(\vec{x}, \vec{y}) = \tilde{\ell}(\vec{x}, \vec{y}) \leq f(\vec{\mu}, \vec{\lambda}(\vec{x}))$$

for any $\vec{\mu} \in K_{\vec{y}}$, and by Lemma 5.20 we can choose $\vec{\mu}$ so that $|\vec{\mu}| \leq K|\vec{y}|$. By (5.20),

$$\ell(\vec{x}, \vec{y}) \leq \sum_{i=1}^{k} \left(\bar{\lambda} + \mu_i(1 + \log \mu_i + L)\right)$$
$$\leq k\left(\bar{\lambda} + K|\vec{y}|(\log |\vec{y}| + L)\right).$$

The bounds follow since $\alpha \log \alpha$ is continuous and unbounded.   ∎

Lemma 5.33. *Assume the* $\log \lambda_i(\vec{x})$ *are bounded and continuous. Then for any* B_1, *the functions* $\tilde{\ell}(\cdot, \vec{y})$ *and* $\ell(\cdot, \vec{y})$ *are continuous in* $\vec{x}$, *uniformly in* $\vec{y}$ *in*

$$S \triangleq \{\vec{y} : \vec{y} \in C, \ |\vec{y}| \leq B_1\}.$$

Proof. This actually follows from the first two inequalities in (5.19) of the proof of Lemma 5.23, but here is another proof using the second representation.

By definition, $f(\vec{\mu}, \vec{\lambda}) \to \infty$ if for any i, the rate $\mu_i \to \infty$. By Lemma 5.20, there is a constant B_2 so that for any $\vec{y}$ in S there exists a $\vec{\mu} \in K_y$ with $|\vec{\mu}| \leq B_2$. Therefore, in computing $\tilde{\ell}$ for any $\vec{y} \in S$, we may restrict our attention to $\vec{\mu}$ such that $|\vec{\mu}| \leq B_3$.

But, by definition, $f(\vec{\mu}, \vec{\lambda}(\vec{x}))$ is continuous in $\vec{x}$, uniformly in $|\vec{\mu}| \leq B_3$. Therefore, for all $\vec{y} \in S$ and any $\vec{\mu} \in K_{\vec{y}}$ with $|\vec{\mu}| \leq B_3$,

$$\tilde{\ell}(\vec{x}', \vec{y}) - \tilde{\ell}(\vec{x}, \vec{y}) \leq f(\vec{\mu}, \vec{\lambda}(\vec{x}')) - \tilde{\ell}(\vec{x}, \vec{y})$$
$$\leq f(\vec{\mu}, \vec{\lambda}(\vec{x})) + \delta - \tilde{\ell}(\vec{x}, \vec{y}).$$

Now choose $\vec{\mu}$ so as to (nearly) minimize $f(\vec{\mu}, \vec{x})$ to establish that

$$\tilde{\ell}(\vec{x}', \vec{y}) - \tilde{\ell}(\vec{x}, \vec{y}) \leq \delta,$$

where δ depends only on $\vec{x}$ and $|\vec{x} - \vec{x}'|$ but not on $\vec{y} \in S$. Repeat the calculation for $\tilde{\ell}(\vec{x}, \vec{y}) - \tilde{\ell}(\vec{x}', \vec{y})$ and the result for $\tilde{\ell}$ follows. The same holds for ℓ since, by Theorem 5.26, the functions are equivalent. ∎

Putting together Lemmas 5.22 and 5.33, we obtain a stronger continuity property for ℓ.

Exercise 5.34. Assume the $\log \lambda_i(\vec{x})$ are bounded and continuous. Then for any B_1, the function $\ell(\vec{x}, \vec{y})$ is continuous in both arguments, uniformly in $|\vec{x}| \le B_1$ and $\vec{y} \in C$ with $|\vec{y}| \le B_1$. Hint: use the triangle inequality and a compactness argument. ♠

Note that the result of this exercise holds if the conditions on $\log \lambda_i$ are satisfied for $|\vec{x}| \le B_1$. If they are globally Lipschitz, then the uniformity is for all $\vec{x}$. We can now establish a "continuity property" for I_0^T, which is useful for approximations.

Theorem 5.35. Assume the $\log \lambda_i(\vec{x})$ are bounded and continuous. Fix $\vec{r} \in C^d[0, T]$ with $I_0^T(\vec{r}) < \infty$. For any $\varepsilon > 0$, there exists a δ so that

$$\sup_{0 \le t \le T} |\vec{q}(t) - \vec{r}(t)| < \delta \quad \text{implies} \quad \left| \int_0^T \ell(\vec{q}(t), \vec{r}\,'(t))\, dt - I_0^T(\vec{r}) \right| \le \varepsilon.$$

Proof. Note that since $\tilde{\ell}(\vec{x}, \vec{y}) = \infty$ for $\vec{y} \notin C$ and $I_0^T(\vec{r})$ is finite, we can ignore any point t for which $\vec{r}\,'(t) \notin C$. By Lemma 5.32 and 5.17, for $B \ge \max\{B_1, B_2\}$,

$$\int_0^T \ell(\vec{q}(t), \vec{r}\,'(t))\mathbf{1}\left[\vec{r}\,'(t) \ge B\right] dt \le \int_0^T C_2 |\vec{r}\,'(t)| \log |\vec{r}\,'(t)| \mathbf{1}\left[\vec{r}\,'(t) \ge B\right] dt$$

$$\le \int_0^T \frac{C_2}{C_1} \ell(\vec{r}(t), \vec{r}\,'(t))\mathbf{1}\left[\vec{r}\,'(t) \ge B\right] dt$$

$$\overset{\triangle}{=} \varepsilon(B).$$

By Lemma 5.18, we have $\varepsilon(B) \to 0$ as $B \to \infty$ since $I_0^T(\vec{r})$ is finite. Fix B so that $\varepsilon(B) \le \varepsilon/4$ and also $\varepsilon(B)C_1/C_2 \le \varepsilon/4$. By Lemma 5.33, there is a δ so that if $|\vec{x} - \vec{x}'| \le \delta$, then

$$|\tilde{\ell}(\vec{x}, \vec{y}) - \tilde{\ell}(\vec{x}', \vec{y})| \le \frac{\varepsilon}{2T}$$

for all $\vec{y}$ with $|\vec{y}| \le B$ and all $\vec{x}, \vec{x}'$ in the compact set

$$\left\{ \vec{x} : \inf_{0 \le t \le T} |\vec{x} - \vec{r}(t)| \le \delta \right\}.$$

Therefore

$$\left| \int_0^T \ell(\vec{q}(t), \vec{r}\,'(t))\, dt - \int_0^T \ell(\vec{r}(t), \vec{r}\,'(t))\, dt \right|$$

$$\leq 2\frac{\varepsilon}{4} + \int_0^T \left| \ell(\vec{q}(t), \vec{r}\,'(t))\, dt - \ell(\vec{r}(t), \vec{r}\,'(t)) \right| \mathbf{1}\left[\vec{r}\,'(t) \leq B \right] dt$$

$$\leq \frac{\varepsilon}{2} + T \cdot \frac{\varepsilon}{2T}. \qquad\blacksquare$$

Semicontinuity is used extensively in the theory of large deviations (it is also used extensively in convex analysis, which is a closely related field). We remind you that a function $f(x)$ is lower semicontinuous if f can only jump down. We shall need a fairly general setup for the upper bound, showing that the approximation methods we employ preserve lower semicontinuity. Some of the definitions below are somewhat complicated by the desire to make them generalize to more complicated processes. This will become clear in Chapter 8.

Definition 5.36. $\quad \overset{\delta}{g}(\vec{x}, \vec{\theta}) \overset{\triangle}{=} \sup_{|\vec{z}-\vec{x}|\leq\delta} \sum_{i=1}^k \lambda_i(\vec{z}) \left(e^{\langle\vec{\theta}, e_i\rangle} - 1 \right).$

Definition 5.37. $\quad \ell^\delta(\vec{x}, \vec{y}) \overset{\triangle}{=} \sup_{\vec{\theta}} \left(\langle\vec{\theta}, \vec{y}\rangle - \overset{\delta}{g}(\vec{x}, \vec{\theta}) \right).$

Definition 5.38. $\quad \overset{\delta}{I}\!{}^T_0(\vec{r}) \overset{\triangle}{=} \int_0^T \ell^\delta(\vec{r}(t), \vec{r}\,'(t))\, dt$ if $\vec{r}$ is absolutely continuous, and is defined to be infinite otherwise.

Lemma 5.39. *Assume* $\lambda_i(\vec{x})$ *are bounded. Then the conclusions of Lemma 5.17 and Lemma 5.18 apply to* ℓ^δ *and* $\overset{\delta}{I}\!{}^T_0$; *that is,* $\ell^\delta(\vec{x}, \vec{y})$ *grows faster than* $|\vec{y}| \log |\vec{y}|$ *uniformly in* $\vec{x}$, *and the functions*

$$\left\{ \vec{r} : \overset{\delta}{I}\!{}^T_0(\vec{r}) \leq K \right\}$$

are uniformly absolutely continuous.

Proof. Identical to the proofs of Lemma 5.17 and Lemma 5.18, respectively. $\quad\blacksquare$

Lemma 5.40 Lower semicontinuity. *If* $\lambda_i(\vec{x})$ *are bounded and continuous, then the function* $\ell^\delta(\vec{x}, \vec{y})$ *is lower semicontinuous in* $(\delta, \vec{x}, \vec{y})$.

Remark. We clearly have $\overset{\delta}{g} \geq g$, so $\ell^\delta \leq \ell$; this would seem to make semicontinuity difficult to show. Nevertheless, here is the

Proof. Define

$$f(\delta, \vec{x}, \vec{y}; \vec{\theta}) \overset{\triangle}{=} \langle\vec{\theta}, \vec{y}\rangle - \sup_{|\vec{z}-\vec{x}|\leq\delta} \sum_{i=1}^k \lambda_i(\vec{z}) \left(e^{\langle\vec{\theta}, e_i\rangle} - 1 \right). \qquad (5.28)$$

In Exercise 5.41 below it is shown that f is jointly continuous in $(\delta, \vec{x}, \vec{y})$, for each $\vec{\theta}$. The result is now a consequence of Exercise A.30. $\quad\blacksquare$

Exercise 5.41. If the $\lambda_i(\vec{x})$ are continuous then f of Equation (5.28) is continuous in $(\delta, \vec{x}, \vec{y})$. Consequently, $\ell^\delta(\vec{x}, \vec{y})$ is right-continuous in δ, i.e.,

$$\lim_{\delta' \downarrow \delta} \ell^{\delta'}(\vec{x}, \vec{y}) = \ell^\delta(\vec{x}, \vec{y})$$

uniformly over $(\vec{x}, \vec{y})$ in bounded sets. In particular, $\ell^\delta(\vec{x}, \vec{y}) \to \ell(\vec{x}, \vec{y})$ as $\delta \to 0$. Hint: f is decreasing in δ. It suffices to show continuity in δ uniformly in $(\vec{x}, \vec{y})$, and continuity in $(\vec{x}, y)$.   ♠

Lemma 5.42. If $\lambda_i(\vec{x})$ are bounded and continuous, then $I_0^T(\vec{r})$ is lower semicontinuous (in the d_d metric). The same holds for $\overset{\delta}{I_0^T}(\vec{r})$.

Proof. We clearly need to consider only sequences of absolutely continuous functions. By Lemma A.62, we can use either of the metrics d_d or d_c: we shall use the latter. Let $\{\vec{r}_n\}$ be a sequence of functions in $C^d[0, T]$ converging (under d_c) to $\vec{r}$. We may assume that $I_0^T(\vec{r}_n)$ is bounded, say by the constant K. By Lemma 5.18, the functions in the set $\{\vec{r}_n\}$ are uniformly absolutely continuous, and therefore $\vec{r}$ is also absolutely continuous. Therefore, given δ we can partition the interval $[0, T]$ into J intervals $[t_j, t_{j+1})$ each of length Δ, such that

$$\max_j \sup_{t_j \leq s < t_{j+1}} |\vec{r}_n(s) - \vec{r}_n(t_j)| < \delta$$

for all n. Then, by Definitions 5.36–5.37 and Lemma 5.15,

$$\int_0^T \ell(\vec{r}_n(t), \vec{r}_n'(t)) \, dt \geq \sum_{j=1}^{J} \int_{t_j}^{t_{j+1}} \ell^\delta(\vec{r}_n(t_j), \vec{r}_n'(t)) \, dt$$

$$\geq \sum_{j=1}^{J} \Delta \cdot \ell^\delta \left(\vec{r}_n(t_j), \frac{\vec{r}_n(t_{j+1}) - \vec{r}_n(t_j)}{t_{j+1} - t_j} \right).$$

Define the function $\vec{r}_J$ by

$$\vec{r}_J(t) = \vec{r}(t_j) \quad \text{for} \quad t_j \leq t < t_{j+1}$$

and let $\vec{r}^J(t) \overset{\Delta}{=} \vec{r}_J(t + \Delta)$ (we have omitted the dependence on n from the notation). By the previous calculation, for each J and δ,

$$\liminf_{n \to \infty} \int_0^T \ell(\vec{r}_n(t), \vec{r}_n'(t)) \, dt \geq \int_0^{T-\Delta} \ell^\delta \left(\vec{r}_J(t), \frac{\vec{r}^J(t) - \vec{r}_J(t)}{\Delta} \right) dt.$$

Now choose a sequence of nested partitions, say $J_i = 2^i$, so that $\Delta_i = T \cdot 2^{-i}$, and a corresponding sequence δ_i that converges to zero. Then

$$\vec{r}_J(t) \to \vec{r}(t) \quad \text{and} \quad \vec{r}^J(t) \to \vec{r}(t) \text{ as } \Delta \to 0,$$

since $\vec{r}$ is absolutely continuous, and for (almost) every $t \in [0, T]$ as $i \to \infty$,

$$\frac{\vec{r}^{J_i}(t) - \vec{r}_{J_i}(t)}{\Delta_i} \to \vec{r}\,'(t).$$

Since ℓ^δ is non-negative, Fatou's Lemma Theorem A.93 and Lemma 5.40 imply

$$\liminf_{i \to \infty} \int_0^{T - \Delta_i} \ell^{\delta_i} \left(\vec{r}_{J_i}(t), \frac{\vec{r}^{J_i}(t) - \vec{r}_{J_i}(t)}{\Delta_i} \right) dt$$

$$\geq \int_0^T \liminf_{i \to \infty} \mathbf{1}\left[t \leq T - \Delta_i \right] \ell^{\delta_i} \left(\vec{r}_{J_i}(t), \frac{\vec{r}^{J_i}(t) - \vec{r}_{J_i}(t)}{\Delta_i} \right) dt$$

$$\geq \int_0^T \ell(\vec{r}(t), \vec{r}\,'(t)) \, dt.$$

This establishes the lower semicontinuity. The proof for $\overset{\delta}{I}_0^T(\vec{r})$ is identical. ∎

Lemma 5.43. *Assume the* $\log \lambda_i$ *are bounded and continuous. Then for any* $\vec{r}$ *with* $I_0^T(\vec{r}) < \infty$ *and any* $\varepsilon > 0$, *there exists a step function* $\vec{\theta}(t)$ *so that*

$$\int_0^T \ell(\vec{\theta}(t), \vec{\lambda}(\vec{r}(t)), \vec{r}\,'(t)) \, dt \geq I_0^T(\vec{r}) - \varepsilon.$$

Similarly, if $\overset{\delta}{I}_0^T(\vec{r}) < \infty$, *then there exists a step function* $\vec{\theta}$ *so that*

$$\int_0^T \ell^\delta(\vec{\theta}(t), \vec{\lambda}(\vec{r}(t)), \vec{r}\,'(t)) \, dt \geq \overset{\delta}{I}_0^T(\vec{r}) - \varepsilon.$$

Proof. Since $I_0^T(\vec{r}) < \infty$ and since $\ell(\vec{\theta}, \vec{\lambda}(\vec{x}), \vec{y}) \leq \ell(\vec{x}, \vec{y})$ for all $\vec{\theta}$, Lemma 5.18 implies that, for B large enough,

$$\int_0^T \mathbf{1}\left[|\vec{r}\,'(t)| \geq B \right] \ell(\vec{\theta}(t), \vec{\lambda}(\vec{r}(t)), \vec{r}\,'(t)) \, dt$$

$$\leq \int_0^T \mathbf{1}\left[|\vec{r}\,'(t)| \geq B \right] \ell(\vec{r}(t), \vec{r}\,'(t)) \, dt \leq \frac{\varepsilon}{4}.$$

Choose $\vec{\theta}_1(t) = \vec{0}$ whenever $|\vec{r}\,'(t)| \geq B$ or $\vec{r}\,'(t) \notin C$. Let $\bar{r} \overset{\Delta}{=} \sup_{0 \leq t \leq T} |\vec{r}(t)|$ which is finite, since $\vec{r}$ is continuous. By Lemma 5.23, for B_1 large enough

$$\sup_{|\vec{\theta}| \leq B_1} \ell(\vec{\theta}, \vec{x}, \vec{y}) \geq \ell(\vec{x}, \vec{y}) - \frac{\varepsilon}{4T}$$

for all $|\vec{y}| \leq B$ in C and all $\vec{x}$ with $|\vec{x}| \leq \bar{r}$. But on the bounded set

$$\left\{ |\vec{\theta}| \leq B_1, \ |\vec{x}| \leq \bar{r}, \ \vec{y} \in C, \ |\vec{y}| \leq B \right\}$$

the function $\ell(\vec{\theta}, \vec{x}, \vec{y})$ is uniformly continuous. Therefore there exist a $\delta > 0$ and a finite collection $\{\vec{\theta}_{ij}, \vec{x}_i, \vec{y}_j\}$ so that

$$\ell(\vec{\theta}_{ij}, \vec{x}, \vec{y}) \geq \ell(\vec{x}, \vec{y}) - \frac{\varepsilon}{2T} \quad \text{whenever} \quad |\vec{x} - \vec{x}_i| + |\vec{y} - \vec{y}_j| \leq \delta.$$

Define the function

$$\vec{\theta}_1(t) = \vec{\theta}_{ij} \quad \text{whenever} \quad |\vec{r}(t) - \vec{x}_i| + |\vec{r}'(t) - \vec{y}_j| \leq \delta$$

with some tie-breaking rule. The function $\vec{\theta}_1$ is a simple function: it takes a finite number of values. However, it may not be constant on *intervals of time*. So we approximate $\vec{\theta}_1(t)$ by a step function. Choose η so that

$$\int_0^T \mathbf{1}\,[t \in A]\,\ell(\vec{r}(t), \vec{r}'(t))\,dt \leq \frac{\varepsilon}{4}$$

whenever the set A has measure (length) less than η. By [Roy Prop. 22 p. 68] we can indeed approximate $\vec{\theta}_1$ by a step function $\vec{\theta}$, so that the functions agree outside a set of measure η. Collecting all approximations gives the result. The proof for ℓ^δ is identical. ∎

Definition 5.44. $\Phi_{\vec{x}}(K) \triangleq \left\{ \vec{r} \in C^d[0, T] : I_0^T(\vec{r}) \leq K, \, \vec{r}(0) = \vec{x} \right\}.$

Definition 5.45. $\overset{\delta}{\Phi}_{\vec{x}}(K) \triangleq \left\{ \vec{r} \in C^d[0, T] : \overset{\delta}{I_0^T}(\vec{r}) \leq K, \, \vec{r}(0) = \vec{x} \right\}.$

Now let C be any compact set in $\mathbb{R}^d$.

Proposition 5.46. *If the* $\lambda_i(\vec{x})$ *are bounded and continuous, then*

$$\bigcup_{\vec{x} \in C} \Phi_{\vec{x}}(K) \quad \text{and} \quad \bigcup_{\vec{x} \in C} \overset{\delta}{\Phi}_{\vec{x}}(K)$$

are compact sets in $C^d[0, T]$, *for any compact* $C \subset \mathbb{R}^d$.

Proof. By the Arzelà-Ascoli Theorem A.51, a set $\mathcal{K}$ in $C^d[0, T]$ is pre-compact (i.e., has compact closure) if and only if the functions in $\mathcal{K}$ are equicontinuous, and the set $\{\vec{r}(0) : \vec{r} \in \mathcal{K}\}$ is compact. By assumption, $\vec{x} \in C$, a compact set. By Lemma 5.18, the functions in these sets are equicontinuous. By Lemma 5.42, $I_0^T(\vec{r})$ is lower semicontinuous, so that by definition, the limit $\vec{r}$ of any convergent sequence satisfies $I_0^T(\vec{r}) \leq K$, and $\vec{r}(0) \in C$; therefore the set is closed, and compactness is established. The proof for the second set is identical. ∎

This compactness allows us to obtain a further semicontinuity property. The large deviations upper bound is of the form

$$I_{\vec{x}}^*(F) \triangleq \inf\{I_0^T(\vec{r}) : \vec{r} \in F, \, \vec{r}(0) = \vec{x}\}.$$

Lemma 5.47. *If the* $\lambda_i(\vec{x})$ *are bounded and continuous, then for each closed set* $F \in C^d[0, T]$, $I_{\vec{x}}^*(F)$ *is lower semicontinuous in* $\vec{x}$.

Proof. We need to establish that if $\vec{x}_n \to \vec{x}$ and $\liminf_n I_{\vec{x}_n}^*(F) = K < \infty$, then $I_{\vec{x}}^*(F) \leq K$. By Proposition 5.46, for any positive ε and δ, the sets

$$F \cap \overset{\delta}{\Phi}_{\vec{x}_n}(K + \varepsilon) \quad \text{and} \quad F \cap \bigcup_{|\vec{x} - \vec{y}| \leq \delta} \overset{\delta}{\Phi}_{\vec{y}}(K + \varepsilon)$$

are compact (intersection of compact and closed sets). By definition they are non-empty (for all n large). By Lemma 5.42, $I_0^T(\cdot)$ is lower semicontinuous, and so by Theorem A.31, there are $\vec{r}_n \in F$ so that $I_{\vec{x}_n}^*(F) = I_0^T(\vec{r}_n)$, at least for large n. Moreover, $\{\vec{r}_n\}$ lies in a compact set, so there is a convergent subsequence of $\{\vec{r}_n\}$ with a limit, say $\vec{r}$. Since F is closed, $\vec{r} \in F$, and clearly $\vec{r}(0) = \vec{x}$. Since I_0^T is lower semicontinuous,

$$I_{\vec{x}}^*(F) \leq I_0^T(\vec{r}) \leq \liminf_n I_0^T(\vec{r}_n) = K. \qquad \blacksquare$$

We now estimate how well $\overset{\delta}{\Phi}_{\vec{x}}(K)$ approximates $\Phi_{\vec{x}}(K)$.

Lemma 5.48. *Let* C *be compact in* $\mathbb{R}^d$, *and assume the* $\log \lambda_i(\vec{x})$ *are bounded and continuous. Given* $K > 0$ *and* $\varepsilon > 0$, *there exists a* $\delta > 0$ *such that*

$$\overset{\delta}{\Phi}_{\vec{x}}(K - \varepsilon) \subset \{\vec{r} : d(\vec{r}, \Phi_{\vec{x}}(K)) \leq \varepsilon\}$$

for all $\vec{x} \in C$ *(here* $d = d_c$*).*

This means that cheap functions in the $\overset{\delta}{I}$ sense are very close to equally cheap functions in the I sense.

Proof. By contradiction. Choose

$$\delta_i \downarrow 0, \quad \vec{x}_i \to \vec{x}, \quad \vec{x}_i \in C, \quad \vec{r}_i \in \overset{\delta_i}{\Phi}_{x_i}(K - \varepsilon) \quad i = 0, 1, \ldots.$$

If the claim is false, then we can make these choices so that

$$d(\vec{r}_i, \Phi_{\vec{x}_i}(K)) > \varepsilon \text{ for all } i.$$

Since ℓ^δ increases as $\delta \downarrow 0$, the $\vec{r}_i$ are contained in the set

$$\bigcup_{\vec{x} \in C} \overset{\delta_0}{\Phi}_{\vec{x}}(K - \varepsilon),$$

which is compact by Proposition 5.46. So, take a subsequence converging to a function $\vec{r}$. Now Definitions 5.36–5.38 imply that for any path $\vec{s}$, if $\delta_j \geq \delta_i$ then

$$\overset{\delta_j}{I}{}_0^T(\vec{s}) \leq \overset{\delta_i}{I}{}_0^T(\vec{s}).$$

This and Lemma 5.40 give, for each j,

$$I_0^T{}^{\delta_j}(\vec{r}) \le \liminf_{i \to \infty} I_0^T{}^{\delta_j}(\vec{r}_i)$$

$$\le \liminf_{i \to \infty} I_0^T{}^{\delta_i}(\vec{r}_i)$$

$$\le K - \varepsilon.$$

Now the monotone convergence theorem A.91(ii) gives

$$I_0^T(\vec{r}) = \lim_{j \to \infty} I_0^T{}^{\delta_j}(\vec{r})$$

$$\le K - \varepsilon.$$

Now fix i (to be chosen later) and set

$$\tilde{\vec{r}}(t) \stackrel{\triangle}{=} \vec{r}(t) + (\vec{x}_i - \vec{x}) \tag{5.29}$$

so that $\tilde{\vec{r}}(0) = \vec{x}_i$. Then since $\vec{r}$ is continuous and $\tilde{\vec{r}}\,'(t) = \vec{r}\,'(t)$, by Theorem 5.35,

$$I_0^T\left(\tilde{\vec{r}}\right) = I_0^T(\vec{r}) + f(|\vec{x} - \vec{x}_i|),$$

where $f(\varepsilon) \to 0$ as $\varepsilon \to 0$. Thus, for i large enough, $I_0^T(\tilde{\vec{r}}) < K$, so $\tilde{\vec{r}} \in \Phi_{\vec{x}_i}(K)$ while $d(\tilde{\vec{r}}, \vec{r}_i) < \varepsilon$, a contradiction. ∎

Proposition 5.49. *Assume the $\log \lambda_i(\vec{x})$ are bounded and continuous. Then, for each $\vec{x}$, $I_0^T(\cdot)$ is a good rate function on $C^d[0, T] \cap \{\vec{r} : \vec{r}(0) = \vec{x}\}$.*

Proof. Since ℓ is non-negative, I_0^T is non-negative and, by Lemma 5.42, it is lower semicontinuous. By Proposition 5.46, its level sets are compact. ∎

Corollary 5.50. *Assume the $\log \lambda_i(\vec{x})$ are bounded and continuous. Then, for each $\vec{x}$, $I_0^T(\cdot)$ is a good rate function on $D^d[0, T] \cap \{\vec{r} : \vec{r}(0) = \vec{x}\}$ under either the metric d_c or d_d.*

Proof. Since $I_0^T(\vec{r})$ is finite only for (absolutely) continuous functions, it suffices to consider only sequences in $C^d[0, T]$: but limits of such sequences under either metric are continuous (cf. Theorem A.58). Moreover, in this case convergence under d_d is equivalent to convergence under d_c (Lemma A.62). Thus I_0^T is lower semicontinuous. The level sets consist of paths in $C^d[0, T]$ and, by Proposition 5.49 are compact in $(C^d[0, T], d_c)$. By A.60 they are thus compact in $(D^d[0, T], d_c)$ and in $(D^d[0, T], d_d)$. ∎

5.3. The Lower Bound

In this section we present and prove the lower bound. There are no new ideas here, but there are many more technicalities than appear in Chernoff's Theorem. To calculate the probability that the jump Markov process $\vec{z}_n(t)$ is near some continuous function $\vec{r}(t)$, we perform the following steps (whose details form the body of this section):

1) Approximate the path $\vec{r}(t)$ by the "fluid limit" path $\vec{z}_\infty(t)$ of a new jump Markov process, with new jump rates that are constant on all intervals of time;

2) Write the change of measure formula for jump Markov processes;

3) Prove that in the limit this formula becomes simple; and

4) Prove the lower bound for $\vec{z}_n$ and $\vec{z}_\infty$ using Kurtz's Theorem.

We show in Theorem 5.26, §5.2 that the lower bound matches the upper bound.

Consider the processes $\vec{z}_n$ with jump directions $\vec{e}_i/n$, and two measures $\mathbb{P}$ and $\tilde{\mathbb{P}}$. Under $\mathbb{P}$ (respectively $\tilde{\mathbb{P}}$) the process $\vec{z}_n$ has jump rates $\{n\lambda_i(\vec{x}),\ i = 1, \ldots, k\}$ (respectively $\{n\mu_i(\vec{x}),\ i = 1, \ldots, k\}$) over a time interval $[0, T]$. We know from (4.19), or Theorem B.6, that (for each n)

$$\frac{d\mathbb{P}}{d\tilde{\mathbb{P}}}(\omega) = \exp\left\{ -\int_0^T n \sum_{i=1}^k (\lambda_i(\vec{z}_n(s)) - \mu_i(\vec{z}_n(s)))\, ds \right.$$

$$\left. + \sum_{j=1}^N \log\left(\frac{\lambda_{l(j)}(\vec{z}_n(t_j^-))}{\mu_{l(j)}(\vec{z}_n(t_j^-))} \right) \right\}. \quad (5.30)$$

Here N is the total number of jumps (transitions) the process $\vec{z}_n$ makes in $[0, T]$ and t_j is the time of the jth jump, so that $\vec{z}_n(t_j^-)$ is the state of the process just before the jump. Finally, $l(j)$ denotes the directions of the jth jump, i.e., $l(j) = l$ implies $x(t_j) - x(t_j^-) = \vec{e}_l$.

Remark. Strictly speaking, (5.30) describes the Radon-Nikodym derivative of $\mathbb{P}$ restricted to the process $\vec{z}_n$, with respect to $\tilde{\mathbb{P}}$ restricted to $\vec{z}_n$. To avoid cumbersome notation, we shall be cavalier about this. This same comment applies to the definition of H below.

Let us first reduce the lower bound to a simpler calculation: the probability that the jump Markov process $\vec{z}_n(t)$ is near some continuous function $\vec{r}(t)$. Suppose we have an open set $G \in D^d[0, T]$ with

$$I^* \triangleq \inf_{\vec{r} \in G} I_0^T(\vec{r}) < \infty.$$

Then we can find a function $\vec{r} \in G$ with $I_0^T(\vec{r}) \leq I^* + \delta$, so that $\vec{r}$ is necessarily absolutely continuous. Then since G is open, by Definition A.7 the set

$$N_\varepsilon(\vec{r}) \triangleq \left\{ \vec{x} \in D^d[0, T] : \sup_{0 \leq t \leq T} |\vec{x}(t) - \vec{r}(t)| < \varepsilon \right\} \quad (5.31)$$

satisfies $N_\varepsilon(\vec{r}) \subset G$ for all ε small enough. Therefore,

$$\mathbb{P}_{\vec{x}}(\vec{z}_n \in G) \geq \mathbb{P}_{\vec{x}}(\vec{z}_n \in N_\varepsilon(\vec{r}))$$

for all ε small enough. So, to prove a lower bound on G, it suffices to prove that

$$\lim_{\varepsilon \downarrow 0} \liminf_{n \to \infty} \frac{1}{n} \log \mathbb{P}\left(\vec{z}_n \in N_\varepsilon(\vec{r})\right) \geq -I_0^T(\vec{r}) \geq -I^* - \delta. \tag{5.32}$$

Here is the idea behind the proof of this inequality. Define H by

$$e^H \stackrel{\triangle}{=} \frac{d\mathbb{P}}{d\tilde{\mathbb{P}}}(\omega).$$

Exactly as in Chernoff's Theorem, we use a change of measure:

$$
\begin{aligned}
\mathbb{P}(\vec{z}_n \in N_\varepsilon(\vec{r})) &= \int_{\vec{z}_n \in N_\varepsilon(\vec{r})} d\mathbb{P}(\omega) \\
&= \int_{\vec{z}_n \in N_\varepsilon(\vec{r})} \frac{d\mathbb{P}}{d\tilde{\mathbb{P}}}(\omega) \, d\tilde{\mathbb{P}}(\omega) \\
&= \int_{\vec{z}_n \in N_\varepsilon(\vec{r})} e^{H(\omega)} \, d\tilde{\mathbb{P}}(\omega) \tag{5.33} \\
&\geq \tilde{\mathbb{P}}(\vec{z}_n \in N_\varepsilon(\vec{r})) \\
&\quad \times \exp\left(\frac{1}{\tilde{\mathbb{P}}(\vec{z}_n \in N_\varepsilon(\vec{r}))} \int_{\vec{z}_n \in N_\varepsilon(\vec{r})} H(\omega) \, d\tilde{\mathbb{P}}(\omega)\right);
\end{aligned}
$$

by version (A.11) of Jensen's inequality. Now, as in Chernoff's Theorem, suppose that we arrange things so that $\tilde{\mathbb{P}}(\vec{z}_n \in N_\varepsilon(\vec{r})) \to 1$ as $n \to \infty$. By Kurtz's Theorem, this amounts to having $\vec{r}(t) = \vec{z}_\infty(t)$ under $\tilde{\mathbb{P}}$, i.e.,

$$\vec{r}'(t) = \sum_{i=1}^k \mu_i(\vec{r}(t)) \, \vec{e}_i. \tag{5.34}$$

If $\tilde{\mathbb{P}}(\vec{z}_n \in N_\varepsilon(\vec{r})) \to 1$ then by (5.33),

$$\liminf_{n \to \infty} \frac{1}{n} \log \mathbb{P}\left(\vec{z}_n \in N_\varepsilon(\vec{r})\right) \geq \liminf_{n \to \infty} \frac{1}{n} \int_{\vec{z}_n \in N_\varepsilon(\vec{r})} H(\omega) \, d\tilde{\mathbb{P}}(\omega). \tag{5.35}$$

This holds for every $\tilde{\mathbb{P}}$, provided the constraint (5.34) holds. Therefore, to get a tight lower bound our problem is reduced to calculating

$$\sup_{\tilde{\mathbb{P}}} \left\{ \liminf_{n \to \infty} \frac{1}{n} \int_{\vec{z}_n \in N_\varepsilon(\vec{r})} H(\omega) \, d\tilde{\mathbb{P}}(\omega) : \sum_{i=1}^k \mu_i(\vec{r}(t)) \, \vec{e}_i = \vec{r}'(t) \right\}, \tag{5.36}$$

where the jump rates μ_i determine $\tilde{\mathbb{P}}$. It is more convenient to do the computations in terms of J as defined in (5.23): in §5.2 we showed that $I = J$.

With this orientation in mind, we proceed to the proofs. Recall the definitions (5.20), (5.22), and (5.23) from §5.2,

$$f(\vec{\mu}, \vec{\lambda}) \triangleq \sum_{i=1}^{k} \lambda_i - \mu_i + \mu_i \log \frac{\mu_i}{\lambda_i}$$

$$K_y \triangleq \left\{ \vec{\mu} : \mu_i \geq 0, \ \sum_{i=1}^{k} \mu_i \vec{e}_i = \vec{y} \right\}$$

$$\tilde{\ell}(\vec{x}, \vec{y}) \triangleq \begin{cases} \inf_{\vec{\mu} \in K_y} f(\vec{\mu}, \vec{\lambda}(\vec{x})) & \text{if } K_y \neq \emptyset; \\ \infty & \text{otherwise.} \end{cases}$$

$$J_0^T(\vec{r}) \triangleq \begin{cases} \int_0^T \tilde{\ell}(\vec{r}(s), \vec{r}'(s)) \, ds & \text{if } \vec{r} \text{ is absolutely continuous;} \\ \infty & \text{otherwise.} \end{cases}$$

The main result of this section is the following theorem.

Theorem 5.51. *Assume that for each i, the function $\log \lambda_i(\vec{x})$ is bounded and Lipschitz continuous. Let G be an open set in $\left(D^d[0, T], d_d\right)$. Then*

$$\liminf_{n \to \infty} \frac{1}{n} \log \mathbb{P}_{\vec{x}}(\vec{z}_n \in G) \geq -\inf\left\{ J_0^T(\vec{r}) : \vec{r} \in G, \ \vec{r}(0) = \vec{x} \right\}$$

uniformly in $\vec{x}$ over compact sets.

By the discussion following (5.31), it suffices to prove that for any path $\vec{r} \in G$ with $\vec{r}(0) = \vec{x}$ and $J_0^T(\vec{r}) < \infty$ and for any $\varepsilon > 0$,

$$\liminf_{n \to \infty} \frac{1}{n} \log \mathbb{P}_{\vec{r}(0)} \left(\sup_{0 \leq t \leq T} |\vec{z}_n(t) - \vec{r}(t)| < \varepsilon \right) \geq -J_0^T(\vec{r}) \qquad (5.37)$$

uniformly over $\vec{x}$ in compact sets.

Remark. Note that since $\vec{r}$ is continuous, the set $N_\varepsilon(\vec{r})$ is open in $\left(D^d[0, T], d_d\right)$.

Let $\tilde{\mathbb{P}}$ be the distribution that corresponds to the same jump directions $\vec{e}_i/n$, but with constant rates $n\mu_i$ (the reason we choose constant rates is discussed below the proof of Corollary 5.53). We write $\tilde{\mathbb{E}}$ for the expectation with respect to $\tilde{\mathbb{P}}$, and write $\vec{y}_\infty(t)$ for the deterministic limit of $\vec{z}_n$ under $\tilde{\mathbb{P}}$ with initial condition $\vec{y}_\infty(0) = \vec{x}$ [see Kurtz's Theorem 5.3 and Equation (5.7)]. Define

$$N_\varepsilon(\vec{y}_\infty) \triangleq \left\{ \vec{x} \in D^d[0, T] : \sup_{0 \leq t \leq T} |\vec{x}(t) - \vec{y}_\infty(t)| < \varepsilon \right\}.$$

The heart of the proof of Kurtz' Theorem is Lemma 5.52, where the limiting form of the change of measure formula (5.30) is established. To establish the appropriate limit of the (logarithm of the) function on the right-hand side of (5.30), we derive a limit formula for general functions $f_i(\vec{x})$, $1 \le i \le k$, $\vec{x} \in \mathbb{R}^d$.

With the notation of (5.30), we have

Lemma 5.52. *Denote by N the number of jumps of $\vec{z}_n$ during $[0, T]$. Then for any bounded continuous functions $\{f_i(\vec{x}), \ 1 \le i \le k\}$ and any $\varepsilon > 0$,*

$$\lim_{n \to \infty} \tilde{\mathbb{E}}_{\vec{x}} \left(\frac{1}{n} \sum_{j=1}^{N} f_{l(j)}(\vec{z}_n(t_j^-)) \right)$$

$$= \lim_{n \to \infty} \tilde{\mathbb{E}}_{\vec{x}} \left(\mathbf{1}\left[\vec{z}_n \in N_\varepsilon(\vec{y}_\infty)\right] \frac{1}{n} \sum_{j=1}^{N} f_{l(j)}(\vec{z}_n(t_j^-)) \right)$$

$$= \int_0^T \sum_{i=1}^{k} \mu_i f_i(\vec{y}_\infty(t)) \, dt,$$

where the convergence is uniform in $\vec{x}$ over bounded sets.

Remark. The middle term here appears in $\frac{1}{n} \int_{N_\varepsilon} H \, d\tilde{\mathbb{P}}$. The lemma states that the expectation has a straightforward limit, reminiscent of Kurtz's Theorem 5.3.

Proof. Fix a direction $\vec{e}_i$ and let M_n be a Poisson random variable with rate $nT\mu_i$. Then for all a,

$$\tilde{\mathbb{P}}_{\vec{x}} \left(\frac{1}{n} \sum_{j=1}^{N} \mathbf{1}\left[l(j) = i\right] > a \right) = \tilde{\mathbb{P}}(M_n > na). \tag{5.38}$$

However, M_n is also a sum of n i.i.d. Poisson random variables with rate $T\mu_i$. By Chernoff's Theorem and the calculation in Example 1.13,

$$\mathbb{P}(M_n > na) \le e^{-n\ell(a)} \tag{5.39}$$

for all $a > T\mu_i$, where $\ell(a)/a \to \infty$ as $a \to \infty$. This establishes that the random variables M_n/n have uniformly bounded second moment (in fact, they have uniformly bounded moments of all orders). In particular, by Definition A.94 and Exercise A.96, they are uniformly integrable and, by (5.38), so are the normalized number of jumps in direction $\vec{e}_i$. Since by Kurtz's Theorem 5.3,

$$\tilde{\mathbb{P}}(\vec{z}_n \in N_\varepsilon(\vec{y}_\infty)) \to 1 \text{ as } n \to \infty, \tag{5.40}$$

we conclude from the comment following Definition A.94 that if one of the limits below exists, then indeed

$$\lim_{n \to \infty} \tilde{\mathbb{E}}_{\vec{x}} \left(\frac{1}{n} \sum_{j=1}^{N} \mathbf{1}\left[l(j) = i\right] \right)$$

$$= \lim_{n \to \infty} \tilde{\mathbb{E}}_{\vec{x}} \left(\mathbf{1} \left[\vec{z}_n \in N_\varepsilon(\vec{y}_\infty) \right] \frac{1}{n} \sum_{j=1}^{N} \mathbf{1} \left[l(j) = i \right] \right). \quad (5.41)$$

Now since (5.41) holds for all i and the f_i are bounded, the first equality of the lemma follows, where the uniformity in $\vec{x}$ is inherited from the uniformity in Kurtz's Theorem 5.3 and (5.38).

To prove the second equality, assume first that f_i are constant (i.e., independent of $\vec{x}$). By Theorem 4.5 the number of jumps $\vec{z}_n$ makes in each direction $\vec{e}_i/n$ is an independent Poisson $n\mu_i T$ random variable, so

$$\tilde{\mathbb{E}}_{\vec{x}} \left(\frac{1}{n} \sum_{j=1}^{N} f_{l(j)}(\vec{z}_n(t_j^-)) \right) = \sum_{i=1}^{k} T \mu_i f_i, \quad (5.42)$$

which completes the proof for constant functions.

To deal with general f_i, we approximate by Riemann sums. We start by restricting attention to $\{\vec{z}_n \in N_\varepsilon(\vec{y}_\infty)\}$:

$$\tilde{\mathbb{E}} \left[\frac{1}{n} \sum_{j=1}^{N} f_{l(j)}(\vec{z}_n(t_j^-)) - \int_0^T \sum_{i=1}^{k} \mu_i f_i(\vec{y}_\infty(t)) \, dt \right] \quad (5.43)$$

$$= \tilde{\mathbb{E}} \left(\mathbf{1} \left[\vec{z}_n \in N_\varepsilon(\vec{y}_\infty) \right] \left[\frac{1}{n} \sum_{j=1}^{N} f_{l(j)}(\vec{z}_n(t_j^-)) \right. \right.$$

$$\left. \left. - \int_0^T \sum_{i=1}^{k} \mu_i f_i(\vec{y}_\infty(t)) \, dt \right] \right)$$

$$(5.44)$$

$$+ \tilde{\mathbb{E}} \left[\mathbf{1} \left[\{\vec{z}_n \in N_\varepsilon(\vec{y}_\infty)\}^c \right] \frac{1}{n} \sum_{j=1}^{N} f_{l(j)}(\vec{z}_n(t_j^-)) \right]$$

$$- \tilde{\mathbb{E}} \left[\mathbf{1} \left[\{\vec{z}_n \in N_\varepsilon(\vec{y}_\infty)\}^c \right] \int_0^T \sum_{i=1}^{k} \mu_i f_i(\vec{y}_\infty(t)) \, dt \right].$$

The last term tends to zero as $n \to \infty$ since f_i are bounded and, by Kurtz's Theorem, $\tilde{\mathbb{P}}_{\vec{x}}(\vec{z}_n \in N_\varepsilon(\vec{y}_\infty)) \to 1$. These facts and the argument in (5.41) imply that the second term on the right of (5.44) tends to zero as $n \to \infty$ as well. Both converge uniformly in $\vec{x}$ over bounded sets. We now estimate the first term on the right of (5.44).

Divide the interval $[0, T]$ into J subintervals of length $\Delta \overset{\triangle}{=} T/J$, and for each j and $\delta > 0$, define

$$B_{\delta j} \overset{\triangle}{=} \left\{ \vec{x} \in \mathbb{R}^d : |\vec{x} - \vec{y}_\infty(t)| < \delta \quad \text{for some} \quad t \in (j\Delta, \, j\Delta + \Delta) \right\}. \quad (5.45)$$

Since by (5.7) the path $\vec{y}_\infty$ is linear on $[0, T]$ and since f_i are continuous, for any $\eta > 0$ we can choose J large and $\delta > 0$ small enough so that

$$\sup_{\vec{x} \in B_{\delta j}} f_i(\vec{x}) - \inf_{\vec{x} \in B_{\delta j}} f_i(\vec{x}) < \eta \tag{5.46}$$

for all i, j. It will be convenient to choose the value of ε as $\varepsilon = \delta$; this is without loss of generality since we have already established that the limits are independent of the value of ε. By Kurtz's Theorem 5.3 and the hypotheses, these choices can be made uniformly in $\vec{x}$ in bounded sets. Consider now the first term in (5.44). Since we are on the set $\{\vec{z}_n \in N_\varepsilon(\vec{y}_\infty)\}$, we can replace each $\vec{z}_n(t_j^-)$ with $\vec{y}_\infty(t_j^-)$, and the resulting error e_0 satisfies

$$\tilde{\mathbb{E}}|e_0| \le \tilde{\mathbb{E}}\frac{N}{n}\eta = \eta T \sum_{i=1}^{k} \mu_i$$

by virtue of (5.46) since $\varepsilon = \delta$. Now break the first sum on the right of (5.44) into the J time intervals. Fix m and let $\Delta_i \vec{z}_n$ denote the number of jumps of $\vec{z}_n$ in direction $\vec{e}_i / n$ during the mth interval. On that interval we have by (5.45) and (5.46) the following upper and lower bounds

$$\mathbf{1}\left[\vec{z}_n \in N_\varepsilon(\vec{y}_\infty)\right]\frac{1}{n}\sum_{i=1}^{k}\Delta_i \vec{z}_n(f_i(\vec{y}_\infty(m\Delta)) - \eta) \tag{5.47}$$

$$\le \mathbf{1}\left[\vec{z}_n \in N_\varepsilon(\vec{y}_\infty)\right]\frac{1}{n}\sum_{j=1}^{N}f_{l(j)}(\vec{y}_\infty(t_j^-))\mathbf{1}\left[m\Delta \le t_j \le m\Delta + \Delta\right]$$

$$\le \mathbf{1}\left[\vec{z}_n \in N_\varepsilon(\vec{y}_\infty)\right]\frac{1}{n}\sum_{i=1}^{k}\Delta_i \vec{z}_n(f_i(\vec{y}_\infty(m\Delta)) + \eta).$$

But as in the proof for the constant functions,

$$\tilde{\mathbb{E}}\frac{1}{n}\sum_{i=1}^{k}\Delta_i \vec{z}_n f_i(\vec{y}_\infty(m\Delta)) = \Delta \sum_{i=1}^{k}\mu_i f_i(\vec{y}_\infty(m\Delta)).$$

Therefore

$$\tilde{\mathbb{E}}\left[\mathbf{1}\left[\vec{z}_n \in N_\varepsilon(\vec{y}_\infty)\right]\frac{1}{n}\sum_{i=1}^{k}\Delta_i \vec{z}_n f_i(\vec{y}_\infty(m\Delta))\right] - \Delta \sum_{i=1}^{k}\mu_i f_i(\vec{y}_\infty(m\Delta)) \to 0$$

as $n \to \infty$. Hence the first sum on the right of (5.43) can be approximated by

$$\sum_{j=0}^{J-1}\sum_{i=1}^{k}\mu_i f_i(\vec{y}_\infty(j\Delta)) \cdot \Delta$$

with an error that is linear in η. Taking $n \to \infty$ we obtain upper and lower Riemann sums for the integral in (5.43). Now η can be made arbitrarily small by taking J large and δ small, and this uniformly in $\vec{x}$ over bounded sets, which concludes the approximation.   ∎

We can now compute a lower bound on the probability that the process $\vec{z}_n$ follows the "fluid limit" of the twisted process. Recall the definition (5.20) of $f(\vec{\mu}, \vec{\lambda})$, which is repeated above Theorem 5.51.

Corollary 5.53. *Assume the μ_i are constant and that the $\log \lambda_i(\vec{x})$ are bounded and continuous. Define $\vec{y}_\infty$ through (5.34). Then*

$$\liminf_{n\to\infty} \frac{1}{n} \log \mathbb{P}_{\vec{x}}(\vec{z}_n \in N_\varepsilon(\vec{y}_\infty)) \geq -\int_0^T f\left(\vec{\mu}, \vec{\lambda}(\vec{y}_\infty(t))\right) dt$$

and the convergence is uniform in $\vec{x}$ over bounded sets.

Proof. Since $\tilde{\mathbb{P}}(\vec{z}_n \in N_\varepsilon(\vec{y}_\infty)) \to 1$, we have $n^{-1} \log \tilde{\mathbb{P}}(\vec{z}_n \in N_\varepsilon(\vec{y}_\infty)) \to 0$. So, by Equations (5.30) and (5.33),

$$\liminf_{n\to\infty} \frac{1}{n} \log \mathbb{P}(\vec{z}_n \in N_\varepsilon(\vec{y}_\infty))$$

$$\geq \liminf_{n\to\infty} \tilde{\mathbb{E}}\left(-\mathbf{1}\left[\vec{z}_n \in N_\varepsilon(\vec{y}_\infty)\right] \int_0^T \sum_{i=1}^k (\lambda_i(\vec{z}_n(s)) - \mu_i(\vec{z}_n(s)) \, ds)\right)$$

$$+ \liminf_{n\to\infty} \tilde{\mathbb{E}}\left(-\mathbf{1}\left[\vec{z}_n \in N_\varepsilon(\vec{y}_\infty)\right] \frac{1}{n} \sum_{j=1}^N \log\left(\frac{\mu_{l(j)}(\vec{z}_n(t_j^-))}{\lambda_{l(j)}(\vec{z}_n(t_j^-))}\right)\right).$$

Now note that if, for some i, the rate μ_i is zero, then under $\tilde{\mathbb{E}}$ no jumps in that direction occur, so that the last term is well defined. In fact, we can simply omit all zero rates from the formulas. But then the convergence of the first term on the right follows from Kurtz's Theorem 5.3 due to the continuity of the λ_i. The convergence of the second term follows from Lemma 5.52, with

$$f_i(\vec{x}) \overset{\triangle}{=} \mathbf{1}\,[\mu_i \neq 0] \log(\mu_i/\lambda_i(\vec{x})).$$

Using the definition (5.20) of $f\left(\vec{\mu}, \vec{\lambda}\right)$, the result follows.   ∎

Remark. It is only for this corollary that we need the $\log \lambda_i$ bounded and continuous. All other results only require the $\lambda_i(\vec{x})$ to be bounded and continuous.

To complete the proof of the lower bound, we would like to set $\vec{y}_\infty(t) = \vec{r}(t)$. This involves finding a nicely behaved set of rates $\mu_i(\vec{x})$ so that

$$\frac{d}{dt}\vec{r}(t) = \sum_{i=1}^k \mu_i(\vec{r}(t)) \vec{e}_i$$

and extending Lemma 5.52 and Corollary 5.53 to non-constant rates μ_i. Although it is possible to extend the results to rates $\mu_i(\vec{x})$ that are bounded and Lipschitz continuous (Definition A.25), there are two problems here. One is that we do not have much control over the smoothness of $r(t)$, so that it is impossible to guarantee that the μ_i are smooth. For example, choose $\vec{r}$ only absolutely continuous, but not continuously differentiable. If μ_i were continuous, then the ODE (5.8) would imply that $\vec{r}\,'$ is continuous, which is a contradiction.

The other problem is that $\vec{r}(t)$ may cross the same point $\vec{x}$ at two different times with two different derivatives, meaning $\sum_{i=1}^{k}\mu_i(\vec{r}(t))\vec{e}_i$ would have to take two different values. We get around both of these difficulties by discretizing $[0, T]$ into J intervals, and taking μ_i to be constant on each interval. It is then a trivial matter to construct $\vec{y}_\infty(t)$ on $[0, T]$ by piecing it together from its piecewise linear parts on the intervals $[jT/J, (jT + T)/J]$. Similarly, all other estimates will be pieced together, and the lemmas that assumed smoothness will be applied on these subintervals. For the purpose of the proof, we will not need these approximating μ_i to converge, and indeed in general they do not.

For a given J, define $\Delta \overset{\triangle}{=} T/J$ and let $t_j \overset{\triangle}{=} j\Delta$. On each interval $[t_j, t_{j+1}] \overset{\triangle}{=} [j\Delta, (j + 1)\Delta]$, define $\Delta\vec{r} = \vec{r}(t_{j+1}) - \vec{r}(t_j)$. Take $\vec{\mu}_j \overset{\triangle}{=} \{\mu_{ij}, i = 1, \ldots, k\}$ so as to satisfy

$$\sum_{i=1}^{k}\mu_{ij}\vec{e}_i = \frac{\Delta\vec{r}}{\Delta}. \tag{5.48}$$

Note that such a choice of μ_{ij} is possible provided $\Delta\vec{r}$ is in the positive cone generated by the $\{\vec{e}_i\}$. If it is strictly outside (say, distance ε), then

$$\mathbb{P}\left(\inf_{0\leq s<t\leq T}|\vec{z}_n(t) - \vec{z}_n(s) - \Delta\vec{r}| < \frac{\varepsilon}{2}\right) = 0$$

for all n. We show below that in this case, $J_0^T(\vec{r}) = \infty$, so that the lower bound is trivial. To this end, it is convenient to use the fact, established in Theorem 5.26, §5.2, that $\tilde{\ell} \equiv \ell$ [Equations (5.20) and (5.2)].

Recall that $\vec{y}_\infty$ is a solution of (5.7) (with μ_{ij} replacing λ_i on the appropriate interval), hence it is piecewise linear. By construction, $\vec{y}_\infty(jT/J) = \vec{r}(jT/J)$ for all j, and therefore for any ε we can choose J large enough so that

$$\sup_{0\leq t\leq T}|\vec{r}(t) - \vec{y}_\infty(t)| < \frac{\varepsilon}{2}. \tag{5.49}$$

Proof of the lower bound, Theorem 5.51. If $J_0^T(\vec{r}) = \infty$ then there is nothing to prove. If it is finite then, since $\vec{r}$ is continuous, Theorem 5.35 implies that for any δ we can choose J large enough so that

$$
\int_0^T \tilde{\ell}(\vec{r}(t), \vec{r}\,'(t))\, dt = \sum_{j=0}^{J-1} \int_{t_j}^{t_{j+1}} \tilde{\ell}(\vec{r}(t), \vec{r}\,'(t))\, dt
$$

$$
\geq \sum_{j=0}^{J-1} \int_{t_j}^{t_{j+1}} \tilde{\ell}\left(\vec{r}\,(t_j), \vec{r}\,'(t)\right) dt - \delta \qquad (5.50)
$$

$$
\geq \sum_{j=0}^{J-1} \Delta \cdot \tilde{\ell}\left(\vec{r}\,(t_j), \frac{\Delta \vec{r}_j}{\Delta}\right) - \delta,
$$

where Lemma 5.16 implies the last inequality. Now $\vec{\mu}_j$ defined in (5.48) is clearly in $K_{\vec{y}}$, where $\vec{y} = \Delta\vec{r}/\Delta$. By the definitions (5.20)–(5.22) of f and $\tilde{\ell}$ we can choose $\vec{\mu}_j$ so that

$$
\tilde{\ell}\left(\vec{r}\,(t_j), \frac{\Delta \vec{r}_j}{\Delta}\right) \geq f\left(\vec{\mu}_j, \vec{\lambda}\left(\vec{r}\,(t_j)\right)\right) - \frac{\delta}{T}.
$$

Thus

$$
\int_0^T \tilde{\ell}(\vec{r}(t), \vec{r}\,'(t))\, dt \geq \sum_{j=0}^{J-1} f\left(\vec{\mu}_j, \vec{\lambda}\left(\vec{r}\,(t_j)\right)\right) \Delta - 2\delta. \qquad (5.51)
$$

Since by (5.49)

$$
\sup_{0 \leq t \leq T} \left|\vec{r}(t) - \vec{y}_\infty(t)\right| \leq \frac{\varepsilon}{2},
$$

we have, by the Markov property,

$$
\mathbb{P}_{\vec{x}}\left(\sup_{0 \leq t \leq T} |\vec{z}_n(t) - \vec{r}(t)| < \varepsilon\right) \geq \mathbb{P}_{\vec{x}}\left(\sup_{0 \leq t \leq T} |\vec{z}_n(t) - \vec{y}_\infty(t)| < \frac{\varepsilon}{2}\right)
$$

$$
\geq \mathbb{P}_{\vec{x}}\left(\sup_{0 \leq t \leq T-\Delta} |\vec{z}_n(t) - \vec{y}_\infty(t)| < \frac{\varepsilon}{2}\frac{J-1}{J}\right)
$$

$$
\times \inf_{v \in B} \mathbb{P}_v\left(\sup_{T-\Delta \leq t \leq T} |\vec{z}_n(t) - \vec{y}_\infty(t)| < \frac{\varepsilon}{2}\right),
$$

where B is the set of initial conditions

$$
B \triangleq \left\{v : |v - \vec{y}_\infty(T-\Delta)| < \frac{\varepsilon}{2}\frac{J-1}{J}\right\}.
$$

But for $v \in B$,

$$
\mathbb{P}_v\left(\sup_{T-\Delta \leq t \leq T} |\vec{z}_n(t) - \vec{y}_\infty(t)| < \frac{\varepsilon}{2}\right)
$$

$$
\geq \mathbb{P}_v\left(\sup_{T-\Delta \leq t \leq T} |[\vec{z}_n(t) - \vec{z}_n(T-\Delta)] - [\vec{y}_\infty(t) - \vec{y}_\infty(T-\Delta)]| < \frac{\varepsilon}{2J}\right).
$$

Therefore, applying Corollary 5.53 over the interval $[T - \Delta, T]$, we obtain

$$
\liminf_{n\to\infty} \frac{1}{n} \log \mathbb{P}_{\vec{x}} \left(\sup_{0\le t\le T} |\vec{z}_n(t) - \vec{r}(t)| < \varepsilon \right)
$$

$$
\ge \liminf_{n\to\infty} \frac{1}{n} \log \mathbb{P}_{\vec{x}} \left(\sup_{0\le t\le T-\Delta} |\vec{z}_n(t) - \vec{y}_\infty(t)| < \frac{\varepsilon}{2}\frac{J-1}{J} \right)
$$

$$
- f\left(\vec{\mu}_j, \vec{\lambda}\left(\vec{r}\left(T - \Delta \right) \right) \right) \cdot \Delta .
$$

Iterating this argument, we obtain by (5.51)

$$
\liminf_{n\to\infty} \frac{1}{n} \log \mathbb{P}_{\vec{x}} \left(\sup_{0\le t\le T} |\vec{z}_n(t) - \vec{r}(t)| < \varepsilon \right) \ge - \sum_{j=0}^{J-1} f\left(\vec{\mu}_j, \vec{\lambda}\left(\vec{r}\left(t_j \right) \right) \right) \cdot \Delta
$$

$$
\ge - \int_0^T \tilde{\ell}(\vec{r}(t), \vec{r}\,'(t))\, dt - 2\delta.
$$

Since δ is arbitrary, the proof is complete.   ■

5.4. The Upper Bound: Orientation

In this section we establish an upper bound for closed sets of paths. Conceptually, the upper bound is more difficult than the lower bound. This is because the lower bound is established by estimating the probability that the scaled process $\vec{z}_n(t)$ is close to a particular path $\vec{r}(t)$. For the upper bound, though, we have to estimate the probability that $\vec{z}_n(t)$ is far from a *set* of paths. This conceptual difficulty is reflected in a more complex proof. Paradoxically, the upper bound can be established (at the present time) in greater generality than the lower bound* (i.e., for a wider class of processes).

Here is a statement of the theorem we will prove below.

Theorem 5.54. *Let the* $\log \lambda_i(\vec{x})$ *be bounded and Lipschitz continuous in* $\mathbb{R}^d$, *and let* F *be closed in* $\left(D^d[0, T], d_d \right)$. *Then*

$$
\limsup_{n\to\infty} \frac{1}{n} \log \mathbb{P}_{\vec{x}}\,(\vec{z}_n \in F) \le - \inf\{I_0^T\,(\vec{r}) : \vec{r} \in F, \; \vec{r}(0) = \vec{x}\}.
$$

This estimate holds also for nearby initial points; see Theorem 5.64.

Since $I_0^T\,(\vec{z}_\infty) = 0$, there is nothing to prove if $\vec{z}_\infty \in F$, so below we exclude this case. Most proofs of upper bounds follow a standard sequence, which we now outline. First define the set of "cheap paths"

$$
\Phi(K) \overset{\Delta}{=} \{\vec{r} : I_0^T\,(\vec{r}) \le K\}.
$$

---

* This situation is unlikely to persist, in our opinion, and simply reflects the relative youth of the theory of large deviations as applied to jump processes.

Take the largest K so that $\Phi(K) \cap F = \emptyset$, or $F \subset \Phi^c(K)$. If we could show that

$$\mathbb{P}(\vec{z}_n \in \Phi^c(K)) \leq e^{-nK+o(n)},$$

then we are done [since our basic space of paths is the space $D^d[0, T]$, the set $\Phi^c(K)$ contains, by definition, those paths in $D^d[0, T]$ that are not in $\Phi(K)$]. However, this cannot be true; for $\vec{z}_n$ is not continuous, so that $I_0^T(\vec{z}_n) = \infty$ (unless $\vec{z}_n$ is constant on $[0, T]$).

To resolve this difficulty, all proofs we know construct a random piecewise linear path $\vec{y}_n$ which is close to $\vec{z}_n$ in the sense that

$$\mathbb{P}(d(\vec{z}_n, \vec{y}_n) > \varepsilon) < e^{-n(K+\delta)} \qquad (5.52)$$

for all n large. The notation $d(\vec{x}, \vec{y}) \overset{\triangle}{=} \sup_{0 \leq t \leq T} |\vec{x}(t) - \vec{y}(t)|$ is used, with the obvious extension to a distance between a path and a set of paths. The key property of $\vec{y}_n$, the derivation of which is described later in this section, is that for all κ,

$$\mathbb{P}(I_0^T(\vec{y}_n) \geq \kappa) \leq e^{-n\kappa}. \qquad (5.53)$$

Now one shows a continuity property for I_0^T, namely that for any $\delta > 0$, there is an ε small enough that

$$\Phi^c(K) \subset \{\vec{r} : d(\vec{r}, \Phi(K - \delta)) > \varepsilon\}.$$

Hence from (5.52) and (5.53) (with $\kappa = K - \delta$),

$$\begin{aligned}
\mathbb{P}(\vec{z}_n \in F) &\leq \mathbb{P}(d(\vec{y}_n, \Phi(K - \delta)) > \varepsilon) + \mathbb{P}(d(\vec{z}_n, \vec{y}_n) > \varepsilon) \\
&\leq \mathbb{P}(d(\vec{z}_n, \vec{y}_n) > \varepsilon) + \mathbb{P}(I_0^T(\vec{y}_n) \geq K - \delta) \\
&\leq 2e^{-n(K-\delta)}.
\end{aligned}$$

This proves the result.

Construction of the random path $\vec{y}_n$ is easy. Divide $[0, T]$ into J intervals $[t_j, t_{j+1})$ for $0 \leq j \leq J - 1$, and let $\vec{y}_n(t)$ be the piecewise linear function that agrees with $\vec{z}_n$ at the endpoints $\{t_j\}$. The same sort of estimates used to prove Kurtz's Theorem are used to show that $\vec{z}_n(t)$ and $\vec{y}_n(t)$ must be close in the sense of (5.52), at least if J is large enough. Now the whole difficulty is in establishing (5.53). Here is where the proof we use differs from "standard" proofs.

In most proofs, one shows (5.53) by an explicit calculation. First approximate the coefficients $\lambda_i(\vec{z}_n(t))$ on the interval $[t_j, t_{j+1})$ by constants $\lambda_i(\vec{z}_n(t_j))$. This new process, with constant coefficients, has a Poisson distribution for the number of jumps in each direction $\vec{e}_i$. Therefore we can obtain explicit estimates for this process. Using the exponential martingale (as in Chernoff's Theorem 1.5 and in Kurtz's Theorem 5.3) we can establish that

$$\mathbb{E}\left(\exp\left(n\tilde{I}_{t_j}^{t_{j+1}}(\vec{y}_n)\right)\right) \leq C,$$

where $\tilde{I}$ is calculated using the constant coefficients. Then by taking J, the number of intervals in the definition of $\vec{y}_n$, to be $o(n)$, we obtain by the Markov property

$$\mathbb{E}\left(\exp\left(n\tilde{I}_0^T(\vec{y}_n)\right)\right) \leq C^J = e^{o(n)}.$$

Hence from Chebycheff's inequality,

$$\mathbb{P}\left(\tilde{I}_0^T(\vec{y}_n) \geq K\right) = \mathbb{P}\left(\exp(n\tilde{I}_0^T(\vec{y}_n)) \geq e^{nK}\right)$$
$$\leq C^J e^{-nK}$$
$$= e^{-nK+o(n)}.$$

Now another estimate shows that $I_0^T(\vec{y}_n) \approx \tilde{I}_0^T(\vec{y}_n)$, completing the proof.

The problem with the proof as outlined above is that it seems to depend in a critical way on the fact that $\vec{y}_n$ is a constant coefficient process. Our ultimate goal is to treat queueing processes, where boundaries occur naturally (since all queue sizes are, by definition, non-negative). A typical one-dimensional queueing process would have service rate μ when $z > 0$ but service rate 0 at $z = 0$. Clearly, when the queue is nearly empty it cannot be approximated well by a constant coefficient process. In order to be able to extend the proofs below to cases with discontinuities of this sort (rather than start anew), we take a different approach. This approach is useful in Chapter 8, where proofs of upper bounds for different processes are nearly identical to the ones in this chapter.

We therefore follow the line of reasoning of [DEW], which does not introduce a new process. The estimation of $\mathbb{E}\left(\exp\left(n I_0^T(\vec{y}_n)\right)\right)$ is done by finding an *approximating functional* $\overset{\delta}{I}_0^T(\vec{y})$, which corresponds in some cases to a constant coefficient process, but is sensitive enough to estimate correctly the effect of the boundary in many cases. We also choose a very fine partition J for the definition of $\vec{y}_n$; in fact, we choose $J = n$. This requires us to show that

$$n \log \mathbb{E}\exp\left(n\tilde{I}_{t_j}^{t_{j+1}}(\vec{y}_n)\right) \leq C < \infty$$

for each j provided n is large enough, so that (by the Markov property)

$$\mathbb{E}\exp\left(n I_0^T(\vec{y}_n)\right) \leq \prod_0^{J-1}\left(\mathbb{E}\exp\left(n\tilde{I}_{t_j}^{t_{j+1}}(\vec{y}_n)\right)\right)$$
$$\leq e^C$$

for n large. (In fact, we show $C = 0$ suffices.) This follows by an argument due to Dupuis and Kushner [DK2].

5.5. Proof of the Upper Bound

We have already indicated that we estimate the probability of a closed set F by estimating the probability of $\Phi^c(K)$, where K is chosen just small enough that $F \subset \Phi^c(K) \, (\subset D^d[0, T])$.

We rely heavily on the following simple consequence of Corollary 5.8, of §5.1.

Corollary 5.55. *If the $\lambda_i(\vec{x})$ are bounded, then there are positive constants c_1 and c_2, independent of $\vec{x}$, such that for any $0 \le t \le t + \Delta \le T$,*

$$
\mathbb{P}_{\vec{x}}\left(\sup_{t \le s \le t+\Delta} |\vec{z}_n(s) - \vec{z}_n(t)| \ge a\right) \le \exp\left(-nac_1 \log\left(\frac{ac_2}{\Delta}\right)\right).
$$

Proof. For $t = 0$ the result follows after simple algebra from Corollary 5.8 and the estimate (5.10), by replacing the interval size T with Δ. The result for $0 \le t < T$ follows by a smoothing argument from the uniformity in $\vec{x}$, due to the Markov property. $\blacksquare$

As in the case of the lower bound, it will be convenient to partition the time interval $[0, T]$ into small pieces. So, given n, define

$$
t_j^n \triangleq j\frac{T}{n}, \quad j = 0, 1, \ldots, n. \tag{5.54}
$$

Definition 5.56. *The piecewise linear interpolation $y_n(t)$ of $\vec{z}_n(t)$:*

$$
\vec{y}_n(t) \triangleq \left(\frac{n}{T}t - j\right)\vec{z}_n\left(t_{j+1}^n\right) + \left((j+1) - \frac{n}{T}t\right)\vec{z}_n\left(t_j^n\right), \quad t \in \left[t_j^n, t_{j+1}^n\right].
$$

In the nomenclature of large deviations, the following result shows that $\vec{y}_n$ is exponentially close to $\vec{z}_n$.

Lemma 5.57. *Assume the $\lambda_i(\vec{x})$ are bounded. Then for each $\delta > 0$ we have, uniformly in $\vec{x} \in \mathbb{R}^d$,*

$$
\limsup_{n \to \infty} \frac{1}{n} \log \mathbb{P}_{\vec{x}}\left(d(\vec{z}_n(t), \vec{y}_n(t)) > \delta\right) = -\infty.
$$

Proof. This follows from Corollary 5.55 with $\Delta = T/n$. Consider some interval $[t_j^n, t_{j+1}^n]$. Since $\vec{z}_n(t)$ and $\vec{y}_n(t)$ agree at the endpoints of these intervals, clearly

$$
|\vec{y}_n(t_{j+1}^n) - \vec{y}_n(t_j^n)| > \frac{\delta}{2} \quad \text{implies} \quad |\vec{z}_n(t_{j+1}^n) - \vec{z}_n(t_j^n)| > \frac{\delta}{2}.
$$

On the other hand, from the triangle inequality,

$$
|\vec{z}_n(t) - \vec{z}_n(t_j^n)| \ge |\vec{z}_n(t) - \vec{y}_n(t)| - |\vec{y}_n(t_{j+1}^n) - \vec{y}_n(t_j^n)|
$$

since $\vec{y}_n$ is piecewise linear and $\vec{y}_n(t_j^n) = \vec{z}_n(t_j^n)$. Therefore if $|\vec{z}_n(t) - \vec{y}_n(t)| > \delta$ for some t in the jth interval, then we must have

$$\sup_{t_j^n \le t \le t_{j+1}^n} |\vec{z}_n(t) - \vec{z}_n(t_j^n)| \ge \frac{\delta}{2}.$$

Applying Corollary 5.55,

$$\mathbb{P}_{\vec{x}}\left(\sup_{t_j^n \le t \le t_{j+1}^n} |\vec{z}_n(t) - \vec{z}_n(t_j^n)| > \delta/2 \right) \le \exp\left(-n\frac{\delta c_1}{2} \log\left(\frac{n\delta c_3}{2} \right) \right),$$

where $c_3 \stackrel{\triangle}{=} c_2/T$. Since the constants c_1 and c_3 do not depend on $\vec{x}$,

$$\mathbb{P}_{\vec{x}}(d(\vec{z}_n(t), \vec{y}_n(t)) > \delta) \le \sum_{j=0}^{n-1} \mathbb{P}_{\vec{x}}\left(\sup_{t_j^n \le t \le t_{j+1}^n} |\vec{z}_n(t) - \vec{y}_n(t)| > \delta \right)$$

$$\le \sum_{j=0}^{n-1} \mathbb{P}_{\vec{x}}\left(\sup_{t_j^n \le t \le t_{j+1}^n} |\vec{z}_n(t) - \vec{z}_n(t_j^n)| > \frac{\delta}{2} \right)$$

$$\le n \exp\left(-n\frac{\delta c_1}{2} \log \frac{n\delta c_3}{2} \right).$$

The result follows since c_1 and c_3 are positive. ∎

Remark. Standard constructions of the process $y_n(t)$ use a fixed number J of intervals, chosen large enough that the estimate

$$\mathbb{P}\left(d(\vec{z}_n, \vec{z}_\infty) > \frac{\delta}{2} \right) \le 2d \exp(-\eta \log Jn)$$

holds for some positive η, and

$$\eta \log J > K \stackrel{\triangle}{=} \inf_{\vec{r} \in F} I_0^T(\vec{r}).$$

We follow the recent trend and estimate all values of K simultaneously.

Our first key observation is that, with exponentially high probability, the processes $\vec{y}_n$ stay in compact sets: in the nomenclature of large deviations, $\{\vec{y}_n\}$ is *exponentially tight*.

Lemma 5.58 (exponential tightness). *Assume the $\lambda_i(\vec{x})$ are bounded. Let $C \subset \mathbb{R}^d$ be a compact set. For each $B < \infty$ there is a compact set $\mathcal{K} \subset C^d[0, T]$ such that for all $\vec{x} \in C$,*

$$\limsup_{n \to \infty} \frac{1}{n} \log \mathbb{P}_{\vec{x}}\left(\vec{y}_n \notin \mathcal{K} \right) \le -B.$$

Proof. We construct an explicit sequence of compact sets that will be used in the remainder of this section as well. Define the (global) modulus of continuity of a function f by

$$V_\delta(f) \triangleq \sup\{|f(t) - f(s)| : 0 \le s \le t \le T, \ |t - s| < \delta\}. \tag{5.55}$$

Now define

$$\mathcal{K}(M) \triangleq \bigcap_{m=M}^{\infty} \left\{ \vec{r} \in C^d[0, T] : \vec{r}(0) \in C, \ V_{2^{-m}}(\vec{r}) \le \frac{1}{\log m} \right\}. \tag{5.56}$$

Each set in the intersection is closed, and so the set $\mathcal{K}(M)$ is closed by Definition A.1. The functions in $\mathcal{K}(M)$ are clearly equicontinuous, with uniformly bounded initial values $\vec{r}(0)$, and therefore the set is compact by the Arzelà-Ascoli Theorem A.51.

Now if $2^{-m} < T/n$, then

$$V_{2^{-m-1}}(\vec{y}_n) = \frac{1}{2} V_{2^{-m}}(\vec{y}_n)$$

since $\vec{y}_n$ is piecewise linear. Therefore, to check whether $\vec{y}_n$ is in $\mathcal{K}(M)$, we only need to consider a finite intersection, for values of m up to

$$M(n) \triangleq \max\left\{ M, \left\lceil \frac{\log(n/T)}{\log 2} \right\rceil \right\},$$

where $\lceil \alpha \rceil$ is the smallest integer larger than α. From Corollary 5.55 and a union bound (Lemma A.115), for any $\vec{x} \in C$ and any n with $M(n) > M$,

$$\mathbb{P}_{\vec{x}}(\vec{y}_n \notin \mathcal{K}(M)) \le \sum_{m=M}^{M(n)} \mathbb{P}_{\vec{x}}\left(V_{2^{-m}}(\vec{y}_n) > \frac{1}{\log m} \right)$$

$$\le \sum_{m=M}^{M(n)} \sum_{j=0}^{n-1} \mathbb{P}_{\vec{x}}\left(\sup_{0 \le t \le 2^{-m}} |\vec{z}_n(t_j^n + t) - \vec{z}_n(t_j^n)| > \frac{1}{\log M} \right)$$

$$\le M(n) \cdot n \exp\left(-n \frac{c_1}{\log M} \log\left(\frac{2^M c_2}{\log M} \right) \right),$$

using Corollary 5.55 in the same manner as in Lemma 5.57, with interval size $\delta = 2^{-m}$ and $a = 1/\log M$. Thus

$$\limsup_{n \to \infty} \frac{1}{n} \log \mathbb{P}_{\vec{x}}\left(\vec{y}_n \notin \mathcal{K}(M) \right) \le -c_4 \frac{M}{\log M}$$

for some positive constant c_4, uniformly in $\vec{x} \in C$.   ∎

The next step is the main local estimate for $\vec{y}_n(t)$:

$$\mathbb{E}_{\vec{x}} \exp\left(\frac{n}{T} \left\langle \vec{y}_n\left(\frac{T}{n} \right) - \vec{y}_n(0), \vec{\theta} \right\rangle \right) \le \hat{g}(\vec{x}, \vec{\theta}).$$

Once this key estimate has been obtained, we use a technique of Dupuis and Kushner [DK2] to conclude that

$$\mathbb{P}_{\vec{x}}\left(\overset{\delta}{I}{}_0^T(\vec{y}_n) \geq K\right) \leq e^{-n(K-\varepsilon)} .$$

The basic idea is

$$\overset{\delta}{I}{}_0^T(\vec{y}_n) \approx \sum_{j=0}^{n-1} \frac{T}{n} \sup_{\vec{\theta}} \left(\frac{\left\langle \vec{y}_n\left(t_{j+1}^n\right) - \vec{y}_n\left(t_j^n\right), \vec{\theta} \right\rangle}{T/n} - \overset{\delta}{g}\left(\vec{y}_n\left(t_j^n\right), \vec{\theta}\right) \right), \quad (5.57)$$

so choosing a maximizing $\vec{\theta}^*$ and using the exponential martingale (5.59),

$$\mathbb{E}\left(\exp n \overset{\delta}{I}{}_0^T(\vec{y}_n)\right) \approx \prod_{j=0}^{n-1} \mathbb{E}\left(\exp\left(n\left\langle \vec{y}(t_{j+1}^n) - \vec{y}(t_j^n), \vec{\theta}^* \right\rangle - T\overset{\delta}{g}\left(\vec{y}_n, \vec{\theta}^*\right)\right)\right)$$

$$\leq 1 \qquad\qquad\qquad\qquad\qquad\qquad\qquad (5.58)$$

by the local estimate and (5.57). Hence

$$\mathbb{P}\left(\overset{\delta}{I}{}_0^T(\vec{y}_n) \geq K\right) = \mathbb{P}\left(\exp\left(n \overset{\delta}{I}{}_0^T(\vec{y}_n)\right) \geq e^{nK}\right)$$

$$\leq e^{-nK}$$

by Chebycheff's inequality. The technicalities are simply to overcome the vagueness in the $\approx$ appearing in Equations (5.57) and (5.58).

Lemma 5.59. *If the $\lambda_i(\vec{x})$ are bounded, then uniformly in $\vec{x}$ in $\mathbb{R}^d$ and in $\vec{\theta}$ in bounded sets,*

$$\limsup_{n\to\infty} \log \mathbb{E}_{\vec{x}} \exp n \left\langle \vec{y}_n\left(\frac{T}{n}\right) - \vec{y}_n(0), \vec{\theta} \right\rangle \leq T\overset{\delta}{g}(\vec{x}, \vec{\theta}).$$

Proof. As usual in upper bounds, the key is the exponential martingale. Since $\vec{y}_n(t_j^n) = \vec{z}_n(t_j^n)$, for any $\vec{\theta}$ we have

$$1 = \mathbb{E}_{\vec{x}} \exp\left(n\left[\left\langle \vec{y}_n\left(\frac{T}{n}\right) - \vec{y}_n(0), \vec{\theta} \right\rangle - \int_0^{T/n} \sum_{i=1}^k \lambda_i(\vec{z}_n(t))(e^{\langle\vec{\theta},\vec{e}_i\rangle} - 1)\, dt\right]\right).$$

$$(5.59)$$

Now fix any $\vec{y}$ with $|\vec{y} - \vec{x}| \leq \delta/2$. Let

$$S_\delta \overset{\triangle}{=} \left\{ \omega : \sup_{0 \leq t \leq T/n} |\vec{z}_n(t) - \vec{x}| < \frac{\delta}{2} \right\}.$$

Then since $e^a \geq 0$ for all a and by Definition 5.36 of $\overset{\delta}{g}$,

$$1 \geq \mathbb{E}_{\vec{x}} \left(\mathbf{1}[S_\delta] \exp n \left(\left\langle \vec{y}_n \left(\frac{T}{n} \right) - \vec{y}_n(0), \vec{\theta} \right\rangle - \frac{T}{n} \overset{\delta}{g} \left(\vec{y}, \vec{\theta} \right) \right) \right)$$

$$= \exp \left(-T \overset{\delta}{g} \left(\vec{y}, \vec{\theta} \right) \right) \mathbb{E}_{\vec{x}} \exp n \left\langle \vec{y}_n \left(\frac{T}{n} \right) - \vec{y}_n(0), \vec{\theta} \right\rangle$$

$$- \exp \left(-T \overset{\delta}{g} \left(\vec{y}, \vec{\theta} \right) \right) \mathbb{E}_{\vec{x}} \left(\mathbf{1}[S_\delta^c] \exp n \left\langle \vec{y}_n \left(\frac{T}{n} \right) - \vec{y}_n(0), \vec{\theta} \right\rangle \right).$$

Hence

$$\mathbb{E}_{\vec{x}} \exp n \left\langle \vec{y}_n \left(\frac{T}{n} \right) - \vec{y}_n(0), \vec{\theta} \right\rangle$$

$$\leq \exp \left(T \overset{\delta}{g}(\vec{y}, \vec{\theta}) \right) + \mathbb{E}_{\vec{x}} \left(\mathbf{1}[S_\delta^c] \exp n \left\langle \vec{y}_n \left(\frac{T}{n} \right) - \vec{y}_n(0), \vec{\theta} \right\rangle \right).$$

Let us estimate the last term: from Corollary 5.55,

$$\mathbb{E}_{\vec{x}} \left(\mathbf{1}[S_\delta^c] \exp \left(n \left\langle \vec{y}_n \left(\frac{T}{n} \right) - \vec{y}_n(0), \vec{\theta} \right\rangle \right) \right)$$

$$\leq \sum_{K=1}^{\infty} \exp \left(n(K+1) \frac{\delta}{2} |\vec{\theta}| \right)$$

$$\times \mathbb{P}_{\vec{x}} \left(\frac{K\delta}{2} \leq \sup_{0 \leq t \leq T/n} |\vec{z}_n(t) - \vec{x}| \leq \frac{(K+1)\delta}{2} \right)$$

$$\leq \sum_{K=1}^{\infty} \exp n \left((K+1) \frac{\delta}{2} B - \frac{K\delta c_1}{2} \log \left(\frac{K\delta c_2 n}{2T} \right) \right)$$

$$\overset{\triangle}{=} e_1(n\delta)$$

for all $\vec{x} \in \mathbb{R}^d$ and $\vec{\theta}$ with $|\vec{\theta}| \leq B$. Hence

$$\mathbb{E}_x \left(\exp n \left\langle \vec{y}_n \left(\frac{T}{n} \right) - \vec{y}_n(0), \vec{\theta} \right\rangle \right) \leq \exp \left(T \overset{\delta}{g} \left(\vec{y}, \vec{\theta} \right) + e_1(n\delta) \right)$$

where $e_1(a) \to 0$ as $a \to \infty$. ∎

We can glean a little more from the argument. Notice that for all $\delta \geq 0$,

$$\overset{\delta}{g}(\vec{x}, \vec{\theta}) \leq G_1(\vec{\theta}) \overset{\triangle}{=} \sum_{i=1}^{k} \left(\sup_{\vec{x} \in \mathbb{R}^d} \lambda_i(\vec{x}) \right) \left(e^{\langle e_i, \vec{\theta} \rangle} - 1 \right)$$

$$\leq G_2(\vec{\theta}) \overset{\triangle}{=} k \left(\sup_{\vec{x}, i} \lambda_i(\vec{x}) \right) e^{|\vec{\theta}| \cdot \sup_i |e_i|}.$$

Hence, without resorting to the set S_δ, we obtain

$$\limsup_{n\to\infty} \log \mathbb{E}_{\vec{x}} \exp\left(n\left\langle \vec{y}_n\left(\frac{T}{n}\right) - \vec{y}_n(0), \vec{\theta}\right\rangle\right) \le T G_2(\vec{\theta}).$$

We use this crude bound below.

To establish the upper bound, we need the estimate

$$\limsup_{n\to\infty} \frac{1}{n} \log \mathbb{P}\left(d\left(\vec{y}_n, \overset{\delta}{\Phi}_{\vec{x}}(K)\right) > \varepsilon\right) \le -(K - \varepsilon).$$

By Lemma 5.58 we can essentially ignore paths outside $\mathcal{K}(M)$, for large enough values of M. Our next step is therefore an estimate for subsets of $\mathcal{K}(M)$. The next two results follow an argument of Dupuis and Kushner [DK2] to knit together the local estimates into a global one.

Definition 5.60. *For a set $K \subset C^d[0, T]$ let $K_x \overset{\triangle}{=} \{\vec{r} \in K : \vec{r}(0) = \vec{x}\}$.*

Lemma 5.61. *Assume the $\lambda_i(\vec{x})$ are bounded and continuous. Let $C \subset \mathbb{R}^d$ be a compact set and fix a step function $\vec{\theta}$. For each $\delta > 0$ and each compact set $K \subset \mathcal{K}(M)$ of functions $\vec{r}$ with $\vec{r}(0) \in C$,*

$$\limsup_{n\to\infty} \frac{1}{n} \log \mathbb{P}_{\vec{x}}\left(\vec{y}_n \in K_x\right) \le - \inf_{\vec{r}\in K_x} \overset{\delta}{I}{}^T_0(\vec{r}, \vec{\theta}),$$

uniformly in $\vec{x} \in C$, where

$$\overset{\delta}{I}{}^T_0(\vec{r}, \vec{\theta}) \overset{\triangle}{=} \int_0^T \langle \vec{r}'(t), \vec{\theta}(t) \rangle - \overset{\delta}{g}\left(\vec{r}(t), \vec{\theta}(t)\right) dt$$

whenever $\vec{r}$ is absolutely continuous, and is defined as $+\infty$ otherwise.

Proof. Since $\vec{y}_n$ is absolutely continuous by construction,

$$\mathbb{P}_{\vec{x}}(\vec{y}_n \in K_x) = \mathbb{P}_{\vec{x}}(\vec{y}_n \in K_x \cap \{\vec{r} : \vec{r} \text{ absolutely continuous }\}),$$

while by definition,

$$\inf_{\vec{r}\in K_x} \overset{\delta}{I}{}^T_0(\vec{r}, \vec{\theta}) = \inf\left\{\overset{\delta}{I}{}^T_0(\vec{r}, \vec{\theta}) : \vec{r} \in K_x, \vec{r} \text{ absolutely continuous }\right\}$$

so that, without further mention, the discussion will be restricted to absolutely continuous functions. For any $\eta > 0$ and $B > 0$, Lemma 5.59 implies that there is an n_0 so that for all $x \in \mathbb{R}^d$ and all $n > n_0$ we have

$$\mathbb{E}_{\vec{x}} \exp\left(n\left\langle \vec{y}_n\left(\frac{T}{n}\right) - \vec{y}_n(0), \vec{\theta}\right\rangle\right) \le \exp(T \overset{\delta}{g}(\vec{x}, \vec{\theta}) + \eta) \qquad (5.60)$$

uniformly in bounded $\vec{\theta}$ [recall $\mathbb{E}_{\vec{x}}$ means $\vec{y}_n(0) = \vec{x}$]. Now assume for notational convenience that the step function $\vec{\theta}(t)$ is right continuous (it is clearly bounded).

For any function $\vec{r}$ define the sum

$$S_n(\vec{r}, \vec{\theta}) \triangleq \sum_{j=0}^{n-1} \left(\left\langle \vec{r}(t_{j+1}^n) - \vec{r}(t_j^n), \vec{\theta}(t_j^n) \right\rangle - \frac{T}{n} \overset{\delta}{g} \left(\vec{r}(t_j^n), \vec{\theta}(t_j^n) \right) \right).$$

The sum $S_n(\vec{r}, \vec{\theta})$ is clearly an approximation to $I_0^T(\vec{r})$, for a suitable choice of $\vec{\theta}$. The Markov property of $\vec{z}_n(t)$ and (5.60) give us

$$\mathbb{E}_{\vec{x}} \exp n S_n(\vec{y}_n, \vec{\theta}) \leq e^{n\eta}$$

for all $n > n_0$. Then since η was arbitrary, we obtain

$$\limsup_{n\to\infty} \frac{1}{n} \log \mathbb{E}_{\vec{x}} \exp(n S_n(\vec{y}_n, \vec{\theta})) \leq 0, \tag{5.61}$$

uniformly over $\vec{x} \in \mathbb{R}^d$.

Since for $\vec{y}_n \in \mathcal{K}_x$ we have

$$S_n(\vec{y}_n, \vec{\theta}) - \inf_{\vec{r} \in \mathcal{K}_x} S_n(\vec{r}, \vec{\theta}) \geq 0 \quad \text{so that} \quad \exp\left(S_n(\vec{y}_n, \vec{\theta}) - \inf_{\vec{r} \in \mathcal{K}_x} S_n(\vec{r}, \vec{\theta}) \right) \geq 1, \tag{5.62}$$

and so

$$\mathbb{P}_{\vec{x}}(\vec{y}_n \in \mathcal{K}_x) \leq \mathbb{E}_{\vec{x}} \exp\left[n \left\{ S_n(\vec{y}_n, \vec{\theta}) - \inf_{\vec{r} \in \mathcal{K}_x} S_n(\vec{r}, \vec{\theta}) \right\} \right]. \tag{5.63}$$

Combining this with (5.61) we have, uniformly over $\vec{x} \in \mathbb{R}^d$

$$\limsup_{n\to\infty} \frac{1}{n} \log \mathbb{P}_{\vec{x}}(\vec{y}_n \in \mathcal{K}_x) \leq -\liminf_{n\to\infty} \left(\inf_{\vec{r} \in \mathcal{K}_x} S_n(\vec{r}, \vec{\theta}) \right). \tag{5.64}$$

(We are not really giving up much here, in the "large deviations" sense. To find the probability that you are in a particular set, you just have to look at the smallest value of I_0^T over that set.)

We now represent the sum on the right-hand side of (5.64) as an integral. Since $\mathcal{K}$ is compact, the functions $\vec{r} \in \mathcal{K}$ are equicontinuous, and their values are bounded. So, let C_1 be a compact set in $\mathbb{R}^d$ so that

$$C_1 \supset \{\vec{x} : \vec{x} = \vec{r}(t) \text{ for some } \vec{r} \in \mathcal{K} \text{ and some } 0 \leq t \leq T\}.$$

Now $\vec{\theta}$ is a step function, so that it is constant on an interval, say $[0, \tau]$. But then [with t_j^n defined in (5.54)]

$$\sum_{j=0}^{n-1} \mathbf{1}\left[t_{j+1}^n \leq \tau\right] \left\langle \vec{r}(t_{j+1}^n) - \vec{r}(t_j^n), \vec{\theta}(0) \right\rangle = \int_0^\tau \left\langle \vec{r}'(t), \vec{\theta}(0) \right\rangle dt + e_2(n), \tag{5.65}$$

where $e_2(n)$ arises since in general, τ may not match any of the t_j^n. But $e_2(n)$ goes to zero uniformly in $\vec{r} \in \mathcal{K}$, since

$$|e_2(n)| \leq |\vec{\theta}(0)| \frac{T}{n} \sup_{\vec{x} \in C_1} |\vec{x}|.$$

Now $\overset{\delta}{g}$ is continuous, so it is bounded and uniformly continuous on C_1. But the functions in $\mathcal{K}$ are uniformly continuous, so

$$\left| \overset{\delta}{g}\left(\vec{r}(t_j^n), \vec{\theta}(t_j^n)\right) - \overset{\delta}{g}\left(\vec{r}(t), \vec{\theta}(t_j^n)\right) \right|, \quad t_j^n \leq t \leq t_{j+1}^n$$

goes to zero with n, uniformly in j and in $\vec{r} \in \mathcal{K}$. Therefore

$$\sum_{j=0}^{n-1} \mathbf{1}\left[t_{j+1}^n \leq \tau\right] \frac{T}{n} \overset{\delta}{g}\left(\vec{r}(t_j^n), \vec{\theta}(t_j^n)\right) = \int_0^\tau \overset{\delta}{g}\left(\vec{r}(t), \vec{\theta}(0)\right) dt + e_3(n), \quad (5.66)$$

with $e_3(n)$ converging to zero uniformly for $\vec{r}$ in $\mathcal{K}$. Repeating the argument on the (finite number of) intervals on which $\vec{\theta}$ is constant and using the uniformity in $\vec{r} \in \mathcal{K}$, we obtain

$$\liminf_{n\to\infty} \inf_{r\in\mathcal{K}_x} \mathcal{S}_n(\vec{r}, \vec{\theta}) = \inf_{\vec{r}\in\mathcal{K}_x} \overset{\delta}{I}_0^T(\vec{r}, \vec{\theta}) \qquad (5.67)$$

uniformly in $\vec{x} \in C$. This together with (5.64) establishes the result. ∎

Proposition 5.62. *Assume the* $\log \lambda_i(\vec{x})$ *are bounded and continuous and fix a compact set* $C \subset \mathbb{R}^d$. *Then for each* $K > 0, \delta > 0$ *and* $\varepsilon > 0$,

$$\limsup_{n\to\infty} \frac{1}{n} \log \mathbb{P}_{\vec{x}}\left(d\left(\vec{y}_n, \overset{\delta}{\Phi}_{\vec{x}}(K)\right) > \varepsilon\right) \leq -(K - \varepsilon)$$

uniformly in $\vec{x}$ *in* C.

Proof. We shall establish this result with δ and ε replaced by 2δ and 2ε. Pick an $\varepsilon > 0$. For each absolutely continuous $\vec{r} \in \mathcal{K}(M)$, let $\vec{\theta}_{\vec{r}}$ be the step function of Lemma 5.43. We claim that there is a neighborhood $N_{\vec{r}}$ of $\vec{r}$ such that for any absolutely continuous $\vec{s} \in N_{\vec{r}}$

$$\overset{\delta}{I}_0^T(\vec{s}, \vec{\theta}_{\vec{r}}) \geq \overset{\delta}{I}_0^T(\vec{r}) - 2\varepsilon. \qquad (5.68)$$

To see this, let $[t_1, t_2)$ be an interval on which $\vec{\theta}_r$ is constant and note that

$$\int_{t_1}^{t_2} \left\langle \vec{s}\,'(t), \vec{\theta}_{\vec{r}}(t_1)\right\rangle dt = \left\langle \vec{s}(t_2) - \vec{s}(t_1), \vec{\theta}_{\vec{r}}(t_1)\right\rangle$$

since $\vec{s}$ is absolutely continuous. Therefore the function $\overset{\delta}{I}_0^T(\vec{s}, \vec{\theta}_{\vec{r}})$ is continuous over absolutely continuous functions $\vec{s}$. The claim now follows from the definition of $\vec{\theta}_{\vec{r}}$ in Lemma 5.43.

Since $\mathcal{K}(M)$ is compact, we can choose a finite subcover $\{N_{\vec{s}_i}\}$ of balls around $\vec{s}_i$ with radii $\varepsilon/2$. Now choose, for each i, an absolutely continuous $\vec{r}_i$ in $N_{\vec{s}_i}$, and let $\{N_{\vec{r}_i}\}$ be a ball around $\vec{r}_i$ with radius ε. We have thus constructed a finite cover

$\{N_{\vec{r}_i}\}$ for $\mathcal{K}(M)$, by balls with radii ε. Let $\{\mathcal{K}_i\}$ denote the collection obtained from this finite subcover, after intersecting with the compact set $\mathcal{K}(M)$ and taking the closure, i.e., $\mathcal{K}_i \overset{\triangle}{=} \overline{N_{\vec{r}_i} \cap \mathcal{K}(M)}$. Now we identify those sets $\mathcal{K}_i$ that are far from $\overset{2\delta}{\Phi_{\vec{x}}}(K)$. Define

$$\mathcal{I} = \left\{ i : d\left(\vec{r}_i, \bigcup_{\vec{x}\in C} \overset{2\delta}{\Phi_{\vec{x}}}(K)\right) \geq \varepsilon\right\}.$$

Then we have, uniformly in $\vec{x} \in C$,

$$\limsup_{n\to\infty} \frac{1}{n} \log \mathbb{P}_{\vec{x}}\left(d\left(\vec{y}_n, \overset{2\delta}{\Phi_{\vec{x}}}(K)\right) \geq 2\varepsilon\right)$$

$$\leq \limsup_{n\to\infty} \frac{1}{n} \log\left[\mathbb{P}_{\vec{x}}(\vec{y}_n \notin \mathcal{K}(M)) + \sum_{i\in\mathcal{I}} \mathbb{P}_{\vec{x}}(\vec{y}_n \in \mathcal{K}_i)\right]$$

$$\leq -\min\left\{c_4 \frac{M}{\log M}, \min_{i\in\mathcal{I}}\left\{\overset{\delta}{I}{}_0^T(\vec{r}_i) - 2\varepsilon\right\}\right\}.$$

Now choose M large enough so that the second term dominates. By Definitions 5.36–5.38 and 5.45 we have $\overset{\delta}{I}{}_0^T(\vec{r}_i) \geq \overset{2\delta}{I}{}_0^T(\vec{r}_i) \geq K$ for $i \in \mathcal{I}$. Therefore,

$$\limsup_{n\to\infty} \frac{1}{n} \log \mathbb{P}_{\vec{x}}\left(d\left(\vec{y}_n, \overset{2\delta}{\Phi_{\vec{x}}}(K)\right) \geq 2\varepsilon\right) \leq -K + 2\varepsilon. \qquad \blacksquare$$

We are now ready to establish an upper bound.

For each point $\vec{x}$ in $\mathbb{R}^d$ and set S in $D^d[0, T]$ define

$$S_{\vec{x}} = \{\vec{r} \in S : \vec{r}(0) = \vec{x}\}$$

$$I_{\vec{x}}(S) = \inf\left\{I_0^T(\vec{r}) : \vec{r} \in S, \ \vec{r}(0) = \vec{x}\right\}.$$

Lemma 5.63. *Assume the $\lambda_i(\vec{x})$ are bounded and continuous. For each closed set F in $D^d[0, T]$ and each $\vec{x}$ in $\mathbb{R}^d$,*

$$\lim_{\varepsilon\downarrow 0} \inf_{|\vec{x}-\vec{y}|<\varepsilon} I_{\vec{y}}(F) = I_{\vec{x}}(F).$$

Proof. It clearly suffices to establish that

$$\lim_{\varepsilon\downarrow 0} \inf_{|\vec{x}-\vec{y}|<\varepsilon} I_{\vec{y}}(F) \geq I_{\vec{x}}(F).$$

But this follows from Lemma 5.47. $\qquad \blacksquare$

Theorem 5.64. *Assume the $\log \lambda_i(\vec{x})$ are bounded and continuous. Then, for each closed set $F \subset D^d[0, T]$ and each $\vec{x} \in \mathbb{R}^d$,*

$$\limsup_{\vec{y}\to\vec{x}, \ n\to\infty} \frac{1}{n} \log \mathbb{P}_{\vec{y}}(\vec{z}_n \in F) \leq -I_{\vec{x}}(F).$$

Remark. Note that this implies that the order of the two limits is immaterial!

Proof. Suppose $I_{\vec{x}}(F) = K < \infty$. By Lemma 5.63, for each ε we can find a δ_ε so that

$$|\vec{y} - \vec{x}| \le \delta \Rightarrow I_{\vec{y}}(F) \ge I_{\vec{x}}(F) - \varepsilon$$

whenever $\delta \le \delta_\varepsilon$. By definition, the compact set

$$S_\delta \overset{\triangle}{=} \bigcup_{|\vec{y} - \vec{x}| \le \delta} \Phi_{\vec{y}}(K - 2\varepsilon)$$

and the closed set

$$F^\delta \overset{\triangle}{=} \{\vec{r} \in F : |\vec{r}(0) - \vec{x}| \le \delta\}$$

do not have any points in common. By Theorem A.19 this implies that there is a minimal distance between them: $\eta_\delta \overset{\triangle}{=} d\left(S_\delta, F^\delta\right) > 0$. Clearly, if $|\vec{y} - \vec{x}| \le \delta$ then for all $\eta \le \eta_\delta$,

$$\mathbb{P}_{\vec{y}}\left(\vec{z}_n \in F\right) = \mathbb{P}_{\vec{y}}\left(\vec{z}_n \in F^\delta\right)$$

$$\le \mathbb{P}_{\vec{y}}\left(d\left(\vec{y}_n, F^\delta\right) < \frac{\eta}{2}\right) + \mathbb{P}_{\vec{y}}\left(d\left(\vec{y}_n, \vec{z}_n\right) \ge \frac{\eta}{2}\right). \quad (5.69)$$

But if $\vec{r}(0) = \vec{y}$,

$$d\left(\vec{r}, F^\delta\right) < \frac{\eta}{2} \quad \text{implies} \quad d\left(\vec{r}, \Phi_{\vec{y}}(K - 2\varepsilon)\right) > \frac{\eta}{2}.$$

Now choose δ (and η) small enough so that by Lemma 5.48

$$d(\vec{r}, \Phi_{\vec{y}}(K - 2\varepsilon)) > \frac{\eta}{2} \quad \text{implies} \quad d\left(\vec{r}, \Phi_{\vec{y}}^{2\delta}\left(K - 2\varepsilon - \frac{\eta}{4}\right)\right) > \frac{\eta}{4}.$$

From this and Proposition 5.62,

$$\limsup_{n \to \infty} \frac{1}{n} \log \mathbb{P}_{\vec{y}}\left(d(\vec{y}_n, F^\delta) < \frac{\eta}{2}\right)$$

$$\le \limsup_{n \to \infty} \frac{1}{n} \log \mathbb{P}_{\vec{y}}\left(d\left(\vec{y}_n, \Phi^{2\delta}\left(K - 2\varepsilon - \frac{\eta}{4}\right)\right) \ge \frac{\eta}{4}\right)$$

$$\le -\left(K - 2\varepsilon - \frac{\eta}{4}\right) \quad (5.70)$$

uniformly in $\vec{y}$ with, say, $|\vec{y} - \vec{x}| \le 1$. On the other hand, by Lemma 5.57,

$$\limsup_{n \to \infty} \frac{1}{n} \log \mathbb{P}_{\vec{y}}\left(d\left(\vec{y}_n, \vec{z}_n\right) \ge \frac{\eta}{2}\right) = -\infty \quad (5.71)$$

uniformly in $\vec{y}$. Now substitute (5.70) and (5.71) into (5.69) to obtain

$$\limsup_{n \to \infty} \frac{1}{n} \log \mathbb{P}_{\vec{y}}\left(\vec{z}_n \in F\right) \le -\left(K - 2\varepsilon - \frac{\eta}{4}\right)$$

whenever $|\vec{y} - \vec{x}| \leq \delta$. Since ε and η can be chosen arbitrarily small, the result follows for the case when $I_{\vec{x}}(F) < \infty$. Exactly the same proof establishes the result in the case $I_{\vec{x}}(F) = \infty$. ∎

Proof of the upper bound, Theorem 5.54. Corollary of Theorem 5.64. ∎

The uniformity that holds for the lower bound does not, in general, hold for the upper bound, as shown in Exercise 5.67 below. However, if $I_{\vec{x}}(F)$ happens to be continuous in $\vec{x}$, uniformity holds.

Corollary 5.65. *Assume the $\log \lambda_i(\vec{x})$ are bounded and continuous, and let F be a closed subset of $D^d[0, T]$. If $I_{\vec{x}}(F)$ is continuous in $\vec{x}$ over a compact set C, then the upper bound is uniform in $\vec{x}$ over the set C.*

Remark. The standard terminology in large deviations in this context is somewhat unfortunate. Uniformity usually refers to the result of Theorem 5.64, while standard terminology in mathematical analysis would suggest that a uniformity of a limit means that n can be chosen large so that the bound almost holds, for all $\vec{x}$ in C. The uniformity referred to in the corollary is interpreted in the latter sense.

Proof. By the continuity and compactness assumptions, $I_{\vec{x}}(F)$ is bounded. Suppose that the uniformity does not hold for the closed set F. Then there are sequences $n_j \to \infty$ and $\vec{x}_j \in C$ with

$$\frac{1}{n_j} \log \mathbb{P}_{\vec{x}_j}(\vec{z}_n \in F) \geq -I_{\vec{x}_j}(F) + \varepsilon \tag{5.72}$$

for some $\varepsilon > 0$. Take a convergent subsequence $\vec{x}_j \to \vec{x}$; by the assumed continuity, $I_{\vec{x}}(F) = \liminf I_{\vec{x}_j}(F)$. But then the continuity and (5.72) imply

$$\liminf_{j \to \infty} \frac{1}{n_j} \log \mathbb{P}_{\vec{x}_j}(\vec{z}_n \in F) \geq -I_{\vec{x}}(F) + \varepsilon.$$

Theorem 5.64, however, yields

$$\limsup_{j \to \infty} \frac{1}{n_j} \log \mathbb{P}_{\vec{x}_j}(\vec{z}_n \in F) \leq -I_{\vec{x}}(F)$$

and this contradiction establishes the corollary. ∎

Exercise 5.66. Extend Corollary 5.65 to the case where $I_{\vec{x}}(F)$ satisfies the following continuity condition. At each $\vec{x}$, either $I_{\vec{x}}(F)$ is continuous, or $I_{\vec{x}}(F) = \infty$ and $\lim_{\vec{y} \to \vec{x}} I_{\vec{y}}(F) \to \infty$. Hint: in this case, for a proof by contradiction, assume that for some $K > 0$ there are sequences $n_j \to \infty$ and $\vec{x}_j \in C$ with

$$\frac{1}{n_j} \log \mathbb{P}_{\vec{x}_j}(\vec{z}_n \in F) \geq \begin{cases} -I_{\vec{x}_j}(F) + \varepsilon & \text{if } I_{\vec{x}_j}(F) \leq K \\ -K + \varepsilon & \text{if } I_{\vec{x}_j}(F) \geq K. \end{cases}$$

♠

Exercise 5.67. Show that when $I_{\tilde{x}}$ is not continuous, then uniformity may fail. Use the following construction. Let z_n be a scaled Poisson process as in Chapter 4. Let $\{x_j\}$ be a real sequence converging to the point x. Define

$$F_j = \left\{ \phi \in D^d[0, T]; \ \phi(0) = x_j, \ \sup_t |\phi(t) - \phi(t^-)| \geq \frac{1}{j} \right\}, \quad j = 1, 2, \ldots .$$

In other words, F_j contains all functions that start at x_j and have at least one jump of size at least $1/j$. Denote

$$F_0 = \left\{ \phi \in D^d[0, T]; \ \phi(0) = x \right\}.$$

Then obviously $I_x(F_0) = 0$. Show that $F = \bigcup_0^\infty F_j$ is closed in $D^d[0, T]$ and that $P_{x_j}(z_n \in F)$ is close to one for all large $n \leq j$. Conclude that

$$I_x(F) = 0 \quad \text{while} \quad I_{x_j}(F) = \infty, \quad j = 1, 2, \ldots$$

and for each $K > 0$ there exists n_K so that

$$\frac{1}{n} \log P_{x_j}(z_n \in F) \leq -K$$

for all $n \geq n_K$, and all j. Thus uniformity does not hold.   ♠

Chapter 6

Freidlin-Wentzell Theory

The theory of large deviations we have developed to this point concerns the behavior of a process over fixed time intervals. The Freidlin-Wentzell theory takes these theorems and bootstraps them into estimates on the behavior of a process over very long time intervals, by splitting time into finite intervals and exploiting the Markov property. The main result is the famous "rare events occur with probability e^{-nI^*}," where I^* is the minimum over the set of paths that cause the event (call it S), so that

$$I^* = I^*(S) \overset{\triangle}{=} \inf \left\{ I_0^T (\vec{r}) : T \geq 0, \ \{\vec{r}(t); \ 0 \leq t \leq T\} \in S \right\}. \tag{6.1}$$

This identification of a probability with e^{-nI^*} is the variational principle of large deviations: probability is estimated by solving a variational problem.

For the next few paragraphs we discuss an example in order to bring out the ideas of the theory. Suppose that we are given a process $\vec{z}_n(t)$, such as a queueing process, where there is a very strong tendency for $\vec{z}_n(t)$ to stay near a point $\vec{q}$. That is, we suppose that $\vec{z}_\infty(t)$ follows a vector field that has a unique global attractor $\vec{q}$. What does $\vec{z}_n(t)$ do over long periods of time? By Kurtz's Theorem 5.3, it almost certainly approaches $\vec{q}$ over any fixed time interval. So fix $T > 0$ and consider the intervals $(0, T); (T, 2T); \ldots; (jT, (j+1)T)$. On each of these intervals Kurtz's Theorem holds, meaning it is very unlikely for $\vec{z}_n(t)$ to wander far from $\vec{q}$. Eventually, though, since it is possible for $\vec{z}_n(t)$ to wander away for a while, it will. So the question is, How long does this take, and what escape route would $\vec{z}_n(t)$ follow?

Let's be a little more specific. Suppose we put a ball of radius one around $\vec{q}$, and ask how long it takes for $\vec{z}_n(t)$ to exit this ball. Consider the set of paths that begin at $\vec{q}$ and end on the boundary of the ball;

$$S \overset{\triangle}{=} \{\vec{r} : \vec{r}(0) = \vec{q}, \ |\vec{r}(T) - \vec{q}| = 1 \text{ for some } T > 0, \ |\vec{r}(t) - \vec{q}| < 1 \text{ for } t < T\}. \tag{6.2}$$

Note that different paths may be defined on different time intervals. Define $I^*(S)$ as in (6.1) and suppose that there is a unique path $\vec{r}^* \in S$ (and therefore unique T^*) with $I_0^{T^*}(\vec{r}^*) = I^*(S)$. Now we can pretend that each interval $(jT^*, (j+1)T^*)$ is an independent effort by the process $\vec{z}_n(t)$ to escape the ball—they are nearly independent by Kurtz's Theorem since most likely $\vec{z}_n(jT^*) \approx \vec{q}$, so that the process on $[jT^*, (j+1)T^*)$ is independent of the process on other intervals. By a large deviations lower bound, in each interval there is a probability of at least e^{-nI^*} for $\vec{z}_n$ to follow the path $\vec{r}^*$ out of the ball. So we expect that after about

e^{nI^*} time intervals (or a geometrically distributed number of these intervals with mean e^{nI^*}), the process $\vec{z}_n(t)$ would follow $\vec{r}^*$ and escape. Once it escapes, by Kurtz's Theorem it is almost certain that $\vec{z}_n(t)$ will follow $\vec{z}_\infty(t)$ back to $\vec{q}$, and begin its Sisyphean escape attempt again.

The preceding paragraph seems to give only a lower bound on the probability of escape in one time interval—we consider only escape attempts along the path $\vec{r}^*$. Here is where the large deviations upper bound comes in. First, using Kurtz's Theorem and the Markov property we can deduce that the probability that escape takes a long time is very small—see Lemma 6.28. But once we restrict our attention to bounded time intervals, we can apply existing theory! By Proposition 5.46 the set $\Phi(I^* + \delta)$ is compact, for each $\delta \geq 0$. Since S is closed and $\vec{r}^*$ is, by assumption, unique, Lemma 2.8 states that, as a consequence of the lower semicontinuity of I_0^T, for every $\varepsilon > 0$ there is a $\delta > 0$ such that $I_0^T(\vec{r}) \geq I^* + \delta$ whenever $|\vec{r} - \vec{r}^*| \geq \varepsilon$ and $\vec{r} \in S$. That is, all the other escape paths (far from $\vec{r}^*$) are in a closed subset of $\Phi^c(I^* + \delta)$. Lemma 2.8 now states that the probability that $\vec{z}_n$ will escape by following *any* path that is more than ε away from $\vec{r}^*$ **before** it escapes by following $\vec{r}^*$ is smaller than $e^{-n\delta}$. This follows from our upper bound, which shows that the probability that $\vec{z}_n(t)$ belongs to the set of paths in S which are more than ε away from $\vec{r}^*$ is less than $e^{-n(I^* + \delta)} = e^{-nI^*}e^{-n\delta}$.

So we see that when I has an unique minimizer $\vec{r}^*$ over a set S, then given that the event in S occurred, the overwhelmingly likely way that it did is by $\vec{z}_n(t)$ being near $\vec{r}^*$. This argument holds for any starting point $\vec{x}$, since by Kurtz's Theorem with very high probability, starting at any $\vec{x}$ the process will first approach $\vec{q}$, at which point the preceding argument applies. To make this slightly more precise, consider the jump-Markov process $\vec{z}_n$ defined by (6.3) below—this is the process considered in Chapter 5. Let τ_n denote the first time of escape; in terms of (6.2),

$$\tau_n \overset{\triangle}{=} \inf\{t > 0 : |\vec{z}_n - \vec{q}| \geq 1\}.$$

Since we only want to describe the path up to the time it actually escaped, it is convenient to shift time by T^*, so that now

$$|\vec{r}(0) - \vec{q}| = 1 \quad \text{and} \quad \vec{r}(-T^*) = \vec{q}.$$

Then by the argument above, there is a law of large numbers for rare events: given any positive $T < T^*$ and $\varepsilon > 0$,

$$\lim_{n\to\infty} \mathbb{P}_{\vec{x}}\left(\sup_{\tau_n - T \leq t \leq \tau_n} |\vec{z}_n(t) - \vec{r}^*(t - \tau_n)| < \varepsilon\right) = 1$$

for every $\vec{x}$ in the domain of attraction of $\vec{q}$. Remember, $\vec{r}^*$ is not random: it solves a deterministic variational problem.

The fact that, generically, rare events happen in only one way has many practical implications. If the rare event in question is undesirable (a frequent happenstance, since if we wanted it to occur we would try to make it frequent, not rare) then auxiliary control on z_n may be in order. We can say, without further study, that the control must affect paths near $\vec{r}\,^*$ to be effective (see §13.7 for an example of a seemingly reasonable control that does not, and hence fails as a control). Also, the speed $|d\vec{r}\,^*/dt|$ with which $\vec{r}\,^*$ moves indicates the requisite speed of a control. Equivalently, it shows how quickly the bad event "builds up" after the extremely long quiescent period.

This brings up another point many people raise when first learning the theory of large deviations. Since we know that the only way an event such as escape can occur is for the path to follow $\vec{r}\,^*$, why don't we try to predict the escape attempts by watching when $\vec{z}_n(t)$ begins to follow the early parts of $\vec{r}\,^*(t)$ (near $\vec{q}$), and when it does this, conclude that $\vec{z}_n(t)$ will now follow $\vec{r}\,^*$ until it exits at the point $\vec{r}\,^*(0)$? The answer is essentially contained in the Markovian nature of $\vec{z}_n(t)$. The vast majority of the time when $\vec{z}_n(t)$ manages to get, say, halfway out by following $\vec{r}\,^*(t)$ part of the way, it will immediately decide to turn and follow $\vec{z}_\infty(t)$ back down to $\vec{q}$. That is, even though the "prediction" would be correct for most successful escapes, we would have an overwhelming incidence of false alarms.

Our heuristic discussion has mixed up the first time an event happens with the steady-state manner in which it happens. This is permissible as long as the process comes back to a small neighborhood of $\vec{q}$ quickly after each excursion (i.e., we need strong recurrence or ergodicity constraints). This will not be a problem in our applications, but it is something to be careful about in general. We also prove, under some constraints, that the steady-state distribution of $\vec{z}_n(t)$ approaches a δ-function at $\vec{q}$ as n becomes large. Similarly, we prove that the steady-state probability of a neighborhood of the set $\{|\vec{z}_n - \vec{q}| = 1\}$ approaches a "conditional" δ-function; that is, if we condition the process to actually be near this set, then almost all the probability mass will be found at $\vec{r}\,^*(0)$. This provides the answer to "How does this unlikely event occur?"

The Freidlin-Wentzell theory applies to much more general types of events than escape from a neighborhood of $\vec{q}$. The arguments go through without modification as long as the event in question can't occur if $\vec{z}_n(t)$ stays near $\vec{q}$, and the set of paths that cause the event is regular enough (see §6.2). For example, in Chapter 13 (starting with §13.2) we will be interested in the event

$$\sup_{0 \leq s \leq T} \int_s^T (\langle \vec{z}_n(t), a \rangle - C)\, dt \geq B$$

for some fixed T, a, B, and C.

One very interesting subject that we do not cover is the question of multiple attractors $\vec{q}_i$. Suppose that there is a fixed set of points $\vec{q}_i$ such that $\lim_{t\to\infty} \vec{z}_\infty(t) = \vec{q}_i$ uniformly for $\vec{z}_\infty(0)$ in some neighborhood of each $\vec{q}_i$, and suppose that the closures of the regions of attraction of the $\vec{q}_i$ cover the state space. Then the process $\vec{z}_n(t)$ behaves as follows: first, $\vec{z}_n(t)$ approaches whichever $\vec{q}_i$ attracts $\vec{z}_n(0)$. It stays in the region for a long time (mean $e^{nI_i^*}$, $I_i^* > 0$). Eventually $\vec{z}_n(t)$ escapes to another region near $\vec{q}_j$ for a long time. Generally, this transition takes place with probability p_{ij} depending only on the points $\vec{q}_i$ and $\vec{q}_j$. This continues. We see that there is an induced Markov chain on the $\vec{q}_i$, depending on which one $\vec{z}_n(t)$ is near. The interested reader may consult [FW, §6.6 p. 198].

Unfortunately, the Freidlin-Wentzell theory is technical, and even more unfortunately, some of the most important things to learn aren't the theorems, but the techniques brought forth in the proofs. We have tried to keep the conditions for the theorems strong, so as to make the proofs simpler and more easily understood. We weaken the conditions through an excessive number of exercises. We strongly recommend that, on your first perusal of this chapter, you just read Exercises 6.23, 6.34, 6.41, and 6.42, and completely ignore all other exercises. The reason for including so many technical exercises is that we need many similar results for the applications chapters, and proving these results is not difficult, once the basics have been understood. However, it is time- and page-consuming to detail the arguments.

We recommend that you begin your study by reading §6.1, including the assumptions but not the exercises, through the paragraph following the statement of Theorem 6.17. Then skip to §6.4 and read the entire section. Then go back to §6.1 and read the lemmas leading up to the proofs of Theorems 6.15 and 6.17, again skipping the exercises, and read the proofs, too. Continue to §6.2. Then, if you are motivated, try reading the entire chapter, and test your prowess on the Exercises.

6.1. The Exit Problem

The first problem we consider in this chapter is the classic exit problem. This problem was the first considered by Freidlin and Wentzell using large deviations. The question is: If a process is given sufficient time to escape from a given set G that contains an "attracting point," how long would it take, and what path would it follow on its escape route? Our setup basically follows that of the general process theory in $\mathbb{R}^d$. The process $\vec{z}_n$ we are interested in is defined by its generator

$$L_n f(\vec{x}) = \sum_{i=1}^{k} n\lambda_i(\vec{x}) \left(f\left(\vec{x} + \frac{\vec{e}_i}{n}\right) - f(\vec{x}) \right). \tag{6.3}$$

We start with a (non-empty) set G in $\mathbb{R}^d$. The set of paths that exit G at time zero is

$$S \triangleq \{\vec{r} : \vec{r}(-T) = \vec{q}, \ \vec{r}(t) \in G^o \text{ for } t < T, \ \vec{r}(0) \in \partial G, \ T > 0\}, \tag{6.4}$$

where $T = \infty$ is acceptable. Note that since none of the quantities of interest depend explicitly on time, we may, whenever convenient, shift everything in time; the only effect this may have is to slightly confuse the reader. For a set of paths S' we use the notation

$$I^*(S') \triangleq \inf \left\{ I^0_{-T}(\vec{r}) : \vec{r} \in S' \right\}, \tag{6.5}$$

where T depends on $\vec{r}$. The following assumptions are used in order to smooth the presentation (by avoiding technicalities), and will be relaxed, mostly through exercises. Figure 6.7 pictorially shows what most of the assumptions mean.

Assumption 6.1. G is bounded in $\mathbb{R}^d$ with $G = \overline{G^o}$; that is, G consists of the closure of its interior.

Assumption 6.2. The log-jump rates $\log \lambda_i(\vec{x})$ are uniformly Lipschitz continuous on a compact set whose interior contains $\overline{G}$.

That is, the jump rates are defined, are smooth, and are bounded away from zero on a set that includes the closure of G in its interior.

Assumption 6.3. The boundary ∂G of G is smooth (Definition A.33), and at every point $\vec{x}$ in ∂G there is an open exterior cone $c(\vec{x})$ so that, for some $\varepsilon > 0$,

$$c(\vec{x}) \cap \{\vec{y} : |\vec{y} - \vec{x}| < \varepsilon\} \cap G = \emptyset.$$

That is, we can find a small neighborhood outside G so that each point in this neighborhood is connected to $\vec{x}$ by a straight line that does not re-enter G.

As in (5.7) we define a vector field

$$\vec{f}(\vec{x}) \stackrel{\triangle}{=} \sum_{i=1}^{k} \lambda_i(\vec{x})\vec{e}_i.$$

Recall that $\vec{f}(\vec{x})$ defines $\vec{z}_\infty(t)$, since $d\vec{z}_\infty(t)/dt = \vec{f}(\vec{z}_\infty(t))$. Let $\vec{n}(\vec{x})$ denote the unit outward normal to the set G, which exists since the boundary is smooth. The next assumption guarantees that whenever $\vec{z}_\infty$ starts on the boundary of G, it will immediately go into the interior; therefore, if $\vec{z}_\infty$ starts inside G, it never escapes G.

Assumption 6.4. There is a $\delta > 0$ so that, at every point $\vec{x}$ in ∂G,

$$\langle \vec{f}(\vec{x}), \vec{n}(\vec{x}) \rangle \leq -\delta < 0.$$

Assumption 6.5. There is a point $\vec{q} \in G$ and, for every $\varepsilon > 0$, there is a $T < \infty$ such that, uniformly over $\vec{x} \in G$, with $\vec{z}_\infty(0) = \vec{x}$,

$$|\vec{z}_\infty(t) - \vec{q}| < \varepsilon \text{ for all } t > T.$$

Assumption 6.6. There exists a unique path $\vec{r}^*(t)$ and corresponding time T^* (possibly infinite!) such that

$$I^*(S) \stackrel{\triangle}{=} \inf_{\vec{r} \in S} I^0_{-T}(\vec{r}) = I^0_{-T^*}(\vec{r}^*).$$

Remark. Note that, in our search for an optimal path, we could include in S paths that leave G before time 0, as long as they start at $\vec{q}$ and are on the boundary at $t = 0$. To see that this would change neither I^* nor $\vec{r}^*$, assume $\vec{r}^*(t_1) \in \partial G$ for some time $t_1 > -T^*$ before 0. Since ℓ is non-negative, the definition of $\vec{r}^*$ and

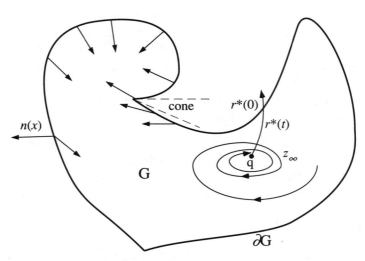

Figure 6.7. The assumptions.

time-homogeneity implies $I_{t_1}^0(\vec{r}^*) = 0$. This in turn implies $\vec{r}^* = \vec{z}_\infty$ on $[t_1, 0]$
by Exercise 5.14. But this contradicts Assumption 6.4, which guarantees that if
$\vec{z}_\infty(t_1) \in \partial G$ then $\vec{z}_\infty(t) \in G^o$ for all $t > t_1$. It also contradicts the uniqueness
assumption.

The assumptions imply that $\vec{q}$ is the unique point in G with $\vec{f}(\vec{q}) = \vec{0}$. We shall
need the following stronger consequence.

Exercise 6.8. Under Assumptions 6.1, 6.2, and 6.5, we have Lyapunov stability;
see e.g., [Hal p. 26]. That is, for any $\varepsilon > 0$ there is an $\varepsilon_s > 0$ so that

$$|\vec{z}_\infty(0) - \vec{q}| < \varepsilon_s \text{ implies } |\vec{z}_\infty(t) - \vec{q}| \le \frac{\varepsilon}{2} \text{ for all } t \ge 0. \qquad (6.6)$$

(The standard definition uses ε rather than $\varepsilon/2$. The reason for our choice will
become clear in the course of the section). Hint: by contradiction. Assume for
some ε that, for a sequence of starting points $\vec{x}_i \to \vec{q}$ the solutions at time t_i are
outside the ε ball around $\vec{q}$. The times are bounded by T of Assumption 6.5. Use
the continuity of solutions with respect to initial conditions (Theorem A.68) to
obtain a contradiction.   ♠

Exercise 6.9. Assumptions 6.2 and 6.5 hold in fact in any (small enough) open
neighborhood of the set G.   ♠

These assumptions probably seem strange, and perhaps strong; here is a remedy. In the Exercises below we restrict $\vec{x}$ to the positive orthant.

Exercise 6.10. Consider the $M/M/\infty$ queue; with the scaling as in Chapter 12, we have a one-dimensional model ($d = 1$), with

$$\vec{e}_1 = 1 \qquad\qquad \lambda_1 = \lambda > 0$$
$$\vec{e}_2 = -1 \qquad\qquad \lambda_2(x) = x.$$

Find a point $q > 0$ and a ball G around q for which Assumptions 6.1–6.5 hold.♠

Exercise 6.11. Consider now d independent $M/M/\infty$ queues, where the rates may be different from queue to queue. Then we obtain

$$\vec{e}_i = (0, \ldots, 0, 1, 0, \ldots, 0) \qquad\qquad \lambda_i(\vec{x}) = \lambda_i > 0$$
$$\vec{e}_{d+i} = (0, \ldots, 0, -1, 0, \ldots, 0) \qquad\qquad \lambda_{d+i}(\vec{x}) = \lambda_{d+i} x_i, \qquad \lambda_{d+i} > 0,$$

$i = 1, 2, \ldots, d$ and where the ± 1 is in the ith position. Find a point $\vec{q}$ in the (strictly) positive orthant and a ball G around $\vec{q}$ for which Assumptions 6.1–6.5 hold. ♠

Exercises 6.10–6.11 are indeed easy—almost trivial. Some coupling between the queues makes the picture more interesting.

Exercise 6.12. Consider the case with $d = 2$ and

$$\vec{e}_1 = (1, 0) \qquad \vec{e}_2 = (-1, 0) \qquad \vec{e}_3 = (0, 1) \qquad \vec{e}_4 = (0, -1)$$
$$\lambda_1 = 3 \qquad \lambda_2(\vec{x}) = 4 + x_1 - 2x_2 \qquad \lambda_3 = 3 \qquad \lambda_4(\vec{x}) = 1 + 2x_1 + x_2.$$

Find a point $\vec{q}$ in the (strictly) positive orthant and a ball G around $\vec{q}$ for which Assumptions 6.1–6.5 hold. Is it true that any closed, convex set contained in G also satisfies 6.1–6.5? Can you choose ε_s in Exercise 6.8 so that $\varepsilon_s = \varepsilon/2$? ♠

Exercise 6.13. Does the uniqueness assumption in 6.6 hold in Exercises 6.10–6.12? ♠

We need some notation for the statements of our theorems. The precise definitions are given below; here is an informal discussion of what we need. The symbol $\varphi_t(\vec{x}, \vec{y})$ represents the cheapest cost (here "cost" is the I-function) for going from $\vec{x}$ to $\vec{y}$ in time t, and $V(\vec{x}, \vec{y})$ represents the cheapest cost when minimizing over time as well. The function $V(\vec{x})$ is $V(\vec{q}, \vec{x})$; this is the cheapest cost for travel from the center of the flow $\vec{z}_\infty$ to a point $\vec{x}$. We show in Theorem 6.89 below that $V(\vec{x})$ measures the cost for the process $\vec{z}_n(t)$ to visit a neighborhood of $\vec{x}$. Recall that G^o is the interior of the set G.

Definitions 6.14.

$$\tau_n(G) \stackrel{\triangle}{=} \inf\{t \geq 0 : \vec{z}_n(t) \notin G^o\}$$
$$N_\varepsilon(\vec{x}) \stackrel{\triangle}{=} \left\{\vec{y} \in \mathbb{R}^d : |\vec{y} - \vec{x}| < \varepsilon\right\}$$
$$G^\delta \stackrel{\triangle}{=} \left\{\vec{y} \in \mathbb{R}^d : |\vec{y} - \vec{x}| < \delta \text{ for some } \vec{x} \in G\right\} = \bigcup_{x \in G} N_\delta(\vec{x})$$

$$\varphi_t(\vec{x}, \vec{y}) \overset{\triangle}{=} \inf\left\{ I^0_{-t}(\vec{r}) : \vec{r}(-t) = \vec{x}, \ \vec{r}(0) = \vec{y}, \ \vec{r}(s) \in G^o \text{ for } -t \le s < 0 \right\}$$

$$V(\vec{x}, \vec{y}) \overset{\triangle}{=} \inf_{t \ge 0} \varphi_t(\vec{x}, \vec{y})$$

$$V(\vec{x}) \overset{\triangle}{=} V(\vec{q}, \vec{x}).$$

Finally, we extend $\vec{r}^*$ so that it is defined for all time, by setting

$$\vec{r}^*(t) \overset{\triangle}{=} \begin{cases} \vec{q} & \text{for } t < -T^* \\ \vec{r}^*(0) + \vec{n}_c(\vec{r}^*(0)) \cdot t & \text{for } t > 0, \end{cases} \tag{6.7}$$

where $\vec{n}_c(\vec{r}^*(0))$ is the center of the exterior cone of Assumption 6.3 at the exit point $\vec{r}^*(0)$.

(We could choose $\vec{n}_c$ to be the outward normal: however, the current definition extends easily to the case of a piecewise-smooth boundary). We now state the main results of this section. For the definition of uniform convergence see Definition A.10.

Theorem 6.15. *Under Assumptions 6.1–6.6, for each $T < T^*$ and $\varepsilon > 0$,*

$$\lim_{n \to \infty} \mathbb{P}_{\vec{x}} \left(\sup_{\tau_n(G)-T \le t \le \tau_n(G)} |\vec{z}_n(t) - \vec{r}^*(t - \tau_n(G))| < \varepsilon, \ \tau_n(G) > T \right) = 1$$

uniformly over $\vec{x}$ in any compact subset of G^o.

That is, the process eventually leaves G, and the last part of the path leading to the exit is near $\vec{r}^*$. The reason this is not an immediate consequence of the previous large deviations bounds is, of course, that the timing of the event of interest is not bounded; as we shall show, $\tau_n(G)$ is unbounded as a function of n. Theorem 6.17 below describes how fast it grows.

Corollary 6.16. *Under Assumptions 6.1–6.6, for any bounded continuous function $u(\vec{x})$,*

$$\lim_{n \to \infty} \mathbb{E}_{\vec{x}} u(\vec{z}_n[\tau_n(G)]) = u(\vec{r}^*(0)).$$

This corollary was one of the original successes of the Freidlin-Wentzell theory. Using the equivalence of escape distributions of diffusion processes and solution of Dirichlet problems (see e.g., B.33), this corollary shows that certain limiting singular Dirichlet problems have constant solutions. See [FW] for more details of this point. We don't make use of this corollary in the remainder of our book.

Theorem 6.17. *Under Assumptions 6.1–6.6, with $I^* = I^*(S)$ given by (6.1) and (6.4), we have uniformly over $\vec{x}$ in any compact subset of G^o,*

(i) *For each $\varepsilon > 0$,* $\displaystyle \lim_{n \to \infty} \mathbb{P}_{\vec{x}} \left(\frac{\log \tau_n(G)}{n} \in (I^*(S) - \varepsilon, I^*(S) + \varepsilon) \right) = 1.$

(ii) $\displaystyle\lim_{n\to\infty}\frac{1}{n}\log\mathbb{E}_{\vec{x}}(\tau_n(G))=I^*(S).$

Theorems 6.15 and 6.17 are based on the same approach. We give a construction and sequence of estimates and lemmas, then use them to establish the theorems. Here are the ideas behind the formulas. By Lemma 6.28, it is very costly for $\vec{z}_n(t)$ to escape G by wandering inside G but away from $\vec{q}$ for a long time. Therefore, we only need to consider a finite-time problem; but for this we already have a theory! Lemma 6.21 shows that the "cost" of leaving G starting near $\vec{q}$ is continuous in the starting point. Thus we can look at consecutive attempts to leave G, starting near $\vec{q}$. The same continuity (and the Markov property) imply that these consecutive attempts to leave G take lengths of time that are "almost i.i.d." (Exercise 6.42), and moreover, by Lemma 6.36, many attempts are required to leave G. Therefore we can apply limit theorems for i.i.d. random variables to obtain Theorem 6.17.

Exercise 6.18. Show that the results of Theorems 6.15 and 6.17 are not uniform in $x\in G$. Hint: the problem is near the boundary. Construct a counterexample using Exercises 6.10–6.12. ♠

We begin the technical exposition with a lemma showing that the assumptions imply that the process occupies all d dimensions of the space.

Lemma 6.19. *Under Assumptions 6.1 and 6.5, the positive cone spanned by the $\{\vec{e}_i\}$ is all of $\mathbb{R}^d$, that is,*

$$\mathcal{C}\overset{\triangle}{=}\left\{\sum_{i=1}^{k}a_i\vec{e}_i\ :\ a_i\geq 0\right\}=\mathbb{R}^d.\qquad(6.8)$$

Proof. The lemma will clearly be established once we show that $N_\varepsilon(0)\subset\mathcal{C}$ for some $\varepsilon>0$. By the assumptions we can find an $\varepsilon>0$ small enough so that

$$\vec{y}\in N_\varepsilon(0)\quad\text{implies}\quad\vec{q}-\vec{y}\in G^o.$$

Fix $\vec{y}\in N_\varepsilon(0)$, and consider the path $\vec{z}_\infty$ with $\vec{z}_\infty(0)=\vec{q}-\vec{y}$. Then

$$\vec{z}_\infty(t)=\vec{q}-\vec{y}+\int_0^t\sum_{i=1}^{k}\vec{e}_i\lambda_i(\vec{z}_\infty(s))\,ds\to\vec{q},$$

since by Assumption 6.5 the point $\vec{q}\in G^o$ attracts each point in G^o. Thus

$$\vec{y}=\lim_{t\to\infty}\sum_{i=1}^{k}\left(\int_0^t\lambda_i(\vec{z}_\infty(s))\,ds\right)\vec{e}_i$$

and so we can approximate $\vec{y}$ by a point in $\mathcal{C}$ as closely as desired. Thus we can approximate any point in $N_\varepsilon(0)$ (in particular, standard basis vectors for $\mathbb{R}^d$ and their negatives) by points in $\mathcal{C}$ with any desired accuracy. This proves $N_\varepsilon(0)\subset\mathcal{C}$ and the lemma is established. ∎

Exercise 6.20. Assume only that $\vec{q} \in G^o$ and that all solutions $\vec{z}_\infty$ starting in G converge to $\vec{q}$. Then (6.8) holds.   ♠

There are pragmatic reasons for embracing the condition that the positive cone C equals $\mathbb{R}^d$. When it does not, there may be some directions that the process $\vec{z}_n(t)$ will take that are "transient," in the sense that the process will only increase in those directions. We do not wish to allow transient directions, since we are interested in processes that have a unique attracting stable point. If $C \neq \mathbb{R}^d$ and there are no such directions, then the dimension of the process can be reduced.

Given $\vec{x}$, $\vec{y}$ and ε let $\vec{r}(t)$ be a straight line between the points

$$\vec{r}(t) = r_{\vec{x},\vec{y}}(t) \overset{\Delta}{=} \vec{x} + (\vec{y} - \vec{x})\,(\varepsilon)^{-1} \cdot t$$

Lemma 6.21 shows that the cost to go between two close points is linear in the distance between the points.

Lemma 6.21. *Assume (6.8) and Assumption 6.2. Then there exists a $C < \infty$ such that for any $\varepsilon > 0$, if $|\vec{x} - \vec{y}| \leq \varepsilon$ and if $\vec{r}(t) \in G$ for $0 \leq t < \varepsilon$, then $\varphi_\varepsilon(\vec{x}, \vec{y}) \leq C\varepsilon$.*

Proof. Since by Assumption 6.2 $\log \vec{\lambda}(\vec{x})$ is bounded in G, Lemma 5.32 of §5.2 shows that, for some $C < \infty$,

$$\ell(\vec{x}, \vec{z}) \leq C \quad \text{whenever} \quad |\vec{z}| \leq 1.$$

Now $|\vec{r}'(t)| \leq 1, \vec{r}(0) = \vec{x}$ and $\vec{r}(\varepsilon) = \vec{y}$ and so

$$\varphi_\varepsilon(\vec{x}, \vec{y}) \leq \int_0^\varepsilon \ell(\vec{r}(t), \vec{r}'(t))\, dt \leq C\varepsilon. \tag{6.9}$$

∎

Remark. The condition that $\vec{r}(t) \in G$ for $0 \leq t \leq \varepsilon$ clearly holds whenever $\vec{x}, \vec{y} \in N_\varepsilon(\vec{z}) \subset G$ for some $\vec{z}$. From the proof it follows that the result extends to points in the compact set of Assumption 6.2 which contains G.

Exercise 6.22. Suppose that in Lemma 6.21, Assumption 6.2 is replaced with the condition that the rates $\lambda_i(\vec{x})$ are bounded above, and for all i, all $\vec{x} \in G^o$ and all $\vec{v}$ with $|\vec{v}| \leq 1$,

$$\int_0^\varepsilon |\log \lambda_i(\vec{x} + t\vec{v})|\, dt \leq c(\varepsilon) \quad \text{where} \quad c(\varepsilon) \to 0 \text{ as } \varepsilon \to 0. \tag{6.10}$$

Then the conclusion of the lemma holds in the form $\varphi_\varepsilon(\vec{x}, \vec{y}) \leq C(\varepsilon)$ where $C(\varepsilon) \to 0$ as $\varepsilon \to 0$. In particular, (6.10) holds whenever the rates $\lambda_i(\vec{x})$ are polynomials in $\vec{x}$. Hint: examine the proof of Lemma 5.32.   ♠

Exercise 6.23. When $d \geq 2$ we can choose a cheap path $\vec{r}$ that stays away from a neighborhood of a point $\vec{z}$, that is,

$$\inf_{0 \leq t \leq \varepsilon} |\vec{r}(t) - \vec{z}| = \min(|\vec{x} - \vec{z}|, |\vec{y} - \vec{z}|) \overset{\Delta}{=} \delta. \tag{6.11}$$

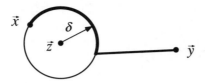

Figure 6.24. The dark line is a path satisfying (6.11).

This would be desirable when the assumptions do not hold near $\vec{z}$ (usually $\vec{q}$), for example if the rates go to zero near $\vec{z}$. Fix some $\eta > 0$, and for each $\vec{x}, \vec{y} \in N_\varepsilon(\vec{z})$, $\delta < \varepsilon < \eta$ construct a path $\vec{r}$ with $\vec{r}(0) = \vec{x}$, $\vec{r}(\varepsilon) = \vec{y}$ that satisfies both (6.9) and (6.11) for a constant $C(\delta)$ independent of ε. Assume (6.8) and that $\log \lambda_i$ are bounded on

$$\left\{ \vec{x} : \vec{x} \in \overline{N}_\eta(\vec{z}), \ \vec{x} \notin N_{\delta/2}(\vec{z}) \right\}.$$

If $\log \lambda_i$ are bounded on $N_\eta(\vec{z})$ then the constant $C(\delta)$ does not depend on δ. On the other hand, if we only assume that

$$\sup_{\delta \leq |\vec{x} - \vec{z}| \leq 1} \delta(\lambda_i(\vec{x}) + |\log \lambda_i(\vec{x})|) \to 0 \quad \text{as} \quad \delta \to 0 \tag{6.12}$$

for each i, then $C(\delta)$ satisfies $\delta C(\delta) \to 0$ as $\delta \to 0$. Hint: for the last part, examine the proof of Lemma 5.32. ♠

Thus it is "cheap" for the process to move between close points. Note that in the one-dimensional case we need special arguments when some rates are zero at $\vec{q}$. Here is an important consequence.

Exercise 6.25. Assume $\log \lambda_i(\vec{x})$ are bounded and continuous. Fix a path $\vec{r}$ and define $\vec{r}_\alpha(t) \stackrel{\triangle}{=} \vec{r}(\alpha t)$: note that $\vec{r}_\alpha$ traverses the same path as $\vec{r}$, except that the time it takes to do so is different. For each T, the function $I_0^{T/\alpha}(\vec{r}_\alpha)$ is continuous in α near $\alpha = 1$. Hint: by Lemmas 5.17 and 5.32, if $|\vec{y}|$ is large then $\ell(\vec{x}, \vec{y})$ is bounded above and below by a constant times $|\vec{y}| \log |\vec{y}|$. As in the proof of Theorem 5.35 show that the set where $\vec{r}'$ is large can be ignored. Now invoke Exercise 5.34. ♠

The next lemma shows that the "price" of avoiding a neighborhood of the stable point $\vec{q}$ increases linearly in time. This is the main ingredient in reducing the infinite-time problem to a bounded time interval. For a given positive T and ε, consider the set of paths that avoid q:

$$S_T(\vec{x}) \stackrel{\triangle}{=} \left\{ \vec{r} \in D^d[0, T] : \inf_{0 \leq t \leq T} |\vec{r}(t) - \vec{q}| > \varepsilon, \right.$$

$$\left. \vec{r}(0) = \vec{x}, \ \vec{r}(t) \in G^o \text{ for all } t \in (0, T) \right\}. \tag{6.13}$$

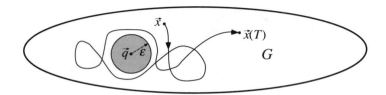

Figure 6.26. A path in S_T.

Let $\mathcal{O}$ be the open subset of G^o that includes all points outside $N_\varepsilon(\vec{q})$; formally, $\mathcal{O} = G^o \setminus \{\vec{x} : d(\vec{x}, \vec{q}) \leq \varepsilon\}$. Define $S_T \overset{\triangle}{=} \cup_{\vec{x} \in \mathcal{O}} S_T(\vec{x})$.

Exercise 6.27. The function $I^*(S_T(\vec{x})) \overset{\triangle}{=} \inf\{I_0^T(\vec{r}) : \vec{r} \in S_T(\vec{x})\}$ is continuous in $\vec{x}$ over the compact set $\overline{\mathcal{O}}$. Hint: combine Exercise 6.25 and Lemma 6.21. ♠

Lemma 6.28. *Under Assumptions 6.1–6.5, for any $\varepsilon > 0$ there are $T_1 > 0$ and $\delta > 0$ such that for all $T > T_1$,*

$$I^*(S_T) \overset{\triangle}{=} \inf_{\vec{x} \in \mathcal{O}} \inf_{\vec{r} \in S_T(\vec{x})} I_0^T(\vec{r}) \geq \delta(T - T_1).$$

Proof. Note that by (6.6), $\varepsilon_s \leq \varepsilon/2$. Let T_S be the longest time (over all initial points in G) any trajectory $\vec{z}_\infty(t)$ stays outside the ball $N_{\varepsilon_s}(\vec{q})$; by Assumption 6.5, $T_S < \infty$. Fix $\vec{x} \in \mathcal{O}$; we establish the result for each such $\vec{x}$, and show that the bound is uniform over $\mathcal{O}$. Since S_T is open in $D^d[0, T]$ (see Exercise A.63), the large deviations lower bound Theorem 5.51 gives

$$\mathbb{P}_{\vec{x}}(\vec{z}_n \in S_T) \geq e^{-(nI^*(S_T(\vec{x})) + o(n))}.$$

By Exercise 6.27 and Corollary 5.65, this bound is uniform in $\vec{x} \in \mathcal{O}$. But, from Kurtz's Theorem 5.3 and the definition of T_S, with $T_1 = T_S + 1$, for large enough n we have

$$\mathbb{P}_{\vec{x}}(\vec{z}_n \in S_{T_1}(\vec{x})) \leq e^{-n2C(\varepsilon_s)},$$

where C is a positive function that does not depend on $\vec{x}$. Hence by taking n large enough, for any $\vec{x} \in \mathcal{O}$,

$$I^*(S_{T_1}(\vec{x})) \geq C(\varepsilon_s) > 0.$$

Now let $\delta' \overset{\triangle}{=} C(\varepsilon_s)$. Then

$$I^*(S_{T_1}(\vec{x})) \geq \delta' = \delta'(T_1 - T_S).$$

Since the rates do not depend explicitly on time, the definition of I gives

$$I^*(S_{2T_1}(\vec{x})) = \inf_{\vec{r} \in S_{2T_1}(\vec{x})} I_0^{2T_1}(\vec{r})$$

$$\geq \inf_{\vec{r} \in S_{T_1}(\vec{x})} I_0^{T_1}(\vec{r}) + \inf_{\vec{y} \in \mathcal{O}} \inf_{\vec{r} \in S_{T_1}(\vec{y})} I_0^{T_1}(\vec{r})$$

$$\geq 2\delta'(T_1 - T_S).$$

Similarly, for each $k > 1$ and $\vec{x} \in \mathcal{O}$, $I^*(S_{kT_1}(\vec{x})) \geq k\delta'(T_1 - T_S) = k\delta'$.

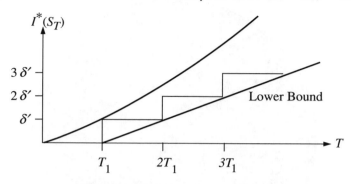

Figure 6.29. Linear lower bound.

Since $I^*(S_T(\vec{x}))$ is increasing in T, the choice $\delta = \delta'(T_1)^{-1}$ establishes the lemma. ∎

Exercise 6.30. Assume that, for some $\eta > \varepsilon > 0$, there is a $T = T(\varepsilon, \eta)$ so that if $\vec{z}_\infty(0) \in G^o \cap N^C_{\varepsilon/2}$, then for some $t \leq T$ either $\vec{z}_\infty(t) \in N_{\varepsilon/2}$ or $\vec{z}_\infty(t) \notin G^\eta$. Assume further that the $\log \lambda_i(\vec{x})$ are uniformly Lipschitz in $G^\eta \cap N^C_{\varepsilon/2}$. Then the conclusion of Lemma 6.28 holds (uniformly in $G^\eta \cap N^C_{\varepsilon/2}$). Hint: follow the proof of Lemma 6.28, and for each point $\vec{x}$ invoke Kurtz's Theorem around the appropriate path $\vec{z}_\infty$. ♠

Now we make a standard Freidlin-Wentzell construction. Look at Figure 6.31 as you read the definitions that follow; it may help explain what's going on. Recall that by the definition of ε_s in Exercise 6.8, $\varepsilon_s = \varepsilon_s(\varepsilon) \leq \varepsilon/2$. Let ε be small enough so that $\inf_{\vec{x} \in \partial G} |\vec{q} - \vec{x}| \geq 2\varepsilon$, and consider the neighborhoods $N_{\varepsilon_s}(\vec{q})$ and $N_\varepsilon(\vec{q})$. We construct a sequence of consecutive (stopping) times so that, at the odd-numbered times the process leaves the ε neighborhood of $\vec{q}$, while at even-numbered times the process either enters the smaller ε_s neighborhood, or leaves G^o entirely. Formally, let $\tau_{-1} \equiv 0$ and for integer $j \geq 0$,

$$\tau_{2j} = \inf\left\{t > \tau_{2j-1} : \vec{z}_n(t) \in \overline{N_{\varepsilon_s}(\vec{q})} \text{ or } \vec{z}_n(t) \notin G^o\right\}$$
$$\tau_{2j+1} = \inf\left\{t > \tau_{2j} : \vec{z}_n(t) \notin N_\varepsilon(\vec{q})\right\}. \qquad (6.14)$$

Obviously, each τ_i depends on ε, on ε_s, and on n, but we shall suppress these dependencies in the interest of brevity. We terminate the sequence at the first time τ_{2j} where $\vec{z}_n(\tau_{2j}) \notin G^o$ [that is, at $\tau_n(G)$]. For notational convenience, we set $\tau_i = \infty$ if the processes exited G^o at some τ_{2j} with $2j < i$, and let $\vec{z}_n(\infty) \stackrel{\Delta}{=} \vec{z}_n(\tau_{2j})$.

The following lemmas show that the distribution of the increments $\tau_{j+1} - \tau_j$ possesses a geometric tail, for all n large. Moreover, the probability that $\vec{z}_n$ leaves

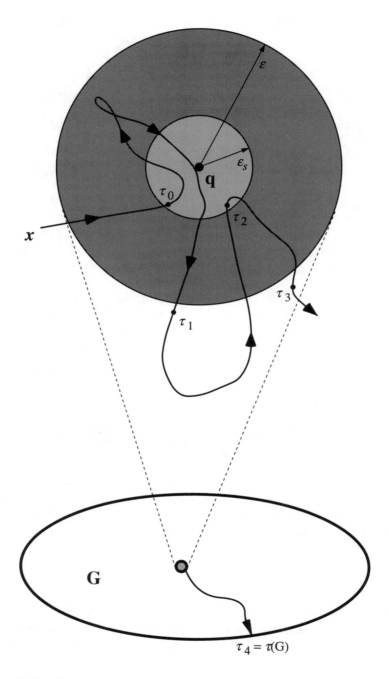

Figure 6.31. The sequence of exit times. Since $\vec{z}_n(t)$ isn't continuous, the $\vec{z}_n(\tau_i)$ aren't exactly on $|\vec{z}_n| = \varepsilon$ or $|\vec{z}_n| = \varepsilon_s$.

G^o at, say, the jth attempt (i.e., at time τ_{2j}) decreases exponentially (in n) with rate roughly $I^*(S)$.

Lemma 6.32. *Under Assumptions 6.1–6.5, for each $\varepsilon > 0$ small enough there is a $T < \infty$, depending only on ε, and there are positive functions C_e and C_o such that if n is large enough then for $k = 1, 2, \ldots$*

$$\mathbb{P}_{\vec{x}}\left(\tau_{j+1} - \tau_j > kT \mid \vec{z}_n(\tau_j) \in G^o\right) \leq \begin{cases} \left(1 - e^{-nC_e(\varepsilon)}\right)^k & j \text{ even,} \\ e^{-nC_o(\varepsilon)k} & j \text{ odd.} \end{cases} \tag{6.15}$$

Moreover, $C_e(\varepsilon) = O(\varepsilon)$.

Proof. For even j, note that if $\vec{z}_n(\tau_j) \in G^o$ then necessarily $\vec{z}_n(\tau_j) \in \overline{N_{\varepsilon_s}(\vec{q})}$, so

$$\mathbb{P}_{\vec{x}}\left(\tau_{j+1} - \tau_j > kT \mid \vec{z}_n(\tau_j) \in G^o\right)$$
$$= \mathbb{P}_{\vec{x}}\left(\vec{z}_n(t) \in N_\varepsilon(\vec{q}), \ \tau_j < t \leq \tau_j + kT \mid \vec{z}_n(\tau_j) \in G^o\right)$$
$$\leq \sup_{|\vec{z}-\vec{q}| \leq \varepsilon_s} \mathbb{P}_{\vec{x}}\left(\vec{z}_n(t) \in N_\varepsilon(\vec{q}), \ \tau_j < t \leq \tau_j + kT \mid \vec{z}_n(\tau_j) = \vec{z}\right)$$
$$= \sup_{|\vec{z}-\vec{q}| \leq \varepsilon_s} \mathbb{P}_{\vec{z}}\left(\vec{z}_n(t) \in N_\varepsilon(\vec{q}), \ 0 < t \leq kT\right)$$

by the (strong) Markov property—Theorems A.137–A.138. Using the Markov property again, conditioning and then replacing the conditioning by maximization as above,

$$\mathbb{P}_{\vec{z}}\left(\vec{z}_n(t) \in N_\varepsilon(\vec{q}), \ 0 < t \leq kT\right)$$
$$\leq \mathbb{E}_{\vec{z}}\left\{\mathbf{1}\left[\vec{z}_n(t) \in N_\varepsilon(\vec{q}), \ 0 < t \leq (k-1)T\right]\right.$$
$$\left. \times \sup_{\vec{y} \in N_\varepsilon(\vec{q})} \mathbb{E}_{\vec{y}}\mathbf{1}\left[\vec{z}_n(t) \in N_\varepsilon(\vec{q}), \ 0 \leq t \leq T\right]\right\}.$$

Now for each $\vec{y}$ in $N_\varepsilon(\vec{q})$, let $\vec{z} \stackrel{\Delta}{=} \vec{y} + 2\varepsilon(\vec{y} - \vec{q})|\vec{y} - \vec{q}|^{-1}$, so that $\vec{z} \notin N_{2\varepsilon}(\vec{q})$. Let $\vec{r}$ be the path constructed in Lemma 6.21, with $\vec{r}(0) = \vec{y}$ and $\vec{r}(\varepsilon) = \vec{z}$, and let S_ε be the set of paths that remain within ε of $\vec{r}$ during $[0, T]$. Then

$$\mathbb{E}_{\vec{y}}\mathbf{1}\left[\vec{z}_n(t) \in N_\varepsilon(\vec{q}), \ 0 \leq t \leq T\right] \leq 1 - \mathbb{P}_{\vec{y}}\left(|\vec{z}_n(t) - \vec{r}(t)| < \varepsilon, \ 0 \leq t \leq T\right)$$
$$\leq 1 - e^{-nI_{\vec{y}}^*(S_\varepsilon) + O(n)}$$
$$\leq 1 - e^{-nC \cdot 2\varepsilon},$$

where the second inequality holds uniformly in $\vec{y}$ in $N_\varepsilon(\vec{q})$ by the lower bound, and the last inequality holds by (6.9) of Lemma 6.21, where C does not depend on $\vec{y}$. Thus with $C_e(\varepsilon) \stackrel{\Delta}{=} C \cdot 2\varepsilon$,

$$\mathbb{P}_{\vec{z}}\Big(\vec{z}_n(t) \in N_\varepsilon(\vec{q}),\ 0 < t \le kT\Big)$$

$$\le \mathbb{P}_{\vec{z}}\Big(\vec{z}_n(t) \in N_\varepsilon(\vec{q}),\ 0 < t \le (k-1)T\Big) \cdot \Big(1 - e^{-nC_e(\varepsilon)}\Big)$$

$$\le \Big(1 - e^{-nC_e(\varepsilon)}\Big)^k$$

uniformly in $\vec{z}$, by iterating the same argument, and C_e is linear in ε for small ε. We see that we may take any $T \ge \varepsilon$ for this bound to hold.

For odd j, we use very similar arguments. Clearly the probability to avoid $N_{\varepsilon_s}(\vec{q})$ for the given duration is larger than the probability of both avoiding $N_{\varepsilon_s}(\vec{q})$ and remaining in G^o. Also, for all n large enough, $\vec{z}_n(\tau_j) \in G^o$ if and only if $\vec{z}_n(\tau_j) \in N_{2\varepsilon}(\vec{q})$. Therefore

$$\mathbb{P}_{\vec{x}}\left(\tau_{j+1} - \tau_j > kT \mid \vec{z}_n(\tau_j) \in G^o\right)$$

$$\le \mathbb{P}_{\vec{x}}\left(\vec{z}_n(t) \notin N_{\varepsilon_s}(\vec{q}),\ \tau_j < t \le \tau_j + kT \mid \vec{z}_n(\tau_j) \in G^o\right)$$

$$\le \sup_{\vec{z} \in N_{2\varepsilon}(\vec{q})} \mathbb{P}_{\vec{x}}\left(\vec{z}_n(t) \notin N_{\varepsilon_s}(\vec{q}),\ \tau_j < t \le \tau_j + kT \mid \vec{z}_n(\tau_j) = \vec{z}\right).$$

By Assumption 6.5, there exists some $T < \infty$ so that, for any initial point $\vec{z} \in G$, the path $\vec{z}_\infty$ starting with $\vec{z}_\infty(0) = \vec{z}$ satisfies $\vec{z}_\infty(t) \in N_{\varepsilon_s/2}(\vec{q})$ for all $t \ge T$. Hence by Kurtz's Theorem 5.3, for some $C(\varepsilon) > 0$ and all n large enough, considering paths that are within $\varepsilon_s/2$ of $\vec{z}_\infty(t)$ yields, for all $\vec{z} \in G^o$,

$$\mathbb{P}_{\vec{z}}\left(\vec{z}_n(t) \notin N_{\varepsilon_s},\ 0 \le t \le T\right) \le e^{-nC(\varepsilon)}.$$

Using the same iteration as in the proof for even j establishes the lemma. ∎

Exercise 6.33. Assume $\vec{q} \in G^o$ and (6.8). If $\log \lambda_i(\vec{x})$ are uniformly Lipschitz in $N_{2\varepsilon}(\vec{q})$ then the conclusion of Lemma 6.32 holds for even j. Under the assumptions of Exercise 6.30, Lemma 6.32 holds for odd j. Hint: see Exercise 6.30. ♠

Thus the times between events have geometric tails. That the time to leave a small neighborhood of $\vec{q}$ cannot be too short is easy to show:

Exercise 6.34. Assume 6.2 and (6.6). For even j, for any T there is a $\delta > 0$ so that

$$\mathbb{P}_{\vec{x}}\left(\tau_{j+1} - \tau_j \le T \mid \vec{z}_n(\tau_j) \in G^o\right) \le e^{-n\delta}. \tag{6.16}$$

Hint: use Kurtz's Theorem 5.3 and the definition of ε_s. ♠

For the models we have in mind, often the point $\vec{q}$ is exactly where the assumptions concerning continuity and boundedness of the log-rates fail; typically either some rates become zero, or they are discontinuous. So, let G be a non-empty set satisfying $G = \overline{G^o}$ and let $\vec{q} \in G^o$. Let $G_1, G_2, \ldots$ be defined by

$$G_i \overset{\triangle}{=} (G \cap N_i(\vec{q})) \setminus N_{1/i}(\vec{q}). \tag{6.17}$$

That is, G_i contains all points in G that are not too close to $\vec{q}$, and are also not too far.

Exercise 6.35. Assume Lyapunov stability (6.6) with $\vec{q} \in G^o$. If Assumption 6.2 holds for each G_i then (6.16) holds for even j. If Assumption 6.2 holds in some $N_{\varepsilon_0}(\vec{q})$ then (6.16) holds for even j provided ε is small enough. Hint for the first claim: for n large enough,

$$
\mathbb{P}_{\vec{x}} \left(\tau_{j+1} - \tau_j \leq T \mid \vec{z}_n(\tau_j) \in G^o \right)
$$
$$
\leq \sup_{\varepsilon_s/2 \leq |\vec{z} - \vec{q}| \leq \varepsilon_s} \mathbb{P}_{\vec{z}} \left(|\vec{z}_n(t) - \vec{q}| \geq \varepsilon, \text{ some } t \leq T, \ \vec{z}_n(s) \notin N_{\varepsilon_s/2} \text{ all } s \leq t \right)
$$

(why?). Now apply the previous argument. ♠

Recall the definitions of T^* and $I^* = I^*(S)$ of Assumption 6.6 and (6.4)–(6.5), and that $\vec{r}^*$ is defined for all t, through (6.7). We now prove our main lemma for the Freidlin-Wentzell theory. This lemma states that the probability that escape occurs on any cycle is about e^{-nI^*}. Furthermore, it shows that if we condition on escape occurring at a particular cycle, then the latter part of $\vec{z}_n(t)$ is close to $\vec{r}^*(t)$.

Lemma 6.36. *Assume 6.1–6.6, let $d \geq 2$, and let j be odd. For any $\delta > 0$ small, if n is large enough, then uniformly in compact subsets of G^o,*

$$
e^{-n(I^*+\delta)} \leq \mathbb{P}_{\vec{x}} \left(\vec{z}_n(\tau_{j+1}) \notin G^o \mid \vec{z}_n(\tau_j) \in G^o \right) \leq e^{-n(I^*-\delta)}. \tag{6.18}
$$

For any $\eta > 0$ and $\delta(\eta) > \delta > 0$ there are $\varepsilon_0 < \eta$, $T' < T^*$, and $T'' < \infty$ such that

$$
\mathbb{P}_{\vec{x}} \left(\sup_{0 \leq s \leq T'} \left| \vec{z}_n(\tau_{j+1} - s) - \vec{r}^*(-s) \right| < \varepsilon, \ \tau_{j+1} \geq \tau_j + T' \right.
$$
$$
\left. \text{and} \quad \vec{z}_n(\tau_{j+1}) \notin G^o \mid \vec{z}_n(\tau_j) \in G^o \right) \geq e^{-n(I^*+\delta)} \tag{6.19}
$$

$$
\mathbb{P}_{\vec{x}} \left(\sup_{0 \leq s \leq T'', \ s \leq \tau_{j+1} - \tau_j} \left| \vec{z}_n(\tau_{j+1} - s) - \vec{r}^*(-s) \right| \geq \eta, \right.
$$
$$
\left. \text{and} \quad \vec{z}_n(\tau_{j+1}) \notin G^o \mid \vec{z}_n(\tau_j) \in G^o \right) \leq e^{-n(I^*+2\delta)} \tag{6.20}
$$

for all $\varepsilon < \varepsilon_0$ and $n > n(\varepsilon)$, uniformly for $\vec{x}$ in compact subsets of G^o. The constant T' is arbitrary in $(0, T^*)$, while T'' is positive, finite, and can be chosen arbitrarily large.

Note that the range of the "sup" in (6.20) is chosen so that $\vec{z}_n(t)$ is well defined throughout the range: we calculate the probability that, while on its journey to leave G^o, the process stays away from $\vec{r}^*$. In fact, one improbable way the process may exit is by doing so in a very short time. The proof is based on Lemma 6.21 and the large deviations bounds.

Proof. For the lower bounds, fix a point $\vec{z}$ in $N_{2\varepsilon}(\vec{q})$ but outside $N_\varepsilon(\vec{q})$; note that (for n large enough) $\vec{z}_n(\tau_j)$ satisfies these two conditions (when $\tau_j < \infty$). Consider the (deterministic!) path $\vec{r}^*(t)$ (of Assumption 6.6) from the time $(-T_1)$ it last exits $N_\varepsilon(\vec{q})$ until it hits ∂G at time zero. By taking ε small we can guarantee that $T_1 > T'$. By Exercise 6.23, there is a path $\vec{r}_1(t)$ with $\vec{r}_1(-T_1 - \alpha) = \vec{z}$, $\vec{r}_1(t) = \vec{r}^*(t)$ for $t \geq -T_1$ so that the path remains outside $N_\varepsilon(\vec{q})$, and $I^0_{-T_1 - \alpha}(\vec{r}_1) \leq I^*(S) + C\varepsilon$ (see Figure 6.37).

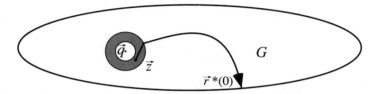

Figure 6.37. The path $\vec{r}^*(t)$ changed on a small interval so as to start at $\vec{z}$.

By Assumption 6.3 and the remark following Lemma 6.21, for ε small enough we can extend the path $\vec{r}_1$ up to time $T_2 > 0$ so that $N_\varepsilon(\vec{r}_1(T_2)) \cap G = \emptyset$, and so that $I^{T_2}_{-T_1 - \alpha}(\vec{r}_1) \leq I^*(S) + (C + C_1)\varepsilon$. Hence by the strong Markov property, for any $\vec{z} \in N_{2\varepsilon}(\vec{q}) \setminus N_\varepsilon(\vec{q})$,

$$\mathbb{P}_{\vec{x}}\left(\vec{z}_n(\tau_{j+1}) \notin G^o \mid \vec{z}_n(\tau_j) = \vec{z}\right)$$

$$\geq \mathbb{P}_{\vec{z}}\left(\sup_{-T_1 - \alpha \leq t \leq T_2} |\vec{z}_n(t + T_1 + \alpha) - \vec{r}_1(t)| < \varepsilon\right)$$

$$\geq e^{-n(I^* + (C + C_1)\varepsilon + \delta')}$$

by the large deviations lower bound Theorem 5.51, for any $\delta' > 0$ and $n \geq n(\delta')$. (That the sets we use in this proof, such as $\{\vec{r} : \sup_{-T_1 - \alpha \leq t \leq T_2} |\vec{r}(t + T_1 + \alpha) - \vec{r}_1(t)| < \varepsilon\}$, are open or closed as required follows from Exercise A.63.) Since this holds uniformly in $\vec{z}$, the lower bound in (6.18) is established. The lower bound (6.19) follows from the last two inequalities, since by construction $\vec{r}_1(t)$ agrees with $\vec{r}^*(t)$ on $[-T_1, 0]$. In particular, by definition of T_1, if $\vec{z}_n$ stays near $\vec{r}^*$ then, for n large enough a choice of small enough ε shows that $\tau_{j+1} - \tau_j \geq T'$.

For the upper bounds (6.18) and (6.20) we use the large deviations upper bound. Using conditioning and the Markov property as above, let $\vec{z}$ be in $N_{2\varepsilon}(\vec{q})$ but outside $N_\varepsilon(\vec{q})$, and define

$$F_T \stackrel{\triangle}{=} \left\{\vec{r} : \vec{r}(s) \notin N_{\varepsilon_s}(\vec{q}) \text{ for } 0 \leq s \leq t, \; \vec{r}(t) \notin G^o \text{ for some } 0 \leq t \leq T\right\}.$$

Then F_T is a closed set, and to establish the upper bound (6.18) we only need to estimate

$$\mathbb{P}_{\vec{z}}\left(\vec{z}_n(\tau_0) \notin G^o\right) \leq \mathbb{P}_{\vec{z}}(\tau_0 \geq T) + \mathbb{P}_{\vec{z}}(\vec{z}_n \in F_T), \qquad (6.21)$$

where T is to be chosen. By Lemma 6.21 and Exercises 6.23 and 6.25, $I^*_{\vec{z}}(F_T)$ (the infimum of I over paths in F_T that start at $\vec{z}$) is continuous in $\vec{z}$. Hence by

Corollary 5.65 the upper bound estimate is uniform in n for $\vec{z}$ in the compact set $N_{2\varepsilon}(\vec{q})$. Thus, for all n large enough and all such $\vec{z}$

$$\mathbb{P}_{\vec{z}}(\vec{z}_n \in F_T) \leq e^{-n(I_{\vec{z}}^*(F_T) - \delta)}.$$

But Lemma 6.21 proves that, for ε small enough,

$$I_{\vec{z}}^*(F_T) \geq I^*(G) - \delta$$

for all T (proof by contradiction is easy). We now bound the first term on the right-hand-side of (6.21), by choosing T large. Using Lemma 6.32, let $\tilde{T}$ be large enough so that

$$\mathbb{P}_{\vec{z}}\left(\tau_0 \geq k\tilde{T}\right) \leq e^{-nC_o(\varepsilon)k}.$$

Thus choosing k so large that $C_0(\varepsilon)k > I^* + 1$, this term is indeed negligible when $T = T'' \overset{\triangle}{=} k\tilde{T}$ (note that we might have $T > T^*$, but this does not affect the result).

For the final upper bound (6.20), by the last calculation it suffices to estimate the probability of $\vec{z}_n \in F_2$, where

$$F_2 \overset{\triangle}{=} F_{T''} \cap \left\{ \vec{r} : \sup_{-t_r \leq t \leq 0} \left| \vec{r}(t + t_r) - \vec{r}^*(t) \right| \geq \eta \right\},$$

where t_r is the first time $\vec{r}$ leaves G^o. Since $\vec{r}^*$ is the unique minimizer of I over S, the path $\{\vec{r}^*(t), -T'' \leq t \leq 0\}$ is also the unique minimizer of $I^0_{-T''}$, among paths starting at $\vec{r}^*(-T'')$ (recall that if $t < -T^*$ then by definition $\vec{r}^*(t) = \vec{q}$). But, as discussed in the introduction to this chapter, using the fact that I^T_0 is a good rate function (lower semicontinuous, with compact level sets) or, using Proposition 5.46, we obtain that for any $\eta > 0$, there is a δ small enough so that

$$I^*_{\vec{r}^*(-T'')}(F_2) \geq I_{\vec{r}^*(-T'')}(\vec{r}^*) + 4\delta,$$

and this holds for any $T'' \geq T_1$. Note that F_2 contains paths on $[0, T]$, where $T > T^*$ is possible: we use $T < \infty$, since by definition $\vec{r}^*$ is optimal among all paths of all lengths. Now choose ε small, and increase T'' if necessary, so that $\vec{r}^*(-T'')$ is in $N_{\varepsilon}(\vec{q})$. Applying Lemma 6.21 to both sides of the last inequality, we obtain

$$I_{\vec{z}}^*(F_2) \geq I^* + 4\delta - \delta = I^* + 3\delta$$

for all $\vec{z}$ in the compact set $\overline{N_{2\varepsilon}}(\vec{q})$. The large deviations upper bound of Corollary 5.65 now gives

$$\mathbb{P}_{\vec{z}}(\vec{z}_n \in F_2) \leq e^{-n(I^* + 2\delta)}$$

for n large enough. The uniformity in $\vec{z}$ over $\overline{N_{2\varepsilon}}(\vec{q})$ follows from the continuity of $I_{\vec{x}}^*(F_2)$ in $\vec{x}$, which is established just like the continuity of $I_{\vec{x}}^*(F)$. Finally, since by decreasing T'' the probability on the left-hand side of (6.20) decreases, we conclude that T'' can also be chosen arbitrarily—it need not be large. ∎

Lemma 6.36 does not extend to $d = 1$; if the optimal path r^* exits G^o, say, to the right of q while $z_n(\tau_j)$ is to the left of q, then the probability in (6.18) corresponds to the optimal way of exiting to the left of q; there is no reason this should be close to I^*. The probability on the right-hand side of (6.19) is zero in this case, so that this bound as well as the lower bound in (6.18) do not hold. However, the basic idea is still valid:

Exercise 6.38. Lemma 6.36 holds in the case $d = 1$ if the conditioning is on $z_n(\tau_{j-1})$. Hint: by choosing ε small we can guarantee that the ratio of the probability of exiting N_{ε_δ} in the "right" direction compared to the "wrong" direction is bounded below (actually, diverges). ♠

The methods of Lemma 6.36 allow us also to compute the probability of exiting the boundary through a prescribed subset of the boundary, which does not include the optimal exit point $\vec{r}^*(0)$. Fix a set $H \subset \partial G$ and assume that it is a "nice" set, that is, H is the closure of its relative interior. Now recall (6.4) and define

$$S_H \triangleq S \cap \{\vec{r} : \vec{r}(-T) = \vec{q}, \ \vec{r}(0) \in H, \ T > 0\}. \tag{6.22}$$

We shall say that $\vec{z}_n$ exited G^o at τ_{j+1} and exit was through H if the straight line between $\vec{z}_n(\tau_{j+1})$ and $\vec{z}_n(\tau_{j+1}^-)$, the position just before exiting, crosses H.

Exercise 6.39. Assume $H \subset \partial G$ is the closure of its relative interior. Under the conditions of Lemma 6.36, for any $\delta > 0$ small enough, n large enough, and uniformly in compact subsets of G^o,

$$e^{-n(I^*(S_H)+\delta)} \leq \mathbb{P}_{\vec{x}}\left(\vec{z}_n(\tau_{j+1}) \notin G^o, \text{ exit was through } H \mid \vec{z}_n(\tau_j) \in G^o\right)$$

$$\leq e^{-n(I^*(S_H)-\delta)}.$$

Extend also to $d = 1$ as in Exercise 6.38. Hint: just as in the proof of Lemma 6.36, the idea is to show that exit must occur in a bounded time, and then apply the theory of Chapter 5. For the lower bound note that if $V(\vec{x}^*) = \inf_{\vec{x} \in H} V(\vec{x})$ then there is a $\vec{y} \in H^o$ near $\vec{x}^*$ with $V(\vec{y})$ near $V(\vec{x}^*)$. Use Lemma 6.21 for approximations and continuity. ♠

Exercise 6.40. The purpose of this exercise is to show that Lemma 6.36 holds without even assuming stability of $\vec{q}$, and under very weak conditions. The key condition is that it is not likely for the process to remain away from $\vec{q}$ but inside G for a long time. Recall the notation (6.17). Let $d \geq 2$. Assume that $\log \lambda_i$ are Lipschitz continuous in each G_i, that the $\vec{e}_i$ span $\mathbb{R}^d$ (6.8) and that the rates do not increase or decrease too fast, so that (6.12) holds. Fix δ and choose $\vec{r}_\delta$ in S so that $\vec{r}_\delta(-T_\delta) = \vec{q}, \vec{r}_\delta(0) \in \partial G$ and

$$I^0_{-T_\delta}(\vec{r}_\delta) < I^*(S) + \delta.$$

(i) If for some $C > 0$ we can extend $\vec{r}_\delta$ for all ε small enough up to time u^ε so that $\vec{r}_\delta(u^\varepsilon)$ is away from the boundary of G, but not too far, or more precisely

$$N_\varepsilon(\vec{r}_\delta(u^\varepsilon)) \cap G = \emptyset \quad \text{and} \quad |\vec{r}_\delta(u^\varepsilon) - \vec{r}_\delta(0)| < C\varepsilon,$$

and if $\log \lambda_i$ are Lipschitz continuous also in a ε-neighborhood of

$$\{\vec{r}_\delta(t) : -T_\delta \leq t \leq u^\varepsilon\},$$

then the lower bounds (6.18)–(6.19) hold, with $\vec{r}_\delta$ replacing $\vec{r}^*$.

(ii) Assume that if the process starts near $\vec{q}$, then it either gets closer or escapes G altogether. More precisely, assume that for some $\eta > \varepsilon > 0$, there is a $T = T(\varepsilon, \eta)$ so that if $\vec{z}_\infty(0) \in N_{2\varepsilon}(\vec{q}) \cap N^C_{\varepsilon/2}(\vec{q})$ then for some $t \leq T$ either $\vec{z}_\infty(t) \in N_{\varepsilon/2}$ or $\vec{z}_\infty(t) \notin G^\eta$. Then the upper bound in (6.18) holds. If in addition $\vec{r}^*$ exists and is unique, then (6.20) holds.

(iii) The results of (i)–(ii) hold in the case $d = 1$, in the form of Exercise 6.38, provided that for ε small enough, the process can exit the ε-ball in the desired direction, i.e.,

$$\mathbb{P}_x \left(|z_n(\tau_j) - r^*(0)| < |q - r^*(0)| \mid z_n(\tau_{j-1}) \in G^o \right) \geq e^{-n(\delta/2)}. \quad (6.23)$$

Hints: for (i), the proof of Lemma 6.36 extends. For (ii) note that for n large, $\vec{z}$ needs only be taken outside $N_\varepsilon(\vec{q})$ but in $N_{2\varepsilon}(\vec{q})$. For the last part of (iii) compute the probability of a sequence of jumps all in the right direction. ♠

Remark. The complicated condition concerning $T(\varepsilon, \eta)$ in Exercise 6.40, part (ii), is implied, for example, by Lyapunov stability. It simply means that there are no limit cycles for $\vec{z}_\infty$ inside G. Equation (6.23) holds if the rates are bounded and (at least one of) the rates of the jumps in the direction from q to $r^*(0)$ satisfies, for some positive η,

$$\lambda_i(q) > 0 \quad \text{and} \quad \int_{q-\eta}^{q+\eta} |\log \lambda_i(x)| \, dx < \infty$$

which holds, in particular, whenever $\lambda_i(x)$ is polynomial in x. On the other hand, if all jumps are of size one and the point $\vec{q}$ is absorbing [that is, $\lambda_i(\vec{q}) = 0$ for all i], then the condition does not hold. Indeed, in this case the lower bound cannot hold.

Lemmas 6.32–6.36 indicate that the segments of the processes between stopping times are almost identically distributed, and in a sense independent. We shall need the following more precise statements in the course of the proof of Theorem 6.17.

Exercise 6.41. Assume the estimates (6.15) of Lemma 6.32 and (6.16) of Exercise 6.34 (or 6.35) hold. Fix $\delta > 0$ and define $b_j \triangleq \mathbb{1}[\tau_{2j} - \tau_{2j-2} \geq T]$— the event that the jth "cycle" is longer than T. Let $\tilde{b}_1, \tilde{b}_2, \ldots$ be a sequence of Bernoulli random variables with $\mathbb{P}\left(\tilde{b}_j = 0\right) = e^{-n\delta}$. There is a positive T so that, if $\tau_{2K} < \infty$,

$$\mathbb{P}_{\tilde{x}}\left(\sum_{j=1}^K b_j \leq M\right) \leq \mathbb{P}_{\tilde{x}}\left(\sum_{j=1}^K \tilde{b}_j \leq M\right)$$

for all M and all n large enough. Hint: use (6.16) to bound the probability that $b_j = 1$. Use induction on K, condition on a sum up to $K - 1$ and on the position at τ_{2K-1}, and use the estimates (6.15) of Lemma 6.32. Remember the position is discrete, for each n, and use the Markov property. ♠

Exercise 6.42. Assume the estimates (6.15) of Lemma 6.32 and (6.16) of Exercise 6.34 (or 6.35) hold. Fix $\delta > 0$ and define $c_j \overset{\triangle}{=} \tau_{2j} - \tau_{2j-2}$—the length of the jth "cycle." Let $\tilde{c}_1, \tilde{c}_2, \ldots$ be a sequence of i.i.d. exponential random variables with mean $e^{n\delta}$. There is a positive T so that, if $\tau_{2K} < \infty$,

$$\mathbb{P}_{\tilde{x}}\left(\sum_{j=1}^{K} c_j \geq M\right) \leq \mathbb{P}_{\tilde{x}}\left(\sum_{j=1}^{K}(\tilde{c}_j + 2T) \geq M\right)$$

for all M and all n large enough. Hint: let first $g_1, g_2, \ldots$ be geometric random variables with mean $e^{n\delta}$. For integer i use

$$\mathbb{P}_{\tilde{x}}\left(\sum_{j=1}^{K} c_j \geq M\right) \leq \sum_i \mathbb{P}\left(M - iT \leq \sum_{j=1}^{K} c_j \leq M - iT + T, \; c_K \geq iT\right)$$

and now use the same conditioning as in Exercise 6.41. Finally, replace the geometric random variables with $T +$ exponential ones. ♠

All the ingredients are now in place to prove the main results of this section. For convenience, let us restate each theorem before its proof.

Theorem 6.15. *Under Assumptions 6.1–6.6, for each finite $T < T^*$ and $\varepsilon > 0$,*

$$\lim_{n \to \infty} \mathbb{P}_{\tilde{x}}\left(\sup_{\tau_n(G)-T \leq t \leq \tau_n(G)} |\vec{z}_n(t) - \vec{r}^*(t - \tau_n(G))| < \varepsilon, \; \tau_n(G) > T\right) = 1.$$

The rate of convergence is exponential, uniformly over $\tilde{x}$ in any compact subset of G^o.

Proof. We provide the proof in the case $d \geq 2$. The special calculations for $d = 1$ are given as Exercise 6.44.

First note that $\tau_n(G)$ is almost surely finite (use a coupling argument as in Examples 4.8–4.9 to show that, eventually, there will be a sequence of jumps in the, say, $\vec{e}_i$ direction, uninterrupted by other jumps, that will cause an exit, independently of the starting position). On the other hand, by Kurtz's Theorem 5.3, for any finite time T, the probability that the process exits G before T (and thus does not follow $\vec{z}_\infty$) $\mathbb{P}_{\tilde{x}}(\tau_n(G) \leq T) \to 0$, uniformly in $\tilde{x}$ in any compact subset of G^o. Therefore we can (and shall) ignore the stipulation $\{\tau_n(G) > T\}$.

We shall use the notation of Lemma 6.36. Fix some $T < T^*$. Since we want to show that events of the type (6.19) dominate, we may as well increase the ε

of (6.19) to equal η and set $\eta = \varepsilon$. With this convention, we define a sequence of events, and for convenience we suppress the dependence on n.

$$R_{2j} \triangleq \left\{ \omega : \sup_{\tau_{2j}-T \le t \le \tau_{2j}} |\vec{z}_n(t) - \vec{r}^*(t - \tau_{2j})| < \varepsilon \quad \text{and} \quad \vec{z}_n(\tau_{2j}) \notin G^o \right\}.$$

R_{2j} is the "desired" event, that the process exits near $\vec{r}^*$, and does so at τ_{2j}. Now let W_{2j} denote the event that the exit is "wild" in that it wanders away from $\vec{r}^*$.

$$W_{2j} \triangleq \left\{ \omega : \sup_{\tau_{2j}-T \le t \le \tau_{2j}} |\vec{z}_n(t) - \vec{r}^*(t - \tau_{2j})| \ge \varepsilon \quad \text{and} \quad \vec{z}_n(\tau_{2j}) \notin G^o \right\}.$$

Note that $R_{2j} \bigcup W_{2j} = \{ \omega : \vec{z}_n(\tau_{2j}) \notin G^o \}$. Recall that we adopted the convention that $\tau_j = \infty$ if $\tau_k = \tau_n(G)$ for some $k < j$. The event we are interested in is the union of the R_{2j}, and clearly

$$R \triangleq \bigcup_{j>0} R_{2j} = \bigcup_{j>0} \{ R_{2j} \cap \{ \tau_{2j-1} < \infty \} \}.$$

Therefore

$$\mathbb{P}_{\vec{x}}(R) = \mathbb{P}_{\vec{x}} \left(\bigcup_{j>0} \{ R_{2j} \cap \{ \tau_{2j-1} < \infty \} \} \right)$$

$$= \sum_{j=1}^{\infty} \mathbb{P}_{\vec{x}} \left(R_{2j} \cap \{ \tau_{2j-1} < \infty \} \right)$$

$$= \sum_{j=1}^{\infty} \mathbb{P}_{\vec{x}} \left(R_{2j} \mid \tau_{2j-1} < \infty \right) \mathbb{P}_{\vec{x}} \left(\tau_{2j-1} < \infty \right)$$

$$\ge e^{-n(I^*+\delta)} \sum_{j=1}^{\infty} \mathbb{P}_{\vec{x}} \left(\tau_{2j-1} < \infty \right)$$

for all n large, by (6.19) of Lemma 6.36. This estimate shows also that the last sum is finite. Similarly (with the obvious definition of W),

$$\mathbb{P}_{\vec{x}}(W) = \mathbb{P}_{\vec{x}} \left(\bigcup_{j>0} \{ W_{2j} \cap \{ \tau_{2j-1} < \infty \} \} \right)$$

$$= \sum_{j=1}^{\infty} \mathbb{P}_{\vec{x}} \left(W_{2j} \mid \tau_{2j-1} < \infty \right) \mathbb{P}_{\vec{x}} \left(\tau_{2j-1} < \infty \right)$$

$$\le e^{-n\delta} e^{-n(I^*+\delta)} \sum_{j=1}^{\infty} \mathbb{P}_{\vec{x}} \left(\tau_{2j-1} < \infty \right)$$

for all n large, by (6.20) of Lemma 6.36. Combining the two estimates

$$\mathbb{P}_{\vec{x}}(R) + \mathbb{P}_{\vec{x}}(W) \le \mathbb{P}_{\vec{x}}(R)\left(1 + e^{-n\delta}\right) \tag{6.24}$$

for all n large. Now let

$$R_0 \overset{\triangle}{=} \{\omega \ : \ \tau_G = \tau_0\}.$$

That is, R_0 is the set of paths that escape right away, before entering $N_{\varepsilon_s}(\vec{q})$ at all. Then it is an easy consequence of Assumption 6.4 and Kurtz's Theorem that, uniformly over $\vec{x}$ in any compact subset of G^o, we have a constant η such that

$$\mathbb{P}_{\vec{x}}(R_0) < e^{-n\eta} \tag{6.25}$$

for all large enough n. Since $\tau_n(G) < \infty$ a.s. and since the sets R_0, R, and W are disjoint,

$$\mathbb{P}_{\vec{x}}\left(R_0 \bigcup R \bigcup W\right) = \mathbb{P}_{\vec{x}}(R_0) + \mathbb{P}_{\vec{x}}(R) + \mathbb{P}_{\vec{x}}(W) = 1,$$

so that (6.24) and (6.25) imply that, for large n,

$$\mathbb{P}_{\vec{x}}(R) \ge 1 - 2e^{-n(\eta \wedge \delta)} \to 1 \text{ as } n \to \infty$$

so that convergence is exponential. Finally, the uniformity in $\vec{x}$ follows since the estimates in Lemma 6.36 are uniform. ∎

Exercise 6.43. Extend Theorem 6.15 to $T \ge T^*$. Hint: since T is finite, only the case $T^* < \infty$ needs be treated. Use the extension to the definition of $\vec{r}^*$ for large (negative) times and the estimates on the probability of remaining near $\vec{q}$. ♠

Exercise 6.44. Extend Theorem 6.15 to $d = 1$. Hint: see Exercise 6.38. ♠

We can now characterize the relative probability of exiting the set G through a specified part H of the boundary ∂G. Recall the definition (6.22) of S_H and the following discussion of the meaning of exiting G^o through H.

Corollary 6.45. *Assume $H \subset \partial G$ is the closure of its relative interior. Under Assumptions 6.1–6.5, uniformly over $\vec{x}$ in any compact subset of G^o,*

$$\lim_{n \to \infty} \frac{1}{n} \log \mathbb{P}_{\vec{x}}\left(\vec{z}_n(t) \text{ exits } G^o \text{ through } H\right) = I^*(S) - I^*(S_H).$$

Suppose now that all the assumptions hold in a larger set $\tilde{G}$, but $\vec{q} \in \tilde{G}^o \setminus G^o$. Then for each $\vec{x} \in G^o$,

$$\lim_{n \to \infty} \frac{1}{n} \log \mathbb{P}_{\vec{x}}\left(\vec{z}_n(t) \text{ exits } G^o \text{ through } H\right)$$

$$= \inf\left\{I^0_{-T}(\vec{r}) : \vec{r}(-T) = \vec{x}, \ \vec{r}(0) \in H, \ T > 0\right\}.$$

The first result gives the (rough) distribution of the location of exit points for the process. Note that we are not assuming 6.6, so that there may be multiple points $x \in \partial G$ where $V(\vec{x})$ is minimized. Some of these points may be in H. In that case, the result does not mean that the exit point is in H with probability approaching one; it only says that the probability decreases at most at a subexponential rate in n. The second result means that if we start in a set G that has $\vec{z}_\infty$ leave in short order, then the probability of leaving any other way can be measured by the I-function. If we replace H by ∂G in the second equation, the minimum will obviously be zero since, when $\vec{q} \notin \partial G$, $\vec{z}_\infty$ leaves G quickly, with zero cost. If $\vec{q} \in \partial G$, then we get to arbitrarily small neighborhoods of $\vec{q}$ with zero cost by following $\vec{z}_\infty$, and then we can choose a straight line path to leave G with cost at most proportional to the size of the neighborhood (as in Lemma 6.21).

Exercise 6.46. Prove Corollary 6.45. Hints: the first result requires that you look at the paths that leave G through H, and compare them to paths that leave G via $\vec{r}\,^*$. This is done in Exercise 6.39: now follow the proof of Theorem 6.15. For the second result, show that cheap paths don't stay in G^o for very long, unless they are approaching $\vec{q}$ (which may be in ∂G). Then consider the cases where $\vec{q} \in H$ and $\vec{q} \notin H$ separately; since H is closed, if $\vec{q} \notin H$ then there is a minimal distance between $\vec{q}$ and H. ♠

Inspection of the proof of Theorem 6.15 establishes the following.

Corollary 6.47. *Let $\vec{z}_n(t)$ be a Markov process, and let $\vec{r}\,^*$ and I^* be respectively a path and a number that make (6.19) and (6.20) of Lemma 6.36 hold. Assume $\tau_n(G) < \infty$ and that $\tau_n(G) \to \infty$. Then the conclusion of Theorem 6.15 holds.*

The asymptotic distribution of the exit time is concentrated around the point $\exp(nI^*(S))$ in the sense of Theorem 6.17, which we recite for convenience.

Theorem 6.17. *Under Assumptions 6.1–6.6 we have, uniformly over $\vec{x}$ in any compact subset of G^o,*

(i) For each $\varepsilon > 0$, $\displaystyle\lim_{n\to\infty} \mathbb{P}_{\vec{x}}\left(\frac{\log \tau_n(G)}{n} \in (I^*(S) - \varepsilon, I^*(S) + \varepsilon)\right) = 1.$

(ii) $\displaystyle\lim_{n\to\infty} \frac{1}{n}\log \mathbb{E}_{\vec{x}}(\tau_n(G)) = I^*(S).$

Proof. We start by showing that the escape time has to be at least as large as specified in (i). Fix any δ small enough and define

$$ j_G \stackrel{\triangle}{=} \inf\{j : \tau_{2j} = \tau_n(G)\} $$

(this is the index of τ_G, the escape time, so $\tau_G = \tau_{2j_G}$). To simplify notation in this proof, let $K_n \stackrel{\triangle}{=} e^{n(I^* - 2\delta)}$. The idea of this proof is that the number of cycles it takes for $\vec{z}_n$ to escape G is approximately geometrically distributed, with mean number e^{nI^*}, and that each cycle takes at least T for some time T.

We take T as defined in Exercise 6.41. Now

$$\mathbb{P}_{\vec{x}}\left(\tau_n(G) < T e^{n(I^* - 3\delta)}\right) = \sum_{i=1}^{\infty} \mathbb{P}_{\vec{x}}\left(j_G = i,\ \tau_n(G) < T K_n e^{-n\delta}\right)$$

$$\leq \sum_{i=1}^{K_n - 1} \mathbb{P}_{\vec{x}}\left(\vec{z}_n(\tau_{2i}) \notin G^o,\ \vec{z}_n(\tau_{2i-1}) \in G^o\right)$$

$$+ \sum_{i=K_n}^{\infty} \mathbb{P}_{\vec{x}}\left(\tau_n(G) < T K_n e^{-n\delta}\ \Big|\ j_G = i\right).$$

For the first summand, if $d \geq 2$ Lemma 6.36 implies

$$\mathbb{P}_{\vec{x}}\left(\vec{z}_n(\tau_{2i}) \notin G^o,\ \vec{z}_n(\tau_{2i-1}) \in G^o\right) \leq \mathbb{P}_{\vec{x}}\left(\vec{z}_n(\tau_{2i}) \notin G^o\ \big|\ \vec{z}_n(\tau_{2i-1}) \in G^o\right)$$

$$\leq e^{-n(I^* - \delta)}.$$

(If $d = 1$, condition on τ_{2i-2} and use Exercise 6.38 to obtain this estimate.) Now, for any non-negative random variables $\sigma_1, \sigma_2, \ldots$ and any $T > 0$, $\sigma_i \geq T \mathbf{1}[\sigma_i \geq T]$, so that for any K and A,

$$\mathbb{P}\left(\sum_{i=1}^{K} \sigma_i < A\right) \leq \mathbb{P}\left(T \sum_{i=1}^{K} \mathbf{1}[\sigma_i \geq T] < A\right).$$

Therefore, using the notation and results of Exercise 6.41, for all n large, and for $i \geq K_n$,

$$\mathbb{P}_{\vec{x}}\left(\tau_n(G) < T K_n e^{-n\delta}\ \Big|\ j_G = i\right) \leq \mathbb{P}_{\vec{x}}\left(T \sum_{j=1}^{i} \tilde{b}_j \leq T K_n e^{-n\delta}\right)$$

$$\leq \mathbb{P}_{\vec{x}}\left(\frac{1}{i} \sum_{j=1}^{i} \tilde{b}_j \leq \frac{1}{3}\right).$$

This puts us in a position to use large deviations, with n fixed, and i becoming large. Note that the rate function $\ell_p(a)$ for Bernoulli random variables is monotone in p (the probability of a one) when a is far from the mean (consult Exercise 1.17). So, with $p = p(n)$,

$$\mathbb{P}_{\vec{x}}\left(\frac{1}{i} \sum_{j=1}^{i} \tilde{b}_j \leq \frac{1}{3}\right) \leq e^{-i\ell_{p(n)}(1/3)} \leq e^{-i\ell},$$

where $\ell = \ell_{2/3}(1/3) > 0$ is the value of the rate function when $p = 2/3$ at $a = 1/3$. Putting the estimates together, for every $T > 0$ we have

$$\mathbb{P}_{\vec{x}}\left(\tau_n(G) < T e^{n(I^* - 3\delta)}\right) \leq K_n e^{-n(I^* - \delta)} + \frac{e^{-\ell K_n}}{1 - e^{-\ell}}.$$

Therefore, since the second term decreases much faster than the first, we obtain

$$\mathbb{P}_{\bar{x}} \left(\frac{\log \tau_n(G)}{n} \geq I^* - 3\delta \right) \geq 1 - \frac{e^{-n\delta}}{2}.$$

This gives a lower bound on $\tau_n(G)$ for part (i) of the theorem. This also provides us with the desired lower bound on the expectation in (ii).

To get an upper bound on $\tau_n(G)$ in (i), define T to be the larger of the T defined in Lemma 6.32 and the T defined in Exercise 6.42 (note that this T makes the conclusion of both hold). Fix $s \geq 2\delta$ and let now $K_n \stackrel{\triangle}{=} e^{n(I^*+s)}$. As in the first part of the proof (and with the same caveat in the case $d = 1$), using now Exercise 6.42 and the notation therein,

$$\mathbb{P}_{\bar{x}} \left(\tau_n(G) > T e^{n(I^*+2s)} \right)$$

$$\leq \mathbb{P}_{\bar{x}} (jG \geq K_n) + \sum_{i=1}^{K_n} \mathbb{P}_{\bar{x}} \left(\sum_{j=1}^{i} c_j \geq T e^{n(I^*+2s)} \right)$$

$$\leq \mathbb{P}_{\bar{x}} \left(\vec{z}_n(\tau_0) \in G^o \right) \prod_{i=1}^{K_n} \mathbb{P}_{\bar{x}} \left(\vec{z}_n(\tau_{2i}) \in G^o \mid \tau_{2i-1} < \infty \right)$$

$$+ \sum_{i=1}^{K_n} \mathbb{P}_{\bar{x}} \left(\frac{1}{i} \sum_{j=1}^{i} \tilde{c}_j \geq \frac{T e^{n(I^*+2s)}}{i} - 2T \right)$$

$$\leq \left(1 - e^{-n(I^*+\delta)} \right)^{K_n} + \sum_{i=1}^{K_n} \exp \left(-i \ell_n \left(\frac{T e^{n(I^*+2s)}}{i} - 2T \right) \right), \quad (6.26)$$

where the first term of the last line of (6.26) is obtained by Lemma 6.32. The second term is obtained from the large deviations upper bound for fixed n, so that ℓ_n is the rate function for i.i.d. exponential random variables $\tilde{c}_1, \tilde{c}_2, \ldots, \tilde{c}_i$ with mean $e^{n\delta}$. Since $s \geq 2\delta$ and $i \leq K_n$, the argument of ℓ_n is larger than $T e^{2n\delta}$, which in turn is larger than the mean $e^{n\delta}$ of $\tilde{c}_j$. Thus ℓ_n is increasing in this region. Now using the scaling property of the rate function for exponential random variables (cf. Exercise 1.20), $\ell_n(\alpha) = \ell_1(\alpha e^{-n\delta})$, where ℓ_1 corresponds to mean one. Therefore,

$$\ell_n \left(\frac{T e^{n(I^*+2s)}}{i} - 2T \right) \geq \ell_1 \left(T e^{n(2s-s-\delta)} - 2T e^{-n\delta} \right)$$

$$\geq C_1 e^{n(s-\delta)} - C_2$$

since ℓ_1 is convex, and $C_1 > 0$ since ℓ_1 is increasing in the range of interest.

Therefore we have for the second term in (6.26)

$$\sum_{i=1}^{K_n} \exp\left(-i\ell_n\left(\frac{Te^{n(I^*+2s)}}{i} - 2T\right)\right) \leq \sum_{i=1}^{K_n} \exp\left(-i\left(C_1 e^{n(s-\delta)} - C_2\right)\right)$$

$$\leq K_n \exp\left(-C_1 e^{n(s-\delta)} + C_2\right)$$

$$= \exp\left(n(I^* + s) - C_1 e^{n(s-\delta)} + C_2\right)$$

$$(6.27)$$

which decreases superexponentially fast with n, uniformly in $s \geq 2\delta$. As for the first term in (6.26), since

$$\lim_{x \to 0} (1 - x)^{1/x} = \frac{1}{e}$$

we have, with $x_n = e^{-n(I^*+\delta)}$

$$\left(1 - e^{-n(I^*+\delta)}\right)^{K_n} = (1 - x_n)^{(\exp n(s-\delta))/x_n}$$

$$\leq (2/e)^{\exp n(s-\delta)} .$$

$$(6.28)$$

Therefore, combining (6.26), (6.27), and (6.28), we see that

$$\mathbb{P}_{\bar{x}}\left(\tau_n(G) \geq t\right) \leq \left(1 - e^{-n(I^*+\delta)}\right)^{K_n} (2/e)^{\exp(n(s-\delta))}$$

$$+ \exp\left(n(I^* + s) - C_1 e^{n(s-\delta)} + C_2\right)$$

$$(6.29)$$

and the upper bound in (i) is established.

To obtain the upper bound for (ii) note that the bound (6.29) is uniform in $s \geq 2\delta$. Let $\alpha = Te^{n(I^*+4\delta)}$. Clearly

$$\mathbb{E}_{\bar{x}}\tau_n(G) \leq \alpha + \int_\alpha^\infty \mathbb{P}_{\bar{x}}\left(\tau_n(G) \geq t\right) dt.$$

With the change of variable $t = e^{n(I^*+s)}$, we get $dt = 2nTe^{n(I^*+2s)} ds$, and using (6.29) we easily find positive numbers C_3 and C_4 so that

$$\int_\alpha^\infty \mathbb{P}_{\bar{x}}\left(\tau_n(G) \geq t\right) dt \leq 2nTe^{nI^*} \int_{2\delta}^\infty \mathbb{P}_{\bar{x}}\left(\tau_n(G) \geq Te^{n(I^*+2s)}\right) e^{2ns} ds$$

$$\leq 2nTe^{nI^*} \int_{2\delta}^\infty \left((2/e)^{\exp n(s-\delta)} + e^{(C_4 - C_3 \exp(ns))}\right) ds,$$

which converges to zero as $n \to \infty$, and the theorem is established. ∎

6.2. Beyond the Exit Problem

While the exit problem is interesting and useful, the reasoning behind the theorems is much more general. Let us examine the reasoning in detail in order to define a more general class of events for which the Freidlin-Wentzell theory holds. To illustrate the idea, consider the following example.

Example 6.48. What is the probability that the process will trace your name? To make this precise, let $d = 2$ and assume that the continuous path $\{\vec{s}(t) : -1 \leq t \leq 0\}$ traces your name. Fix some positive ε, assume that the point $\vec{q}$ is stable and is far from points on $\vec{s}$, and define

$$S_g \triangleq \left\{ \vec{r} : \sup_{-1 \leq t \leq 0} |\vec{r}(t) - \vec{s}(t)| < \varepsilon \right\}.$$

Thus S_g contains all paths that trace your name to within ε, and do so with a prescribed speed. The analogue of Equation (6.4) would then be the set

$$\left\{ \vec{r} : \vec{r}(-T) = \vec{q} \text{ for some } T \geq 1, \ \{\vec{r}(s) : -1 \leq s \leq 0\} \in S_g \right\}.$$

The questions are then, What is the probability that your name is ever traced? How would the path arrive to $s(-1)$ starting from $\vec{q}$? and How long will you have to hold your breath until this happens?

The answer is that Theorems 6.15 and 6.17 apply: we just have to redefine $\tau_n(G)$ and S. Some additional care may be required since we no longer work with bounded processes. Let us now obtain such results for a slightly more general case, but do keep the example in mind.

We start with a set of paths $S_e \subset D^d[0, u]$ for some arbitrary positive, finite u. This set describes the events of interest. Now let

$$S \triangleq \{\vec{r} : \vec{r}(-T) = \vec{q} \text{ for some } T > 0, \ \{\vec{r}(s), -u \leq s \leq 0\} \in S_e\}. \quad (6.30)$$

Note that this does not prevent us from defining events that take a very short time to happen (shorter than u), since we can extend the path backwards in time by making it stay at $\vec{q}$. In fact, it will be convenient to consider S as a subset of $D^d[0, \infty)$, by precisely such an extension. Thus the set S is such that the last portion of each path is in S_e.

Compare this to Example 6.48 and Equation (6.4): in the special case $u = 0$ and $S_e = \{\vec{r} : \vec{r}(0) \in \partial G\}$ Equation (6.30) reduces to (6.4). We make the following assumptions.

Assumption 6.49. There exist $\varepsilon_0 > 0$ so that if $\vec{r} \in S_e$ and $\vec{r}(t) \in N_{\varepsilon_0}$ for some t then any path $\vec{v}$ that satisfies $\vec{v}(s) = \vec{r}(s)$, $s \geq t$ is also in S_e. Moreover, $|\vec{r}(u) - \vec{q}| \geq \varepsilon_0$.

That is, whether or not a path is in S_e can be determined from its behavior after the last time it left $N_\varepsilon(\vec{q})$. Recall the definition of I^* in Equation (6.5).

Assumption 6.50. S is a continuity set, and every point in S is the limit of points in the interior of S; that is,

$$S \subset \overline{S^o} \quad \text{and} \quad I^*(S^o) = I^*(\overline{S}).$$

Assumption 6.51. There is some K so that $V(\vec{y}) > I^* + 1$ whenever $\vec{y}$ is outside $N_K(\vec{q})$.

Assumption 6.52. Assumption 6.5 holds with $N_K(\vec{q})$ replacing G. Assumption 6.2 holds with G replaced by the set of all points that $\vec{z}_\infty$ can reach with $\vec{z}_\infty(0) \in N_K(\vec{q})$.

We need a technical device, intended to restrict the analysis to a bounded domain. Let

$$S_{eK} \stackrel{\triangle}{=} S_e \bigcup \{\vec{r} \in D^d[0, u] : \vec{r}(u) \notin N_K(\vec{q})\}$$

and let S_K be defined through S_{eK} as in (6.30). S_{eK} is the set of paths that either cause the event of interest, or escape to $|\vec{x} - \vec{q}| > K$. Since we have chosen K to be large enough that the event of interest will very likely occur before escape to $|\vec{x} - \vec{q}| > K$, nothing is lost, and we gain compactness in the paths we need to consider. From here on, unless explicitly stated, we consider the events defined by S_{eK}. Here are the appropriate modifications of the previous definitions. As in Definition 6.14, let

$$\tau_n(S_{eK}) \stackrel{\triangle}{=} \inf\{t \geq 0 : \{\vec{z}_n(s), \ t - u \leq s \leq t\} \in \bar{S}_{eK}\}.$$

Similarly, the stopping times of Equation (6.14) are redefined by adopting the convention

$$\text{replace} \quad \vec{z}_n(t) \notin G^o \quad \text{by} \quad \{\vec{z}_n(s), \ t - u \leq s \leq t\} \in \bar{S}_{eK}. \tag{6.31}$$

These stopping times are well defined for all $0 < \varepsilon < \varepsilon_0$. From the definitions it easily follows that

Exercise 6.53. Under Assumption 6.51, if S satisfies Assumption 6.50 then so does S_K. Moreover, $I^*(S) = I^*(S_K)$. ♠

We can now follow the proofs of the exit problem, with very little change. We shall not bother with extensions, e.g., to an irregular point $\vec{q}$, although again the same arguments apply. We shall not repeat the statements of previous results, since they remain unchanged under the convention (6.31).

Lemma 6.54. *Under Assumptions 6.49–6.52, conclusion (6.15) of Lemma 6.32 holds, as does conclusion (6.16) of Exercise 6.34.*

Exercise 6.55. Verify that the proof of Lemma 6.32 applies under the setup of Lemma 6.54. Ditto for Exercise 6.34. ♠

Lemma 6.56. *Under Assumptions 6.6 and 6.49–6.52, the conclusion of Lemma 6.36 holds for $d \geq 2$.*

Exercise 6.57. Prove Lemma 6.56 and extend it to $d = 1$. Hints: note that the conditioning is, in fact, on $\vec{z}_n(\tau_j) \in N_{2\varepsilon}(\vec{q})$. The only delicate point concerns the construction of the path $\vec{r}_1$—but remember S_{eK} contains no isolated points:

every point is on the boundary of a ball inside S_{eK}. Since S is a continuity set, some paths in the interior of S near $\vec{r}\,^*$ have almost the same cost.   ♠

Exercise 6.58. Extend Lemma 6.56 to the case where S_e is used, and not S_{eK}, including the case $d = 1$.   ♠

Theorem 6.59. *Under Assumptions 6.6 and 6.49–6.52, the conclusion of Theorem 6.15 and conclusion (i) of Theorem 6.17 hold for the events defined through S_{eK}, as well as for the events defined through S_e, uniformly in $\vec{x} \in N_\varepsilon(\vec{q})$, for some $0 < \varepsilon < \varepsilon_0$.*

Exercise 6.60. Prove Theorem 6.59. Hints: consider first the events defined by S_{eK}, in which case the proofs of Theorem 6.15 and Theorem 6.17(i) go through. The condition $\tau_n(S_{eK}) < \infty$ follows since S_{eK} contains an open neighborhood. This implies $\mathbb{P}\left(\tau_n(S_{eK}) = \tau_n(S_e)\right) \to 1$ and the conclusions for S_e follow.   ♠

Exercise 6.61. Extend Theorem 6.59 to an arbitrary starting point $\vec{x}$, under the condition that

$$\left\{ \bigcup_{s \geq 0} \{\vec{z}_\infty(s+t) : 0 \leq t \leq u\} \right\} \cap \bar{S}_e = \emptyset,$$

where $\vec{z}_\infty(0) = \vec{x}$; that is, it is not likely that S_e will be entered before reaching $N_\varepsilon(\vec{q})$.   ♠

Extending part (ii) of Theorem 6.17 is not as simple; although it is likely that the process does not wander far before entering S_e, it is possible that when it does, it spends a very large time away, and furthermore it might not ever come back. Thus estimating the expectation is more tricky, and some additional recurrence conditions are necessary.

Returning to Example 6.48, if a stable process is chosen to trace your name, clearly it makes sense to choose a process such that the point $\vec{q}$ is actually on the path $\vec{s}$! Indeed, Assumption 6.49 is not necessary. Here is an illustration.

Exercise 6.62. Let $\{\vec{s}(t) : 0 \leq t \leq y\}$ be a continuous path going through $\vec{q}$ exactly once. Establish Theorem 6.59 for the events defined by S_g.   ♠

6.3. Discontinuities

In many applications, some of the rates go to zero near the stable point $\vec{q}$, in contradiction of Assumption 6.2 on the boundedness of $\log \lambda_i(x)$. Or, the rates may be simply discontinuous near $\vec{q}$. In particular, one-dimensional processes such as the $M/M/1$ queue have a discontinuity at $q = 0$. In other applications, there are discontinuities at other points: in the basic AMS model of Chapter 13, for example, some rates go to zero at the point zero, while $q > 0$. Nonetheless, we may extend the Freidlin-Wentzell theory to some of these cases, as was hinted at throughout the exercises. Other assumptions, such as 6.4 or the uniqueness in 6.6, were imposed in order to simplify the presentation, but in practice may not hold, or may be difficult to verify.

On the other hand, general conditions under which the two main theorems hold may be too restrictive for our applications. This motivates the various extensions, carried out in the exercises, as well as the conditions in the corollaries below: the technical conditions can be verified separately in each application. The proofs follow directly (although not instantaneously) from previous results, and are therefore relegated to further exercises.

Corollary 6.63. *Assume 6.1, 6.3, 6.4, and 6.6. Let the point $\vec{q}$ be stable in the sense of Lyapunov so that (6.6) holds. Assume there are sets $G_1, G_2, \ldots$ satisfying (6.17) so that the log rates $\log \lambda_i(\vec{x})$ are Lipschitz continuous on each G_i, and that (6.12) holds. If $d = 1$ assume that (6.23) holds. Assume $\vec{r}^*$ exists and is unique. Finally, assume*

$$\infty > \tau_n(G) \to \infty \text{ a.s. as } n \to \infty.$$

Then the conclusions of Theorem 6.15 hold.

Exercise 6.64. Prove Corollary 6.63. Hint: this follows from Corollary 6.47 and Exercise 6.40. Observe that the condition in Exercise 6.40 (ii) is implied by Lyapunov stability, and that Lyapunov stability implies that the $\vec{e}_i$ span $\mathbb{R}^d$. ♠

Corollary 6.65. *Assume the conditions of Corollary 6.63. Assume further that for even j, there is a $C(\varepsilon) > 0$ such that*

$$P(\tau_{j+1} - \tau_j > kT) \le \left(1 - e^{-nC(\varepsilon)}\right)^k.$$

Then the conclusions of Theorem 6.17 hold.

Note that if some jump rates go to zero, the conclusion (6.15) of Lemma 6.32 might not hold for even j. A direct verification is thus required.

Exercise 6.66. Prove Corollary 6.65. ♠

Using these extensions we may obtain the Freidlin-Wentzell estimates for processes that have various sorts of discontinuities. Since one-dimensional processes are both the easiest to analyze and are the most common in applications (perhaps because they are easy to analyze ...), it is worthwhile to work a little more on these processes.

Exercise 6.67. Let z_n be a one dimensional process, let $G \stackrel{\triangle}{=} (a, b)$ with $a < q < b$. Assume q is stable in the sense of Lyapunov, so that (6.6) holds. Assume that the log rates $\log \lambda_i(\vec{x})$ are bounded above, are Lipschitz continuous except perhaps at q, and satisfy (6.12) (that is, they do not decrease too fast). If (at least) one of the rates is bounded below and if

$$\inf \left\{ I_0^T(r) : r(0) = q, \ r(T) = a, \ T > 0 \right\}$$

$$\neq \inf \left\{ I_0^T(r) : r(0) = q, \ r(T) = b, \ T > 0 \right\},$$

then the conclusions of Theorems 6.15 and 6.17 hold. Note that with the value of a held fixed, there is exactly one value of b for which the costs of exiting to either side are the same, so that the last condition is not restrictive. Hints: the last conditions in Corollaries 6.63 and 6.65 respectively can be established by coupling arguments (see Definition 4.7 and following exercises). Consider processes that are the same except that some rates are increased, and some jump directions are eliminated; see Examples 4.8–4.9. For the existence of r^* use the compactness implied by Lemma 5.18. For its uniqueness use the strict convexity of ℓ.   ♠

Exercise 6.68. Let z_n be a one dimensional process with non-negative values; that is, the rates of jump to the left are all zero when the process is at (or to the left of) the point zero. Let $G \stackrel{\triangle}{=} (-1, b)$ with $0 < q < b$, so that exits can occur only at b. Assume q is stable in the sense of Lyapunov, so that (6.6) holds. Assume that $\log \lambda_i(\vec{x})$ are bounded above, are Lipschitz continuous except perhaps at zero, and satisfy (6.12) (that is, they do not decrease too fast). If (at least) one of the rates of jump to the right is bounded below then the conclusions of Theorems 6.15 and 6.17 hold.   ♠

Nonsmooth domains and rates.

We now extend Theorem 6.17 to processes on nonsmooth domains with nonsmooth jump rates. The idea behind the extension is that we only need certain estimates on the behavior of $\vec{z}_n(t)$, such as its propensity to stay near $\vec{q}$; the assumptions we made on the smoothness of $\lambda_i(\vec{x})$ in order to prove these properties are irrelevant so long as the properties hold. We suppose that $\vec{z}_n(t)$ is defined in a domain G that is piecewise smooth, but might have corners or edges (see Assumption 6.71).

Here is the new set of assumptions that we use to replace 6.1–6.6. We follow their statements by a set of lemmas that show how to extend the reasoning of §6.1 to some nonsmooth problems.

Assumption 6.69. There is a path $\vec{z}_\infty(t)$ satisfying Equation (6.6) so that the conclusion of Kurtz's Theorem (Theorem 5.3) holds.

We use this assumption instead of Lipschitz continuity of the $\lambda_i(\vec{x})$ since the jump rates may not be continuous when there are boundaries, but the conclusion of Kurtz's Theorem might hold anyway.

Assumption 6.70. Assumption 6.5 holds for every bounded subset G containing the unique stable point $\vec{q}$.

Assumption 6.71. The domain G over which $\vec{z}_n(t)$ is defined is the closure of its interior, and has the following property. There is a constant K such that for any two points $\vec{x}$ and $\vec{y} \in G^o$ there is a $T < K|\vec{x} - \vec{y}|$ and an associated absolutely continuous path $\vec{r}(t)$ with $\vec{r}(0) = \vec{x}, \vec{r}(T) = \vec{y}$, and the path satisfies $|\vec{r}'(t)| \leq 1$.

Assumption 6.72. The rates $\lambda_i(\vec{x})$ are bounded above for $\vec{x}$ in bounded regions, and (6.10) holds.

Assumption 6.73. There is a function $f(\varepsilon)$ with $f \to 0$ as $\varepsilon \to 0$ such that given any $\vec{x}$ and $\vec{y} \in G$, the path $\vec{r}$ of Assumption 6.71 can be chosen in such a way that $I_0^T(\vec{r}) \leq f(|\vec{x} - \vec{y}|)$.

This enables us to use Exercise 6.22 to conclude that small distances can be covered in small time with small cost. This assumption also implies that $\vec{q}$ is not an absorbing point.

Assumption 6.74. The large deviations principle holds for the process $\vec{z}_n(t)$.

Exercise 6.75. Check that Assumptions 6.69–6.74 imply that the conclusion of Lemma 6.28 holds for any bounded set G that does not contain $\vec{q}$. ♠

Assumption 6.76. For every $\varepsilon > 0$ there is a $\delta > 0$ and $n_0 > 0$ such that for each $n > n_0$ and each $\vec{x}$ less than δ from ∂G,

$$\mathbb{P}_{\vec{x}}\left(\vec{z}_n(t) \notin G^o \text{ for some } t \in [0, 1]\right) \geq \exp(-n\varepsilon).$$

This assumption obviates the need to extend the process $\vec{z}_n(t)$ outside of G. We simply assume that $\vec{z}_n(t)$ might exit G whenever it is close to the boundary. It is related to Assumption 6.73, but we decided to keep it separate since it is easy enough to check.

Theorem 6.77. *Suppose that Assumptions 6.69–6.76 are satisfied. Define I^* as in (6.1) with S defined by (6.4). Then uniformly over $\vec{x}$ in any compact subset of G^o,*

(i) For each $\varepsilon > 0$, $\displaystyle\lim_{n\to\infty} \mathbb{P}_{\vec{x}}\left(\frac{\log \tau_n(G)}{n} \in (I^*(S) - \varepsilon, I^*(S) + \varepsilon)\right) = 1$.

(ii) $\displaystyle\lim_{n\to\infty} \frac{1}{n} \log \mathbb{E}_{\vec{x}}(\tau_n(G)) = I^*(S)$.

This theorem is proved by checking through the sequence of lemmas that lead up to the proof of Theorem 6.17 and seeing that the new assumptions are indeed sufficient. Here is a sequence of statements that the reader is invited to check that lead to the proof of Theorem 6.77.

First, Lemma 6.19 that the positive cone spanned by the $\vec{e}_i$ is all of $\mathbb{R}^d$ holds under Assumptions 6.71 and 6.70, since these assumptions are strictly stronger than Assumption 6.1 and 6.5, which are the assumptions of the lemma. Secondly, Assumption 6.73 takes the place of Lemma 6.21.

For the next lemma, recall the definition of $S_T(\vec{x})$, the paths that start at $\vec{x}$ and stay at least ε from $\vec{q}$:

$$S_T(\vec{x}) \stackrel{\triangle}{=} \left\{ \vec{r} \in D^d[0, T] : \right.$$

$$\left. \inf_{0 \leq t \leq T} |\vec{r}(t) - \vec{q}| > \varepsilon, \ \vec{r}(0) = \vec{x}, \ \vec{r}(t) \in G^o \text{ for all } t \in (0, T) \right\}.$$

Also recall the definition $S_T \stackrel{\triangle}{=} \cup_{\vec{x} \in \mathcal{O}} S_T(\vec{x})$ for given ε and $\mathcal{O}$.

Lemma 6.78. *Under Assumptions 6.69, 6.70, 6.72, and 6.74, the conclusion of Lemma 6.28 holds for bounded subsets of G. Namely, for any $\varepsilon > 0$ and any bounded open set $\mathcal{O}$ with $N_\varepsilon(\vec{q}) \cap \mathcal{O} = \emptyset$, there are $T_1 > 0$ and $\delta > 0$ such that for all $T > T_1$,*

$$I^*(S_T) \stackrel{\triangle}{=} \inf_{\vec{x} \in \mathcal{O}} \inf_{\vec{r} \in S_T(\vec{x})} I_0^T(\vec{r}) \geq \delta(T - T_1).$$

Exercise 6.79. Prove Lemma 6.78. ♠

Lemma 6.80. *Under Assumptions 6.69–6.74, the conclusion of Lemma 6.32 obtains; namely, for each $\varepsilon > 0$ small enough there is a $T < \infty$, depending only on ε, and there are positive functions C_e and C_o such that if n is large enough, then for $k = 1, 2, \ldots$*

$$\mathbb{P}_{\vec{x}} \left(\tau_{j+1} - \tau_j > kT \mid \vec{z}_n(\tau_j) \in G^o \right) \leq \begin{cases} \left(1 - e^{-nC_e(\varepsilon)}\right)^k & j \text{ even}, \\ e^{-nC_o(\varepsilon)k} & j \text{ odd}. \end{cases}$$

Moreover, $C_e(\varepsilon) \to 0$ as $\varepsilon \to 0$.

Exercise 6.81. Prove Lemma 6.80. ♠

Lemma 6.82. *Under Assumptions 6.69–6.76, the first conclusion of Lemma 6.36 [Equation (6.18)] obtains; namely, let $d \geq 2$ and let j be odd. For any $\delta > 0$ small, if n is large enough, then uniformly in compact subsets of G^o,*

$$e^{-n(I^*+\delta)} \leq \mathbb{P}_{\vec{x}} \left(\vec{z}_n(\tau_{j+1}) \notin G^o \mid \vec{z}_n(\tau_j) \in G^o \right) \leq e^{-n(I^*-\delta)}.$$

Exercise 6.83. Prove Lemma 6.82. ♠

Exercise 6.84. Prove Theorem 6.77. Hint: it's just like Theorem 6.17, using Lemma 6.82 instead of Lemma 6.36, and using the large deviations principle where needed. We didn't assume that $\vec{r}^*(t)$ exists or is unique; if needed, in its place, simply use an approximately minimal path. ♠

We conclude this section with the following conjecture:

When $\vec{q}$ is a regular point (that is, the $\log \lambda_i(\vec{x})$ are Lipschitz near $\vec{q}$), then $T^* = \infty$.

This implies that if $T^* < \infty$, then $\vec{q}$ is irregular in one of the senses mentioned. The reason such a result is to be expected is as follows. The time to escape a small neighborhood of $\vec{q}$ is roughly the inverse of the derivative $\vec{r}'(t)$. Therefore the cost to escape with derivative of size y is about $\ell(q, y)/y$. But $\ell(q, 0) = 0$ and $\ell(q, y)$ is strictly convex in y. Therefore the function $\ell(q, y)/y$ is minimal at $y = 0$ (draw a picture, or compute a Taylor expansion), corresponding to an infinite time to escape.

6.4. Convergence of Invariant Measures

We have seen that the times between rare events can be estimated by calculating the large deviations rate associated with that event. In particular, if a set D does not contain the point $\vec{q}$, then we expect that $\vec{z}_n(t)$ enters D at times τ that are exponentially far apart (Theorem 6.17):

$$\frac{1}{n} \log \tau \to I^*(D).$$

This leads us to expect that the steady-state probability $\pi_n(D)$ that $\vec{z}_n(t) \in D$ is approximately $e^{-nI^*(D)}$. That is, the probability of a rare event should be approximately one over the time between occurrences of the event. This section contains the mathematical justification of this heuristic.

The following additional assumptions are used in this section.

Assumption 6.85. The jumps $\vec{e}_i$ take integer values in each direction.

Assumption 6.86. The process $\vec{z}_n$ is positive recurrent.

Assumption 6.85 means that the process $\vec{z}_n$ takes values in the d-dimensional lattice with mesh size $1/n$. This assumption is imposed in order to simplify proofs; for a derivation of the more general case, see e.g., [FW]. Assumption 6.86 is essential: in fact, we shall even strengthen it in Assumption 6.88 below. It means that any point in the lattice can be reached from any other point, and the time to return, say from zero back to zero, has finite mean (see, e.g., Çinlar [Ci] where the older terminology "non-null" is used in lieu of "positive recurrent"). As a consequence, the following ergodic properties hold [Ci (5.11) p. 264, (5.26)–(5.29) pp. 268–269]. For each n there exists a probability measure π_n so that, for any set D,

$$\lim_{t \to \infty} \mathbb{P}(\vec{z}_n(t) \in D) = \int_D d\pi_n(x) \tag{6.32}$$

$$\lim_{t \to \infty} \frac{1}{t} \int_0^t \mathbf{1}\left[\vec{z}_n(s) \in D\right] ds = \pi_n(D) \quad \text{w.p. 1.} \tag{6.33}$$

The measure π_n is the (unique) invariant distribution of the process: if $\vec{z}_n(t)$ is distributed according to π_n at time t, then $\vec{z}_n(s)$ will be distributed according to π_n for all $s \geq t$.

To carry out our program we investigate the behavior of the process on excursions away from a neighborhood of the stable point $\vec{q}$. Fix ε and $\varepsilon_s = \varepsilon_s(\varepsilon)$. Define $G = \mathbb{R}^d$ and recall the definition of the stopping times (6.14). These stopping times are all (w.p. 1) finite, due to Assumption 6.86. Define the discrete-time Markov chain $m_0, m_1, \ldots$ through

$$m_j \overset{\triangle}{=} \vec{z}_n(\tau_{2j}), \ j = 0, 1, \ldots.$$

Then it is easy to verify [Ci] that $m_0, m_1, \ldots$ is an ergodic Markov chain taking a finite number of values (the points on the lattice inside N_{ε_s}). Denote the invariant distribution of this chain by μ_n, and let $\mathbb{E}_{\mu_n}$ denote expectation starting with the initial distribution μ_n [that is, $m(0)$ is distributed according to μ_n]. In Exercise 6.87 below it is shown that for all sets D,

$$\pi_n(D) = \left(\mathbb{E}_{\mu_n}\tau_2\right)^{-1} \mathbb{E}_{\mu_n} \int_0^{\tau_2} \mathbf{1}\left[\vec{z}_n(s) \in D\right] ds. \tag{6.34}.$$

Exercise 6.87. Assume 6.85 and 6.86 and derive (6.34). Hint: see references to [Ci] above. Use (6.33), choose τ_{2j} as the increasing times. Assume first that N_{ε_s} contains exactly one point—say $\vec{z}$. Then the cycle times $\tau_{2j+2} - \tau_{2j}$ and the amount of times spent in D over these cycles form i.i.d. sequences. Multiply and divide by j and use the law of large numbers for each. Now recall the number of points in N_{ε_s} is finite and do the calculation separately for each starting point in N_{ε_s}. Look at the time spent in D during cycles (τ_{2j}, τ_{2j+2}) that start at some fixed $\vec{z}$. The fraction of such cycles converges to $\mu_n(\vec{z})$, and the amounts of time spent in D over such cycles are i.i.d. random variables. ♠

Let $\tau_\varepsilon(n) \overset{\triangle}{=} \inf\{t : |\vec{z}_n(t) - \vec{q}| < \varepsilon\}$. Assume

Assumption 6.88. For each N and ε, and all n,

$$\sup_{|\vec{z}-\vec{q}|\leq N} \mathbb{E}_{\vec{z}}\tau_\varepsilon(n) \leq C_{\varepsilon,N} < \infty.$$

Theorem 6.89. *Assume that $\lambda_i(\vec{x})$ are uniformly Lipschitz continuous in some neighborhood of $\vec{q}$. Let $\vec{q}$ be stable in the sense of Lyapunov so that (6.6) holds and let Assumptions 6.85–6.88 hold. Then for any $\varepsilon > 0$,*

$$\lim_{n\to\infty} \int_{N_\varepsilon(\vec{q})} d\pi_n = 1. \tag{6.35}$$

Let $D \subset \mathbb{R}^d$ be a smooth, bounded open set, let $S_e = \{\vec{r} : \vec{r}(0) \in \bar{D}\}$ and define S through Equation (6.30). Define I^* as in Assumption 6.50 and let Assumptions 6.51–6.52 hold as well, for some K, where $D \subset N_K(\vec{q})$. Then

$$\lim_{n\to\infty} \frac{1}{n} \log \pi_n(D) = - \inf_{\vec{x}\in D} V(\vec{x}), \tag{6.36}$$

i.e., for any $\delta > 0$ *and* n *large enough,*

$$- \inf_{\vec{x} \in D} V(\vec{x}) - \delta \leq \frac{1}{n} \log \pi_n(D) \leq - \inf_{\vec{x} \in D} V(\vec{x}) + \delta.$$

The first claim is that, as $n \to \infty$, the steady-state distribution approaches a δ-measure at the point $\vec{q}$. This is very reasonable: as $n \to \infty$, it becomes increasingly difficult for $\vec{z}_n(t)$ to escape from neighborhoods of $\vec{z}_\infty(t)$ which, in turn, converges to $\vec{q}$ as t becomes large. The second claim is that V, the minimal I-function, really measures how difficult it is to get to a set D, or how often $\vec{z}_n(t)$ will be there. The first claim is proved using only ergodicity and Kurtz's Theorem. The second one uses the Freidlin-Wentzell construction, with a bit more arguing.

Proof. Fix some ε and let $D = \{\vec{x} : |\vec{x} - \vec{q}| > \varepsilon\}$. To apply the representation (6.34), we estimate the two expectations in (6.34). For each $\vec{z} \in N_{\varepsilon_s}(\vec{q})$, the argument of Exercise 6.34 or Exercise 6.35 implies that $\mathbb{E}_{\vec{z}} \tau_2 \to \infty$ as $n \to \infty$. On the other hand, $\mathbb{E}_{\vec{z}}(\tau_2 - \tau_1) \leq C_{\varepsilon,N}$ (positive recurrence), and so by (6.34), $\pi_n(D) \to 0$ and (6.35) is established.

To establish the second claim, we use Lemma 6.36 or more precisely its extensions Lemma 6.56 and Exercises 6.57–6.58. We define the set S to be the set of paths starting at $\vec{q}$ and ending in D. Then for any δ, for ε small and n large

$$\mathbb{P}_{\vec{z}}\left(\int_0^{\tau_2} \mathbf{1}\left[\vec{z}_n(s) \in D\right] ds > 0 \right) = \mathbb{P}_{\vec{z}}(\tau_n(D) < \tau_2) \leq e^{-n(I^* - \delta)}$$

for each $\vec{z} \in N_{\varepsilon_s}$, by (6.18). But

$$\mathbb{E}_{\vec{z}}\left(\int_0^{\tau_2} \mathbf{1}\left[\vec{z}_n(s) \in D\right] ds \right) \leq \mathbb{P}_{\vec{z}}(\tau_n(D) < \tau_2) \cdot \max_{\vec{x} \in D} \mathbb{E}_{\vec{x}} \tau_0$$

which provides the desired upper bound, since the second term is bounded by Assumption 6.88. Now $\vec{z}_n$ is a jump process whose rate inside D is bounded above by

$$n \cdot \bar{\lambda} \stackrel{\triangle}{=} n \cdot \max_i \sup_{\vec{x} \in \bar{D}} \lambda_i(\vec{x}).$$

Therefore, conditioned on entering D, the expected time to stay inside is bounded below by $(n\bar{\lambda})^{-1}$, so that a similar approximation using (6.18) gives

$$\mathbb{E}_{\vec{z}}\left(\int_0^{\tau_2} \mathbf{1}\left[\vec{z}_n(s) \in D\right] ds \right) \geq e^{-n(I^* + \delta)} \cdot \frac{1}{n\bar{\lambda}}$$

and the proof is concluded.   ∎

This result holds for more general processes, although we have to be careful in stating a theorem since recurrence is no longer such a simple notion. For a more general theory, consult, e.g., [FW]. It is easy to relax the assumption that D is bounded: we only need that the rates are bounded below at entrance points to D, so that once inside D, the time spent there is bounded below. We will also need to relax the conditions on the rates, especially in the one-dimensional case.

Exercise 6.90. With the terminology of Exercise 6.35 assume Lyapunov stability (6.6) and let Assumption 6.2 hold *for each* G_i. Then under Assumptions 6.85–6.88 the conclusion (6.35) holds. ♠

Exercise 6.91. Consider the one-dimensional case, and let Assumptions 6.85–6.88 hold. Let z_n be a process with non-negative values. Assume Lyapunov stability (6.6) with either $q = 0$ or $q > 0$. Let Assumption 6.2 hold *for each* G_i, but allow the rate of jump to the left to vanish at zero (either decrease or jump down at 0). If (6.23) holds than (6.36) holds. Hint: see Exercise 6.40 and the remark following it, where a sufficient condition for (6.23) is given. ♠

Exercise 6.91 covers most of our applications.

Our final result in this section is an extension of Theorem 6.15 to the invariant measure problem when the jump rates $\lambda_i(\vec{x})$ are bounded but are not necessarily smooth (such as when there are boundaries). It is a natural result, showing that steady state is achieved essentially by upcrossings. That is, if we know that $\vec{z}_n(t)$ is in a set D in steady state, then we are pretty sure that the way $\vec{z}_n(t)$ got to D was by following $\vec{r}^*$, whenever there is a unique $\vec{r}^*$ that goes from $\vec{q}$ to D.

Since we are not going to assume smoothness of the rates $\lambda_i(\vec{x})$ or smoothness of the domain of the process, we need some new assumptions. We first assume that the domain of $\vec{z}_n(t)$ is defined by a finite number of linear inequalities on $\mathbb{R}^d$; it can be a half space, or an intersection of any finite number of half spaces. We do not assume that it is bounded. For statements of the assumptions we use, look at the text above Theorem 6.77.

Theorem 6.92. *Suppose that Assumptions 6.69–6.74, 6.85 and 6.86 are satisfied. Let $D \subset \mathbb{R}^d$ be a smooth, bounded open set, let $S_e = \{\vec{r} : \vec{r}(0) \in \bar{D}\}$, and define S through Equation (6.30). Define I^* as in Assumption 6.50 and let Assumptions 6.51–6.52 hold as well, for some K, where $D \subset N_K(\vec{q})$. Suppose that there is a unique $\vec{r}^*$ such that $I_0^T(\vec{r}^*) = I^*$. Suppose further that $\vec{r}^*$ intersects D at exactly one point even when extended beyond time zero by the path $\vec{z}_\infty$ [this is equivalent to the assumption that $V(\vec{x}) = I^*$ for $\vec{x} \in D$ for a unique $\vec{x} = \vec{r}^*(0)$]. Then for all $T < \infty$ and any $\varepsilon > 0$,*

$$\lim_{n \to \infty} \mathbb{P}\left(\sup_{0 \leq s \leq T} |\vec{z}_n(\tau - s) - \vec{r}^*(-s)| < \varepsilon \;\middle|\; \vec{z}_n(\tau) \in D \right) = 1.$$

Exercise 6.93. Prove Theorem 6.92. Hint: see Theorem 6.15. We have assumed enough to make the reasoning simple. ♠

Chapter 7

Applications and Extensions

In this chapter we derive a few consequences and extensions of the theory developed in Chapters 1, 5, and 6. The first two sections contain estimates similar to laws of large numbers, applied to some simple jump Markov processes. In §7.1 we give rough exponential estimates for the amount of time a finite Markov process spends in each state, and how often each transition occurs. In §7.2 we obtain similar bounds for the time spent at zero by certain processes on an infinite state space (the reals or the positive integers). Section §7.4 has something completely different. There we show that the optimal path $\vec{r}\,^*$ arising in a variational problem corresponds, in some sense, to a local change of measure: when $\vec{z}_n(t)$ is following $\vec{r}\,^*(t)$, its jumps in direction $\vec{e}_i$ occur at about rate $n\lambda_i e^{\langle \vec{\theta}(t), \vec{e}_i \rangle}$, where $\vec{\theta}(t)$ is the maximizing $\vec{\theta}$ in the definition of $\ell(\vec{r}\,^*(t), \vec{r}\,^{*\prime}(t))$. This happens only in a certain average sense, but the sense is strong enough for many applications. There are usually many ways that the process *could* jump so as to follow a particular path; the result is that the way it actually jumps is nearly deterministic.

The results of §7.1 and §7.2 are pretty simple—we include them mostly because they are needed for the theory of Chapter 8, not because of any novelty. (We hope that you find our techniques to be appealing, too.) The results of these two sections are extended to a large deviations regime in Appendix D. Section §7.4 contains a new approach to a simplified version of an advanced problem: when a rare event occurs, how do the jumps of the process behave? This is very close to a "Level III" result, which we don't even mention in this book.

7.1. Empirical Distributions of Finite Markov Processes

In this section we study the limits and rates of convergence associated with empirical distributions of finite-state Markov processes. The limits we consider are on the order of a law of large numbers, or of Kurtz's Theorem; they are not about rare events. We do not attempt to derive sharp bounds. Rather, we derive bounds on the rate of convergence that are uniform over a class of Markov processes.

We start with notation. Let $x(t)$ be a finite state Markov process with states $\{1, \ldots, D\}$. Fix the given set of allowed transitions $m \to j$ by defining an *incidence matrix*, that is a $D \times D$ matrix $\mathcal{I}$ whose entries are zero or one; $\mathcal{I}_{mj} = 1$ means that transition from state m to state j is allowed. Let λ_{mj} denote the rate of transition of $x(t)$ from m to j, and let $\vec{\lambda}$ denote the matrix of all rates of allowed transitions (set $\lambda_{mj} = 0$ if $\mathcal{I}_{mj} = 0$). Note that, contrary to the conventions in the theory of continuous time Markov processes and contrary to our convention concerning the diagonal terms in the definition (4.5) of the generator, here $\lambda_{mm} \geq 0$ with possibly a strict inequality. This corresponds to a jump from m to m. In Exercise 7.8 the results of this section will be extended further, to allow several different jumps between a pair of states. The reason for these quirks will become clear in §8.4. Throughout this section we impose the following assumption.

Assumption 7.1. The *transition structure* $\mathcal{I}$ is ergodic, that is, the finite state Markov process with transition rates $\{\lambda_{mj} = \mathcal{I}_{mj}\}$ is ergodic. (That is, we assume that between any two states m and $j \in \{1, \ldots, D\}$ there is a sequence of allowed transitions leading from m to j.)

We let $\vec{\pi}^\lambda = (\pi_1^\lambda, \ldots, \pi_D^\lambda)$ denote the unique stationary distribution of $x(t)$—it is unique by the ergodicity assumption (this is a simple consequence of Lemma 7.4 below). For each T let $\pi_m(T)$ denote the proportion of time in $[0, T]$ that $x(t)$ spends in state m; that is,

$$\pi_m(T) \overset{\triangle}{=} \frac{1}{T} \int_0^T \mathbf{1}\,[x(t) = m]\, dt. \tag{7.1}$$

Let $R_c(\delta)$ be a collection of allowed rates $\vec{\lambda}$ that are bounded above and below:

$$R_c(\delta) \overset{\triangle}{=} \left\{ \vec{\lambda} : \mathcal{I}_{mj} = 1 \text{ implies } \delta \leq \lambda_{mj} \leq \frac{1}{\delta}, \text{ and } \mathcal{I}_{mj} = 0 \text{ implies } \lambda_{mj} = 0 \right\}. \tag{7.2}$$

We begin with a simple result showing that, under Assumption 7.1, $\vec{\pi}(T)$ approaches $\vec{\pi}^\lambda$ exponentially quickly, uniformly in the jump rates over $R_c(\delta)$. But first some technical stuff. Denote

$$\lambda_m \overset{\triangle}{=} \sum_{j=1}^D \lambda_{mj}. \tag{7.3}$$

Exercise 7.2. If $\vec{\lambda} \in R_c(\delta)$ and $\mathcal{I}_{mj} = 1$ then $\mathbb{P}_m\big(x(t) = j \text{ for some } t \in [0, 1]\big) \geq \delta^3/D\,e$. Hint: what's the probability that the first jump occurs before time one, and that it is to state j? ♠

Fix a state m and define $\tau_{-1} = 0$ and (for $i \geq 0$)

$$\tau_{2i} \overset{\triangle}{=} \inf \{t > \tau_{2i-1} : x(t) = m\}$$
$$\tau_{2i+1} \overset{\triangle}{=} \inf \{t > \tau_{2i} : \text{ a transition occurs at } t\}. \tag{7.4}$$

These times should remind you of the τ_i from the Freidlin-Wentzell theory. They are cycle times into and out of a state m. The only subtlety is that we might have $\tau_{2i+1} = \tau_{2i+2}$, if a transition occurs at that time from state m to itself. We also define the cycle lengths, the differences of the cycle times:

$$y_i = \tau_i - \tau_{i-1} \quad \text{for} \quad i \geq 0. \tag{7.5}$$

Thus y_{2i+1} is the time $x(t)$ spends at m until its next jump, while y_{2i} is the time it spends away from m (which equals zero in case that the transition in the definition of y_{2i-1} is from m to m). All of the y_{2i+1} are identically distributed, and all the y_{2i} are identically distributed, except perhaps y_0.

Exercise 7.3. Under Assumption 7.1, $\pi_m^\lambda = \mathbb{E}y_1/(\mathbb{E}y_1 + \mathbb{E}y_2)$. Hint: as in Exercise 6.87, use (6.33) and take the limit at points τ_{2n}.   ♠

Lemma 7.4. *Let $\mathcal{I}$ be ergodic. For each δ, ε, and $T > 0$ there exists a $C > 0$ and n_0 such that for all starting points $x \in \{1, \ldots, D\}$, all $n > n_0$, and all jump rates $\vec{\lambda} \in R_c(\delta)$,*

$$\frac{1}{n} \log \mathbb{P}_x \left(\left| \vec{\pi}(nT) - \vec{\pi}^\lambda \right| > \varepsilon \right) < -C.$$

The idea behind this lemma is simple. As in the Freidlin-Wentzell theory, cycle times have exponential tails. Chernoff's Theorem then says that over long periods of time there will be lots of cycles. Furthermore, the average amount of time in each type of cycle will be very close to its mean, with probability exponentially close to one. The estimates are uniform over $\lambda \in R_c(\delta)$ since Chernoff's Theorem can be made uniform there.

Proof. We consider each component $\pi_m(T)$ separately. We use the recurrence cycles defined in (7.5) above. Let us first show that the y_i have exponential tails. By definition, y_{2i+1} is an exponential random variable with parameter λ_m (recall (7.3)). Clearly y_{2i}, $i = 1, 2, \ldots$ are i.i.d. random variables. Now since there are only D states, it is possible to reach m from any state $j \neq m$ in at most D steps. On the other hand, since $\vec{\lambda} \in R_c(\delta)$, the probability that each transition along this path actually occurs (given that the preceding one did), and does so within a time interval of length one is, by Exercise 7.2, bounded below, say by $p > 0$ (Exercise 7.2 says that we can take $p = \delta^3/De$). But then by the Markov property, for integer $M \geq 1$,

$$\min_{1 \leq j \leq D} \mathbb{P}_j \left(x(t) = m \quad \text{for some} \quad t \in [0, D] \right) \geq p_{\vec{1}}^D$$

since we can always make a sequence of transitions from j to m that has no more than D steps. Therefore,

$$\max_{1 \leq j \leq D} \mathbb{P}_j \left(x(t) \neq m \quad \text{for} \quad t \in [0, MD] \right) \leq \left(1 - p^D \right)^M$$

and so y_{2i} has an exponential tail; this argument applies also to y_0.

Exercise 1.8 now implies that there is a positive function $h(\alpha)$ that vanishes only at $\alpha = 0$ such that for any $\alpha > 0$,

$$\mathbb{P}\left(\left| \frac{y_1 + y_3 + \cdots + y_{2n-1}}{n} - \mathbb{E}y_1 \right| > \alpha \right) \leq e^{-nh(\alpha)}$$

$$\mathbb{P}\left(\left| \frac{y_0 + y_2 + \cdots + y_{2n}}{n} - \mathbb{E}y_2 \right| > \alpha \right) \leq e^{-nh(\alpha)}$$

for all n and all processes satisfying the hypotheses. Let $y = y_1 + y_2$ and let $M = M(n) = \lfloor nT/\mathbb{E}y \rfloor$. Then, for any state m and any positive η,

$$
\mathbb{P}_x \left(\left| \pi_m(nT) - \pi_m^\lambda \right| > \varepsilon \right)
$$

$$
\leq \mathbb{P}_x \left(\left| \pi_m(nT) - \pi_m^\lambda \right| > \varepsilon, \; |y_0 + \ldots + y_{2M} - M\mathbb{E}y| \leq M\eta \right)
$$

$$
+ \mathbb{P}_x \left(|y_0 + \ldots + y_{2M} - M\mathbb{E}y| > M\eta \right).
$$

(7.6)

The second term is bounded above by $e^{-Mh(\eta)}$. Now for any ε', if n (or M) is large and η is chosen small, then $|y_0 + \ldots + y_{2M} - M\mathbb{E}y| \leq M\eta$ implies

$$
\left| \frac{1}{nT} \int_0^{nT} \mathbf{1}[x(t) = m] \, dt - \frac{1}{nT} \sum_{i=1}^{M} y_{2i-1} \right| < \varepsilon'
$$

(approximate both by taking the upper limit of the integral to be $M\mathbb{E}y$). Thus we obtain, using Exercise 7.3, that

$$
\mathbb{P}_x \left(\left| \pi_m(nT) - \pi_m^\lambda \right| > \varepsilon \right) \leq \mathbb{P}_x \left(\frac{1}{\mathbb{E}y} \left| \frac{\mathbb{E}y}{nT} \sum_{i=1}^{M} y_{2i-1} - \mathbb{E}y \, \pi_m^\lambda \right| > \frac{\varepsilon}{2} \right) + e^{-Mh(\eta)}
$$

$$
\leq \mathbb{P}_x \left(\left| \frac{1}{M} \sum_{i=1}^{M} y_{2i-1} - \mathbb{E}y_1 \right| > \mathbb{E}y \, \frac{\varepsilon}{2} \right) + e^{-Mh(\eta)}
$$

for all n large. The result now follows from Exercise 1.8, applied to $\{y_1, y_3, \ldots\}$. ∎

Definition 7.5. *Let $n_m(T)$ denote the number of jumps from state m in $[0, T]$ (including jumps to m). Let $n_{mj}(T)$ denote the number of (direct) jumps from state m to state j in $[0, T]$. In particular,*

$$
n_m(T) = \sum_j n_{mj}(T).
$$

The next lemma shows that $n_m(nT)/(nT)$ approaches its mean $1/\mathbb{E}y$ exponentially quickly, and furthermore that $n_{mj}(nT)/(nT)$ approaches its mean $\pi_m^\lambda \lambda_{mj}$ exponentially quickly. It is a simple consequence of Chernoff's Theorem.

Lemma 7.6. *Let $\mathcal{I}$ be ergodic. For each $\delta > 0$, ε and $T > 0$ there exists a $C > 0$ and n_0 such that for all starting points $x \in \{1, \ldots, D\}$, and all jump rates $\vec{\lambda} \in R_c(\delta)$,*

$$
\frac{1}{n} \log \mathbb{P}_x \left(\left| \frac{n_m(nT)}{nT} - \frac{1}{\mathbb{E}y} \right| > \varepsilon \right) < -C \quad \text{for all} \quad n \geq n_0,
$$

$$
\frac{1}{n} \log \mathbb{P}_x \left(\left| \frac{n_{mj}(nT)}{nT} - \pi_m^\lambda \lambda_{mj} \right| > \varepsilon \right) < -C \quad \text{for all} \quad n \geq n_0.
$$

Proof. Since the number of jumps from m differs from the number of cycles by at most one, we shall prove this result with $n_m(T)$ as the number of cycles. Define

$$M_+(n) \triangleq \left\lfloor nT\left(\frac{1}{\mathbb{E}y} + \varepsilon\right)\right\rfloor \qquad M_-(n) \triangleq \left\lceil nT\left(\frac{1}{\mathbb{E}y} - \varepsilon\right)\right\rceil.$$

Then by definition

$$\mathbb{P}_{\bar{x}}\left(\left|\frac{n_{mj}(nT)}{nT} - \frac{1}{\mathbb{E}y}\right| > \varepsilon\right)$$

$$\leq \mathbb{P}_{\bar{x}}\Big(n_m(nT) > M_+(n) \quad \text{or} \quad n_m(nT) < M_-(n)\Big)$$

$$\leq \mathbb{P}_{\bar{x}}\left(\sum_{i=1}^{M_+(n)}(y_{2i-1}+y_{2i}) < nT\right) + \mathbb{P}_{\bar{x}}\left(\sum_{i=1}^{M_-(n)}(y_{2i-1}+y_{2i}) > nT\right).$$

Consider the probability of the first event in the last term. For n large enough, clearly $nT \leq M_+(n)\mathbb{E}y - M_+(n)\varepsilon/2$. By Lemma 7.4 the random variables $y_{2i-1} + y_{2i}$ have an exponential tail, so that by Exercise 1.8

$$\mathbb{P}_{\bar{x}}\left(\frac{1}{M_+(n)}\sum_{i=1}^{M_+(n)}(y_{2i-1}+y_{2i}) < \mathbb{E}y - \varepsilon/2\right) \leq 2e^{-M_+(n)h(\varepsilon/2)}$$

for some function h, which depends only on δ. Exactly the the same calculation applies to the second term, using M_-, so the first result is established.

To get the second result, by the triangle inequality,

$$\left|\frac{n_{mj}(nT)}{nT} - \pi_m^\lambda \lambda_{mj}\right| \leq \frac{n_{mj}(nT)}{n_m(T)}\left|\frac{n_m(T)}{nT} - \pi_m^\lambda \lambda_m\right|$$

$$+ \pi_m^\lambda \lambda_m\left|\frac{n_{mj}(nT)}{n_m(T)} - \frac{\lambda_{mj}}{\lambda_m}\right|.$$

But by definition, $n_{mj}(nT) \leq n_m(T)$, and so

$$\mathbb{P}_{\bar{x}}\left(\left|\frac{n_{mj}(nT)}{nT} - \pi_m^\lambda \lambda_{mj}\right| > \varepsilon\right) \leq \mathbb{P}_{\bar{x}}\left(\left|\frac{n_m(T)}{nT} - \pi_m^\lambda \lambda_m\right| > \frac{\varepsilon}{2}\right)$$

$$+ \mathbb{P}_{\bar{x}}\left(\left|\frac{n_{mj}(nT)}{n_m(T)} - \frac{\lambda_{mj}}{\lambda_m}\right| > \frac{\varepsilon}{2\pi_m^\lambda \lambda_m} \text{ and } \left|\frac{n_m(T)}{nT} - \pi_m^\lambda \lambda_m\right| \leq \frac{\varepsilon}{2}\right).$$

By Exercise 7.7 below, $\pi_m^\lambda \lambda_m = 1/\mathbb{E}y$, so that by the first part of the lemma the first term decreases exponentially fast, as desired. For the second term, recall from Propositions 4.3 and 4.5 that we can construct n_{mj} as follows. Each time the process $x(t)$ enters state m, the duration it stays there is an independent exponential random variable with parameter λ_m. Let $\{b_\ell, \ \ell = 1, 2, \ldots\}$ be a sequence of coin throws, i.e., i.i.d. Bernoulli random variables with $\mathbb{P}(b_\ell = 1) = p = \lambda_{mj}/\lambda_m$.

Then, at the ℓth time x visits state m, it will next go to j if and only if $b_\ell = 1$. Let $M_-(n) = \lfloor nT (\pi_m^\lambda \lambda_m - \varepsilon) \rfloor$ and $M_+(n) = \lceil nT (\pi_m^\lambda \lambda_m + \varepsilon) \rceil$. Then

$$\mathbb{P}_{\bar{x}} \left(\left| \frac{n_{mj}(nT)}{n_m(T)} - \frac{\lambda_{mj}}{\lambda_m} \right| > \frac{\varepsilon}{2\pi_m^\lambda \lambda_m} \,\Big|\, \left| \frac{n_m(T)}{nT} - \pi_m^\lambda \lambda_m \right| \le \frac{\varepsilon}{2} \right)$$

$$\le \sum_{\ell = M_-(n)}^{M_+(n)} \mathbb{P} \left(\left| \frac{b_1 + \ldots + b_\ell}{\ell} - p \right| > \frac{\varepsilon}{2\pi_m^\lambda \lambda_m} \right)$$

$$\le 2(nT\varepsilon + 1) \exp \left(-M_-(n) h_b \left(\frac{\varepsilon}{2\pi_m^\lambda \lambda_m} \right) \right)$$

by Exercise 1.8 applied to the Bernoulli random variables.   ∎

Exercise 7.7. For a finite, ergodic Markov process, $\lambda_m \pi_m^\lambda = 1/\mathbb{E} y$. Hints: see Exercise 7.3. Note that $n_m(t)$, restricted to the intervals when the process is at m is, by definition, a Poisson process with rate λ_m. Show that

$$M(t) \overset{\triangle}{=} n_m(t) - \int_0^t \lambda_m \mathbf{1} \, [x(s) = m] \, ds$$

is a martingale (extend Theorem 4.13 to the unbounded function $f(x) \equiv x$). Its mean is clearly zero and, as shown in the first part of Lemma 7.6, $M(nT)/nT$ converges in probability to a constant. Show that the expectation of the limit is zero due to uniform integrability.   ♠

Imagine now the Markov process $x(t)$ as moving on a graph, whose nodes are the states. Index the arcs leaving state m as e_i, $i = 1, \ldots, k(m)$. There may be several arcs connecting the pair m, j and there may be none; there may be arcs connecting a state m to itself. Let $\lambda_i(m)$ be the rate at which the process moves (from m) along the arc e_i. The motivation for studying this process (and for the notation) is made clear in §8.4. Denote the state that is reached from m by following arc e_i as $m'(m; i)$, and set

$$\lambda_{mj} = \sum_{i=1}^{k(m)} \lambda_i(m) \mathbf{1} \left[j = m'(m; i) \right] \quad \text{and} \quad \mathcal{I}_{mj} = \mathbf{1} \left[j = m'(m; i) \text{ for some } i \right].$$

Let $n_i(m; T)$ denote the number of jumps along the arc e_i by time T.

Exercise 7.8. Let $\mathcal{I}$ be ergodic. Then the conclusions of Lemmas 7.4 and 7.6 hold for the process described above. If moreover $\lambda_i(m) \ge \delta$ then

$$\frac{1}{n} \log \mathbb{P}_x \left(\left| \frac{n_i(m; nT)}{nT} - \pi_m^\lambda \lambda_i(m) \right| > \varepsilon \right) < -C \quad \text{for all} \quad n \ge n_0.$$

Note that the estimate is uniform over $R_c(\delta)$, which restricts λ_{mj}, but not $\lambda_i(m)$. Hint: use Theorem 4.5, as in the proof of Lemma 7.6.   ♠

7.2. Simple Jump Processes

Let us see how our results from previous chapters enable us to analyze some transition times and some portion of the empirical distributions of certain simple Markov processes. Specifically we examine the time a process on $\mathbb{R}$ or on $\mathcal{Z}$ takes to cross from a positive state to zero or to a negative state. Using Kurtz's Theorem or using martingale techniques, we obtain simple uniform estimates on the transition time. For processes on $\mathcal{Z}^+$ we also obtain estimates on the observed time spent at state zero over long periods; this is similar to the results of the previous section, but we don't look at the whole empirical distribution, just the time at zero. We are interested in these questions for use in Chapter 8 in connection with large deviations theory for processes with a boundary. But some of the results below will also be used in the applications chapters.

Our first model is a one-dimensional jump Markov process with jump directions and constant jump rates $\{(e_i, \lambda_i), \ i = 1, \ldots, k\}$. The generator of the process is thus

$$(Lf)(x) = \sum_{i=1}^{k} \lambda_i (f(x + e_i) - f(x)), \tag{7.7}$$

for any real-valued function f. The drift of the process is defined as

$$\beta \overset{\Delta}{=} \sum_{i=1}^{k} \lambda_i e_i. \tag{7.8}$$

From Kurtz's Theorem 5.3 and (5.7) we know that β is the average drift of the process. We are interested in the behavior of the process until the time

$$\tau \overset{\Delta}{=} \inf\{t > 0 : x(t) \leq 0\},$$

that is, until the first time it goes below zero. Let us first get a rough estimate on τ.

Exercise 7.9. Assume $\lambda_i \leq \overline{\lambda} < \infty$ and $\beta < -\delta < 0$. Then there are functions f_1 and f_2, depending only on δ and $\overline{\lambda}$, so that for all $t \geq 1$, all $x \leq t\delta/2$ and all $\alpha > 0$,

$$\mathbb{P}_x(\tau \geq t) \leq e^{-tf_1(\delta)} \quad \text{and} \quad \mathbb{P}_x(x(t) \geq \alpha t) \leq e^{-tf_2(\alpha)},$$

where $f_1(\delta) > 0$, and $f_2(\alpha) > 0$. Hint: this follows from Kurtz's Theorem 5.3, except for the uniformity in $\overline{\lambda}$. Scale the process to have jump rate t and jump size $1/t$. Since we are only looking at the probability that the sample path drifts higher than $z_{\infty} + \delta/2$, we do not get a factor of 2 in front of the estimate. For a more elementary proof, use Exercise 1.8. Note that the first term is bounded by $\mathbb{P}(x(t) \geq 0)$, and represent $x(t)$ as a sum of t i.i.d. random variables $x(t) = x_1 + \ldots + x_t$. ♠

Exercise 7.10. Let $h(x) \equiv x$. Show that

$$M(t) \overset{\Delta}{=} h(x(t)) - \int_0^t Lh(x(s)) \, ds \tag{7.9}$$

is a martingale. Hint: this would hold by Theorems 4.12 and 4.13, if h were bounded. Approximate by a process that freezes if it reaches N, and show using Theorem 5.1(i) that the approximation error decreases exponentially in N. (This is a standard result in martingale theory, and a standard technique.) ♠

The next lemma uses a martingale technique to estimate the mean time to hit zero. The general approach goes by the name Dynkin's formula and is given in §B.4—see especially (B.33)–(B.37).

Lemma 7.11. *Assume $\beta < 0$. Let $\bar{e} \triangleq \max_i \{-e_i\}$ denote the largest downward jump size of the process. Then*

$$\frac{-x}{\beta} \le \mathbb{E}_x \tau \le \frac{-x - \bar{e}}{\beta}.$$

Proof. By Exercise 7.10, $M(t)$ is a martingale. By Exercise A.130 (or Theorem A.129) applied to the martingale M of Exercise 7.10 we have

$$\mathbb{E}_x M(\tau) = \mathbb{E}_x \left(h(x(\tau)) - \int_0^\tau L h(x(s)) \, ds \right) = \mathbb{E}_x M(0). \qquad (7.10)$$

(We are justified in using τ as a stopping time since for any $\alpha > 0$

$$\mathbb{E}_x \left(x(t) \mathbf{1} [\tau \ge t] \right) \le \alpha t \, \mathbb{P}_x (\tau \ge t) + \mathbb{E}_x \left(x(t) \mathbf{1} [\vec{x}(t) \ge \alpha t] \right),$$

so that Exercise 7.9 implies that $\mathbb{E}_x (x(t) \mathbf{1} [\tau \ge t]) \to 0$ as $t \to \infty$.) But by definition, $-\bar{e} \le h(x(\tau)) \le 0$, $L h(x) \equiv \beta$ and $M(0) = x$. Put these into (7.10). ∎

The reason that we had to justify the martingale argument with Exercise 7.10 is that h is an unbounded function over an unbounded domain. Here is another, simpler example. Let

$$\tau_M \triangleq \inf\{t > 0 : x(t) = M \text{ or } x(t) = 0\}.$$

Exercise 7.12. Assume $\beta < 0$. Fix $x > 0$ and show that

$$p_x(M) \triangleq \mathbb{P}_x (x (\tau_M) = M) = e^{-M\theta^* + o(M)},$$

where θ^* is the unique strictly positive solution of

$$\sum_{i=1}^k \lambda_i \left(e^{\theta^* e_i} - 1 \right) = 0.$$

Extend to $x = o(M)$. Hint: existence and uniqueness of θ^* follow from convexity; compute derivatives, in particular at $\theta = 0$. Assume first that $e_i = \pm 1$ or 0, for all i. For $h(x) = e^{\theta x}$, show that $M(t \wedge \tau_M)$ is a martingale by considering a process that "freezes" at zero and M, and follow the argument of Lemma 7.11, to obtain an exact expression for $p_x(M)$. Now repeat for general e_i. ♠

Although the method used in Exercise 7.12 is important, it is more in keeping with our theme to use the Freidlin-Wentzell ideas to show this result.

Exercise 7.13. Repeat Exercise 7.12 using large deviations methods. Hint: consider $z_M(t)$ and use Corollary 6.45 to obtain a variational problem. Consider the cheapest path from zero to one over all times T. By Lemmas 5.13 and 5.16 $r^*(t) = ct$ for some c and $I^* = T\ell(1/T)$. Optimize over T and use (C.6) to get $I^* = \theta^*$.   ♠

Next we obtain some uniform estimates for τ.

Lemma 7.14. *Assume $\beta < -\delta < 0$ and $\lambda_i \leq \bar{\lambda} < \infty$, and denote $T_\varepsilon \overset{\triangle}{=} -(1+\varepsilon)/\beta$. There exists a function g depending only on δ and $\bar{\lambda}$ so that for each $\varepsilon > 0$, $g(\varepsilon) > 0$ and*

$$\mathbb{P}_n \left(\tau > (1+\varepsilon)\mathbb{E}_n\tau \right) < e^{-ng(\varepsilon)}, \tag{7.11}$$

$$\mathbb{P}_n \left(\tau < (1-\varepsilon)\mathbb{E}_n\tau \right) < e^{-ng(\varepsilon)}, \tag{7.12}$$

$$\mathbb{P}_n \left(\tau \geq \alpha \cdot nT_\varepsilon \right) \leq e^{-n(\alpha-1)g(\varepsilon)} \quad \text{for all} \quad \alpha > 0. \tag{7.13}$$

Proof. The proof of (7.11)–(7.12) is done in Exercise 7.15 below. To get (7.13) we simply have to rescale the previous estimate (7.11). Obviously, for any $\alpha \geq 1$,

$$\mathbb{P}_n \left(\tau \geq \alpha \cdot nT_\varepsilon \right) \leq \mathbb{P}_{\alpha n} \left(\tau \geq \alpha \cdot nT_\varepsilon \right)$$

and (7.13) follows.   ■

Exercise 7.15. Establish (7.11)–(7.12). Hint: use

$$\mathbb{P}_n \left(\tau > nT_\varepsilon \right) \leq \mathbb{P}_n \left(x(nT_\varepsilon) - (n + bnT_\varepsilon) > n\varepsilon \right)$$

and follow the same idea as in Exercise 7.9.   ♠

We now modify $x(t)$ to create an ergodic process on the positive integers. We impose the following assumption.

Assumption 7.16. The process $x(t)$ takes positive integer values. Its generator is given by Equation (7.7) as long as $x > 0$. The e_i are integers, bounded below by (-1), and $e_i = -1$ for some i. When $x(t) = 0$ the jump directions and jump rates are $\{(e_i(0), \lambda_i(0)), i = 1, \ldots, k(0)\}$, where the $e_i(0)$ are positive integers.

The process described in Assumption 7.16 is a "skip-free to the left lattice process." (It is called skip-free to the left since it skips no state when moving to the left.) Note that for such a process, using the notation of Lemma 7.11, $\mathbb{E}_x\tau = -x/\beta$, since $h(x(\tau)) = 0$.

The drift β in the interior is still defined through (7.7), using the data in the interior $x > 0$. Let $\pi_0(T)$ denote the fraction of time $x(t) = 0$ in the interval $[0, T)$ [$\pi_0(T)$ is the empirical distribution at zero defined in (7.1)], and let π_0^λ denote the unique probability that $x(t) = 0$ in steady state.

Exercise 7.17. Under Assumption 7.16, if $\beta < 0$ then the process $x(t)$ is ergodic, and its (unique) invariant distribution $\vec{\pi}^\lambda$ satisfies

$$\pi_0 \sum_{i=1}^{k(0)} \lambda_i(0) e_i(0) + (1 - \pi_0)\beta = 0. \tag{7.14}$$

Hint: use Lemma 7.11 for ergodicity. To obtain (7.14) take the expectation of (7.9) with $x(0)$ [and hence $x(t)$] distributed according to $\vec{\pi}^\lambda$. ♠

Lemma 7.18. *Assume 7.16, fix $\delta > 0$, and assume that the drift (7.7) satisfies $\beta < -\delta$, and that λ_i and $\lambda_i(0)$ are all bounded above by δ^{-1}. If $e_i(0) \geq 1$ for some i, then the process $x(t)$ is ergodic, so that π_0^λ exists and is unique. Moreover, for every ε and $T > 0$ there exists a $C > 0$ and n_0, depending only on δ such that uniformly over x in bounded sets and for any $n > n_0$,*

$$\frac{1}{n} \log \mathbb{P}_{\vec{x}} \left(|\pi_0(nT) - \pi_0^\lambda| > \varepsilon \right) < -C.$$

Exercise 7.19. Prove Lemma 7.18. All conclusions except ergodicity continue to hold if $e_i(0) = 0$ for all i: in that case, all states except zero are transient. You can also let x grow at any prescribed speed that is $o(n)$. Hint: ergodicity follows from the definitions and Lemma 7.11. The estimates follow from (7.13) of Lemma 7.14 with the same proof as in Lemma 7.4. ♠

We also need to estimate the number of jumps of each type in any time interval $[0, nT]$. For each i let $n_i(0, T)$ be the number of jumps $x(t)$ makes in direction $\vec{e}_i(0)$ in time $[0, T]$, and let $n_1(1, T)$ be the number of jumps $x(t)$ makes in direction $\vec{e}_i(1)$ in time $[0, T]$.

Lemma 7.20. *Under the same assumptions as Lemma 7.18, there is a C such that for each i we have*

$$\frac{1}{n} \log \mathbb{P}_{\vec{x}} \left(\left| \frac{n_i(0, nT)}{nT} - \pi_0^\lambda \lambda_i(0) \right| > \varepsilon \right) < -C$$

and for each i we also have

$$\frac{1}{n} \log \mathbb{P}_{\vec{x}} \left(\left| \frac{n_i(1, nT)}{nT} - \left(1 - \pi_0^\lambda\right) \lambda_i(1) \right| > \varepsilon \right) < -C.$$

Exercise 7.21. Prove Lemma 7.20. Hint: this goes exactly like Lemma 7.6. Use Lemmas 7.11 and 7.14 to estimate the lengths of the recurrence cycles. ♠

7.3. The Free M/M/1 Process

This section contains some calculations that do the same thing in three different ways. We look at a simple process $y(t)$ called the *free M/M/1 process*, and estimate the probability of a certain rare event in three different ways. The point is to build confidence in the large deviations estimates, to demonstrate our ability to obtain explicit expressions, and to obtain some estimates that will be used in many application chapters.

The free process is an $M/M/1$ queue without a boundary. That is, it is the difference of a Poisson (λ) and an independent Poisson (μ) process. The transition structure of $y(t)$ is

$$\begin{aligned} e_1 &= +1 & \lambda_1(x) &= \lambda \\ e_2 &= -1 & \lambda_2(x) &= \mu. \end{aligned} \tag{7.15}$$

It is obvious that this process satisfies the conditions of Kurtz's Theorem and of the large deviations theorems of Chapter 5. At any fixed t, the random variable $y(t)$ is the difference of Poisson (λt) and Poisson (μt) random variables. Then $z_n(t) \stackrel{\triangle}{=} n^{-1} y(nt)$ is the usual scaled process (check the generator!). Our question is: Given any $a > \mathbb{E}(z_n(1))$, what is the probability that the process $z_n(t)$ exceeds a at time one; that is, what is $\mathbb{P}_0(z_n(1) \geq a)$?

To begin, we have to know what we are deviating from. By (5.7) or (5.8),

$$z_\infty(t) = z_\infty(0) + (\lambda - \mu)t,$$

and by Kurtz's Theorem the process $z_n(t)$ converges to $z_\infty(t)$ as $n \to \infty$ in a rather strong sense. From (5.2) we have

$$\ell(x, y) = \ell(y) = \sup_\theta \left\{ \theta y - \lambda(e^\theta - 1) - \mu(e^{-\theta} - 1) \right\}. \tag{7.16}$$

Solving this for θ (see Exercise 7.24) we find

$$\ell(y) = y \log \left(\frac{y + \sqrt{y^2 + 4\lambda\mu}}{2\lambda} \right) + \lambda + \mu - \sqrt{y^2 + 4\lambda\mu}. \tag{7.17}$$

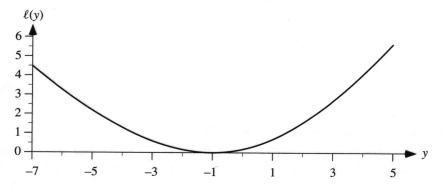

Figure 7.22. The ℓ-function for the free process with $\lambda = 1$, $\mu = 2$. Note that $\ell(y) = 0$ at $y = \lambda - \mu = -1$.

Let's calculate $\mathbb{P}(z_n(1) \geq a)$ in several different ways, for $a > \lambda - \mu = \mathbb{E}(z_n(1))$. This should support our confidence that the abstract derivations of Chapter 5 are correct, provide some intuition, and show some connections between methods.

1. Processes. Using Theorem 5.1 and Definitions (5.2)–(5.5) of Chapter 5, we expect

$$\mathbb{P}_0(z_n(1) \geq a) = e^{-nI^* + o(n)},$$

where

$$I^* = \inf_{r \in F} \int_0^1 \ell(r(t), r'(t)) \, dt \tag{7.18}$$

$$F = \{r : r(0) = 0, \ r(1) \geq a\} .$$

By Lemma 5.16, the cheapest path from 0 to b is a straight line, with cost $\ell(r, b) = \ell(b)$. Since by Exercise 7.24 below $\ell(b)$ is convex increasing for $b > \lambda - \mu$,

$$I^* = \ell(a) .$$

Exercise 7.23. Establish (7.18) by proving that F is a continuity set for $I_0^T(\cdot)$. Since F is the closure of an open set G (in $D^d[0, T] \cap \{r : r(0) = 0\}$), you simply need to establish that

$$\inf_{r \in G} \int_0^1 \ell(r(t), r'(t)) \, dt = \inf_{r \in F} \int_0^1 \ell(r(t), r'(t)) \, dt.$$

Show that this holds for any value of λ and μ.   ♠

2. Random variables. Let $w_1, w_2, \ldots$ be i.i.d. random variables, where w is the difference of two independent Poisson random variables

$$w \overset{\mathcal{L}}{=} \text{Pois}(\lambda) - \text{Pois}(\mu) .$$

Then $n z_n(1)$ is a sum of n such random variables. By Chernoff's Theorem, for every $\varepsilon > 0$ there is an n_0 such that if $n > n_0$ then

$$e^{-n(\ell(a)+\varepsilon)} \leq \mathbb{P}_0(z_n(1) \geq a) \leq e^{-n\ell(a)}.$$

By the definition of w and (1.4),

$$\ell(a) = \sup_\theta \{\theta a - \log E(e^{\theta w})\}$$

$$= \sup_\theta \{\theta a - \lambda(e^\theta - 1) - \mu(e^{-\theta} - 1)\}$$

so that again the value of ℓ is given by (7.17).

3. Direct (formal) approximations. Let X and Y be independent Poisson random variables with parameters $n\lambda$ and $n\mu$, respectively. Then

$$\mathbb{P}_0(z_n(1) \geq a) = \sum_{j=0}^\infty \mathbb{P}(X - j \geq na)\mathbb{P}(Y = j). \tag{7.19}$$

Now let $j = nr$, and apply Stirling's formula (A.1) to get

$$\mathbb{P}(Y = nr) = e^{-n\mu} \frac{(n\mu)^{rn}}{(rn)!}$$

$$\approx e^{-n\mu} (n\mu)^{rn} \left[\sqrt{2\pi rn} (rn)^{rn} e^{-rn} \right]^{-1}.$$

Using this approximation, and applying Chernoff's Theorem (or Stirling's formula, as in Example 1.13) to the first term in (7.19) we obtain from Example 1.13

$$\mathbb{P}_0(z_n(1) \geq a) \approx \sum_r \exp -n \left[(a + r) \left(\log \frac{a + r}{\lambda} - 1 \right) + \lambda \right] \tag{7.20}$$

$$\times \exp \left[-n\mu + rn \log n\mu - rn \log rn + rn \right],$$

where r takes the values i/n for $i \geq 0$. Laplace's method, described in (1.3) says that the value of the right-hand side of (7.20) is about equal to the maximal term. So, write the exponent in (7.20) as

$$-n \Bigg[(a + r) \log(a + r) - (a + r) \log \lambda - (a + r)$$

$$+ \lambda + \mu + r \log \left(\frac{r}{\mu} \right) - r \Bigg]. \tag{7.21}$$

Setting the derivative equal to zero, we obtain

$$\log(a + r) + 1 - \log \lambda - 1 + \log r - \log \mu + 1 - 1 = 0.$$

This gives

$$r(a + r) = \lambda \mu$$

or

$$r^2 + ar - \lambda \mu = 0. \tag{7.22}$$

Since $r > 0$ we obtain that the maximizing r is

$$r = \frac{-a + \sqrt{a^2 + 4\lambda\mu}}{2}.$$

To obtain an expression for r^{-1} divide (7.22) by r^2:

$$1 + \frac{a}{r} - \frac{\lambda\mu}{r^2} = 0$$

so that

$$\frac{1}{r} = \frac{a + \sqrt{a^2 + 4\lambda\mu}}{2\lambda\mu}.$$

Therefore we have

$$(a + r) \log \frac{a + r}{\lambda} = \frac{a + \sqrt{a^2 + 4\lambda\mu}}{2} \log \frac{a + \sqrt{a^2 + 4\lambda\mu}}{2\lambda}$$

$$r \log \frac{r}{\mu} = -r \log \frac{\mu}{r} = \frac{a - \sqrt{a^2 + 4\lambda\mu}}{2} \log \frac{a + \sqrt{a^2 + 4\lambda\mu}}{2\lambda};$$

so evaluating the exponent in (7.21),

$$(a+r)\log\frac{a+r}{\lambda} + r\log\frac{r}{\mu} - (a+r) + \lambda + \mu - r$$

$$= a\log\frac{a+\sqrt{a^2+4\lambda\mu}}{2\lambda} + \lambda + \mu - (2r+a),$$

which yields (7.17).

So, we obtain exactly the same results with any of these approaches, and the calculations are fairly simple. Things get more complicated when we ask questions about the process, not just its value at time one. This is why we had to develop the theory in Chapter 5.

We conclude this section with some calculations concerning the function ℓ. These will be of use in Chapter 11. Note that any calculation concerning the behavior of $\ell(y)$ of (7.16) immediately extends to the behavior of $\ell(x, y)$ as a function of y (with x held fixed). Thus, the calculations apply to any one-dimensional jump Markov process with jumps of size $+1$ and (-1). This variation will be utilized in several applications chapters.

Exercise 7.24. Recall the Definition (7.16) of $\ell(y)$. Show that the supremum is achieved at $\theta^*(y)$ where

$$e^{\theta^*(y)} = \frac{y+\sqrt{y^2+4\lambda\mu}}{2\lambda}. \tag{7.23}$$

Calculate also $e^{-\theta^*(y)}$. (Hint: get quadratic equations in e^θ and $e^{-\theta}$, respectively. Note that one solution of the resulting quadratic equation corresponds to a minimum, at $\theta = 0$.) Using these, establish (7.17). Using the explicit expression (7.17) verify the calculation of (C.6) which states that $d\ell(y)/dy = \theta^*(y)$. Show that consequently, the minimum of $\ell(y)$ occurs at the same point $y_m = \lambda - \mu$ that $\theta^*(y) = 0$. Using this show that

$$\ell(y) \geq 0 \quad \text{with equality if and only if} \quad y = \lambda - \mu.$$

Note that (7.23) implies that θ^* is monotone in y. Conclude that ℓ is strictly convex, and is increasing for $y > \lambda - \mu$. Finally, show that

$$\ell''(\lambda - \mu) = \frac{1}{\mu+\lambda}, \tag{7.24}$$

as follows. By the calculations above,

$$\frac{\partial^2\ell(y)}{\partial^2 y} = \frac{\partial\theta^*(y)}{\partial y}.$$

By definition of ℓ and θ^*,

$$y - \lambda e^{\theta^*(y)} + \mu e^{-\theta^*(y)} = 0.$$

Take derivatives with respect to y of both sides and solve for $\partial\theta^*/\partial y$, noting that $\theta^*(\lambda - \mu) = 0$.   ♠

7.4. Meaning of the Twisted Distribution

The main result in this section, Theorem 7.26, is a fairly weak result that turns out to be very useful for applications. Furthermore, its proof is simple. Here is the idea behind the result. From (1.8) we know that the way rare events occur is by following a "twisted distribution." So, if $\vec{z}_n(t)$ follows a path $\vec{r}(t)$, then the jumps of $\vec{z}_n$ in the $\vec{e}_i$ direction should occur approximately as a Poisson process with rate $\lambda_i e^{\langle \vec{\theta}(t), \vec{e}_i \rangle}$. Here $\vec{\theta}(t)$ comes from the definition (5.2), (5.4) of ℓ with $\vec{r}(t) = \vec{x}$ and $\vec{r}'(t) = \vec{y}$:

$$g(\vec{x}, \vec{\theta}) \triangleq \sum_{i=1}^{k} \lambda_i(\vec{x}) \left(e^{\langle \vec{\theta}, \vec{e}_i \rangle} - 1 \right)$$

$$\ell(\vec{x}, \vec{y}) \triangleq \sup_{\vec{\theta}} \left(\langle \vec{\theta}, \vec{y} \rangle - g(\vec{x}, \vec{\theta}) \right)$$

$$= \left(\langle \vec{\theta}(t), \vec{y} \rangle - g(\vec{x}, \vec{\theta}(t)) \right). \tag{7.25}$$

We won't really prove that the jumps occur in a Poisson fashion, but we will prove that certain average jump rates converge on large enough intervals. This result is close to what is often termed "Level III" large deviations, or at least "Level II." Therefore we will often refer to the result as "Level 1 1/2" large deviations.

To establish the result, we define the following extension of our process:

$$\vec{w}_n(t) \triangleq (\vec{z}_n(t), \vec{y}_n(t)), \quad \text{where}$$

$$\vec{y}_n^i(t) \triangleq \frac{1}{n} \left(\text{no. of jumps of } \vec{z}_n(s), \ s \in (0, t) \text{ in direction } \vec{e}_i \right); \quad 1 \leq i \leq k.$$

$$\tag{7.26}$$

Then $\vec{w}_n(t)$ is a jump Markov process on $\mathbb{R}^{d+k}$, with jump directions $\vec{\epsilon}_i/n$ and jump rates L_i defined as

$$\vec{\epsilon}_i \triangleq (\vec{e}_i, \underbrace{0, 0, \ldots, 0}_{i-1 \text{ zeros}}, 1, \underbrace{0, \ldots, 0}_{k-i \text{ zeros}})$$

$$L_i(\vec{w}) \triangleq \lambda_i(w^1, \ldots, w^d).$$

The process $\vec{w}_n(t)$ tracks not only $\vec{z}_n(t)$, but also how many jumps of each type $\vec{z}_n(t)$ makes. Analogously, for paths $\vec{r}(t)$ for which there is a solution $\theta(t)$ to (7.25), we define the extended path $\vec{w}r(t) \in \mathbb{R}^{d+k}$ by

$$\vec{w}r^i(t) \triangleq \begin{cases} r_i(t) & 1 \leq i \leq d, \\ \displaystyle\int_0^t \lambda_j(\vec{r}(s)) e^{\langle \vec{\theta}(s), \vec{e}_j \rangle} \, ds & i = d+j, 1 \leq j \leq k, \end{cases}$$

where $\vec{\theta}(s)$ is a solution of (7.25). To prevent confusion, we shall occasionally append (or prepend) a subscript $_w$ to variables and functions associated with the process $\vec{w}_n$ so that, for example, $\vec{x}_w$ is a vector in $\mathbb{R}^{d+k}$ and ℓ_w is defined as ℓ above, but for the process $\vec{w}_n$. Finally, we define G_ε, the set of extensions to $\vec{r}(t)$ that are far from $\vec{w}r(t)$, as

$$G_\varepsilon \triangleq \{(\vec{r}(t), \vec{u}(t)) \ : \ \vec{u}(0) = 0, \ d(\vec{w}r, (\vec{r}, \vec{u})) \geq \varepsilon\}.$$

Theorem 7.26 below is virtually a corollary of the following lemma.

Lemma 7.25. *Assume that* $\log \lambda_i(\vec{x})$ *is bounded and Lipschitz continuous for each* i. *Let* $\vec{r}$ *be a path with* $I_0^T(\vec{r}) < \infty$. *Then*

(i) $\vec{w}r$ *is well-defined and* $_wI_0^T(\vec{w}r) = I_0^T(\vec{r})$.

(ii) *for each* $\varepsilon > 0$,

$$\inf_{\vec{v} \in G_\varepsilon} {}_wI_0^T(\vec{v}) > I_0^T(\vec{r}).$$

That is, for each $\varepsilon > 0$, *there is a* $\delta > 0$ *with*

$$_wI_0^T(\vec{v}) \geq I_0^T(\vec{r}) + \delta$$

for every $\vec{v} \in G_\varepsilon$.

(iii) *For each* $\gamma > 0$ *there are positive* ε *and* η *so that if* $v_w = (\vec{s}, \vec{u})$ *satisfies*

$$d(\vec{r}, \vec{s}) < \varepsilon \quad \text{and} \quad d(\vec{w}r, v_w) > \gamma$$

then

$$_wI_0^T(v_w) > I_0^T(\vec{r}) + \eta.$$

Remark. Note that this lemma is purely an analytic statement about the rate function. It is not a probabilistic statement.

The idea behind the proof is simple: using the change of measure that gives us $I_0^T(\vec{r})$, we show that $_wI_0^T(\vec{w}r) = I_0^T(\vec{r})$. Then we show that if $\vec{v}$ is an extension (to $\mathbb{R}^{d+k}$) of $\vec{r}$ that is far from $\vec{w}r$, we must have $_wI_0^T(\vec{v}) > I_0^T(\vec{r})$. This follows from the essential uniqueness of the change of measure [although there may not be a finite $\vec{\theta}(t)$ that solves (7.25)]. The rest follows from the large deviations principle for the process $\vec{w}_n$.

Proof. For the first claim we use the fact that $I_0^T(\vec{r}) = J_0^T(\vec{r})$, given in Theorem 5.26. Let $(\vec{x}, \vec{y}) = (\vec{r}(t), \vec{r}'(t))$ for some t and let $\vec{x}_w = (\vec{x}, x^{d+1}, \ldots, x^{d+k})$ with the same relation between $\vec{y}_w$ and $\vec{y}$. Using the notation of (5.20)–(5.23), if $\vec{\theta}(t)$ is finite, then

$$\tilde{\ell}(\vec{x}, \vec{y}) = \tilde{\ell}(\vec{\theta}(t), \vec{\lambda}(\vec{x}), \vec{y})$$
$$= f(a^*, \vec{\lambda}(\vec{x}))$$

by Theorem 5.26, where $a^* = a^*(t)$ is given by

$$a_i^* = \lambda_i(\vec{x})e^{\langle\vec{\theta}(t), \vec{e}_i\rangle}. \tag{7.27}$$

By Exercise 5.29, the left-hand side is always defined and unique, and hence we can define the right-hand side uniquely by this equality, and therefore, $\vec{w}r$ is well defined even when there is no finite maximizer $\vec{\theta}$. Now

$$\tilde{\ell}_w(\vec{x}_w, \vec{y}_w) \geq \tilde{\ell}(\vec{x}, \vec{y})$$

since in the latter case the infimum in (5.22) is taken over a larger set (there are more values of $\vec{\mu}$ that satisfy the equality constraint). Since by definition, $f_w(\vec{a}, \vec{\lambda}) = f(\vec{a}, \vec{\lambda})$ for any $\vec{a}$, we have established that

$$\tilde{\ell}_w(\vec{x}_w, \vec{y}_w) \geq f_w(a^*, \vec{\lambda}(\vec{x})).$$

If we now establish that

$$\vec{y}_w = \sum_{i=1}^{k} a_i^* \vec{\epsilon}_i \quad \text{for} \quad (\vec{x}_w, \vec{y}_w) = (\vec{w}r(t), \vec{w}r'(t)), \tag{7.28}$$

then Lemma 5.25 would establish that

$$\tilde{\ell}_w(\vec{x}_w, \vec{y}_w) = f_w(a^*, \vec{\lambda}(\vec{x})) = \tilde{\ell}(\vec{x}, \vec{y})$$

which implies the first claim. But since $\vec{y}_w = \vec{w}r'(t)$,

$$\vec{y}_w = \left(\vec{y}, \left\{ \lambda_i(\vec{x}) e^{\langle \vec{\theta}(t), \vec{e}_i \rangle}, \ 1 \leq i \leq k \right\} \right)$$

which, together with (7.27), establishes (7.28).

To establish the second claim, let $\vec{v}$ by any point in G_ε and let t be a point such that

$$\frac{d}{dt} \vec{v}(t) \neq \frac{d}{dt} \vec{w}r(t).$$

Let us first show that necessarily

$$\ell_w(\vec{v}(t), \vec{v}'(t)) > \ell_w(\vec{r}(t), \vec{r}'(t)).$$

Let a_w^* be the minimizer of $f_w(\vec{a}_w, \vec{\lambda}(\vec{x}))$ over $_wK_{\vec{v}'(t)}$, which is unique by Exercise 5.24. We have already established that, under Definition (7.28),

$$\ell_w(\vec{v}(t), \vec{v}'(t)) = f_w(a_w^*, L(\vec{v}(t)) \geq f(a^*, \vec{\lambda}(\vec{x})).$$

If equality holds, then by definition a_w^* minimizes f over K_y. But such a minimizer is unique by Exercise 5.24, so that necessarily $a_w^* = a^*$. This in turn implies that $\vec{v}'(t) = \vec{w}r'(t)$, contradicting our assumption.

Now suppose that the second claim were not true. We have already established that for any $\vec{v}$ in G_ε, $_wI_0^T(\vec{w}r) \geq I_0^T(\vec{r})$. So, suppose we have a sequence of paths $\vec{v}_i \in G_\varepsilon$ with $_wI_0^T(\vec{v}_i) \to I_0^T(\vec{r})$. By Proposition 5.46, this means that the $\vec{v}_i$ are in a compact set (at least for all i large). Let $\vec{v}$ be a limit of the $\vec{v}_i$; then $\vec{v}$ is necessarily absolutely continuous with $\vec{v}(0) = 0$. By lower semicontinuity of $_wI_0^T$, it follows that $_wI_0^T(\vec{v}) \leq I_0^T(\vec{r})$. Since $\vec{v}$ is an extension of $\vec{r}$, the first claim of the lemma now implies that $_w I_0^T (\vec{v}) = I_0^T (\vec{r})$. But this and the previous arguments imply that $\tilde{\ell}_w(\vec{v}(t), \vec{v}'(t)) = \tilde{\ell}(\vec{w}r(t), \vec{w}r'(t))$, and consequently $\vec{v}'(t) = \vec{w}r'(t)$ for (almost) all t. Since $\vec{v}$ is absolutely continuous and $\vec{v}(0) = \vec{w}r(0) = 0$, this in turn implies $\vec{v}(t) = \vec{w}r(t)$ for all t, contradicting the fact that $\vec{v} \in G_\varepsilon$.

The proof of (iii) is similar: assume the statement is false, and take a sequence $\vec{v}_w^i$ and constants ε_i and η_i, both decreasing to zero, so that

$$d(\vec{r}, \vec{s}^i) < \varepsilon_i \quad \text{and} \quad _w I_0^T (v_w^i) \leq I_0^T (\vec{r}) + \eta_i$$

while

$$d(\vec{w}r, v_w^i) > \gamma.$$

Then there is a converging subsequence, and necessarily its limit $\vec{v}_w$ is in G_γ. By lower semicontinuity, $_w I_0^T (\vec{v}_w) \leq I_0^T (\vec{r})$, contradicting (ii). $\blacksquare$

There is a delicate measurability issue here: although $a^*(t)$ is measurable, this is not easy to establish and relies on measurable selection theorems. But we do not need this: we only need that functions such as $\ell(\vec{r}(t), \vec{r}'(t))$ are measurable functions of t: this follows from standard results, since, for example, ℓ is the maximum of the continuous function $\ell(\vec{\theta}, \vec{\lambda}, \vec{y})$. The same comments apply to $\vec{\theta}(t)$.

The utility of this lemma is the following theorem. Given ε, n, T, and $\Delta > 0$, define $k = T/\Delta$, and

$$D_n(\Delta, t) \triangleq \frac{\vec{y}_n(t + \Delta) - \vec{y}_n(t)}{\Delta}$$

$$R_n(\Delta, t) \triangleq \sum_{i=1}^{k} \left| D_n^i(\Delta, t) - \lambda_i(\vec{r}(t)) e^{\langle \vec{\theta}(t), \vec{e}_i \rangle} \right|$$

and denote the probability and expectation conditioned on staying near $\vec{r}$ by

$$\mathbb{P}^\varepsilon (A) \triangleq \mathbb{P}(A \mid d(\vec{z}_n, \vec{r}) < \varepsilon)$$

$$\mathbb{E}^\varepsilon (a) \triangleq \mathbb{E}(a \mid d(\vec{z}_n, \vec{r}) < \varepsilon).$$

Here D_n represents the observed rate of jumps of each type over intervals of size Δ, and R_n represents the difference between this and the "theoretical" jump rate $\lambda_i(\vec{r}(t)) e^{\langle \vec{\theta}(t), \vec{e}_i \rangle}$. The next theorem shows that the maximal deviation of the observed and theoretical jump rates goes to zero in an appropriate limit.

Theorem 7.26. *Assume that $\log \lambda_i (\vec{x})$ is bounded and Lipschitz continuous for each i. Let $\vec{r}$ be a path with $I_0^T (\vec{r}) < \infty$, and so that $a^*(t)$ of (7.27) is continuous. Then for each $\delta > 0$,*

$$\lim_{\Delta \downarrow 0} \lim_{\varepsilon \downarrow 0} \lim_{n \to \infty} \mathbb{P}^\varepsilon \left(\sup_{0 \leq t \leq T - \Delta} R_n(\Delta, t) < \delta \right) = 1, \tag{7.29}$$

$$\lim_{\Delta \downarrow 0} \lim_{\varepsilon \downarrow 0} \lim_{n \to \infty} \mathbb{E}^{\varepsilon} \left(\sup_{0 \leq t \leq T - \Delta} R_n(\Delta, t) \right) = 0. \tag{7.30}$$

This theorem supports the claim that, conditioned on $\vec{z}_n(t)$ staying close to $\vec{r}(t)$, the jumps of $\vec{z}_n(t)$ in direction $\vec{e}_i$ occur as if the jump rate were $n\lambda_i e^{\langle \vec{\theta}, \vec{e}_i \rangle}$. However, the sense in which this holds is pretty weak: we only obtain

$$\frac{1}{n} \text{no. of jumps} \to \lambda_i e^{\langle \vec{\theta}, \vec{e}_i \rangle}$$

in mean and in probability. We don't obtain any information that would enable us to distinguish the jump process from any other point process with the same average rate. That is, we can't claim that $\lambda_i e^{\langle \vec{\theta}, \vec{e}_i \rangle}$ represents a Poisson rate. However, for many applications these theorems are good enough.

Proof. Fix δ and some arbitrary $\gamma > 0$ and consider first those paths $\vec{w}_n$ so that $d(\vec{w}_n, \vec{w}r) \leq \gamma$. Then by definition and by Equation (7.27),

$$\frac{1}{\Delta} \left(\int_t^{t+\Delta} a_i^*(u)\, du - 2\gamma \right) \leq \frac{\vec{y}_n^i(t + \Delta) - \vec{y}_n^i(t)}{\Delta} \leq \frac{1}{\Delta} \left(\int_t^{t+\Delta} a_i^*(u)\, du + 2\gamma \right).$$

Since by assumption $a^*(t)$ is continuous, if Δ is small enough then

$$|a^*(v) - a^*(u)| < \delta/2 \quad \text{for all} \quad 0 \leq u \leq v \leq T, \ v - u < \Delta.$$

For such a value of Δ, choose γ so that $2\gamma/\Delta < \delta/2k$. We conclude that, for all Δ small enough there is a small γ so that $d(\vec{w}_n, \vec{w}r) \leq \gamma$ implies

$$\left| a_i^*(t) - \frac{\vec{y}_n^i(t + \Delta) - \vec{y}_n^i(t)}{\Delta} \right| < \frac{\delta}{k}$$

for all t in $[0, T - \Delta]$. Thus to establish (7.29) it suffices to show that for all γ and all ε, $\mathbb{P}^{\varepsilon}(d(\vec{w}_n, \vec{w}r) > \gamma) \to 0$ as $n \to \infty$. However,

$$\mathbb{P}^{\varepsilon}(d(\vec{w}_n, \vec{w}r) > \gamma) = \frac{\mathbb{P}\left(d(\vec{w}_n, \vec{w}r) > \gamma, \ d(\vec{z}_n, \vec{r}) < \varepsilon \right)}{\mathbb{P}(d(\vec{z}_n, \vec{r}) < \varepsilon)}. \tag{7.31}$$

We now apply the large deviations upper bound to the numerator of (7.31). By Lemma 7.25(iii), there is some $\eta > 0$ so that

$$w I_0^T(v_w) > I_0^T(\vec{r}) + \eta,$$

for all $v_w = (\vec{s}, \vec{u})$ with $d(v_w, \vec{w}r) \geq \gamma$ and $d(\vec{s}, \vec{r}) \leq \varepsilon$. Therefore, for all n large enough,

$$\mathbb{P}(d(\vec{w}_n, \vec{w}r) \geq \gamma, \ d(\vec{z}_n, \vec{r}) \leq \varepsilon) \leq e^{-n\left(I_0^T(\vec{r}) + \eta - \eta/4 \right)}, \tag{7.32}$$

and this upper bounds the numerator. On the other hand, since obviously

$$\inf \left\{ I_0^T(\vec{s}) : d(\vec{s}, \vec{r}) < \varepsilon \right\} \leq I_0^T(\vec{r})$$

the large deviations upper bound implies

$$\mathbb{P}\left(d(\vec{z}_n, \vec{r}) < \varepsilon\right) \geq e^{-n\left(I_0^T(\vec{r}) + \eta/4\right)} \tag{7.33}$$

for all n large. Putting the estimates (7.32) and (7.33) into (7.31), we have

$$\mathbb{P}^\varepsilon\left(d(\vec{w}_n, \vec{w}r) > \gamma\right) \leq e^{-n\eta/2},$$

and (7.29) is established.

To establish (7.30) fix Δ small enough as in the previous argument. Since $a^*(t)$ is continuous, it is bounded and so, for all K large, $R_n(\Delta, t) \geq K$ only if $|D_n(\Delta, t)| \geq K/2$. Let

$$A_{nK} \stackrel{\Delta}{=} \left\{ \omega : \sup_{0 \leq t \leq T - \Delta} |D_n(\Delta, t)| \leq K/2 \right\}.$$

By (7.29),

$$\mathbb{E}^\varepsilon\left(\sup_{0 \leq t \leq T - \Delta} R_n(\Delta, t)\mathbf{1}\left[A_{nK}\right] \right) \to 0$$

as $n \to \infty$. But Corollary 5.8 implies that

$$
\mathbb{P}^\varepsilon\left(\sup_{0 \leq t \leq T - \Delta} |D_n(\Delta, t)| \geq \frac{K}{2} \right)
$$

$$
= \frac{\mathbb{P}\left(\sup_{0 \leq t \leq T - \Delta} |D_n(\Delta, t)| \geq \frac{K}{2},\ d(\vec{z}_n, \vec{r}) < \varepsilon \right)}{\mathbb{P}\left(d(\vec{z}_n, \vec{r}) < \varepsilon\right)}
$$

$$
\leq \frac{\exp -nC\left(\Delta K/2\right)}{\exp -n(I_0^T(\vec{r}) + \rho)}
$$

for some ρ, and where $C(x)/x \to \infty$ as $x \to \infty$. Thus for K large enough the probability that R_n is large has an exponentially decreasing tail, which decreases with n. This implies that

$$\mathbb{E}^\varepsilon\left(\sup_{0 \leq t \leq T - \Delta} R_n(\Delta, t)\mathbf{1}\left[A_{nK}^c\right] \right) \to 0$$

with K, and (7.30) is established.   ∎

7.5. End Notes

Ellis [Ell] has a much more complete theory for discrete-time Markov processes. Some of the results we develop can be recovered from standard sources (e.g., Chung [Chu]); however, we felt it instructive (and possibly easier) to obtain the results as a consequence of our previous development. Moreover, with our approach the uniformity of the bounds over $\lambda \in R_c(\delta)$ comes almost for free, while it might not be obvious from most other methods.

Level III large deviations are described in [Ell] (we believe Ellis is the originator of the notions of Levels I, II, and III). The theory was first developed in [DV4]. Our result is much weaker than those in that we do not attempt to look at deviations in the space of processes. Nevertheless, we obtain an estimate of the behavior of the process during a rare event using almost no extra theory. Our approach may well extend to the case where $a^*(t)$ is not continuous, by some sort of local time average, but we have not done this extension.

Chapter 8

Boundary Theory

This chapter establishes the large deviations principle for two special classes of jump processes. The *flat boundary process* is a jump Markov process of the type we dealt with in Chapter 5, except that it is restricted to a half-space, say $x_0 \geq 0$. Moreover, jump directions as well as jump rates *on the boundary* $x_0 = 0$ may be completely different from those in the interior $x_0 > 0$. In particular, the jump rates are not continuous at the boundary.

The *finite levels process* differs from our standard $\mathbb{R}^d$-valued jump Markov process in that, at any given time, there is a state, or level, associated with the process. For each such level there is a set of jump directions and jump rates for the $\mathbb{R}^d$ part of the process. Transitions between levels are also Markovian, and may (or may not) be associated with jumps in the $\mathbb{R}^d$ valued process.

In both cases, the jump *directions* depend to some extent on position, so these processes are outside the scope of Chapter 5. The method of analysis is similar for both types of processes, and so the theories are developed together as much as possible. The analysis is an extension of the arguments given in Chapter 5; we refer the reader back to that chapter for motivation and an outline of the ideas. In fact, Chapter 5 was set up so that we could follow the same trail, providing additional proofs when necessary.

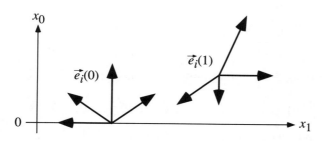

Figure 8.1. A flat boundary model.

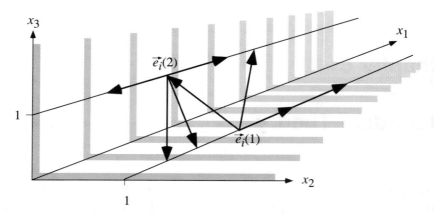

Figure 8.2. A finite levels model, with $d = 1$ and $D = 2$.

We now proceed to a formal definition of the two processes. We fix two positive integers d and D. For a vector $\vec{v} \in \mathbb{R}^{d+D}$ we use the notation

$$\vec{v} = (v_1, v_2, \ldots, v_{d-1}, v_d, v_{d+1} \ldots, v_{d+D}) = (v^d, v^D),$$

where $v^d = (v_1, v_2, \ldots, v_{d-1}, v_d)$ and $v^D = (v_{d+1} \ldots, v_{d+D})$.

Definition 8.3. *The finite levels process is a jump Markov process with values in* $\mathbb{R}^d \times \{1, \ldots, D\}$. *When at state* $(\vec{x}^d, m)$ *the rate of jump to* $(\vec{x}^d + \vec{e}_i^d(m), m'(m; i))$ *is* $\lambda_i(\vec{x}^d; m)$. *Here* $\vec{e}_i^d(m) \in \mathbb{Z}^d$, $1 \leq m \leq D$, $1 \leq i \leq k(m)$.

Note that each pair $(m; i)$ determines m'. We may have $\vec{e}_i^d(m) = \vec{0}$ for some m and i, corresponding to a change in the level without a change in position. Similarly, we may have $m' = m$, corresponding to a change in position without a change in the level. By definition, $k(m)$ is the number of possible (distinct) jumps while at level m.

It will be convenient to give this process a different representation, one that is closer to the processes we have encountered before. The new process $\vec{x}(t)$ has values in $\mathbb{R}^{d+D}$, and

$$(\vec{x}^d, \underbrace{0, \ldots, 0}_{m-1 \text{ zeros}}, 1, 0 \ldots, 0) \quad \text{corresponds to} \quad (\vec{x}^d, m). \tag{8.1}$$

The corresponding definition of jump directions is

$$\vec{e}_i(m) \overset{\triangle}{=} (\vec{e}_i^d(m), 0, \ldots, 0, \alpha, 0, \ldots, 0, -\alpha, 0, \ldots, 0), \qquad \alpha = 0 \text{ or } 1. \tag{8.2}$$

An $\alpha = 1$, with α in coordinate $d + m'$ together with a $(-\alpha)$ in coordinate $d + m$ corresponds to a change from level m to level m', associated with a jump $\vec{e}_i^d(m)$ (possibly zero) in the first component. An $\alpha = 0$ corresponds to no change in the level. We shall henceforth use the representation (8.1)–(8.2), but will retain the notation $\vec{e}_i(m)$, with an explicit dependence on the level m. Note that (8.1) allows us to recover the level m from $\vec{x}$. Similarly for the jump rates

$$\lambda_i(\vec{x}; m) \overset{\triangle}{=} \lambda_i(\vec{x}^d; m), \qquad i = 1, \ldots, k(m). \tag{8.3}$$

We use either of the notations in (8.3), that is, with the first argument of λ_i of dimension d or $d+D$, as convenient. With the scaled process in mind, we generalize (8.1) and obtain the values of $\vec{x}$ under consideration as

$$(\vec{x}^d, \underbrace{0, \ldots, 0}_{m-1 \text{ zeros}}, \alpha, 0 \ldots, 0), \ \alpha > 0 \quad \text{corresponds to} \quad (\vec{x}^d, m). \tag{8.4}$$

Since this process is "modulated" by a Markov chain, we distinguish associated functions by a subscript "c." With this notation, the development of §4.2, Definition 4.6 and Equation (4.5) apply, so that the generator of this process is defined for all $\vec{x} \in \mathbb{R}^{d+D}$ taking the form (8.4) with $\alpha = 1$. For real-valued functions f on $\mathbb{R}^{d+D}$ we define the generator L_c by

$$L_c f(\vec{x}) = \sum_{i=1}^{k(m)} \lambda_i(\vec{x}; m)(f(\vec{x} + \vec{e}_i(m)) - f(\vec{x})).$$

The scaled process $\vec{z}_n(t)$ is obtained, as before, by increasing transition rates by a factor of n, while decreasing step sizes by the same factor. Thus the generator of the scaled process is given by

$$L_{c,n} f(\vec{z}) = \sum_{i=1}^{k(m)} n \cdot \lambda_i(\vec{z}; m) \left(f \left(\vec{z} + \frac{\vec{e}_i(m)}{n} \right) - f(\vec{z}) \right)$$

and is defined for all $\vec{z} \in \mathbb{R}^{d+D}$ taking the form (8.4) with $\alpha = 1/n$.

We concentrate on $\vec{z}_n^d(t)$, the scaled finite levels process where we ignore the level. While $\vec{z}_n^d(t)$ is not a Markov process, we will prove both the upper and lower bounds in terms of $\vec{z}_n^d(t)$, not $\vec{z}_n(t)$.

We make the following assumptions on the finite levels process.

Assumption 8.4. For each $\vec{x}^d$, the continuous-time Markov chain with states $\{1, \ldots, D\}$ and rates $\lambda_i(\vec{x}; m)$ of transition from m to $m'(m; i)$ is ergodic.

Assumption 8.5. The $\log \lambda_i(\vec{x}; m)$ are bounded and Lipschitz continuous in $\vec{x}^d$. [Continuity is in the first d coordinates; see (8.3).]

Exercise 8.6. Under Assumption 8.5, if the ergodicity in Assumption 8.4 holds for one value of $\vec{x}$, then it holds for all values. Hint: ergodicity depends only on whether the rates are zero or not. ♠

What can we expect from this process, in terms of its large deviations behavior? The last D coordinates clearly converge to zero. Since the rates are smooth (with m fixed), the process does not change much statistically as its $\vec{x}^d$ part moves a little. On the other hand, it does change as it jumps from level to level. But as the number of such jumps per unit time is increased, the "levels part" of the process will have time to settle down to its invariant distribution before the rest of the process moves much. Therefore, the large deviations behavior should be determined from an "averaged levels process." This is indeed the form of the most likely path—see Equation (8.6). To estimate unlikely events, we shall find the "twisted rates," then compute a related invariant measure, and again average the process; see Theorem 8.19.

Here is the definition of the other process of interest.

Definition 8.7. *The flat boundary process is a jump Markov process with state $\vec{x} = (x_0, \vec{x}^d)$, where $x_0 \geq 0$ and $\vec{x}^d \in \mathbb{R}^d$. At $x_0 = 0$ its jump directions and corresponding rates are $\{\vec{e}_i(0), \lambda_i(0; \vec{x}); \ i = 1, \ldots, k(0)\}$, while if $x_0 > 0$ then its jump directions and corresponding rates are $\{\vec{e}_i(1), \lambda_i(1; \vec{x}); \ i = 1, \ldots, k(1)\}$.*

Thus this process has one set of jump directions and rates in the interior, and a different set on the boundary $x_0 = 0$. We distinguish functions associated with the boundary process by a subscript "b." The generator for this process is then given by

$$
L_b f(\vec{x}) = \begin{cases} \displaystyle\sum_{i=1}^{k(0)} \lambda_i(0; \vec{x})(f(\vec{x} + \vec{e}_i(0)) - f(\vec{x})) & \text{if } x_0 = 0 \\ \displaystyle\sum_{i=1}^{k(1)} \lambda_i(1; \vec{x})(f(\vec{x} + \vec{e}_i(1)) - f(\vec{x})) & \text{if } x_0 > 0. \end{cases}
$$

The scaled process $\vec{z}_n(t)$ and its generator are obtained by multiplying the rates by n, and dividing the jump sizes by n.

Let e_i^0 denote the component of $\vec{e}_i$ in the direction x_0. We make the following assumptions on the flat boundary process:

Assumption 8.8. The jump directions satisfy

(i) $e_i^0(m) \in \mathbb{Z}$ for $m = 0, 1$, $i = 1, \ldots, k(m)$, i.e., the jumps towards or away from the boundary are integer-valued.

(ii) $e_i^0(1) \geq -1$ for $i = 1, \ldots, k(1)$, with equality for at least one i. That is, from the interior the process cannot jump "down" by more than one unit, and some jump actually does go down.

(iii) $e_i^0(0) \geq 0$ for $i = 1, \ldots, k(0)$. That is, the process cannot jump down from the boundary.

(iv) $e_i^0(0) \geq 1$ for some i. That is, the process can make transitions from the boundary to the interior.

This assumption guarantees that the process does not jump beyond the boundary, and in addition the process will, with positive probability, move between the boundary and the interior.

Assumption 8.9. For each i, $\log \lambda_i(0; \vec{x})$ is bounded and Lipschitz continuous in $\vec{x} = (0, \vec{x}^d)$. Similarly, $\log \lambda_i(1; \vec{x})$ is bounded and Lipschitz continuous in $\vec{x} \in [0, \infty) \times \mathbb{R}^d$.

There are a few processes that can be modeled as either flat boundary or finite levels processes. For example, consider in $\mathbb{R}^2$ a process that lives on $-\infty < x < \infty$, $y = 0$ or $y = 1$ as in Figure 8.10.

Figure 8.10. A process that can be modeled as either a flat boundary process or a finite levels process.

8.1. The Rate Functions

Let us now define the likely path $\vec{z}_\infty$ and the rate function for the finite levels and the flat boundary processes. Recall that quantities related to these processes are distinguished by a subscript "c" for the Markov chain (levels) process or "b" for the flat boundary process.

The finite levels process.

For large n, the scaled finite levels process cannot move away from $\vec{x}^D = \vec{0}^D$. We therefore define its likely behavior only in terms of its first d coordinates. Let $\{v_i(m) : m = 1, \ldots D, \ i = 1, \ldots, k(m)\}$ be a set of rates for the jump directions $\{\vec{e}_i(m)\}$; that is, $v_i(m)$ is a set of constant coefficient jump rates for the finite levels process (no dependence on $\vec{x}^d$). Define $\vec{\pi}(\vec{v}) = \{\pi(\vec{v}; 1), \ldots, \pi(\vec{v}; D)\}$ to be a distribution on $\{1, \ldots, D\}$ satisfying

$$\sum_{m=1}^{D} \pi(\vec{v}; m) \sum_{i=1}^{k(m)} v_i(m) \vec{e}_i^D = \vec{0}^D. \tag{8.5}$$

Comparing this with the definition A.141 of an invariant measure, Definition 8.3 and (8.1)–(8.2) describing the $\vec{x}^D$-part of the process, it follows that the distribution $\vec{\pi}(\vec{v})$ is invariant for the continuous-time Markov chain with rates $\vec{v}$. By Theorem A.142 such a distribution always exists, and moreover for fixed $\vec{x}$, $\vec{\pi}(\vec{\lambda}(\vec{x}))$ is actually unique since under Assumptions 8.4 and 8.5 the rates $\vec{\lambda}$ define an ergodic chain. By Lemma 7.4 the probability $\pi(\vec{v}; m)$ is proportional to the amount of time this chain spends at state m. [Recall that $\lambda_i(\vec{x}; m)$ is defined by the right-hand side of (8.3), so that we need not worry about the value of $\vec{x}^D$.]

We define our scaled process $\vec{z}_n(t)$ as the first d coordinates of the chain $\vec{x}(t)$ scaled in the usual way:

$\vec{x}(t)$ has jump rates $n\lambda_i(\vec{x}(t); m)$ in direction $\vec{e}_i(m)/n$

$$\vec{z}_n(t) \stackrel{\triangle}{=} \vec{x}^d(t).$$

As we establish in Theorem 8.71 below, the most likely path $_c\vec{z}_\infty$ for the finite levels process is the solution of

$$\frac{d}{dt}{}_c\vec{z}_\infty(t) = \sum_{m=1}^{D} \pi\left(\vec{\lambda}({}_c\vec{z}_\infty(t)); m\right) \sum_{i=1}^{k(m)} \lambda_i({}_c\vec{z}_\infty(t); m) \cdot \vec{e}_i^{\,d}(m). \qquad (8.6)$$

Exercise 8.11. Show that when the $\log \lambda_i(\vec{x}; m)$ are bounded and Lipschitz continuous, then $\pi(\vec{\lambda}(\vec{x}); m)$ are Lipschitz continuous. Conclude that (8.6) is a well-posed differential equation; that is, it possesses a unique solution that is continuous in the initial conditions. ♠

For any $\vec{x}^d, \vec{y}^d \in \mathbb{R}^d$ define the local rate function for the finite levels process

$$\ell_c(\vec{x}^d, \vec{y}^d) \stackrel{\triangle}{=} \sup_{\vec{\theta} \in \mathbb{R}^{d+D}} \left(\langle \vec{\theta}^d, \vec{y}^d \rangle - \max_m \sum_{i=1}^{k(m)} \lambda_i(\vec{x}^d; m) \left(e^{\langle \vec{\theta}, \vec{e}_i(m) \rangle} - 1 \right) \right). \qquad (8.7)$$

[See Theorem 8.19 and Lemma 8.20 for alternate forms of ℓ_c.] Recall that we are interested only in the behavior of the first d coordinates of this process. This however is affected by the levels structure, as is evident from a comparison of (8.7) and (5.2).

Remark 8.12. Note that the rate function defined by (8.7) is not a standard Legendre transform: the variable $\vec{\theta}$ is in $\mathbb{R}^{d+D}$, whereas the direction $\vec{y}$ is in $\mathbb{R}^d$. This distinction is what enables the rate $\ell_c(\vec{x}, \vec{y})$ to have nice properties, and in particular is what makes it equal to the natural rate function for the lower bound.

Let us compute a quick example to show that this rate function makes sense. Consider a process with $d = 1$ and $D = 2$ with the following jump rates and directions:

$$\vec{e}_1(1) = (1, -1, 1) \qquad \vec{e}_1(2) = (-1, 1, -1)$$
$$\lambda_1(1) = 1 \qquad \lambda_1(2) = 1.$$

This process shuttles back and forth between the points $(0, 1, 0)$ and $(1, 0, 1)$ with jump rate one. Therefore the rate function should equal

$$\ell(x, y) = \begin{cases} 0 & \text{if } y = 0, \\ \infty & \text{if } y \neq 0. \end{cases}$$

Let us see that this is so.

If $y > 0$, then consider $\vec{\theta}$ of the form $(2a, a, -a)$ for positive a. We have

$$\ell_c(x, y) = \sup_{\vec{\theta} \in \mathbb{R}^3} \left(y\theta_1 - \max\left(e^{\theta_1 - \theta_2 + \theta_3} - 1, e^{-\theta_1 + \theta_2 - \theta_3} - 1 \right) \right)$$
$$\geq ya - \max(0, 0) \to \infty$$

as $a \to \infty$; similarly, choose $a \to -\infty$ if $y < 0$. Finally, if $y = 0$, then notice that

$$
\begin{aligned}
\ell_c(x, y) &= \sup_{\vec{\theta} \in \mathbb{R}^3} \left(-\max \left(e^{\theta_1 - \theta_2 + \theta_3} - 1, \, e^{-\theta_1 + \theta_2 - \theta_3} - 1 \right) \right) \\
&= \sup_{a \geq 0} \left(-\max(a - 1, 1/a - 1) \right) \\
&= 0;
\end{aligned}
$$

so the formula works.

Finally, the rate function for the finite levels process is defined for functions $\vec{r}$ with range in $\mathbb{R}^d$:

$$
{}_cI_0^T(\vec{r}) =
\begin{cases}
\displaystyle \int_0^T \ell_c\left(\vec{r}(s), \vec{r}'(s)\right) \, ds & \text{if } \vec{r} \text{ is absolutely continuous;} \\
\infty & \text{otherwise.}
\end{cases}
\tag{8.8}
$$

The flat boundary process.

We now develop appropriate definitions for $\vec{z}_\infty(t)$ and $\ell(\vec{x}, \vec{y})$ for the flat boundary process. The process $\vec{z}_n(t)$ is defined simply as the scaled version of $\vec{x}(t)$:

$$
\vec{z}_n(t) \text{ has jump rates}
\begin{cases}
n\lambda_i(1; \vec{x}) & z_n^0 > 0 \\
n\lambda_i(0; \vec{x}) & z_n^0 = 0
\end{cases}
\text{ in direction }
\begin{cases}
\vec{e}_i(1)/n & z_n^0 > 0 \\
\vec{e}_i(0)/n & z_n^0 = 0.
\end{cases}
$$

The drift (7.8) in the direction of x_0 (normal to the boundary) is

$$
\beta(\vec{x}) \overset{\triangle}{=} \sum_{i=1}^{k(1)} \lambda_i(1; \vec{x}) e_i^0(1).
\tag{8.9}
$$

Now define $\pi_0(\vec{x})$ for $\vec{x} = (0, \vec{x}^d)$ as the solution of

$$
\pi_0(\vec{x}) \sum_{i=1}^{k(0)} \lambda_i(0; \vec{x}) e_i^0(0) + (1 - \pi_0(\vec{x})) \beta(\vec{x}) = 0, \quad \beta(\vec{x}) < 0
\tag{8.10}
$$

$$
\pi_0(\vec{x}) = 0, \quad \beta(\vec{x}) \geq 0.
$$

Then by Exercise 7.17, π_0 is the probability of $x_0(t) = 0$ under the invariant distribution of the flat boundary process, but where jump rates are frozen at $\lambda_i(m, \vec{x})$, and $\vec{x}^d$ is fixed. That is, the process is allowed to move only in the x_0 direction, with the rates frozen at their boundary or near-boundary values. By Lemma 7.18 π_0 is therefore the proportion of time the process with frozen rates spends on the boundary, and $1 - \pi_0$ is the proportion of time the process spends in the interior $x_0 > 0$. We would expect the most likely path $_b\vec{z}_\infty(t) \in \mathbb{R}^{d+1}$ to be the solution of the differential equation

$$\frac{d}{dt}\, {}_b\vec{z}_\infty(t)$$

$$= \begin{cases} \displaystyle\sum_{i=1}^{k(1)} \lambda_i\left(1;\, {}_b\vec{z}_\infty(t)\right) \vec{e}_i(1) & \text{if } {}_b z_\infty^0(t) > 0 \\ & \text{or } \beta({}_b\vec{z}_\infty(t)) \geq 0; \\[4mm] \displaystyle\pi_0({}_b\vec{z}_\infty(t)) \sum_{i=1}^{k(0)} \lambda_i\left(0;\, {}_b\vec{z}_\infty(t)\right) \vec{e}_i(0) & \text{if } {}_b z_\infty^0(t) = 0 \\ & \text{and } \beta({}_b\vec{z}_\infty(t)) \leq 0. \\[2mm] \quad + \left(1 - \pi_0({}_b\vec{z}_\infty(t))\right) \displaystyle\sum_{i=1}^{k(1)} \lambda_i\left(1;\, {}_b\vec{z}_\infty(t)\right) \vec{e}_i(1) & \end{cases}$$

$$(8.11)$$

It might seem that the first thing to check is that ${}_b\vec{z}_\infty$ exists and is unique, because when the drift β is negative at $x_0 = 0$, the right-hand side of (8.11) is not even continuous. However, we really won't have occasion to use such a result—we only need a much weaker one, which is established in Exercise 8.13 and used in Lemma 8.57 below. The exercise establishes that (8.11) possesses a solution (which is the most likely path) when the jump rates do not depend on $\vec{x}^d$. The result is established (along with much more) in [DI].

Exercise 8.13. If the $\lambda_i(m; \vec{x})$ do not depend on $\vec{x}^d$ and are Lipschitz continuous in x_0, then (8.11) is a well-posed differential equation, that is, it possesses a unique solution that is continuous in the initial conditions. Hint: consider what happens when a path ${}_b z_\infty^0(t)$ starts at a point $\vec{x}$ with $x_0 > 0$. The only difficulty occurs at the time ${}_b z_\infty^0(t)$ hits the boundary $x_0 = 0$.   ♠

Exercise 8.14. If the $\log \lambda_i(m; \vec{x})$ are bounded and Lipschitz continuous, then $\pi_0(\vec{x})$ is Lipschitz continuous.   ♠

For $\vec{x}$ and $\vec{y}$ in $\mathbb{R}^{d+1}$, define the local rate function for the flat boundary process as follows.

$\ell_b(\vec{x}, \vec{y})$

$$
= \begin{cases}
\displaystyle\sup_{\vec{\theta} \in \mathbb{R}^{d+1}} \left(\langle \vec{\theta}, \vec{y} \rangle - \sum_{i=1}^{k(0)} \lambda_i(1; \vec{x}) \left(e^{\langle \vec{\theta}, \vec{e}_i(1) \rangle} - 1 \right) \right) & \text{if } x_0 > 0 \text{ or } y_0 > 0 \\[2em]
\displaystyle\sup_{\vec{\theta} \in \mathbb{R}^{d+1}} \left(\langle \vec{\theta}, \vec{y} \rangle - \max_{m=0,1} \right. & \\
\qquad\qquad\qquad \left. \sum_{i=1}^{k(m)} \lambda_i(m; \vec{x}) \left(e^{\langle \vec{\theta}, \vec{e}_i(m) \rangle} - 1 \right) \right) & \text{if } x_0 = 0 \text{ and } y_0 = 0 \\[2em]
\infty & \text{if } x_0 = 0 \text{ and } y_0 < 0 \\
\infty & \text{if } x_0 < 0.
\end{cases}
$$

(8.12)

This local rate function is the usual one, except that on the boundary $x_0 = 0$ the rate may be lower than the rate in the interior for directions parallel to the boundary. For alternate ways of writing the rate function, see Theorem 8.19 and Lemma 8.20. Finally, the rate function for the flat boundary process is defined for functions $\vec{r}$ with range in $\mathbb{R}^{d+1}$:

$$
{}_b I_0^T(\vec{r}) = \begin{cases} \displaystyle\int_0^T \ell_b\left(\vec{r}(s), \vec{r}'(s)\right) ds & \text{if } \vec{r} \text{ is absolutely continuous;} \\ \infty & \text{otherwise.} \end{cases}
$$

(8.13)

Note that the local rate function for a flat boundary process, when restricted to stay on the boundary, coincides with that of a finite levels process with two levels: compare the middle term in (8.11) to (8.6), and the second term in (8.12) to (8.7). In order to exploit this similarity, we adopt the convention that, for the boundary process,

$$
\sum_{m=1}^{D} \overset{\triangle}{=} \sum_{m=0}^{1} \quad \text{and} \quad \{m = 1, \ldots, D\} \overset{\triangle}{=} \{m = 0, 1\}.
$$

(8.14)

We are now ready to state the main results of this chapter. The proofs are given in §8.3 and 8.5. The large deviations principle we establish for the finite levels model is the following.

Theorem 8.15. *Let Assumptions 8.4 and 8.5 hold for the finite levels process. Then $_cI_0^T$ is a good rate function in the space $\left(D^d[0, T], d_d\right)$, and*

(i) *For every closed set $F \in D^d[0, T]$, uniformly over initial levels $m = 1, \ldots, D$,*

$$\limsup_{\vec{y} \to \vec{x}, \, n \to \infty} \frac{1}{n} \log \mathbb{P}_{\vec{y}} \, (\vec{z}_n \in F) \leq - \inf \left\{ _cI_0^T (\vec{r}) : \vec{r} \in F, \, \vec{r}(0) = \vec{x} \right\}.$$

(ii) *For every open set $G \in D^d[0, T]$, uniformly over initial levels $m = 1, \ldots, D$ and starting position $\vec{x}^d$ in compact subsets of $\mathbb{R}^d$,*

$$\liminf_{n \to \infty} \frac{1}{n} \log \mathbb{P}_{\vec{x}} \, (\vec{z}_n \in G) \geq - \inf \left\{ _cI_0^T (\vec{r}) : \vec{r} \in G, \, \vec{r}(0) = \vec{x} \right\}.$$

For the flat boundary model, we obtain the usual large deviations principle:

Theorem 8.16. *Let Assumptions 8.8 and 8.9 hold for the flat boundary process. Consider points $\vec{x}$ and $\vec{y}$ in $[0, \infty) \times \mathbb{R}^d$. Then $_bI_0^T$ is a good rate function in the space $\left(D^{d+1}[0, T], d_d\right)$, and*

(i) *For every closed set $F \in D^{d+1}[0, T]$,*

$$\limsup_{\vec{y} \to \vec{x}, \, n \to \infty} \frac{1}{n} \log \mathbb{P}_{\vec{y}} \, (\vec{z}_n \in F) \leq - \inf \left\{ _bI_0^T (\vec{r}) : \vec{r} \in F, \, \vec{r}(0) = \vec{x} \right\}.$$

(ii) *For every open set $G \in D^{d+1}[0, T]$, uniformly over starting position $\vec{x}$ in compact subsets,*

$$\liminf_{n \to \infty} \frac{1}{n} \log \mathbb{P}_{\vec{x}} \, (\vec{z}_n \in G) \geq - \inf \left\{ _bI_0^T (\vec{r}) : \vec{r} \in G, \, \vec{r}(0) = \vec{x} \right\}.$$

The remainder of the chapter is organized as follows. In §8.2 we establish some properties of the rate functions, roughly along the lines of §5.2. In §8.3 we derive the upper bound for our processes; the proof is nearly identical to that of §5.5 and consists essentially of checking that the same steps are valid. Then we establish, in §8.4, some facts about constant coefficient jump Markov processes, including Kurtz's Theorem for the constant coefficient case, and conclude with proofs of the lower bounds and Kurtz's Theorem for the two types of processes (§8.5).

A flat boundary process would seem to be a limit of finite levels processes as the number of levels becomes infinite. We would naturally approximate a flat boundary process with a finite levels process by having one level represent the boundary $x_0 = 0$, another represent $x_0 = 1$, etc., up to $x_0 = D - 1$, and then making D large. (A different approximation would be needed for each scaling of the process, so that justifying or using this approximation is not obvious). There is, however, an underlying geometry in the flat boundary process that is missing from the finite levels process. We allow jumps in the finite levels process to be quite arbitrary. But by definition of the flat boundary process, the jump directions at each level above zero are the same. In particular, the distance any one jump can take in terms of the number of levels it could cross is strictly bounded, and is independent of the approximating D. Let us compare how a process might appear in the two presentations. With the flat boundary process, we know that during a large deviation each jump $\vec{e}_i$ has its rate adjusted by $\exp(\langle \vec{\theta}, \vec{e}_i \rangle)$, where $\vec{\theta}$ is the optimal "twisting" coefficient. This means, for example, that if two jump directions $\vec{e}_i$ and $\vec{e}_j$ are related by $\vec{e}_i = C\vec{e}_j$ for some constant C, then their twists will be related also. This would not be at all clear in a finite levels model of the same process, where there is no geometry; we would never have any two jump directions that go between levels being proportional, since all levels are mutually orthogonal. Thus, when we compare the flat boundary theory to the finite levels theory, we are comparing theories where the difficulties are in different places. The flat boundary theory is difficult because of the potentially infinite state space: $x_0 \in \{0, 1, 2, \ldots\}$. But the jump rules at each level above $x_0 = 0$ are exactly the same. In the finite levels process we add a new dimension for each level. This generally makes calculation more difficult as the number of levels increases.

Exercise 8.17. Show that if Assumption 8.4 (Assumption 8.8) does not hold, then the function defined in (8.8) [(8.13) respectively] cannot be the large deviations rate function. Hint: show that this function may be too small, so that the corresponding upper bound is too large. Use the following example. Take a standard Poisson process that moves right in Level one, moves left in Level 2, and has no transitions between Levels one and two. Compute the upper bound given by (8.8) for moving right in Level two. Similarly for the boundary process, let the process move right in the interior and left on the boundary. ♠

8.2. Properties of the Rate Function

In this section we restate all the main results from §5.2, with new proofs whenever the previous proofs do not hold. Many of the statements in §5.2 hold for all bounded $\lambda_i(m)$ without requiring continuity, so that they do not have to be reproved. But first we establish the new key Theorem 8.19 for our processes: the equivalence between the natural rate functions for the upper and lower bounds. This analogue of Theorem 5.26 is useful in deriving some of the required results.

Corollary 8.18. *For each $\vec{x}$, the function $\ell_c(\vec{x}, \vec{y})$, defined in (8.7), and the function $\ell_b(\vec{x}, \vec{y})$, defined in (8.12), are non-negative and lower semicontinuous in $\vec{y}$. Moreover, $\ell_c(\vec{x}, \vec{y})$ is convex in $\vec{y}$, and $\ell_b(\vec{x}, \vec{y})$ is convex whenever $x_0 \neq 0$. Finally, if $x_0 = 0$ then the function $\ell_b(\vec{x}, (0, \vec{y}^d))$ is convex in $\vec{y}^d \in \mathbb{R}^d$.*

Recall that for ℓ_c, $\vec{x}$ and $\vec{y}$ are in $\mathbb{R}^d$ while for ℓ_b, $\vec{x}$ and $\vec{y}$ are in $\mathbb{R}^{d+1}$. In general, $\ell_b(\vec{x}, \vec{y})$ may be non-convex when $x_0 = 0$, since it may jump down at $y_0 = 0$.

Proof. For ℓ_c and for ℓ_b with $x_0 > 0$, repeat the steps of Proposition 5.10 to obtain non-negativity, convexity, and lower semicontinuity. Exactly the same proof applies to $\ell_b(\vec{x}, (0, \vec{y}^d))$, $\vec{y}^d \in \mathbb{R}^d$ with $x_0 = 0$. Setting $\vec{\theta} = \vec{0}$ in the definitions proves the non-negativity in all cases.

It remains to prove lower semicontinuity of ℓ_b when $x_0 = 0$. By Exercise A.29 it clearly suffices to consider sequences $\vec{y}_n$ whose limit $\vec{z}$ satisfies $z_0 = 0$; otherwise, the previous argument applies. But then the only possible jump of the function ℓ_b is down (due to the "max"), so that the function is lower semicontinuous at this point.   ∎

Theorem 8.19. *Let $\lambda_i(m)$ be positive. For each $\vec{y} \in \mathbb{R}^d$ the following are equal.*

$$(i) \quad \sup_{\vec{\theta} \in \mathbb{R}^{d+D}} \left(\langle \vec{\theta}^d, \vec{y} \rangle - \max_{1 \le m \le D} \sum_{i=1}^{k(m)} \lambda_i(m) \left(e^{\langle \vec{\theta}, \vec{e}_i(m) \rangle} - 1 \right) \right),$$

$$(ii) \quad \inf_{\vec{\mu} \in \mathcal{B}} \inf_{\vec{v} \in \mathcal{S}(\vec{\mu})} \sum_{m=1}^{D} \mu_m \left(\sum_{i=1}^{k(m)} v_i(m) \log \frac{v_i(m)}{\lambda_i(m)} - v_i(m) + \lambda_i(m) \right),$$

where

$$\mathcal{B} \triangleq \left\{ \vec{\mu} \in \mathbb{R}^D \; : \; \mu_m \ge 0, \; \sum_{m=1}^{D} \mu_m = 1 \right\},$$

and

$$\mathcal{S}(\vec{\mu}) \triangleq \left\{ \vec{v} \; : \; v_i(m) \ge 0, \; \sum_{m=1}^{D} \mu_m \sum_{i=1}^{k(m)} v_i(m) \vec{e}_i(m) = \left(\vec{y}, \vec{0}^D \right) \right\}.$$

Furthermore, (i) is exactly (8.7), so that the quantities (i)–(ii) represent the local rate function of the finite levels process. Similarly, (i) becomes (8.12) when restricted to $x_0 = 0$ and $y_0 = 0$, so that the quantities (i)–(ii) represent the local rate function of the flat boundary process when it is moving along the boundary.

Remark. *(i)* is the natural local rate function for the upper bound. *(ii)* is the natural local rate function for the lower bound. *(iii)* below is the natural local rate function that arises in a different approach to the upper bound. If you haven't already read Remark 8.12, we recommend that you do so before continuing with this section.

We prove the equivalence of *(i)* and *(ii)* by deriving three additional forms.

Lemma 8.20. *Let $\lambda_i(m)$ be positive. For each $\vec{y} \in \mathbb{R}^d$ the following are equal.*

$$(iii) \quad \inf_{\vec{\mu} \in \mathcal{B}} \sup_{\vec{\theta} \in \mathbb{R}^d} \left(\langle \vec{\theta}, \vec{y} \rangle - \sum_{m=1}^{D} \mu_m \sum_{i=1}^{k(m)} \lambda_i(m) \left(e^{\langle \vec{\theta}, \vec{e}_i^d(m) \rangle} - 1 \right) + \ell_{\vec{\theta}}(0; \vec{\mu}) \right)$$

where

$$\ell_{\vec{\theta}}(0; \vec{\mu}) \triangleq \sup_{\vec{\gamma} \in \mathbb{R}^D} \left(-\sum_{m=1}^{D} \mu_m \sum_{i=1}^{k(m)} \lambda_i(m) e^{\langle \vec{\theta}, \vec{e}_i^d(m) \rangle} \left(e^{\langle \vec{\gamma}, \vec{e}_i^D(m) \rangle} - 1 \right) \right).$$

(iv) $\displaystyle \sup_{\vec{\theta} \in \mathbb{R}^{d+D}} \ \inf_{\vec{\mu} \in \mathcal{B}} \left(\left\langle \vec{\theta}^d, \vec{y} \right\rangle - \sum_{m=1}^{D} \mu_m \sum_{i=1}^{k(m)} \lambda_i(m) \left(e^{\langle \vec{\theta}, \vec{e}_i(m) \rangle} - 1 \right) \right).$

(v) $\displaystyle \inf_{\vec{\mu} \in \mathcal{B}} \ \inf_{\{\vec{b}_m\} \in \mathcal{T}(\vec{\mu})} \sum_{m=1}^{D} \mu_m$

$$\left\{ \sup_{\vec{\theta}_m \in \mathbb{R}^{d+D}} \left(\left\langle \vec{\theta}_m, \vec{b}_m \right\rangle - \sum_{i=1}^{k(m)} \lambda_i(m) \left(e^{\langle \vec{\theta}_m, \vec{e}_i(m) \rangle} - 1 \right) \right) \right\}$$

where

$$\mathcal{T}(\vec{\mu}) \triangleq \left\{ \{\vec{b}_1, \ldots, \vec{b}_D\} \ : \ \vec{b}_m \in \mathbb{R}^{d+D}, \ \sum_{m=1}^{D} \mu_m \vec{b}_m = (\vec{y}, \vec{0}^D) \right\}.$$

Remark. We don't use form *(iii)* at all in this book. It is included because it is a natural upper bound based on a different proof than the one we elected to use, and for historical reasons: it is the one we derived first.

Proof. It is immediate that if $\lambda_i(m) = 0$ for some i and m, then we can simply omit the corresponding term from all expressions. We shall therefore assume that the $\lambda_i(m)$ are strictly positive.

To show *(iii)* = *(iv)*, note that the assumptions of Theorem A.44 are satisfied, so we can rewrite *(iv)* as

$$\inf_{\vec{\mu} \in \mathcal{B}} \ \sup_{\vec{\theta} \in \mathbb{R}^{d+D}} \left(\left\langle \vec{\theta}^d, \vec{y} \right\rangle - \sum_{m=1}^{D} \mu_m \sum_{i=1}^{k(m)} \lambda_i(m) \left(e^{\langle \vec{\theta}, \vec{e}_i(m) \rangle} - 1 \right) \right). \qquad (8.15)$$

Then for each $\vec{\mu} \in \mathcal{B}$,

$$\sup_{\vec{\theta} \in \mathbb{R}^{d+D}} \left(\left\langle \vec{\theta}^d, \vec{y} \right\rangle - \sum_{m=1}^{D} \mu_m \sum_{i=1}^{k(m)} \lambda_i(m) \left(e^{\langle \vec{\theta}, \vec{e}_i(m) \rangle} - 1 \right) \right)$$

$$= \sup_{\vec{\alpha} \in \mathbb{R}^d, \ \vec{\gamma} \in \mathbb{R}^D} \left(\langle \vec{\alpha}, \vec{y} \rangle - \sum_{m=1}^{D} \mu_m \sum_{i=1}^{k(m)} \lambda_i(m) \left(e^{\langle \vec{\alpha}, \vec{e}_i^d(m) \rangle} e^{\langle \vec{\gamma}, \vec{e}_i^D(m) \rangle} - 1 \right) \right)$$

$$= \sup_{\vec{\alpha} \in \mathbb{R}^d} \left(\langle \vec{\alpha}, \vec{y} \rangle - \sum_{m=1}^{D} \mu_m \sum_{i=1}^{k(m)} \lambda_i(m) \left(e^{\langle \vec{\alpha}, \vec{e}_i^d(m) \rangle} - 1 \right) \right.$$

$$\left. + \sup_{\vec{\gamma} \in \mathbb{R}^D} \left(-\sum_{m=1}^{D} \mu_m \sum_{i=1}^{k(m)} \lambda_i(m) \left(e^{\langle \vec{\alpha}, \vec{e}_i^d(m) \rangle} \left(e^{\langle \vec{\gamma}, \vec{e}_i^D(m) \rangle} - 1 \right) \right) \right) \right)$$

so that *(iii)* = *(iv)*.

To establish $(iv) = (v)$, we use the representation (8.15) for (iv). It suffices to show that for each $\vec{\mu} \in \mathcal{B}$ we have

$$\inf_{\{\vec{b}_m\} \in \mathcal{T}(\vec{\mu})} \sum_{m=1}^{D} \mu_m \left\{ \sup_{\vec{\theta}_m \in \mathbb{R}^{d+D}} \left(\langle \vec{\theta}_m, \vec{b}_m \rangle - \sum_{i=1}^{k(m)} \lambda_i(m) \left(e^{\langle \vec{\theta}_m, \vec{e}_i(m) \rangle} - 1 \right) \right) \right\}$$

$$= \sup_{\vec{\theta} \in \mathbb{R}^{d+D}} \left(\langle \vec{\theta}^d, \vec{y} \rangle - \sum_{m=1}^{D} \mu_m \sum_{i=1}^{k(m)} \lambda_i(m) \left(e^{\langle \vec{\theta}, \vec{e}_i(m) \rangle} - 1 \right) \right).$$

$$(8.16)$$

For each fixed $\{\vec{b}_m\}$ in $\mathcal{T}(\vec{\mu})$ clearly

$$\sum_{m=1}^{D} \mu_m \sup_{\vec{\theta} \in \mathbb{R}^{d+D}} \left(\langle \vec{\theta}, \vec{b}_m \rangle - \sum_{i=1}^{k(m)} \lambda_i(m) \left(e^{\langle \vec{\theta}, \vec{e}_i(m) \rangle} - 1 \right) \right) \qquad (8.17a)$$

$$\geq \sup_{\vec{\theta} \in \mathbb{R}^{d+D}} \sum_{m=1}^{D} \mu_m \left(\langle \vec{\theta}, \vec{b}_m \rangle - \sum_{i=1}^{k(m)} \lambda_i(m) \left(e^{\langle \vec{\theta}, \vec{e}_i(m) \rangle} - 1 \right) \right) \quad (8.17b)$$

$$= \sup_{\vec{\theta} \in \mathbb{R}^{d+D}} \left(\langle \vec{\theta}, (\vec{y}, \vec{0}^D) \rangle - \sum_{m=1}^{D} \sum_{i=1}^{k(m)} \mu_m \lambda_i(m) \left(e^{\langle \vec{\theta}, \vec{e}_i(m) \rangle} - 1 \right) \right)$$

and therefore $(v) \geq (iv)$. The result will follow if we can find some $\{\vec{b}_m\} \in \mathcal{T}(\vec{\mu})$ so that equality holds in (8.17). We assume that (8.17b) is finite, and without loss of generality that $\mu_m > 0$ for all m. Consider now (8.17b) as the ℓ-function of a process with (constant) jump rates $\{\mu_m \lambda_i(m)\}$ and jump directions $\{\vec{e}_i(m)\}$, where $i = 1, \ldots, k(m)$ and $m = 1, \ldots, D$. Then by the remark following Lemma 5.21, assuming ℓ is finite implies that $\vec{y}$ lies in the positive cone $\mathcal{C}$ generated by all the jump directions $\{\vec{e}_i(m)\}$ (Definition 5.19). In addition, by Lemma 5.21 there is a maximizing sequence $\vec{\theta}_1^*, \vec{\theta}_2^*, \ldots$ so that at these values (8.17b) converges to its supremum, and $\langle \vec{\theta}_k^*, \vec{y} \rangle$ and $\exp(\langle \vec{\theta}_k^*, \vec{e}_i(m) \rangle)$ both tend to finite limits, which we denote $\langle \vec{\theta}^*, \vec{y} \rangle$ and $\exp(\langle \vec{\theta}^*, \vec{e}_i(m) \rangle)$. (Recall that $\vec{\theta}_1^*, \vec{\theta}_2^*, \ldots$ need not converge—in fact, the sequence need not even be bounded.) We define

$$\vec{b}_m \triangleq \lim_{k \to \infty} \sum_{i=1}^{k(m)} \lambda_i(m) e^{\langle \vec{\theta}_k^*, \vec{e}_i(m) \rangle} \vec{e}_i(m). \qquad (8.18)$$

Then by (5.24) and the proof of Theorem 5.26 we have

$$\sum_{m=1}^{D} \mu_m \vec{b}_m = (\vec{y}, \vec{0}^D) \quad \text{meaning} \quad \{\vec{b}_m\} \in \mathcal{T}(\vec{\mu}). \qquad (8.19)$$

By definition, $\vec{b}_m \in \mathcal{C}(m)$ for all $m = 1, \ldots, D$, where $\mathcal{C}(m)$ is the positive cone generated by $\{\vec{e}_i(m) : i = 1, \ldots, k(m)\}$. For these given $\vec{b}_m$, consider

$$\sup_{\vec{\theta} \in \mathbb{R}^{d+D}} \left(\langle \vec{\theta}, \vec{b}_m \rangle - \sum_{i=1}^{k(m)} \lambda_i(m) \left(e^{\langle \vec{\theta}, \vec{e}_i(m) \rangle} - 1 \right) \right)$$

which is finite for each m. Let $\vec{\theta}_1^*(m)$, $\vec{\theta}_2^*(m)$, ... be a maximizing sequence. Then by the first line of (5.24) we have

$$\vec{b}_m = \lim_{k \to \infty} \sum_{i=1}^{k(m)} \lambda_i(m) e^{\langle \vec{\theta}_k^*(m), \vec{e}_i(m) \rangle} \vec{e}_i(m). \tag{8.20}$$

Now we claim that (8.18) and (8.20) together imply that as $k \to \infty$, $\exp\langle \vec{\theta}_k^*, \vec{e}_i(m) \rangle$ and $\exp\langle \vec{\theta}_k^*(m), \vec{e}_i(m) \rangle$ have the same limits. That is,

$$\lim_{k \to \infty} \langle \vec{\theta}_k^*, \vec{e}_i(m) \rangle = \lim_{k \to \infty} \langle \vec{\theta}_k^*(m), \vec{e}_i(m) \rangle \quad \text{for all} \quad i, m \tag{8.21}$$

in the sense that either both converge to the same finite limit, or both diverge to $(-\infty)$. By Theorem 5.26 and Exercise 5.29, the limits on both sides are unique. Now using our notation for the limits, we have two representations of $\vec{y}$:

$$\vec{y} = \sum_{m=1}^{D} \sum_{i=1}^{k(m)} \mu_m \lambda_i(m) e^{\langle \vec{\theta}^*, \vec{e}_i(m) \rangle} \vec{e}_i(m)$$

$$= \sum_{m=1}^{D} \sum_{i=1}^{k(m)} \mu_m \lambda_i(m) e^{\langle \vec{\theta}^*(m), \vec{e}_i(m) \rangle} \vec{e}_i(m).$$

By Lemma 5.21 we have, for each i and m

$$\langle \vec{\theta}_k^*, \vec{e}_i(m) \rangle \to -\infty \quad \text{implies} \quad e^{\langle \vec{\theta}^*(m), \vec{e}_i(m) \rangle} = 0,$$

since it is a coefficient in a representation of $\vec{y}$, and hence $\langle \vec{\theta}_k^*(m), \vec{e}_i(m) \rangle \to -\infty$. The same argument applied to $\vec{b}_m$ with the representations (8.18) and (8.20) shows the converse implication, so that if either side of (8.21) diverges for some i and m, so does the other. Now

$$\vec{0} = \vec{b}_m - \vec{b}_m$$

$$= \lim_{k \to \infty} \sum_{i=1}^{k(m)} \lambda_i(m) \left(e^{\langle \vec{\theta}_k^*, \vec{e}_i(m) \rangle} - e^{\langle \vec{\theta}_k^*(m), \vec{e}_i(m) \rangle} \right) \vec{e}_i(m)$$

$$= \sum_i \lambda_i(m) \left(e^{\langle \vec{\theta}^*, \vec{e}_i(m) \rangle} - e^{\langle \vec{\theta}^*(m), \vec{e}_i(m) \rangle} \right) \vec{e}_i(m),$$

where in the last sum we include only those indices i for which the limit in (8.21) is finite. Taking inner product with $\vec{\theta}_k^*$ and with $\vec{\theta}_k^*(m)$ and subtracting,

$$\lim_{k \to \infty} \sum_i \lambda_i(m) \left(e^{\langle \vec{\theta}_k^*, \vec{e}_i(m) \rangle} - e^{\langle \vec{\theta}_k^*(m), \vec{e}_i(m) \rangle} \right) \left(\langle \vec{\theta}_k^*, \vec{e}_i(m) \rangle - \langle \vec{\theta}_k^*(m), \vec{e}_i(m) \rangle \right)$$

$$= \sum_i \lambda_i(m) \left(e^{\langle \vec{\theta}^*, \vec{e}_i(m) \rangle} - e^{\langle \vec{\theta}^*(m), \vec{e}_i(m) \rangle} \right) \left(\langle \vec{\theta}^*, \vec{e}_i(m) \rangle - \langle \vec{\theta}^*(m), \vec{e}_i(m) \rangle \right)$$

$$= 0,$$

where again we include only those indices i for which the limit in (8.21) is finite. Since $(e^x - e^y)(x - y) \geq 0$, with equality if and only if $x = y$, we conclude that $\langle \vec{\theta}^*(m), \vec{e}_i(m) \rangle = \langle \vec{\theta}^*, \vec{e}_i(m) \rangle$ for all i and m.

Substituting these $\vec{b}_m$ into (8.17) and replacing the sup over $\vec{\theta}$ with the limits, the equality in (8.21) implies equality in (8.17), and the proof of the lemma is complete.   ∎

Proof of Theorem 8.19. The proof follows from Lemma 8.20 by showing that $(i) = (iv)$ and $(ii) = (v)$. It is immediate that if $\lambda_i(m) = 0$ for some i and m, then we can simply omit the corresponding term from all expressions. We therefore assume that the $\lambda_i(m)$ are strictly positive.

Since the expression in parentheses in (iv) is linear in the μ_m, the infimum can be attained with $\mu_m = 1$ for some m, and $(i) = (iv)$ follows.

To show $(ii) = (v)$ define, as in (5.21)

$$K_{\vec{b}}(m) \triangleq \left\{ \vec{v}(m) : v_i(m) \geq 0, \ \sum_{i=1}^{k(m)} v_i(m) \vec{e}_i(m) = \vec{b} \right\}.$$

By definition of $\mathcal{S}$,

$$\mathcal{S}(\vec{\mu}) = \bigcup_{\{\vec{b}_m\} \in \mathcal{T}(\vec{\mu})} \left\{ \vec{v} : \vec{v}(m) \in K_{\vec{b}_m}(m), \ m = 1, \ldots, D \right\}.$$

Therefore $(ii) = (v)$ will follow once we establish that, for each fixed $\vec{\mu}$, m and $\vec{b} \in \mathbb{R}^{d+D}$,

$$\sup_{\vec{\theta} \in \mathbb{R}^{d+D}} \left(\langle \vec{\theta}, \vec{b} \rangle - \sum_{i=1}^{k(m)} \lambda_i(m) \left(e^{\langle \vec{\theta}, \vec{e}_i(m) \rangle} - 1 \right) \right)$$

$$= \inf_{\vec{v}(m) \in K_b(m)} \left(\sum_{i=1}^{k(m)} v_i(m) \log \frac{v_i(m)}{\lambda_i(m)} - v_i(m) + \lambda_i(m) \right).$$

But this last equality is established in Theorem 5.26, and $(ii) = (v)$ follows.

The final claims are obvious from the definitions.   ∎

We now proceed along the path of §5.2, and verify each of the relevant results for the local rate functions ℓ_c and ℓ_b. Some of the results will apply only to the restriction of ℓ_b to either the boundary or the interior. This was already the case in Corollary 8.18, which established a version of Proposition 5.10 and Corollary 5.12.

An important difference between the theory of this chapter and that for smooth processes is that here we cannot expect strict convexity of the local rate function. One immediate consequence is that, in the constant coefficient case, straight lines are no longer the unique optimal paths.

Exercise 8.21. Consider a finite levels model with $d = 1$ and $D = 2$ where there are two jump directions and corresponding rates:

$$\lambda_1(1) = 1 \qquad \vec{e}_1(1) = (1, 0, 0)$$
$$\lambda_1(2) = 1 \qquad \vec{e}_1(2) = (2, 0, 0)$$

(The process never switches between levels; the projection on $\mathbb{R}^D$ is a constant. Therefore this process violates Assumption 8.4.) Show that $\ell(y) = 0$ for any $\vec{y} \in [1, 2]$. This proves that the rate function is not strictly convex. Hint: this is easy using form *(ii)* of the rate function. Just divide up the mass among the two levels so the mean drift is correct. ♠

Exercise 8.22. Under Assumption 8.4, $\ell_c(\vec{x}, \vec{y}) = 0$ if and only if for some $\vec{\mu} \in \mathcal{B}$,

$$\vec{y} = \sum_{m=1}^{D} \mu_m \sum_{i=1}^{k(m)} \lambda_i(m)\vec{e}_i^d(m) \quad \text{and} \quad \sum_{m=1}^{D} \mu_m \sum_{i=1}^{k(m)} \lambda_i(m)\vec{e}_i^D(m) = \vec{0}^D. \qquad (8.22)$$

Under Assumption 8.8, the same holds for $\ell_b\left((0, \vec{x}^d), (0, \vec{y}^d)\right)$ with $\vec{y} = \vec{y}^d$, and moreover $\vec{\mu} = (\pi_0, 1 - \pi_0)$ [Equation (8.9)]. Hint: see the definition of $\mathcal{S}$ and observe that for $x > 0$, the function $a \log(a/x) - a + x$ is continuous, positive, and vanishes if and only if $a = x$. ♠

Note that under Assumption 8.4 there exists a unique $\vec{\mu}$ satisfying (8.22).

Lemma 8.23. *Assume $\lambda_i(\vec{x}; m)$ do not depend on $\vec{x}^d$. Then for all $\vec{r}$ absolutely continuous and all T,*

$$_c I_0^T(\vec{r}) \triangleq \int_0^T \ell_c(\vec{r}(t), \vec{r}\,'(t))\, dt \geq T \cdot \ell_c\left(\vec{r}(0), \frac{\Delta\vec{r}}{T}\right), \qquad (8.23)$$

where $\Delta\vec{r} \triangleq \vec{r}(T) - \vec{r}(0)$. The same holds for $_b I_0^T(\vec{r})$ provided either $r_0(t) = r_0'(t) = 0$ for all t, or $r_0(t) > 0$ for all t.

Proof. The same as for Lemma 5.16. ∎

Remark. Unlike Lemma 5.16, we do not claim that equality holds in (8.23) if and only if $\vec{r}\,'(t)$ is a constant, since the rate function might not be strictly convex. This means that we are not guaranteed that every minimal path for a constant coefficient process is a straight line; there might be other minimal paths. However, uniqueness clearly holds in the flat boundary model away from the boundary, since the process is of the type considered in Chapter 5.

Lemma 8.24. *If $\lambda_i(m)$ are bounded, then there exist constants C_1 and B_1 so that for all $\vec{x}$ and all $|\vec{y}| \geq B_1$,*

$$\ell_c(\vec{x}, \vec{y}) \geq C_1|\vec{y}| \log|\vec{y}| \quad \text{and} \quad \ell_b(\vec{x}, \vec{y}) \geq C_1|\vec{y}| \log|\vec{y}|.$$

Proof. As in Lemma 5.17, but put $\vec{\theta} = (\vec{\theta}^d, \vec{0}^D)$ or $(0, \vec{\theta}^d)$ respectively.   ∎

For a more detailed statement of the following, see Lemma 5.18.

Lemma 8.25. *Assume the $\lambda_i(m)$ are bounded and consider $I_0^T = {}_c I_0^T$ or $I_0^T = {}_b I_0^T$. For each K the functions in $\{\vec{r} : I_0^T(\vec{r}) \leq K\}$ are uniformly equicontinuous. Moreover, for each $\varepsilon > 0$ there is some $B = B(K)$ so that*

$$\int_0^T \mathbf{1}\left[|\vec{r}\,'(t)| \geq B\right] dt \leq \varepsilon.$$

Proof. This follows from Lemma 8.24: see proof of Lemma 5.18.   ∎

The definition of a cone that is appropriate for the finite levels processes and for the flat boundary process around $x_0 = 0$ (cf. Definition 5.19) is

Definition 8.26. *The positive cone C_f associated with $\{\vec{e}_i(m)\}$ is*

$$C_f \stackrel{\triangle}{=} \left\{ \vec{y} \in \mathbb{R}^d : (\vec{y}, \vec{0}^D) = \sum_{m=1}^{D} \sum_{i=1}^{k(m)} a_i(m)\vec{e}_i(m) \text{ for some } \vec{a} \text{ with } a_i(m) \geq 0 \right\}.$$

Lemma 8.27. *For any $B_1 > 0$ there is a B_2 so that $\vec{y} \in C_f$ and $|\vec{y}| \leq B_1$ implies that there exists a representation $\vec{a}$ of $\vec{y}$ so that*

$$(\vec{y}, \vec{0}^D) = \sum_{m=1}^{D} \sum_{i=1}^{k(m)} a_i(m)\,\vec{e}_i(m), \quad a_i(m) \geq 0 \quad \text{and} \quad |\vec{a}| \leq B_2,$$

where $|\vec{a}| \stackrel{\triangle}{=} \max_{i,m} |a_i(m)|$. In fact, we can choose $B_2 = K B_1$ for some constant K.

Proof. This follows immediately from Lemma 5.20.   ∎

Define

$$\ell_c(\vec{\theta}, \vec{\lambda}, \vec{y}) \stackrel{\triangle}{=} \langle \vec{\theta}, \vec{y} \rangle - \max_{1 \leq m \leq D} \sum_{i=1}^{k(m)} \lambda_i(m) \left(e^{\langle \vec{\theta}, \vec{e}_i(m) \rangle} - 1 \right). \tag{8.24}$$

Define $\ell_b(\vec{\theta}, \vec{\lambda}, \vec{y})$ similarly, through (8.24) if $x_0 = y_0 = 0$ and through (5.16) if $x_0 > 0$ or $y_0 > 0$.

Let $\vec{y}$ have the representation $\vec{y} = \sum_{i,m} a_i(m)\vec{e}_i(m)$.

Lemma 8.28. *Assume $\log \lambda_i(\vec{x}; m)$ are bounded. Given any c_1 there exists a constant c_2 so that for all $\vec{y} \in C_f$ with $|\vec{y}| \leq c_1$,*

$$\ell_c(\vec{\theta}, \vec{\lambda}, \vec{y}) \geq -1 \quad \text{implies} \quad \langle \vec{\theta}, \vec{e}_i(m) \rangle \leq c_2 \text{ for all } i, m.$$

Moreover, if $\lim_{j\to\infty} \ell_c(\vec{\theta}_j, \vec{\lambda}(\vec{x}), \vec{y}) = \ell_c(\vec{x}, \vec{y})$ *and* $\liminf_{j\to\infty}\langle\vec{\theta}_j, \vec{e}_i(m)\rangle = -\infty$ *for some* i, m, *then necessarily* $a_i(m) = 0$ *in every representation of* $\vec{y}$. *The same conclusions hold for* ℓ_b.

Proof. Let $K \overset{\triangle}{=} \sum_{m=1}^{D} k(m)$. Clearly

$$\max_{1\le m\le D} \sum_{i=1}^{k(m)} \lambda_i(m)\left(e^{\langle\vec{\theta},\vec{e}_i(m)\rangle} - 1\right) \ge K^{-1} \sum_{m=1}^{D} \sum_{i=1}^{k(m)} \lambda_i(m)\left(e^{\langle\vec{\theta},\vec{e}_i(m)\rangle} - K\right).$$

Therefore,

$$\ell_c(\vec{\theta}, \vec{\lambda}, \vec{y}) \le \sum_{m=1}^{D} \sum_{i=1}^{k(m)} \left(a_i(m)\langle\vec{\theta}, \vec{e}_i(m)\rangle - K^{-1}\lambda_i(m)\left(e^{\langle\vec{\theta},\vec{e}_i(m)\rangle} - K\right)\right).$$

The proof is now exactly as in Lemma 5.21. The proof for ℓ_b is identical. ∎

As in the remark following the statement of Lemma 5.21, it follows that we can assume that $\langle\vec{\theta}_j, \vec{e}_i(m)\rangle$ converges to some constants C_{im}.

Lemma 8.29. *Assume* $\log \lambda_i(\vec{x}; m)$ *are bounded. Then* $\ell_c(\vec{x}, \vec{y})$ *is bounded for* $\vec{y}$ *in bounded subsets of* C_f, *uniformly in* $\vec{x}$. *Moreover, for each* x, $\ell_c(\vec{x}, \cdot)$ *is continuous on* C_f. *The same conclusions apply to* ℓ_b *when* $x_0 > 0$ *or, if* $x_0 = 0$, *provided* ℓ_b *is restricted either to* $y_0 = 0$ *or to* $y_0 > 0$.

Proof. From the proofs of Lemmas 5.21 and 8.28 it follows that ℓ_c is finite at each point in C_f. Boundedness follows by the arguments of Lemma 5.22. By Corollary 8.18 the function ℓ_c is convex and lower semicontinuous. Theorem A.45 now implies the continuity, since for all B, the set $C_f \cap \{\vec{y} : |\vec{y}| \le B\}$ is a polytope. Finally, since C_f is closed, ℓ_c is uniformly continuous on bounded subsets.

For a more explicit proof, follow the proof of Lemma 5.22. The proof for ℓ_b is identical. ∎

Lemma 8.30. *Assume* $\log \lambda_i(\vec{x}; m)$ *are bounded and let the ergodicity Assumption 8.4 hold. Then for each* $\varepsilon > 0$ *and* c_1 *there exists a bound* B *so that for all* $\vec{y} \in C_f$ *with* $|\vec{y}| \le c_1$,

$$\sup_{|\vec{\theta}|\le B} \ell_c\left(\vec{\theta}, \vec{\lambda}(\vec{x}), \vec{y}\right) \ge \ell_c\left(\vec{\lambda}(\vec{x}), \vec{y}\right) - \varepsilon.$$

Under Assumption 8.8 this conclusion holds also for ℓ_b.

Proof. The proof of Lemma 5.23 applies, with the following changes. Let π be the invariant measure for the levels process, which exists due Assumption 8.4. Set

$$\vec{z}(\vec{y}, \delta) \overset{\triangle}{=} \vec{y} + \sum_{m=1}^{D} \sum_{i=1}^{k(m)} \delta\pi(m)\vec{e}_i(m),$$

and restrict to the smallest affine set Aff C_f containing C_f [Definition A.37(i)]. Interpreting open sets as open relatively to Aff C_f, the previous part of the proof where $\vec{x}$ is held fixed applies. Replacing (5.18) with

$$\left| \ell_c\left(\vec{\theta}, \vec{\lambda}, \vec{y}\right) - \ell_c\left(\vec{\theta}, \vec{\lambda}', \vec{y}\right) \right| \leq \max_{1 \leq m \leq D} \sum_{i=1}^{k(m)} \left| \lambda_i(m) - \lambda_i'(m) \right| \left| e^{\langle \vec{\theta}, \vec{e}_i(m) \rangle} - 1 \right|$$

$$\leq K e^{c_2 \delta}$$

the rest of the proof of Lemma 5.23 applies as well. The conclusion for ℓ_b follows by applying either the same argument, or Lemma 5.22, depending on the region. ∎

Exercise 8.31. Prove Lemma 8.30 when Assumption 8.4 is replaced by the assumption that there are no transient levels. Hint: there exists an invariant measure with positive probability at all levels. ♠

Lemma 8.32. *For each $\vec{y} \in C_f$ the infimum in Theorem 8.19(ii) is attained. That is, there is a $\vec{\mu} \in \mathcal{B}$ and a $\vec{v} \in \mathcal{S}(\vec{\mu})$ such that*

$$\sup_{\vec{\theta} \in \mathbb{R}^{d+D}} \left(\left\langle \vec{\theta}^d, \vec{y} \right\rangle - \max_{1 \leq m \leq D} \sum_{i=1}^{k(m)} \lambda_i(m) \left(e^{\langle \vec{\theta}, \vec{e}_i(m) \rangle} - 1 \right) \right)$$

$$= \sum_{m=1}^{D} \mu_m \left(\sum_{i=1}^{k(m)} v_i(m) \log \frac{v_i(m)}{\lambda_i(m)} - v_i(m) + \lambda_i(m) \right).$$

Proof. See Exercise 8.33 below. ∎

Exercise 8.33. Prove Lemma 8.32. Hint: take a sequence of points $\left(\vec{\mu}^{(k)}, \vec{v}^{(k)} \right)$, $k = 1, \dots$ that lead to the infimum, and take a subsequence along which every coordinate either converges, or diverges monotonically to ∞. If $v_i^{(k)}(m)$ diverges, set $v_i(m) \stackrel{\triangle}{=} \lambda_i(m)$, and otherwise give $v_i(m)$ and $\mu(m)$ the value of the limits. Then $\vec{\mu}, \vec{v}$ attains the infimum. ♠

Let $v_i(m)$ be another set of rates for the jump directions $\vec{e}_i(m)$. Suppose that the invariant measure π_v is unique (this means that Assumption 8.4 holds). Define

$$f_c(\vec{v}, \vec{\lambda}) = \sum_{m=1}^{D} \pi_v(m) \left(\sum_{i=1}^{k(m)} \lambda_i(m) - v_i(m) + v_i(m) \log \frac{v_i(m)}{\lambda_i(m)} \right). \tag{8.25}$$

Lemma 8.34. *Let $\vec{\theta} \in \mathbb{R}^d$, and let $\lambda_i(m)$ and $a_i(m)$ be positive numbers. Then*

$$\vec{y} = \sum_{m=1}^{D} \sum_{i=1}^{k(m)} a_i(m) \vec{e}_i(m) \quad \text{implies} \quad \ell_c(\vec{\theta}, \vec{\lambda}, \vec{y}) \leq f_c(\vec{a}, \vec{\lambda}).$$

Proof. Follows from Theorem 8.19. ∎

Similarly, the analogue of Theorem 5.26 follows from the representation in Theorem 8.19 *(ii)*

$$\ell_c(\vec{x}, \vec{y}) \overset{\Delta}{=} \inf_{\vec{v}} \left\{ f_c(\vec{v}, \vec{\lambda}(\vec{x})) : \sum_{m=1}^{D} \left(\pi_v(m) \sum_{i=1}^{k(m)} v_i(m) \vec{e}_i^d(m) \right) = \vec{y} \right\}, \quad (8.26)$$

since the last D coordinates in the definition of $\mathcal{S}(\mu)$ imply that μ is an invariant measure. The same comment applies to the flat boundary process with $x_0 = y_0 = 0$, except that there m in (8.25) takes the values zero and one. In this case the invariant measure is on the "states" zero (boundary) and one (interior).

Lemma 8.35. *Suppose Assumptions 8.4 (ergodicity) and 8.5 (Lipschitz continuity) hold. Then for any constant B_1, the function $\ell_c(\cdot, \vec{y})$ is continuous, uniformly over the set*

$$\{\vec{y} : \vec{y} \in \mathcal{C}_f, \ |\vec{y}| \le B_1\}.$$

Under Assumption 8.9 (Lipschitz continuity) the same holds for $\ell_b(\vec{x}, \vec{y})$ when $\vec{x}$ is restricted to $\{x_0 > 0\}$, and when in addition Assumption 8.8 holds, the same holds when $\vec{x}$ is restricted to $\{x_0 = 0\}$.

Proof. This is established in Equation (5.19): see the proof of Lemma 8.30. ∎

Lemma 8.35 is clearly false for the flat boundary process if we allow x_0 to approach the boundary. For example, suppose that there is a jump direction $\vec{e}_j(0)$ such that $\langle \vec{e}_j(0), \vec{e}_i(1) \rangle < 0$ for all i. Then $\ell_b(\vec{x}, \vec{e}_j(0)) = \infty$ for $x_0 > 0$, but $\ell_b(\vec{x}, \vec{e}_j(0)) < \infty$ when $x_0 = 0$.

Lemma 8.36. *If $\log \lambda_i(\vec{x}; m)$ are bounded then there exist C_2 and B_2 so that for all $\vec{y} \in \mathcal{C}_f$ and all $\vec{x}$,*

$$\ell_c(\vec{x}, \vec{y}) \le \begin{cases} C_2 & \text{if } |\vec{y}| \le B_2, \\ C_2|\vec{y}| \log |\vec{y}| & \text{if } |\vec{y}| \ge B_2. \end{cases}$$

Proof. The same as the proof of Lemma 5.32, since obviously any component of an invariant measure is bounded by one. ∎

Theorem 5.35 depends only on Lemmas 5.17, 5.18, and 5.33, and so it holds for the finite levels process. We need to modify the statement of the Theorem to have it hold for the flat boundary process, since Lemma 5.33 does not hold there.

Theorem 8.37. *Let Assumptions 8.4–8.5 hold both for the original $\lambda_i(\vec{x}; m)$ as well as for some rates $\mu_i(\vec{x}; m)$. Then the conclusions of Theorem 5.35 hold for the finite levels process. Fix $\vec{r} \in C^d[0, T]$ with $_c I_0^T(\vec{r}) < \infty$. Define $_\mu I_0^T(\vec{r})$ as* the function $_cI_0^T(\vec{r})$ with the $\mu_i(\vec{x}; m)$ replacing the $\lambda_i(\vec{x}; m)$ in the definition of ℓ_c. For any $\varepsilon > 0$ there exists a δ so that

$$\max_{i,m} \sup_{0 \le t \le T} |\mu_i(\vec{r}(t), m) - \lambda_i(\vec{r}(t), m)| < \delta \quad \text{implies} \quad \left|_\mu I_0^T(\vec{r}) - {}_cI_0^T(\vec{r})\right| \le \varepsilon.$$
(8.27)

Conclusion (8.27) holds for $_bI_0^T(\vec{r})$ with an identical definition of $_\mu I_0^T(\vec{r})$, but under Assumptions 8.8–8.9.

Proof. Exactly the same as the proof of Theorem 5.35.   ∎

The definition 5.36 of $\overset{\delta}{g}$ takes the following form in the present case.

$$\overset{\delta}{g}(\vec{x}, \vec{\theta}) \overset{\triangle}{=} \sup_{\vec{x} \,:\, |\vec{z}^d - \vec{x}^d| \le \delta} \sum_{i=1}^{k(m)} \lambda_i(\vec{z}^d; m) \left(e^{\langle \vec{\theta}, \vec{e}_i(m)\rangle} - 1\right)$$

$$= \sup_{|\vec{z}^d - \vec{x}^d| \le \delta} \max_m \sum_{i=1}^{k(m)} \lambda_i(\vec{z}^d; m) \left(e^{\langle \vec{\theta}, \vec{e}_i(m)\rangle} - 1\right),$$
(8.28)

where in the middle term, $m = m(\vec{x})$. With this new definition of $\overset{\delta}{g}$ we define, for $\vec{y} \in C_f$

$$\ell_c^\delta(\vec{\theta}, \vec{x}, \vec{y}) \overset{\triangle}{=} \left\langle \vec{\theta}^d, \vec{y}^d \right\rangle - \overset{\delta}{g}(\vec{x}, \vec{\theta})$$

$$\ell_c^\delta(\vec{x}, \vec{y}) \overset{\triangle}{=} \sup_{\vec{\theta} \in \mathbb{R}^{d+D}} \left(\left\langle \vec{\theta}^d, \vec{y}^d \right\rangle - \overset{\delta}{g}(\vec{x}, \vec{\theta}) \right).$$

These are also the definitions of ℓ_b^δ when $x_0 = y_0 = 0$. Otherwise, the definition of ℓ_b^δ is given through Definitions 5.36–5.37. As before, we set $\ell^\delta(\vec{x}, \vec{y}) = \infty$ whenever $\vec{y} \notin C_f$.

Lemma 8.38 Lower semicontinuity. *If the $\lambda_i(\vec{x}; m)$ are bounded and continuous, then the function $\ell_c^\delta(\vec{x}, \vec{y})$ is lower semicontinuous in $(\delta, \vec{x}, \vec{y})$. The same holds for $\ell_b^\delta(\vec{x}, \vec{y})$.*

Proof. Identical to the proof of Lemma 5.40, except in the case of ℓ_b with $x_0 = 0$. In this case the function f of (5.28) is only lower semicontinuous in $\vec{x}$, and the rest of the proof is unchanged.   ∎

Exercise 8.39. If the $\lambda_i(\vec{x}; m)$ are continuous then $\ell_c^\delta(\vec{x}, \vec{y})$ is right continuous in δ, that is,

$$\lim_{\delta' \downarrow \delta} \ell^{\delta'}(\vec{x}, \vec{y}) = \ell^\delta(\vec{x}, \vec{y})$$

uniformly in $(\vec{x}, \vec{y})$ in bounded sets. The same holds for ℓ_b^δ. In particular,

$$\ell_c(\vec{x}, \vec{y}) = \lim_{\delta \downarrow 0} \ell_c^\delta(\vec{x}, \vec{y})$$

$$= \sup_{\vec{\theta} \in \mathbb{R}^{d+D}} \left(\left\langle \vec{\theta}^d, \vec{y}^d \right\rangle - \lim_{\delta \downarrow 0} \overset{\delta}{g}(\vec{x}, \vec{\theta}) \right).$$

Hint: see Exercise 5.41.

♠

Lemma 8.40. *If* $\log \lambda_i(\vec{x}; m)$ *are bounded and continuous, then both* $_cI_0^T(\vec{r})$ *and* $_c\overset{\delta}{I}_0^T(\vec{r})$ *are lower semicontinuous (in the* d_d *metric). The same holds for* $_bI_0^T(\vec{r})$ *and* $_b\overset{\delta}{I}_0^T(\vec{r})$.

Proof. The same as the proof of Lemma 5.42.
■

Lemma 8.41. *Let Assumptions 8.4–8.5 hold. Then for any* $\vec{r}$ *with* $_cI_0^T(\vec{r}) < \infty$ *and any* $\varepsilon > 0$ *there exists a step function* $\vec{\theta}(t)$ *so that*

$$\int_0^T \ell_c(\vec{\theta}(t), \vec{\lambda}(\vec{r}(t)), \vec{r}\,'(t))\, dt \geq {_cI_0^T}(\vec{r}) - \varepsilon.$$

The same conclusion holds if $(\ell_c, {_cI_0^T})$ *is replaced with* $\left(\ell_c^\delta, {_c\overset{\delta}{I}_0^T} \right)$ *or, under Assumptions 8.8–8.9, with* $(\ell_b, {_bI_0^T})$ *or with* $\left(\ell_b^\delta, {_b\overset{\delta}{I}_0^T} \right)$.

Proof. Note that $\ell_c^\delta\left(\vec{\theta}, \vec{\lambda}(\vec{x}), \vec{y} \right)$ is continuous in $\left(\vec{\theta}, \vec{x}, \vec{y} \right)$, uniformly on compact sets. The proof for the finite levels model is therefore the same as that of Lemma 5.43. For the flat boundary model, divide time into $\{ t : r^0(t) = 0 \}$ and $\{ t : r^0(t) > 0 \}$ for the purpose of defining the simple function $\vec{\theta}_1(t)$. The rest of the argument from Lemma 5.43 goes through as before.
■

Recall Definitions 5.44–5.45 of the level sets $\Phi_{\vec{x}}(K)$ of $I_0^T(\vec{r})$ and $\overset{\delta}{\Phi}_{\vec{x}}(K)$ of $\overset{\delta}{I}_0^T(\vec{r})$. We omit the subscripts $_c$ and $_b$ below, where no confusion may arise.

Proposition 8.42. *If* $\log \lambda_i(\vec{x}; m)$ *are bounded and continuous, then for any compact* $C \subset \mathbb{R}^d$

$$\bigcup_{\vec{x} \in C} \Phi_{\vec{x}}(K) \quad and \quad \bigcup_{\vec{x} \in C} \overset{\delta}{\Phi}_{\vec{x}}(K)$$

are compact sets in $C^d[0, T]$, *for both the finite levels and the flat boundary processes.*

Proof. Identical to the proof of Proposition 5.46, with Lemma 8.25 replacing Lemma 5.18. ∎

Recall the notation $I_{\vec{x}}^*(F) \overset{\triangle}{=} \inf\{I_0^T(\vec{r}) : \vec{r} \in F,\ \vec{r}(0) = x\}$.

Lemma 8.43. *If* $\log \lambda_i(\vec{x})$ *are bounded and continuous, then for each closed set* $F \in C^d[0, T]$, $_cI_{\vec{x}}^*(F)$ *and* $_bI_{\vec{x}}^*(F)$ *are lower semicontinuous in* $\vec{x}$.

Proof. Identical to the proof of Lemma 5.47, with Proposition 8.42 replacing Proposition 5.46. ∎

Finally, Lemma 5.48 holds, with a slightly different proof.

Lemma 8.44. *Let* C *be compact and let Assumptions 8.4–8.5 (finite levels) or Assumptions 8.8–8.9 (flat boundary) hold. Given* $K, \varepsilon > 0$, *there exists a* $\delta > 0$ *such that*

$$\Phi_{\vec{x}}^{\delta}(K - \varepsilon) \subset \left\{\vec{r} : d\left(\vec{r}, \Phi_{\vec{x}^d}(K)\right) \le \varepsilon\right\} \quad \text{for all} \quad \vec{x}^d \in C.$$

Proof. There is no difference in the proof for the finite levels process. For the flat boundary process the proof goes exactly as before up to the definition (5.29) of $\tilde{\vec{r}}(t)$. We are no longer able to take $\tilde{\vec{r}}(t) = \vec{r} + (\vec{x}_i - \vec{x})$ since, if $x_{i,0} \ne x_0$, this would change the times the path spends on the boundary (or even "penetrate" the boundary!) so that the hypotheses of Theorem 8.37 would not hold. The appropriate approximation $\tilde{\vec{r}}$ is derived in Exercise 8.45 below. With this $\tilde{\vec{r}}$, the proof is concluded exactly as before. ∎

Exercise 8.45. Derive an approximating $\tilde{\vec{r}}$ for Lemma 8.44, so that for i large

$$\sup_{0 \le t \le T} \left|\tilde{\vec{r}} - \vec{r}_i\right| < \varepsilon, \quad {}_bI_0^T\left(\tilde{\vec{r}}\right) < K \quad \text{and} \quad \tilde{\vec{r}} \in \Phi_{x_i}(K).$$

Hints: if $\inf_{0 \le t \le T} r_0(t) > 0$ or if $x_{i,0} = x_0$ for infinitely many i, the previous construction works. Otherwise, show that it suffices to consider separately the two cases: $x_{i,0} > x_0$ for all i, or $x_{i,0} < x_0$ for all i. In the first case, take a path that starts at $\vec{x}_i$, follows any (interior) jump direction with a negative component in the zero direction until its distance from the boundary is x_0, and continue following $\vec{r}$ (shifted in time). This creates a path with the same time spent on the boundary (except for a small initial and small terminal interval). In the second case, begin with $\vec{r}$ shifted to start at $\vec{x}_i$ until the first time this path hits the boundary. Continue by following the segment of $\vec{r}$ starting at the point $\vec{r}$ first hits the boundary (shifted in time, and shifted in space to match the first segment). Use continuity of the path to establish the desired properties. ♠

Note that Lemma 8.44 fails in the flat boundary case if the boundary cannot be reached. For an example, take any two jump processes on $\mathbb{R}^d$ whose jump directions $\vec{e}_j^1$, $\vec{e}_j^2$ span disjoint cones. Let one process define $\vec{z}_n(t)$ for $z_n^0 = 0$, and the other define $\vec{z}_n(t)$ for $z_n^0 > 0$. Then if $x_i^0 > 0$, $x^0 = 0$, we see that we cannot approximate $\vec{r}_{\vec{x}}(t)$ by $\vec{r}_{\vec{x}_i}(t)$. Of course, if the boundary cannot be reached we do not need a boundary theory ...

8.3. Proof of the Upper Bound

We now move on to the proof of the upper bound. Here is where the hard work of §5.5 bears fruit: almost nothing needs to be changed in order that the same theorems apply to the boundary and finite level processes. Recall that the scaled process $\vec{z}_n$ does not include the levels: see the definitions in §8.1. We also omit the subscript c and b. We shall prove

Theorem 8.46 (Finite Levels). *Suppose Assumptions 8.4 (ergodicity) and 8.5 (Lipschitz continuity) hold. Then for each $T > 0$, each closed set $F \subset D^d[0, T]$ and each $\vec{x} \in \mathbb{R}^d$, uniformly in $j \in [1, \ldots, D]$,*

$$\limsup_{\vec{y} \to \vec{x}, n \to \infty} \frac{1}{n} \log \mathbb{P}_{\vec{y}, j} (\vec{z}_n \in F) \leq -I_{\vec{x}}(F).$$

Theorem 8.47 (Flat Boundary). *Suppose Assumptions 8.8 (irreducibility) and 8.9 (Lipschitz continuity) hold. Then for each $T > 0$, each closed set $F \subset D^{d+1}[0, T]$ and each $\vec{x} \in \mathbb{R}^{d+1}$ with $x_0 \geq 0$,*

$$\limsup_{\vec{y} \to \vec{x}, n \to \infty} \frac{1}{n} \log \mathbb{P}_{\vec{y}} (\vec{z}_n \in F) \leq -I_{\vec{x}}(F)$$

provided $y_0 \geq 0$.

Proving these theorems is simply a matter of checking that the argument given in §5.5 goes through. We refer to that section for motivation and some of the proofs. We now check the steps one by one for the two processes.

For the finite levels process define $\vec{y}_n(t)$ as the linear interpolation of $\vec{z}_n^d(t)$:

$$\vec{y}_n(t) \triangleq \left(\frac{n}{T}t - j\right) \vec{z}_n^d\left(t_j^n\right) + \left((j+1) - \frac{n}{T}t\right) \vec{z}_n^d\left(t_{j+1}^n\right), \ t \in \left[t_j^n, t_{j+1}^n\right].$$

For the flat boundary process we take $\vec{y}_n(t)$ exactly as in Definition 5.56:

$$\vec{y}_n(t) \triangleq \left(\frac{n}{T}t - j\right) \vec{z}_n\left(t_j^n\right) + \left((j+1) - \frac{n}{T}t\right) \vec{z}_n\left(t_{j+1}^n\right), \ t \in \left[t_j^n, t_{j+1}^n\right].$$

To simplify the notation, we will often omit the dependence on the initial level j.

The first four claims in §5.5 do not assume continuity of the $\lambda_i(\vec{x}; m)$. Therefore they hold for both the finite levels process and the flat boundary process, and the only changes are that in the finite level case we ignore D coordinates. We now check Corollary 5.55.

Corollary 8.48. *If the* $\lambda_i(\vec{x}; m)$ *are bounded, then there are positive constants* c_1 *and* c_2 *such that for any* $0 \leq t \leq t + \Delta \leq T$,

$$\mathbb{P}_{\vec{x}} \left(\sup_{t \leq s \leq t+\Delta} |\vec{z}_n(s) - \vec{z}_n(t)| \geq a \right) \leq \exp\left(-nac_1 \log\left(\frac{ac_2}{\Delta} \right) \right)$$

uniformly in $\vec{x}$.

Proof. Corollary 5.55 did not require continuity of the $\lambda_i(\vec{x})$, so its proof holds. We can replace $\vec{z}_n$ by $\vec{z}_n^d$ for the finite levels process, and the result still holds since

$$\left| \vec{z}_n^d(s) - \vec{z}_n^d(t) \right| \leq |\vec{z}_n(s) - \vec{z}_n(t)| . \qquad \blacksquare$$

Lemma 8.49. *For the flat boundary process with bounded* $\lambda_i(\vec{x}; m)$,

$$\limsup_{n \to \infty} \frac{1}{n} \log \mathbb{P}_{\vec{x}} \left(d(\vec{z}_n(t), \vec{y}_n(t)) > \delta \right) = -\infty$$

uniformly in $\vec{x} \in \mathbb{R}^{d+1}$. *For the finite levels process with bounded* $\lambda_i(\vec{x}; m)$,

$$\limsup_{n \to \infty} \frac{1}{n} \log \mathbb{P}_{\vec{x}} \left(d(\vec{z}_n^d(t), \vec{y}_n(t)) > \delta \right) = -\infty$$

uniformly in $\vec{x} \in \mathbb{R}^d$.

Proof. This follows from Corollary 8.48, exactly as Lemma 5.57 followed from Corollary 5.55. $\qquad \blacksquare$

Recall the definition (5.56) of the compact set $\mathcal{K}(M)$:

$$\mathcal{K}(M) \stackrel{\Delta}{=} \bigcap_{m=M}^{\infty} \left\{ \vec{r} \in C^d[0, T] : \vec{r}(0) \in C, \ V_{2-m}(\vec{r}) \leq \frac{1}{\log m} \right\},$$

where V_δ was defined in (5.55) as the modulus of continuity.

Lemma 8.50. *Assume the* $\lambda_i(\vec{x}; m)$ *are bounded. Let* C *be a compact set, where* $C \subset \mathbb{R}^d$ *for the finite levels process, and* $C \subset \mathbb{R}^{d+1}$ *for the flat boundary process. For each* $B < \infty$ *there is a compact* $\mathcal{K} \subset C^d[0, T]$ *(for the finite levels model, while* $\mathcal{K} \subset C^{d+1}[0, T]$ *for the flat boundary process) such that for all* $\vec{x} \in C$,

$$\limsup_{n \to \infty} \frac{1}{n} \log \mathbb{P}_{\vec{x}} \left(\vec{y}_n \notin \mathcal{K} \right) \leq -B$$

for the flat boundary process. For the finite levels model, for all $j \in [1, \ldots, D]$,

$$\limsup_{n \to \infty} \frac{1}{n} \log \mathbb{P}_{\vec{x},j} \left(\vec{y}_n \notin \mathcal{K} \right) \leq -B.$$

Proof. Lemma 5.58 applies, as no continuity was assumed. $\qquad \blacksquare$

Lemma 8.51. *If the $\lambda_i(\vec{x}; m)$ are bounded, then uniformly over $\vec{x} \in \mathbb{R}^d$ (over $\vec{x} \in \mathbb{R}^{d+1}$ for the flat boundary process) and in bounded $\vec{\theta}$,*

$$\limsup_{n\to\infty} \log \mathbb{E}_{\vec{x}} \left(\exp \left(n \left\langle \vec{y}_n \left(\frac{T}{n} \right) - \vec{y}_n(0), \vec{\theta} \right\rangle \right) \right) \leq T \overset{\delta}{g}(\vec{x}, \vec{\theta}).$$

Proof. For the flat boundary process, the proof is the same as that of Lemma 5.59. For the finite levels process, simply note that we only use $\vec{\theta}$ with $\theta^D = 0$. The estimation goes as before, since now $\langle \vec{y}_n, \vec{\theta} \rangle = \langle \vec{y}_n^d, \vec{\theta}^d \rangle$ and the rest of the estimates have also been reduced to $\mathbb{R}^d$. ∎

We have to rework the proof of Lemma 5.61, since continuity of $\lambda_i(\vec{x})$ was assumed.

Lemma 8.52. *Let Assumptions 8.5 or 8.9 (Lipschitz continuity) hold. Let C be a compact set, where $C \subset \mathbb{R}^d$ for the finite levels process, and $C \subset \mathbb{R}^{d+1}$ for the flat boundary process. Fix a step function $\vec{\theta}$. For each $\delta > 0$ and each compact set $\mathcal{K} \subset \mathcal{K}(M)$ we have*

$$\limsup_{n\to\infty} \frac{1}{n} \log \mathbb{P}_{\vec{x}} \left(\vec{y}_n \in \mathcal{K}_x \right) \leq - \inf_{\vec{r}\in\mathcal{K}_x} \overset{\delta}{I}{}_0^T (\vec{r}, \vec{\theta}),$$

uniformly in $\vec{x} \in C$, where

$$\overset{\delta}{I}{}_0^T (\vec{r}, \vec{\theta}) \triangleq \int_0^T \langle \vec{r}\,'(t), \vec{\theta}(t) \rangle - \overset{\delta}{g}(\vec{r}(t), \vec{\theta}(t)) \, dt$$

whenever $\vec{r}$ is absolutely continuous, and is defined as $+\infty$ otherwise. The starting point can be any level $j \in [0, \dots, D]$ for the finite levels model.

Proof. For the finite levels model, the proof is exactly the same as the proof of Lemma 5.61, since the function $\overset{\delta}{g}$ is continuous in $\vec{x}$ ($= \vec{x}^d$). For the flat boundary model this function is not continuous; there may be a jump when $x_0 = \delta$. However, the argument up to (5.64) holds; that is, for the flat boundary model we have

$$\limsup_{n\to\infty} \frac{1}{n} \log \mathbb{P}_{\vec{x}}(\vec{y}_n \in \mathcal{K}_x) \leq - \liminf_{n\to\infty} \left(\inf_{\vec{r}\in\mathcal{K}_x} S_n(\vec{r}, \vec{\theta}) \right), \tag{5.64}$$

where

$$S_n(\vec{r}, \vec{\theta}) \triangleq \sum_{j=0}^{n-1} \left(\left\langle \vec{r}(t_{j+1}^n) - \vec{r}(t_j^n), \vec{\theta}(t_j^n) \right\rangle - \frac{T}{n} \overset{\delta}{g}(\vec{r}(t_j^n), \vec{\theta}(t_j^n)) \right).$$

To continue, note that for any $\delta > 0$ and any step function $\vec{\theta}(t)$, there exists an n_0 such that for all $n \geq n_0$, all $\vec{r} \in \mathcal{K}$, and all $t \in [t_j^n, t_{j+1}^n]$ with no jump of $\vec{\theta}(t)$ in this interval,

$$\overset{\delta}{g}(\vec{r}_j, \vec{\theta}_j) \leq \overset{2\delta}{g}(\vec{r}(t), \vec{\theta}(t)).$$

This is because all $\vec{r} \in \mathcal{K}$ have the same modulus of continuity V. Therefore for each step function $\vec{\theta}(t)$, if τ is the time of the first jump of $\vec{\theta}$,

$$\sum_{j=0}^{n-1} \mathbf{1}\left[t_{j+1}^n \le \tau\right] \frac{T}{n} \overset{\delta}{g}(\vec{r}(t_j^n), \vec{\theta}(t_j^n)) \le \int_0^\tau \overset{2\delta}{g}\,(\vec{r}(t), \vec{\theta}(0))\,dt + e_3(n). \qquad (8.29)$$

The error term $e_3(n)$ arises because the time τ might be in the interior of an interval $[t_j^n, t_{j+1}^n]$. Although $\overset{2\delta}{g}$ may not be continuous, it is bounded on the compact set

$$C_1 = \{\vec{r}(t) : \vec{r} \in \mathcal{K},\ \vec{r}(0) \in C,\ 0 \le t \le T\},$$

and therefore $e_3(n) \to 0$ as $n \to \infty$ uniformly in $\vec{r} \in \mathcal{K}$ and $\vec{x} \in C$.

Equation (8.29) is a generalization of (5.66). Also, (5.65) holds verbatim:

$$\sum_{j=0}^{n-1} \mathbf{1}\left[t_{j+1}^n \le \tau\right] \left\langle \vec{r}(t_{j+1}^n) - \vec{r}(t_j^n), \vec{\theta}(0) \right\rangle = \int_0^\tau \left\langle \vec{r}'(t), \vec{\theta}(0) \right\rangle dt + e_2(n),$$

where the error $e_2(n)$ is uniform, since C_1 is compact. So we obtain for each $\vec{\theta}$ and $\delta > 0$,

$$\liminf_{n \to \infty} \inf_{\vec{r} \in \mathcal{K}_x} S_n(\vec{r}, \vec{\theta}) \ge \inf_{\vec{r} \in \mathcal{K}_x} \int_0^T \langle \vec{r}'(t), \vec{\theta}(t) \rangle - \overset{2\delta}{g}\,(\vec{r}(t), \vec{\theta}(t)) \, dt.$$

And finally we obtain for any $\delta > 0$ (replacing δ by $\delta/2$ if necessary)

$$\limsup_{n \to \infty} \frac{1}{n} \log \mathbb{P}_{\vec{x}} \left(\vec{y}_n \in \mathcal{K}_x \right) \le - \inf_{\vec{r} \in \mathcal{K}_x} \overset{\delta}{I}_0^T (\vec{r}, \vec{\theta})$$

uniformly in $x \in C$. $\blacksquare$

The proof of Proposition 5.62 uses continuity of $\overset{\delta}{g}$ and the previous results, so that it goes through practically verbatim for finite levels processes. Since an approximating function $\vec{\theta}(\cdot)$ is required, ergodicity assumptions appears.

Proposition 8.53. *Let Assumptions 8.5–8.4 or 8.9–8.8 hold. Let C be a compact set, where $C \subset \mathbb{R}^d$ for the finite levels process, and $C \subset \mathbb{R}^d \times [0, \infty)$ for the flat boundary process. Then, for each K, δ, and $\varepsilon > 0$, uniformly in $\vec{x}$: For the finite levels process, uniformly in $j \in [1, \dots, D]$,*

$$\limsup_{n \to \infty} \frac{1}{n} \log \mathbb{P}_{\vec{x}, j} \left(d\left(\vec{y}_n, \overset{\delta}{\Phi}_{\vec{x}}(K) \right) > \varepsilon \right) \le -(K - \varepsilon).$$

For the flat boundary process,

$$\limsup_{n \to \infty} \frac{1}{n} \log \mathbb{P}_{\vec{x}} \left(d\left(\vec{y}_n, \overset{\delta}{\Phi}_{\vec{x}}(K) \right) > \varepsilon \right) \le -(K - \varepsilon).$$

Proof. For the finite levels process, the proof is the same as that for Proposition 5.62. For the flat boundary process, as we show in Exercise 8.54 below, given any $\vec{\theta}$ and $\vec{r}$, there exists a neighborhood $N(\vec{r})$ such that for any $\vec{s} \in N(\vec{r})$,

$$\overset{\delta}{I}{}_0^T(\vec{s}, \vec{\theta}) \geq \overset{\delta}{I}{}_0^T(\vec{r}, \vec{\theta}) - \varepsilon. \tag{8.30}$$

Then the rest of the proof goes through exactly as before. ∎

Exercise 8.54. Establish (8.30). Hint: the usual continuity arguments apply when $\vec{s}(t)$ and $\vec{r}(t)$ are both on the boundary or both in the interior. Moreover, if $\vec{r}(t)$ is within δ of the boundary but $\vec{s}(t)$ is not, then the inequality is in the right direction. Show that the amount of time the reverse happens can be made small (uniformly in $\vec{s}$!) by choosing the neighborhood to be small. ♠

Finally, Lemma 5.63 of §5.5 was a consequence of lower semicontinuity—Lemma 5.47. But we established lower semicontinuity in Lemma 8.43 in §8.2. So, we have the following result.

Lemma 8.55. *Let Assumptions 8.5 or 8.9 (Lipschitz continuity) hold. Then for each pair of closed sets $F \in D^d[0, T]$ and $\vec{x} \in \mathbb{R}^d$ for the finite levels process, or $F \in D^{d+1}[0, T]$ and $\vec{x} \in \mathbb{R}^{d+1}$ for the flat boundary process, we have*

$$\lim_{\varepsilon \downarrow 0} \inf_{|\vec{x} - \vec{y}| < \varepsilon} I_{\vec{y}}(F) = I_{\vec{x}}(F).$$

We now have all the ingredients we need for proving the large deviations upper bound for both types of boundary processes.

Proof of the upper bounds Theorem 8.46—finite level processes, and Theorem 8.47—flat boundary processes. Identical to the proof of Theorem 5.64. ∎

Corollary 8.56. *Under the assumptions of either Theorem 8.46 or Theorem 8.47, if $I_{\vec{x}}(F)$ is continuous in $\vec{x}$ over a compact set C, then the upper bound is uniform in $\vec{x}$ over the set C.*

Proof. Identical to the proof of Corollary 5.65. ∎

The reader will note that the sequence of lemmas we have given, and hence the upper bound, holds for more general processes than we have considered. See [DEW] for a more general statement.

8.4. Constant Coefficient Processes

As in §5.3, our proof of the lower bound uses an approximation by a locally constant coefficient process. However, our processes change their character as the finite levels process moves from level to level, or as the flat boundary process moves along the boundary, staying part of the time on the boundary and part of the time just inside. Therefore we need to estimate the amount of time these processes spend at the various levels or on the boundary, respectively. We saw in §7.1 that in the limit, the fraction of the time the process spends at each level equals an invariant measure and, moreover, that fraction of time converges to the measure exponentially fast. This will allow us to prove the lower bound assuming that the "levels part" of the process is always in "steady state." The only point requiring extra attention is that we need the fact that our estimates hold uniformly among all processes with jump rates (coefficients) whose logarithms satisfy a given bound. This will eventually enable us to prove the lower bound uniformly in a strong sense.

We start with notation. For a finite levels process $\vec{x}(t)$ of Definition 8.3, we consider only $\vec{x}^D(t)$; that is, the level. So, to simplify the notation, we take $x(t) \in \{1, \ldots, D\}$.

Fix the given set of allowed transitions $m \to j$ by defining the *incidence matrix*

$$\mathcal{I}_{mj} = \mathbf{1}\left[j = m'(m; i) \text{ for some } i\right]. \tag{8.31}$$

Let λ_{mj} denote the rate of transition of $x(t)$ from m to j, that is, in the notation of Definition 8.3,

$$\lambda_{mj} = \sum_{i=1}^{k(m)} \lambda_i(m)\mathbf{1}\left[j = m'(m; i)\right]. \tag{8.32}$$

Both $\lambda_i(m)$ and λ_{mj} are independent of the position $\vec{x}^d$ since we assume that the processes have constant coefficients. As before, $\vec{\lambda} \in \mathbb{R}^D$ is the vector of all rates of allowed transitions. Note that, contrary to the conventions in the theory of continuous-time Markov chains and contrary to our convention concerning the diagonal terms in the definition (4.5) of the generator, here $\lambda_{mm} \geq 0$ with possibly a strict inequality. This corresponds to the jumps in the $\vec{x}^d$ part of the process without an associated jump in the $\vec{x}^D$ part.

We let $\vec{\pi}^\lambda = (\pi_1^\lambda, \ldots, \pi_D^\lambda)$ denote the unique stationary distribution of $x(t)$—it is unique since by Assumption 8.4 $x(t)$ is ergodic. For each T let $\pi_m(T)$ denote the proportion of time in $[0, T]$ that $x(t)$ spends in state m; that is,

$$\pi_m(T) \stackrel{\triangle}{=} \frac{1}{T} \int_0^T \mathbf{1}\left[x(t) = m\right] dt.$$

For the flat boundary process we consider only the component perpendicular to the boundary; that is, we let $x(t) = x_0(t) \in \{0, 1, 2, \ldots\}$. Recall that we are interested in the constant coefficient case only and, for the flat boundary process, in the case where the drift of (8.9) satisfies $\beta < 0$. Under these conditions, by

Exercise 7.17 the Markov chain (with countable state space) $x(t)$ is ergodic, and we let π_0 denote the invariant probability that $x(t) = 0$ (i.e., $\vec{x}$ is on the boundary); see (8.10).

Let $\pi_0(T)$ be the observed proportion of time that x is zero in $[0, T]$; that is,

$$\pi_0(T) \stackrel{\triangle}{=} \frac{1}{T} \int_0^T \mathbf{1}\left[x(t) = 0\right] dt.$$

Thus we have constructed two continuous-time Markov chains with finite and countable state spaces, respectively. In §7.1 and §7.2 we studied the rates with which they settle into steady state, and those results are the key to the proofs below. Most of the results concerning those Markov chains can be recovered from the monograph [Chu] by Chung; however, it is instructive (and sometimes easier) to obtain the results as a consequence of our previous development; moreover, the uniformity results appear more naturally.

Now we prove an analogue of Kurtz's Theorem for the finite levels constant coefficient process $\vec{x}(t) = (\vec{x}^d, \vec{x}^D)(t)$. Recall that $\vec{z}_n(t) = \frac{1}{n}\vec{x}^d(nt)$, and with $\vec{\pi}$ as defined above, (8.6) takes the form

$$\vec{z}_\infty(t) = \vec{z}_\infty(0) + t \cdot \sum_{m=1}^{D} \pi_m^\lambda \sum_{i=1}^{k(m)} \lambda_i(m)\vec{e}_i^d(m). \tag{8.33}$$

Note that since $\vec{\pi}^\lambda$ is invariant, $\sum_{m=1}^{D} \pi_m^\lambda \sum_{i=1}^{k(m)} \lambda_i(m)\vec{e}_i^D(m) = \vec{0}^D$.

Lemma 8.57. *Assume the jump structure $\mathcal{I}$ is ergodic (Assumption 7.1, implied by Assumption 8.4), and let $\vec{z}_\infty(0) = \vec{x}$. For any δ, ε, and $T > 0$, there are positive constants C_1 and C_2 such that for all jump processes satisfying $\delta \leq \lambda_i(m) \leq \delta^{-1}$ for all i and m, for all starting points $\vec{x}$ and all positive n,*

$$\mathbb{P}_{\vec{x}}\left(\sup_{0 \leq t \leq T} |\vec{z}_n(t) - \vec{z}_\infty(t)| > \varepsilon\right) < C_1 e^{-nC_2}.$$

Proof. This follows from Lemma 7.6 and its extension, Exercise 7.8. Let M be a bound on the total jump rate the process $\vec{x}$ undergoes, and $\bar{e}$ a bound on the jump size:

$$M \stackrel{\triangle}{=} \sum_{m=1}^{D}\sum_{i=1}^{k(m)} \pi_m^\lambda \lambda_i(m) \quad \text{and} \quad \bar{e} \stackrel{\triangle}{=} \max_{i,m} |\vec{e}_i^d(m)|. \tag{8.34}$$

We break time $[0, T]$ into intervals of length $\Delta = \varepsilon/6M\bar{e}$, and let $\eta = \varepsilon/3\bar{e}kDT$, where $k \stackrel{\triangle}{=} \max_m k(m)$ is the maximum number of jump directions at any level. Let $n_i(m; n\Delta)$ be the number of jumps in direction $\vec{e}_i^d(m)$ during the interval $[0, n\Delta)$. Since the process $\vec{x}(t)$ is constant coefficient, $n_i(m; n\Delta)$ corresponds also to the number of jumps of the finite-state Markov chain described in the text above Exercise 7.8. According to Lemma 7.6 and Exercise 7.8, there is a

$C = C(\Delta) > 0$ and an $n_0 = n_0(\Delta)$ such that for any $n > n_0$,

$$\mathbb{P}\left(\max_{i,m}\left|\frac{n_i(m;n\Delta)}{n\Delta} - \pi_m^\lambda \lambda_i(m)\right| > \eta\right) \le e^{-nC}. \tag{8.35}$$

Now consider the first interval of time $[0, \Delta]$. If for all i, m

$$\left|\frac{n_i(m;n\Delta)}{n\Delta} - \pi_m^\lambda \lambda_i(m)\right| \le \eta \tag{8.36}$$

then by Equation (8.33),

$$|\vec{z}_n(\Delta) - \vec{z}_n(0) - [\vec{z}_\infty(\Delta) - \vec{z}_\infty(0)]|$$

$$= \left|\sum_{m=1}^{D}\sum_{i=1}^{k(m)}\left(\frac{n_i(m;n\Delta)}{n} - \Delta\pi_m^\lambda \lambda_i(m)\right)\vec{e}_i^d(m)\right|$$

$$\le \Delta\bar{e}\sum_{m=1}^{D}\sum_{i=1}^{k(m)}\left|\frac{n_i(m;n\Delta)}{n\Delta} - \pi_m^\lambda \lambda_i(m)\right|$$

$$\le \Delta\bar{e}kD\eta,$$

which is bounded by $\varepsilon\Delta/3T$ by definition of η. Moreover, under (8.36) the maximal change in an interval of size Δ is, by a similar calculation

$$\sup_{0\le t\le\Delta}|\vec{z}_n(t) - \vec{z}_n(0)| + |(\vec{z}_\infty(t) - \vec{z}_\infty(0))|$$

$$\le \sum_{m=1}^{D}\sum_{i=1}^{k(m)}\left(\bar{e}\left(\pi_m^\lambda \lambda_i(m) + \eta\right)\Delta + \bar{e}\pi_m^\lambda \lambda_i(m)\Delta\right) \tag{8.37}$$

$$= \bar{e}\left(2M + kd\eta\right)\Delta \le \frac{2\varepsilon}{3}$$

by definition of M, where we assume $\Delta < 1$ and $T \ge 1$. Let now $n_i(m; n\Delta)$ denote the number of $\vec{e}_i(m)$ jumps in an interval, say $[\ell\Delta, (\ell+1)\Delta)$. Exactly the same calculations show that, under (8.36),

$$|\vec{z}_n(\ell\Delta + \Delta) - \vec{z}_n(\ell\Delta) - [\vec{z}_\infty(\ell\Delta + \Delta) - \vec{z}_\infty(\ell\Delta)]| \le \frac{\varepsilon\Delta}{3T}$$

$$\sup_{\ell\Delta\le t\le(\ell\Delta+\Delta)}|\vec{z}_n(t) - \vec{z}_n(\ell\Delta) - [\vec{z}_\infty(t) - \vec{z}_\infty(\ell\Delta)]| \le \frac{2\varepsilon}{3}$$

for each $\ell \le T/\Delta$. This means that under (8.36), at an endpoint of a subinterval, the distance between the two paths could have increased by at most $\varepsilon\Delta/3T$ from the distance at the starting point of that subinterval, so that if (8.36) holds for all ℓ then

$$\max_{1\le\ell\le T/\Delta}|\vec{z}_n(\ell\Delta) - \vec{z}_\infty(\ell\Delta)| \le \frac{\varepsilon}{3}.$$

Using the triangle inequality with (8.37) establishes that in this case

$$\sup_{0 \le t \le T} |\vec{z}_n(t) - \vec{z}_\infty(t)| \le \varepsilon.$$

Since the events (8.36) on different subintervals are independent we have from (8.35)

$$\mathbb{P}\left(\sup_{0 \le t \le T} |\vec{z}_n(t) - \vec{z}_\infty(t)| \le \varepsilon \right) \ge \left(1 - e^{-nC} \right)^{\frac{T}{\Delta}}.$$

A little algebra shows that this proves the lemma.   ∎

Exercise 8.58. Extend Lemma 8.57 to the case where $\vec{\lambda} \in R_c(\delta)$, but possibly $\lambda_i(m) < \delta$ for some i, m [$R_c(\delta)$ is defined in Equation (7.2)]. Note that the point is the uniformity of the result in $\vec{\lambda}$. Hint: the proof of Lemma 8.57 holds verbatim for $\vec{\lambda} \in R_c(\delta) \cap \{\lambda_i(m) \notin (0, \delta)\}$. Choose $\delta' \ll \delta$ and approximate any process satisfying $\vec{\lambda} \in R_c(\delta)$, as well as the limit $\vec{z}_\infty$ of the process by increasing $\lambda_i(m)$ to δ' whenever $\lambda_i(m) < \delta'$, while keeping λ_{mj} fixed.   ♠

Let $\mathcal{X}$ denote a set of arcs connecting the states $G \overset{\triangle}{=} \{1, \dots, D\}$ of the Markov chain. With a slight abuse of notation (that will pay off handsomely), we denote the ith arc leaving state m as $\vec{e}_i(m)$, with $i = 1, \dots, k(m)$. This arc leads to state $m'(m; i)$—see Definition 8.3. In terms of our standard notation, we simply ignore the first d coordinates of $\vec{e}_i(m)$ and concentrate on the last D, describing the finite state Markov chain. Note that $\mathcal{X}$ is ergodic (Assumption 7.1) if and only if the arcs make a (directed) path between any state i and any other state j; in short, $\mathcal{X}$ connects G. Let $|\mathcal{X}|$ denote the number of arcs in $\mathcal{X}$.

Lemma 8.59. *Given a set of arcs $\mathcal{X}$ that connect G and a probability distribution $\vec{\pi}$ on G with $\pi(m) \ge \alpha > 0$, there is a set of jump rates $\vec{\mu}$ such that $\vec{\pi} = \vec{\pi}_\mu$ is the (unique) invariant measure under $\vec{\mu}$ and $\alpha|\mathcal{X}|^{-1} \le \mu_i(m) \le 1$ for all $\vec{e}_i(m) \in \mathcal{X}$.*

Note that we can take any subset $\tilde{\mathcal{X}}$ of $\mathcal{X}$ that connects G, and insist that $\mu_i(m) > 0$ only when $\vec{e}_i(m) \in \tilde{\mathcal{X}}$, and that the resulting Markov chain be ergodic.

Proof. This is trivial when $D = 1$, so assume $D \ge 2$. We construct the rates $\vec{\mu}$ by regarding the arcs in $\mathcal{X}$ as carrying fluid flow, where the rate of flow along an arc $\vec{e}_i(m)$ is $\pi(m)\mu_i(m)$. Note that by Definition A.141 [but now allowing arcs $\vec{e}_i(m)$ with $m'(m; i) = m$, so that the term $j = m$ is included in both sums], if we construct such a flow then the balance equations hold, and so $(\vec{\mu}, \vec{\pi}_\mu)$ are indeed a set of rates and their invariant measure. Start by renumbering the states so that $0 < \alpha \le \pi(1) \le \pi(2) \cdots \le \pi(D) < 1$. Take any arc $\vec{e}_i(1)$ out of state 1 and set $\mu_i^1(1) = 1$. This makes the flow along the arc equal to $\pi(1)$. Continue connecting arcs end-to-end in a loop-free manner until you arrive back at state one. There is always such a circuit since $\mathcal{X}$ connects the states, and we can remove any extra loops along the circuit without destroying its connectivity. We choose the rate along each arc $\vec{e}_j(n)$ in the circuit to be $\mu_j^1(n) = \pi(1)/\pi(n)$, so that the total amount of flow $\pi(n)\mu_j^1(n)$ is preserved at the level $\pi(1)$. By our ordering of the states, $\alpha \le \mu_i^1(m) \le 1$ for all arcs along this circuit.

If this circuit reaches every arc then we are done. If not, set $\mu_i^1(m) = 0$ for every arc $\vec{e}_i(m) \in \mathcal{X}$ not covered by this circuit. Take an arc $\vec{e}_i(m)$ that has not been covered yet, where m is the smallest (and hence of smallest probability $\pi(m)$) among states where untraversed arcs originate. Make a new loop-free circuit of arcs from $m'(m; i)$ back to m (possibly overlapping some arcs that have already been covered). Along each of the arcs $\vec{e}_j(n)$ along this circuit define $\mu_j^2(n)$ by

$$\alpha \le \mu_j^2(n) = \mu_j^1(n) + \frac{\pi(1)}{\pi(n)} \le 2.$$

This operation obviously preserves the total flow at level $\pi(1)$ along the new arcs, and it is easy to verify that the local balance equations (amount of flow entering a state equals flow leaving the state) remain valid, including the case where a state or an arc is shared by the two circuits. Continue in this manner until all arcs have been traversed, and let the resulting rates be $\mu_i^k(m)$. The smallest flow along any arc is at least α, and the largest is at most $|\mathcal{X}|$, which bounds the number of iterations required to cover all arcs [recall we allow an arc from m to itself, which forms a loop, as well as multiple arcs between a pair of states (m, n)]. Setting $\mu_i(m) = \mu_i^k(m)|\mathcal{X}|^{-1}$ establishes the lemma. $\blacksquare$

If we assume that between any two states there is at most one arc in each direction, then clearly $|\mathcal{X}| < D(D + 1)/2$, and we obtain a universal bound depending only on α and D. For our finite levels model, $|\mathcal{X}| = \sum_{m=1}^{D} k(m)$.

Definition 8.60. *Given any number $K > 0$ and any set of arcs $\mathcal{X}$, define*

$$M(K) \stackrel{\triangle}{=} \left\{ \vec{\mu}, \vec{\pi}_\mu : \vec{\pi}_\mu \text{ is invariant for } \vec{\mu}, \ \mu_i(m) \le K \text{ for all } \vec{e}_i(m) \in \mathcal{X} \right\},$$

$$M(\delta, K) \stackrel{\triangle}{=} \left\{ (\vec{\mu}, \vec{\pi}_\mu) \in M(K) : \vec{\pi}_\mu \text{ is unique and } \mu_i(m) \notin (0, \delta) \right\},$$

where "$\vec{\pi}_\mu$ is unique" means $\vec{\mu}$ makes an ergodic chain.

Note that if $\vec{\mu}$ is a set of strictly positive rates and if $\mathcal{X}$ connects G then necessarily the chain is ergodic so that $\vec{\pi}_\mu$ is unique. In this case the definition of $\vec{y}_\mu$

$$\left(\vec{y}_\mu, \vec{0}^D \right) \stackrel{\triangle}{=} \sum_{m=1}^{D} \pi_\mu(m) \sum_{i=1}^{k} \mu_i(m) \vec{e}_i(m) \tag{8.38}$$

is well posed. For brevity, we shall use the notation $\vec{y}_\mu$ also when there is no uniqueness of $\vec{\pi}_\mu$, but where no ambiguity arises. For an arbitrary set of rates μ and a probability π on G define also

$$f(\vec{\mu}, \vec{\pi}) \stackrel{\triangle}{=} \sum_{m=1}^{D} \pi(m) \left(\sum_{i=1}^{k(m)} \lambda_i(m) - \mu_i(m) + \mu_i(m) \log \frac{\mu_i(m)}{\lambda_i(m)} \right). \tag{8.39}$$

Given $\vec{y} \in C_f$, Lemma 8.32 implies that there exist $(\vec{\eta}, \pi_\eta)$ so that

$$\vec{y} = \vec{y}_\eta \quad \text{and} \quad f(\vec{\eta}, \vec{\pi}_\eta) = \ell(\vec{x}, \vec{y}). \tag{8.40}$$

Lemma 8.61. *Fix $\mathcal{X}$ that connects G and a point $\vec{x}$. For every positive ε and L there is a $\delta > 0$ for which the following hold. Let $\ell(\vec{x}, \vec{y}) \leq L$ and define $\vec{\eta}, \vec{\pi}_\eta$ through (8.40). Then there is a pair $(\vec{\mu}, \vec{\pi}_\mu) \in M(\delta, \delta^{-1})$ so that*

$$\left| \pi_\mu(m) - \pi_\eta(m) \right| < \varepsilon \quad \text{for all} \quad 1 \leq m \leq D \tag{8.41a}$$

$$\left| \pi_\mu(m)\mu_i(m) - \pi_\eta(m)\eta_i(m) \right| < \varepsilon \quad \text{for all} \quad \vec{e}_i(m) \in \mathcal{X} \tag{8.41b}$$

$$\left| \vec{y} - \vec{y}_\mu \right| < \varepsilon \tag{8.41c}$$

$$f(\vec{\mu}, \vec{\pi}_\mu) < \ell(\vec{x}, \vec{y}) + \varepsilon. \tag{8.41d}$$

Proof. Assume without loss of generality that $\varepsilon < 1 < L$. The proof consists of showing that the left-hand sides of (8.41a)–(8.41d) are bounded by functions $\varepsilon_i(\alpha, s)$ that go to zero as α and s go to zero in an appropriate way. Fix $0 < \alpha < D^{-1}$, to be specified later, and break the states in G into two classes named B and S:

$$m \in B \quad \text{means} \quad \pi_\eta(m) > \alpha$$
$$m \in S \quad \text{means} \quad \pi_\eta(m) \leq \alpha.$$

S might be empty, but since $\alpha < D^{-1}$ the set B is not empty. Within B break up the states into disjoint communicating classes: choose a state k in B, and let b_1 consist of all states m in B so that there is a circuit from k to m and back, with arcs in $\mathcal{X}$ passing only through states in B. Choose a state in B but outside b_1 to construct the set b_2 and repeat. By construction, between any two states in b_k there is a path going only through states in b_k, and b_k cannot be enlarged without either violating this property or going through S. Note that all the b_k might be singletons or, at the other extreme, all of B might be one class. Define

$$h(x) \overset{\triangle}{=} \min_{i,m} \left\{ x \log \frac{x}{\lambda_i(m)} - x + \lambda_i(m) \right\}.$$

Then by assumption $\pi_\eta(m) h(\eta_i(m)) \leq L$ and so

$$\pi_\eta(m)\eta_i(m) \leq L \frac{\eta_i(m)}{h(\eta_i(m))} \downarrow 0 \quad \text{monotonically as} \quad \eta_i(m) \uparrow \infty.$$

The rates $\mu_i(m)$ we shall construct will be bounded by $\overline{\eta} = \overline{\eta}(\alpha) > 0$, defined through

$$\alpha \overline{\eta} \sqrt{\log \overline{\eta}} = 1. \tag{8.42}$$

Note that, as $\alpha \downarrow 0$ we have $\overline{\eta}(\alpha) \uparrow \infty$ and $\alpha \overline{\eta}(\alpha) = (\log \overline{\eta})^{-1/2} \downarrow 0$. Therefore, for any $m \in S$,

$$\pi_\eta(m)\eta_i(m) \leq \max \left\{ \alpha \overline{\eta}, L \frac{\overline{\eta}}{h(\overline{\eta})} \right\} \overset{\triangle}{=} \varepsilon_1(\alpha)$$

for all α small, where we substituted $\bar{\eta} = \bar{\eta}(\alpha)$ to obtain ε_1 as a function of α alone. Therefore, the total flow between S and B is at most $|\mathcal{X}|\,\varepsilon_1(\alpha)$. We claim that if m and n are not both in the same class b_k, then the flow along any arc $\vec{e}_i(m)$ with $n = m'(m; i)$ is also at most $|\mathcal{X}|\,\varepsilon_1(\alpha)$. To see this, note that any flow that goes into a set b_k from outside either comes from a state in S directly, or must come from another b_m. But there can be no loops among flows between different sets b_m, since the sets were defined to be maximal sets that communicate. Therefore if we trace any flow backward it must eventually come from S, and is therefore bounded by $|\mathcal{X}|\,\varepsilon_1(\alpha)$.

We now wish to set all flows outside the b_ks to zero, maintaining the invariant measure $\vec{\pi}_\eta$, without making a large change to any flow. It is clear how to do this: consider all the flow through the network that runs through S. Some of it may run through some b_ks, but the total flow is small, so simply cut it off. The resulting transition rates, call them $\tilde{\mu}_i(m)$, satisfy $\tilde{\mu}_i(m) = 0$ unless both m and $m'(m; i)$ are in the same b_k. To make $\vec{\pi}_\eta$ invariant under the rates $\tilde{\mu}_i(m)$, we need to modify the flows. Each state in S and each set b_k is now isolated from all others, so that they can be assigned any probability, and in particular, the one resulting from $\vec{\pi}_\eta$. We only have to verify that the flows within each b_k can be changed by changing only the rates, and so that the balance equations hold for $(\{\tilde{\mu}_i(m)\}, \vec{\pi}_\eta)$. We only need to compensate for a decrease in flow, which is due to our cutting off the flow from S. This can be done with a change in the flow of at most $|\mathcal{X}|\,\varepsilon_1(\alpha)$, effected by making the appropriate $\tilde{\mu}_i(m)$ smaller than the original $\eta_i(m)$. Under this construction, if $\eta_i(m) \geq \bar{\eta}$ than $\pi_\eta(m) \leq \alpha$ so that $m \in S$, and therefore $\tilde{\mu}_i(m) = 0$. Consequently $\tilde{\mu}_i(m) \leq \eta_i(m)$ for all arcs $\vec{e}_i(m)$.

To define $\vec{\pi}_\mu$, we increase the probability for states in S to α, and decrease for states in B by a factor:

$$\pi_\mu(m) = \alpha \quad \text{for all} \quad m \in S$$
$$\pi_\mu(m) = C\pi_\eta(m) \quad \text{for all} \quad m \in B,$$

where $C \leq 1$ is chosen to make the probabilities sum to one:

$$C \sum_{m \in B} \pi_\eta(m) + |S|\,\alpha = 1.$$

C is close to one since

$$C = \frac{1 - |S|\,\alpha}{\sum_{m \in B} \pi_\eta(m)} \geq 1 - |S|\,\alpha.$$

Since $|S| < D \leq |\mathcal{X}|$, (8.41a) holds, with $\varepsilon_a(\alpha) \overset{\triangle}{=} \alpha\,|\mathcal{X}|$ on the right.

Note that $\vec{\pi}_\mu$ is invariant for the rates $\tilde{\mu}_i(m)$. This follows since each $m \in S$ does not communicate with any of the other states, so any measure is invariant, and each $m \in b_k$ communicates only with its fellow members of b_k, so the various b_ks can have any total amount of probability.

As our final step in the construction of $\vec{\mu}$, note that $\mathcal{X}$ and $\vec{\pi}_\mu$ satisfy the assumptions of Lemma 8.59. So, define $\gamma_i(m)$ as the jump rates that arise from

Lemma 8.59 that make $\vec{\pi}_\mu$ invariant, and let $\mu_i(m) = \tilde{\mu}_i(m) + s\gamma_i(m)$. The small parameter s depends on α, and will be specified in (iia) below. Since $\vec{\pi}_\mu$ is invariant for the $\tilde{\mu}_i(m)$ as well as for the $\gamma_i(m)$, it is invariant for $\vec{\mu}$. From the lower bound on $\gamma_i(m)$, given in Lemma 8.59, we obtain the lower bound $s\alpha|\mathcal{X}|^{-1} \le \mu_i(m)$. Since $\tilde{\mu}_i(m) \le \overline{\eta}$ and $\gamma_i(m) \le 1$, we have the upper bound $\mu_i(m) \le \overline{\eta}(\alpha)+s$. Thus the rates $\vec{\mu}$ are strictly positive, and $(\vec{\mu}, \vec{\pi}_\mu) \in M(\delta, \delta^{-1})$ for small enough δ (which depends on α).

It remains to establish (8.41b)–(8.41d). The flows $\pi_\mu(m)\mu_i(m)$ differ from $\pi_\eta(m)\eta_i(m)$ by

$$\left|\pi_\eta(m)\eta_i(m) - \pi_\mu(m)\mu_i(m)\right|$$
$$\le \left|\pi_\eta(m)\eta_i(m) - \pi_\eta(m)\tilde{\mu}_i(m)\right| + \left|\pi_\eta(m)\tilde{\mu}_i(m) - \pi_\eta(m)\mu_i(m)\right|$$
$$+ \left|\pi_\eta(m)\mu_i(m) - \pi_\mu(m)\mu_i(m)\right|$$
$$< |\mathcal{X}|\,\varepsilon_1(\alpha) + s + \overline{\eta}\varepsilon_a(\alpha) \overset{\triangle}{=} \varepsilon_b(\alpha, s).$$

By definition of h and $\overline{\eta}$ it follows that $\varepsilon_1(\alpha) = \alpha\overline{\eta}$ for all α small enough. Therefore, $\varepsilon_b(\alpha, s) \le s + c_1\alpha(\overline{\eta} + 1) \to 0$ as $\alpha \to 0$ and $s \to 0$. The change in $\vec{y}$ is now easy to bound:

$$|\vec{y} - \vec{y}_\mu| = \sum_{m=1}^{D}\sum_{i=1}^{k} \left(\pi_\eta(m)\eta_i(m) - \pi_\mu(m)\mu_i(m)\right) \vec{e}_i(m)$$
$$\le \overline{e}\,|\mathcal{X}|\,\varepsilon_b(\alpha, s),$$

where $\overline{e}$ is a bound on the maximal jump size.

Finally, we establish (8.41d). By (8.39)–(8.40), both $f(\vec{\mu}, \vec{\pi}_\mu)$ and $\ell(\vec{x}, \vec{y})$ are defined as sums [the latter in terms of $(\vec{\eta}, \vec{\pi}_\eta)$] over m and i. We find an upper bound for each of the summands. Fix i and m and, for brevity, omit the indices i and m from $\pi_\mu(m)$, $\pi_\eta(m)$, $\eta_i(m)$, $\mu_i(m)$, and $\lambda_i(m)$. Now

$$\pi_\mu \left(\mu \log \frac{\mu}{\lambda} - \mu + \lambda\right) - \pi_\eta \left(\eta \log \frac{\eta}{\lambda} - \eta + \lambda\right)$$
$$\le \left|\pi_\mu\mu - \pi_\eta\eta\right| + \left|\pi_\mu - \pi_\eta\right|\lambda + \left|\pi_\mu\mu - \pi_\eta\eta\right||\log\lambda| + \pi_\mu\mu \log\mu - \pi_\eta\eta \log\eta$$
$$\le \varepsilon_2(s, \alpha) + \pi_\mu\mu \log\mu - \pi_\eta\eta \log\eta$$

where $\varepsilon_2 = c_2(\varepsilon_a + \varepsilon_b)$ for some constant c_2, depending only on λ. Denote the remaining two terms by ε_ℓ. We check that ε_ℓ can be made small using different bounds over different regions of η and π_η.

(i) $\pi_\eta \le \alpha$, that is, $m \in S$.

In this case $\pi_\mu = \alpha$ and $\mu \le s$. Therefore, for s small,

$$\varepsilon_\ell \le \alpha\,|s \log s| - \alpha \inf_{0\le x\le 1} x \log x, \tag{8.43}$$

since $\eta \log \eta > 0$ for $\eta > 1$. So, in this case we have a bound that is linear in α and decreasing in s. Note that this covers the case $\eta \ge \overline{\eta}$.

(ii) $\pi_\eta > \alpha$, that is, $m \in B$.

   (iia) $1 < \eta < \overline{\eta}$.

$$\varepsilon_\ell \le \pi_\mu \mu \log(\eta + s) - \pi_\eta \eta \log \eta$$
$$\le \pi_\mu \mu \log \eta + \pi_\mu \mu \frac{s}{\eta} - \pi_\eta \eta \log \eta.$$

But note that the only increase in flow in constructing the μ appears due to the addition of the rates γ. Therefore $\pi_\mu \mu \le \pi_\eta \eta + s$, and since $\log \eta \ge 0$,

$$\varepsilon_\ell \le s \log \overline{\eta} + 2s.$$

We now set $s = s(\alpha)$ as

$$s(\alpha) = (\log[\overline{\eta}(\alpha)])^{-2}.$$

With this choice, $s \to 0$ and $s \log \overline{\eta} \to 0$ as $\alpha \to 0$.

   (iib) $s \le \eta \le 1$.

$$\log \mu \le \log \eta + \frac{\mu - \eta}{\eta}$$

and so

$$\varepsilon_\ell \le \pi_\mu \mu \left(\log \eta + \frac{\mu - \eta}{\eta} \right) - \pi_\eta \eta \log \eta$$
$$\le |\pi_\mu \mu - \pi_\eta \eta| \, |\log s| + \pi_\mu \frac{\mu}{\eta} s$$
$$\le \varepsilon_b(\alpha, s) \, |\log s| + 2s.$$

We need to show that $\varepsilon_b(\alpha, s)| \log s| \to 0$ as $\alpha \to 0$. But

$$\overline{\eta}(\alpha)\alpha| \log s| = \frac{|\log s|}{\sqrt{\log[\overline{\eta}(\alpha)]}} \to 0 \quad \text{as} \quad \alpha \to 0, \tag{8.44}$$

for our choice of $s(\alpha)$ (by an application of l'Hospital's rule). Finally,

   (iic) $s > \eta$.

This is easy, since $\mu \le \eta + s \le 2s$, so that $\varepsilon_\ell \le 2 |2s \log 2s|$, and we are done.

   We conclude that if we choose α small enough, (8.41) holds, and then $\delta > 0$ is chosen so that $\delta \overline{\eta} \le 1$ and $\delta |\mathcal{X}| \le \alpha s(\alpha)$. ∎

Exercise 8.62. Show that if the $\log \lambda_i(\vec{x}; m)$ are bounded, then Lemma 8.61 holds uniformly in $\vec{x}$. Moreover, instead of making all rates bounded below, we could specify a connecting subset of $\mathcal{X}$ so that arcs outside the subset have rate zero, and keep the same constants. Hint: just follow the proof of the Lemma. ♠

Exercise 8.63. Show that, given any $\varepsilon > 0$ there is a $\delta > 0$ such that for any pair $(\vec{\eta}, \vec{\pi}_\eta)$ with drift $\vec{y}_\eta$ there is a pair $(\vec{\mu}, \vec{\pi}_\mu) \in M(\delta, \delta^{-1})$ that has drift $\vec{y}_\mu$ satisfying

$$|\vec{y}_\eta - \vec{y}_\mu| < \varepsilon$$
$$\ell(\vec{x}, \vec{y}_\mu) \le \ell(\vec{x}, \vec{y}_\eta) + \varepsilon.$$

Hint: $\ell(\vec{x}, \vec{y}_\eta) \leq f(\vec{\eta}, \vec{\pi}_\eta)$. Now use Lemma 8.61.   ♠

The corresponding result for the flat boundary model is easier to derive. In this case we would like to avoid very large rates, and also rates for which, when the process is on the boundary, it is indifferent between staying there and moving inside. Here are the precise definitions [cf. (8.10)]. Recall that our underlying Markov chain has the positive integers as state space.

Recall the definitions of the drift $\beta(\vec{x})$ (8.9) and the invariant probability $\pi_0(\vec{x})$ (8.10) for the process to be on the boundary $x_0 = 0$.

Definition 8.64. *Given a process with jump rates $\vec{\mu}$ and with an associated invariant measure $\vec{\pi}_\mu$ on $m = 0, 1, \ldots$ define the net drift at zero by*

$$\vec{y}_\mu(0) \stackrel{\triangle}{=} \pi_\mu(0) \sum_{i=1}^{k(0)} \mu_i(0)\vec{e}_i(0) + (1 - \pi_\mu(0)) \sum_{i=1}^{k(1)} \mu_i(1)\vec{e}_i(1).$$

As in (8.9), define the drift associated with the rates $\vec{\mu}$ as

$$\beta_\mu \stackrel{\triangle}{=} \sum_{i=1}^{k(1)} \mu_i(1)e_i^0(1).$$

The set of rates $M'(\delta, K)$ is defined by

$$M'(\delta, K) = \left\{ \vec{\mu} : \delta \leq |\mu_i(m)| \leq K, \ \beta_\mu \notin (-\delta, 0] \right\}. \tag{8.45}$$

Note that the symbol $\vec{y}_\mu$ has a different interpretation in the context of the finite levels process. We only need the value of $\vec{\pi}$ at zero for these calculations. The proofs of the following lemmas are left to Exercise 8.67.

Lemma 8.65. *For every ε and K with $0 < \varepsilon < K$ there is a $\delta > 0$ such that for any process satisfying Assumption 8.8 with jump rates $\vec{\eta}$ and an invariant measure $\vec{\pi}_\eta$ with $(\vec{\eta}, \pi_\eta(m)) \in M(K)$, there exists a $\vec{\mu} \in M'(\delta, K)$ such that, for $m = 0, 1$ and for all i,*

$$\left| \pi_\eta(0) - \pi_\mu(0) \right| < \varepsilon$$
$$\left| \pi_\eta(m)\eta_i(m) - \pi_\mu(m)\mu_i(m) \right| < \varepsilon.$$

Lemma 8.66. *For every ε and K with $0 < \varepsilon < K$ there is a $\delta > 0$ such that for any process satisfying Assumption 8.8 with jump rates $\vec{\eta}$ and an invariant measure π_η with $(\vec{\eta}, \pi_\eta(m)) \in M(K)$ and $\beta_\eta \in (-\delta, 0]$, there exists a $\mu_i(m) \in M'(\delta, K)$ such that for any $\vec{x}$ with $x_0 = 0$,*

$$\left| \vec{y}_\eta - \vec{y}_\mu \right| \leq \varepsilon$$
$$\ell_b(\vec{x}, \vec{y}_\mu) \leq \ell_b(\vec{x}, \vec{y}_\eta) + \varepsilon.$$

Recall that ℓ does not depend on $\vec{x}$ except through x_0, since the jump rates are constant in each region.

Exercise 8.67. Prove Lemmas 8.65 and 8.66. Hint: use the idea of flows of Lemma 8.61. We need only to adjust two drifts in order for β_μ to avoid $(-\delta, 0]$: one from the boundary to the interior, the other from the interior to the boundary.♠

8.5. The Lower Bound

This section contains proofs of the lower bounds for the finite levels process and the flat boundary process. The lower bound for the finite levels process is simpler than the bound for the flat boundary process, so we'll prove it first. Here are the statements of the theorems: the proofs appear after Lemma 8.72 later in this section, and at the end of the chapter respectively.

Theorem 8.68. *Let Assumptions 8.4 (ergodicity) and 8.5 (continuity) hold. Then uniformly over $\vec{x}$ in compact subsets of $\mathbb{R}^d$, and over $m \in \{1, \ldots, D\}$, for each open set $G \subset D^d[0, T]$,*

$$\liminf_{n\to\infty} \frac{1}{n} \log \mathbb{P}_{\vec{x},m} \left(\vec{z}_n^d \in G \right) \geq - \inf \left\{ {}_c I_0^T (\vec{r}) : \vec{r} \in G, \ \vec{r}(0) = \vec{x} \right\}.$$

Theorem 8.69. *Let Assumptions 8.8 (jump structure) and 8.9 (continuity) hold. Then uniformly over $\vec{x}$ in compact subsets of $[0, \infty) \times \mathbb{R}^d$, for each open set $G \subset D^{d+1}[0, T]$,*

$$\liminf_{n\to\infty} \frac{1}{n} \log \mathbb{P}_{\vec{x}} (\vec{z}_n \in G) \geq - \inf \left\{ {}_b I_0^T (\vec{r}) : \vec{r} \in G, \ \vec{r}(0) = \vec{x} \right\}.$$

The proof of the bound for finite levels processes goes very much along the lines of §5.3. The theorems are proved by showing that, for each fixed path $\vec{r} \in G$, with $I^* \triangleq \inf \left\{ I_0^T (\vec{r}) : \vec{s} \in G, \ \vec{s}(0) = \vec{x} \right\}$, for each $\zeta > 0$,

$$\lim_{\varepsilon \downarrow 0} \liminf_{n\to\infty} \frac{1}{n} \log \mathbb{P} (\vec{z}_n \in N_\varepsilon(\vec{r})) \geq -I_0^T (\vec{r}) \geq -I^* - \zeta, \tag{8.46}$$

where $I_0^T (\vec{r})$ represents either ${}_b I_0^T (\vec{r})$ or ${}_c I_0^T (\vec{r})$ as appropriate [this is the same as (5.32)]. There are two new technicaliúes. The first has already been discussed: the equivalence of the local rate functions for the upper and lower bounds. This is more difficult to prove in this setting, but was established in Theorem 8.19. The second is a new problem in obtaining uniformity in the convergence of $\vec{z}_n$ to $\vec{z}_\infty$ for constant coefficient processes. After we divide time into intervals of length Δ, we locally change the measure to a constant coefficient process. We don't have much control over the coefficients of the new process; they simply solve a minimization problem. The process $\vec{z}_n(t)$ is then supposed to follow the local drift $\vec{z}_\infty(t)$ of the new process. According to Lemma 7.4, the rate of convergence of $\vec{z}_n(t)$ to $\vec{z}_\infty(t)$ depends on having a lower bound on the jump rates in the new process. This is where Lemma 8.61 comes in. It enables us to approximate any jump process by one from a family that mixes uniformly quickly. This gives us a bound that holds over all the Δ-intervals and over all relevant changes of measure.

The lower bound for the flat boundary process requires a few more technicalities. One is related to the derivation of Equation (5.50): the cheapest path between point $\vec{x}$ and point $\vec{y}$ in time T for a constant coefficient flat boundary process is *not* necessarily a straight line. Since the local rate function ℓ can be smaller along the boundary than in the interior, it might be cheaper for a path to go from $\vec{x}$ to the boundary, travel along the boundary for a while, and then travel to $\vec{y}$. This invalidates (5.50), which was the starting point for proving the lower bound in §5.3. This difficulty is resolved by noting that a cheapest path might not be a straight line, but it can be composed of at most three straight line segments: one down to the boundary, one along the boundary, and one back from the boundary. (Because the rate function is not strictly convex on the boundary, the straight line path along the boundary might not be the only minimal cost path, but convexity assures us that it is a minimal cost path, so we can use it for constructing a cheapest path.) This enables us to do all the calculations almost as before. There is also a difficulty in proving uniform mixing, as for the finite levels model, but it is almost trivial to fix this technicality for the flat boundary model, since we only need to make sure that the mixing occurs in the x_0 direction. The final technical point for the flat boundary process is the question of whether the process starts exactly on the boundary or not at any particular time. We make our estimates for the case $z_{n,0}(0) = 0$, but when $\vec{z}_n(t)$ is following a path $\vec{r}(t)$ we don't know exactly how high $z_{n,0}(t)$ will be. This is fixed by changing the path over a small interval of time with small cost so that it either leaves the boundary or goes to the boundary.

We begin our proof of the lower bound for the finite levels process with the change of measure formula. Let the measure $\mathbb{P}$ correspond to jump rates $\lambda_i(m)$, let $\tilde{\mathbb{P}}$ correspond to jump rates $\mu_i(m)$, and let $\vec{u}_m \in \mathbb{R}^D$ be the unit vector in the mth coordinate direction. As in Equation (5.30), let N denote the total number of jumps (transitions) the process $\vec{z}_n$ makes in $[0, T]$, let t_j be the time of the jth jump, and let $l(j)$ denote the direction of the jth jump. We shall use the notation $\lambda_i(\vec{z}_n)$ or $\lambda_i(m, \vec{z}_n^d)$ as convenient. Then we have

$$\frac{d\mathbb{P}}{d\tilde{\mathbb{P}}}(\omega) = \exp\left[-\int_0^T n \sum_{m=1}^D \mathbf{1}\left[\vec{z}_n^D = \vec{u}_m \right] \sum_{i=1}^{k(m)} \left(\lambda_i(m, \vec{z}_n^d(s)) - \mu_i(m, \vec{z}_n^d(s)) \right) ds \right.$$

$$\left. + \sum_{j=1}^N \log\left(\frac{\lambda_{l(j)}(\vec{z}_n(t_j^-))}{\mu_{l(j)}(\vec{z}_n(t_j^-))} \right) \right] \stackrel{\triangle}{=} e^{H(\omega)}. \quad (8.47)$$

This follows from Theorem B.6. Now given any continuous path $\vec{r} \in C^d[0, T]$ and any $\varepsilon > 0$ we write, as in (5.33),

$$\mathbb{P}_{\vec{x},m}\left(\vec{z}_n^d \in N_\varepsilon(\vec{r})\right) = \int\limits_{\vec{z}_n^d \in N_\varepsilon(\vec{r})} d\mathbb{P}(\omega)$$

$$= \int\limits_{\vec{z}_n^d \in N_\varepsilon(\vec{r})} e^{H(\omega)} d\tilde{\mathbb{P}}(\omega)$$

$$\geq \tilde{\mathbb{P}}\left(\vec{z}_n^d \in N_\varepsilon(\vec{r})\right)$$

$$\cdot \exp\left[\frac{1}{\tilde{\mathbb{P}}\left(\vec{z}_n^d \in N_\varepsilon(\vec{r})\right)} \int\limits_{\vec{z}_n^d \in N_\varepsilon(\vec{r})} H(\omega)\, d\tilde{\mathbb{P}}(\omega)\right],$$

$$(8.48)$$

where initial conditions are not denoted explicitly. Therefore if $\tilde{\mathbb{P}}$ is chosen so that $\tilde{\mathbb{P}}\left(\vec{z}_n^d \in N_\varepsilon(\vec{r})\right) \to 1$ as $n \to \infty$ then

$$\liminf_{n\to\infty} \frac{1}{n} \log \mathbb{P}_{\vec{x},m}\left(\vec{z}_n^d \in N_\varepsilon(\vec{r})\right) \geq \liminf_{n\to\infty} \frac{1}{n} \int\limits_{\vec{z}_n^d \in N_\varepsilon(\vec{r})} H(\omega)\, d\tilde{\mathbb{P}}(\omega).$$

We now develop the analogue of Lemma 5.52. Fix a set of jump rates $\mu_i(m)$ that do not depend on $\vec{x}$ and that make $\vec{z}_n^D$ an ergodic process. Let $\tilde{\mathbb{P}}_{\vec{x},m}$ denote the probability and $\tilde{\mathbb{E}}_{\vec{x},m}$ the expectation with respect to the scaled jump process $\vec{z}_n(t)$ with jump rates $\mu_i(m)$, starting from an initial position $(\vec{x}, \vec{u}_m)$ where $\vec{u}_m$ is the unit vector in the mth coordinate direction. As usual $\pi_\mu(m)$ is the invariant measure, which is unique since under these rates $\vec{z}_n^D$ is ergodic. Let $\vec{y}_\infty(t)$ be the limit $\vec{z}_\infty^d(t)$ for the process with rates $\mu_i(m)$. By (8.33), $\vec{y}_\infty(t)$ is a straight line, and since for these rates $\vec{z}_n^D$ is ergodic, $\vec{y}_\infty(t)$ is independent of the initial level m. Suppose we are given a set of bounded continuous functions $f_{i,m}(\vec{x}^d)$ where the index m ranges over the levels ($1 \leq m \leq D$) and the index i ranges over the jump directions on each level ($1 \leq i \leq k(m)$). Let $l(j)$ denote the index (i, m) of the jth transition of $\vec{z}_n(t)$, where again i is the index of the jump direction $\vec{e}_i(m)$ and m is the level at the start of the jump.

Lemma 8.70. *Assume $\vec{\mu}$ makes the Markov chain ergodic. For any fixed $T > 0$ let N denote the number of jumps of $\vec{z}_n$ in $[0, T]$. Then for $f_{i,m}(\vec{x}^d)$ bounded continuous, uniformly over over starting levels $m \in \{1, \ldots, D\}$ and $\vec{x}^d$ in compact sets, for each $\varepsilon > 0$ we have*

$$\lim_{n\to\infty} \tilde{\mathbb{E}}_{\vec{x},m}\left(\frac{1}{n}\sum_{j=1}^{N} f_{l(j)}(\vec{z}_n(t_j^-))\right)$$

$$= \lim_{n \to \infty} \tilde{\mathbb{E}}_{\vec{x},m} \left(\mathbf{1} \left[\vec{z}_n^d \in N_\varepsilon(\vec{y}_\infty) \right] \frac{1}{n} \sum_{j=1}^N f_{l(j)}(\vec{z}_n(t_j^-)) \right)$$

$$= \int_0^T \sum_{m'=1}^D \pi_\mu(m') \sum_{i=1}^{k(m')} \mu_i(m') f_{i,m'}(\vec{y}_\infty(t)) \, dt.$$

In order to prove this lemma we first need to prove Kurtz's Theorem for finite levels processes with constant coefficients.

Theorem 8.71. *Given any* K, δ, T, *and* $\varepsilon > 0$ *there exist positive* $C_1, C_2(\varepsilon)$ *and* n_0 *such that for all* $\vec{x}^d \in \mathbb{R}^d$, *all* $m \in \{1, \ldots, D\}$, *all* $(\mu, \pi_\mu) \in M(\delta, K)$, *and any* $n \geq n_0$,

$$\tilde{\mathbb{P}}_{\vec{x},m} \left(\sup_{0 \leq t \leq T} \left| \vec{z}_n(t) - \vec{y}_\infty(t) \right| > \varepsilon \right) \leq C_1 e^{-nC_2(\varepsilon)}. \tag{8.49}$$

Furthermore,

$$\lim_{\varepsilon \to \infty} \frac{C_2(\varepsilon)}{\varepsilon} = \infty. \tag{8.50}$$

Proof. To prove (8.50), let $x(t)$ be a Poisson process with rate

$$M \stackrel{\triangle}{=} T \sum_{m=1}^D k(m)K.$$

Then $|\vec{z}_n(t)|$ is bounded by $n^{-1}x(nT)$ in probability; that is,

$$\tilde{\mathbb{P}}_{\vec{x},m} \left(\sup_{0 \leq t \leq T} |\vec{z}_n(t)| > \varepsilon - \sup_{0 \leq t \leq T} |\vec{y}_\infty(t)| \right) \leq \mathbb{P} \left(\frac{x(nT)}{n} > \varepsilon - \sup_{0 \leq t \leq T} |\vec{y}_\infty(t)| \right).$$

But from Example 1.13, the rate function for this random variable goes to infinity superlinearly, and so (8.50) holds.

The estimate (8.49) follows directly from Exercise 7.8. Recall that $\bar{e}$ is the maximum of $|\vec{e}_i(m)|$. Given ε and K, divide time into intervals of length Δ such that

$$\Delta K \sum_{m=1}^D k(m)\bar{e} < \frac{\varepsilon}{4}.$$

Also take $\eta = \varepsilon \Delta/(2T \, \bar{e})$ and let $J \stackrel{\triangle}{=} T/\Delta$ and $s_j \stackrel{\triangle}{=} j\Delta$.

By a union bound,

$$
\tilde{\mathbb{P}}_{\vec{x},m}\left(\sup_{0\le t\le T}\left|\vec{z}_n(t)-\vec{y}_\infty(t)\right|>\varepsilon\right)
$$

$$
\le \tilde{\mathbb{P}}_{\vec{x},m}\left(\max_{0\le j\le J}\left|\vec{z}_n(s_j)-\vec{y}_n(s_j)\right|\le\eta,\ \max_{0\le j\le J}\sup_{s_j\le t\le s_{j+1}}\left|\vec{z}_n(t)-\vec{z}_n(s_j)\right|>\varepsilon/2\right)
$$

$$
+\tilde{\mathbb{P}}_{\vec{x},m}\left(\max_{0\le j\le T/\Delta}\left|\vec{z}_n(s_j)-\vec{y}_n(s_j)\right|>\eta\right)
$$

$$
\le \frac{T}{\Delta}\tilde{\mathbb{P}}_{\vec{x},m}\left(\sup_{0\le t\le\Delta}\left|\vec{z}_n(t)-\vec{z}_n(0)\right|>\varepsilon/2\right)+\sum_{j=0}^{J}\tilde{\mathbb{P}}_{\vec{x},m}\left(\left|\vec{z}_n(s_j)-\vec{y}_n(s_j)\right|>\eta\right).
$$

The first term on the last line is bounded, by the same argument as before, by a Poisson random variable with mean at most $\varepsilon/4$, and so it decays exponentially in n. Each term in the last sum is bounded, by Exercise 7.8, by $e^{-nC_3(j)}$ for some functions $C_3(j)$ that are all positive. Since Δ is fixed, so is J, and the result follows. ∎

Proof of Lemma 8.70. Using Theorem 8.71, this goes exactly as Lemma 5.52. ∎

As in (5.20), define

$$
f\left(\vec{\mu},\vec{\lambda}\right)\triangleq\sum_{m=1}^{D}\pi(m)\left(\sum_{i=1}^{k(m)}\lambda_i(m)-\mu_i(m)+\mu_i(m)\log\frac{\mu_i(m)}{\lambda_i(m)}\right) \tag{8.51}
$$

[cf. (8.39)] where, since we assume $\vec{\mu}$ makes the Markov chain ergodic, there is a unique invariant measure $\vec{\pi}_{\vec{\mu}}$, which we simply denote $\pi(m)$.

Lemma 8.72. *Assume that $\vec{\mu}$ makes the Markov chain ergodic, and that the parameters $\log\lambda_i(m)$ are bounded and continuous. Then for any $\varepsilon>0$,*

$$
\liminf_{n\to\infty}\frac{1}{n}\log\mathbb{P}_{\vec{x},m}\left(\vec{z}_n\in N_\varepsilon(\vec{y}_\infty)\right)\ge-\int_0^T f(\vec{\mu},\vec{\lambda}(\vec{y}_\infty(t)))\,dt
$$

and the convergence is uniform over $\vec{x}$ in compact sets and over $m\in\{1,\dots,D\}$.

Proof. Exactly as the proof of Corollary 5.53. ∎

Proof of the lower bound for finite levels processes, Theorem 8.68. This is left to the reader, as Exercise 8.73. ∎

Exercise 8.73. Prove Theorem 8.68. Hints: follow the proof of Theorem 5.51. Since the mesh Δ is fixed, we have a bound on ℓ and hence on $\frac{\Delta r}{\Delta}$. Use Lemma

8.61 to choose new $\vec{\mu}$ in each interval so that $\vec{\mu} \in M(\delta, \delta^{-1})$, while preserving (5.49). These $\mu_i(m)$ satisfy the conditions of Lemma 8.72.   ♠

We proceed to the proof of the lower bound for flat boundary processes. We start by fixing an absolutely continuous path $\vec{r}(t)$ for which the cost $I_0^T(\vec{r})$ is finite. Our first step is the construction of a piecewise linear path $\vec{w}(t)$ that is used to approximate $\vec{r}(t)$.

Given $\varepsilon > 0$ and T, choose Δ so that the change of $\vec{r}$ over any interval of length Δ is less than $\varepsilon/4$; that is,

$$\sup_{|t-s|<\Delta} |\vec{r}(t) - \vec{r}(s)| < \varepsilon/4.$$

This is possible since $\vec{r}$ is continuous and T is finite. Furthermore, take Δ smaller than ε, and small enough that Δ divides T. With the notation $\Delta\vec{r}(t) = \vec{r}(t+\Delta) - \vec{r}(t)$ we obviously have the bound $|\Delta\vec{r}(t)|/\Delta < \varepsilon/4\Delta$. Let $J_0 \stackrel{\Delta}{=} T/\Delta$ and

$$s_j \stackrel{\Delta}{=} j\Delta, \ 0 \le j \le J_0.$$

We now define a piecewise linear approximation $\vec{w}(t)$ of $\vec{r}$. Fix an interval $[s_j, s_{j+1}]$. Let

$$\ell_j(\vec{x}, \vec{y}) \stackrel{\Delta}{=} \ell(\vec{r}(s_j), \vec{y})$$
$$I_j(\vec{u}) \stackrel{\Delta}{=} \int_{s_j}^{s_{j+1}} \ell_j(\vec{u}(t), \vec{u}'(t)) \, dt$$

for absolutely continuous $\vec{u}$. That is, the local cost function ℓ_j is defined in the usual way, but with the rates $\lambda_i(1; \vec{x})$ fixed at $\lambda_i(1; \vec{r}(s_j))$, and $\lambda_i(0; \vec{x})$ fixed at $\lambda_i(0; (0, \vec{r}^d(s_j)))$ [recall that $(0, \vec{r}^d(s_j))$ is the projection of r onto the boundary]. The cost function I_j is the usual one, but on the interval $[s_j, s_{j+1}]$. Consider a cheapest path that starts at $\vec{r}(s_j)$ and ends at $\vec{r}(s_{j+1})$. By Lemma 5.16 we can choose this path to have one of two forms. It could be a straight line joining $\vec{r}(s_j)$ and $\vec{r}(s_{j+1})$, which we denote by $\vec{v}$. Or, it is composed of at most three straight line segments, one of which runs along the boundary, and the others connect this segment to the endpoints; we call the cheapest such path $\vec{u}$. [If only one of $\vec{r}(s_j)$ or $\vec{r}(s_{j+1})$ lies on the boundary then $\vec{u}$ consists of at most two segments. If both lie on the boundary then $\vec{u} = \vec{v}$.]

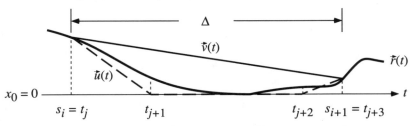

Figure 8.74. Construction of $\vec{w}(t)$ for the flat boundary model.

Definition 8.75. *Given Δ, define $\vec{w}$ on each interval $[s_j, s_{j+1}]$ by*

$$\vec{w}(t) = \begin{cases} \vec{v}(t) & \text{if } \vec{r}_0(s) > 0 \text{ for all } s_j \le s \le s_{j+1}; \\ \vec{v}(t) & \text{if } I_j(\vec{v}) \le I_j(\vec{u}); \\ \vec{u}(t) & \text{otherwise.} \end{cases}$$

Note that $\vec{w}$ is piecewise linear, and so absolutely continuous, though it is not everywhere differentiable. We define t_k as the left endpoint of the kth interval, $k \ge 0$, in the definition of $\vec{w}$, and let J be the index of the time T. (Clearly $J = J_0$ only if $\vec{w} = \vec{v}$ on each subinterval, and then $t_j = s_j$ for each j. In general $J_0 \le J < 3J_0$.)

Lemma 8.76. *Recall that $\vec{r}$ is fixed and $I_0^T(\vec{r}) < \infty$. Under Assumption 8.8, for each positive δ and K there exists a constant $\gamma > 0$ such that for any set of jump rates $\vec{\mu}$ with $\delta \le \mu_i(m) \le K$, all $0 < \varepsilon \le 1$ and all Δ as above,*

$$\sup_{0 \le t \le T} |\vec{w}(t) - \vec{r}(t)| < \gamma \varepsilon.$$

Proof. See Exercise 8.77.   ∎

Exercise 8.77. Prove Lemma 8.76. Hints: only the case $\vec{w} = \vec{u}$ requires proof. Fix δ and K, and let ℓ_μ be the rate function corresponding to the *constant rates* $\vec{\mu}$. Use Lemma 8.36 to show that for some constants c_m and c_M,

$$c_m (1 + |\vec{y}| \log |\vec{y}|) \stackrel{\triangle}{=} \ell_m(\vec{y}) \le \ell_\mu(\vec{x}, \vec{y}) \le \ell_M(\vec{y}) \stackrel{\triangle}{=} c_M (1 + |\vec{y}| \log |\vec{y}|)$$

for all $\vec{\mu} \in M'(\delta, K)$. Conclude that $I_0^T(\vec{r}) < \infty$ implies $I_0^T(\vec{v}) < c_r < \infty$ for all Δ. Now fix an interval $[t_k, t_{k+1}]$ with $s_j = t_k$ for some j and denote $r' = |\vec{r}'(s_j)|$ and $w' = |\vec{w}'(s_j)|$. If w' is small or $w' < r'$ then we are done. Otherwise, if r' is large

$$(t_{k+1} - t_k)c_1 w' \log r' \le I_j(\vec{w}) \le I_j(\vec{v}) \le \Delta c_2 r' \log r'$$

from which a bound on $(t_{k+1} - t_k)w'$ follows. Now do the case where r' is small. A similar argument works on the intervals $[t_{k+m}, t_{k+m+1}]$, $m = 1, 2$.   ♠

The next approximation lemma shows that it is not too unlikely for $\vec{z}_n(t)$ to stay near $\vec{w}(t)$. This is the counterpart of Corollary 5.53, the main result used in proving the lower bound. For fixed $\vec{r}$ and Δ, let

$$\overline{w}' \stackrel{\triangle}{=} \max_{0 \le t \le T} \left| \frac{d}{dt} \vec{w}(t) \right| \tag{8.52}$$

$$\overline{e} \stackrel{\triangle}{=} \max_{i,m} |\vec{e}_i(m)| \tag{8.53}$$

$$A \stackrel{\triangle}{=} \overline{w}' + \overline{e} + 1 \tag{8.54}$$

$$\mathcal{D} \stackrel{\triangle}{=} \left\{ i : e_i^0(1) = -1 \right\} \tag{8.55}$$

$$\tau_{\mathcal{D}} \stackrel{\triangle}{=} \frac{1}{\min_{i \in \mathcal{D}} \lambda_i(1; \vec{x})} \tag{8.56}$$

$$\lambda_{\mathcal{D}} \stackrel{\triangle}{=} \sum_{i \in \mathcal{D}} \lambda_i(1; \vec{x}) \tag{8.57}$$

$$\overline{\lambda} \stackrel{\triangle}{=} \max_{\vec{x}} \sum_{i=1}^{k(0)} \lambda_i(0; \vec{x}) + \max_{\vec{x}} \sum_{i=1}^{k(1)} \lambda_i(1; \vec{x}) \tag{8.58}$$

Clearly $\overline{w}'$ is finite but may diverge as $\Delta \downarrow 0$. For $\eta > 0$ and $0 \le j \le J$ define

$$S_j(\eta) \stackrel{\triangle}{=} \left\{ \vec{x} \in \mathbb{R}^{d+1} : x_0 \ge 0, \ |\vec{x} - \vec{w}(t_j)| < 2^j A\eta \right\}. \tag{8.59}$$

Lemma 8.78. *Suppose that we have a flat boundary process satisfying Assumptions 8.8 (structure of the flat boundary process) and 8.9 (smooth jump rates). For any $\varepsilon > 0$ there exists a $\Delta > 0$ such that with $\{\vec{w}(t), \{s_j\}, \{t_j\}\}$ as defined in Definition 8.75 and A defined in (8.54) there is an η_0 such that for any $\eta < \eta_0$, if $|\vec{z}_n(t_j) - \vec{w}(t_j)| < \eta$ then*

$$\inf_{\vec{x} \in S_j(\eta)} \liminf_{n \to \infty} \frac{1}{n} \log \mathbb{P}_{\vec{x}} \left(\sup_{t_j \le t \le t_{j+1}} |\vec{z}_n(t) - \vec{w}(t)| < A\eta \right)$$

$$\ge -I_{t_j}^{t_{j+1}}(\vec{w}) - \varepsilon(t_{j+1} - t_j)/2. \tag{8.60}$$

Furthermore, the same estimate holds when we restrict the endpoints to be on or off the boundary when $\vec{w}(t)$ is. That is, the same limit holds if we consider the event

$$\left\{ \sup_{t_j \le t \le t_{j+1}} |\vec{z}_n(t) - \vec{w}(t)| < A\eta, \ \mathbf{1}\left[z_{n,0}(t_{j+1}) = 0\right] = \mathbf{1}\left[w_{n,0}(t_{j+1}) = 0\right] \right\}$$

$$\tag{8.61}$$

instead. We can even restrict t_j to have $\mathbf{1}\left[z_{n,0}(t_j) = 0\right] = \mathbf{1}\left[w_{n,0}(t_j) = 0\right]$ and have the result (8.60).

Remark. This lemma is very nearly the lower bound. It says that we can approximate the path $\vec{w}$ to within $A\eta$ with cost that is within ε of $I_0^T(\vec{r})$.

Lemma 8.78 is proved by considering the four cases determined by whether or not each starting point $\vec{w}(t_j)$ and ending point $\vec{w}(t_{j+1})$ is on or off the boundary.

Figure 8.79. The four types of line segments of $\vec{w}$.

We break the calculations into bite-size chunks, starting with a useful estimate.

Exercise 8.80. Let x_λ be a Poisson random variable with parameter λ. For any $\lambda \geq 1$ prove that

$$\mathbb{P}\,(x_\lambda > \lambda) \geq \frac{1}{2} - \frac{1}{e} > \frac{1}{8}.$$

Hint: Show that $\mathbb{P}\,(x_\lambda > \lambda)$ is increasing in λ when λ is not an integer. Then for integer λ and integer $k > 0$ show that

$$\mathbb{P}\,(x_\lambda = \lambda - 1 - k) < \mathbb{P}\,(x_\lambda = \lambda + k)$$

since

$$\frac{\lambda^{\lambda+k}}{\lambda} = \frac{\lambda^\lambda}{\lambda!}\frac{\lambda}{\lambda+1}\cdots\frac{\lambda}{\lambda+k} \quad \text{and} \quad \frac{\lambda^{\lambda-1-k}}{(\lambda-1-k)!} = \frac{\lambda^\lambda}{\lambda!}\frac{\lambda}{\lambda}\frac{\lambda-1}{\lambda}\cdots\frac{\lambda-1-k}{\lambda}$$

and note that for any positive n and m that

$$\frac{n-m}{n} < \frac{n}{n+m}.$$

Then since $\mathbb{P}\,(x_\lambda = \lambda - 1) = \mathbb{P}\,(x_\lambda = \lambda)$, we obtain

$$\mathbb{P}\,(x_\lambda > \lambda) \geq \frac{1}{2} - \mathbb{P}\,(x_\lambda = \lambda).$$

For integer values of λ, $\mathbb{P}\,(x_\lambda = \lambda) \leq 1/e$. This concludes the exercise... but can you show that our bound is exactly half the optimal lower bound of $1 - 2/e > 1/4$? ♠

The next lemmas use some of the constants defined in (8.52)–(8.58).

Lemma 8.81. *Let Assumptions 8.8 and 8.9 hold. For any $\eta > 0$ and any $\vec{x}$ with $x_0 = \eta$, any $\tau \geq \tau_{\mathcal{D}}\eta$ and any $n \geq \tau_{\mathcal{D}}$ with $n\eta \in \mathcal{Z}$,*

$$\mathbb{P}_{\vec{x}}\left(z_{n,0}(\tau) = 0, \quad \sup_{0 \leq t \leq \tau} |\vec{z}_n(t) - \vec{x}| \leq \bar{e}\eta\right) \geq \frac{1}{8}e^{-n\tau\bar{\lambda}}$$

Proof. The left hand side is lower bounded by the probability that $\vec{z}_n(t)$ has no jumps except for those in $\mathcal{D}$ while $z_{n,0} > 0$, and has no jumps at all once it hits the boundary. By Exercise 8.80, the probability that a Poisson (λ) process makes at least λt jumps in time t is at least $\frac{1}{8}$ when $\lambda t \geq 1$. The probability that the process makes no other kinds of jumps in time τ is obviously bounded by $e^{-n\tau\bar{\lambda}}$. ∎

Lemma 8.82. *Let Assumptions 8.8 and 8.9 hold. For any $\eta > 0$ and any $\vec{x}$ with $x_0 = 0$, any $\tau \geq \tau_{\mathcal{D}}\eta$ and any $n \geq \tau_{\mathcal{D}}$ with $n\eta \in \mathcal{Z}$,*

$$\mathbb{P}_{\vec{x}}\left(z_{n,0}(\tau) \geq \eta, \quad \sup_{0 \leq t \leq \tau} |\vec{z}_n(t) - \vec{x}| \leq \bar{e}\eta\right) \geq \frac{1}{8}e^{-n\tau\bar{\lambda}}.$$

Proof. This is just the reverse of Lemma 8.81. Consider jumps of $\vec{z}_n(t)$ away from the boundary rather than towards it. The new point is that the process might overshoot the appropriate level, since the jumps away from the boundary have integer sizes that are not necessarily one. If the process overshoots, it does so by at most a constant amount. ∎

For any set of rates $\vec{\mu}$ in $M'(\delta, K)$ let $\tilde{\mathbb{P}}_{\vec{x}}$ ($\tilde{\mathbb{E}}_{\vec{x}}$) denote the probability distribution (respectively expectation) of the scaled jump process $\vec{z}_n(t)$ with jump rates $\vec{\mu}$, with $\vec{z}_n(0) = \vec{x}$. Let $\vec{y}_\infty$ denote the induced drift process $\vec{z}_\infty$ with rates $\vec{\mu}$. Recall that rates in $M'(\delta, K)$ have a uniformly negative drift, and under Assumption 8.8 they mix the interior and boundary uniformly well. Moreover, the process $z_n^0(t)$ is necessarily an ergodic Markov chain.

Lemma 8.83. *Under Assumption 8.8, given any K, δ, T, and $\varepsilon > 0$ there exist positive C_1, $C_2(\varepsilon)$, and n_0 such that for all $\vec{x} \in \mathbb{R}^{d+1}$ with $x_0 = 0$ and all $\vec{\mu} \in M'(\delta, K)$,*

$$\tilde{\mathbb{P}}_{\vec{x}}\left(\sup_{0 \le t \le T} |\vec{z}_n(t) - \vec{y}_\infty(t)| > \varepsilon \right) \le C_1 e^{-n C_2(\varepsilon)}, \tag{8.62}$$

where $\vec{y}_\infty(t)$ is, as usual, the most likely path under the measure $\vec{\mu}$:

$$\frac{d}{dt}\vec{y}_\infty(t) = \pi_0 \sum_i \mu_i(0)\vec{e}_i(0) + (1 - \pi_0) \sum_i \mu_i(1)\vec{e}_i(1) \tag{8.63}$$

$$\vec{y}_\infty(0) = \vec{x},$$

where π_0 is the stationary measure of $x_0 = 0$ under the law $\vec{\mu}$ (see (7.14)). Furthermore, the function $C_2(\varepsilon)$ grows superlinearly at ∞:

$$\lim_{\varepsilon \to \infty} \frac{C_2(\varepsilon)}{\varepsilon} = \infty.$$

Proof. This follows from Lemma 7.20 in exactly the same way that Theorem 8.71 followed from Exercise 7.8. ∎

Consider the path $\vec{y}_\infty(t)$. If $x_0 = 0$ then it is easy to see that $\vec{y}_\infty(t)$ is a straight line. Suppose we are given a set of bounded continuous functions $f_{i,m}(\vec{x})$ where the index m is zero or one depending on whether $x_0 = 0$ or $x_0 = 1$, and the index i ranges over the jump directions on each level ($1 \le i \le k(m)$). Let $l(j)$ denote the index (i, m) of the jth transition of $\vec{z}_n(t)$, where again i is the index of the jump direction $\vec{e}_i(m)$, and m indicates whether or not the process is on the boundary at the start of the jump. As before, N is the number of jumps of $\vec{z}_n$ on $[0, T]$. As usual $\pi_\mu(0)$ is the invariant measure of the boundary, and $\pi_\mu(1) \overset{\triangle}{=} 1 - \pi_\mu(0)$ is the invariant measure of the interior, which is unique if we assume that $z_n^0(t)$ is ergodic.

Lemma 8.84. *Let Assumptions 8.8 and 8.9 hold and assume the process $z_n^0(t)$ is ergodic under the rates $\vec{\mu}$. Then for $f_{i,m}(\vec{x})$ bounded continuous, uniformly over $\vec{x}$ in compact sets with $x_0 = 0$, for each $\varepsilon > 0$ we have*

$$\lim_{n\to\infty} \tilde{\mathbb{E}}_{\vec{x}}\left(\frac{1}{n}\sum_{j=1}^{N} f_{l(j)}(\vec{z}_n(t_j^-))\right)$$

$$= \lim_{n\to\infty} \tilde{\mathbb{E}}_{\vec{x}}\left(\mathbf{1}\left[\vec{z}_n \in N_\varepsilon(\vec{y}_\infty)\right]\frac{1}{n}\sum_{j=1}^{N} f_{l(j)}(\vec{z}_n(t_j^-))\right)$$

$$= \int_0^T \sum_{m=0}^{1}\pi_\mu(m)\sum_{i=1}^{k(m)}\mu_i(m)f_{i,m}(\vec{y}_\infty(t))\,dt.$$

Proof. This goes exactly as Lemma 5.52, using Lemma 8.83.   ∎

For the next result recall the definition (8.51) of the function f that defines the local rate function:

$$f\left(\vec{\mu},\vec{\lambda}\right) \triangleq \sum_{m=1}^{D}\pi(m)\left(\sum_{i=1}^{k(m)}\lambda_i(m) - \mu_i(m) + \mu_i(m)\log\frac{\mu_i(m)}{\lambda_i(m)}\right).$$

Corollary 8.85. *Let Assumption 8.8 hold and suppose that $\vec{\mu} \in M'(\delta, K)$ for some δ and K. Then uniformly over $\vec{x}$ in compact sets with $x_0 = 0$, for any $\varepsilon > 0$,*

$$\liminf_{n\to\infty}\frac{1}{n}\log\mathbb{P}\left(\vec{z}_n \in N_\varepsilon(\vec{y}_\infty)\right) \geq -\int_0^T f(\vec{\mu},\vec{\lambda}(\vec{y}_\infty(t)))\,dt.$$

Remark. Note that the assumption $\vec{\mu} \in M'(\delta, K)$ implies that the path $\vec{y}_\infty(t)$ lies on the boundary, that is, $y_{\infty,0}(t) = 0$. So this corollary is useful for examining paths that start and end on the boundary. Furthermore, since $\vec{\mu} \in M'(\delta, K)$, we can get a uniform lower bound on the probability that $z_n^0(T) = 0$.

Proof. Exactly as in the proof of Corollary 5.53.   ∎

In order to prove Lemma 8.78, we choose constants in the following order.

1. Choose a $\zeta > 0$; this is the bound on the error in the cost $I_0^T(\vec{r})$.

2. Choose an $\varepsilon_0 > 0$ small enough with respect to ζ that the following inequalities hold. First take $\varepsilon = \zeta/4$ in Theorem 8.37, and let δ be the resulting δ so that (8.27) holds with $_bI_0^T(\vec{r})$ instead of $_cI_0^T(\vec{r})$. Let $\varepsilon_0 = \delta$. We will perform estimates for $\varepsilon < \varepsilon_0$.

3. Choose Δ small enough so that for each $s \in [0, T - \Delta]$,

$$\max_{i,m}\ \sup_{s\leq t\leq s+\Delta}\ |\lambda_i(\vec{r}(s), m) - \lambda_i(\vec{r}(t), m)| < \varepsilon_0.$$

Furthermore, choose Δ small enough that

$$\sup_{0\leq t\leq T}|\vec{w}(t) - \vec{r}(t)| < \varepsilon_0.$$

where $\vec{w}(t)$ is constructed by Definition 8.75.

4. With $\vec{w}(t)$ fixed, find the following:

$$d \stackrel{\triangle}{=} \min_j t_{j+1} - t_j$$

$$Q \stackrel{\triangle}{=} \min_j \{ w_0(t_j) \ : \ w_0(t_j) > 0 \}$$

$$R \stackrel{\triangle}{=} \max_j \{ w_0(t_j) \ : \ w_0(t_{j-1}) = 0 \text{ or } w_0(t_{j+1}) = 0 \}$$

Then choose η_0 small with respect to Q, R/d, and Q/Δ. This also gives a bound on how close the drift of $\vec{y}_{\vec{\mu}}$ is to $\vec{w}$. This also gives rise to the constants δ and K so that $\vec{\mu} \in M'(\delta, K)$.

5. Finally, we let $n \to \infty$ for each fixed choice of all the constants involved in the construction.

Proof of Lemma 8.78. We will only outline the proof of this lemma, since all the pieces have already been given, and there is nothing new in the reasoning. The reader is invited to fill in the details. As illustrated in Figure 8.79, the lemma is proved by dividing considerations into four cases. When the starting and ending positions are both on or both off the boundary, the lemma reduces to Corollary 8.85 (both on the boundary) or Corollary 5.53 (both off the boundary). When one endpoint is on and the other is off, we use Lemmas 8.81 and 8.82 to show that the path can be pushed on or off the boundary at the endpoints as needed in small time and with small cost, and then estimate the behavior of the path during the remainder of the interval.

All four cases begin with the construction of a change of measure $\vec{\mu}$ that has its center (see (8.63)) equal to $\vec{w}(t)$ on the interval $[t_j, t_{j+1}]$, and also achieves the minimal cost with respect to jump rates fixed at time t_j. Such a measure exists by Lemma 8.32. Then, if the interval $[t_j, t_{j+1}]$ has both endpoints $\vec{w}(t_j)$ and $\vec{w}(t_{j+1})$ on the boundary, we use Lemma 8.66 to change $\vec{\mu}$ so that its center is no more than η from $\vec{w}(t)$ and so that $\vec{\mu}$ is in $M'(\delta, K)$. Here δ and K are functions of η given by Lemma 8.66.

When the left-hand endpoint $\vec{w}(t_j)$ is on the boundary ($w_0(t_j) = 0$) but the right-hand endpoint is not, we perform the following construction. First we use Lemma 8.82 with $\tau = \tau_D$ to use up a little time (τ) and a little probability $(\exp(-n\tau\lambda))$ but pushing the process $\vec{z}_n(t)$ away from the boundary. Then we can make sure that $\vec{z}_n(t)$ follows parallel to $\vec{w}(t)$ for the remainder of the interval. We make sure that it doesn't hit the boundary again by setting the neighborhood size to $\eta/2$. The discrepancy between $\vec{z}_n(t)$ and $\vec{w}(t)$ grows by the following mechanisms. The initial interval τ has $\vec{z}_n(t)$ move in a direction that has nothing to do with $\vec{w}(t)$. But we know that $\vec{z}_n(\tau)$ is no more than $\bar{e}\eta$ from its starting point, and $\vec{w}$ has gone no more than $\tau\vec{w}'$ during this time by (8.52). Therefore they move apart no more than a constant times η. Now they move parallel and grow apart no more than $\eta/2$ during the remaining time.

Now suppose that the right-hand endpoint is on the boundary but the left-hand endpoint is not. We need to see the earliest time that the path $\vec{z}_n(t)$ might hit the boundary. The initial discrepancy between $\vec{z}_n(t)$ and $\vec{w}(t)$ is η, and the discrepancy grows by no more than η over the interval $[t_j, t_{j+1}]$. Suppose that $2\eta < Q$ (recall Q is the smallest distance an endpoint of $\vec{w}$ can be above the boundary if it is nonzero). Then the earliest that the process $\vec{z}_n(t)$ might reach the boundary is $t_{j+1} - \Delta 2\eta/Q$ (the interval $t_{j+1} - t_j \leq \Delta$). Now the highest the process might be at this time is 4η. Then we use Lemma 8.81 to push the process down to the boundary, moving no more than $\bar{e}4\eta$ in the process, while $\vec{w}$ moves by no more than $\overline{w}'\Delta 2\eta/Q$. This finishes the proof.   ∎

Proof of the lower bound for flat boundary processes, Theorem 8.69.
Direct from Lemma 8.78 along the lines of the proof of Theorem 8.68.   ∎

8.6. End Notes

As pointed out in the text, the approach to the upper bound follows [DEW]. The lower bound follows the usual procedure for both types of processes considered, but the key Lemma 8.61 for the finite levels process is new, as is the finite levels process itself.

Others have approached this subject with a variety of techniques. The new book by Dupuis and Ellis [DE2] is a general approach to proving the existence of large deviations principles for a wide variety of processes with discontinuous rates. Previously, these authors [DE1] gave a proof of the large deviations principle for a discrete-time process with a discontinuity in the interior of the state space (or, if you like, the state space is two half-spaces joined at the boundary). The large deviations principle for the flat boundary process was developed independently by Ignatyuk and Scherbakov [IS].

Applications

This half of the book deals with applications of the theory of large deviations to queues, communication systems, and computing. The chapters are meant to be largely independent of each other, except that several chapters are based on calculations and ideas from Chapter 11, the $M/M/1$ queue, especially §11.5, and to a lesser extent §11.4 and 11.6. Even though these are applications, there is some theoretical discussion, usually under the rubric "justification" in most chapters, because the theory has to be modified, extended, or simplified for most of these applications. This is despite our having written the first half of the book specifically for these applications! This is a general lament of those trying to apply large deviations (or most other "applied mathematics" techniques): the theory just never quite seems to fit. We hope that the grounding you get in the present book will enable you to approach your own applications successfully.

Despite the diversity of applications and requisite theory, almost all the applications are approached the same way. Here is how we suggest you approach problems, either in this book or in studying your own application:

1. Make a Markov model of the process (this already rules out many applications).

2. Scale the problem, if necessary, so that jump rates are of order n and jumps are of order $1/n$ for some large parameter n.

3. Calculate $z_\infty(t)$. This step alone will usually provide you with most of the insight needed for your problem. Analyzing how the system will most likely behave, calculating the attracting points of $z_\infty(t)$, and seeing how it behaves as a function of the problem's parameters is often enough information in practice. This is usually easy to do, at least numerically, since it involves solving a system of ordinary differential equations.

4. See if the issue of convergence to $z_\infty(t)$ (Kurtz's Theorem) for this problem is covered by the theory as developed in this book or another source. If so, great! If not, try to extend the theory to cover the problem at hand.

5. Based on your analysis from step 3, and on the problem at hand, write down the appropriate large deviations rate function. Set up the questions you want answered, and write (formally!) the corresponding variational problems.

6. Solve each variational problem: analytically, or numerically, if necessary. See how the solution behaves as a function of the problem's parameters.

7. See if the problem is covered by the theory as developed in this book or another source. If not, try to extend the theory to cover the problem at hand.

Note that in some cases Kurtz's Theorem is established as a consequence of a large deviations principle, in which case step 4 becomes a part of step 7.

We strongly suggest leaving steps 4 and 7, the justifications, to the end. First, if the answer isn't very informative, then what have you lost if it's not justified? And if you can't solve the variational problem anyway, then why bother trying to justify it? And if the solutions *do* tell you something, then you will be much more motivated to try to justify them, so much more likely to succeed. This is also consistent with the way most variational problems should be approached. It is usually quite difficult to justify every step in the derivation of an analytic solution. However, if by some mysterious way you have arrived at a solution that you suspect to be correct, then a *verification theorem* such as Theorem C.1 can be used to check that this is indeed a kosher solution, and often even to certify its uniqueness. Some intuition into variational problems, as well as statements of the theorems that we use, are collected in Appendix C. The application chapters should be read with this philosophy in mind: the purpose of most calculations is to show how to go about *guessing* a solution to the variational problem. There is no point in trying to be rigorous at this stage, since it is much easier to check that the solution we have guessed is indeed the right one.

There are many tricks that we and others have developed for solving the variational problems that arise in large deviations problems. Most of these tricks work best on one-dimensional problems; see §11.5. Often, the key to solving a high dimensional problem is to reduce it to one-dimensional problem(s). Sometimes some insight can be gained by simplifying the problem in other ways, too. In short, large deviations is not a panacea, and you will have to be as clever and diligent as you would when using any other technique in applied mathematics.

The first two application chapters, "allocating independent subtasks" (on parallel processors) and "parallel algorithms: rollback," require no extra theory, and are independent of all other applications. This is why they are given first. All subsequent applications use techniques and calculations from Chapter 11. This is the chapter you should read first if you are interested in performance analysis.

Erlang's model is the basis of all circuit-switched communication. The Anick-Mitra-Sondhi model is the basis of all packet-switched communication. And the Aloha model is the basis of all multiple-access channels. These three chapters, then, cover all the basic models of communication networks. Information theory is outside the scope of this book, although these subjects have also benefitted recently from large deviations theory; see [Bu, DZ] and references therein.

The other chapters largely concern models that arise in parallel computation. None of them are as basic as the communication models, possibly excepting the model of rollback as developed in Chapter 10. They may be scattershot, but also may be useful; take your pick.

The degree to which this part is independent of the first part of the book depends on how much you are willing to trust our word. Familiarity with the processes and the scaling that we use is necessary: these are covered in Chapter 4. Some familiarity with the statements of the large deviations theorems is also necessary: this is covered in the introductory parts of Chapters 5 and 6, and in §6.4. Finally, at least the intuitive part of the calculus of variations, as sketched in Appendix C is indispensable. And now, time to dive in, head first!

Chapter 9

Allocating Independent Subtasks

In this chapter we touch the surface of an important problem: How should jobs be allocated on parallel processors? The subject is much too broad and deep even to be introduced properly here. We content ourselves with the presentation of a very simple model, which is nonetheless important and moreover is amenable to analysis via large deviations. We need nothing more from large deviations theory than Chernoff's Theorem 1.5 or 1.10. We introduce the useful notions of characteristic maximum and increasing failure rate, and we bring out a correspondence between Chernoff's Theorem and the central limit theorem.

We consider a computer consisting of a number of processors, memory modules, and a communication network between processors and memory. On such machines, one is often confronted with solving a task composed of many independent subtasks (or jobs) where it is necessary to synchronize the processors after all of the subtasks have been processed. This models, for instance, a single fork and join.

Let p be the number of processors and n be the number of subtasks. We assume that assigning one or more subtasks to a processor entails some overhead. Overhead may arise from communication delays, memory delays, or contention between processors for various resources. We assume that the overhead is independent of the number of subtasks assigned. This could be the case, for example, if there is a central job queue at one of the memory modules, and the main delay is getting to the job queue, not the process of acquiring jobs once there. The time may be deterministic or an independent random variable.

One way to allocate the subtasks would be to assign $K = n/p$ to each processor. The problem here is that the processors may finish at wildly different times, and many processors will be idled until the last one finishes. Another way is to assign them one at a time; this would seem to result in the least difference between finishing times, but might incur excessive delay due to the overhead of performing many job assignments for each processor. Finally, one can imagine an intermediate scheme where subtasks are dynamically assigned in "batches," i.e., several at a time, with the idea of keeping the finishing times fairly even but not adding too much overhead. We evaluate the efficiency of parceling out all the tasks at once, and show that this algorithm is not too bad compared to an optimal allocation. Clearly, in practice things may be much more complicated. For instance, K could vary with time, the subtasks may be statistically dependent, and they may not be identically distributed. For an overview and some results on models with synchronization constraints, see [BaM].

To summarize, our model is as follows. There are p processors, initially idle. At $t = 0$ they each take $K = n/p$ subtasks from a job queue, each experiencing a delay v in that access. The processors then continue to run independently until all the jobs are done. Running times of the subtasks are independent identically distributed (i.i.d.) random variables possessing a moment generating function (so that Chernoff's Theorem can be applied). The time we are trying to estimate is

$$T = \max\{Y_1, \ldots, Y_p\},$$

where Y_i represents the running time of processor i and is the sum of $K + 1$ independent random variables: one representing the overhead, and K representing the processing time of each subtask. T is sometimes called the makespan.

The precise assumptions are given in Corollary 9.13 below. The corollary is an immediate consequence of Lemmas 9.6 and 9.9 and Theorems 9.7 and 9.8, which provide estimates of T when there is no overhead. The most restrictive assumption is that the distribution function of running times (and overhead) has an increasing failure rate (IFR, Definition 9.1 below). This assumption excludes many pathological distributions. However, it is general enough to include the exponential, gamma, Weibull, uniform, deterministic, and truncated normal distributions. The reader will note that we actually make use of a much weaker assumption. However, the IFR assumption is essential in a more complete analysis, and we introduce it here more for pedagogic purposes than mathematical necessity.

9.1. Useful Notions

In the section we present the notion of IFR distributions, which are used extensively in reliability theory [BP]. We then present the notion of characteristic maximum and motivate its usefulness.

Definition 9.1 [BP Chapter 3, Definition 1.1]. *The distribution function $G(x)$ is said to have an increasing failure rate (abbreviated IFR) if $G(0) = 0$ (i.e., it is the distribution function of a positive random variable) and if for all $t > 0$,*

$$\frac{1 - G(x + t)}{1 - G(x)} \quad \text{is monotone decreasing in } x.$$

When G has a density g then this is equivalent to

$$\frac{g(x)}{1 - G(x)} \quad \text{is monotone increasing in } x.$$

Note that our definition includes the borderline case of exponential distributions, which have constant failure rate. Intuitively, an IFR random variable shows aging. Here are some properties of IFR distributions.

Lemma 9.2.

1. *The sum of independent IFR random variables is IFR [BP Chapter 4 Theorem 4.2].*

2. *The following are IFR distributions [BP Chapter 3]:*
   a. *Exponential*
   b. *Gamma with* $\mu/\sigma \geq 1$, *where* μ *is the mean,* σ *the standard deviation*
   c. *Weibull with rate* ≥ 1
   d. *Truncated normal (i.e., normal constrained to be positive)*
   e. *Uniform on the interval* $(0, C)$ *for any* $C > 0$
   f. *Constant* $= C$ *for any* $C > 0$.

Our analysis is also based on the notion of the characteristic maximum. This is a good estimate of the expected value of the maximum of a large number of independent random variables. We shall introduce this notion, show how it relates to the distribution of the maximum, and then devote ourselves to methods of obtaining estimates of the characteristic maximum.

Let $Y_1, \ldots, Y_p$ be random variables with common distribution function

$$G(x) \overset{\triangle}{=} \mathbb{P}(Y \leq x).$$

Definition 9.3. *The characteristic maximum of a random variable* Y *with distribution function* G *is* $m_p \overset{\triangle}{=} \inf\{x : 1 - G(x) \leq 1/p\}$.

If $G(x)$ is continuous then obviously m_p is a solution of the equation

$$1 - G(m_p) = \frac{1}{p}. \tag{9.1}$$

However, this equation may have many solutions, in which case m_p is the smallest. If G is not continuous, then (9.1) may not have a solution. This is likely to be the case for discrete distributions.

Exercise 9.4. For an exponential distribution function $G(x) = 1 - e^{-\lambda x}$, show that $m_p = \log p/\lambda$.   ♠

Exercise 9.5. Let G be a standard normal distribution:

$$G(x) = \Phi(x) = \int_{-\infty}^{x} \frac{1}{\sqrt{2\pi}} e^{-t^2/2} \, dt.$$

From the well-known inequalities [Mc, p. 4]

$$\frac{e^{-x^2/2}}{\sqrt{2\pi}\,(x + x^{-1})} < 1 - \Phi(x) < \frac{e^{-x^2/2}}{\sqrt{2\pi}\,x} \tag{9.2}$$

obtain the inequalities [LR1, p. 287]

$$\sqrt{2\log p - \log\log p - 3} < m_p < \sqrt{2\log p - \log\log p} \quad \text{for} \quad p \geq 5. \quad ♠$$

The utility of the characteristic maximum will be brought out in four statements. The first two are immediate consequences of the definition of m_p, and the latter are strong asymptotic estimates of the mean of the maximum in terms of m_p. We assume that $G(x)$ is continuous: this is convenient simply because (9.1) holds. The reader is invited to explore the extensions at his peril.

Lemma 9.6. *Let $Y_1, Y_2, \ldots$ be identically distributed with a continuous distribution function G. Then*

$$\mathbb{E}(\text{number of } Y_1, \ldots, Y_p \text{ that exceed } m_p) = 1 \qquad (9.3)$$

regardless of their statistical dependence. If they are also independent then

$$\lim_{p \to \infty} \mathbb{P}(\max\{Y_1, \ldots, Y_p\} \le m_p) = 1/e. \qquad (9.4)$$

Proof. Let $X_i = \mathbf{1}\left[Y_i > m_p\right]$. Then

$$\mathbb{E}(\text{number of } Y_1, \ldots, Y_p \text{ that exceed } m_p) = \mathbb{E}\left(\sum_{i=1}^{p} X_i\right)$$

$$= \sum_{i=1}^{p} \mathbb{E}(X_i)$$

$$= p\mathbb{P}(Y > m_p)$$

$$= p \cdot 1/p = 1.$$

If they are independent then $\mathbb{P}(\max\{Y_1, \ldots, Y_p\} \le m_p) = (1 - 1/p)^p \to 1/e.$ ∎

Theorem 9.7 [LR1, p. 287]. *Suppose that G is continuous, and*

$$\lim_{x \to \infty} \frac{1 - G(x + c)}{1 - G(x)} = 0 \quad \text{for all} \quad c > 0.$$

Then, no matter how the $Y_1, Y_2, \ldots$ are dependent,

$$\lim_{p \to \infty} \mathbb{E}(\max\{Y_1, \ldots, Y_p\}) - m_p \le 0. \qquad (9.5)$$

If $Y_1, Y_2, \ldots$ are independent then equality holds in (9.5).

Theorem 9.8 [LR1, p. 287]. *Suppose that G is continuous, and*

$$\lim_{x \to \infty} \frac{1 - G(cx)}{1 - G(x)} = 0 \quad \text{for all} \quad c > 1.$$

Then, no matter how $Y_1, Y_2, \ldots$ are dependent,

$$\lim_{p \to \infty} \frac{\mathbb{E}(\max\{Y_1, \ldots, Y_p\})}{m_p} \le 1. \qquad (9.6)$$

If $Y_1, Y_2, \ldots$ are independent then equality holds in (9.6).

The first fact of Lemma 9.6 states that the expected number of observations $Y_1, \ldots, Y_p$ that exceed m_p is equal to one, regardless of how $Y_1, \ldots, Y_p$ are dependent. The second states that, for independent random variables, m_p is approximately the $(1/e)$th quantile of the distribution of the maximum. The two theorems state that, under certain conditions on the tails of the distribution $G(x)$, the characteristic maximum is a good estimate of the mean of the maximum. Theorem 9.7 says that m_p is very close to $\mathbb{E}(\max\{Y_1, \ldots, Y_p\})$, whereas Theorem 9.8 only states that m_p grows faster than $\mathbb{E}(\max\{Y_1, \ldots, Y_p\}) - m_p$. The condition on $G(x)$ in Theorem 9.7 is that the tail of G at x is much larger than the tail at x plus a fixed distance, whereas in Theorem 9.8 the tail at x is large compared to the tail at x times a fixed amount.

Only Theorem 9.8 will be used in our analysis; the other facts are to give the reader a feel for the utility of m_p. Theorem 9.7 could be used in special cases for better bounds than we obtain with Theorem 9.8. These theorems are proved also in [LR2 p. 100], and we will not prove them here. For more information, see also Galambos [Gal, Chapter 1, Exercises 15 and 16].

9.2. Analysis

We now present a simple lemma showing that the hypothesis of Theorem 9.8 is satisfied for continuous IFR distributions.

Lemma 9.9. *If $G(x)$ is IFR then*

$$\lim_{x \to \infty} \frac{1 - G(cx)}{1 - G(x)} = 0 \quad \text{for all} \quad c > 1. \tag{9.7}$$

Proof. From the definition of IFR, $[1 - G(x + t)]/[1 - G(x)]$ is a decreasing function of x for each $t > 0$, and is clearly less than one for large values of x. Fix $c > 1$ and $t > 0$ and let x be much larger than t. Denote

$$k(x) \stackrel{\triangle}{=} \max\left\{i \text{ integer } : i \le \frac{cx - x}{t}\right\}. \tag{9.8}$$

Then $k(x) \to \infty$ as $x \to \infty$. Hence

$$\frac{1 - G(cx)}{1 - G(x)} \le \frac{1 - G(x + t)}{1 - G(x)} \cdot \frac{1 - G(x + 2t)}{1 - G(x + t)} \cdots \frac{1 - G(x + k(x)t)}{1 - G(x + k(x)t - t)}$$

$$\le \left(\frac{1 - G(x + t)}{1 - G(x)}\right)^{k(x)},$$

which converges to zero as $x \to \infty$ for each $c > 1$ and $t > 0$. ∎

Returning to the problem of estimating the makespan T, consider first the case without overhead, so that

$$T \stackrel{\triangle}{=} \max\{Y_i, \ldots, Y_p\} \quad \text{and} \quad Y_i = x_{i1} + \cdots + x_{iK}, \tag{9.9}$$

where the x_{ij} are independent subtasks with distribution function $F(x)$. We have the following straightforward procedure for estimating $\mathbb{E}(T)$. Find $G(x)$ as the K-fold convolution of F, and then find m_p. Then Theorem 9.8 states that

$$\frac{\mathbb{E}(T)}{m_p} \approx 1 \quad \text{for } p \text{ large.}$$

A similar procedure would allow also to incorporate the (independent) overhead: this is the content of Corollary 9.13 below. Unfortunately this procedure is difficult to carry out except for some special cases, because it may be difficult to calculate $G(x)$. We shall therefore sketch a general method of estimating m_p for K large. This estimate requires that the tail of $F(x)$ has an exponential bound, and that K grows faster than $\log p$.

Exercise 9.10. Let $Y = x_1 + \cdots + x_K$, where the x_i are independent with continuous distribution F, mean μ, and standard deviation σ. Suppose that for some $\epsilon > 0$

$$\int_0^\infty e^{\epsilon x}\, dF(x) < \infty. \tag{9.10}$$

Show by formal calculation that

$$\limsup_{p \to \infty,\ K/\log p \to \infty} \left| \frac{m_p - K\mu - \sigma\sqrt{2K \log p}}{\log p} \right| < \infty. \tag{9.11}$$

That is,

$$m_p = K\mu + \sigma\sqrt{2K \log p} + O(\log p). \tag{9.12}$$

Hint: define

$$M(a, \theta) = e^{-\theta a} \int e^{\theta x} dF(x) = \int e^{\theta(x-a)} dF(x), \tag{9.13}$$

$$D(a, \theta) = \frac{\partial}{\partial \theta} M(a, \theta) = \int (x - a) e^{\theta(x-a)} dF(x). \tag{9.14}$$

Recall that

$$\mathbb{P}\left(\frac{1}{K} \sum_i^K x_i \geq a \right) \approx \left(\inf_\theta M(a, \theta) \right)^K. \tag{9.15}$$

By definition, m_p is the solution to the equation

$$\mathbb{P}\left(\sum_i^K x_i \geq m_p \right) = \frac{1}{p}. \tag{9.16}$$

Therefore,

$$\left[\mathbb{P}\left(\sum_i^K x_i \geq m_p \right) \right]^{1/K} = \left(\frac{1}{p} \right)^{1/K} \approx 1 - \frac{\log p}{K}, \tag{9.17}$$

since $\log p/K$ is small. So the problem has essentially been reduced to finding the number a that satisfies

$$\inf_{\theta} M(a, \theta) = 1 - \frac{\log p}{K}, \tag{9.18}$$

yielding an estimate $m_p = aK$. Expand M about the point $(\mu, 0)$ in a two-term Taylor series with remainder, and find

$$m_p = K(\mu + \Delta a + O(\delta a^2))$$

$$= K\mu + \sigma\sqrt{2K \log p} + K \cdot O\left(\frac{\sigma^2 \log p}{K}\right) \tag{9.19}$$

$$= K\mu + \sigma\sqrt{2K \log p} + O(\log p).$$

Equivalently, you can expand the Chernoff rate $\ell(a)$ near $a = \mu$ as

$$\ell(a) \approx (a - \mu)^2/\sigma^2. \qquad \spadesuit$$

Since $K/\log p$ is large, the error term in 9.10 is $O(\log p) < \sqrt{K \log p}$.

Corollary 9.11. *If $F(x)$ is IFR, then the estimate of Exercise 9.10 implies*

$$E(T) = K\mu + \sigma\sqrt{2K \log p} + O(\log p). \tag{9.20}$$

Proof. This is immediate from Exercise 9.10 and Lemma 9.9. ∎

Comparing this with the result of Exercise 9.5, we see that in this limit, the Normal law derived from the central limit theorem and the estimate from Chernoff's Theorem agree to the lowest order. (In the jargon of matched asymptotic expansions, the inner expansion of large deviations matches the outer expansion of the central limit theorem). This formal calculation can be made precise: this is done, for example, in [Ne1, Ne2] and [Ree]. The delicate point here is the errors in the approximations, since generally the location where a function takes a value is very sensitive to changes in the function.

There are some circumstances where $K/\log p$ might not be large even though K is large. We can still use the theory of large deviations to estimate $\mathbb{E}(T)$ in any particular case, although there does not seem to be a general statement corresponding to Exercise 9.10.

Exercise 9.12. Show that for the exponential distribution, when $\log p/K$ is large then

$$m_p \approx \frac{\log p}{\lambda} + \frac{K}{\lambda} \log \frac{\log p}{K}. \qquad \spadesuit$$

The analysis of the model with random overhead is subsumed by our previous analysis, in the following way. Suppose that it takes processor i time $v(i)$ to access the job queue. The makespan is then given by (9.9), but with $\tilde{Y}_i \stackrel{\Delta}{=} Y_i + v(i)$ replacing Y_i. Let $\tilde{m}_p$ be the characteristic maximum of $\tilde{Y}_i$.

Corollary 9.13. *Let the processing times $Y_1, Y_2, \ldots$ be i.i.d. IFR random variables with a continuous distribution function. Let $v(i)$ be i.i.d. IFR random variables, independent of the processing times Y_i, with a continuous distribution function. Then the conclusions of Lemmas 9.6 and 9.9 and Theorems 9.7 and 9.8 hold (with $\tilde{Y}_i$ replacing Y_i and $\tilde{m}_p$ replacing m_p).*

Proof. Immediate; just check that all the assumptions hold.   ■

The formal approximation of Exercise 9.10 can be extended to cover the case with overhead. To do that, we need the following definition.

Definition 9.14. *A random variable z has an infinitely divisible distribution if, for each positive integer n, there are i.i.d. random variables $\{z_1^n, \ldots, z_n^n\}$ so that $z = z_1^n + \cdots + z_n^n$.*

Exercise 9.15. Assume in addition that $v(i)$ is infinitely divisible with mean μ_v and standard deviation σ_v, and that $v_j^K(i)$ satisfies (9.10). Compute (formally!) the appropriate extension of Equation (9.12). Hint: let $\tilde{x}_j = x_j + v_j^K(i)$. The standard deviation of $\tilde{x}_j$ is then $\sigma + (\sigma_v/\sqrt{K})$.

Examples of infinitely divisible distributions are: any constant, and the gamma or normal distributions. For instance, the system in which jobs are (normally) distributed $N(\mu, \sigma^2)$, the batch size is K, and overhead is a constant v, is equivalent to the system with jobs distributed $N(\mu + v/K, \sigma^2)$, batch size K, and no overhead. The system in which jobs are distributed $N(\mu, \sigma^2)$, the batch size is K, and the overhead is distributed $N(v, s^2)$, is equivalent to the system with jobs distributed $N(\mu + v/K, \sigma^2 + s^2/K)$, batch size K, and no overhead.

The foregoing analysis can be extended to find an approximate value of the expected makespan when $K < n/p$, that is, when each processor takes a number of batches. There is an initial period when all processors are busy, repeatedly taking jobs, K at a time from the central queue, and then finishing the batches one at a time. The last batch to be taken starts the final phase of operations, when all the processors simply finish whatever they have at the moment. Under the assumption that the jobs have IFR service requirements, we expect that the distribution of remaining times is in some sense smaller than their original distribution. It is therefore straightforward to come up with bounds and approximations for the expected lengths of each of the phases of operation. See [KrW] for details.

Now consider the more general problem of finding an optimal allocation of the jobs among the processors. We could assign the jobs, say, K at a time. It seems clear that $K = 1$ is optimal when $v = 0$, and $K = n/p$ is optimal when v is very large. We shall now show that $K = n/p$ may be acceptable even in the worst case, i.e., when $v = 0$. Let $T(p)$ be the makespan (time it takes to solve the problem) using p processors and batches of size $K = n/p$. In parallel processing the standard measures of quality are speedup and efficiency, which are closely related. The efficiency is defined to be $T(1)/pT(p)$ and is always between zero and one. It measures how efficiently each processor is being used relative to a sequential machine (uniprocessor). An efficiency of around one half is usually considered to be very good. Since our setup is random, we will define efficiency by $\mathbb{E}(T(1))/p\mathbb{E}(T(p))$. When $v = 0$, $K = n/p$, and $K/\log p$ is large, our analysis showed that

$$\mathbb{E}T \approx K\mu + \sigma\sqrt{2K \log p}$$

$$= \frac{n}{p}\mu + \sigma\sqrt{2n\frac{\log p}{p}}. \tag{9.21}$$

Our definition of efficiency therefore gives

$$\text{efficiency} \overset{>}{\sim} \frac{n\mu}{p\left(\frac{n}{p}\mu + \sigma\sqrt{2n\frac{\log p}{p}}\right)}$$

$$= \frac{1}{1 + \frac{\sigma}{\mu}\sqrt{2p\frac{\log p}{n}}}. \tag{9.22}$$

For many applications with positive random variables, σ/μ is likely to be less than one. (If the jobs are distributed normally, they could be positive with high probability only if σ/μ is no larger than $1/4$). The analysis holds if $n/(p \log p)$ is large, which has already been assumed. So, even if $n/(p \log p)$ is about one,

$$\text{efficiency} \overset{>}{\sim} \frac{1}{1 + \sqrt{2}} \approx 0.4.$$

9.3. End Notes

Much more is known about the problem. See [Ri] and [WeG] for some recent results and references. There are surprises for everyone in this subject. For example, [Gr] has constructed a sequence of 20 job lengths that finish more quickly when allocated one at a time and in order to two processors than to three, and four processors take longer yet! This chapter is essentially an abbreviated version of [KrW].

Chapter 10

Parallel Algorithms: Rollback

This chapter contains a description and analysis of a model arising in algorithms for parallel computing. The mathematical model is a branching random walk; this model is studied in Chapter 3.

10.1. Rollback Algorithms

As computer chips approach physical limits, practitioners turn to parallel computation for increased speed. This section is concerned with a parallel algorithm for parallel execution of event-driven simulation, such as calls in a telephone network. The idea is also applicable to scientific computations, business programs, etc.

To introduce the idea behind the algorithm, consider a discrete-time system consisting of a computer with two processors. At odd times ($t = 1, 3, 5, \ldots$) odd-numbered programs are routed to processor 1, and at even times ($t = 2, 4, 6, \ldots$) even-numbered programs are routed to processor 2. Execution of each program takes one unit of time. This routing clearly does not lead to an efficient use of resources, as each processor is idle half the time. If we could route the programs to each processor as soon as the processor becomes available, we would increase the parallelism of the computation, and the computation time would be halved. Doing this makes each processor have its own idea of what the (program) time is; this is called its "virtual time."

Now suppose this algorithm is implemented, and that program 3 computes the value of some variable, say x, at the end of its execution (time 4), and that this value is needed by program 4 about the middle of its execution (time 4.5). But programs 3 and 4 are now executed concurrently, so that when processor 2 is at (its virtual) time 4.5 and needs to read the value of x, processor 1 is at (his virtual) time 3.5, and has not yet computed the final value of x. Thus, program 4 reads (and uses) the value of x before its final evaluation, so that the value used may be incorrect.

We come to the conclusion that disposing of the (time) synchronization between the processors leads to faster execution, but the result may be erroneous. This is not very useful . . . but don't despair! Here is a way out.

To each computation of a value of a variable we associate the time that it was computed (in terms of the execution time of the original, not the speeded-up program, or equivalently in terms of the virtual time); call this the "time-stamp" of that computation. In addition, whenever a processor computes a value of a variable that is to be used by another processor, it sends that value together with its computation time. In our speeded-up example, processor 1 (as part of program 3) would send the value x to processor 2, with the time-stamp 4, which is the time it was supposed to be computed. Now processor 2 receives, at the end of execution of program 4 (which is its virtual time 5), the value of x with a time-stamp 4. But since program 4 already used a value for x (at virtual time 4.5), it becomes clear that a mistake was made.

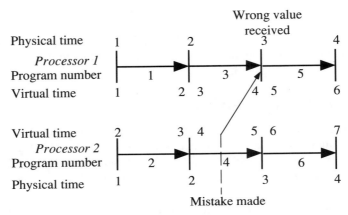

Figure 10.1. Schematic of parallel execution of processes.

This communications and time-stamping mechanism provides a way to detect errors. In order to correct these faults, each processor needs to keep track of its past states, and restart the computation from the point the error was made. In addition, each processor keeps track of variables it sends to other processors, with their respective time-stamps. The mechanism of restarting, which involves moving the virtual-time **backwards**, is called rollback.

But things are not so simple; for suppose in addition that program 5 starts by using a variable, say y, whose value is computed slightly before the end of program 4. Then this value is computed prior to the discovery of the error, and upon discovery of its error, program 4 should notify program 5 (at processor 1) that the value of y may be incorrect. Thus a secondary rollback occurs. In general, we may get a cascade of such secondary rollbacks.

Now that we have mastered the ideas of virtual time, time-stamps and rollback, let us go beyond the two-processor system. Imagine a huge system consisting of many processors, each running according to its own (virtual-time) clock, and employing a time-stamp—rollback algorithm to correct errors. It seems clear that this would be efficient when few rollbacks occur, but could be wasteful if a lot of time is spent on recovery from errors.

To figure out how many rollbacks are to be expected from a particular system, consider the growth of a single rollback. Assume processor 2 discovers at virtual time 5 that it has made a mistake at virtual time 4.5. It would then be prudent to assume that all computations done during the interval [4.5, 5] are in error. In particular, any value of a variable that was computed during that interval and used by other processors is suspect. Therefore, processor 2 needs to send "cancellation messages," or warning messages, to any such processor regarding any such computation. These correction messages need to carry the time-stamp of the corrected variable.

Now repeat the same reasoning for each of the recipient processors. You obtain a tree (actually, a graph), with a node at each processor that participates in this process. Note that when a processor sends a correction message to another processor, that second processor is affected by the correction only if the time-stamp of the correction is in the processor's (virtual) past, for otherwise the information that is corrected was not used as yet.

In order to analyze this phenomenon we need, of course, to describe both the pullulation of the graph, as well as the size of the rollback generated at each processor. Since a precise analysis, or even a construction of a precise model, seems difficult, we create a model that upper bounds the damage caused by rollback. We assume that the system is large, and that communication is local. That is, each processor communicates with (and therefore can initiate rollback in) a fixed, small (random) number of other processors. We make the following assumptions in our upper bounding model:

A1. The network is infinite in size, and homogeneous in structure.

A2. The graph is a tree, i.e., there are no loops.

A3. The number of branches of this tree at each point, which is the number of processors any one typical processor is "connected to," is a random variable with mean $b > 1$, independent of all other random variables around.

A4. The change in the size of rollback in one step is a random variable with negative mean, independent of all other random variables around.

Assumptions A1 and A2 clearly increase the number of processors involved in any one rollback "snowball;" A1 increases the number of nodes, and A2 says that each rollback message finds a fresh processor to attack, even if in the original model several messages are sent to the same processor. A3 restricts the class of systems to which our analysis applies, but large systems tend to have such restrictions in order to avoid excessive overhead due to internal communications. A4 states that the design of the system was careful enough, so that rollback in one step does not induce (statistically) increased rollback in the next step. To complete the description of the resulting process, assume a rollback of size x has started somewhere. From Assumptions A1–A3 the process that goes through all processors that could potentially be involved in the rollback is a standard branching process (see §3.2). From Assumption A4, going from the root processor along any sequence of branches of this tree, we obtain a random walk (§3.1). This combined process is called a "Branching random walk;" for a formal definition and construction, see §3.2. However, our process is slightly different, since not every communication results in a rollback. As mentioned above, if the time-stamp of the correction message is in the (virtual) future of a processor, it need not roll back. But if a processor does not roll back, it does not propagate the rollback process further. Thus our process is obtained by "trimming" a branching random walk tree: whenever the total rollback obtains a negative value, that node of the branching process (processor) and all subsequence nodes and branches are killed. We call the process resulting from this modification a "branching random walk with a barrier." This process was constructed and analyzed in §3.2. We now apply these results to our model. The reader interested in more information about the rollback mechanism and its analysis should consult [LSW].

10.2. Analysis of a Rollback Tree

Theorem 3.17 gives us the tools to identify the conditions under which rollback is guaranteed to expire in finite time. Since the analysis applies to our upper bound, the conditions are conservative. Experimental results suggest, however, that for some systems our bounds are tight.

Denote by F the distribution of the change in rollback size, and let b denote the mean number of offspring of a node. Define ℓ through (1.4), where x_1 is distributed according to F.

Theorem 10.2. *Assume A1–A4, and that $F(0) < 1$.*
If $\log b < \ell(0)$ then
  (i) with probability one, every rollback eventually dies.
  (ii) Moreover, the total number of nodes involved in a single rollback has finite mean.
If $\log b > \ell(0)$, then
  (iii) each rollback tree has a positive probability of surviving indefinitely. Moreover, the expected number of nodes at generation n that are involved grows exponentially fast.

Proof. (i) and (iii) are immediate consequences of Theorem 3.17. To obtain (ii), fix $0 < \varepsilon < \ell(0) - \log b$ so that $r \overset{\triangle}{=} \log b - \ell(0) + \varepsilon < 0$. From Theorem 3.17(i), for some $n_0 = n_0(\varepsilon)$,

$$\sum_{n=n_0}^{\infty} \mathbb{E}z(n, x) \le \sum_{n=n_0}^{\infty} e^{nr} \le \frac{1}{1 - e^r}.$$

Since $\mathbb{E}z(n, x)$ is finite for each n we have (ii). ∎

Remark. Theorem 10.2(iii) implies that if rollbacks are generated often, then with high probability, eventually there will be a rollback that will survive, and hence involve an unbounded number of nodes.

Since this is an upper bounding model, it can give conditions that guarantee stability of rollback for a class of systems. However, in order to analyze the efficiency of such algorithms we need also to analyze the number of rollbacks that arise. In order to analyze their memory requirements we need to analyze the size of rollbacks. This is done in [LSW], where conditions for stability are given. In

[BLS] it is shown that if the initial rollback is large, then (asymptotically) all roll-backs initialized by this rollback will be no larger.

Let us now apply this theorem to get concrete information about the stability of a simple system. Suppose we can model the changes in rollback size as random variables that take the values 1 and $-K$ with probability q and $(1-q)$ respectively. The mean change in rollback size is $q - K(1-q)$, which is negative whenever $q < \frac{K}{K+1}$.

Example 10.3. For $K = 1$, we have

$$\ell(0) \overset{\triangle}{=} \sup_{\theta} \left[-\log(qe^{\theta} + (1-q)e^{-\theta}) \right].$$

Differentiating $qe^{\theta} + (1-q)e^{-\theta}$ to locate the maximum, we get a quadratic equation in e^{θ} and substituting the solution $e^{\theta*} = \sqrt{1-q}/\sqrt{q}$ yields

$$\ell(0) = -\log\left(2\sqrt{q(1-q)}\right).$$

This implies that for $b = 2$ the critical value of q that separates the stable and unstable regions is

$$q = \frac{2 - \sqrt{3}}{4} \approx 0.066987.$$

Example 10.4. For $q = 1/2$,

$$\ell(0) = -\log\left(\frac{1}{2} K^{\frac{1}{K+1}} (1 + frac1K)\right).$$

The stability condition for $b = 2$ is therefore

$$K^{\frac{1}{K+1}} (1 + 1/K) < 1.$$

However, this condition does not hold for any finite value of K. Therefore, it is not possible to obtain stability when $q = 1/2$!

Exercise 10.5. Verify the calculations of Examples 10.3 and 10.4. Compute the critical values of q for $b = 2$, as a function of K. Compare to the critical value obtained from the naïve requirement that the mean step size is negative. ♠

Example 10.6. For $K = 1$, the critical value of q as a function of b is

$$q^* = \frac{1}{2}\left(1 - \sqrt{b^2 - 1}/b\right).$$

The following qualitative statements are immediate:

(i) q^* increases to $1/2$ as b decreases to one.

(ii) q^* increases to $1/2$ much more slowly than b decreases to one.

(iii) For b large, $q^* \approx 1/(4b^2)$.

It is also possible to obtain some explicit expressions for $\sum_{i=1}^{\infty} \mathbb{E}z(n, x)$; see [BLS].

Chapter 11

The M/M/1 Queue

This chapter contains a partial study of the $M/M/1$ queue from the point of view of large deviations. This is a *partial* study because nearly any question one can ask of the $M/M/1$ queue can be answered, and we ask only a few. While many of our results have been obtained before by different methods, we hope that you will find the present approach natural and simple. The $M/M/1$ queue has been a workhorse of communication engineers since at least the time of Erlang (1917). Its steady-state and transient distributions are known explicitly (see, e.g., [Mas]). Nearly all queueing theory can be viewed as perturbations of the basic $M/M/1$ model: $M/M/n$, $M/G/1$, $GI/M/1$, Jackson networks, etc. It is almost certain that any quantity that is not obtainable for the $M/M/1$ queue is unobtainable for every other queue (although some questions are interesting for other queues that are not interesting for the $M/M/1$ queue).

We use this chapter to develop some of the basic ideas and tools that will be utilized in later chapters. Some generic calculations are also carried out here, for example, in §11.5. The ideas from §§11.4 and 11.6 are also used in later chapters, but to a lesser extent.

11.1. The Model

The $M/M/1$ queue is a jump Markov process $x(t)$ on $\mathcal{Z}^+$, with the following jump directions and rates:

$$
\begin{aligned}
e_1 &= +1 & \lambda_1(x) &= \lambda \\
e_2 &= -1 & \lambda_2(x) &= \begin{cases} \mu & x = 1, 2, 3, \ldots, \\ 0 & x = 0. \end{cases}
\end{aligned}
\tag{11.1}
$$

That is, customers arrive at the queue according to a Poisson process with rate λ. The service times are exponential with parameter μ. Our usual scaling (5.1) gives us a process $z_n(t)$, which jumps $+1/n$ with rate $n\lambda$, and jumps $(-1/n)$ with rate $n\mu$ when $z_n > 0$. In the parlance of queueing theory, λ denotes the arrival rate to a queue, μ denotes the service rate, and $x(t)$ denotes the number of customers in the queue. In order for the queue to be stable (have a nondegenerate steady state) we must have

$$
\rho \overset{\triangle}{=} \frac{\lambda}{\mu} < 1.
\tag{11.2}
$$

Throughout this chapter we will assume $\rho < 1$, unless specifically stated otherwise. Our process $z_n(t)$ is obviously identical in distribution to $\frac{1}{n}x(nt)$. Note that the $M/M/1$ process (and hence its scaled version) is closely related to the free

$M/M/1$ process, which was the subject of §7.3. (This is an $M/M/1$ process but without the boundary, so that $\lambda_2 \equiv \mu$.) In order to avoid excessive repetition, we shall assume that you have mastered the ideas of §7.3.

Here is how the rest of the chapter is organized. In §11.2 we give formal derivations of most of the results of this chapter. This is in keeping with our point of view as expounded in the Introduction to the Applications: derive first, justify later! Most of the remainder of the chapter is concerned with justification (that is, proofs). Sections §11.3 and §11.4 give a derivation of Kurtz's Theorem and a derivation of the large deviations principle for the $M/M/1$ queue, respectively. This is mostly for pedagogic purposes, since Theorems 8.47 and 8.69 establish the principle. However, we don't suppose that everyone who reads this chapter will have gone through all the previous chapters, so we include a fairly simple and straightforward approach. Once we have the large deviations principle established, we proceed to justify the Freidlin-Wentzell theory for the $M/M/1$ process. Then, after justifying using these calculations to calculate the probability of hitting isolated points, we finish with a new calculation: how long are the busy periods of an $M/M/1$ queue, and how large can the process get during long busy periods?

11.2. Heuristic Calculations

We now follow the steps for analyzing a system using large deviations as given in the Introduction to the Applications. We already have a Markov model of the system at hand, and have done the appropriate scaling. The next step is to calculate the system's likely behavior by calculating $z_\infty(t)$. We have by (5.7) or more correctly by Exercise 8.13,

$$\frac{d}{dt} z_\infty = \begin{cases} \lambda - \mu & \text{if } z_\infty > 0 \\ 0 & \text{if } z_\infty = 0. \end{cases} \tag{11.3}$$

The second line in (11.3) comes from (8.11); it can be understood intuitively as a statement that a stable $M/M/1$ queue tends to stay bounded once it becomes empty. Therefore,

$$z_\infty(t) = \max\left(z_\infty(0) - t(\mu - \lambda), 0\right), \tag{11.4}$$

or equivalently, with $T^* = z_\infty(0)/(\mu - \lambda)$ and $z_\infty(0) \geq 0$,

$$z_\infty(t) = \begin{cases} z_\infty(0) - (\mu - \lambda)t & \text{if } t \leq T^* \\ 0 & \text{if } t \geq T^*. \end{cases} \tag{11.5}$$

That is, the scaled $M/M/1$ process tends to drift to zero, and once there it stays.

The next step is to formulate some large deviations problems. The local rate function ℓ for the $M/M/1$ queue is, by (8.12) or by Theorem 11.13,

$$\ell(x, y) = \begin{cases} \sup_\theta \left\{\theta y - \lambda(e^\theta - 1) - \mu(e^{-\theta} - 1)\right\} & \text{for } x > 0 \text{ or } y > 0; \\ 0 & \text{for } x = 0 \text{ and } y = 0; \\ \infty & \text{for } x < 0 \text{ or } x = 0 \text{ and } y < 0. \end{cases} \tag{11.6}$$

Using the calculation in Exercise 7.24 of §7.3 we see that for $x > 0$ or $y > 0$ the local rate function has the explicit form

$$\ell(x, y) = y \log \left(\frac{y + \sqrt{y^2 + 4\lambda\mu}}{2\lambda} \right) + \lambda + \mu - \sqrt{y^2 + 4\lambda\mu}. \qquad (11.7)$$

Now let's calculate the probability (in steady state) that $z_n(t) \geq a$ for any $a > 0$. [Of course, an exact expression for this probability is easy to obtain since $\mathbb{P}_{ss}(x(t) = j) = (\lambda/\mu)^j$. But let us derive this through our favorite method.] Since zero is the unique global attracting point, it would seem that we can use the Freidlin-Wentzell theory to calculate the escape times from $z_n \leq a$ and to calculate $\mathbb{P}_{ss}(z_n \in G)$ for any nice set G; in particular, we should be able to calculate $\mathbb{P}_{ss}(a \leq z_n \leq b)$ for any $a < b$. We simply have to calculate

$$V(a) \overset{\triangle}{=} \inf_{F(a)} I_0^T(r)$$

$$F(a) \overset{\triangle}{=} \{r, T : r(0) = 0, r(T) = a, T > 0\}. \qquad (11.8)$$

We have a complete solution of this variational problem in §C.3. Also, we give an alternate discussion for one-dimensional birth-death processes in §11.5. For a direct derivation, note that the function $\ell(x, y)$ is independent of x as long as $x > 0$. Therefore the Euler equation (C.2),

$$\frac{\partial \ell}{\partial r} - \frac{d}{dt} \frac{\partial \ell}{\partial r'} = 0$$

simplifies to

$$r'' \frac{\partial^2 \ell}{(\partial r')^2} = 0.$$

However, the function ℓ is strictly convex, so that the second derivative is strictly positive. Therefore, in the region $x > 0$, any solution $r^*(t)$ of Euler's equation must be a straight line: $r^*(t) = ct, 0 \leq t \leq a/c$. (This obviously follows from Lemma 5.16). We simply have to minimize the cost of this path:

$$V(a) \overset{\triangle}{=} \inf_{c>0} \int_0^{a/c} \ell(x, c)\, dt$$

$$= \inf_{c>0} \frac{a}{c} \left(c \log \frac{c + \sqrt{c^2 + 4\lambda\mu}}{2\lambda} + \lambda + \mu - \sqrt{c^2 + 4\lambda\mu} \right).$$

Differentiating with respect to c to obtain the minimum we obtain

$$\frac{\partial}{\partial c} \ell(x, c) = \log \frac{c + \sqrt{c^2 + 4\lambda\mu}}{2\lambda}$$

$$V(a) = a \log \frac{\mu}{\lambda}$$

(remember that by (C.6), $d\ell/dy = \theta^*$ and use Exercise 7.24). That is, we have found that

$$\mathbb{P}_{ss}\,(z_n \in (a, b)) = e^{-nV(a)+o(n)},$$

or put another way,

$$\mathbb{P}_{ss}\,(x(t) \in (na, nb)) = \left(\frac{\lambda}{\mu}\right)^{(na+o(n))}.$$

By Exercise 6.91 we can set $b = \infty$ and get

$$\mathbb{P}_{ss}(x(t) \geq na) = \left(\frac{\lambda}{\mu}\right)^{(na+o(n))}.$$

(The astute reader will note that the $o(n)$ term is actually quite small for the $M/M/1$ queue.)

Another immediate consequence of this analysis and of Theorem 6.17 is the following estimate of escape times. Let $\tau(n; a)$ be the first time that $z_n(t) \geq a$.

Corollary 11.1. *For a stable $M/M/1$ queueing process and any $a > 0$ and $\delta > 0$ we have, uniformly over $z_n(0) \in [0, a - \delta)$,*

$$\lim_{n \to \infty} \mathbb{P}_{z_n(0)} \left(\frac{\log \tau(n; a)}{n} \in (V(a) - \delta, V(a) + \delta)\right) = 1.$$

We can also compute transient distributions for $z_n(t)$. For example, what is $\mathbb{P}_x\,(z_n(t) = y)$? That is, what is the probability that there are ny customers in the queue after time nt, if the queue starts with nx customers? Clearly we need to calculate

$$\varphi_t(x, y) \stackrel{\triangle}{=} \inf_{F(x,y,t)} I_0^T(r)$$

(see Definitions 6.14) where

$$F(x, y, t) = \{r \;:\; r(0) = x, \; r(t) = y, \; t > 0\}.$$

A moment's thought will show that there are two types of paths that go from x to y: paths that avoid the boundary $z = 0$, and paths that hit the boundary. For the first type of path we know from Lemma 5.16 that the cheapest way to go from x to y is in a straight line. However, for the second type of path, there are several possibilities. We can easily see that optimal paths that touch the boundary must be "U" shaped: a straight line segment from x to zero, possibly a segment where the path remains at zero, then a straight line segment from zero to y. This is because the first and last segments have costs identical to the cost of the free process of §7.3, for which all optimal paths are straight line segments. Furthermore, the time spent at zero costs nothing, while the cost of making an excursion away from zero would have positive cost, since $\rho < 1$.

Now consider the lowest cost path from x to y that touches the boundary, where we place no restrictions on the time t the path takes. We can complete the segment from x to zero with no cost, by following z_∞. The lowest cost path from zero to y is the solution to (11.8). We found that the path is linear, and we show in (11.22) that the optimal slope is $\mu - \lambda$. Therefore, the amount of time it takes to reach y is $y(\mu - \lambda)^{-1}$. Hence a lowest cost path from x to y reaching zero in between is given by

$$
r(s) = \begin{cases} x + (\lambda - \mu)s & 0 \le s \le \dfrac{x}{\mu - \lambda} \\ \left(s - \dfrac{x}{\mu - \lambda}\right)(\mu - \lambda) & \dfrac{x}{\mu - \lambda} \le s \le \dfrac{x}{\mu - \lambda} + \dfrac{y}{\mu - \lambda} \stackrel{\triangle}{=} t_1 \end{cases} \tag{11.9}
$$

for which, by (11.23) the cost is

$$
I_0^{t_1}(r) = y \log \frac{\mu}{\lambda}.
$$

Other paths with the same I-function but that take more time to go from x to y are constructed by allowing the path to remain at zero once it reaches zero; by (11.18) their cost is indeed the same. Note that the path from 0 to y is exactly the time reversal of z_∞!

Thus we have identified the paths with minimal I-function that start at x, pass through zero and take at least $t_1 \stackrel{\triangle}{=} (x + y)(\mu - \lambda)^{-1}$ to reach from x to y. Let us denote by $I_{f,0}^t$ and by ℓ_f the rate and local rate functions for the free process of §7.3. Now for $t = t_1$, the path r does not remain at zero for a positive amount of time; it only touches zero. From (11.18) is follows for this path that

$$
I_{f,0}^t(r) = I_0^t(\bar{r}), \tag{11.10}
$$

i.e., the cost is the same as for the free process. But Lemma 5.16 shows that for the free process the lowest-cost path is the straight line from x to y (in fact, due to the strict convexity of ℓ, this is *the only* optimal path). However, for the path (11.9) (indeed, for any path that does not spend a positive amount of time on the boundary), (11.10) holds as well. Therefore, for $t = t_1$ the straight line path is the only optimal path.

As t increases, we have for the straight line path

$$
I_{f,0}^t(r) = t\ell_f \left(\frac{y - x}{t}\right) \sim t\ell_f(0) \to \infty,
$$

since ℓ_f is continuous and positive at zero. Therefore, for some $t^* > t_1$ we will have

$$
I_0^{t^*}(r) = y \log \frac{\mu}{\lambda}.
$$

For every $t > t^*$, we shall see that the lowest cost path uses the boundary. Note that at $t = t^*$ there are two optimal paths r^*: the straight line from x to y in time t^*, and the "U"-shaped path that follows z_∞ from x to zero, and follows the time reverse of z_∞ from zero to y at a later time.

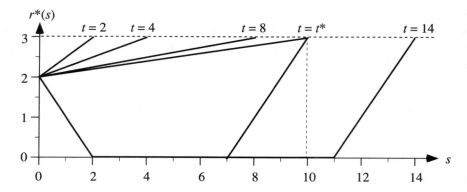

Figure 11.2. Critical paths for travel from x to y in time t for $x = 2$, $y = 3$, $\lambda = 1$, and $\mu = 2$. We have sketched the paths for $t = 2, 4, 8, t^* \approx 10, 14$.

We summarize this result in the form of a theorem. The proof is given below in §11.7.

Theorem 11.3. *Define $t_1 \triangleq (x + y)(\mu - \lambda)^{-1}$. For the stable $M/M/1$ queue, there is a $t^* > t_1$ so that the optimal cost and path from x to y in time t are given by*

$$
\varphi_t(x, y) =
\begin{cases}
y \log \frac{\mu}{\lambda} & \text{if } t \geq t^*; \\[2mm]
t\ell_f\left(\dfrac{y - x}{t}\right) & \text{if } t \leq t^*.
\end{cases}
$$

$$
r^*(s) =
\begin{cases}
x + (y - x)s/t & \text{if } t < t^*; \\
x - s(\mu - \lambda) & \text{if } t > t^* \text{ and } s < \frac{x}{\mu - \lambda}; \\
0 & \text{if } t > t^* \text{ and } \frac{x}{\mu - \lambda} \leq s \leq t - \frac{y}{\mu - \lambda}; \\
y + (s - t)(\mu - \lambda) & \text{if } t > t^* \text{ and } t - \frac{y}{\mu - \lambda} \leq s \leq t.
\end{cases}
$$

Moreover, t^* is the (unique) larger positive solution of

$$
t\ell_f\left(\frac{y - x}{t}\right) = y \log \frac{\mu}{\lambda}.
$$

The utility of this theorem is the following result, which is suggested by the large deviations principle, but whose proof is in §11.6 and §11.7.

Theorem 11.4. *If $\lambda < \mu$ then for any positive t, x and y,*

$$
\lim_{n \to \infty} \frac{1}{n} \log \mathbb{P}_x\left(y - 1/2n < z_n(t) \leq y + 1/2n\right) = \varphi_t(x, y).
$$

This is the probability that $z_n(t)$ hits the point y, to within the resolution of the grid. We discuss the implications of this result in §11.8.

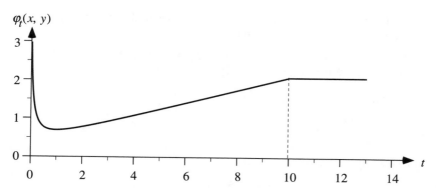

Figure 11.5. $\varphi_t(x, y)$ for the $M/M/1$ queue with $x = 2$, $y = 3$, $\lambda = 1$, $\mu = 2$.

This concludes our heuristic analysis of the $M/M/1$ queue via large deviations. We now begin the justifications and more detailed analysis of the system.

11.3. Most Probable Behavior

In this section we prove that Kurtz's Theorem holds for the $M/M/1$ queue, with $z_\infty(t)$ defined by any of (11.3)–(11.5). If you read Chapter 8 you can, of course, skip this section. We prove the theorem holds by comparing the $M/M/1$ process with the free process. The free process was defined in §7.3. It is the $M/M/1$ queue without a boundary. That is, it is the difference of a Poisson (λ) and an independent Poisson (μ) process. We denote the free process by $\zeta(t)$. The transition structure of $\zeta(t)$ is given by

$$
\begin{array}{ll}
e_1 = +1 & \lambda_1(x) = \lambda \\
e_2 = -1 & \lambda_2(x) = \mu.
\end{array}
\tag{11.11}
$$

It is obvious that this process satisfies the conditions of Kurtz's Theorem and of the large deviations theorems of Chapter 5. At any fixed t, the random variable $\zeta(t)$ is the difference of Poisson (λt) and Poisson (μt) random variables. In §7.3 we showed that the rate function for the free process is

$$
\ell(x, y) = \ell(y) = y \log \frac{y^2 + \sqrt{y + 4\lambda\mu}}{2\lambda} + \lambda + \mu - \sqrt{y^2 + 4\lambda\mu}.
$$

We make the large deviations scaling

$$
\zeta_n(t) \overset{\triangle}{=} \frac{1}{n}\zeta(nt).
$$

[We call the free process $\zeta_n(t)$ so as not to confuse it with the scaled $M/M/1$ process $z_n(t)$.]

Here is an outline of the argument we use:

1. The time it takes for $z_n(t)$ to go from an initial point $x > 0$ to zero is about $x\,(\mu - \lambda)^{-1}$. We obtain the rate function for deviations from this expected time.

2. Until time $x(\mu - \lambda)^{-1}$, $z_n = \zeta_n$ with very high probability. Therefore Kurtz's Theorem holds until this time, since it holds for ζ_n.

3. $\sup_{0 \le t \le T} z_n(t) \le \varepsilon$ with overwhelming probability, if $z_n(0) = 0$.

We establish point 2 first since the proof is the shortest.

Theorem 11.6. *There is a function $C(\varepsilon, T) > 0$, with*

$$\lim_{\varepsilon \downarrow 0} \frac{C(\varepsilon, T)}{\varepsilon^2} \in (0, \infty) \quad and \quad \lim_{\varepsilon \to \infty} \frac{C(\varepsilon, T)}{\varepsilon} = \infty,$$

such that for any x, T and $\varepsilon > 0$ with

$$z_\infty(0) = x > T(\mu - \lambda) + \varepsilon,$$

the process $\vec{z}_n$ satisfies

$$\mathbb{P}_x \left(\sup_{0 \le t \le T} |z_n(t) - z_\infty(t)| > \varepsilon \right) < 2e^{-nC(\varepsilon, T)}.$$

Proof. If $|z_n(t) - z_\infty(t)| \le \varepsilon$ for all $t \in [0, T]$, then $z_n(t) > 0$ for all these t. Hence on this set $z_n(t)$ is identical to the free process—a process without boundary, and so Kurtz's Theorem 5.3 applies. ∎

We can also get useful bounds on the hitting time

$$\tau_n \overset{\triangle}{=} \inf\{t : z_n(t) = 0\}.$$

Theorem 11.6 shows that $\tau_n \approx x(\mu - \lambda)^{-1}$. But we can be more precise than this. Recall the definition (7.16) of $\ell_f(y)$. Fix x and define $v(y)$ as y plus the expected time to hit zero. Then $f(y)$ is the cost of the straight line path from x to zero in time $v(y)$.

$$v(y) \overset{\triangle}{=} \frac{x}{\mu - \lambda} + y, \qquad f(y) \overset{\triangle}{=} v(y)\ell \left(-\frac{x}{v(y)} \right). \tag{11.12}$$

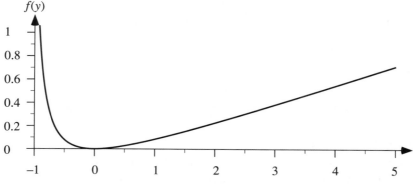

Figure 11.7. The function $f(y)$ for $\lambda = 1$, $\mu = 2$, $x = 1$.

Exercise 11.8. Prove that f is convex and $f(0) = 0$. Hint: in the region where it is finite, ℓ is convex and smooth; compute second derivatives.   ♠

Theorem 11.9. *If $\rho < 1$ then for any $x > 0$ and any y,*

$$\lim_{n\to\infty} \frac{1}{n} \log \mathbb{P}_x \left(\tau_n \le \frac{x}{\mu - \lambda} + y \right) = -f(y) \quad y < 0$$

$$\lim_{n\to\infty} \frac{1}{n} \log \mathbb{P}_x \left(\tau_n \ge \frac{x}{\mu - \lambda} + y \right) = -f(y) \quad y > 0.$$

This shows that $f(y)$ is the correct rate function for estimating the probability that τ_n differs from its expected value by y. The proof is a very straightforward application of the large deviations principle for the free process—you might well want to do it yourself. We include the proof since it brings up a common point: we can estimate level crossing times by large deviations, since we can always move small distances in small time with fairly high probability.

Proof. We relate the behavior of τ_n to the behavior of the free process $\zeta_n(t)$, which is the same as $z_n(t)$ but has no boundary: for ζ_n,

$$\lambda_1(x) = n\lambda, \qquad \lambda_2(x) = n\mu$$

for all x. Clearly we can construct z_n and ζ_n so that

$$z_n(t) = \zeta_n(t), \quad 0 \le t \le \tau_n$$

(see Definition 4.7 and subsequent discussion). Therefore, for $y < 0$, the definition of τ_n gives

$$\left\{ \omega : \tau_n \le \frac{x}{\mu - \lambda} + y \right\} = \left\{ \omega : \inf \left\{ \zeta_n(t) : 0 \le t \le \frac{x}{\mu - \lambda} + y \right\} \le 0 \right\}$$

so that, with

$$T \overset{\triangle}{=} v(y) \overset{\triangle}{=} \frac{x}{\mu - \lambda} + y, \tag{11.13}$$

we have by Theorem 5.1

$$\frac{1}{n} \log \mathbb{P}_x \left(\tau_n \le T \right) = \frac{1}{n} \log \mathbb{P}_x \left(\inf \{ \zeta_n(t) : 0 \le t \le T \} \le 0 \right)$$

$$\le - \inf_{r \in F} I_0^T(r) + o(1),$$

where the closed set F referred to in Theorem 5.1 is given by

$$F = \left\{ r : r(0) = x, \; r(t) = 0, \; \text{for some } t \le T \right\}.$$

In this case we can easily compute the optimal path. Let t_r be the first time t (after zero) that $r(t) = 0$. By the definition (5.5) of I_0^T, we need to consider only functions r that follow a slope $\lambda - \mu$ between t_r and T, since then $\ell_f(r'(t)) = 0$

over this interval (remember that we are doing the computation for the free process ζ_n!). Lemma 5.16 now implies that, for such functions,

$$
\begin{aligned}
I_0^T(r) &= \int_0^{t_r} \ell_f(r'(s))\, ds \\
&\geq T\ell_f\left(\frac{\Delta r}{T}\right),
\end{aligned}
\tag{11.14}
$$

where $\Delta r = -x - (T - t_r)(\mu - \lambda)$. Since ℓ_f is convex and vanishes at $\lambda - \mu$, the minimum of the right-hand side of (11.14) is obtained at $t_r = T$. The optimal value is thus

$$
\inf_{r\in F} I_0^T(r) = T\ell_f(-x/T) = \left(\frac{x}{\mu - \lambda} + y\right)\ell_f\left(-\frac{x}{\frac{x}{\mu-\lambda} + y}\right) = f(y)
$$

and the optimal path r^* is a straight line with $r(T) = 0$. This establishes the upper bound for the case $y < 0$.

To get a lower bound, let y_ε denote the linear function with $y_\varepsilon(0) = x$ and $y_\varepsilon(T) = -\varepsilon$, and continue to use T as defined in (11.13). Then from Theorem 5.1 applied to the free process

$$
\begin{aligned}
\mathbb{P}_x\big(\inf\{\zeta_n(t) : 0 \leq t \leq T\} \leq 0\big) &\geq \mathbb{P}_x\left(\sup_{0\leq t\leq T}|\zeta_n(t) - y_\varepsilon(t)| < \varepsilon\right) \\
&\geq \exp\left(-nT\ell_f\left(-\frac{x+\varepsilon}{T}\right)\right),
\end{aligned}
$$

since this set is open, and since the previous calculation shows that the value of $I_0^T(y_\varepsilon)$ is $T\ell_f(-(x+\varepsilon)/T)$. Now ℓ_f is continuous, so taking $\varepsilon \to 0$ establishes the lower bound and concludes the proof in the case $y < 0$.

For $y > 0$,

$$
\{\omega : \tau_n > T\} = \{\omega : \inf\{\zeta_n(t) : 0 \leq t \leq T\} > 0\}.
$$

Since the times between jumps have a density (exponential), $\mathbb{P}(\tau_n = T) = 0$. Therefore

$$
\begin{aligned}
\frac{1}{n}\log\mathbb{P}_x(\tau_n \geq T) &= \frac{1}{n}\log\mathbb{P}_x(\tau_n > T) \\
&= \frac{1}{n}\log\mathbb{P}_x\big(\inf\{\zeta_n(t) : 0 \leq t \leq T\} > 0\big) \\
&\geq -\inf\{I_0^T(r) : r \in G, r(0) = x\} + o(1),
\end{aligned}
$$

where the open set G is

$$
G = \left\{r : \inf_{0\leq t\leq T} r(t) > 0\right\}.
$$

Then exactly the same arguments as in the case $y < 0$ establish the desired lower bound, except that the optimal path is not in G. To obtain an upper bound note that

$$\{\omega : \inf\{\zeta_n(t) : 0 \leq t \leq T\} > 0\} \subset \{\omega : \inf\{\zeta_n(t) : 0 \leq t \leq T\} \geq 0\}$$

and apply the same argument as in the case $y < 0$ to conclude that the optimal path is a straight line, with $r(T) = 0$. ∎

Exercise 11.10. Use the relation

$$\left\{\omega : \inf\left\{\zeta_n(t) : 0 \leq t \leq \frac{x}{\mu - \lambda} + y\right\} > 0\right\} \subset \{\omega : \zeta_n(T) > 0\}$$

to establish the upper bound in the case $y > 0$. Use a similar argument to obtain the lower bound for the case $y < 0$. Hint: apply Chernoff's Theorem to the random variable ζ_n. ♠

Now the question is, once $z_n(t)$ becomes zero, how do we proceed in the calculation? Once $z_\infty(t) = 0$ it remains there since $\mu - \lambda < 0$ (Chapter 8). From the point of view of Chapter 6, zero is an irregular attracting point, so we do not actually need boundary theory to examine the $M/M/1$ process.

But we can avoid both of these advanced chapters, and argue directly that

$$\mathbb{P}_0\left(\sup_{0 \leq t \leq T} z_n(t) < \varepsilon\right) = e^{o(n)},$$

using the fabled reflection map (defined immediately below in §11.4). Let $\zeta_n(t)$ be the free process.

Lemma 11.11.

$$\mathbb{P}_0\left(\sup_{0 \leq t \leq T} z_n(t) \leq \varepsilon\right) \geq 1 - C_1 e^{-nC_2(\varepsilon/2)}.$$

Proof. By Kurtz's Theorem, for each $\varepsilon > 0$,

$$\mathbb{P}\left(\sup_{0 \leq t \leq T} |\zeta_n(t) - (\lambda - \mu)t| > \varepsilon\right) \leq C_1 e^{-nC_2(\varepsilon)}. \tag{11.15}$$

Now if $\sup_{0 \leq t \leq T} |\zeta_n(t) - (\lambda - \mu)t| \leq \varepsilon$, then

$$\zeta_n(t) - \inf_{0 \leq s \leq t} \zeta_n(s) \leq 2\varepsilon .$$

Therefore, by (11.16), the lemma is established. ∎

We can now conclude that the most probable behavior of z_n is to follow a straight line (z_∞) until it becomes zero. After that time, by the Markov property, we can ignore the question of how it behaved until then: the most probable behavior after $z_n(t)$ becomes zero is to stay near zero.

11.4. Reflection Map

In order to analyze the $M/M/1$ queue via large deviations we obviously need a large deviations principle for the process. This principle is established in Chapter 8. However, this strikes us as killing a fly with a sledgehammer. So we show in this section how to establish the large deviations principle for the $M/M/1$ queue by using the contraction principle on the "free process." The mapping we use for this contraction is called the reflection map.

There is an obvious way to obtain an $M/M/1$ process from the free process $\zeta_n(t)$ (which was defined in §11.3): simply "delete" all transitions down whenever the queue is empty. This can be formalized via the fabled reflection map. Assume $\zeta_n(0) \geq 0$ and define

$$z_n(t) = \zeta_n(t) - \inf_{0 \leq s \leq t} \zeta_n(s) . \tag{11.16}$$

Then z_n is a (scaled) $M/M/1$ process starting with $z_n(0) = 0$.

> To see that $z_n(t)$ defined by (11.16) is in fact an $M/M/1$ queue, couple a scaled $M/M/1$ process and $\zeta_n(t)$ to have the same arrivals and departures, except if the $M/M/1$ queue is empty and a departure occurs from ζ_n, then ζ_n changes but the queue does not. On a sample path basis, then, by induction on arrivals and departures, we see that the queue (call it $z_n(t)$) satisfies
>
> $$z_n(j^{\text{th}} \text{ event}) = \zeta_n(j^{\text{th}} \text{ event}) - \inf_{i \leq j} \zeta_n(i) .$$
>
> The service distribution remains exponential, due to the memoryless property of the exponential distribution. This mapping is explained in more detail in any book with the word "queueing" in the title, or any that mentions "reflecting Brownian motion."

This construction allows us to obtain a large deviations principle for z_n despite the fact that the rates (11.1) are not continuous. To do that, denote by R the reflection map. More formally, fix an interval $[0, T]$ and consider the map, defined in (11.16), which takes functions ζ_n on $[0, T]$ to functions z_n on $[0, T]$.

Lemma 11.12. *The map R takes $D[0, T]$ into $D[0, T]$ and is continuous.*

Proof. Since $D[0, T]$ is a topological vector space A.22, it is closed under addition, and addition is continuous. Therefore it suffices to show that the map

$$Ax(t) \stackrel{\triangle}{=} \inf_{0 \leq s \leq t} x(s) \tag{11.17}$$

satisfies the statements of the theorem. Now if x is right continuous with left limits, then it is easy to see that $(Ax)(t)$ is also right continuous with left limits. Thus A takes $D[0, T]$ into $D[0, T]$.

To establish continuity, we use the definition of the metric on $D[0, T]$. If $d(x, y) < \varepsilon$ then, for some time transformation λ (cf. (A.2) and Theorem A.58) we have

$$\sup_{0 \le t \le T} |x(t) - y(\lambda(t))| \le 2\varepsilon.$$

But then, for each t (with the same λ),

$$\left| \inf_{0 \le s \le t} x(s) - \inf_{0 \le s \le \lambda(t)} y(s) \right| \le 2\varepsilon$$

so that

$$d(Ax, Ay) \le \sup_{0 \le t \le T} \left| \inf_{0 \le s \le t} x(s) - \inf_{0 \le s \le \lambda(t)} y(s) \right| \le 2\varepsilon.$$

Therefore, the map A (and hence R) is continuous. ∎

We now invoke the contraction principle (Theorem 2.13) to establish that z_n possesses exponential upper and lower bounds, and to identify the rate function. Recall that ℓ_f and $I_{f,0}^T$ correspond to the free process; ℓ_f is given in (11.7), and as usual $I_{f,0}^T$ is the integral of ℓ_f.

Theorem 11.13. *The $M/M/1$ process with $\rho < 1$ satisfies a large deviations principle with the good rate function*

$$I_0^T(\vec{r}) = \begin{cases} \displaystyle\int_0^T \ell(r(t), r'(t))\, dt & \text{if } r \text{ is absolutely continuous;} \\ \infty & \text{otherwise,} \end{cases}$$

(11.18)

$$\ell(x, y) = \begin{cases} \ell_f(y) & \text{if } x > 0 \text{ or } y \ne 0 ; \\ 0 & \text{if } x = 0 \text{ and } y = 0 ; \\ \infty & \text{if } x < 0. \end{cases}$$

Proof. Note that Lemma 11.12 establishes the continuity of the reflection map R from $C[0, T]$ to $C[0, T]$. Thus the contraction principle (Theorem 2.13) implies that the $M/M/1$ process satisfies a large deviations principle, and the rate function I_0^T satisfies

$$I_0^T(r) = \inf \left\{ I_{f,0}^T(u) : R(u) = r \right\}.$$

By (11.16), if $R(u) = r$ then

$$u'(t) = r'(t) \quad \text{whenever} \quad r(t) > 0 \quad \text{or} \quad r'(t) > 0.$$

So, from the definition of $I_{f,0}^T$ and for r absolutely continuous,

$$\inf \left\{ I_{f,0}^T(u) : R(u) = r \right\} \ge \inf \left\{ \int_0^T \ell_f(u'(t)) \mathbf{1}\,[r(t) > 0]\, dt : R(u) = r \right\}$$

$$= \int_0^T \ell_f(r'(t)) \mathbf{1}\,[r(t) > 0]\, dt$$

with equality if we choose u to be linear with slope $\lambda - \mu$ whenever $r(t) = 0$. The measure of the set of times where $r(t) = 0$ but $r'(t) \neq 0$ is zero, so that the definition of ℓ at these points is immaterial. Finally, if $r(t) < 0$ for some t then there is no appropriate u and we obtain the representation (11.18).   ■

Exercise 11.14. To deal with a process that starts at a non-zero point, define

$$z_n(t) = \zeta_n(t) - \min \left(0, \inf_{0 \leq s \leq t} \zeta_n(s) \right). \tag{11.19}$$

Show that Lemma 11.12 and Theorem 11.13 hold verbatim with this new map R.♠

11.5. The Exit Problem and Steady State

We have seen that $z_n(t)$ goes linearly to zero. Once it hits zero it tends to stay nearby, as discussed in Chapter 6; basically, if it tried to wander away, it would immediately have to follow $z_\infty(t)$ again back down to zero. Now we discuss the frequency and manner of the excursions of $z_n(t)$ away from zero.

A generic calculation.

The derivation of the level crossing rate $V(x)$ given in §11.2 was not completely justified; the Euler equation is a necessary but not sufficient condition for a minimal path. There is a special one-dimensional justification in Appendix C, but we also want a more standard solution of the variational problem that can be understood and applied as a standard method.

We solve the variational problem in a generic way, so that we can use the solution for other one-dimensional problems (for which exact solutions may not be available). The results of §C.3 are tailored for this case. The class of models we cover are birth-death processes. These are one-dimensional processes for which the jump directions are $e_1 = 1$ and $e_2 = -1$. Recall the Definitions (5.2)–(5.4)

$$\ell(x, y) \overset{\triangle}{=} \sup_{\theta} \{\theta y - g(x, \theta)\}$$

$$g(x, \theta) \overset{\triangle}{=} \sum_{i=1}^{k} \lambda_i(x) \left(e^{\theta e_i} - 1 \right),$$

which, for the free process associated with the $M/M/1$ queue takes the form (7.16)

$$\ell_f(x, y) = \sup_{\theta} \left\{ \theta y - \lambda(e^{\theta} - 1) - \mu(e^{-\theta} - 1) \right\}. \tag{11.20}$$

The formula for a general birth-death process is identical, except that the rates may depend on position, so that $\lambda = \lambda(x)$ and $\mu = \mu(x)$. We shall often omit the x variable from the notation, but only when it does not affect the calculations.

Theorem 11.15. *Consider the variational problem of (11.8) for a birth-death process. Assume that both $\lambda(x)$ and $\mu(x)$ are continuous in x, are strictly positive in $[0, a]$, and that $\lambda(x)/\mu(x) \leq \rho < 1$. Then*

$$V(a) = \int_0^a \log\left(\frac{\mu(x)}{\lambda(x)}\right) dx, \qquad (11.21)$$

and the unique optimal path is given by the solution to

$$r'(t) = \mu(r(t)) - \lambda(r(t)), \quad r(0) = 0. \qquad (11.22)$$

Proof. The uniqueness of $r(t)$ is established in Exercise 11.16. The explicit calculation (7.17) clearly applies to this case, so that ℓ is given by (11.7), but where λ and μ depend on x. Under our assumptions on the rates we conclude that $\ell(x, y)$ is jointly continuous in (x, y), and continuously differentiable in y. Therefore, all the assumptions imposed in §C.3 hold. The assumption that $\lambda(x)/\mu(x) \leq \rho < 1$ implies that, starting at each point x in $[0, a]$, the path $z_\infty(t)$ is strictly decreasing. This means that the variational problem indeed corresponds to an unlikely event.

We therefore conclude from Lemma C.5 that we should be looking for monotone increasing solutions. From Lemmas C.6 and C.8 we conclude that there is a solution v to the variational problem, and from Lemmas C.9 that the value $V(a)$ can be computed via (C.12), where θ^* solves (C.11):

$$g(r, \theta) = \lambda(e^\theta - 1) + \mu(e^{-\theta} - 1) = 0$$

$$\Rightarrow \qquad \lambda\left(e^\theta\right)^2 - (\lambda + \mu)e^\theta + \mu = 0$$

$$\Rightarrow \qquad e^\theta = \frac{\lambda + \mu \pm \sqrt{(\lambda + \mu)^2 - 4\lambda\mu}}{2\lambda}$$

$$= \frac{\lambda + \mu \pm (\mu - \lambda)}{2\lambda}$$

$$= 1, \text{ or } \frac{\mu}{\lambda}.$$

The choice $e^\theta = 1$ gives $\theta = 0$ so that the Definition (7.16) or (11.20) of ℓ gives $\ell(r, r') = 0$. But $\ell = 0$ means $r = z_\infty$ (Corollary 5.12 and Exercise 5.27. This is a generic calculation: $\theta = 0$ corresponds to $r = z_\infty$, which is probable behavior, not improbable). This means $r' < 0$ by the assumption on z_∞, and is not the right solution since it cannot satisfy the boundary conditions. Thus we have the explicit formula for θ^*

$$e^{\theta^*(x)} = \frac{\mu(x)}{\lambda(x)}$$

or $\theta = \log \mu/\lambda$. This establishes (11.21). Now use the fact that ℓ must be maximized by this θ. From the definition (11.20) of ℓ, equating the derivative to zero,

$$0 = r' - \lambda e^\theta + \mu e^{-\theta}$$

$$= r' - \lambda \frac{\mu}{\lambda} + \mu \frac{\lambda}{\mu}.$$

This establishes (11.22).   ∎

Exercise 11.16. Prove that (11.22) has a unique solution. Hint:

$$\int_0^{r(t)} \frac{dr}{\mu(r) - \lambda(r)} = t.$$
♠

Exercise 11.17. Show that the conclusions of Theorem 11.15 hold for $M/M/1$ queues. Hint: the only issue is that μ is not continuous at zero.
♠

Using Exercise 11.17 and last calculation of Theorem 11.15, we see that the $M/M/1$ queue has a minimal path $r(t) = t(\mu - \lambda)$, which is indeed monotone and smooth, and by (C.12),

$$V(a) = a \log \frac{\mu}{\lambda}. \tag{11.23}$$

In this case the optimal time T^* equals $a(\mu - \lambda)^{-1}$.

Our calculation of $V(a)$ enables us to conclude that, in steady state,

$$\mathbb{P}(z_n \geq a) \approx \exp\left(-na \log \frac{\mu}{\lambda}\right).$$

Unscaling $z_n(t)$ to $nx(t/n)$, where $x(t)$ is the number in the queue at time t, we have that in steady state

$$\mathbb{P}(x(t) \geq na) \approx \exp\left(-na \log \frac{\mu}{\lambda}\right) = \left(\frac{\lambda}{\mu}\right)^{na}. \tag{11.24}$$

This is exact! That is, the error in the calculation is zero whenever na is an integer.

Our solution of the variational problem provides much more information. Note, since $r^*(t) = t(\mu - \lambda)$ for $0 \leq t \leq a(\mu - \lambda)^{-1}$, that the most likely way the queue became big is by following, over that time interval, exactly the reverse of the most likely behavior—z_∞. More precisely, $r(t) = z_\infty(-t)$ (perhaps plus initial conditions). But this is true in a more detailed way: by Theorem 7.26, during this time interval the arrival process has rate $\lambda e^{\theta^*} = \mu$, and the service process has rate $\mu e^{-\theta^*} = \lambda$. That is, *the arrival and service rates are interchanged during the excursion from zero up to a*. The exact statement is given in the next theorem, and is formulated in terms of the (unscaled) arrival process $a(t)$ and departure, or service process $s(t)$ associated with the $M/M/1$ queue.

Theorem 11.18. *Consider the $M/M/1$ in steady state. Conditioned on $z_n(0) = a > 0$, over any interval of length t contained in $(-a/(\mu - \lambda), 0)$, the total num-*

ber of arrivals $a(t)$ and services $s(t)$ satisfy

$$\lim_{n\to\infty} \mathbb{P}\left(\left|\frac{1}{n}a(nt) - \mu t\right| < \varepsilon\right) = 1$$

$$\lim_{n\to\infty} \mathbb{P}\left(\left|\frac{1}{n}s(nt) - \lambda t\right| < \varepsilon\right) = 1$$

$$\lim_{n\to\infty} \frac{1}{n}\mathbb{E}(a(nt)) = \mu t$$

$$\lim_{n\to\infty} \frac{1}{n}\mathbb{E}(s(nt)) = \lambda t .$$

The proof is found in §7.4, Theorem 7.26. Consequences of this fact are elaborated in Chapter 16.

The dwell time in neighborhoods of zero follows easily from the calculation of $V(a)$. Applying Theorem 6.17 we have the following:

Theorem 11.19. *Given $a > 0$ and $z_\infty(0) = x < a$, let*

$$\tau_n \overset{\triangle}{=} \inf_{t>0}\{z_n(t) \geq a\}.$$

Then

$$\lim_{n\to\infty} \frac{1}{n} \log \mathbb{E}_x(\tau_n) = V(a) = a \log \frac{\mu}{\lambda},$$

and for every $\delta > 0$,

$$\lim_{n\to\infty} \mathbb{P}_x\left(\tau_n > e^{n(V(a)+\delta)}\right) = \lim_{n\to\infty} \mathbb{P}_x\left(\tau_n < e^{n(V(a)-\delta)}\right) = 0.$$

11.6. The Probability of Hitting a Point

The object of this section is to prove Theorem 11.4.

Theorem 11.4. *If $\rho < 1$ then for any positive t, x, and y,*

$$\lim_{n\to\infty} \frac{1}{n} \log \mathbb{P}_x\left(y - 1/2n < z_n(t) \leq y + 1/2n\right) = \varphi_t(x, y).$$

The point of this lemma is that we estimate what the probability is of hitting a point. Since the associated set (in the variational problem) is not open, the large deviations lower bound is not immediate.

Proof. The probability we are trying to estimate

$$p_{x,y}^n(t) \overset{\triangle}{=} \mathbb{P}_x\left(y - 1/2n < \vec{z}_n(t) \leq y + 1/2n\right) \tag{11.25}$$

is clearly positive, for every n. Estimating $p_{x,y}^n(t)$ using large deviations, the process-level upper bound gives, for each positive ε as $n \to \infty$,

$$p_{x,y}^n(t) \leq \mathbb{P}_x\left(y - \varepsilon \leq \vec{z}_n(t) \leq y + \varepsilon\right)$$
$$\leq e^{-nI_\varepsilon^* + o(n)},$$

where the value of I_ε^* is obtained by minimizing I_0^t over a closed set

$$I_\varepsilon^* = \inf\{I_0^t(r) : r(0) = x,\ y - \varepsilon \le r(t) \le y + \varepsilon\}.$$

By Lemma 6.21 and Exercise 6.25,

$$\lim_{\varepsilon\downarrow 0} I_\varepsilon^* = \varphi_t(x, y) \stackrel{\triangle}{=} \inf\{I_0^t(r) : r(0) = x,\ r(t) = y\}. \tag{11.26}$$

This establishes the upper bound for $p_{x,y}^n(t)$. But the lower bound does not apply to closed sets! It gives us only

$$\lim_{\varepsilon\downarrow 0}\lim_{n\to\infty}\inf \frac{1}{n}\log \mathbb{P}_x\big(z_n(t) \in N_\varepsilon(y)\big) \ge -\varphi_t(x, y),$$

where $N_\varepsilon = \{w : |w - y| < \varepsilon\}$. This is not good enough to give a bound on $p_{x,y}^n(t)$, since the number of values in $N_\varepsilon(y)$ the process $\vec{z}_n$ can take grows (linearly) with n, whereas by definition the desired probability concerns exactly one point. However, we may use a general argument to establish the desired lower bound on this probability, which we expect to be

$$p_{x,y}^n(t) \ge e^{-n\varphi_t(x,y)+o(n)}. \tag{11.27}$$

We give the argument for the $M/M/1$ process, and then outline the extension to the general case in Exercise 11.20.

Fix a small ε. Clearly, by the Markov property,

$$p_{x,y}^n(t) \ge \mathbb{P}_x\big(z_n(t - \varepsilon) \in N_\varepsilon(y)\big) \cdot \inf_w \mathbb{P}_w(z_n(\varepsilon) = y),$$

where the infimum is over all points in $N_\varepsilon(y)$ that are of the form k/n, where k is an integer. But the lower bound gives

$$\mathbb{P}_x\big(z_n(t - \varepsilon) \in N_\varepsilon(y)\big) \ge e^{-nI_\varepsilon^*+o(n)},$$

where here I_ε^* is defined as

$$I_\varepsilon^* = \inf\{I_0^{t-\varepsilon}(r) : r(0) = x,\ r(t - \varepsilon) \in N_\varepsilon(y)\} \to \varphi_t(x, y) \text{ as } \varepsilon \downarrow 0,$$

where the convergence to I^* follows, as before, from arguments given in the proof of Lemma 6.21 and in Exercise 6.25. Now suppose we show that there is a $C < \infty$ such that, uniformly over $w \in N_\varepsilon(y)$,

$$\mathbb{P}_w(z_n(\varepsilon) = y) \ge e^{-nC\varepsilon} \tag{11.28}$$

(whenever w and y have the form k/n for integer k). Then, since ε is arbitrary, the lower bound (11.27) follows.

To establish (11.28), we use a very crude (but generally useful) bound. In the notation of Theorem 11.18, if $w < y$,

$$\mathbb{P}_w(z_n(\varepsilon) = y) \ge \mathbb{P}_{y-\varepsilon}(a(n\varepsilon) = n\varepsilon) \cdot \mathbb{P}_{y-\varepsilon}(s(n\varepsilon) = 0),$$

using the independence of the arrival and service processes. The second probability is just $e^{-n\mu\varepsilon}$. For the first one, use Stirling's formula or Chernoff's theorem to obtain the bound. The case $w > y$ is handled in the same way. ∎

Exercise 11.20. Consider a jump-Markov process where the jump-directions $\vec{e}_i$ have integer components. Assume that the positive cone generated by the collection $\{\vec{e}_i\}$ equals $\mathbb{R}^d$. Generalize the arguments given above. Hint: establish that the number of jumps the process must take to go between points has an upper bound that is linear in the distance. This establishes the equivalent of Lemma 5.20, but for the integer case. Then calculate a probability of the process following one particular sequence of jumps to go from one point to another. ♠

This result was used extensively in the proof of the large deviations principle for processes with a flat boundary (Chapter 8).

11.7. Transient Behavior

This section contains a proof of Theorem 11.3 which we restate for convenience.

Theorem 11.3. *Let* $t_1 \stackrel{\triangle}{=} (x + y)(\mu - \lambda)^{-1}$. *For the stable* $M/M/1$ *queue, there is a* $t^* > t_1$ *so that the optimal cost and path from* x *to* y *in time* t *are*

$$\varphi_t(x, y) = \begin{cases} y \log \frac{\mu}{\lambda} & \text{if } t \geq t^*; \\ t\ell_f\left(\dfrac{y - x}{t}\right) & \text{if } t \leq t^*. \end{cases}$$

$$r^*(s) = \begin{cases} x + (y - x)s/t & \text{if } t < t^*; \\ x - s(\mu - \lambda) & \text{if } t > t^* \text{ and } s < \frac{x}{\mu - \lambda}; \\ 0 & \text{if } t > t^* \text{ and } \frac{x}{\mu - \lambda} \leq s \leq t - \frac{y}{\mu - \lambda}; \\ y + (s - t)(\mu - \lambda) & \text{if } t > t^* \text{ and } t - \frac{y}{\mu - \lambda} \leq s \leq t. \end{cases}$$

Moreover, t^* *is the (unique) larger positive solution of*

$$t\ell_f\left(\frac{y - x}{t}\right) = y \log \frac{\mu}{\lambda}. \tag{11.29}$$

The proof of Theorem 11.3 will be given below.

Recall the discussion of §11.2. In summary, we have identified the paths with minimal I-function that start at x, pass through zero, and take an amount of time at least equal to $t_1 \stackrel{\triangle}{=} (x + y)(\mu - \lambda)^{-1}$ to go from x to y. Now for $t = t_1$, the path r does not remain at zero for a positive amount of time; it only touches zero. From (11.18) it follows that this path satisfies (11.10),

$$I_{f,0}^t(r) = I_0^t(r),$$

i.e., the cost is the same as for the free process. But Lemma 5.16 shows that for the free process the lowest cost path is the straight line from x to y (in fact, due to the strict convexity of ℓ_f (see Exercise 11.8), this is *the only* optimal path). However, for such paths (11.10) holds as well. Therefore, for $t = t_1$ the straight line path is optimal.

The same argument also establishes that for $t < t_1$, the straight line path is optimal. First, any path that touches the boundary but doesn't spend time there will be more expensive than the straight line path by the above argument. Secondly, it is clearly not optimal to touch the boundary, leave it, and return, since the excursion from zero to zero can be deleted, thus reducing the cost. Now consider a path that spends a time interval $[s_1, s_2]$ at zero. By the same considerations as above, this path consists of three straight lines. Since $t < t_1$, necessarily either the slope of the decreasing path is more negative than $\lambda - \mu$, or the slope of the increasing path is larger than $\mu - \lambda$ (or both). The convexity of $t\ell_f(x/t)$ (see Exercise 11.8) now implies that the cost is reduced by changing both slopes so that $s_1 = s_2$, and the argument is complete.

As t increases, we have for the straight line path

$$I_{f,0}^t(r) = t\ell_f \left(\frac{y-x}{t} \right) \sim t\ell_f(0) \to \infty$$

since ℓ_f is continuous from the right and positive at zero. Therefore, for some $t^* > t_1 = (x+y)(\mu - \lambda)^{-1}$ we will have

$$I_{f,0}^t(r) = y \log \frac{\mu}{\lambda}.$$

For every $t > t^*$, then, the lowest cost path uses the boundary. [Note that by Exercise 11.8, the function $t\ell_f(a/t)$ is convex for $t > 0$ and is increasing for $t \geq |a|/(\mu - \lambda)$.]

We now formalize the statements and arguments. Assume x, y are strictly positive.

Proof of Theorem 11.3. The first claim follows from the preceding arguments, which also establish that

$$t_1 \ell_f \left(\frac{y-x}{t_1} \right) < y \log \frac{\mu}{\lambda}. \tag{11.30}$$

Since $t\ell_f(a/t)$ is continuous, there is at least one solution to the stated equality. But from the note above, $t\ell_f(a/t)$ is strictly increasing for $t > t_1$, so that the solution is unique.   ■

Exercise 11.21. For $x = 2$, $y = 3$, $\lambda = 1$, and $\mu = 2$, calculate t^* numerically. How close is it to 10?   ♠

Exercise 11.22. What is the form of Theorem 11.3 in the case where x or y (or both) are zero?   ♠

11.8. Approach to Steady State

We have seen that, as far as large deviations are concerned, there is a definite time t^*, depending on x and y, after which $p^n_{x,y}(t)$ does not change; that is, t^* *is the time when steady state is reached.* What can this mean? In fact, steady state is never fully "reached"; initial conditions never die out completely. However, in our scaling, they become asymptotically negligible after a fixed finite time t^*, as $n \to \infty$. For the purpose of intuition, consider $p_{00}(t)$. Starting at state zero, the queue has busy/idle cycles. These have i.i.d. lengths. So after time $t = n\varepsilon$, for any $\varepsilon > 0$, we should have $|p_{00}(t) - p_{00}(\infty)| < e^{-Cn\varepsilon}$ for some $C > 0$. That is, for any $\varepsilon > 0$, in our scaling the transition from zero to zero achieves steady state in time less than ε.

Unscaling the process.

We have estimated $p^n_{x,y}(t)$ for the process $z_n(t)$. What does this mean for the original problem, the $M/M/1$ queueing process $x(t)$?

$$e^{-n\varphi_t(x,y)} \asymp \mathbb{P}\left(z_n(t) = y \mid z_n(0) = x\right) = \mathbb{P}\left(x(nt) = ny \mid x(0) = nx\right),$$

so we have roughly

$$\mathbb{P}\left(x(t) = ny \mid x(0) = nx\right) \asymp e^{-nI^*(t/n)}.$$

We find that after time $nt^*(nx, ny)$, the process $x(t)$ virtually achieves steady state for transitions from nx to ny. This is not exactly a standard notion of steady state. Perhaps it is more natural to ask "How long will it take $x(t)$ to have all transition probabilities $p_{nx,ny}(t)$ within e^{-nC} of their ultimate value?" For this, we simply have to find t_2 such that either

$$t^*(nx, ny) \leq t_2 \tag{11.31}$$

or

$$I^*(t_2) \geq C \tag{11.32}$$

for all $y \geq 0$. If $t^*(nx, ny) < t_2$, then we are guaranteed that

$$\frac{1}{n}\left|\log \frac{p_{nx,ny}(t_2)}{p_{nx,ny}(\infty)}\right| \to 0 \text{ as } n \to \infty,$$

which may not be the same as our requirement, but it's the best we can do. If $I^*(t_2) > C$ then $p^n_{x,y}(t) < e^{-nC}$ for all $t \geq t_2$. So we are left with the question of existence of a t_2 that satisfies either (11.31) or (11.32) for all $y \geq 0$.

Exercise 11.23. Prove that, for any $x > 0$, there always exists a $t_2 < \infty$ satisfying either (11.31) or (11.32) for all $y \geq 0$. ♠

Explicit asymptotics.

There are simple cases when the transition probability $p_{x,y}^n(t)$ can be estimated. Suppose $y < x$ and $z_\infty(t) = y$ (i.e., $t = (x - y)(\mu - \lambda)^{-1}$). Then we know that $p_{x,N_\varepsilon(y)}(t) \approx 1$ from Kurtz's Theorem. That is, for $t < x(\mu - \lambda)^{-1}$, the transition probability looks like a delta function around $z_\infty(t) = x - \mu t + \lambda t$. Furthermore, when $t > x(\mu - \lambda)^{-1}$ the transition function is close to a delta function around zero.

Now let's estimate how quickly the transition function approaches steady state. For each x and y, we see that $I^* = $ constant when $t > t^*$, where t^* solves

$$t^* \ell_f \left(\frac{y - x}{t^*} \right) = y \log \frac{\mu}{\lambda} .$$

There are a few cases where we can calculate explicit asymptotics for t^*:

1) $|y - x|$ small relative to $|y|$. Then since $t^* > t_1$ and t_1 is of the order of $x + y$,

$$t \ell_f \left(\frac{y - x}{t} \right) \approx t \left(\ell_f(0) + \frac{y - x}{t} \ell_f'(0) \right).$$

By a simple calculation [see (11.7)], $\ell_f(0) = \left(\sqrt{\mu} - \sqrt{\lambda} \right)^2$. From (C.6) and the definition of ℓ_f,

$$\ell_f'(0) = \theta^*(0) = \log \sqrt{\frac{\mu}{\lambda}} = \frac{1}{2} \log \frac{\mu}{\lambda}.$$

Substituting in the above approximation we obtain

$$t \ell_f \left(\frac{y - x}{t} \right) \approx t \left(\sqrt{\mu} - \sqrt{\lambda} \right)^2 + \frac{y - x}{2} \log(\mu/\lambda)$$

$$t^* \approx \frac{\frac{y + x}{2} \log \frac{\mu}{\lambda}}{(\sqrt{\mu} - \sqrt{\lambda})^2}.$$

This is exact if $y = x$.

2) y is small relative to x. In that case, $t_1 \approx x(\mu - \lambda)^{-1}$. But then

$$\ell_f \left(\frac{y - x}{t_1} \right) \approx \ell_f(\lambda - \mu) = 0.$$

Let us expand ℓ_f near $\lambda - \mu$. This is a minimum, so $\ell_f'(\lambda - \mu) = 0$. Denote $w \triangleq \ell_f''(\lambda - \mu)$ and let $t = (x - y)(\mu - \lambda)^{-1} + \Delta$ and define

$\varepsilon \overset{\triangle}{=} \Delta(x - y)^{-1}$. Then

$$t\ell_f\left(\frac{y-x}{t}\right) = \left(\frac{x-y}{\mu-\lambda} + \Delta\right)\ell_f\left((\lambda-\mu)\left(1 + \frac{\Delta(\lambda-\mu)}{x-y} + O\left(\varepsilon^2\right)\right)\right)$$

$$= (x-y)\left(\frac{1}{\mu-\lambda} + \varepsilon\right)\frac{w}{2}\left(\varepsilon^2(\lambda-\mu)^4 + O\left(\varepsilon^3\right)\right)$$

$$= \frac{w}{2}\frac{(\mu-\lambda)^3}{x-y}\Delta^2 + O\left(\varepsilon^2\right).$$

Using the explicit expression $w = (\lambda + \mu)^{-1}$ from (7.24) of Exercise 7.24, and the expression for t^* from Theorem 11.3 we find

$$t^* \approx \frac{x-y}{\mu-\lambda} + \sqrt{\frac{(x-y)y\log(\mu/\lambda)}{(w/2)(\mu-\lambda)^3}}$$

$$\approx \frac{x}{\mu-\lambda} + \frac{\sqrt{y}}{\mu-\lambda}\sqrt{2x\frac{(\mu+\lambda)}{(\mu-\lambda)}\log\frac{\mu}{\lambda}}.$$

11.9. Further Extensions

Long busy periods.

The $M/M/1$ queue, like most queues, has busy periods (when $x(t) \geq 1$) and idle periods [when $x(t) = 0$]. The idle periods are exponentially distributed with mean $1/\lambda$. What do the busy periods look like?

We investigate the probability that a busy period is longer than n. Since during a busy period the $M/M/1$ process coincides with the free process, we are immediately led to consider

$$I^* = \inf\left\{\int_0^1 \ell_f(r, r')\,dt : r(0) = 0,\ r(t) > 0,\ 0 < t < 1\right\}. \qquad (11.33)$$

Since $\ell_f(r, r') = \ell_f(r')$ is increasing for $r' > \mu - \lambda$ (or from Lemma 5.16), we see that I^* is minimized for $r(t) = 0, 0 \leq t \leq 1$, which just barely misses satisfying the boundary conditions. Also by (7.17), $\ell_f(r, 0) = (\sqrt{\mu} - \sqrt{\lambda})^2$. So we are led to conjecture the following

Theorem 11.24. *For the stable $M/M/1$ queue,*

$$\lim_{n\to\infty}\frac{1}{n}\log \mathbb{P}_0\left(busy\ period \geq n\right) = -\left(\sqrt{\mu} - \sqrt{\lambda}\right)^2. \qquad (11.34)$$

Theorem 11.25. *For the stable $M/M/1$ queue,*

$$\lim_{n\to\infty}\mathbb{P}_0\left(\sup_{0\leq t\leq n}\frac{x(t)}{n} \geq \varepsilon \,\middle|\, busy\ period\ is \geq n\right) = 0. \qquad (11.35)$$

[The second theorem comes from noticing that $r^* = 0$ for all t, so the maximal excursion is $o(n)$.]

Exercise 11.26. Prove Theorem 11.24. Hint: by the remarks above it suffices to consider the free process. Now show that the set of interest is a continuity set. ♠

We give two additional proofs of the lower bound for Theorem 11.24, one requiring nothing more than Chernoff's Theorem and the Ballot Theorem 3.3, the other a straightforward process proof. We also give a simple upper bound—using Chernoff's Theorem directly.

Proof 1. A busy period always starts with $x(0) = 1$.

$$\mathbb{P}_1 \left(\inf_{0 \le t \le n} x(t) = 1 \right)$$

$$= \sum_{j=0}^{\infty} \mathbb{P}_1 \left(\inf_{0 \le t \le n} x(t) = 1 \;\middle|\; \# \text{ transitions } = j \right) \cdot \mathbb{P}(\# \text{ transitions } = j)$$

$$\ge \sum_{j=0}^{\infty} \frac{1}{j} \mathbb{P} \left(x(n) \ge 1 \;\middle|\; \# \text{ transition } = j \right) \cdot \mathbb{P}(\# \text{ transitions } = j)$$

by the Ballot Theorem 3.3. Since the "free" process is identical to $x(t)$ on the set where $\inf_{0 \le t \le n} x(t) = 1$, each transition is $+1$ with probability $\lambda (\lambda + \mu)^{-1}$, and -1 with probability $\mu (\lambda + \mu)^{-1}$. Hence

$$\mathbb{P} \left(x(n) \ge 1 \;\middle|\; \# \text{ transitions } = j \right) = \mathbb{P} \left(x_1 + \ldots + x_j \ge \frac{j}{2} \right), \qquad (11.36)$$

where $x_1, x_2, \ldots$ are i.i.d. binomial random variables with parameter $p = \frac{\lambda}{\lambda + \mu}$. Now Chernoff's Theorem 1.5 and Exercise 1.17 show that

$$\mathbb{P} \left(x_1 + \ldots + x_j \ge \frac{j}{2} \right) = e^{-j \ell_b \left(\frac{1}{2} \right) + o(j)},$$

where

$$\ell_b(a) = a \log \frac{a}{p} + (1 - a) \log \frac{1 - a}{1 - p}.$$

Also,

$$\mathbb{P}(\# \text{ transitions } = j) = e^{-(\lambda + \mu)n} \frac{((\lambda + \mu)n)^j}{j!}.$$

So we obtain

$$\mathbb{P} \left(\inf_{0 \le t \le n} x(t) = 1 \right)$$

$$\ge \sum_{j} e^{-n(\lambda + \mu)} \frac{(n(\lambda + \mu))^j}{j! \, j} \exp \left(-j \left[\frac{1}{2} \log \frac{1}{2p} + \frac{1}{2} \log \frac{1}{2(1 - p)} \right] + o(j) \right).$$

Now take $j = n(\lambda + \mu)$ and use Stirling's formula to obtain

$$\mathbb{P}\left(\inf_{0 \leq t \leq n} x(t) = 1\right) \geq e^{-n\left(\sqrt{\mu} - \sqrt{\lambda}\right)^2 + o(n)}. \tag{11.37}$$

The upper bound is even easier. By Chernoff's Theorem applied to the random variable w with distribution Poisson(λ) $-$ Poisson(μ),

$$\mathbb{P}_1(x(n) \geq 0) \leq e^{-n\ell_w\left(-\frac{1}{n}\right)},$$

where $\ell_w(0) = \left(\sqrt{\mu} - \sqrt{\lambda}\right)^2$ and $\ell_w(x)$ is continuous. (See §7.3 for details of this calculation.)   ■

Proof 2. For time $n\varepsilon$, let no service occur: this occurs with probability $e^{-n\varepsilon\mu}$. Then $x(n\varepsilon) \overset{\mathcal{L}}{=} \text{Pois}(n\varepsilon\lambda)$, and therefore

$$\mathbb{P}\left(\frac{n\varepsilon\lambda}{2} < x(n\varepsilon) \text{ and } x(t) \geq 1,\ 0 \leq t \leq n\varepsilon\right) > \frac{1}{2}e^{-n\varepsilon\mu}.$$

Then the probability that $x(t)$ deviates by less than $n\varepsilon\lambda/2$ from the path $r(t) \equiv \varepsilon$ for $\varepsilon n \leq t \leq n$ is at least $e^{-n(1-\varepsilon)\left(\sqrt{\mu} - \sqrt{\lambda}\right)^2 + o(n)}$, by the large deviations lower bound. That is,

$$\mathbb{P}\left(x(t) \geq 1 \text{ for } 0 \leq 1 \leq n\right) \geq \frac{1}{2}\exp\left(-\frac{n\mu\varepsilon}{2} - n(1-\varepsilon)\left(\sqrt{\mu} - \sqrt{\lambda}\right)^2 + o(1)\right).$$
                                                                                                                    ■

Proof of Theorem 11.25. From the second proof of Theorem 11.24,

$$\mathbb{P}\left(1 \leq \sup_{0 \leq t \leq n} x(t) \leq n\varepsilon\right) \geq e^{-n\left(\sqrt{\mu} - \sqrt{\lambda}\right)^2 + O(n\varepsilon) + o(n)}. \tag{11.38}$$

But from the process upper bound,

$$\mathbb{P}\left(1 \leq \inf_{0 \leq t \leq n} x(t),\ \sup_{0 \leq t \leq n} x(t) \geq n\varepsilon\right) \leq e^{-nI_\varepsilon^* + o(n)}$$

$$I_\varepsilon^* = \inf\left(I_0^1(r) : r(0) = 0,\ r(t) \geq 0,\ \sup_{0 \leq t \leq 1} r(t) \geq \varepsilon\right) > \left(\sqrt{\mu} - \sqrt{\lambda}\right)^2.$$

You may verify that the path $r^*(t)$ for this problem consists of a straight line from zero to ε and then back down to zero. The reason that this must be the form of any minimal cost path is that the path has to reach from zero to ε, and if it does so in time s then the lowest cost way of doing it is a straight line. Similarly, it is clearly cheaper for $r(1)$ to be zero than any positive value, so the cheapest way to go from ε to zero in time $1 - s$ is again by a straight line. Now we just calculate a value of s that makes $I(r)$ minimal. Now show that the cost for this path is $I^* = \left(\sqrt{\mu} - \sqrt{\lambda}\right)^2 + O\left(\varepsilon^2\right)$ where the additional term is positive.   ■

Finite M/M/1 queue.

Another important question is the effect of a boundary at a level M for some $M > 0$. That is, in practice a queue has a finite capacity, and we'd like to know how this finiteness affects the queue's statistics. From the large deviations point of view, this inclusion is almost a triviality. We simply put a barrier at $z_n = M$ (equivalently at $x = nM$), for example, by setting

$$\lambda(x) = \begin{cases} \lambda & \text{if } x < nM; \\ 0 & \text{if } x = nM, \end{cases}$$

$$\mu(x) = \begin{cases} \mu & \text{if } x > 0 \\ 0 & \text{if } x = 0. \end{cases}$$

Then all the calculations we have done do not change at all so long as $\mu < \lambda$. We still have

$$\mathbb{P}_{ss}(z_n \geq a) = e^{-nV(a)+o(n)}, \tag{11.39}$$

where $V(a) = a \log \frac{\mu}{\lambda}$, as long as $a < M$. Furthermore we have

$$\mathbb{P}_x(z_n(t) = y) = e^{-n\varphi_t(x,y)+o(n)}, \tag{11.40}$$

so long as x and y are both smaller than M, where $\varphi_t(x, y)$ is given by Theorem 11.3 for $x, y \in [0, M]$. We can even extend the analysis to the case $\lambda > \mu$ when we have a barrier at M: simply consider the process $y(t) = M - x(t)$, and we see that $y(t)$ is a stable $M/M/1$ queue with a barrier at M.

Exercise 11.27. What happens if $\lambda = \mu$ when there is a barrier at M? (Answer: $V(a) = 0$ for all $a \in [0, M]$, but $\varphi_t(x, y)$ has the same expression as given in Theorem 11.3, but with $t^* = \infty$. This is because $\varphi_t(x, y) \to 0$ as $t \to \infty$.) ♠

11.10. End Notes

The $M/M/1$ queue is called, confusingly enough, an Erlang system in the older literature. There is a voluminous literature on it and its close relatives, for both steady state and transient analysis. For a classical approach to the $M/M/1$ queue, Takács [Ta1] may still be the best reference. For a modern point of view see Massey [Mas]. For more than you ever wanted to know about it and its relatives, see Cohen [C1]. Other books that have worthwhile discussions include [Co, Kl], and most likely any book with "queueing" in the title probably has something to say on the subject.

Our result on the transient behavior of the $M/M/1$ queue, Theorem 11.3, may be extracted from [AKW], once the appropriate question is asked. Also, our result on long busy periods can be extended and improved using either more advanced large deviations or other methods. For example, the distribution of the queue length (conditioned on being in a busy period that started a long time ago) tends to a limit that is calculable. The fact that $M/M/1$ queues fill by switching their arrival rate and service rate (Theorem 11.18) can be made more precise by considering the time reversed process: see e.g. [SW]. This has many other implications, too, some of which are explored in later chapters (in particular, in Chapter 16).

We hope that the experienced reader has enjoyed a discussion of the $M/M/1$ queue that had neither a derivation of its steady-state distribution nor a transient analysis rooted in Laplace transforms or special functions. It's amazing how many ways there are of looking at the same thing, isn't it?

Chapter 12

Erlang's Model

This chapter contains an analysis of the transient behavior of Erlang's model, the basic model of circuit-switched traffic. Consider two points (switches) connected by n wires (trunks, communication channels). Customers arrive at the system according to a Poisson process with rate λ. Each customer attempts to hold one trunk for a period of time that is exponentially distributed with mean one, and holding times are independent. (The choice of mean one defines our time unit. The time unit is often taken to be 3 to 5 minutes for voice calls in the United States.) If all n trunks are busy when a customer arrives, he departs forever (this model with no retries is called "blocked calls cleared"). In the parlance of queueing theory, the system just described is an $M/M/n/n$ queue: Poisson arrivals, a capacity–n queue, n exponential servers. There are obviously several deficiencies in this model. Nevertheless, we must understand it in order to have a chance of understanding more realistic and complex models.

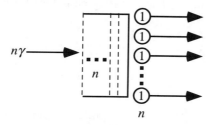

Figure 12.1. The basic Erlang model.

One of the most important performance measures of the system is the *blocking probability*: the probability that an arriving customer finds all trunks occupied. That is, we try to estimate $\mathbb{P}(x(t) = n)$. In modern systems, this is an (increasingly) rare event. This probability is easy to calculate in steady-state; Erlang's treatment from 1917 still holds up, and anyone who has studied birth-death processes can write down the steady state probability by inspection. However, until quite recently [MW], the transient behavior of the blocking probability was hidden behind "the Laplace veil"; that is, the transform was known, but the behavior of the function itself was obtainable only through numerical calculation. We aim to rectify that situation in this chapter.

Once we have a handle on the transient behavior of Erlang's basic model, we extend the results in several ways. One natural extension is to finite population; after all, there is usually not an infinite number of potential customers attempting to use a link, even with a substantial population of teenagers. Also, modern

289

services such as Picturephone$^©$ might need more than one trunk per connection; we are able to analyze the effect of this type of "multirate" connection in certain cases. We hope that by the end of this chapter, the interested reader will be able to develop his or her own model and moreover, based on our techniques, be able to solve it.

12.1. Scaling and Preliminary Calculations

Unlike the $M/M/1$ process, in this case we have a scale parameter "built in": it is the number of trunks. We need to rephrase things so that we can apply the theory.

If n is the number of trunks, then for the process $x(t) \overset{\triangle}{=}$ the number of trunks occupied at time t,

$$e_1 = +1 \qquad \lambda_1(x) = \begin{cases} \lambda & x < n \\ 0 & x \geq n \end{cases}$$

$$e_2 = -1 \qquad \lambda_2(x) = x, \ 0 \leq x \leq n.$$

In order to get a scaling to which we can apply our theory, (5.1) suggests that we use

$$z_n(t) \overset{\triangle}{=} \frac{1}{n}x(t), \qquad \lambda \overset{\triangle}{=} n\gamma$$

for some positive constant γ. The resulting generator for z_n is of the form (5.1):

$$L_n f(z) = n\gamma \mathbf{1}\,[z < 1]\left(f\left(z + \frac{1}{n}\right) - f(z)\right)$$

$$+ nz\mathbf{1}\,[z > 0]\left(f\left(z - \frac{1}{n}\right) - f(z)\right). \quad (12.1)$$

The case $\lambda = n + O\left(n^{1/2}\right)$ is also an important regime, but we do not analyze it here; it is better suited to analysis by diffusion techniques (see e.g., [KnM]).

Before we perform the calculations, let us discuss whether the process $z_n(t)$ satisfies a large deviations principle. There are two possible problems with the process, namely the boundary at $z = 0$ and the boundary at $z = 1$. For the former we have $\lambda_2(0) = 0$, so $\log \lambda_2(x)$ is not bounded near $x = 0$. Therefore the process $z_n(t)$ does not satisfy the assumptions of either Chapter 5 or Chapter 8. Also, $\lambda_1(x)$ has a jump at $x = 1$, but this falls within the domain of the theory of Chapter 8. Nevertheless the point zero is a removable irregularity in the sense of the Freidlin-Wentzell theory. We provide a justification of the large deviations principle for Erlang's model in §12.6. For now, we simply continue with our standard program of analysis.

The main quantity we are interested in is $\mathbb{P}_x(z_n(t) = 1)$, for various values of x and t. We perform the calculation for $x = 0$ in §12.2, for $x = 1$ in §12.3, and finally in §12.5 for general $x \in [0, 1]$.

As usual, the first item of business is to calculate $z_\infty(t)$. Note that Kurtz's Theorem 5.3 applies as long as we stay away from $z = 1$: it is established for Erlang's model in Theorem 12.17. From (5.7) then,

$$\frac{d}{dt}z_\infty(t) = \gamma - z_\infty(t) \qquad 0 \le z_\infty(t) < 1$$

$$z_\infty(t) = \min\left\{\gamma(1 - e^{-t}) + e^{-t}z_\infty(0), 1\right\}$$

(12.2)

so that the differential equation holds until the first time (if any) that $z_\infty(t) = 1$.

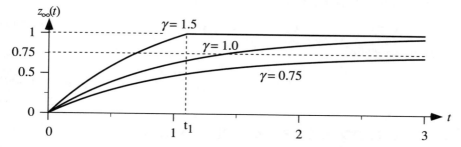

Figure 12.2. The path $z_\infty(t)$ for $\gamma = 0.75$, $\gamma = 1$, and $\gamma = 1.5$. Note that the last case has a corner at t_1.

The time when z_∞ reaches one starting from an empty system is therefore

$$t_1 \overset{\triangle}{=} \begin{cases} \infty & \gamma \le 1 \\ \log\frac{\gamma}{\gamma - 1} & \gamma > 1. \end{cases}$$

This is the longest time that we can expect an Erlang system, with any initial size, to stay away from the boundary at one. We see that there are three cases to consider:

1) $\gamma < 1$. Then $z_\infty(t) \to \gamma$ as $t \to \infty$. We call this "light traffic."

2) $\gamma > 1$. Then $z_\infty(t) = 1$ for some finite $t \le t_1$, and the differential equation for $z_\infty(t)$ breaks down thereafter. In fact, by the results of Chapter 8, after that time $z_\infty(t)$ is identically equal to one. We call this "heavy traffic."

3) $\gamma = 1$. Here $z_\infty(t) \to 1$ as $t \to \infty$. We call this "moderate traffic."

The next item of business is to set up the question of interest, the blocking probability, as a variational problem. In Theorem 12.18 of §12.6 we establish the large deviations principle for Erlang's model, and in §12.6 we show that the blocking probability satisfies

$$\mathbb{P}_x(z_n(t) = 1) = e^{-nI_x^*(t) + o(n)},$$

(12.3)

where $I_x^*(t)$ is the infimum, over absolutely continuous functions:

$$I_x^*(t) = \inf\left\{\int_0^t \ell(r(s), r'(s))\, ds : r(0) = x,\ r(t) = 1\right\}.$$

(12.4)

For fixed y and for $0 < x < 1$, comparing (12.1) with (7.15) we see that the calculation of ℓ agrees with that of the free $M/M/1$ process of §7.3, with arrivals rate γ and departure rate x. Thus from (7.17),

$$\ell(x, y) = y \log \frac{y + \sqrt{y^2 + 4\gamma x}}{2\gamma} + \gamma + x - \sqrt{y^2 + 4\gamma x}. \tag{12.5a}$$

We only need to extend the definition on the boundary to ensure that the process stays within $0 \leq z_n(t) \leq 1$. The appropriate definition (established in §12.6, or more generally in Chapter 8) is

$$\ell(x, y) = \infty \quad \text{if} \quad x = 0, \ y < 0 \quad \text{or} \quad x = 1, \ y > 0, \tag{12.5b}$$

and finally we define

$$\ell(1, 0) = \begin{cases} 0 & \text{if } \gamma \geq 1 \\ \text{as in (12.5a)} & \text{if } \gamma < 1 . \end{cases} \tag{12.5c}$$

12.2. Starting with an Empty System

To compute the probability in (12.3) with $x = 0$, we need to compute the optimal path for the variational problem (12.4), with $x = 0$. This calculation will be useful in light, moderate, and heavy traffic. We shall first solve the variational problem (12.4) corresponding to a model where $\lambda_2(x) = x$ for all $x \geq 0$, i.e., without the upper boundary. Therefore, ℓ is defined through (12.5a) only. Due to (12.5b), if this path does not cross above one before t, it must be optimal for the problem with the boundary in place. This will be the case if $\gamma \leq 1$, and with a little more arguing, we can obtain a solution for $\gamma > 1$. The justification for these formal calculations is given in Lemma 12.6.

Since ℓ does not depend explicitly on time we have from (C.3) along any extremal path r

$$\ell(r, r') - r' \frac{\partial \ell}{\partial r'} = -K \tag{12.6}$$

for some constant K whose value will be determined later (the minus sign is for convenience). By Theorem C.2,

$$\ell(x, y) - y \frac{\partial \ell(x, y)}{\partial y} = -g(x, \theta^*), \tag{12.7}$$

where θ^* is extremizing at x, y in the definitions (5.2)–(5.4). Therefore along extremal paths Theorem C.3 implies that, for some K,

$$g(r, \theta^*(r, r')) = K. \tag{12.8}$$

Exercise 12.3. Show that for this model,

$$\theta^*(x, y) = \log \left(\frac{y + \sqrt{y^2 + 4\gamma x}}{2\gamma} \right) \tag{12.9}$$

$$-g(x, \theta^*(x, y)) = \gamma + x - \sqrt{y^2 + 4\gamma x}. \tag{12.10}$$

Hint: See Exercise 7.24 and hints therein.   ♠

Combining (12.8) with (12.10) we have

$$\gamma + r - \sqrt{(r')^2 + 4\gamma r} = -K$$

and after some algebra,

$$r' = \pm\sqrt{(r + (K - \gamma))^2 + 4K\gamma}. \tag{12.11}$$

Exercise 12.4. For $\gamma \leq 1$ and any extremizing path of (12.4)–(12.5) show that $r'(t) \geq 0$ (for a.e. t). Derive the result also when $\gamma > 1$. Hint: for $\gamma \leq 1$ take $u < v$ with $\int_u^v r'(s)\,ds = 0$. Make a new path by deleting this part of r and inserting a section of length $v - u$ at $r(t) = \gamma$ where the new path is identically γ. For $\gamma > 1$ the new path is at one, due to Equation (12.5).   ♠

Exercise 12.5. If r is optimal for (12.4)–(12.5) then $r(s) \geq z_\infty(s)$, $0 \leq s \leq t$. Hence when $x = 0$, $r'(0) \geq \gamma$, and so $K \geq 0$, and $r'(t) > 0$ for all t. Hint: for the first claim, if $r(s_1) = z_\infty(s_1)$ for some $0 < s_1 < t$, then

$$r_1(u) = \begin{cases} z_\infty(u) & 0 \leq u \leq s_1 \\ r(u) & s_1 \leq u \leq t \end{cases}$$

has lower cost, and satisfies the boundary conditions. For the other claims use Equations (12.8)–(12.11).   ♠

Lemma 12.6. *If $t \leq t_1$ then the solution to the variational problem (12.4)–(12.5) with boundary conditions $r(0) = 0$ and $r(t) = 1$ is given by*

$$r(s) = K(e^s - 1) + \gamma\left(1 - e^{-s}\right)$$

$$K = \frac{1 - \gamma\left(1 - e^{-t}\right)}{e^t - 1} > 0. \tag{12.12}$$

If $\gamma > 1$ and $t \geq t_1$ then $I_0^* = 0$ and $r = \vec{z}_\infty$.

Proof. Since by Exercise 12.4 any extremizing path is increasing, we can assume that ℓ is given by (12.5a) for all x, provided $t < t_1$. The validity of the calculations (12.6)–(12.11) follows from §§C.2 and C.4. By Exercise 12.5, $K \geq 0$ and r' satisfies Equation (12.11) with a positive sign. Since the right-hand side of Equation (12.11) is Lipschitz continuous in r, this equation has a unique solution with $r(0) = 0$, which is therefore the solution of the variational problem by the argument of Theorem C.1. Although we can verify that our proposed solution of the differential equation is correct by substitution, let us develop this formula.

Equation (12.11) defines a hyperbolic sine function. To see this, consider

$$x(s) = A\sinh(s - B) + C$$

$$\Rightarrow \qquad \frac{dx}{ds} = A\cosh(s - B)$$

$$= \sqrt{(x - C)^2 + A^2}$$

since $\cosh^2 y = 1 + \sinh^2 y$. Since K is positive let

$$A^2 = 4K\gamma$$
$$C = \gamma - K.$$

By Exercise 12.5, $r'(0) > 0$ so A is strictly positive. Since

$$r(0) = 0 = C - A \sinh B \quad \text{and} \quad r'(0) > 0$$

we have

$$\sinh B = \frac{C}{A} = \frac{\gamma - K}{\sqrt{4\gamma K}}; \quad \cosh B = \frac{\gamma + K}{\sqrt{4\gamma K}} .$$

Now using the identity

$$\sinh(s - B) = \sinh s \cosh B - \sinh B \cosh s$$

we have

$$\begin{aligned}
r(s) &= \sqrt{4\gamma K} \left[\frac{\gamma + K}{\sqrt{4\gamma K}} \sinh s - \frac{\gamma - K}{\sqrt{4\gamma K}} \cosh s \right] + \gamma - K \\
&= \frac{\gamma + K}{2} \left(e^s - e^{-s}\right) + \frac{K - \gamma}{2} \left(e^s + e^{-s}\right) + \gamma - K \\
&= K \left(e^s - 1\right) + \gamma \left(1 - e^{-s}\right) .
\end{aligned}$$

Finally use $r(t) = 1$ to obtain (12.12).

If $\gamma > 1$ and $t \geq t_1$ then it is trivial to verify that the optimal path is z_∞, with $I_0^* = 0$. ■

Thus we have obtained a solution of the Euler-Lagrange equations that satisfies the boundary conditions and, moreover, established that this is the solution to the variational problem. Let us consider each of the cases, depending on the value of γ, when the system is initially empty.

Exercise 12.7. $I_0^*(t)$ is strictly decreasing in t for $t < t_1$, and bounded below if $\gamma \leq 1$. If $\gamma > 1$ then $I_0^*(t) = 0$ for $t \geq t_1$. Hint: use the patching idea of Exercise 12.4 and the fact that $r'(t) > 0$. ♠

1. Light traffic.

Since

$$z_\infty(t) = \gamma \left(1 - e^{-t}\right) < \gamma < 1,$$

the event $z_n(t) = 1$ is indeed rare. We have

Lemma 12.8. In light traffic, i.e., $\gamma < 1$, the minimal cost path from zero to one in time t is

$$r^*(s) = z_\infty(s) + K \left(e^s - 1\right) = \gamma \left(1 - e^{-s}\right) + K \left(e^s - 1\right) ,$$

where K is given by (12.12) to make $r(t) = 1$. The optimal cost $I^* = I^*(t)$ is given by

$$I^*(t) = \int_0^1 \log \frac{\sqrt{(r - (\gamma - K))^2 + 4\gamma K} + \gamma + r + K}{2\gamma}\, dr - Kt.$$

Remark. We see that $r'(s) > 0$, but as $t \to \infty$, the path $r(s)$ breaks up into two segments: the part that is very nearly $z_\infty(s)$, for s far from t, and the final exponential rise to one near $s = t$.

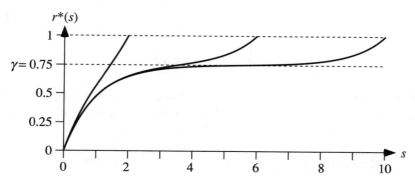

Figure 12.9. The path $r^*(s)$ for $\gamma = 0.75$, $t = 2$, $t = 6$, and $t = 10$.

Proof. The first claim follows from Lemma 12.6. Next, by (12.6), and by (C.6), Theorem C.2,

$$\int_0^t \ell(r(s), r'(s))\, ds = \int_0^t r'(s) \frac{\partial \ell(r(s), r'(s))}{\partial r'}\, ds - Kt$$

$$= \int_0^t r'(s)\theta^*\big(r(s), r'(s)\big)\, ds - Kt.$$

But from (12.9) and (12.11)

$$\theta^*(r, r') = \log\left(\frac{r' + \sqrt{r'^2 + 4r\gamma}}{2\gamma}\right)$$

$$= \log \frac{\sqrt{(r - (\gamma - K))^2 + 4\gamma K} + \gamma + r + K}{2\gamma}. \qquad (12.13)$$

That is, θ^* can be expressed as a function of r alone. We conclude that, since r is monotone,

$$\int_0^t \ell(r(s), r'(s))\, ds = \int_0^1 \theta^*(r)\, dr - Kt.$$

The expression for I_0^* follows from this and (12.13). ∎

We now compute some asymptotics for the optimal cost as t becomes large. Write

$$K = \frac{1 - \gamma \left(1 - e^{-t}\right)}{e^t - 1} = e^{-t} \left(\frac{1}{1 - e^{-t}} - \gamma\right) = e^{-t}(1 - \gamma) + O\left(e^{-2t}\right).$$
(12.14)

As $t \to \infty$, we see that $K \to 0$ and $Kt \to 0$, and so

$$\begin{aligned}
I_0^*(\infty) &= \int_0^1 \log \frac{2\max(\gamma, r)}{2\gamma}\, dr \\
&= \int_\gamma^1 \log \frac{r}{\gamma}\, dr \\
&= \gamma - 1 - \log \gamma.
\end{aligned}$$
(12.15)

We could have derived this from a simpler argument: the cost from zero to γ is zero, since $z_\infty(t)$ goes from 0 to γ. Then for any birth-death process with birth rate $\lambda(x)$ and death rate $\mu(x) > \lambda(x)$, Theorem 11.15 shows

$$\inf\left\{I_0^T(r) : r(0) = a < r(T) = b,\ T > 0\right\} = \int_a^b \log \frac{\mu(x)}{\lambda(x)}\, dx$$

provided $\lambda(x)$ and $\mu(x)$ are continuous and bounded away from zero. In this case the variational problem converges to a limit (as $t \to \infty$)!

Improving the accuracy of the estimate.

Our large deviations calculation of $I_0^*(t)$ enables us to estimate $\mathbb{P}_0\left(z_n(t) = 1\right)$ for every $t > 0$, and it also enables us to estimate the steady-state probability $\mathbb{P}_{ss}\left(z_n(t) = 1\right)$, since by Theorem 6.89 we expect

$$\mathbb{P}_{ss}\left(z_n(t) = 1\right) \approx \exp\left(-n \inf_t I_0^*(t)\right) = \exp\left(-n I_0^*(\infty)\right).$$
(12.16)

Now suppose that through some other means we have an estimate of the steady-state probability; for one approach to this problem, see [Jag]. Let us call the more accurate expression $\mathbb{P}_g(z_n = 1)$, the g standing for "good." Then we can improve our estimate of $\mathbb{P}_0(z_n(t) = 1)$ by the following expedient:

$$\mathbb{P}_0(z_n(t) = 1) \approx e^{-n\left(I_0^*(t) - I_0^*(\infty)\right)} \mathbb{P}_g(z_n = 1)$$
(12.17)

and have our answer be guaranteed accurate as t gets large. This equation is a bit of a cheat, since we have no estimate on the term $o(n)$ in

$$\mathbb{P}_x(z_n(t) = 1) = e^{-n I_0^*(t) + o(n)}.$$
(12.18)

That is, we do not have a proof that (12.17) is more accurate at any particular values of n and t than (12.18). Nevertheless, it is obviously more accurate at large enough values of t, and we have found it to be a good heuristic method for improving the accuracy of large deviations approximations.

We conclude the discussion of the light-traffic case with an explicit approximation for (12.17), as t becomes large. From Lemma 12.8 and (12.15) we have

$$I_0^*(t) - I_0^*(\infty) = -Kt$$

$$+ \int_0^1 \log \left(\frac{\sqrt{K^2 + 2K(\gamma + r) + (r - \gamma)^2} + \gamma + r + K}{2 \max(\gamma, r)} \right) dr. \quad (12.19)$$

We break up the integral into $\int_0^\gamma + \int_\gamma^1$. Then using

$$\int \sqrt{(x^2 + a^2)}\, dx = \frac{1}{2} x \sqrt{x^2 + a^2} + \frac{a^2}{2} \log \left(\frac{x}{a} + \sqrt{\frac{x^2}{a^2} + 1} \right)$$

$$\int \frac{dx}{\sqrt{x^2 + a^2}} = \log \left(x + \sqrt{x^2 + a^2} \right)$$

$$= \log \left(\frac{x}{a} + \sqrt{\frac{x^2}{a^2} + 1} \right)$$

$$\sqrt{1 + x} = 1 + \frac{x}{2} - \frac{1}{8} x^2 + O\left(x^3\right)$$

$$\log(1 + x) = x - \frac{x^2}{2} + O\left(x^3\right)$$

and a lot of algebra, we find that when t is large (so K is small)

$$I_0^*(t) - I_0^*(\infty) \approx K + K \log \frac{1}{K} + K \log(1 - \gamma) - Kt + O\left(K^{3/2}\right).$$

Now substituting (12.14) for K, we obtain

$$I_0^*(t) - I_0^*(\infty) \approx (1 - \gamma)e^{-t}.$$

Therefore,

$$\mathbb{P}_0(z_n(t) = 1) \approx e^{-n(1-\gamma)e^{-t}} \mathbb{P}_g(z_n = 1)$$

$$= \exp \left(-e^{-(t - \log(n(1-\gamma)))} \right) \mathbb{P}_g(z_n = 1).$$

That is, the transient behavior follows the function $e^{-e^{-t}}$, with a time shift of $\log(n(1 - \gamma))$. Thus, using (12.15) as our estimate for $\mathbb{P}_g(z_n = 1)$,

$$\mathbb{P}_0(z_n(t) = 1) \approx \exp \left(-e^{-(t - \log(n(1-\gamma)))} \right) \exp(-n(\gamma - 1 - \log \gamma)). \quad (12.20)$$

Exercise 12.10. Can you evaluate the integral 12.19 explicitly? How accurate is our asymptotic evaluation of this integral? (See [Kn, pp. 769–770].) ♠

2. Moderate traffic ($\gamma = 1$).

The transient analysis is quite similar here to the case of light traffic. We have

$$z_\infty(t) = \gamma \left(1 - e^{-t}\right)$$
$$= 1 - e^{-t} \qquad 0 \le t < \infty$$

and so $I_0^*(\infty) = 0$, since $z_\infty(\infty) = 1$. The variational problem and solution are the same as in the light traffic case, so the optimal path is

$$r(s) = K(e^s - 1) + 1 - e^s, \qquad 0 \le s \le t$$

but

$$K = \frac{1 - \left(1 - e^{-t}\right)}{e^t - 1} = \frac{e^{-t}}{e^t - 1} = e^{-2t} + O\left(e^{-3t}\right).$$

The cost $I_0^*(t)$ is a bit different: we no longer break up the integral into $\int_0^\gamma + \int_\gamma^1$, since already $\gamma = 1$. We find

$$I_0^*(t) = \frac{K}{2} + \frac{K}{2} \log \frac{1}{K} - Kt + O\left(K^{3/2}\right),$$

so $I_0^*(t) \approx \frac{1}{2} e^{-2t}$, and $I_0^*(\infty) = 0$. Therefore, using (12.17) we find

$$\mathbb{P}_0(z_n(t) = 1) \approx \exp\left(-\frac{n}{2} e^{-2t}\right) \mathbb{P}_g(z_n = 1)$$
$$= \exp\left(-e^{-2\left(t - \frac{1}{2} \log \frac{n}{2}\right)}\right) \mathbb{P}_g(z_n = 1).$$

There is a noticeable difference between (12.17) and (12.18) in this case. Using (12.18) we find

$$\mathbb{P}_{ss}(z_n(t) = 1) = e^{-nI_0^*(\infty) + o(n)} = e^{o(n)}.$$

That is, using large deviations alone to estimate the steady-state blocking probability, we simply have to say that it does not shrink exponentially quickly in n, but we have no further information. But steady-state analysis (see, e.g., [Jag]) gives us

$$\mathbb{P}_g(z_n = 1) = \frac{C}{\sqrt{n}} + O\left(\frac{1}{n}\right).$$

In any case, we find the functional form for the approach to steady state, namely $e^{-e^{-t}}$, but this time with a time shift of $\frac{1}{2} \log \frac{n}{2}$ and an acceleration factor (multiplier of t) of 2.

3. Heavy traffic.

When $\gamma > 1$, $z_\infty(t)$ reaches one in finite time:

$$z_\infty(t_1) = \gamma \left(1 - e^{-t_1}\right) = 1$$

at $t_1 = \log \frac{\gamma}{\gamma-1}$. Then for $t < t_1$, $I_0^*(t) > 0$ and for $t > t_1$, $I_0^*(t) = 0$. Therefore *without further calculation* we can see that the approach to steady state is nearly a step function at t_1:

$$\mathbb{P}_0(z_n(t) = 1) \approx \begin{cases} e^{-nI_0^*(t)} \mathbb{P}(z_n(\infty) = 1) \approx 0 & t < t_1 \\ \mathbb{P}_{ss}(z_n(t) = 1) \ (\approx 1 - \frac{1}{\gamma}) & t \geq t_1 \ . \end{cases} \tag{12.21}$$

(More discussion of this equation appears below.) We can write (12.21) in terms of the function $e^{-e^{-t}}$ as follows:

$$\mathbb{P}(z_n(t) = 1) \approx e^{-e^{-\infty(t-t_1)}} \mathbb{P}(z_n(\infty) = 1) \ .$$

We can be a little more accurate than this using some calculation. Since $\gamma > 1$ we have $\gamma - r > 0$ so that, using a direct approximation for K small and $t < t_1$,

$$\begin{aligned} I_0^*(t) &= \int_0^1 \log \frac{\sqrt{K^2 + 2K(r+\gamma) + (\gamma-r)^2} + \gamma + r + K}{2\gamma} \, dr - Kt \\ &\approx \int_0^1 \log \left(1 + \frac{K}{2\gamma} \left(1 + \frac{\gamma + r}{\gamma - r}\right)\right) dr - Kt \\ &\approx K \left[\log \frac{\gamma}{\gamma - 1} - t\right] \ . \end{aligned}$$

Now K is small when t is near t_1. Specifically, if $t = t_1 - \delta$,

$$K = \frac{1 - \gamma\left(1 - e^{-(t_1-\delta)}\right)}{e^{t_1 - \delta} - 1} = \frac{1 - \gamma + \gamma e^{\delta} \frac{\gamma-1}{\gamma}}{e^{-\delta} \frac{\gamma}{\gamma-1} - 1} \approx \delta(\gamma - 1)^2$$

(using $e^{\delta} \approx 1 + \delta$), so, for $t \approx t_1, t < t_1$,

$$I_0^*(t) \approx \delta^2(\gamma - 1)^2 = (t - t_1)^2(\gamma - 1)^2,$$

$$\mathbb{P}_0(z_n(t) = 1) \approx e^{-n(\gamma-1)^2(t-t_1)^2} \mathbb{P}_{ss}(z_n(t) = 1).$$

That is, the "step function" looks like a Gaussian with width of order $1/\sqrt{n}$.

12.3. Starting with a Full System in Light Traffic

We saw in the last section that when $\gamma > 1$ and $z_n(0) = 1$, the system instantly achieves steady state, at least as far as $\mathbb{P}_1(z_n(t) = 1)$ is concerned. Large deviations is also too crude to give any detail about $\mathbb{P}_1(z_n(t) = 1)$ when $\gamma = 1$; diffusion approximations are more appropriate there (see, e.g., [Kn, Har]). So in this section we concentrate exclusively on $\gamma < 1$. We carry out formal calculations first, relying on heuristics and guesses, and obtain explicit results. The justification for these steps is given in Theorem 12.13.

To begin, by Theorem 12.17 we have

$$z_\infty(t) = \left(1 - e^{-t}\right)\gamma + e^{-t}, \tag{12.22}$$

so having $z_n(t) = 1$ is indeed a rare event. By Theorem 12.18 we have

$$\mathbb{P}_1(z_n(t) = 1) \approx e^{-nI_1^*(t)}$$

$$I_1^*(t) = \inf \left\{ \int_0^t \ell(r(t), r'(t))\, dt : r(0) = 1,\ r(t) = 1 \right\}$$

with $\ell(x, y)$ as defined in (12.5). The variational problem is almost the same as before, but we no longer have a monotone function $r^*(t)$, since $r(0) = r(t) = 1$. We have, as in the previous section,

$$\ell\left(r, r'\right) - r'\frac{\partial \ell}{\partial r'} = K \tag{12.23}$$

for some constant K, where we have chosen to put $+K$ (compare to Equation (12.6)) as we anticipate the right-hand side of Equation (12.23) being positive. As in (12.7)–(12.11) we obtain

$$\ell - r'\frac{\partial \ell}{\partial r'} = \gamma + r - \sqrt{(r')^2 + y\gamma r}$$

$$r' = \pm\sqrt{(r - (K + \gamma))^2 - 4K\gamma}\,. \tag{12.24}$$

This is the equation of a hyperbolic cosine:

$$x = A \cosh(s - B) + C$$

$$\frac{dx}{ds} = A \sinh(s - B) = \pm\sqrt{(x - C)^2 - A^2}\,, \tag{12.25}$$

where the sign is determined by the parameters. Now the function $\cosh s$ is convex and symmetric around zero. To keep the path below one we need to start down at $s = 0$, and return to one at $s = t$. This dictates $A > 0$ and $B > 0$. So let

$$A = \sqrt{4K\gamma},\quad C = K + \gamma\,. \tag{12.26}$$

Then setting $r(0) = 1$ gives

$$A \cosh B + C = 1$$

$$\cosh B = \frac{1 - C}{A}\,.$$

The condition $r(t) = 1$ and the symmetry of $\cosh s$ make $t = 2B$. Then using $2(\cosh x)^2 = \cosh 2x + 1$,

$$\cosh B = \cosh\frac{t}{2} = \sqrt{\frac{\cosh t + 1}{2}} = \frac{1 - C}{A} = \frac{1 - K - \gamma}{\sqrt{4K\gamma}}. \qquad (12.27)$$

With $D = (1 + \cosh t)/2$, we obtain a quadratic equation for K, with solution

$$K = 1 - \gamma + 2\gamma D - \sqrt{(1 - \gamma + 2\gamma D)^2 - (1 - \gamma)^2}, \qquad (12.28)$$

where the $(-)$ sign is chosen since the left-hand side is positive so that necessarily $K < 1 - \gamma$.

We see that K is a decreasing function, converges to zero as $t \to \infty$ and that $K = (1 - \sqrt{\gamma})^2 + O(t^2)$ for small t. The minimum of the curve, achieved at $s = B = t/2$, is $C + A$, and it converges to γ as $t \to \infty$.

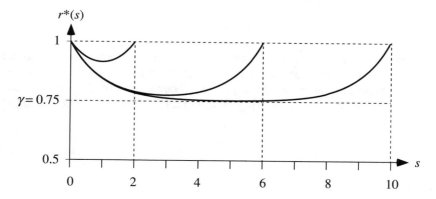

Figure 12.11. The path $r^*(t)$ for $\gamma = 0.75$, $t = 2$, $t = 6$, and $t = 10$.

The value of the I-function must be computed carefully. Using (12.23) and breaking the path into monotone parts,

$$\begin{aligned}
I_0^t(r^*) &= \int_0^t \ell(r, r')\, dt \\
&= \int_0^t r'\frac{\partial \ell}{\partial r'}\, dt + Kt \qquad (12.29) \\
&= \int_1^{C+A} \theta_-\, dr + \int_{C+A}^1 \theta_+\, dr + Kt.
\end{aligned}$$

Here θ_- is obtained from (12.9) by taking $r' < 0$ in (12.24), and θ_+ by taking

$r' > 0$. By (12.9) we have

$$\theta_\pm = \log \frac{r' + \sqrt{(r')^2 + 4\gamma r}}{2\gamma}$$

$$= \log \frac{\pm\sqrt{(r - (K + \gamma))^2 - 4K\gamma} + \sqrt{(r - (K + \gamma))^2 - 4K\gamma + 4\gamma r}}{2\gamma}$$

$$= \log \frac{\pm\sqrt{(r - (K + \gamma))^2 - 4K\gamma} + r + \gamma - K}{2\gamma}$$

$$\text{(12.30)}$$

since $r + \gamma - K \geq C + A + \gamma - K = 2\gamma + \sqrt{4K\gamma} > 0$. To summarize, for given $t > 0$, we find $K(t)$ by (12.28), then calculate A and C from (12.26). We have $r^*(s) = A \cosh(s - t/2) + C$, and I_1^* given by (12.29). Note that the minimum value of $r^*(t)$, $A + C$, decreases as t increases, and that for large t, r^* nearly decomposes into $z_\infty(s)$ near $s = 0$, and the time reversal of that near $s = t$.

Exercise 12.12. Show that $I^*(t) \approx t \left(1 - \sqrt{\gamma}\right)^2$ as $t \to 0$ and, as $t \to \infty$,

$$I^*(t) - I^*(\infty) \approx e^{-t} t \frac{(1 - \gamma)^2}{\gamma} \left(1 + \frac{1 - \gamma}{2}\right).$$

Hints: you don't need to do any calculation for this as soon as you realize that $r^*(s)$ is very nearly one for all $s \in [0, t]$ when t is small. Find an expression for K as a function of t for large t from (12.28). Then plug into (12.29) in a manner similar to the calculations appearing below (12.19). You may need the approximation

$$C + A \approx \gamma + 2(1 - \gamma)e^{-t/2} + \frac{(1 - \gamma)^2}{\gamma} e^{-t} + O\left(t^{-3/2}\right). \qquad \text{(12.31)}$$

Don't forget that $\log(1 + x) \approx 1 + x - x^2/2$!   ♠

12.4. Justification

There are two unrelated items that need justification in this (and any application) chapter.

1) The probability in question [here $\mathbb{P}(z_n(t) = 1)$] is approximable by the solution to a variational problem.

2) Our calculations produced the solution to the variational problem.

We provide the justification of 1) in §12.6. This section contains a proof of the following theorem, which justifies 2):

Theorem 12.13. *Consider the variational problem of obtaining I_1^*, where ℓ is given in (12.5a)–(12.5c). Then the unique optimal path is given by Equations (12.25)–(12.28), and I_1^* is given by (12.29)–(12.30).*

To prove this theorem we must show that

1) Solutions of the variational problem satisfy the differential equation (12.23).

2) Our stated solution of the differential equation is the correct (minimal) solution of the variational problem.

Our approach to the last point is to show that the solutions to the differential equation (12.23) are of a very limited variety, and to examine each possible solution. Let's get to it.

The main new idea is to examine the problem by removing the upper boundary (at $x = 1$). That is, we consider a modified process that has arrival rate γ for all positive x, and has service rate x for all positive x. (We actually modify the service rate so that it is bounded for $x > 2$, but this is just another technicality.) We show that our solution is minimal for the modified process. Therefore it must be optimal for the actual process $z_n(t)$. The advantage of the modified process is that our theorems of Appendix C are specifically tailored for processes without boundaries, so we modify the process in order to prove that the solution to the variational problem must satisfy certain differential equations.

To show that optimal paths for the modified process are given by our solution to (12.23), we first show that every optimal path from one to one in time t never reaches the level $2 - \gamma$. This is because, by the argument of Theorem 11.15, the cheapest path from 1 to $2 - \gamma$ has cost

$$I = \int_1^{2-\gamma} \log \frac{x}{\gamma} \, dx > \int_\gamma^1 \log \frac{x}{\gamma} \, dx. \qquad (12.32)$$

(The inequality is obvious since the two integrals are over the same distance, but the second integrand is pointwise smaller than the first.) Now the cost of an optimal path from one to one in time t is smaller than the cost of the cheapest path from γ to one in time t, since an optimal path must lie above z_∞ (going lower would cost something, but would have no benefit, since the path must eventually cross back above z_∞ to reach level one by time t). But $z_\infty(s) > \gamma$ for all $s > 0$ when $z_\infty(0) = 1$. Also, by Theorem 11.15, the cheapest path from γ to one has cost $\int_\gamma^1 \log \frac{x}{\gamma} \, dx$. Thus we have shown that for the modified process, all cheapest paths from one to one over a time t lie below the level $2 - \gamma$.

Now Theorems C.13 and C.18 apply to the modified process, since this process has smooth bounded jump rates. These theorems state that the DuBois-Reymond condition holds; that is, that the differential equation (12.23) holds over every interval where the optimum $r^*(s)$ satisfies $\frac{d}{ds} r^*(s) \neq 0$, and furthermore that $\frac{d}{ds} r^*(s)$ is continuous. By the calculations following (12.23), each segment of $r^*(s)$ must be a hyperbolic cosine, hyperbolic sine, or interval where $r' = 0$. The segments can be pieced together only at points where $r' = 0$. It is thus clear that any path from one to one cannot have any segments composed of hyperbolic sinusoids, since these paths are monotone and never have $r' = 0$. This means that K must be positive. Similarly, segments of hyperbolic cosines must have their axis of symmetry between zero and t, or else these too would be monotone and would not have $r' = 0$ at any intermediate point. Thus we have reduced our considerations to hyperbolic cosines that have an axis of symmetry in $(0, t)$ and to straight line segments where $r' = 0$. We eliminate hyperbolic cosines that lie above the level $r = 1$ by showing that any locally optimal path above $r = 1$ actually costs more than the path $r \equiv 1$. Then we show that optimal paths cannot have any straight line segments. The idea here is pretty obvious: Euler's equation shows that straight lines are not optimal, so we just compare the straight line to a nearby curve that is more like the hyperbolic cosine. Now we continue with the statements and proofs outlined here.

Claim. *If* $r'(0) > 0$, *then* r *is not optimal.*

Proof. We show that the hyperbolic cosine $r(s)$ that starts at one and curves downward ($r'(0) > 0, r'' < 0$) is more costly than the straight line path $r_1(s) \equiv 1$, $0 \le s \le t$. First note that, since $\theta^*(x, y)$ is monotone in y for every x (see Appendix C),

$$I_0^t(r) = \int_1^{C+A} (\theta_+ - \theta_-)(r) \, dr + Kt > Kt.$$

Now the straight line path $r_1(s)$ has cost $(1 - \sqrt{\gamma})^2$. The hyperbolic cosine path has [by (12.28), noting that we must have the positive sign on the square root] $K \ge 1 + \gamma$. Therefore

$$I_0^t(r) > Kt > (1 + \gamma)t > I_0^t(r_1). \qquad \blacksquare$$

Claim. *No optimal path has* $r' = 0$ *over an interval of time.*

Proof. There are three cases to consider: $r > 1, r = 1$, and $r < 1$. The first case costs more than the second for any time interval. Therefore it cannot be worthwhile to hold at any level above one, since it would have been cheaper to spend the time holding at one. We analyze the second and third cases together. First note that every candidate optimal path remains above γ, so we really only have to consider $\gamma < r \le 1$. Now consider the problem of the cheapest cost path from x to x in time t, where $\gamma < x \le 1$. We show that $r \equiv x$ is not optimal by showing that changing the path in the direction of the hyperbolic cosine path results in lower cost. That is, the straight line does not satisfy the Euler equations, and we make a variation in a direction we suspect will have a lower cost, and lo and behold, it does:

$$\frac{d}{d\varepsilon} \int_0^t \ell\Big(x + \varepsilon(x - r(s)), [x + \varepsilon(x - r(s))]'\Big) \, ds \bigg|_{\varepsilon=0}$$

$$= \int_0^t (x - r(s)) \left(\frac{\partial}{\partial r} \ell(x, 0) - \frac{d}{ds} \frac{\partial}{\partial r'} \ell(x, 0) \right) ds$$

$$= \int_0^t (x - r(s)) \left(\frac{(\sqrt{x} - \sqrt{\gamma})}{\sqrt{x}} + 0 \right) ds$$

$$< 0$$

since $x - r(s) < 0$ for $r \in (0, t)$ and $x > \gamma$. Therefore optimal paths do not hold at any level above γ.   ∎

This concludes our proof of the validity of our solution of the variational problem for paths that go from one to one in time t.

12.5. Erlang's Model: General Starting Point

Suppose now that $z_\infty(0) = x \in (0, 1)$ and we are given a $T > 0$ with the task of estimating $\mathbb{P}_x(z_n(T) = 1)$. The thing to notice is that if, for some previously chosen t with its associated path r^*, we have

$$r^*(s_1) = x$$
$$r^*(t) = 1$$
$$t - s_1 = T$$

and r^* is the unique minimizer of the problem on $[0, t]$, then

$$r(s) = r^*(s + s_1) \tag{12.33}$$

is the (unique) optimal path for the problem of minimizing I over paths that start at x and reach one in time T. This is a special application of the general principle of optimality: solve one variational problem, then see if any part of your extremal fits another problem. See Theorem C.12.

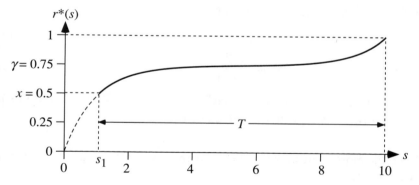

Figure 12.14. A minimal path from zero to one, with the bold portion the minimal path from x to one in time T. Here $x = 0.5$, $\gamma = 0.75$, and $t = 10$, implying that $s_1 = 1.0979$ and so $T = t - s_1 = 8.9021$.

For Erlang's model we are fortunate enough to have a complete solution at hand: the extremals from $x = 0$ and $x = 1$ cover the rest. Let's see how.

1) $\gamma > 1$.

The longest time extremal path here is $z_\infty(t)$;

$$z_\infty(t) = \gamma \left(1 - e^{-t}\right) \qquad 0 \le t \le \log \frac{\gamma}{\gamma - 1}$$

so for given $x \in (0, 1)$, $z_\infty(t) = x$ at $t_x = \log \frac{\gamma}{\gamma - x}$. Therefore, given $x \in (0, 1)$, we will be able to find an extremal that goes from x to one in time t for any $t < \log \frac{\gamma}{\gamma - 1} - \log \frac{\gamma}{\gamma - x} = \log \frac{\gamma - x}{\gamma - 1}$.

We have
$$r^*(s) = \gamma \left(1 - e^{-s}\right) + K \left(e^s - 1\right)$$
$$K(t) = \frac{1 - \gamma \left(1 - e^{-t}\right)}{e^t - 1}.$$

Now $r^*(s_1) = x$ can be solved for $s_1(x, t)$, and then we have to find t such that
$$T = t - s_1(x, t). \tag{12.34}$$

As t shrinks to zero, K increases, and the interval between when $r^*(s) = x$ and $r^*(s) = 1$ shrinks monotonically to zero. This implies that there is a unique $t(T)$ solving (12.34). Given t, we find $I^* = \int_{s_1}^{t} \ell \left(r^*, \frac{d}{ds} r^*\right) ds$.

2) $\gamma = 1$.

Again the set of extremals from the initial condition $x = 0$ covers the rest of the cases. We may choose any $t \in (0, \infty)$, and find that for each $x \in (0, 1)$, the time $T = t - s(x, t)$ is monotone increasing in t. We have for each $x \in (0, 1)$,
$$\lim_{t \downarrow 0} T(t) = 0, \qquad \lim_{t \uparrow \infty} T(t) = \infty.$$

Therefore there is a unique extremal of the form
$$r^*(s) = \left(1 - e^{-s}\right) + K \left(e^s - 1\right) \tag{12.35}$$
$$K = \frac{e^{-t}}{e^t - 1} \tag{12.36}$$

that goes from x (at time s_1) to one (at time t) with $t - s_1 = T$. For this path, $I_1^* = \int_x^1 \theta(r) dr - KT$, with $\theta(r)$ given by (12.9).

3) $\gamma < 1$.

This is the most interesting situation. There are several cases, with a bifurcation of the solution!

A. $x \leq \gamma$.

Here the extremals from $r^*(0) = 0$ cover the cases. The calculations are the same as the previous two cases.

B. $x \geq \gamma$.

1. $T < \log \frac{1-\gamma}{x-\gamma}$.

Here again the extremals from $r^*(0) = 0$ cover the cases. The break point $T = \log \frac{1-\gamma}{x-\gamma}$ is obtained by setting $z_\infty(T) = x$ with $z_\infty(0) = 1$. This is the longest time an upcrossing from x to one can take with $r(0) = 0$, since the limiting case of $t \to \infty$ for the problem of crossing from zero to one in time t gives a path

$r^*(s)$ that decouples into a rise from zero to γ along $z_\infty(s)$, then a final rise near t that is the time-reversal of $z_\infty(s)$ with z_∞ starting at one.

2. $T > \log \frac{1-\gamma}{x-\gamma}$.

Here the extremals from $r^*(0) = 1$ cover all the cases. There is only one thing to notice. Let $r_x(t)$ refer to that path that achieves its minimum at x (i.e., in the notation of Section 12.3, $C + A = x$ for this path). If $T \le t/2$, then we use a strictly increasing path. If $T > t/2$, we use a path that first decreases, then increases. See Figure 12.15 below for the various types of paths. The interested student might wish to find explicit expressions for r^* as a function of T, and formal expressions for I^* as a function of this r^*.

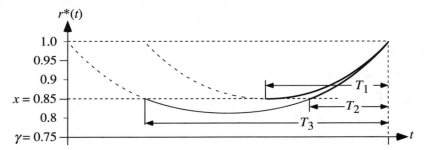

Figure 12.15. $r^*(t)$ for $\gamma = 0.75$, $x = 0.85$. $T_1 = 1.5877$, the bifurcation value. T_2 and T_3 are 1.0213 and 3.1541, respectively, for the curve $t = 4.1754$.

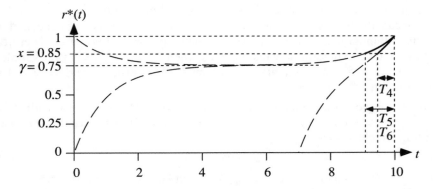

Figure 12.16. $r^*(t)$ for $\gamma = 0.75$, $x = 0.85$. $T_4 = 0.5279$ for $t = 3$. T_5 and T_6 are 0.9155 and 0.91655 for the curves starting at zero and one respectively. These two curves have $t = 10$; in the limit $t \to \infty$ their final portions agree with each other and with $z_\infty(-t)$, with $T = 0.91629 = \log\left(\frac{1-\gamma}{x-\gamma}\right)$.

12.6. Large Deviations Theory

In this section we show that both Kurtz's Theorem 5.3 and the large deviations principle apply to Erlang's model, and we identify the rate function. The reason we have to work at this is that the assumptions of Chapter 5 are not satisfied by our model at the points zero (where $\lambda_2(x) = 0$) and one (where $\lambda_1(x)$ is discontinuous). Both theorems are justified by using very similar methods. We construct processes for which the theorems are known to hold, and use them to prove the theorems for Erlang's model by approximation arguments. The discontinuity at one can be treated via the methods of Chapter 8, and there are fairly general methods that can handle some discontinuities of this type. However, no general methods exist to deal with rates that go to zero, and we work hard to verify the results, using methods that are specific to this model.

Kurtz's Theorem.

Define z_∞ through (12.2) with $z_\infty(0) = x$.

Theorem 12.17. *Given any $\varepsilon > 0$, $T > 0$, and $x \in [0, 1]$, there are numbers $C_1 > 0$ and $C_2 > 0$ such that*

$$\mathbb{P}_x \left(\sup_{0 \le t \le T} |z_n(t) - z_\infty(t)| > \varepsilon \right) < C_1 e^{-nC_2}.$$

Proof. Kurtz's Theorem 5.3 applies for $0 \le x < 1$, as long as z_∞ is not close to one. Clearly when $\gamma \le 1$ the process $z_\infty(t)$ never hits one if $z_\infty(0) < 1$. This means the only cases not covered by Theorem 5.3 are when $\gamma > 1$ or $x = 1$. In fact, we can always break the path z_∞ into two parts, one away from one, and one near that boundary, and prove the result separately for the two parts. Therefore, we simply have to determine what happens when $z_\infty = 1$. This is very simple to calculate: for $\gamma \ge 1$ we compare z_∞ with an $M/M/1$ process, and for $\gamma < 1$ we compare with a free process.

For $\gamma \ge 1$ we compare the process $1 - z_n(t)$ to an $M/M/1$ queue with arrival rate one and service rate γ; we call a scaled version of this process $w_n(t)$. Using coupling we clearly have

$$1 - z_n(t) \le w_n(t), \qquad (12.37)$$

since $1 - z_n(t)$ has arrival rate $\lambda = z_n(t) \le 1$ and has service rate γ. Now we use the $M/M/1$ version of Kurtz's Theorem, 11.6, to conclude that

$$\mathbb{P}_0 \left(\sup_{0 \le t \le T} w_n(t) \ge \varepsilon \right) \le C_1 e^{-nC_2}. \qquad (12.38)$$

This immediately implies that

$$\mathbb{P}_1 \left(\inf_{0 \le t \le T} z_n(t) \le 1 - \varepsilon \right) \le C_1 e^{-nC_2}. \qquad (12.39)$$

For $\gamma < 1$ we can use similar coupling arguments to show that the process must leave a neighborhood of $x = 1$ quickly and very near z_∞; the details are left to the reader. This completes the proof of Kurtz's Theorem. ∎

The large deviations principle.

While the statement and proof of the large deviations principle apply only to the basic Erlang's model of §12.1, you will be able to use the idea to prove the principle for other models developed in later sections and later chapters.

Theorem 12.18. *Fix $T > 0$ and define I_0^T through Equations (12.5) and (5.5). Then I_0^T is a good rate function. For any open set of paths $G \subset D^1[0, T]$, and any closed set of paths $F \subset D^1[0, T]$, and any $x \in [0, 1]$ we have*

$$\limsup_{n \to \infty} \frac{1}{n} \log \mathbb{P}_x \left(z_n \in F \right) \leq - \inf \left\{ I_0^T(r) : r \in F, \ r(0) = x \right\}.$$

Furthermore, uniformly over $x \in [0, 1]$ we have

$$\liminf_{n \to \infty} \frac{1}{n} \log \mathbb{P}_x \left(z_n \in G \right) \geq - \inf \left\{ I_0^T(r) : \vec{r} \in G, \ \vec{r}(0) = \vec{x} \right\}.$$

Proof. Let $_1 I_0^T$ denote the rate function defined through Equation (12.5a), with $\ell_1(x, y) \stackrel{\triangle}{=} \ell(1, y)$ for $x > 1$. Then

$$\left\{ r : I_0^T(r) \leq \alpha \right\} = \left\{ r : {}_1 I_0^T(r) \leq \alpha \right\} \bigcap \left\{ r : \sup_{0 \leq t \leq T} r(t) \leq 1 \right\}.$$

The first set on the right is compact by Proposition 5.46, while the second is closed. Therefore the intersection is a compact set, so that I_0^T has compact level sets. In particular, the level sets are closed, so that I_0^T is lower semicontinuous. So, it is a good rate function.

The upper and lower bounds are established in Chapter 8 for sets of paths that stay away from $x = 0$. To prove the upper bound for sets that include $x = 0$, consider a modified process $z_n^\varepsilon(t)$ defined by the following jump directions and rates:

$$e_1 = 1 \qquad \lambda_1^\varepsilon(x) = \begin{cases} \gamma & x < 1; \\ 0 & x = 1. \end{cases}$$

$$e_2 = -1 \qquad \lambda_2^\varepsilon(x) = \begin{cases} x & x \geq \varepsilon; \\ \varepsilon & 0 < x \leq \varepsilon; \\ 0 & x = 0. \end{cases}$$ (12.40)

We construct a coupling between the processes $z_n(t)$ and $z_n^\varepsilon(t)$ as follows. Suppose $z_n(t)$ is defined as before. We also define an independent Poisson process $N(t)$ with rate ε. Now z_n^ε [as well as $z_n(t)$] is defined to start at x, and to have the same jumps as $z_n(t)$. We then modify z_n^ε as follows. If $N(t)$ makes a jump at time t, and if $z_n^\varepsilon(t)$ is below ε, then with probability $\varepsilon - z_n^\varepsilon(t)$, we induce a jump down. This makes the jump rates of z_n^ε as specified in (12.40). From the time of the first jump of $N(t)$, the processes are no longer coupled, and we can construct $z_n^\varepsilon(t)$ any way we like (this part of the construction doesn't figure into the bound we use).

The process $z_n^\varepsilon(t)$ has the following properties. It is a jump Markov process for which the log rates are bounded and Lipschitz continuous in $(0, 1)$, and so the theory of Chapter 8 applies. The process z_n^ε agrees pathwise with $z_n(t)$ at least until the first time that $N(t)$ makes the first jump. Therefore, for any $x \in [0, 1]$ we have (with probability one)

$$\mathbb{P}_x\left(z_n(t) = z_n^\varepsilon(t),\ 0 \le t \le T \mid z_n(t),\ 0 \le t \le T\right) \ge e^{-nT\varepsilon}$$

since the event that $N(t)$ makes no jumps in $[0, T]$ is independent of $\{z_n(t),\ 0 \le t \le T\}$, and its probability is at least $e^{-nT\varepsilon}$. This immediately implies that for any set of paths S,

$$\mathbb{P}\left(z_n^\varepsilon \in S\right) \ge e^{-nT\varepsilon}\mathbb{P}\left(z_n \in S\right). \tag{12.41}$$

Therefore, we have the following upper bound, for each ε:

$$\limsup_{n\to\infty} \frac{1}{n}\log \mathbb{P}_x\left(z_n \in F\right) \le -\inf\left\{{}_\varepsilon I_0^T(r) : r \in F,\ r(0) = x\right\} + \varepsilon T, \tag{12.42}$$

where ${}_\varepsilon I_0^T$ is the rate function for z_n^ε, defined in the usual way through the local rate function ℓ_ε. That is, the large deviations upper bound for $z_n^\varepsilon(t)$ provides a large deviations upper bound for $z_n(t)$. To see that (12.42) gives the correct upper bound, we need to show that

$$\limsup_{\varepsilon\to 0} \inf\left\{{}_\varepsilon I_0^T(r) : r \in F,\ r(0) = x\right\} \ge \inf\left\{I_0^T(r) : r \in F,\ r(0) = x\right\}. \tag{12.43}$$

We show that (12.43) is correct by constructing a sequence of functions $v_\varepsilon(t)$ that have a limit $v(t)$ as $\varepsilon \to 0$, and also have

$$\limsup_{\varepsilon\to 0} \inf\left\{{}_\varepsilon I_0^T(r) : r \in F,\ r(0) = x\right\} \ge \limsup_{\varepsilon\to 0} {}_\varepsilon I_0^T(v_\varepsilon)$$

$$\ge \inf\left\{I_0^T(r) : r \in F,\ r(0) = x\right\}.$$

Since ${}_\varepsilon I_0^T$ is a good rate function, there are r_ε that achieve the infimum in (12.42) at each ε. We construct the functions v_ε from any such sequence $r_\varepsilon(t)$, as follows. We want $v_\varepsilon(t) \ge \varepsilon$ for all t. First we modify $r_\varepsilon(t)$ near the endpoints ($t = 0$ and $t = T$) to obtain the intermediate function u_ε. If $x < \varepsilon$ then let $u_\varepsilon(0) = \varepsilon$ and let u_ε go down with slope $(-\gamma)$ until it intersects r_ε. Similarly, if $r_\varepsilon(T) < \varepsilon$ then set $u_\varepsilon(T) = \varepsilon$ and let u_ε go up linearly from r_ε to ε with slope γ.

To obtain an upper bound on the error due to this modification, we use the representation (5.20)–(5.22) of ℓ_ε, established in Theorem 5.26. For the initial segment, where $y = \gamma$, choose $\mu_1 = \gamma$ and $\mu_2 = 0$ to obtain

$$\ell_\varepsilon(x, y) \le \gamma + \varepsilon - \gamma + \gamma \log \frac{\gamma}{\gamma} + 0 = \varepsilon$$

for all $x \le \varepsilon$. For the final segment, where $y = -\gamma$, choose $\mu_1 = 0$ and $\mu_2 = \gamma$ to obtain

$$\ell_\varepsilon(x, y) \le \gamma + \varepsilon - \gamma + 0 + \gamma \log \frac{\gamma}{\varepsilon} = \varepsilon + \gamma \log \frac{\gamma}{\varepsilon}$$

for all $x \le \varepsilon$. Since each modification is on an interval of length at most ε/γ, we have

$$\varepsilon I_0^T (r_\varepsilon) \ge \varepsilon I_0^T (u_\varepsilon) + 2\frac{\varepsilon}{\gamma} \left(\varepsilon + \gamma \log \frac{\gamma}{\varepsilon} \right),$$

so that for all ε small the error is bounded by $\epsilon(\varepsilon) \stackrel{\triangle}{=} C\varepsilon | \log \varepsilon |$, for some C, and clearly $\epsilon(\varepsilon)$ goes to zero as $\varepsilon \to 0$.

Finally, define $v_\varepsilon(t) \stackrel{\triangle}{=} \max\{\varepsilon, u_\varepsilon(t)\}$. Since εI_0^T is associated with a constant coefficient process for $0 \le x \le \varepsilon$, any segment of $r_\varepsilon(t)$ that starts at ε, goes down, and then crosses back up to ε, has a higher cost than the straight line path that remains at ε. Therefore

$$\varepsilon I_0^T (r_\varepsilon) \ge \varepsilon I_0^T (v_\varepsilon) + \epsilon(\varepsilon).$$

By Exercise 12.21, $\{r_\varepsilon\}$ are uniformly absolutely continuous, and hence so are $\{v_\varepsilon\}$ and moreover, $v_\varepsilon(0) \le \max\{x, \varepsilon\}$. Therefore there is a converging subsequence of the v_ε, with limit v, and $v(0) = x$. Note that $v \in F$ since by construction, $\sup_{0 \le t \le T} |v_\varepsilon(t) - r_\varepsilon(t)| \le \varepsilon$ and $r_\varepsilon \in F$ which is closed. Since $v_\varepsilon(t) \ge \varepsilon$ for all t we have $\varepsilon I_0^T (v_\varepsilon) = I_0^T (v_\varepsilon)$ and we finally conclude that

$$\limsup_{\varepsilon \to 0} \varepsilon I_0^T (r_\varepsilon) \ge \limsup_{\varepsilon \to 0} \varepsilon I_0^T (v_\varepsilon) = \limsup_{\varepsilon \to 0} I_0^T (v_\varepsilon). \tag{12.44}$$

On the other hand, by the lower semicontinuity of I_0^T, Lemma 5.42, and since $v \in F$ and $v(0) = x$,

$$\limsup_{\varepsilon \to 0} I_0^T (v_\varepsilon) \ge I_0^T (v) \ge \inf \left\{ I_0^T (r) : r \in F, \; r(0) = x \right\} \tag{12.45}$$

and (12.43) is established, and with it the upper bound.

To establish the lower bound it suffices to show that for each (absolutely continuous) path r with $r(0) = x$,

$$\mathbb{P}_x \left(\sup_{0 \le t \le T} |z_n(t) - r(t)| \le \varepsilon \right) \ge \exp \left(-n I_0^T (r) + n \epsilon_1(\varepsilon) + o(n) \right), \tag{12.46}$$

where $\epsilon_1(\varepsilon) \to 0$ as $\varepsilon \to 0$. Assume $r(0) > 0$: the case $r(0) = 0$ is left to Exercise 12.22. Fix r and let

$$r_\varepsilon(t) \overset{\triangle}{=} \max\{r(t), 2\varepsilon\}. \tag{12.47}$$

With the process z_n^ε constructed as for the upper bound,

$$\mathbb{P}_x \left(\sup_{0 \le t \le T} |z_n(t) - r(t)| \le 3\varepsilon \right)$$

$$\ge \mathbb{P}_x \left(\sup_{0 \le t \le T} |z_n(t) - r(t)| \le 3\varepsilon, \; \inf_{0 \le t \le T} z_n(t) > \varepsilon \right)$$

$$= \mathbb{P}_x \left(\sup_{0 \le t \le T} \left| z_n^\varepsilon(t) - r(t) \right| \le 3\varepsilon, \; \inf_{0 \le t \le T} z_n^\varepsilon(t) > \varepsilon \right)$$

$$\ge \mathbb{P}_x \left(\sup_{0 \le t \le T} \left| z_n^\varepsilon(t) - r_\varepsilon(t) \right| \le \varepsilon \right)$$

$$\ge \exp \left(-n \cdot {}_\varepsilon I_0^T (r_\varepsilon) + n O(\varepsilon) + o(n) \right).$$

where the last inequality follows from the lower bound applied to the process z_n^ε. Therefore the lower bound will follow once we establish that

$$\lim_{\varepsilon \to 0} {}_\varepsilon I_0^T (r_\varepsilon) \le I_0^T (r).$$

In Exercises 12.19 and 12.20 we show that for all x, y,

$$\ell_\varepsilon(x, y) - \ell(x, y) \le \begin{cases} 0 & \text{if } y \le 0 \\ f(\varepsilon) & \text{if } y > 0, \end{cases}$$

where $f(\varepsilon) \to 0$ as $\varepsilon \to 0$. Therefore

$$_\varepsilon I_0^T (r) \le I_0^T (r) + T f(\varepsilon). \tag{12.48}$$

On the other hand, as in the proof of the upper bound, we can modify r near zero and T, to obtain functions v_ε with

$$v_\varepsilon(0) = \max\{\varepsilon, r_\varepsilon(0)\} \quad \text{and} \quad v_\varepsilon(T) = \max\{\varepsilon, r_\varepsilon(T)\},$$

and v_ε agrees with r on $[\varepsilon/\gamma, T - \varepsilon/\gamma]$ with a change in the cost $_\varepsilon I_0^T$ by at most $\epsilon(\varepsilon)$. But below ε the functional $_\varepsilon I_0^T$ corresponds to a constant coefficient process. Therefore optimal paths, and in particular paths from ε to ε, are straight lines. This implies that

$$_\varepsilon I_0^T (r_\varepsilon) \le {}_\varepsilon I_0^T (r) + \epsilon(\varepsilon).$$

This together with (12.48) establishes the lower bound.   ∎

Exercise 12.19. If $y \le 0$ then $\ell_\varepsilon(x, y) < \ell(x, y)$. Hint: compute both at $y = 0$. Now use

$$\partial \ell / \partial y = \theta^*$$

and the explicit expression (12.5a) for $\ell(x, y)$. ♠

Exercise 12.20. If $y \ge 0$ then $\ell_\varepsilon(x, y) < \ell(x, y) + f(\varepsilon)$ where $f(\varepsilon) \to 0$ as $\varepsilon \to 0$. Hint: use the explicit expression (12.5a) for $\ell(x, y)$, derive a relation when $y \ge \sqrt{4\gamma\varepsilon}$ and a different relation when it is positive but smaller. ♠

Exercise 12.21. Show that $\{r_\varepsilon\}$, defined in the proof of the upper bound, Theorem 12.18, are uniformly absolutely continuous. Hint: use the definition of r_ε and the preceding exercises to obtain a uniform bound through

$$\varepsilon I_0^T(r_\varepsilon) \le \varepsilon I_0^T(r) \le f(\varepsilon)T + I_0^T(r).$$

Now use the proof of Theorem 5.18. ♠

Exercise 12.22. Complete the case $r(0) = x = 0$ in Theorem 12.18. Hint: use Kurtz's Theorem 5.3 to show that with high probability, $z_n(2\varepsilon/\gamma) \ge \varepsilon$, and approximate r to keep them close. ♠

The last piece of justification concerns the probability of hitting a point. The large deviations bounds concern closed and open sets respectively. However, we are interested in the probability of hitting a point—which corresponds to a closed set. To obtain the probability as a solution to a variational problem we have to show that the lower bound holds for this set. This was, however, discussed in §11.6. Since the arguments are identical, we make this into an

Exercise 12.23. Establish Equation (12.3). ♠

12.7. Extensions to Erlang's Model

In this section we describe extensions of various sorts to Erlang's basic model. These extensions are in various stages of completion. The point is not to provide answers, but to show how to use what we've learned in more complex settings, and to show how the large deviations point of view may be used.

Non-Poisson arrivals.

There are several reasons why the arrival process to a trunk group might not be Poisson. We quickly examine some of them.

1. Finite population.

Suppose that there are nM potential customers who might use a line. To avoid trivialities, we assume that $M > 1$, so $\mathbb{P}($ the number of trunks used $= n) > 0$. Each customer waits an independent, exponentially distributed amount of time with mean M/γ, then attempts to use a line. If successful, the customer holds the line for an independent exponentially distributed amount of time with mean one. If unsuccessful, the customer simply becomes idle again.

Now let

$$z_n(t) = \frac{1}{n}(\text{ number of trunks occupied at time } t). \tag{12.49}$$

We have

$$e_1 = +1 \qquad \lambda_1(x) = \frac{M-x}{M}\gamma, \quad 0 \le x < 1$$
$$e_2 = -1 \qquad \lambda_2(x) = x;$$

that is, if nx trunks are being held, the rate at which they are released is nx, and the rate of demand for trunks is (available population) $\cdot$ (rate/person), or $(nM - nx) \cdot \frac{\gamma}{M}$.

Now that we have specified our model, our analysis can go through almost exactly as before. The fluid limit is

$$\frac{d}{dt}z_\infty(t) = \left(1 - \frac{z_\infty}{M}\right)\gamma - z_\infty = \gamma - \left(1 + \frac{\gamma}{M}\right)z_\infty,$$

$$z_\infty(t) = \left(1 - e^{-\left(1+\frac{\gamma}{M}\right)t}\right)\frac{\gamma}{1+\gamma/M} + e^{-\left(1+\frac{\gamma}{M}\right)t}z_\infty(0). \tag{12.50}$$

The three cases are then the following:

$$\frac{\gamma}{1+\gamma/M} > 1 \qquad \text{heavy traffic,}$$

$$\frac{\gamma}{1+\gamma/M} = 1 \qquad \text{moderate traffic,}$$

$$\frac{\gamma}{1+\gamma/M} < 1 \qquad \text{light traffic.}$$

The analysis goes through exactly as in the basic Erlang model from here. The details are left to the reader.

Exercise 12.24. Show that if $r(0) = 0$, then for the finite population model we have

$$r^*(t) = A \sinh(B(t - D)), \tag{12.51}$$

and if $r(0) = 1$ and traffic is light, then

$$r^*(t) = A \cosh(B(t - D)), \tag{12.52}$$

where A, B, and D are constants. Find an expression for $I^*(t)$ as in the basic Erlang model. Show that, in light traffic,

$$\lim_{t\to\infty,\, n\to\infty} \frac{1}{n} \log \mathbb{P}_x(z_n(t) = 1) = -\int_{\frac{\gamma}{1+\gamma/M}}^{1} \log \frac{\lambda_2(x)}{\lambda_1(x)}\, dx \,. \quad \spadesuit$$

2. Batch arrivals.

Batch arrivals are relatively easy to analyze. Suppose that customers arrive in "waves" or "clumps." We can model this by having the number entering the system at once be j with probability p_j, for $1 \le j \le M$. Services are still single; that is, an arrival is a cluster of calls that depart the system independently. Then a one-dimensional model suffices to analyze the system. As before we let $x(t)$ be the number of trunks occupied at time t, and the transition rates and directions become

$$e_j = j \qquad\qquad \lambda_j(x) = n\gamma p_j, \qquad j = 1\ldots M, \quad 0 \le x < n - j,$$
$$e_{M+1} = -1 \qquad \lambda_{M+1}(x) = x.$$

We set $\lambda_j(x) = 0$ when $x > n - j$ since we don't want any batches to arrive that would force the system to have more than n customers. There are other ways to model this requirement, such as having $\lambda_{n-x}(x) = \sum_{j=n-x}^{M} n\gamma p_j$ whenever $x > n - M$; this would mean that as much of a batch as possible is admitted, so the whole batch isn't blocked when there is room for any part of it. Which model is better depends on the particulars of a system, and cannot be decided on purely mathematical grounds. They are equally easy to analyze. In fact, they are indistinguishable once we make our usual scaling of jump rates and space:

$$\lambda_j(x) = \gamma p_j, \qquad j = 1\ldots M, \quad 0 \le x < 1,$$
$$\lambda_{M+1}(x) = x$$

in both cases except for a neighborhood of size M/n of $x = 1$. Note that this is a model with a "thick flat boundary." It *does not* fall under the domain of the theorems of Chapter 8. However, light traffic can be analyzed using the familiar "small movement over small time periods takes small cost" result (cf. Lemma 6.21).

We are interested in the probability of blocking. To begin our analysis, we calculate the drift:

$$\frac{d}{dt} z_\infty(t) = \gamma \sum_{j=1}^{M} j p_j - z_\infty(t), \qquad 0 < z_\infty < 1. \qquad (12.53)$$

If we set $\eta \overset{\triangle}{=} \gamma \sum_{j=1}^{M} j p_j$, then we see that z_∞ has the same equation (12.2) as before, with η replacing γ. Therefore we can immediately define light, moderate, and heavy traffic as having η be less than one, equal to one, or greater than one, respectively. Furthermore we can define t_1 exactly as before, and have our transient behavior qualitatively depend on whether or not t is greater than t_1, in which case steady-state will have been reached already, or if t is less than t_1 we know we are still in the transient regime.

The local rate function is defined by an algebraic equation that cannot, in general, be solved analytically. Specifically,

$$\ell(x, y) \overset{\triangle}{=} \sup_{\theta} \left(\theta y - \gamma \sum_{j=1}^{M} p_j e^{j\theta} - x e^{-\theta} \right), \qquad (12.54)$$

and by setting $w = e^{\theta}$ and taking derivatives, we obtain an $M + 1^{\text{th}}$ degree equation in w, which is not usually solvable when $M > 3$. However, it is not hard to show that there is a unique solution with $w > 1$, that $\ell(x, y)$ has the convexity and continuity we have come to expect, and indeed the qualitative behavior of the solutions is exactly the same as in Erlang's basic model. For more caveats on this model, see the next section.

Exercise 12.25. For $M = 4$ and $p_j = 0$, $j < 4$, $p_4 = 1$, calculate the value of γ that gives moderate traffic for the batch arrival model. Find an expression for the steady-state blocking probability in light traffic. Numerically evaluate the rate function I that governs the steady-state blocking probability for $\gamma = 1/5$. (We obtain $I \approx 0.00912$.)   ♠

3. Multirate service.

Suppose that there are two classes of customers in our system. One class requires one trunk per customer; the other requires M trunks per customer. This models, for example, advanced services such as low bitrate video or moderate capacity data channels. We suppose that the "fat" customers have arrival rate $n\nu$ and service rate η, while the normal customers have arrival rate $n\gamma$ and service rate one.

Each class has a different blocking probability; the fat customers are blocked whenever there are fewer than M trunks available, while the normal customers are blocked only when no trunks are available. What is the transient response of this system?

We let $\vec{z}_n(t) = ($ number of normal $/n,$ number of fat $/n)$ present in the system at time t. Then

$$e_1 = (1, 0) \qquad \lambda_1(\vec{x}) = \begin{cases} \gamma & \text{if } x_1 + M x_2 < 1 \\ 0 & \text{if } x_1 + M x_2 \geq 1 \end{cases}$$

$$e_2 = (-1, 0) \qquad \lambda_2(\vec{x}) = \begin{cases} 1 & \text{if } x_1 > 0 \\ 0 & \text{if } x_1 = 0 \end{cases}$$

$$e_3 = (0, 1) \qquad \lambda_3(\vec{x}) = \begin{cases} \nu & \text{if } x_1 + M x_2 < 1 \\ 0 & \text{if } x_1 + M x_2 \geq 1 \end{cases}$$

$$e_4 = (0, -1) \qquad \lambda_4(\vec{x}) = \begin{cases} 1 & \text{if } x_2 > 0 \\ 0 & \text{if } x_2 = 0 \,. \end{cases}$$

We see that our scaling cannot distinguish between normal and fat customers for the purpose of blocking analysis. Because of this, the model is inadequate for the representation of moderate or heavy traffic or for analyzing long-term over-loads (we discuss this in more detail below). However, for light traffic, the calculation follows from the basic model's analysis in a very straightforward fashion, as outlined for the following exercises.

We begin with a derivation of $\vec{z}_\infty(t)$.

$$\frac{d}{dt}\vec{z}_\infty(t) = \gamma\vec{e}_1 + v\vec{e}_3 - \langle\vec{e}_2, \vec{z}_\infty(t)\rangle - \eta\langle\vec{e}_4, \vec{z}_\infty(t)\rangle \qquad (12.55)$$

whenever $\vec{z}_\infty(t)$ is away from the boundary; that is, when

$$0 < \langle\vec{e}_1, \vec{z}_\infty\rangle$$
$$0 < \langle\vec{e}_3, \vec{z}_\infty\rangle$$
$$0 < \langle\vec{e}_1 + M\vec{e}_3, \vec{z}_\infty\rangle.$$

Notice that as long as $\vec{z}_\infty(t)$ is away from a boundary, the two components of $\vec{z}_\infty(t)$ decouple. We use this observation to solve not only for $\vec{z}_\infty(t)$, but for $\vec{r}^*$ and I^* as well. We are now in position to say what "light," "medium," and "heavy" traffic means for our multirate model. We say that the system is in light traffic if the point $\vec{q}$ where $\frac{d}{dt}\vec{z}_\infty = 0$ is strictly in the interior of the region

$$x_1 + Mx_2 \leq 1, \quad x_1 \geq 0, \quad x_2 \geq 0. \qquad (12.56)$$

Exercise 12.26. Find $\vec{q}$ as a function of γ, v, and η. ♠

Exercise 12.27. Find the ℓ- and I-functions for the multirate model as explicitly as you can. ♠

Exercise 12.28. In light traffic, write an estimate for the probability of blocking (overflow) as the solution of a variational problem. ♠

Exercise 12.29. Solve the variational problem of the previous exercise. You can do this easily once you note that the two components of $\vec{r}^*(t)$ decouple. The only possible complication is finding the point where $\vec{r}^*(t)$ hits the boundary $x_1 + Mx_2 = 1$. Find the (unique) path from $\vec{q}$ to this boundary by minimizing the cost (I function) over all extremals that hit this boundary. That is, find

$$\inf\{I(\vec{r}) : 0 \leq s \leq 1, \ \vec{r} \in G(s)\}$$
$$G(s) \triangleq \{\vec{r}(t), \ T : \vec{r}(0) = \vec{q}, \ \vec{r}(T) = (s, (1-s)/M)\}. \qquad (12.57)$$

Show that the optimal value of s lies between γ and $1 - v$ and satisfies

$$v\left(\frac{s}{\gamma}\right)^M + s - 1 = 0. \qquad ♠$$

Exercise 12.30. How does the frequency of buffer overflow behave as a function of M? Could you have predicted this functional form heuristically from the solution to the plain Erlang model? You will notice that the definition of "blocking as a function of M" needs to be done carefully; simply increasing M will also increase the steady-state load on the system. To make a fair comparison, we must scale $\lambda_3(\vec{x})$ by $1/M$. This gives us a measure of the cost of burstiness, or equivalently the cost of largeness (distance from a fluid limit). Recall from Chernoff's theorem that probabilities should increase geometrically as a function of the chunk size; is this what you found?   ♠

12.8. Transient Behavior of Trunk Reservation

In the AT&T circuit-switched (voice) network (and in others, for all we know) there are two classes of calls: *direct*, and *via*. Between nearly every pair of the 140 or so (as of 1994) main switches there is a direct connection; that is, the graph of connections is nearly complete. When a call from, say, Cleveland to Buffalo is being set up, the network attempts to place the call on the direct link between the two cities (assuming that there is just one switch serving each city). However, if this link is full, then there are 138 vias to choose from: Cleveland to Rochester to Buffalo, Cleveland to Newark to Buffalo, etc. We shall not describe the algorithm by which the via is chosen; the interested reader can examine the voluminous and growing literature on the subject of good algorithms (DNHR, ALBA, DAR, RTNR).

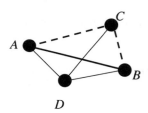

Figure 12.31. Calls between A and B may be carried on the direct link (dark line), or on a via such as C (dashed line).

There is a problem with many algorithms for choosing vias. Each via call takes two links, so potentially reduces the carrying capacity of the network. In fact, just as in Aloha (see Chapter 14), there are often two operating points for the system: one where most traffic is carried on vias, and one where most traffic is direct.

We won't prove the existence of this bistability here, but for plausibility consider the following argument. Suppose that the network is heavily loaded, and that nearly every call is a via. Let the fraction of links that are not full be $p \ll 1$. When a new call arrives, it has a probability p of being routed directly (we assume that if the direct link is available it will be used). With a high probability $(1 - p)$ the call will need to use a via, if it is accepted at all. If there are m nodes in the network then (assuming that the links are statistically independent) the probability that at least one via is available is

$$1 - \left(1 - p^2\right)^{m-2} \approx \begin{cases} (m-2)p^2 & \text{for } p^2 \ll 1/m \\ 1 & \text{for } p^2 \gg 1/m \\ 1 - e^{-p^2(m-2)} & \text{for } p^2 = O\left(1/m\right). \end{cases}$$

So we see that if a call is blocked only when there is no via available, the probability that an arriving call will be carried on a via can be close to one (that is, about equal to p) if m is large and p is small but $p^2 > 1/m$. In this situation the total carried traffic is only about half the network capacity; the network would actually perform better (have lower blocking and higher throughput) if no vias were allowed.

In order to prevent this problem (and for other benefits as well), network de-signers incorporate a scheme called *trunk reservation* into most effective routing strategies. Trunk reservation allows a via to use a link only if there are at least C trunks free on the link at the time of call setup. Direct calls are carried if there are any free trunks. The parameter C is adjusted to make the various blocking probabilities conform to the designer's goals. Empirically, any positive value of C seems to keep a network free of bistability under any reasonable routing strat-egy.

We now present and analyze a horrendously simplified model of trunk reserva-tion. We consider a single link of capacity n. We suppose that there is a Poisson arrival process of direct calls with parameter $n\lambda$, a trunk reservation parameter Cn, and we model the requests (from the rest of the network) for use of the link as a via as a Poisson arrival process with rate $n\gamma$. Each accepted call holds the link for an i.i.d. time distributed exponentially with rate μ. Then the number of calls carried on the link is a Markov process with the following structure:

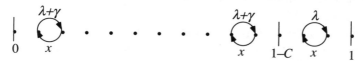

Figure 12.32. Birth-death structure of trunk reservation.

Taking $z_n(t) = n^{-1} \cdot$ (number of calls in progress at t), we have

$$e_1 = 1 \qquad \lambda_1(z) = \begin{cases} \lambda + \gamma & \text{for } 0 \le z < 1 - C \\ \lambda & \text{for } 1 - C \le z < 1 \\ 0 & \text{for } z \ge 1 \end{cases}$$

$$e_2 = -1 \qquad \lambda_2(z) = \mu z.$$

This process does not have Lipschitz continuous rates, so our theory does not ap-ply. Yet we could modify the rate $\lambda_1(z)$ quite easily to obtain a Lipschitz contin-uous model by linear interpolation, viz.

$$\lambda_1(z) = \begin{cases} \lambda + \gamma & \text{for } 0 \le z < 1 - C - \varepsilon \\ \lambda + \frac{\gamma}{2} + \frac{1 - C - z}{\varepsilon} \frac{\gamma}{2} & \text{for } 1 - C - \varepsilon \le z \le 1 - C + \varepsilon \\ \lambda & \text{for } 1 - C + \varepsilon \le z < 1 \\ 0 & \text{for } z \ge 1. \end{cases} \qquad (12.58)$$

In some ways this model might be as realistic as the previous one: decisions are often based on stale information, so calls may be blocked or accepted when they shouldn't be.

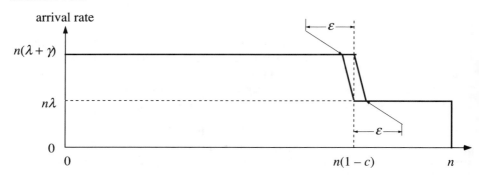

Figure 12.33. The arrival rate $\lambda_1(x)$ for both an upper bounding model and a lower bounding model. Note that the drifts and rate functions of these models are identical except in an ε neighborhood of $1 - C$.

We can analyze either an upper bounding or a lower bounding model via Kurtz's Theorem and via large deviations. The results will have an ε in the calculations. It is clear that we may take $\varepsilon = 0$ even before doing the calculations, though, and just be cognizant of the fact that statements about what is happening near the point $1 - C$ might be off. Note that this singularity can be handled rigorously using results in [DE2].

There are no further theoretical difficulties in analyzing this model. There are at least two quantities of interest: the probability that $z_n(t) = 1$, and the probability that $z_n(t) \geq 1 - C$. We follow our usual procedure in estimating these quantities.

The first item of business is to examine $z_\infty(t)$. We plot the possible character-istics of the flow in Figure 12.34. We see that there are four main cases, and three critical cases. Case 1 has light traffic from the point of view of both the direct traf-fic and the via traffic. Case 2 has light direct traffic, but critical via traffic. Case 3 has heavy via traffic but light direct traffic. And Case 4 has heavy traffic for both types. There are three transitional cases also: where via traffic is barely crit-ical and direct traffic is light (between Cases 1 and 2), where via traffic is barely heavy and direct traffic is light (between Cases 2 and 3), and where via traffic is heavy and where direct traffic is critical (between Cases 3 and 4).

We examine only case 2 in detail; the others are, if anything, easier to analyze, and are left for the interested student. Our approach is the following. We eval-uate the solution of the variational problem for the case $\varepsilon = 0$. It is not hard to see that the variational problem for any $\varepsilon > 0$ has solutions that are close to the solution for our case. Furthermore, I^* is close for the two cases. There is no need to prove the large deviations principle for the case $\varepsilon = 0$; by comparison with an upper bounding model and a lower bounding model as in Figure 12.33, the prin-ciple holds well enough to enable us to calculate the quantities of interest for this problem.

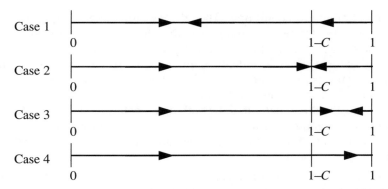

Figure 12.34. Possible patterns of drift using trunk reservation.

Consider the problem of estimating the blocking of direct traffic:

$$\mathbb{P}_x(z_n(t) = 1) = \exp\left(-nI^*(x, t) + o(n)\right), \tag{12.59}$$

$$I^*(x, t) = \inf\left\{\int_0^t \ell(r(s), r'(s))\, ds : r(0) = x,\ r(t) = 1\right\}. \tag{12.60}$$

For each $x \in [0, 1]$, the path z_∞ with $z_\infty(0) = x$ reaches $1 - C$ in finite time $T(x)$; that is, there is a bounded function $T(x)$ with

$$z_\infty(t) \neq 1 - C \quad \text{for all} \quad t \in [0, T(x)),\ z_\infty(0) = x$$
$$z_\infty(t) = 1 - C \quad \text{for all} \quad t \geq T(x),\ z_\infty(0) = x.$$

Furthermore, the minimal cost path from $1 - C$ to 1 is the time reversal of $z_\infty(t)$, so that it also takes finite time. Therefore, for each $x \leq 1 - C$ and $t \geq T(x) + T(1)$, $I^*(x, t)$ is a constant equal to

$$I^* \overset{\triangle}{=} \inf\left\{\int_0^T \ell(r(s), r'(s))\, ds : r(0) = 1 - C,\ r(T) = 1\right\}$$
$$= \int_{1-C}^1 \log\left(\frac{\mu x}{\lambda}\right) dx \tag{12.61}$$
$$= C\left(\log(1 - C) + \log\left(\frac{\mu}{\lambda}\right) - 1\right) - \log(1 - C).$$

This is because the minimal cost path (over all time) from x to $1 - C$ is $z_\infty(t)$, with cost zero; the cost of remaining at $1 - C$ is zero; and the minimal cost path (over all time) from $1 - C$ to one has cost I^*.

For $x > 1 - C$ we perform an analysis reminiscent of the $M/M/1$ queue. There is a time below which a direct path from x to one (which stays above the level $1 - C$) has the lowest cost, and above which the cost is I^*. Here is an outline of the calculation. To compute the cost of a direct path, we pretend that there is no change of arrival rate at $1 - C$. The problem is reduced to the transient behavior of the plain Erlang model, starting at a point x. We know that, for small time, the critical path is strictly increasing, and that as time goes on the path will initially dip before climbing to height one. The cost, as a function of time, is unimodal, starting at infinity at time zero, going to a minimum, and gradually climbing back to the steady-state value I^*(model). Clearly we have I^*(model) $> I^*$, since the behavior of the two systems is identical until a path hits the level $1 - C$, and then it is easier for the present system to reach the level 1 afterwards. Hence there is a unique time T^* when I^*(model)$(T^*) = I^*$. At this time there are two paths with the same cost: the direct path, and the path that reaches the level $1 - C$, remains there for a while, and then climbs back up to one. The only possible question is whether the direct path would have reached level $1 - C$ before T^*, leading to an inconsistency.

Exercise 12.35. Why can't the direct path touch level $1 - C$ before T^* without having higher cost than the other extremal path? ♠

Exercise 12.36. Analyze cases 1, 3, and 4. You only need to use the solution of Erlang's model for λ and $\lambda + \gamma$ arrival rates. Find the optimal paths $r^*(t)$ and the behavior of the cost function $I^*(t)$. Is there a finite time T^* in these models? Why or why not? In the cases when there is no finite time T^*, show that there is a discontinuity in the derivative of $I^*(t)$ as a function of t, and analyze the behavior of r^* that causes it. ♠

There are better models for the behavior of a link under various routing policies utilizing trunk reservation. For example, the rate with which the link is used as a via might depend on the state of the link. This comes about in "Least Busy Alternative" routing, where vias are chosen not at random, but by selecting the least busy among the $n - 2$ possible vias. Least busy could mean several things; a typical measure is to have the busyness measured by the most heavily loaded of the two links on a via, and load is measured in terms of free trunks below the reservation level $1 - C$.

The rate at which vias attempt to use a given link should be a monotone decreasing function of the link occupancy. In "fixed point" steady state calculations, the function is calculated numerically [MGH]. To make a more realistic transient analysis, one might assume that these rates hold in our model, too. We would then use them to calculate the transient behavior of the scheme. The only difference between that calculation and the one from the previous section is that γ is now $\gamma(x)$. The explicit solution of Erlang's model is not available, but quantitative, numerical, and qualitative results are derivable.

Exercise 12.37. Suppose that $\gamma(x) = \gamma \cdot \left(1 - e^{-(C-x)}\right)$. Examine the approach to steady-state of the quantity $\mathbb{P}_x(z_n(t) = 1)$. ♠

The other quantity of interest is $\mathbb{P}_x(z_n(t) \geq 1 - C)$. This quantity is a bit more delicate because of the same reason we don't necessarily have a large deviations principle for the process. Are vias allowed in when $z_n = 1 - C$ or not? The question is whether the arrival rate is right- or left-continuous at the point $1 - C$. For this reason the calculation of the quantity in Case 2 is a bit problematic. The other cases reduce directly to simple Erlang model calculations, and are left to the reader.

Exercise 12.38. Show that in Case 1,

$$\lim_{n\to\infty,\, t\to\infty} \frac{1}{n} \log \mathbb{P}_x(z_n(t) \geq 1 - C) = -\int_q^{1-C} \log \frac{\mu(x)}{\lambda(x)}\, dx \ .$$

Why does the Freidlin-Wentzell theory apply to this case?   ♠

12.9. End Notes

Like the $M/M/1$ queue, Erlang's model has a voluminous literature. We cannot possibly do justice to other authors, so we content ourselves with a very brief discussion of some papers that are directly applicable to the types of results we obtained.

The steady-state solution of Erlang's model is a truncated Poisson distribution with parameter $n\gamma$. Therefore Chernoff's Theorem may be used to derive asymptotic expansions of the blocking probability when $\gamma < 1$ or $\gamma > 1$. The central limit theorem may be used when $\gamma = 1$. It is also straightforward to derive expansions based purely on analytic expressions. See [Jag] for details.

The Laplace transforms of $\mathbb{P}_0(z_n(t) = 1)$ and $\mathbb{P}_1(z_n(t) = 1)$ were derived by Takàcs [Ta1] and Beneš [Be]. These were inverted asymptotically in Mitra and Weiss [MW], giving higher order terms in the expansion. Knessl [Kn] has recently obtained more accurate expansions based on the Fokker-Plank (forward) equations. Our transient results for the extensions of Erlang's model are all new.

Multirate Erlang systems have been analyzed before, of course. For example, J.S. Kaufman [Ka] and J.W. Roberts [Rob] independently found a way to calculate the steady-state distribution of the model. Marty Reiman [Re1] analyzed it in the interesting case of critical loading. Peter Key [Ke] has a clean analysis, and Gazdzicki, Lambadaris, and Mazumdar [GLM] have a thorough large deviations analysis of the model than appears here.

Our model of trunk reservation is weakest in its scaling. Marty Reiman [Re2] has shown that the optimal scaling of the size of the reserved region is generally not Cn, but depending on the parameters λ and γ and the desired blocking probabilities, is either $C\sqrt{n}$, $C\log n$, or simply C.

Chapter 13

The Anick-Mitra-Sondhi Model

Just as Erlang's model is the basic model of circuit-switched communication, the Anick-Mitra-Sondhi model (AMS model) is the basic model of packet-based communication. It captures the essential feature of packet systems (statistical multiplexing), and more realistic models must contain the AMS model as a subsystem or as a limiting case.

The model consists of n statistically independent and identical traffic sources feeding into a buffer (concentrator). The packet sources are modeled as sources of fluid. This scaling makes sense if either packets are very small, so that only large groups of packets make significant demands on resources, or if sources may create a large number of packets during periods of high activity. To allow sources to have several levels of activity, we model them as finite state (d-state) Markov processes. Each state $i \in \{1, \ldots, d\}$ has an associated "activity" a_i, which represents the rate of production of fluid for sources in that state. We can incorporate a fixed number of classes of sources, each with their own statistics, by allowing the Markov process to be decomposable (Definition A.140) as in Figure 13.1. The fraction of sources in each class is fixed by the initial condition. We suppose that there are no transient states in the source model. This excludes, for example, the touchy case of one transient source leading into all the subclasses, so that the number of members of each subclass is random.

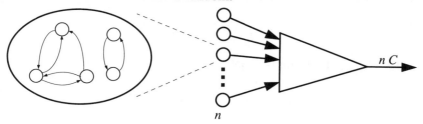

Figure 13.1. An AMS model.

Here is the detailed source model. Suppose that each source has d possible states. Then a source may be identified with a vector in $\mathcal{Z}^d$ whose components are all zero except for component i if the source is in state number i; this component is equal to one. That is, the one points to the index of the state the source is in. Then adding all n vectors representing the n sources, we obtain a vector

$$\vec{x}(t) = (\text{number of sources in state } 1, \ldots, \text{ number of sources in state } d).$$

The vector $\vec{x}(t)$ is a Markov process. We calculate the rates of its transitions as follows. Suppose that each source goes from state i to state j at rate λ_{ij}. That is, transitions in direction

$$\vec{e}_{ij} \stackrel{\triangle}{=} (0, \ldots, 0, +1, 0, 0, \ldots, 0, -1, 0, \ldots, 0)$$

due to a change in the state of a specific source occur at rate λ_{ij}, where $\vec{e}_{ij}$ has a $(+1)$ in the jth component and a (-1) in the ith component. Then for the composite process $\vec{x}(t)$, both jump directions and jump rates are indexed by the double index ij, and $\lambda_{ij}(\vec{x}) = \lambda_{ij} \cdot x_i$, since there are x_i independent sources with potential transitions (of the Markov chain) from i to j. Now defining

$$\vec{z}_n(t) \stackrel{\triangle}{=} \frac{1}{n} \vec{x}(t)$$

we see that $\vec{z}_n(t)$ satisfies the usual scaling (5.1). It has jumps $\vec{e}_{ij}/n$ and associated rates $n\lambda_{ij}(\vec{z})$, where

$$\lambda_{ij}(\vec{z}) = \lambda_{ij} \cdot z_i. \tag{13.1}$$

The buffer is modeled as an infinite-capacity fluid reservoir having an output pump with rate nC. When the aggregate input rate of fluid is, say, $nA > nC$, the buffer content increases at rate $n(A - C)$. When the input rate is $nA < nC$, the buffer drains at rate $n(C - A)$. Since each source in state i produces fluid at rate a_i, the buffer content $b(t)$ satisfies

$$\frac{d}{dt}b(t) = \begin{cases} \langle \vec{a}, n\vec{z}_n(t) \rangle - nC & \text{if } b(t) > 0, \text{ or if } b(t) = 0 \text{ and } \langle \vec{a}, \vec{z}_n \rangle > C \\ 0 & \text{otherwise,} \end{cases}$$
$$\tag{13.2}$$

where $\vec{a} \stackrel{\triangle}{=} (a_1, \ldots, a_d)$. The scaled buffer size is $b_n(t) = b(t)/n$. We now derive some representations for the buffer size, to be used below. Let $u(t)$ be the last time before t that the buffer was empty. Then from (13.2)

$$b_n(t) = \int_{u(t)}^{t} (\langle \vec{a}, \vec{z}_n(s) \rangle - C)\, ds.$$

For any $u < u(t)$, (13.2) implies that

$$\int_{u}^{u(t)} (\langle \vec{a}, \vec{z}_n(s) \rangle - C)\, ds \le 0 \tag{13.3}$$

since b_n is always positive. On the other hand, for all $s < t$,

$$\int_{s}^{t} (\langle \vec{a}, \vec{z}_n(s) \rangle - C)\, ds \le b_n(t) - b_n(s).$$

Therefore, if $b_n(t) = 0$ for some t, then for any $T > t$ we have

$$b_n(T) = \sup_{0 \le u \le T} \int_{u}^{T} (\langle \vec{a}, \vec{z}_n(s) \rangle - C)\, ds$$

$$= \sup_{t \le u \le T} \int_{u}^{T} (\langle \vec{a}, \vec{z}_n(s) \rangle - C)\, ds. \tag{13.4}$$

Our goal is to study the behavior of the buffer. For example, how often is it non-empty? How often does it exceed a level nB? How effective are certain controls for reducing the frequency of large buffers, or overflows in finite buffers? Equivalently, how large does the buffer have to be to have a loss rate (proportion of information that "overflows") less than $\exp(-nA)$ for some given A?

It is easy to see, at least intuitively, that the occurrence of a non-empty buffer should be a large deviation whenever the system is stable. When the number of sources is large, the strong law of large numbers implies that the proportion of sources in state i should be close to the average value

$$q_i \overset{\triangle}{=} \lim_{n\to\infty} \mathbb{E}_{ss}\left(z_{n,i}(t)\right). \tag{13.5}$$

[For a more explicit form of $\vec{q} = (q_1,\ldots,q_d)$ see (13.9) below.] But then the rate of fluid production should be near $np = n\langle\vec{q},\vec{a}\rangle$ almost all the time. Thus p is the average rate of production for one source and $\vec{q}$ is the average state for the scaled process $\vec{z}_n$. The buffer fills only when this rate exceeds nC, and any stable system will have $C > p$. Therefore, the steady-state probability that the buffer is non-empty is about $\exp(-nI^*)$, where

$$I^* = \inf_G I_0^T(\vec{r}) \tag{13.6}$$
$$G = \{\vec{r}, T : \vec{r}(0) = \vec{q},\ \langle\vec{r}(T),\vec{a}\rangle = C\}. \tag{13.7}$$

This is a standard Freidlin-Wentzell level crossing problem. Of course, by definition $0 \le z_{n,i} \le 1$, so that we need consider only paths $\vec{r}$ satisfying this condition.

In §13.1 we examine the simplest model in detail: sources have only two states, called "on" and "off." We will also examine one case of more complex sources, and the effects of various types of controls on the statistics of the buffer. Usually we are unable to solve the resulting variational problems completely, so we examine certain limiting cases (large and small buffers) to get both analytic formulas and a feel for how the solutions depend on the various parameters of the models. In the cases where we don't solve the variational problems analytically we try to provide enough information to make the resulting numerical problems easy to solve. We are often able to turn qualitative information on the solutions into quantitative statements; for example, see §13.7 on the effectiveness of controls.

It is sometimes convenient to lower the dimension of the state space by one. Since $z_{n,1} + \ldots + z_{n,d} = 1$, we can replace $\vec{z}_n(t)$ by the $(d-1)$-dimensional process $(z_{n,1}(t),\ldots,z_{n,d-1}(t))$ where there are now transitions $\vec{e}_{ij}$ with only a $(+1)$ or a (-1), corresponding to transitions from or to state d. For example, in a two-state system with $\lambda_{1,2} = \mu$, $\lambda_{2,1} = \lambda$ we can consider $z_n(t)$ as the number in state one alone, and we obtain

$$\begin{aligned} e_1 &= +1 & \lambda_1(x) &= \lambda - \lambda x \\ e_2 &= -1 & \lambda_2(x) &= \mu x. \end{aligned} \tag{13.8}$$

From (5.7), we can calculate $\vec{z}_\infty(t)$ for the AMS model:

$$\frac{d}{dt}\vec{z}_\infty(t) = \sum_{i,j} \lambda_{ij}(\vec{z}_\infty(t))\vec{e}_{ij},$$

where the sum is over all possible transitions i to j. By (13.1), $\lambda_{ij}(\vec{z}_\infty)$ is affine in $\vec{z}_\infty$; hence

$$\frac{d}{dt}\vec{z}_\infty = A\vec{z}_\infty + \vec{B}$$

for a matrix A and a vector $\vec{B}$. In general, then, $\vec{z}_\infty(t)$ relaxes exponentially to a constant $\vec{q}$ satisfying

$$A\vec{q} = -\vec{B}. \tag{13.9}$$

[This $\vec{q}$ is the same one defined in (13.5) by Theorem 6.89.] For example, in the simple on/off system,

$$\frac{d}{dt}z_\infty(t) = \lambda\,(1 - z_\infty(t)) - \mu z_\infty(t)$$

$$z_\infty(t) = \left(z_\infty(0) - \frac{\lambda}{\lambda + \mu}\right)\exp(-(\lambda + \mu)t) + \frac{\lambda}{\lambda + \mu}. \tag{13.10}$$

13.1. The Simple Source Model

This is the model studied extensively by Anick, Mitra and Sondhi [AMS], Kosten [Ko] and others. Because it is the simplest it is simultaneously the model about which we can say the most, and which is the least realistic. Nevertheless, as with any simple model, it has features that we use for insight into the behavior of more complicated models.

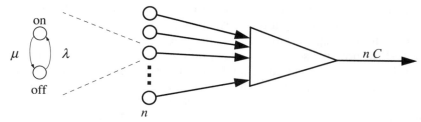

Figure 13.2. The simple AMS model with $d = 2$, $a_1 = 1$, and $a_2 = 0$.

Each source is assumed to go from on to off at rate μ, and from off to on at rate λ. This means that the long-run (steady-state) probability that a source is on is $\lambda/(\lambda + \mu)$. As in (13.5) or (13.9) and (13.10) let

$$q \triangleq \frac{\lambda}{\lambda + \mu}. \tag{13.11}$$

While a source is on we suppose that it pumps fluid at rate one. Hence we expect the buffer to be non-empty whenever more than nC sources are on. To make the problem nontrivial we make the following assumption, which is enforced without further mention throughout this chapter.

Assumption 13.3. $q < C < 1$; equivalently, $\mu C > \lambda(1 - C)$.

Since the sources are independent, the long-run probability that more than nC sources are on is the probability that the sum of n Bernoulli(q) random variables exceeds nC. Using Chernoff's Theorem for Bernoulli random variables as in Example 1.15, we see that this is about equal to $\exp(-n\ell(C))$, where

$$\ell(C) = \sup_\theta \left(\theta C - \log\left(qe^\theta + 1 - q\right)\right) = C \log \frac{C}{q} + (1 - C)\log \frac{1 - C}{1 - q}.$$
(13.12)

So now that we know the answer, let's do the proper calculation and prove it!

Our process $z_n(t)$ has an associated local rate function

$$\ell(x, y) = y \log \frac{y + \sqrt{y^2 + 4\lambda(1 - x)\mu x}}{2\lambda(1 - x)} + \lambda(1 - x) + \mu x - \sqrt{y^2 + 4\lambda(1 - x)\mu x}$$
(13.13)

provided that $0 < x < 1$ or, if $x = 0$ or 1, that y points inwards. As explained below (11.20), this follows from (7.16)–(7.17), since our process is a birth-death process (with jumps of $1/n$) with birth rate $\lambda(1 - x)$ and death rate μx (restricted to [0, 1]). Now to calculate the frequency of a non-empty buffer, use the general principle that steady-state probabilities have the same asymptotic form as upcrossing probabilities. We haven't proved this yet for the AMS model. We'll justify the calculation in §13.6: for now, let's proceed formally. The upcrossing problem is associated with the variational problem of calculating

$$I^* = \inf\left\{I_0^T(r) : (r, T) \in G\right\}$$
$$G = \{r, T : r(0) = q, \ r(T) = C\}$$
(13.14)

where, from (13.11), q is the stationary probability of the process $z_n(t)$.

We have solved the level crossing problem for birth-death processes before in §11.5. We recapitulate the solution given in Theorem 11.15. We have

$$c_1 \stackrel{\triangle}{=} I^* = \int_q^C \log \frac{\mu x}{\lambda(1 - x)} \, dx = C \log \frac{C}{q} + (1 - C)\log \frac{1 - C}{1 - q}.$$
(13.15)

Moreover, the path $r^*(t)$ that achieves this cost (shifted so that $r(0) = C$) is the time reversal of $z_\infty(t)$, namely

$$r^*(t) = q + (C - q)e^{(\lambda + \mu)t}, \quad -\infty \le t \le 0.$$
(13.16)

So we have not only shown that our original heuristic calculation of the frequency of non-empty buffers is correct, but we now know the most likely way that the buffer becomes non-empty.

Exercise 13.4. Prove that indeed (13.15)–(13.16) solve the variational problem (13.14). Hint: show that the conclusions of Theorem 11.15 hold despite the fact that the hypotheses do not. ♠

So, once we justify the relations between the variational problem and the calculation of probabilities, the results (Theorems 6.17 and 6.59) of Chapter 6 will imply the following.

Theorem 13.5. *Define I^* and r^* through (13.15)–(13.16). Under the stability condition $q < C$, for any ε and $T > 0$,*

$$\lim_{n \to \infty} \mathbb{P}_{ss} \left(\sup_{-T < t < 0} \left| z_n(t) - r^*(t) \right| < \varepsilon \ \middle| \ b_n(0) > 0 \right) = 1.$$

Let τ be the first time the buffer becomes non-empty. For each $x < C$ and $\delta > 0$,

$$\mathbb{P}_x \left(e^{n(I^* - \delta)} < \tau < e^{n(I^* + \delta)} \right) \to 1 \quad as \quad n \to \infty.$$

13.2. Buffer Statistics

We wish to estimate the steady-state probability that the buffer $b_n(t)$ exceeds any given level B. We argue that this can be estimated as

$$\mathbb{P}_{ss} \left(b_n(t) > B \right) = e^{-nI^*(B) + o(n)} \tag{13.17}$$

$$I^*(B) \triangleq \inf \left\{ I_0^T(r) : (r, T) \in G(B) \right\} \tag{13.18}$$

$$G(B) \triangleq \left\{ r, T : r(0) = q, \ b^r(T) \geq B \right\} \tag{13.19}$$

and we define $b^r(T)$ in (13.19) by (13.4) with r replacing $\vec{z}_n$ and $b^r(0) = 0$.

We explain this equation below; but first some notation. We need to extend the map that takes $z_n(t)$ to $b_n(t)$ as in (13.2) and (13.4) for a general continuous path $r(t)$ and starting quantity $b(0)$. We take (13.2) as our basic definition, with $r(t)$ replacing $\vec{z}_n(t)$ (the paths are one-dimensional here):

$$\frac{d}{dt} b^r(t) = \begin{cases} r(t) - C & \text{if } b^r(t) > 0, \text{ or if } b^r(t) = 0 \text{ and } r(t) > C; \\ 0 & \text{otherwise.} \end{cases} \tag{13.20}$$

We attempt to generalize (13.4). For any $b \triangleq b^r(0)$ and any $T > 0$ define

$$m(T) \triangleq \inf_{0 \leq t \leq T} \int_0^t \left(r(s) - C \right) ds. \tag{13.21}$$

Clearly there is a time $t \in [0, T]$ where $b^r(t) = 0$ if and only if $m(T) \leq -b$ [note that the dependence of $m(T)$ and τ on the chosen path r is suppressed in our notation]. Therefore

$$b^r(T) = \begin{cases} \displaystyle\sup_{0 \leq t \leq T} \int_t^T \left(r(s) - C \right) ds & \text{if } m(T) \leq -b \\[2ex] \displaystyle b + \int_0^T \left(r(s) - C \right) ds & \text{if } m(T) > -b. \end{cases} \tag{13.22}$$

If we define $\tau(T)$ as the last time the buffer $b^r(t)$ is empty, then $\tau(T) \in [0, T]$ if and only if $m(T) \le -b$. We have the following equivalent definition of $b^r(t)$, where recall that we are considering only $T > 0$ and $b^r(0) = b \ge 0$:

$$b^r(T) = \begin{cases} \displaystyle\int_{\tau(T)}^{T} \left(r(s) - C \right) ds & \text{if } m(T) \le -b \\[2ex] \displaystyle b + \int_{0}^{T} \left(r(s) - C \right) ds & \text{if } m(T) > -b . \end{cases} \tag{13.23}$$

We write $b^r = \mathbf{B}(r)$ when $b^r(t)$ and $r(t)$ are related by any of the equivalent representations (13.20), (13.22), or (13.23). The initial point b and the initial time (which we took to be zero) do not appear explicitly in the notation.

We can now rewrite our variational problem with a form of G slightly different than that given in (13.19):

$$G(B) \stackrel{\triangle}{=} \left\{ r, T : r(0) = q, \ \mathbf{B}(r)(T) \ge B \right\} \tag{13.24}$$

It is not hard to see that the process $b_n(t)$ must be zero quite often; below we bound it by a random walk with negative drift and a boundary at zero, showing that it becomes zero quickly starting from any positive value.

Here is why we expect Equation (13.17) to hold. The Freidlin-Wentzell theory tells us that we expect most steady-state quantities associated with rare events to be estimatable by variational problems, as argued in §6.2. In fact, Theorem 6.59 was designed precisely for the present case. We show that the conclusion of Theorem 6.59 holds for the function $\mathbf{B}(r)$ in §13.6. Therefore we concentrate on solving the variational problem (13.18). First we give some simplifications that enable us to solve it, then we perform the calculations. The main difficulty in solving (13.18) is that $\mathbf{B}(r)$ is defined in terms of an infimum, or in terms of an unknown time $\tau(t)$. So our next goal is to reduce the variational problem to a standard one.

If a path $r(t)$ has minimal cost and makes $b^r(T) \ge B$, then we claim that we can split time into two distinct intervals: $0 \le t \le T_1$, and $T_1 \le t \le T$. For $t \in [0, T_1)$ we have $r(t) < C$, while for $t \in (T_1, T)$ we have $\mathbf{B}(r)(t) = \int_{T_1}^{t} (r(s) - C) \, ds$. This implies that we may replace the set G of (13.19) by

$$\tilde{G}(B) \stackrel{\triangle}{=} \left\{ r, T_1, T : r(0) = q, \ r(T_1) = C, \ \int_{T_1}^{T} (r(t) - C) \, dt = B \right\} \tag{13.25}$$

without changing the solution. There are a few things to explain about this definition. In Exercise 13.7 we show that for any solution of the new variational problem, $r(t) < C$ for $0 \le t \le T_1$, and that the change of "$\ge$" sign to an "$=$" sign is also justified. Finally, the simplification we achieve is to split the variational problem into two distinct and standard variational problems: the cheapest way to go from q to C, and then the cheapest way of making an integral equal to B starting from $r = C$.

It is easy to see why (13.25) gives the correct solution to the variational problem. First, we are concerned with paths that start with $b_n(0) = 0$ and $z_n(0) = q$.

Therefore the first item of business must be to get $z_n(t)$ to above level C; until it does, $b_n(t)$ remains zero. Let us call T_1 the time when the buffer is first non-empty (the time when r first crosses the level C). Next, suppose that the path r that achieves a buffer $\mathbf{B}(r)(T) = B$ has $\mathbf{B}(r)(t) = 0$ at some time $t \in (T_1, T)$. Let t denote the largest such time. Then we clearly have $r(t) = C$. Furthermore, $I^t_{T_1}(r) > 0$ since $r(T_1) = r(t)$, so r cannot follow the most likely path z_∞. Now construct a new path r^* from r by deleting the time from T_1 to t:

$$r^*(s) = \begin{cases} r(s) & s \le T_1 \\ r(s + t - T_1) & s \ge T_1 . \end{cases}$$

Then it is easy to see that r^* is absolutely continuous if r is, $I^T_0(r^*) < I^T_0(r)$, and furthermore r^* makes $\mathbf{B}(r^*)(T - (t - T_1)) = \mathbf{B}(r)(T)$.

This proves that we can restrict our attention to paths r for which $\mathbf{B}(r)$ is never zero from the time r first crosses the level C until the time the buffer reaches the prescribed level B; that is, the map $\mathbf{B}(r)$ is given by the integral of $r - C$ over the whole interval. By a similar "deletion" argument, paths with $r(t) < C$ for $t > T_1$ can be ignored, and we conclude that (13.25) gives the same solution to the variational problem (13.18) as (13.19).

We have reduced our variational problem to the following.

$$I^*(B) = \inf_{G_1} I^{T_1}_0(r) + \inf_{G_2(B)} I^T_0(r) \tag{13.26}$$

$$G_1 \overset{\triangle}{=} \left\{ r, T_1 : r(0) = q, \ r(T_1) = C \right\} \tag{13.27a}$$

$$G_2(B) \overset{\triangle}{=} \left\{ r, T : r(0) = C, \ \int_0^T (r(t) - C)\, dt = B \right\}. \tag{13.27b}$$

[We could, of course, have shifted time by T_1 in the definition of $G_2(B)$ so that the two pieces of the path $r(t)$ would match up.] Now our variational problem has split into two nearly standard variational problems. Indeed, we already found $\inf_{G_1} I^{T_1}_0(r)$ in (13.15). We now turn our attention to the solution of the second variational problem in (13.26).

The second minimization in (13.26) is constrained:

$$\text{minimize} \quad I^T_0(r) \quad \text{subject to} \quad \int_0^T (r(t) - C)\, dt = B. \tag{13.28}$$

Let us use some "soft" (meaning not based on calculation) arguments to show that any solution $r(t)$ to the variational problem (13.28) must satisfy $r(T) = C$. If $r(T) < C$ then $r(s) = C$ at some earlier time $s < T$, and the buffer was larger. Therefore there was an earlier time t when the buffer was equal to B, and $r(t) > C$. So suppose that $r(T) > C$. For s smaller than T, let r_s be the path that follows r until time s, then follows z_∞ until it reaches C (say at T_C). In particular, r_T extends r to times larger than T. This path is well-defined at least for s larger than the last time $s_C < T$ that $r(s_C) = C$. If the buffer size at $t = s_C$ is greater than B then r is not optimal, since $I^T_{s_C}(r) > 0$. So, assume the buffer size at s_c is smaller than B. Now the path r_s, which splits from r at $s \in [s_c, T]$, gives a continuous value for the buffer size at T_C, as a function of s. Since the buffer size corresponding to r_s is below B at $s = s_c$ and above B at $s = T$, there is an intermediate time s_i with buffer size B. Furthermore, the cost of this new path is strictly lower than the cost of r, since the two agree on $[0, s_c]$, but do not agree on the entire interval $(s_c, T]$ (since they end up in different places at time T), and the only path with zero cost on a time interval is z_∞. Therefore we have proved that without loss of generality we may take $r(T) = C$. Let us summarize our discussion concerning the variational problem: for proofs see Exercises 13.7–13.8.

Lemma 13.6. *Any solution of the variational problem (13.26)–(13.27) solves (13.18)–(13.19).*

Exercise 13.7. Show that we may restrict G_1 to paths satisfying $r(t) < C$ for $t < T_1$, and restrict $G_2(B)$ to paths satisfying $r(T) = C$, $\mathbf{B}(r)(T) = B$ and $r(t) \geq C$, $0 \leq t \leq T$. Hint: see the preceding paragraph. ♠

Exercise 13.8. Prove Lemma 13.6. Note that you cannot assume that there exists a solution to (13.18)–(13.19). ♠

One way to solve the constrained optimization problem (13.28) is to introduce a Lagrange multiplier K (for explanation and justification, see the detour below and §13.6) and to consider the extreme points of the functional

$$\int_0^T \left(\ell(r(t), r'(t)) - K(r(t) - C) \right) dt.$$

Therefore, the Lagrange problem we need to solve is

$$I_K^* \triangleq \inf \left\{ \int_0^T \left(\ell(r(t), r'(t)) - K(r(t) - C) \right) dt : (r, T) \in G \right\} \quad (13.29)$$

$$G \triangleq \{r, T : r(0) = r(T) = C\}. \quad (13.30)$$

We then need to choose K to satisfy the buffer size constraint in (13.28).

Intuitively, this is a continuous-time version of the problem of minimizing a function $L(\vec{r})$ subject to a constraint $M(\vec{r}) = C$. We map the functional $I_0^T(r)$ to a function $L(\vec{r})$ by considering the value of r and r' at a mesh of N equally spaced points on $[0, T]$, and letting

$$L(\vec{r}) \triangleq \frac{T}{N} \sum_{i=1}^N \ell(r(Ti/N), r'(Ti/N))$$

$$M(\vec{r}) \triangleq \frac{T}{N} \sum_{i=1}^N (r(Ti/N) - C).$$

This is a standard problem in calculus; if there is a minimum, then it should occur at a critical point for the function $L(\vec{r}) - K M(\vec{r})$.

We proceed to solve the Euler equation for the Lagrange problem, and postpone the technical justifications to §13.6. Since the integrand in (13.29) is independent of t, the DuBois-Reymond equation (C.3) is

$$\ell(r, r') - r'\frac{\partial \ell}{\partial r'}(r, r') - K(r - C) = \text{constant}; \tag{13.31}$$

since T is free, the transversality condition (C.4) implies that the constant in (13.31) is zero. Now using the definition of ℓ in (13.12), recalling (C.6) [or simply copying the argument given for the free $M/M/1$ process (7.23)] and using the general relation $\ell - y \partial \ell / \partial y = -g$ we obtain after some algebra the explicit expression for (13.31)

$$\lambda(1-r(t))+\mu r(t)-\sqrt{(r'(t))^2 + 4\lambda \mu r(t)(1-r(t))}-K(r(t)-C) = 0. \tag{13.32}$$

Squaring and isolating r' we obtain the equation of a hyperbolic cosine:

$$(r')^2 = \big(\mu r + \lambda(1-r) - K(r-C)\big)^2 - 4\lambda \mu r(1-r). \tag{13.33}$$

Taking derivatives with respect to time in (13.33) we obtain the Euler equation

$$r'' = (\mu - \lambda - K)\big(\mu r + \lambda(1-r) - K(r-C)\big) - 2\lambda \mu(1-2r). \tag{13.34}$$

Exercise 13.9. Prove that (13.34) is indeed the Euler equation. Hint: use (C.6), write $\theta^* = f_1(r, r') + f_2(r)$, and let $\Delta = \Delta(r, r')$ denote the square root in (13.13). Delay the expansion of derivatives of θ^* and Δ as much as possible. ♠

We are interested in solutions of the first- and second-order equation with initial conditions $r(0) = C$ and with $r'(0) = \mu C - \lambda(1 - C) > 0$, as determined from (13.33). But Equation (13.34) for r'' is linear in r, and therefore by Theorem A.67 has a unique solution for every initial condition. This seems to establish uniqueness of the solution to the variational problem. However, life is not that easy: a careful inspection of the statements in §C.4 shows that the DuBois-Reymond and the Euler equations need hold only a.e. and that there is no guarantee in general that the solutions be continuously differentiable. But then Theorem A.67 does not apply! Fortunately, in our case, this line of argument can be made rigorous, although this is not trivial. Exercise 13.46 resolves this issue: however, you should not attempt this exercise before reading most of §13.6.

It will be convenient to search for the correct solution by studying the first-order equation. Naturally, we are interested only in those values of K for which the solution r_K satisfies $r_K(T) = C$ for some $T > 0$. Expanding the right-hand side of (13.33), collecting terms in r, and completing the square, we find

$$(r')^2 = a((r-b)^2 + c), \tag{13.35}$$
$$a = (\lambda + K - \mu)^2 + 4\lambda \mu \tag{13.36a}$$
$$b = \big(CK^2 - CK\mu + \lambda^2 + K\lambda + CK\lambda + \mu\lambda\big)/a \tag{13.36b}$$
$$c = a^{-1}(KC + \lambda)^2 - b^2. \tag{13.36c}$$

We see that (13.33) is the equation of either a hyperbolic cosine or a hyperbolic sine, depending on whether c is negative or positive. But the boundary conditions $r(0) = r(T)$ eliminate the hyperbolic sine from consideration, so $c < 0$.

Exercise 13.10. Check that (13.35) gives either a hyperbolic sine or a hyperbolic cosine. Prove that when $q < C$, $c < 0$ implies that $K > \frac{\mu}{1-C} - \frac{\lambda}{C} > 0$. Hint: substitute and use $\cosh^2 x - \sinh^2 x = 1$. ♠

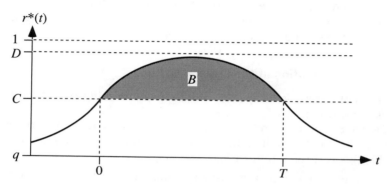

Figure 13.11. $r^*(t)$ for a given value of B. See §13.4 for the meaning of D.

Now denote $r_K(t) = x \cosh(yt + z) + w$. Since cosh is symmetric, the conditions $r(0) = r(T) = C$ imply $y > 0$ and $z < 0$, and we have

$$r_K(t) = x \cosh(yt + z) + w$$
$$y = \sqrt{a} \qquad\qquad\qquad w = b$$
$$x = -\sqrt{-c} \qquad\qquad\qquad T = -2z/y \qquad (13.37)$$
$$\cosh(z) = w - C/(-x) \quad \text{and} \quad z < 0.$$

Given K we can now solve (13.37) for the constants y, w, x, z, T in that order (provided K is such that $c < 0$) to obtain the optimal path r_K.

Our next goal is to find K. We can obtain an explicit expression for

$$\mathbf{B}(r_K)(T) = \int_0^T (r_K(t) - C)\, dt = \frac{2z}{y}(C - w) - \frac{2}{y}\sqrt{(w - C)^2 - x^2}. \quad (13.38)$$

Although this provides an explicit expression, it is too messy to be useful (or to be worth writing down ...). However, we can perform a numerical search for the proper value of K which is determined by $\mathbf{B}(r_K)(T) = B$ (see Figure 13.22). This is particularly simple because of the following properties of $\mathbf{B}(r_K)(T)$ as a function of K, which we establish below:

1. It is monotone decreasing to 0 as $K \to \infty$.
2. It reaches infinity for a fixed positive value of K (see Exercise 13.10 and Theorem 13.16).

We summarize the results we have established concerning the solution of the Lagrange variational problem.

Lemma 13.12. *Assume $q < C$. There exists a unique function r_K solving the DuBois-Reymond and the Euler equations (13.33)–(13.34) with initial conditions $r(0) = C$ and $r'(0) = \mu C - \lambda(1 - C) > 0$. If $KC(1 - C) > r'(0)$ then the unique solution r_K is given by (13.36)–(13.37), and satisfies $r_K(0) = r_K(T) = C$ for some $T > 0$. In this case the solution is symmetric around $T/2$, and its buffer size $\mathbf{B}(r_K)(T)$ is given by (13.38).*

Note that there are multiple solutions to the DuBois-Reymond equation, since we may stay indefinitely at the point r such that $r' = 0$. This ambiguity, however, is resolved by the second-order Euler equation.

Properties of the solution of the Euler equation.

The analysis of our model as well as of its extensions relies on properties of the solutions of Equations (13.33)–(13.34) with initial condition $r(0) = C$. Note that the following results are not restricted to the case $c < 0$, so that a hyperbolic sine is a possible solution. Throughout the section we enforce the condition $q < C < 1$ (remember this is Assumption 13.3).

We now establish a lemma that is surprisingly difficult to prove. It yields an easy proof of Theorem 13.16, which would otherwise be very difficult to establish. The lemma says that as K increases, the time when $r_K(t)$ reaches its maximum (if there is such a time) is strictly decreasing. We develop asymptotics for this decrease in §13.3 and §13.4. You might think to prove this lemma from the explicit formula for T given in (13.37), but we were unable to do so.

Definition 13.13. $K_{\min} \stackrel{\triangle}{=} \dfrac{\mu}{1 - C} - \dfrac{\lambda}{C}$, $T_K \stackrel{\triangle}{=} \dfrac{T}{2} \stackrel{\triangle}{=} \dfrac{z}{y}$, $r_{\max} \stackrel{\triangle}{=} w + x$, and

$$f(r, K) \stackrel{\triangle}{=} \sqrt{(\mu r + \lambda(1 - r) - K(r - C))^2 - 4\lambda\mu r(1 - r)}$$

where T, w, x, y, and z are defined in (13.37), and $f(r, K) = r'_K$ by (13.33).

Note that $r_{\max}$ is the maximum of $r_K(t)$, since $x < 0$ and $\cosh(u) \geq 1$ for any u. Also $K_{\min}$ is the lower bound on K, by Exercise 13.10.

Exercise 13.14. Show that, when $r'_K > 0$, it is strictly decreasing in K for a given r, and it is strictly increasing when $r'_K < 0$. Conclude that $r_{\max}$ is decreasing in K. Hint: take derivatives in (13.33) and note that $r(1 - r) > 0$ and by assumption $\left(r'_K\right)^2 > 0$.   ♠

Lemma 13.15. *T_K strictly decreases from ∞ to zero as K increases from $K_{\min}$ to ∞.*

In order not to interrupt the flow of our narrative, we defer the proof of this lemma to §13.6. The interested reader might want to try to prove it himself or herself; our proof is long and involved, and there may well be a simpler one.

Theorem 13.16. *For each $t \neq 0$ the solutions $r_K(t)$ of (13.33)–(13.34) with initial conditions $r(0) = C, r'(0) = \mu C - \lambda(1 - C)$ are continuous and strictly monotone decreasing in K. Furthermore, for each $t \neq 0$,*

$$\lim_{K \to \infty} r_K(t) = -\infty$$
$$\lim_{K \to -\infty} r_K(t) = \infty.$$

Remark. This theorem states that the paths $r_K(t)$ form a *pencil.* A pencil is a set of paths that radiate from one point and cover the plane without any other intersections. This is a useful notion in the calculus of variations, since it shows that the paths have no conjugate points (we won't go into these notions; see any book on the calculus of variations for a discussion).

Proof. Since the right-hand side of (13.34) is continuous in K and r' is continuous in K by (13.33), $r_K(t)$ is continuous in K. Consider the paths $r_K(t)$ when $r'_K > 0$. Equation (13.33) shows that larger values of K have smaller absolute values of r'_K for the same value of r_K. This means that two different paths $r_K(t)$ cannot intersect while both are increasing. The theorem will therefore be proved if we can show that there are no intersections where one path is increasing and the other is decreasing, or where both are decreasing. But Lemma 13.15 shows that the time when the paths $r_K(t)$ achieve their maxima are strictly ordered: smaller K means a larger time. The paths are symmetric about this point, too, since they are hyperbolic cosines. Therefore since they cannot intersect on the way up, they cannot intersect on the way down.

We also have to examine the monotonicity when $r_K(t) < C$. This is easy, too, considering the order of the paths near the point C. Larger values of K lead to larger values of r'_K, so the paths are strictly ordered. If a path has a minimum there is something more to show, but we leave this to the reader (it is exactly the same as the case when the path has a maximum). ∎

Corollary 13.17. *For any $T > 0$ and $x \in [0, 1]$ there is a unique $K = K(x, T)$ such that $r_K(T) = x$. Furthermore, $0 < r_K(s) < 1$ for each $s \in (0, T)$ and $r'_K(t) < 0$ whenever $r_K(t) < C$.*

Proof. Fix $T \neq 0$. Theorem 13.16 implies that there is a unique K such that $r_K(T) = x$. We only have to check the second statement. Since $x \in [0, 1]$, $r(0) = C \in (0, 1)$ and $r'(0) > 0$, if we show that either r is strictly monotone or that the place where $r' = 0$ is in $(0, 1)$ then we will be done. But (13.33) shows that the only place where $r' = 0$ is where

$$\big(\mu r + \lambda(1 - r) - K(r - C)\big)^2 = 4\lambda \mu r (1 - r).$$

The left-hand side is positive, which is possible only for the right-hand side when $r \in [0, 1]$. Since $r'(0) > 0$, the only solution that is not strictly monotone is a (negative) hyperbolic cosine. Therefore, it remains to exclude the possibility that $r'(1) = 0$. However, if this were true than from (13.33) we get $K(1 - C) = \mu$.

Substituting into (13.34) we obtain $r'' > 0$, that is, a minimum. This contradicts $r(0) = C < 1$. This also establishes the last claim. ∎

Corollary 13.18. *For each $T > 0$ the function $\int_0^T \left(r_K(t) - C\right) dt$ is strictly monotone decreasing from ∞ to $(-\infty)$ as K increases from $-\infty$ to ∞.*

Proof. Immediate from Theorem 13.16. ∎

Lemma 13.19. *For each $x \in [0, 1]$, $B(x, T)$ is strictly monotone increasing in T, $\lim_{T \to \infty} B(x, T) = \infty$ and $\lim_{T \to 0} B(x, T) = 0$ where*

$$B(x, T) \overset{\triangle}{=} \int_0^T \left(r_{K(x,T)}(t) - C\right) dt.$$

Proof. Consider $K(x, T)$ as T increases. By Corollary 13.17, if $x \leq C$ then $r'(T) < 0$. Theorem 13.16 then shows that $K(x, T)$ is monotone decreasing as T increases. Therefore $r_{K(x,T)}(t)$ is monotone increasing as T increases for every $t > 0$, and hence so is $B(x, T)$. The same argument proves the monotonicity of K with respect to T if $x > C$ and $r'(T) < 0$.

If $x > C$ and $r'(T) > 0$ then $r(t)$ is strictly monotone increasing in t for $t \in [0, T]$ [since $r'(t)$ cannot equal zero more than once]. Therefore Theorem 13.16 shows that $r_{K(x,T)}(t)$ is monotone decreasing in T. This means that $K(x, T)$ is monotone increasing in T. Taking the derivative with respect to K in (13.33),

$$2r_K' \frac{dr_K'}{dK} = -2\left(\mu r + \lambda(1 - r) - K(r - C)\right)(r - C).$$

Since $r'(0) = \mu C + \lambda(1 - C) > 0$ the right-hand side is negative, at least near $r = C$. Moreover, for it to become negative (as r increases) it is necessary that $\mu r + \lambda(1 - r) - K(r - C)$ passes through zero. But due to (13.33), this can only happen if $r = 1$. Since $r_K' > 0$ we conclude that $dr_K'/dK < 0$, or that r_K', for a fixed value of r, is decreasing in K. Therefore, the larger K, the smaller the slope of r_K at each level r.

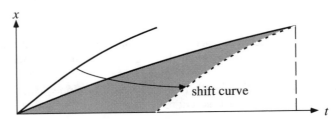

Figure 13.20. Comparing the areas for different values of K.

Now shift the "upper" curve as in Figure 13.20 so that its endpoint meets the "lower" curve, and we see that the total area is decreasing. Finally, the case $r'(T) = 0$ and $x > C$ is handled by one of the two arguments, depending on whether $r_K(t)$ is monotone in t or has a maximum at T.

To see that $B \rightarrow \infty$ as $T \rightarrow \infty$, simply note that K is monotone in T, and so for each $t < T$, $r(t)$ is monotone increasing in T. To see that $B \rightarrow 0$ as $T \rightarrow 0$, note that $r(t) \in (0, 1)$. ∎

Corollary 13.21. *For any $x \in [0, 1]$ and $B > 0$ there is a unique $T = T(x, B)$ such that $B(x, T) = B$.*

Proof. Direct from Lemma 13.19. ∎

Given B, we can clearly find a K such that $x = r(T) = C$ and $B(K) = B$.

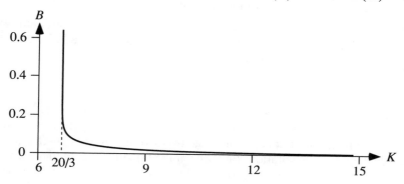

Figure 13.22. $B(K)$ for $\lambda = 1$, $\mu = 2$, and $C = 3/4$. The minimum of K is 6 2/3.

Now given a value of K we can compute the cost $I_0^T(r)$ of the optimal path, starting at C, of getting a buffer size B. From (13.31)

$$I_0^T(r) = \int_0^T \ell(r, r')\, dt = \int_0^T \frac{\partial \ell(r, r')}{\partial r'} r'\, dt + KB. \qquad (13.39)$$

To evaluate the integral on the right of (13.39) consider first the interval $[0, T/2]$ where the optimal path r is strictly increasing. We change the variable of integration from t to r, to obtain an integral of $\partial \ell(r, r'(r))/\partial r'$ with respect to r. Using (C.6) and (7.23) as in the discussion leading from (13.31) to (13.33), and defining

$$\zeta \overset{\triangle}{=} \mu r + \lambda(1 - r) - K(r - C), \qquad (13.40)$$

we obtain the following form for the integrand $\partial \ell / \partial r'$:

$$\log \frac{r' + \sqrt{(r')^2 + 4\lambda \mu r (1 - r)}}{2\lambda(1 - r)} = \log \frac{\sqrt{\zeta^2 - 4\lambda \mu r (1 - r)} + \zeta}{2\lambda(1 - r)},$$

where we use (13.33) to eliminate r' from the equation, taking the positive square root since r' is positive, and we use (13.32) to simplify the square root on the left. This part of the path starts at $r = C$ and ends at $r = w + x$ [the maximum of the cosh: see (13.37)]. The same reasoning gives a similar expression for the path from $w + x$ to C, except that we choose the negative value for r' in (13.33), so that the last square root appears with a minus sign. Finally, the cost, starting at $r(0) = q$, of reaching a buffer size B is the cost c_1 of reaching level C [given in (13.15)], plus the cost computed in (13.39). Collecting all terms, we obtain

$$I^*(B) = c_1 + KB + \int_C^{w+x} \log \frac{\zeta + \sqrt{\zeta^2 - 4\lambda\mu r(1 - r)}}{\zeta - \sqrt{\zeta^2 - 4\lambda\mu r(1 - r)}} \, dr. \qquad (13.41)$$

These integrals can be evaluated explicitly, but there seems little point in giving the evaluations, since they are quite complicated.

Even without having a complete closed form solution to this problem, we can glean some information without much extra work. Consider the Euler equation at the point $r(t) = C$; that is, at $t = 0$ and at $t = T$. We have

$$|r'(t)| = \mu C - \lambda(1 - C) ;$$

that is, the slope does not depend on B, and is equal to the slope of the most likely way of reaching C, and of the most likely way of getting back down to q from C. So even though the variational problem had two distinct regimes, there is a smooth connection between the solutions (at least for the function and its first derivative). This is not entirely coincidental; see the Principle of Smooth Fit for Queueing Problems 13.63.

We collect our results in the form of a theorem. The proof of this theorem relies on the correctness of the large deviations principle for the process $z_n(t)$, the applicability of the Freidlin-Wentzell theory for the process $b_n(t)$ viewed as a function of $z_n(t)$, and the validity of our solution of the variational problem (13.18). These results are established in §13.6.

Theorem 13.23. *Let $z_n(t)$ be described by (13.8), and define $b_n(t)$ by (13.4). Suppose that $q < C < 1$. Then for each $B \geq 0$ we have*

$$\lim_{n \to \infty} \frac{1}{n} \log \mathbb{P}_{ss} (b_n(t) \geq B) = -I^*(B),$$

where $I^*(B)$ is given by (13.41), and where K is the unique number greater than $\mu(1 - C)^{-1} - \lambda/C$ that makes the quantity (13.38) equal to B.

This is where we leave the exact calculation for the buffer of the two-state model. To develop some insight into how the solution behaves, we consider the asymptotic regimes of B small and B large. The first case is interesting because buffers are expensive, and designers may try to minimize the buffers in a system. The second case is possibly even more interesting because of its peculiar mathematical properties. Also, both are interesting for investigating limits taken in different orders (i.e., investigating the effect of limits that don't interchange, namely $n \to \infty$ and $B \to 0$ or ∞).

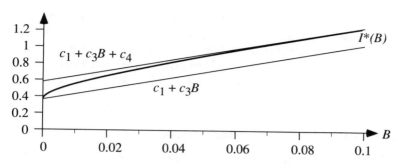

Figure 13.24. $I^*(B)$ for $\lambda = 1$, $\mu = 2$, and $C = 3/4$, and linear upper and lower approximations derived in the next two sections. The lower bound on I^* is easier to calculate than the upper bound.

13.3. Small Buffer

There are at least two approaches to the problem of calculating I^* as B approaches zero. First, we could take our solution to the variational problem, examine limits as $K \to \infty$, and find the consequences. The second approach is more direct, and in any case shows some new ideas, so we follow it instead. We note that buffers are filled by having the function $\vec{r}^*(t)$ follow a concave function with initial derivative equal to $\mu C - \lambda(1 - C)$. By Exercise 13.25, the function cannot get very high before making a buffer of size B. That is, we find that when B is small, $|\vec{r}^*(t) - C|$ is also small during $0 \le t \le T$.

Exercise 13.25. Show that if $r(t)$ is concave, $r(0) = 0$, $r'(0) = a$, $r(t) \ge 0$ for $0 \le t \le T$, and $\int_0^T r(t)\, dt = B$, then $\sup_{0 \le t \le T} r(t) \le \sqrt{2aB}$. Hint: a triangle makes the height largest. ♠

This suggests that we consider the problem of finding the likelihood that a *constant coefficient* process makes a buffer of size B. That is, we replace the rates $\lambda(1 - x)$ and μx by

$$L \stackrel{\triangle}{=} \lambda(1 - C) \quad \text{and} \quad M \stackrel{\triangle}{=} \mu C. \tag{13.42}$$

This provides an approximation of the variational problem, as established in Exercise 13.26 below. So, we consider the variational problem

$$\text{minimize} \quad {}_c I_0^T(r) \stackrel{\triangle}{=} \int_0^T \ell_c(r, r')\, dt$$

$$\text{over the set} \quad G_c \stackrel{\triangle}{=} \left\{ (r, T) : r(0) = C, \ \int_0^T (r(t) - C)\, dt = B \right\},$$

where the local rate function

$$\ell_c(x, y) \stackrel{\triangle}{=} y \log \frac{y + \sqrt{y^2 + 4LM}}{2L} + L + M - \sqrt{y^2 + 4LM} \tag{13.43}$$

is now a function of y alone (the subscript c is used to distinguish the constant coefficients case). We can now use exactly the same arguments that were used in the solution of the variational problem (13.28). We can restrict G_c to paths r that stay above C and satisfy $r(T) = C$. Following the same Lagrange multiplier technique, exactly the same steps show that in this case (13.33) takes the form

$$(r')^2 = (L + M - K(r - C))^2 - 4LM \tag{13.44}$$

which is again the equation of a hyperbolic cosine.

Let $v \triangleq (L + M - K\,(r(t/K) - C))\left(2\sqrt{LM}\,\right)^{-1}$; then

$$\left(\frac{d}{dt}v\right)^2 = v^2 - 1.$$

So $v(t) = \pm \cosh(t - \tilde{z})$ for some $\tilde{z}$. Setting $z_c = K\tilde{z}$, $r_c(0) = r_c(T) = C$,

$$r_c(t) = \left(C + \frac{L + M}{K}\right) - \frac{2\sqrt{LM}}{K}\cosh(Kt - z_c) \tag{13.45}$$

and we have the new expressions

$$z_c = \cosh^{-1}\frac{L + M}{2\sqrt{LM}} = \frac{1}{2}\log\frac{M}{L} > 0$$

$$T_c = \frac{2z_c}{K} \tag{13.46}$$

$$B = \frac{1}{K^2}\left((L + M)\log\frac{M}{L} - 2(M - L)\right).$$

Therefore we obtain the explicit expressions

$$K_c = \left((L + M)\log\frac{M}{L} - 2(M - L)\right)^{1/2}\frac{1}{\sqrt{B}} \triangleq \frac{c_2}{\sqrt{B}} \tag{13.47}$$

$$T_c = \frac{2z_c}{c_2}\sqrt{B}, \tag{13.48}$$

where c_2 is defined by (13.47). Given this value of K_c we can compute I_c^* as in (13.39)–(13.41), with the new rates as in (13.42). Using the elementary integration formulas

$$\int \log\left(x + \sqrt{x^2 - 1}\right)\,dx = x\log\left(x + \sqrt{x^2 - 1}\right) - \sqrt{x^2 - 1}$$

$$\int \log\left(x - \sqrt{x^2 - 1}\right)\,dx = x\log\left(x - \sqrt{x^2 - 1}\right) + \sqrt{x^2 - 1},$$

the fact that

$$\max_{0 \le t \le T} r(t) = C + \frac{1}{K_c}\left(\sqrt{M} - \sqrt{L}\right)^2$$

and the obvious change of variable $u = L + M - K_c(r - C)$, we obtain after a fair amount of (we recommend computer-aided) algebra

$$\int_0^T r'(t) \frac{\partial \ell(r(t), r'(t))}{\partial r'} \, dt = \frac{1}{K_c} \left((L + M) \log \frac{M}{L} - 2(M - L) \right)$$

$$= c_2 \sqrt{B},$$

(13.49)

where c_2 is defined in (13.47). Therefore, substituting (13.47) and (13.49) into (13.39) we find

$$I_c^*(B) = 2c_2 \sqrt{B}.$$

(13.50)

Exercise 13.26. Assume that the paths r^* and r_c^* start and end at C, stay above C, yield buffer size B, and optimize I_0^T and $_c I_0^T$, respectively. Show that

$$\left| I_0^T (r^*) - _c I_0^T (r_c^*) \right| = O(B).$$

Hint: let r denote either of the two paths. Substitute the estimate

$$\sup_{0 \le t \le T} \left| \ell \left(x, r'(t) \right) - \ell_c \left(x, r'(t) \right) \right| \le c_3(x - C), \quad C \le x \le 1$$

(13.51)

in the definitions of I_0^T and $_c I_0^T$ and note that each path makes an area B. To establish (13.51), show first $|dr(t)/dt| \le |z'_\infty(t)|$ (this is obvious if r is decreasing, and use symmetry). This implies (13.51) since ℓ and ℓ_c agree at $r = C$ (check the derivatives). ♠

Theorem 13.27. Assume $q < C < 1$. Then r_c^* and $_c I_0^T (r_c^*)$ approximate r^* and $I_0^T (r^*)$ as follows. For any constant $\tau > 0$,

$$\left| I_0^T (r^*) - _c I_0^T (r_c^*) \right| = O(B)$$

$$\sup_{0 \le t \le \tau \sqrt{B}} \left| r_c^*(t) - r^*(t) \right| = o \left(\sqrt{B} \right).$$

Proof. The first claim is established in Exercise 13.26 above. Define

$$\tilde{K} \triangleq \frac{K}{\sqrt{B}} \qquad \tilde{r}(t) \triangleq \frac{r \left(\sqrt{B} \cdot t \right) - C}{\sqrt{B}}$$

$$\tilde{K}_c \triangleq \frac{K_c}{\sqrt{B}} \qquad \tilde{r}_c(t) \triangleq \frac{r_c \left(\sqrt{B} \cdot t \right) - C}{\sqrt{B}}.$$

Straightforward calculation shows that $\tilde{r}$ and $\tilde{r}_c$ satisfy

$$\left(\frac{d}{dt}\tilde{r}_c(t)\right)^2 = \left(L + M - \tilde{K}_c\tilde{r}_c(t)\right)^2 - 4LM$$

$$\left(\frac{d}{dt}\tilde{r}(t)\right)^2 = \left(L + M + \sqrt{B}(\mu - \lambda)\tilde{r}(t) - \tilde{K}\tilde{r}(t)\right)^2$$

$$- 4LM\left(1 + \frac{\sqrt{B}}{C}\tilde{r}(t)\right)\left(1 - \frac{\sqrt{B}}{1-C}\tilde{r}(t)\right).$$

Now it is easy to see that as $B \to 0$ we must have $K_c \to K$. This is because the coefficients of the two differential equations approach each other. So by the continuity of solutions with respect to parameters, Theorem A.68, the solutions approach each other over any finite time interval, and furthermore the pencil property (Theorem 13.16) shows that the solutions are associated with values of B that are close. Therefore the values of K must be close also. ∎

Therefore, for small values of B we have

$$\mathbb{P}_{ss}(b_n(t) \geq B) = \exp\left(-nc_1 - n2c_2\sqrt{B} + O(nB) + o(n)\right). \qquad (13.52)$$

We also have the following qualitative behavior of the solution as B tends to zero, directly from Theorem 13.27.

1. $T = O(\sqrt{B})$.
2. $r'(0) = M - L = O(1)$. (This is true regardless of B and K.)
3. The maximal value of r satisfies $r_{max} - C = O\left(\sqrt{B}\right)$.

We see that as $B \to 0$, the curvature of $r(t)$ tends to infinity; in fact, as B changes, the function $r(t)$ is simply magnified uniformly in the space and time domains. This is a consequence of the nearly constant-coefficient behavior of the birth-death process near $r(t) = C$. This is not true for large B; for example, in that case, the function $r(t)$ is always bounded by one.

Exercise 13.28. For $\lambda = 1$, $\mu = 2$, $C = 1/2$, take $K = 10$ and using (13.36) and (13.37) compute B, $r(t)$, and I. For this B, using (13.45)–(13.47) find the approximating $r(t)$ and use (13.50) to find the approximating value of I. ♠

Exercise 13.29. Derive the asymptotic formula (13.52) from the exact formula (13.41) by expanding the quantities w, x, y, z, and T for large K. ♠

13.4. Large Buffer

The case of large values of B again may be handled by going back to the variational problem, or by analyzing the exact solution and taking asymptotics; as in the previous section, we perform the former. First note that large values of B are obtained by a function $r(t)$ that is concave and bounded; therefore we must have $r(t) \approx$ constant for long periods of time. So we are led to the question: Which constant is it? Let us call the constant D. The cost per unit time of holding at D is

$$\ell(D, 0) = \left(\sqrt{\mu D} - \sqrt{\lambda(1 - D)} \right)^2, \qquad (13.53)$$

and the growth of the buffer per unit time is $D - C$. Then we minimize

$$\frac{\text{cost}}{\text{buffer growth}} = \frac{\text{cost/unit time}}{\text{buffer growth/unit time}} = \frac{\left(\sqrt{\mu D} - \sqrt{\lambda(1 - D)} \right)^2}{D - C}. \qquad (13.54)$$

Since $q < C$, the relevant minimum satisfies $D > C$ and so occurs where the derivative vanishes. A little algebra gives

$$D = \frac{\mu C^2}{\mu C^2 + \lambda(1 - C)^2} < 1 \qquad (13.55)$$

with a (cost/buffer growth) ratio of

$$\frac{\mu}{1 - C} - \frac{\lambda}{C} \stackrel{\triangle}{=} c_3. \qquad (13.56)$$

That is, $I^*(B) \approx B c_3$ for large values of B.

We can obtain a more accurate approximation for $I^*(B)$, as well as additional information, by calculating r^* for B large. First, from (13.39)–(13.41)

$$I^*(B) = \int \ell(r(t), r'(t)) \, dt$$

$$= \int r'(t) \frac{\partial \ell(r(t), r'(t))}{\partial r'} \, dt + K \int (r(t) - C) \, dt + c_1 \qquad (13.57)$$

$$= \int r'(t) \frac{\partial \ell(r(t), r'(t))}{\partial r'} \, dt + K B + c_1.$$

The last integral turns out to be a bounded function (we will show that it approaches a constant), so we see that for large values of B, $K \approx c_3$. This is another way of seeing that $K \to$ constant as $B \to \infty$. We saw this in Exercise 13.10 by noting that when $K < c_3$ then the curve $r^*(t)$ is a hyperbolic sine, not a hyperbolic cosine.

Now let us evaluate the integral in (13.57) as B becomes large. First, since $r^*(t)$ is a hyperbolic cosine, it is nearly equal to an exponential for t far from the axis of symmetry. Since by Corollary 13.18 and Exercise 13.10 B is monotone in K and is finite for each $K > c_3$, we must have $K \to c_3$ as $B \to \infty$. Algebraically, from the exact solution (13.37), substituting c_3 for K into the equation for y we see that near $t = 0$

$$r^*(t) \approx D - (D - C) \exp\left(-t\left(\lambda\frac{1-C}{C} + \mu\frac{C}{1-C}\right)\right) \qquad (13.58)$$

$$r'(t) \approx \left(\lambda\frac{1-C}{C} + \mu\frac{C}{1-C}\right)(D - r(t)), \qquad (13.59)$$

while near $t = T$ we have

$$r^*(t) \approx D - (D - C) \exp\left((t-T)\left(\lambda\frac{1-C}{C} + \mu\frac{C}{1-C}\right)\right) \qquad (13.60)$$

$$r'(t) \approx \left(\lambda\frac{1-C}{C} + \mu\frac{C}{1-C}\right)(r(t) - D).$$

Hence as B becomes large we have

$$
\begin{aligned}
c_4 &\stackrel{\Delta}{=} \int r'(t)\frac{\partial\ell(r(t), r'(t))}{\partial r'}\,dt \\
&= \int_C^D \log\frac{r'(r) + \sqrt{(r'(r))^2 + 4\lambda\mu r(1-r)}}{2\lambda(1-r)}\,dr \qquad (13.61) \\
&\quad + \int_D^C \log\frac{r'(r) + \sqrt{(r'(r))^2 + 4\lambda\mu r(1-r)}}{2\lambda(1-r)}\,dr,
\end{aligned}
$$

where we use Equation (13.58) for r' as a function of r in the first integral on the right-hand side of (13.61), and (13.60) for the second. These integrals can be explicitly evaluated, but they are so cumbersome that there is little point in presenting the solution.

Now we claim that $K(B)$ tends to c_3 exponentially fast (see Exercise 13.31), so that the factor KB in (13.57) may be replaced by c_3B with negligible error. So, by a combination of exact calculations and asymptotic analyses, we arrive at the conclusion that, as B becomes large,

$$I^*(B) = c_3B + c_1 + c_4 + o(1). \qquad (13.62)$$

The asymptotic path $r^*(t)$ is an exponential increase to the level C followed by an exponential approach to the level D, then a long holding time at D (whose length depends only on the value of B and increases linearly with B), and a symmetrical exponential path away from D back to C and then the final exponential approach to q.

Exercise 13.30. Check that (13.59) gives $r'(0) = \lambda(1 - C) + \mu C$, as we know critical paths for all values of B do. ♠

Exercise 13.31. Show that $|\log(K(B) - c_3)| = O(B)$ as $B \to \infty$. Hint: $x \to 0$ algebraically as $K \downarrow c_3$ in (13.37). Also, $z = O(B)$ as $B \to \infty$ since $y = O(1) > 0$ as $B \to \infty$. ♠

We now prove that the approximation that we have developed is valid. Consider a path $r_c(t)$ defined by (13.58), and then extended symmetrically about the time that makes the path r_c have area $B/2$. Recall from (13.37) that T is the time when the curve r crosses C down.

Theorem 13.32. $\lim\limits_{B \to \infty} \sup\limits_{0 \leq t \leq T} |r_c(t) - r(t)| = 0.$

Proof. The differential equations for the two paths approach each other since we have $K \downarrow c_3$ as $B \to \infty$. Therefore (Theorem A.68) the paths approach each other over any finite time interval. But we know the path $r(t)$ is a hyperbolic cosine making an area B, so is concave and symmetric. Therefore from time of order one to $T/2$ (which is of order B) the path is increasing, and we also know that it never exceeds D, so is even closer to $r_c(t)$ for these times than from a time of order one. To finish the proof we need to show that the time when the curve r_c crosses the level C is close to T. We can do this by showing that the curves are exponentially close in B for times $t \leq T/2$. Using Exercise 13.31 we know that K approaches c_3 exponentially quickly as $B \to \infty$. A straightforward application of Gronwall's Lemma 5.4 shows that there is a constant $\eta > 0$ such that for any $A > 0$,

$$\sup_{0 \leq t \leq A} |r(t) - r_c(t)| \leq e^{-\eta B + tG}, \tag{13.63}$$

where $G \triangleq 2(\mu - \lambda - c_3)^2 + 4\lambda\mu$ is an upper bound on the Lipschitz constant for the differential equation (13.34) when K is close to c_3. It is easy to show from (13.63) that the difference in area between $r_c(t)$ and $r(t)$ over a period of time of length $\sqrt{B}$, say, is decreasing to zero as $B \to \infty$, and that the difference in area during the time $(\sqrt{B}, T/2)$ is also decreasing to zero as $B \to \infty$. From this and the symmetry of the solutions we obtain the theorem. ∎

Corollary 13.33. $\lim\limits_{B \to \infty} \sup\limits_{0 \leq t \leq T} \left| \dfrac{d}{dt} r_c(t) - \dfrac{d}{dt} r(t) \right| = 0.$

Proof. Both functions are concave and are nearly equal. ∎

13.5. Consequences of the Solution

Based on the asymptotic solution, we can calculate a variety of interesting statistics for the system. We will present the calculations without justification. The reader is invited to prove whichever ones interest him or her.

1. Mean buffer.

$$
\begin{aligned}
\mathbb{E}(b_n) &\approx \int_0^\infty \exp(-nI^*(B))\, dB \\
&\approx \int_0^\infty \exp\left(-n\left(c_1 + 2c_2\sqrt{B}\right)\right)\, dB \\
&= \exp(-nc_1)\frac{1}{2n^2 c_2^2},
\end{aligned}
$$

where c_1 is defined in (13.15) and c_2 is defined in (13.47). The second approximation follows from the fact that $I^*(B)$ is monotone increasing in B, so nearly all the relevant portion of the integral is near $B = 0$. The factor $1/2n^2 c_2^2$ should be neglected since the error term is $O(\exp[o(n)])$. That is, we have the following result:

Theorem 13.34. $\displaystyle \lim_{n\to\infty} n^{-1} \log \mathbb{E}_{ss}(b_n) = -c_1.$

2. Transient behavior of the buffer.

How long does the buffer take to drain after it attains a value nB ? Since we know that $r^*(T^*) = C$, and that after T the process $\vec{z}_n(t)$ will follow $\vec{z}_\infty$, we can say that with probability approaching one, $b_n(t) \approx b^*(t)$, where

$$
\begin{aligned}
b^*(t) &\triangleq B + \int_0^t (z_\infty(s) - C)\, ds \\
&= B - \left(C - \frac{\lambda}{\lambda + \mu}\right)\left(t - \frac{1 - \exp(-(\lambda + \mu)t)}{\lambda + \mu}\right)
\end{aligned}
\tag{13.64}
$$

for all t such that the right-hand side is positive, and $b^*(t) = 0$ otherwise. That is, we have the following result.

Theorem 13.35. *For any $T > 0$ and $\varepsilon > 0$,*

$$
\lim_{n\to\infty} \mathbb{P}_{ss}\left(\sup_{0\le t\le T} \left|b_n(t) - b^*(t)\right| < \varepsilon \;\middle|\; b_n(0) = B\right) = 1.
$$

3. Universal shape parameters.

Newell [New] mentions that the transient behavior of a buffer will exhibit a "universal shape," given that it reaches a high level. He states that the buffer begins to fill in a quadratic manner: buffer $(t) \approx C_f t^2$ if it begins to fill at $t = 0$. Furthermore, it empties linearly: buffer $(t) \approx C_e(T - t)$ if it empties at $t = T$. Our analysis for large B shows that

$$b_n(t) \approx (\mu C - \lambda(1 - C))t^2 \qquad \text{for } t \text{ near zero}$$

$$b_n(t) \approx \left(C - \frac{\lambda}{\lambda + \mu} \right)(t - T) \quad \text{for } t \text{ near } T.$$

That is, we have calculated C_f and C_e for our model.

4. Size of buffer during a busy period.

Suppose that we wish to calculate the probability that the buffer will exceed nB in a given busy period (period of non-empty buffer). Then we can immediately write down that

$$\mathbb{P}(b_n(t) \geq B \text{ in a given busy period }) \approx \exp\left(-nc_2\sqrt{B}\right) \qquad B \text{ small}$$

$$\mathbb{P}(b_n(t) \geq B \text{ in a given busy period }) \approx \exp(-nc_3 B - nc_4) \quad B \text{ large}.$$

That is, conditioning on the existence of a busy period simply eliminates the factor $\exp(-nc_1)$ from the buffer statistics.

5. Conditional length of busy period.

Conversely, suppose that we know that a given busy period had a maximum buffer content of nB. How long was the busy period? The length was, with overwhelming probability, $T(B) +$ however long the buffer took to drain from B; see Equation (13.64). $T(B)$ is defined by Equation (13.37) where we find $K(B)$ as described by Figure 13.22.

6. Duration of long busy periods.

We can calculate the probability that a given busy period will exceed T in length. The answer turns out to be the same as the $M/M/1$ queue's busy period, with arrival rate $n\lambda(1 - C)$ and departure rate $n\mu C$:

$$\lim_{n \to \infty} \frac{1}{n} \log \mathbb{P}(\text{ busy period } \geq T) = -T \cdot \left(\sqrt{\mu C} - \sqrt{\lambda(1 - C)}\right)^2. \qquad (13.65)$$

Here is the calculation. Suppose that $\int_0^T (r(t) - C)\, dt = 0$ for some smooth path $r(t)$ with $r(0) = C$. Then we can reorder time so that $\int_0^a (r(t) - C)\, dt \geq 0$, $0 \leq a \leq T$; just put all the time intervals where $r(t)$ is positive at the beginning. We find that the cheapest path r that satisfies $\int_0^T (r(t) - C)\, dt = 0$, $r(0) = C$ extremizes

$$\int_0^T [\ell(r(t), r'(t)) + K(r(t) - C)]\, dt. \qquad (13.66)$$

That is, by our previous calculations, $r(t)$ is an exponential, hyperbolic cosine, or hyperbolic sine. Since $r(0) = C, r'(0) \geq 0$, and $r(T) \leq C$, we must have that r is a hyperbolic cosine (it is the only member of that family that is not monotone). Now look at the cheapest hyperbolic cosine that makes a given positive area in a fixed amount of time, and minimize the cost over all areas. That is, find out how much positive area is generated before the curve dips below C. We obtain that the minimal cost curve makes zero area; that is, the curve is a degenerate one, a constant identically equal to C. There are at least three ways of obtaining this result: by straightforward but tedious calculation, by dividing time into intervals where the curve is above or below C and noting that the curvature must be nonnegative in the portion below C but nonpositive in the portion above, or by noting that it is cheaper for the curve to hold at any fixed level below C than any level above C and so we may lower cost by using the convexity of $\ell(x, 0)$ with respect to x. The details are left to the interested reader.

7. Maximum activity in a long busy period.

Just as in the $M/M/1$ queue, we can show that the most likely way that the buffer has a long busy period is for $z_n(t) \approx C$ for the duration of the period. Furthermore, this means that conditioned on a busy period being long, the maximum level of the buffer is $o(n)$, so conditioning on high buffer content is very different than conditioning on a long busy period. Here is a precise statement of the result.

Theorem 13.36. *Given any $\varepsilon > 0$ and $T > 0$ we have*

$$\lim_{n \to \infty} \mathbb{P}_{ss}\left(\max_{0 \le t \le T} b_n(t) \ge \varepsilon \ \middle|\ \min_{0 \le t \le T} b_n(t) > 0 \right) = 0.$$

13.6. Justification

In this section we check that a large deviations principle holds for $z_n(t)$. We also verity that the Freidlin-Wentzell theory applies to the calculation of fluctuations of $b_n(t)$. We prove Lemma 13.15, and finally, we justify the use of a Lagrange multiplier in calculating the minimal cost path that makes a given buffer B.

We begin with the large deviations principle for $z_n(t)$. Our process does not satisfy the hypotheses of Chapters 5 and 8, since there are two boundaries, at zero and at one, and at each of these places the logarithms of the jump rates $\log \lambda_i(x)$ are not bounded: at zero the service rate μx goes to zero, and at one the arrival rate $\lambda(1 - x)$ goes to zero. This is exactly the type of singularity encountered in Erlang's model at $x = 0$, and the same argument will prove the large deviations principle in this case as well. We follow the argument of §12.6, subsection "The large deviations principle." The idea is to construct a process $z_n^\varepsilon(t)$ that is identical to $z_n(t)$ whenever z_n is more than ε from a boundary, but satisfies a large deviations principle with a rate function $_\varepsilon I_0^T(r)$ that is nearly the same as $I_0^T(r)$. The process we use has the following jump rates and directions:

$$e_1 = 1 \qquad \lambda_1^\varepsilon(x) = \begin{cases} \lambda(1 - x) & x < 1 - \varepsilon \\ \lambda(1 - \varepsilon) & 1 - \varepsilon \le x < 1 \\ 0 & x = 1 \end{cases}$$

$$\tag{13.67}$$

$$e_2 = -1 \qquad \lambda_2^\varepsilon(x) = \begin{cases} \mu x & x \ge \varepsilon \\ \mu \varepsilon & 0 < x \le \varepsilon \\ 0 & x = 0 \end{cases}$$

From here the argument precisely follows that of §12.6, subsection "The Large Deviations Principle;" the interested reader may fill in the details. For the record, here is a statement of the result we now claim holds.

Theorem 13.37. *The process $z_n(t)$ satisfies a large deviations principle with good rate function $I_0^T(r)$.*

The joint process $(z_n(t), b_n(t))$ does not fall into the theory we developed in Chapter 6. Our extension in §6.2 is close to what we need, but some points need to be checked. Instead of trying to simply quote results, we will outline a proof of the Freidlin-Wentzell theory for the process $b_n(t)$ from scratch. We divide time into two types of intervals.

Definition 13.38. *Given an $\varepsilon > 0$, Type 1 intervals of time are those for which $b_n(t) = 0$ and $z_n(t) < C - \varepsilon$. Type 2 intervals are the complement (everything else).*

Let us examine how much time is spent in each interval, and how the processes behave. During a Type 1 interval, Kurtz's Theorem shows that $z_n(t)$ will most likely approach q (if it is not already near q), and therefore that b_n will most likely remain zero, and we expect this interval to be long. Now any Type 2 interval begins (for n large enough) with $C - \varepsilon \le z_n \le C - \varepsilon/2$ and $b_n = 0$. Standard estimates we now outline show that the duration τ of a Type 2 interval has an exponential tail, so that (13.69) holds. This is proved in much the same way as Lemma 6.28. First observe that if b_n remains zero throughout the interval then obviously τ satisfies (13.69). So, we make the following simple estimate to bound the excursions of the new element b_n. By Kurtz's Theorem for $z_n(t)$ we can find a $T < \infty$ such that independent of the starting point $x \in [0, 1]$

$$\mathbb{P}_x\big(b_n(T) > \max\{b_n(0) - 1, 0\}\big) < e^{-n}. \tag{13.68}$$

Obviously b_n cannot increase by more than $T(1 - C)$ in any interval of length T. Therefore, independently of z_n, we can bound b_n over intervals of length T by an asymmetric random walk with strong negative drift (at least until b_n becomes zero). Once b_n reaches zero, we know that z_n is at most C, and by Kurtz's Theorem it is most likely that b_n remains zero and z_n remains below C for the rest of the interval of length T. We conclude from the results for random walks, for example Theorem 3.4, that for T defined above, there is a $g > 0$ and n_0 such that for any $n > n_0$,

$$\mathbb{P}(\text{ a Type 2 interval lasts longer than } KT) < e^{-nKg}. \tag{13.69}$$

Equation (13.69) gives us the tool we need to finish the proof of the applicability of the Freidlin-Wentzell theory to $b_n(t)$. It says that we only have to look at finite times with negligible loss of accuracy—that is, it is the counterpart of Lemma 6.28. Once we have this, then as in Chapter 6 we can prove the results below by simply considering the various types of paths that might occur after the start of a Type 2 interval. The path $z_n(t)$ might cause b_n to exceed B by a time

$T(B)$ [we specify $T(B)$ in a few more sentences]. It might go back towards q, leaving b_n smaller than B. Or it might vacillate, so that it would take longer than $T(B)$ to find out whether or not b_n is going to reach B. We can easily find a path that makes buffer B with a finite cost $I_0^T(r)$ in time T: simply take a straight line from q to one, then hold at one for time $T \approx B/(1 - C)$. This has cost about $\mu B/(1 - C)$. According to (13.69), we can now choose $T(B)$ so that the event of vacillating has a larger exponential rate than the event of causing a buffer of B by time $T(B)$. This means that all the elements of the Freidlin-Wentzell theory are in place. Following the arguments of §6.1 we arrive at the following results.

Theorem 13.39. $\lim\limits_{n \to \infty} \dfrac{1}{n} \log \mathbb{P}_{ss}(b_n(t) \geq B) = -I^*(B)$, where

$$I^*(B) = \inf_G I_0^T(r)$$
$$G = \{r, T : r(0) = q, \mathbf{B}(r)(T) = B\}.$$

Theorem 13.40. Let $\tau_\varepsilon(B)$ denote the time between the beginning of neighboring Type 2 intervals that cause a buffer of at least B. Then for any $\delta > 0$ there is an $\varepsilon_0 > 0$ such that if $\varepsilon < \varepsilon_0$ then for any $B > 0$,

$$\lim\limits_{n \to \infty} \mathbb{P}_{ss}\left(I^*(B) - \delta < \frac{\log \tau_\varepsilon}{n} < I^*(B) + \delta\right) = 1.$$

Theorem 13.41. If there is a unique solution r_B^* to the variational problem of Theorem 13.39, then for any δ and $T > 0$

$$\lim\limits_{n \to \infty} \mathbb{P}_{ss}\left(\sup_{-T \leq t \leq 0} \left|z_n(t) - r_B^*(t)\right| < \delta \,\middle|\, b_n(0) = B\right) = 1.$$

Note that because of the order in which we chose parameters we can take a limit as $B \downarrow 0$ in Theorem 13.40, and obtain a limiting behavior for the occurrences of non-empty buffers ("busy periods"). The buffer may become non-empty and then have a few short busy periods while $z_n(t)$ is nearly equal to C. However, once z_n drops below C by any fixed amount ε, then it is very unlikely that it goes back up to C again quickly. That is, the busy periods occur in bursts, and the Freidlin-Wentzell theory estimates the distribution of time between the bursts, not within the bursts. The limit as $B \downarrow 0$ gives the distribution between bursts, since the two types of time intervals were designed to provide this separation. This finishes our discussion of the Freidlin-Wentzell theory for $b_n(t)$.

Here is a restatement and proof of Lemma 13.15. We use the notation of Definition 13.13.

Lemma 13.15. T_K is strictly decreasing from ∞ to zero as K increases from $K_{\min}$ to ∞.

Proof. Since r increases from C to $r(T_K) = r_{\max}$,

$$T_K = \int_0^{T_K} dt$$

$$= \int_{r=C}^{r_{\max}(K)} \frac{1}{dr/dt}\, dr. \qquad (13.70)$$

$$= \int_{x=0}^{r_{\max}(K)-C} \frac{1}{f\left(r_{\max}(K) - x,\, K\right)}\, dx$$

by a change of variable $x = r_{\max} - r$ and using $r'_K = f(r, K)$. By Exercise 13.14, strict monotonicity will be established once we show that the function $g(x, K) \triangleq (f(r_{\max}(K) - x, K))^2$ is increasing in K for every x in the interval $[0, r_{\max}(K) - C)$. That is, our goal is to show that

$$\frac{\partial g}{\partial K} > 0 \quad \text{for} \quad x \in (0, r_{\max}(K) - C). \qquad (13.71)$$

Since $(f(x, K))^2$ is quadratic in x and $g(0, K) = 0$, (13.35)–(13.36) imply

$$g(x, K) = Q(K)x^2 + L(K)x, \qquad (13.72)$$

$$Q(K) = a + (\lambda + K - \mu)^2 + 4\lambda\mu \qquad (13.73a)$$

$$L(K) = 2a(b - r_{\max}). \qquad (13.73b)$$

Therefore we obtain after quite a bit of algebra that

$$\frac{\partial g}{\partial K} = \frac{\partial Q}{\partial K}x^2 + \frac{\partial L}{\partial K}x \qquad (13.74)$$

$$\frac{\partial L}{\partial K} = \frac{2\sqrt{\lambda\mu}\left[2C(1-C)K + (1-C)\lambda - C\mu\right]}{\sqrt{C(1-C)K^2 + K\left((1-C)\lambda - C\mu\right)}} \qquad (13.75)$$

$$\frac{\partial Q}{\partial K} = 2(\lambda + K - \mu). \qquad (13.76)$$

To derive (13.75), note that the derivative of r is zero at $r_{\max}$ and so

$$0 = ar_{\max}^2 - 2abr_{\max} + (CK + \lambda)^2,$$

$$r_{\max} = b - \sqrt{\frac{-(CK + \lambda)^2 a + a^2 b^2}{a^2}}.$$

Now substitute this expression into (13.73), and eventually obtain

$$L(K) = 4\sqrt{\mu\lambda K}\sqrt{C(1-C)K - C\mu + (1-C)\lambda}.$$

From here the differentiation is tedious but straightforward.

Next we develop some inequalities that will help our analysis. First, it is clear that $\partial L/\partial K > 0$, since

$$K > K_{\min} \overset{\triangle}{=} \frac{\mu}{1-C} - \frac{\lambda}{C} > 0. \tag{13.77}$$

Therefore the only way (13.71) might not hold is if $\partial Q/\partial K < 0$. By (13.76), this only obtains when $\mu > K + \lambda$. This and (13.77) imply

$$\frac{\mu}{\lambda} < \left(\frac{1-C}{C}\right)^2. \tag{13.78}$$

On the other hand, since $C > \lambda/(\lambda + \mu)$,

$$\frac{\mu}{\lambda} > \frac{1-C}{C}. \tag{13.79}$$

Obviously, there will be a non-empty set of μ to examine only if (13.79) and (13.78) hold simultaneously; this can happen only if

$$C < \frac{1}{2} \quad \text{and} \quad \frac{\mu}{\lambda} \in \left(\frac{1-C}{C}, \left(\frac{1-C}{C}\right)^2\right). \tag{13.80}$$

We are trying to show that

$$\frac{\partial L}{\partial K} > x\left(\frac{-\partial Q}{\partial K}\right) \quad \text{for} \quad 0 \le x \le r_{\max}. \tag{13.81}$$

We simplify notation a bit by dividing through by λ, replacing all instances of μ/λ and K/λ by μ and K, respectively. Our goal (13.81), after substituting (13.75) and (13.76), reduces to showing

$$\sqrt{\mu}\left[C(1-C)K + (1-C) - C\mu + C(1-C)K\right]$$
$$> (r_{\max} - C)(\mu - K - 1)\sqrt{K}\sqrt{C(1-C)K + (1-C) - C\mu}. \tag{13.82}$$

Note that

$$C(1-C)K + (1-C) - C\mu = C(1-C)(K - K_{\min}). \tag{13.83}$$

Let $K = mK_{\min}$ for some $m \ge 1$. Our goal, (13.82), becomes

$$\sqrt{\mu}\sqrt{C(1-C)}\,(2m-1)K_{\min} > (r_{\max} - C)(\mu - mK_{\min} - 1)\sqrt{m}\sqrt{m-1}K_{\min}. \tag{13.84}$$

We use the following two simple reductions:

$$r_{\max} - C < D - C < D = \frac{\mu C^2}{\mu C^2 + (1-C)^2} < \frac{\mu C^2}{(1-C)^2} \tag{13.85}$$

$$\mu - mK_{\min} - 1 < \mu - mK_{\min}. \tag{13.86}$$

Therefore we are reduced to showing the following inequality:

$$\frac{(1-C)^2\sqrt{1-C}(2m-1)}{\sqrt{m}\sqrt{m-1}} > \sqrt{\mu C}\,(\mu C - mK_{\min}C) \qquad (13.87)$$

where the parameters satisfy the constraints

$$C \in (0, 1/2) \qquad\qquad m > 1$$

$$CK_{\min} = \frac{\mu C}{1-C} - 1 \qquad\qquad \mu C > 1 - C. \qquad (13.88)$$

Let $w = \mu C$. Then the right-hand side of (13.87) equals

$$\sqrt{w}\left(w - m\left(\frac{w}{1-C} - 1\right)\right) = \sqrt{w}\left(m - w\left(\frac{m}{1-C} - 1\right)\right). \qquad (13.89)$$

The right-hand side of (13.89) is a concave function of w and is maximized at

$$w = \frac{m}{\frac{3m}{1-C} - 3}. \qquad (13.90)$$

There are two cases to consider in order to prove (13.87).

1. Suppose that $m \cdot \left(\frac{3m}{1-C} - 3\right)^{-1} \le 1 - C$. Then the condition on μC in (13.88) shows that (13.89) is maximized at $w = 1 - C$. At that point (13.89) is equal to $(1-C)^{3/2}$ [remember that this is the right-hand side of (13.87)]. But then (13.87) holds, since

$$(1-C)^{5/2}\frac{2m-1}{\sqrt{m}\sqrt{m-1}} > (1-C)^{3/2} \text{ for all } m > 1,\ C < 1/2$$

since

$$\frac{2m-1}{\sqrt{m}\sqrt{m-1}} > 2 > \frac{1}{1-C}.$$

2. Now suppose $m \cdot \left(\frac{3m}{1-C} - 3\right)^{-1} \ge 1 - C$. So, we need to check whether

$$(1-C)^2\sqrt{1-C}\frac{2m-1}{\sqrt{m}\sqrt{m-1}} > \sqrt{\frac{m}{\frac{3m}{1-C} - 3}}\frac{2m}{3}. \qquad (13.91)$$

The condition on m is equivalent to $1 \le m \le 3(1 - C)/2$, which also implies that $C \le 1/3$. We claim that the right-hand side of (13.91) is increasing in m, and the left-hand side is decreasing, so (13.91) holds if it holds for $m = 3(1 - C)/2$. Using this value of m, the left-hand side of (13.91) is $(1 - C)^{3/2}$ times

$$(1-C)\frac{3(1-C) - 1}{\sqrt{\frac{3}{2}(1-C)}\sqrt{\frac{3}{2}(1-C) - 1}} > \frac{2 \cdot 2}{3} > 1$$

since (as straightforward but tedious algebra shows) the left-hand side has no minimum in $(0, 1/3)$ so that the minimum is at $C = 0$. The right-hand side is

$$\sqrt{1 - C} \, \frac{2m}{3} = (1 - C)^{3/2}.$$

This proves that (13.91) holds.

We need to check that the right-hand side of (13.91) is monotone increasing in m for $1 \le m \le \frac{3}{2}(1 - C), 0 < C \le 1/3$. This is a simple exercise we leave to the reader below.

We have shown that T_K is monotone decreasing in K. It is not hard to show that $T_K \to 0$ as $K \uparrow \infty$, or that $T_K \to \infty$ as $K \downarrow K_{\min}$, either directly from the formula (13.37) for T or following the asymptotics of §13.3 and §13.4. ∎

Exercise 13.42. Show that the right-hand side of (13.91) is monotone increasing in m for $1 \le m \le 3(1 - C)/2, 0 < C \le 1/3$. (You might just want to check that the derivative is positive.) ♠

Now we sketch a proof that the Lagrange multiplier rule applies to the AMS model. That is, we justify (13.31), the key equation in our analysis of the model. We would have liked to use standard results in the calculus of variations. Unfortunately, we could not find one that pertained to this model, since there are boundaries at $x = 0$ and $x = 1$ where $\ell(x, y)$ is infinite for finite values of y.

Theorem 13.43 Lagrange Multiplier Rule. *Given a positive T and B, consider the variational problem (13.28) for the basic AMS model. There exists a $K > 0$ so that the solution of the variational problem satisfies (13.31).*

We prove the rule by the following sequence of steps.

1. Show that an auxiliary variational problem $\min_{r \in G} H$ has a solution.

2. Show that the solution of this problem satisfies the Euler equation.

3. Show that there is a subclass of solutions that satisfy $\int_0^T (r(t) - C) \, dt = B$.

4. Show that the minimum of the solutions over all T also satisfies the transversality condition.

5. Show that any solution to the auxiliary problem minimizes the original problem.

Our argument is based on the following technical lemma. Define the vector $\vec{r}$ as $(r, 1 - r)$, which is the vector of "on" and "off" sources (we have been using r in our analysis since $\vec{r}$ is easily derived from r). Further define $\ell(\vec{r}, \vec{r}\,')$ and $I_0^T(\vec{r})$ as the same as the values of the one-dimensional path $r(t)$. You can check that this is the natural definition as given in Chapter 5. For any path $\vec{r}(t)$ and time $T > 0$

define

$$\bar{r}(T) \triangleq \frac{1}{T} \int_0^T \vec{r}(t)\, dt$$

$$\bar{r}'(T) \triangleq \frac{1}{T} \int_0^T \vec{r}'(t)\, dt = \frac{\vec{r}(T) - \vec{r}(0)}{T}.$$

Our next lemma states that the average cost $\frac{1}{T} I_0^T(\vec{r})$ of a path $\vec{r}(t)$ is greater than the cost $\ell(\bar{r}(T), \bar{r}'(T))$ at its average point $(\bar{r}(T), \bar{r}'(T))$.

Lemma 13.44. $I_0^T(\vec{r}) \geq T\ell\left(\bar{r}(T), \bar{r}'(T)\right).$

Proof.

$$T\ell\left(\bar{r}(T), \bar{r}'(T)\right) = \sup_{\vec{\theta}} \left(\langle \vec{\theta}, T\,\bar{r}'(T) \rangle - \sum_i \lambda_i T\, \bar{r}_i(T) \left(e^{\langle \vec{\theta}, \vec{e}_i \rangle} - 1 \right) \right)$$

$$= \sup_{\vec{\theta}} \left(\langle \vec{\theta}, \int_0^T \vec{r}'(t)\, dt \rangle - \sum_i \lambda_i \int_0^T r_i(t)\, dt \left(e^{\langle \vec{\theta}, \vec{e}_i \rangle} - 1 \right) \right)$$

$$\leq \int_0^T \sup_{\vec{\theta}} \left(\langle \vec{\theta}, \vec{r}'(t) \rangle - \sum_i \lambda_i r_i(t) \left(e^{\langle \vec{\theta}, \vec{e}_i \rangle} - 1 \right) \right)$$

$$= I_0^T(\vec{r}).$$
∎

Note that the only property of our model we used in the proof is the linearity of the jump rates $\lambda_i(\vec{r})$ with respect to r_i. We also need the following definitions.

$$T(B) \triangleq \frac{B}{1 - C}; \quad \text{the minimum time a path takes to make area } B.$$

$$D \triangleq \inf_{r > C} \frac{\ell(r, 0)}{r - C} = \frac{\mu C^2}{\mu C^2 + \lambda(1 - C)^2}; \qquad \text{see also (13.55)}.$$

$$m \triangleq \frac{\ell(D, 0)}{D - C} = \frac{\mu}{1 - C} - \frac{\lambda}{C} > 0; \qquad \text{see also (13.56)}.$$

$$H(r, K, T, B) \triangleq \int_0^T \ell(r(t), r'(t))\, dt - K \left(\int_0^T (r(t) - C)\, dt - B \right).$$

$$G(T) \triangleq \{ r \in AC[0, T] : r(0) = r(T) = C \}.$$

Exercise 13.45. Show that if $K > 0$ then there exists an $\varepsilon = \varepsilon(K)$ so that the following hold. If r is any path in $G(T)$ and $r(t) > 1 - \varepsilon$ for some t in $(0, T)$ then there exists a path r_1 in $G(T)$ that never exceeds $1 - \varepsilon$ and also satisfies $H(r_1, K, T, B) < H(r, K, T, B)$. Similarly, if r goes below q it can be replaced with some r_1 that does not. Hints: the second claim is easy—just replace the segment below q with a segment that stays at q. For the first part, assume $r(t) > 1 - \varepsilon$ for $a \leq t \leq b$ and set for notational convenience $a = 0$, $b = T$, and $C = B = 0$.

Using Lemma 13.44 show that

$$H(r, K, T, B) \geq T \left(\ell(\bar{r}, 0) - K\bar{r} \right),$$
$$\frac{d\ell(x, 0)}{dx} \uparrow \infty \quad \text{as} \quad x \uparrow 1 \tag{13.92}$$

and since $\bar{r} > 1 - \varepsilon$ conclude that for all ε small,

$$\frac{\ell(\bar{r}, 0) - \ell(1 - \varepsilon, 0)}{\bar{r} - (1 - \varepsilon)} \geq \frac{d\ell(1 - \varepsilon, 0)}{dx} > K$$
$$\ell(\bar{r}, 0) - K\bar{r} > \ell(1 - \varepsilon, 0) - K(1 - \varepsilon).$$

Therefore the path that remains at $1 - \varepsilon$ is better.   ♠

Exercise 13.46. Prove that there is indeed a unique solution to the variational problem, given through the solution of the DuBois-Reymond and the Euler ODEs. Hint: use the arguments of Exercise 13.45 to avoid singular points. Use arguments as in Lemma 13.47 to show existence of solutions. Use Theorem C.12, Theorem C.18, and the argument preceding this exercise to show uniqueness up to the point where $r'(t) = 0$. Now show an optimal solution cannot stay constant at its maximal value for any interval of time.   ♠

Lemma 13.47. *For each fixed $K, T > 0$, and B, there exists a function $r^*(t)$ in $G(T)$ such that*

$$\inf_{r \in G(T)} H(r, K, T, B) = H(r^*, K, T, B) < \infty.$$

Proof. H is bounded below, since $r(t) \in [0, 1]$, and $H(r, K, T, B) < \infty$ for $r(t) \equiv C$. Now by Proposition 5.46, $\{r : r \in G(T), H(r, K, T, B) \leq A\}$ is compact for each finite A. By Exercise 13.45, we can restrict to paths r that do not approach either zero or one. But then Lemma 5.42 implies that H is lower semicontinuous and the result follows from Theorem A.31.   ∎

Lemma 13.48. *For each fixed $K > 0, T > 0$, and B, the function $r^*(t)$ of Lemma 13.47 has the following properties:*

(i) *$C \leq r^*(t) < 1$ for all $t \in [0, T]$.*

(ii) *$r^*(t)$ satisfies the Euler equation (13.34).*

(iii) *$r^*(t)$ satisfies the Dubois-Reymond equation (13.33) with C replaced by a different constant C_1.*

Proof. Part (i) is established in Exercise 13.45. We prove the rest by using Theorem C.13: it states that any solution of a minimization problem of the form

$$\min \left\{ \int_0^T f(r, r') \, dt : r \in AC, \; r(0) = r(T) = C \right\}$$

will satisfy the Euler equation

$$\frac{\partial f}{\partial r} - \frac{d}{dt}\frac{\partial f}{\partial r'} = 0$$

if f is smooth in r, convex in r', and bounded for r' bounded. Our problem has

$$f(r, r') = \ell(r, r') - K(r - C), \qquad (13.93)$$

which satisfies these conditions when r is bounded away from zero and one. By (i) any solution r^* (which necessarily exists by Lemma 13.47) is uniformly bounded away from zero and one. Therefore r^* satisfies the Euler equation.

Since r satisfies the Euler equation, it satisfies the Dubois-Reymond equation. We write this equation as follows:

$$\lambda(1 - r(t)) + \mu r(t) - \sqrt{(r'(t))^2 + 4\lambda\mu r(t)(1 - r(t))} - K(r(t) - C) = C_2,$$

where C_2 arises since we don't know that the transversality condition holds (T is not free in this problem). We can obviously absorb C_2 into C when $K \neq 0$, resulting in the new equation

$$\lambda(1 - r(t)) + \mu r(t) - \sqrt{(r'(t))^2 + 4\lambda\mu r(t)(1 - r(t))} - K(r(t) - C_1) = 0 \quad (13.94)$$

with $C_1 = C - C_2/K$. This completes the proof of the lemma. ∎

Lemma 13.49. *For each $B > 0$ and $T > T_{min}(B) \overset{\triangle}{=} B/(1 - C)$, there exists a $K > 0$ and C_1 such that the solution $r(t)$ of (13.94) satisfies $r(0) = r(T) = C$ and also $Br(T) = B$.*

Proof. We will not provide a complete proof of this lemma. It is based on calculations similar to those that established Lemma 13.15. These are involved and not particularly illuminating, so we just sketch the ideas.

For each value $r_{max} > 1/2$ and each K larger than a minimal value (which can be calculated but there's no point in giving an explicit value) there is a C_1 such that the solution to (13.94) with $r(0) = C$ has r_{max} as its maximum. It is easy to see that the positive time T_K with $r(T_K) = C$ is decreasing in K and increasing in r_{max}. Therefore $Br(T)$ is increasing in K if we adjust C_1 so that $T_K = T$. There are some complications in that there might be small values of K for which a solution exists, also, but it is easy to see that this solution is monotone in K as well.

There might be some problems with monotonicity for certain small values of C and r_{max}. However, it is easy to show that $Br(T) \to \infty$ as $K \to \infty$, and $Br(T) \to 0$ as K approaches an appropriate minimum. Therefore we are always able to find at least one solution, which is all the lemma claims. ∎

Lemma 13.50. *For each fixed $B > 0$, let $r(t)$ minimize H over the set G with T free, and with $K(B, T)$ defined by Lemma 13.49. Then $r(t)$ satisfies the transversality condition. Specifically,*

$$\frac{d}{dt}r(0) = -\frac{d}{dt}r(T) = \mu C - \lambda(1 - C). \tag{13.95}$$

Proof. We prove the condition assuming that K is a smooth function of T. The proof of smoothness is again a detailed calculation along the lines of the proof of Lemma 13.15, and we omit it for lack of space and motivation.

$$0 = \frac{d}{dT}H(r, K, T, B)$$
$$= \ell(r(T), r'(T)) - K(r(T) - C)$$
$$+ \int_0^T \left(\frac{\partial \ell(r, r')}{\partial r} \frac{\partial r(t)}{\partial T} + \frac{\partial \ell(r, r')}{\partial r'} \frac{\partial r'(t)}{\partial T} \right) dt$$
$$- \frac{dK}{dT} \left(\int_0^T (r(t) - C) dt - B \right) - K \int_0^T \frac{\partial r(t)}{\partial T} dt$$

using Lemma 13.49, integrating by parts, and noting that $\frac{\partial r}{\partial T}(0) = 0$

$$= \ell(r(T), r'(T))$$
$$+ \int_0^T \left(\frac{\partial \ell(r, r')}{\partial r} - K - \frac{d}{dt}\frac{\partial \ell(r, r')}{\partial r'} \right) \frac{\partial r(t)}{\partial T} dt + \frac{\partial \ell}{\partial r'}\frac{\partial r}{\partial T}(T)$$

using the Euler equation

$$= \ell(r(T), r'(T)) + \frac{\partial \ell(r, r')}{\partial r'}\frac{\partial r}{\partial T}(T)$$
$$= \ell(r(T), r'(T)) - r'(T)\frac{\partial \ell(r, r')}{\partial r'}(T),$$

since $\frac{dr}{dT}(T) = -r'(T)$ in order that $r(T) = C$. We use the symmetry of $r(t)$ about $T/2$ to see the result at time zero. ∎

Proof of Theorem 13.43. Given B, minimize H over T with $K = K(B, T)$, obtaining $r(t)$ and T. This pair clearly minimizes $I_0^T(r)$ over paths $r(t)$ satisfying $r(0) = r(T) = C$ and $\int_0^T (r(t) - C) dt = B$. Furthermore, the path satisfies the Euler equation (13.31) with boundary condition (13.33). This concludes the proof of the Lagrange multiplier rule. ∎

13.7. Control Schemes

People continually propose new control schemes for packet networks. Many of them are designed to overcome the large latency (delay-bandwidth product) inherent in long-haul high-speed data transport. Our analysis specifically neglects the problem of delay. Instead, we look at some simple paradigms for control that can be analyzed easily using the calculations we have just performed. Our controls are supposed to reduce the frequency of buffer overflows. We find the steady-state probability that an infinite buffer exceeds a level B. This has the same asymptotic form as the steady-state overflow probability of a finite buffer of size B, since steady-state probabilities and upcrossing probabilities have the same form. We obtain both quantitative and qualitative features of various control schemes.

Our purpose here is to illustrate the use of large deviations techniques and to obtain some insight, rather than to prove every statement as we would if this were a pure mathematics book. We shall therefore leave some calculations unjustified; some for lack of space and motivation, some for lack of ability.

1. Cutoff control.

One of the simplest control schemes is to refuse more than nL sources being active at the same time. That is, the control enforces the inequality $z_n(t) \leq L$. How effective is this at reducing buffer overflows? The precise model here is the following. We have

$$
\begin{aligned}
e_1 &= 1 & \lambda_1 &= \begin{cases} \lambda(1-x) & x < L \\ 0 & x \geq L \end{cases} \\
e_2 &= -1 & \lambda_2 &= \mu x.
\end{aligned}
\tag{13.96}
$$

This model has a standard flat boundary at $x = L$ as well as the nonstandard flat boundary at $x = 0$ that we have analyzed before. Now let's see what effect the level L has on the buffer statistics.

Consider first that $L = 1$ is no control at all, and $L = C$ allows no overflow at all. We might expect that there is a smoothly varying function $f(L, B)$ with

$$
\mathbb{P}(b_n > B) \approx \exp(-nf(L, B))
$$

with $f(1, B) = I^*(B)$ and $f(C, B) = \infty$. But we know that the most likely way large buffers are formed obeys $z_n(t) \leq D$ [see (13.55) for the definition of D]. Therefore we know that

$$
f(L, B) = I^*(B), \quad D \leq L \leq 1.
$$

That is, the control scheme is completely ineffective in reducing the buffer whenever $L \geq D$, at least to the leading order term in n. This holds for all B.

In fact, each fixed buffer level B has an associated $r^*(t)$ for the uncontrolled problem $(L = 1)$ that reaches a maximum $m(B) < D$. [The function $m(B)$ is $w + x$ in the notation of Equation (13.37), where we solve (13.37) for K in terms of B and then substitute to find w and x.] The cutoff control with any level L

above $m(B)$ is thus ineffective in reducing the frequency of buffer overflows; that is, $f(L, B) = I^*(B)$ for $L > m(B)$.

We can easily estimate the effect of the cutoff control for large B and $L < D$. The cost per unit time of remaining at level L is $\left(\sqrt{\mu L} - \sqrt{\lambda(1-L)}\right)^2$ (see Exercise 13.52), and the amount of time needed to fill to level B is about $B/(L-C)$, so the total cost is

$$B\frac{\left(\sqrt{\mu L} - \sqrt{\lambda(1-L)}\right)^2}{L - C},$$

as opposed to

$$B\frac{\left(\sqrt{\mu D} - \sqrt{\lambda(1-D)}\right)^2}{D - C}.$$

Since D minimizes this function, clearly

$$\left.\frac{\partial \text{ cost}}{\partial L}\right|_{L=D} = 0.$$

Therefore we don't expect the cutoff to help much until we are well away from D.

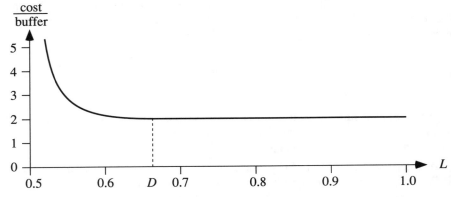

Figure 13.51. Cost/buffer for large B, $\lambda = 1$, $\mu = 2$, $C = 1/2$.

Exercise 13.52. The cost of holding at level L equals $\left(\sqrt{\mu L} - \sqrt{\lambda(1-L)}\right)^2$. Why? Clearly if there were no barrier at L that would be the cost, but the control places a boundary at L. Calculate the cost using the theory of Chapter 8. Alternatively, give a soft argument for why the cost couldn't go up, and this boundary can only make the cost higher, so the cost doesn't change.   ♠

The preceding discussion involved no new calculation, and hence holds for a wide class of models: if there is a unique point $\vec{D}$ where $\{\vec{z}_n(t) \approx \vec{D}$ for a long time$\}$ minimizes a cost function, then we can say:

- Restricting the state space without getting rid of this function won't change the cost.

- Getting rid of this set changes the cost.
- Just barely getting rid of this set doesn't do much if the cost function is smooth.

2. Bit dropping.

In packet voice or video communication the desiderata are different than for data. In particular, faulty packets can be tolerated, whereas in data transmission all packets must be received correctly, so that dropped or garbled packets must be retransmitted or reconstructed. Clever coding schemes have been developed that allow portions of a voice or video packet to be dropped without undue degradation of the quality of the perceived signal. This allows graceful degradation of performance in the event of network congestion: the packets are trimmed down in size, and the network load decreases without dropping any packets.

We now specify three models of bit dropping.

Single watermark. This model has a watermark level B_1 for the buffer. When $b(t) > B_1$ arriving packets are trimmed to a fraction $\rho < 1$ of their original size. When $b(t) < B_1$ no trimming is done. Buffer evolution is described by

$$\frac{d}{dt}b_n(t) = \begin{cases} z_n(t) - C & b_n(t) < B_1, \text{ or } b_n(t) = B_1 \text{ and } z_n(t) < C \\ 0 & b_n(t) = 0 \text{ and } z_n(t) \leq C \\ 0 & b_n(t) = B_1 \text{ and } C \leq z_n(t) \leq C/\rho \\ \rho z_n - C & b_n(t) > B_1, \text{ or } b_n(t) = B_1 \text{ and } z_n(t) > C/\rho. \end{cases}$$
(13.97)

This determines the map
$\mathbf{B} : b_n(t) = \mathbf{B}(z_n)(t)$ for the single watermark model.

Dual watermark. This model has two watermark levels $B_2 < B_1$ for the buffer. The watermarks define two modes of operation for the buffer. Mode 1 is when arriving packets are not trimmed, and Mode 2 is when packets are trimmed to a fraction $\rho < 1$ of their original size. The system changes from Mode 1 to Mode 2 when the buffer reaches level B_1. It changes back to Mode 1 when the buffer reaches level B_2. Therefore the system exhibits hysteresis: when the buffer is between B_2 and B_1, the mode of the system is determined by the history of the system, and not only by its current state. The equations that describe the evolution of the buffer are the following:

$$\frac{d}{dt}b_n(t) = \begin{cases} z_n(t) - C & \text{Mode 1 with } b_n(t) > 0 \text{ or } z_n(t) > C \\ 0 & b_n(t) = 0 \text{ and } z_n(t) \leq C \\ \rho z_n - C & \text{Mode 2.} \end{cases}$$
(13.98)

The mode changes from 1 to 2 when $b_n(t) \geq B_1$, *and changes from 2 to 1 when* $b_n(t) \leq B_2$.

Equation (13.98) and the description of mode changing determine the map $\mathbf{B} :$ $b_n(t) = \mathbf{B}(z_n)(t)$ for the dual watermark model.

The single-watermark model is a limit of the dual-watermark model as $B_2 \uparrow B_1$. You might wish to determine in which space this limit occurs. (It's not in C^1, since the derivatives near B_1 don't converge.)

Full trim. This model has a single watermark B_1, but is similar to the dual-watermark model with $B_2 = 0$. It not only has incoming packets trimmed by a factor $\rho < 1$ when $b_n(t)$ reaches B_1, but all packets in the buffer are also trimmed. The trimming continues until the buffer reaches zero. In other words, the two modes Mode 1 and Mode 2 obtain just as in the dual-watermark model. The equations of evolution of the full trim model are

$$\frac{d}{dt}b_n(t) = \begin{cases} z_n(t) - C & \text{Mode 1 with } b_n(t) > 0 \text{ or } z_n(t) > C \\ 0 & b_n(t) = 0 \text{ and } z_n(t) \leq C \\ \rho z_n - C & \text{Mode 2.} \end{cases} \tag{13.99}$$

The mode changes from 1 to 2 when $b_n(t) \geq B_1$, and changes from 2 to 1 when $b_n(t) = 0$. Furthermore, if the mode changes from 1 to 2 at time T then

$$b_n(T^+) = \rho b_n(T^-). \tag{13.100}$$

Equation (13.99), the description of mode changing, and (13.100) determine the map $\mathbf{B} : b_n(t) = \mathbf{B}(z_n)(t)$ for the full-trim model.

Our goal in this section is to find the asymptotics of the distribution of the buffer when various bit dropping strategies are in effect. We are therefore going to consider the problem

$$\inf \left\{ I_0^T(r) : (r, T) \in G(B) \right\} \tag{13.101}$$

$$G(B) \triangleq \left\{ r, T : r \in AC[0, T],\ r(0) = q = \frac{\lambda}{\lambda + \mu},\ \mathbf{B}(r)(T) \geq B \right\}, \tag{13.102}$$

where $B > 0$ and the bit dropping strategy are given. The functional $\mathbf{B}(\cdot)$ depends on the bit dropping strategy employed as described above. In practice we take functions $r \in G$ to have $r(-\infty) = q$ and $r(0) = C$. We know that the buffer remains zero while $r < C$, and so the first portion of any minimal path must be a minimal cost trajectory to the level C. Thereafter we need to solve a variational problem involving the buffer.

Generic calculations.

There are some common features in the analyses of the bit dropping models. Each model is identical to the original simple AMS model for intervals of time. Therefore it should not be surprising that the solutions of the models are composed of segments of the solutions to original AMS models, pieced together, over these intervals of time. We will provide complete details of the analysis for one model only, namely the single-watermark model. We solve all the models by considering which types of intervals can appear. To avoid trivialities we assume that

Assumption 13.53. $C < \rho < 1$.

If $\rho < C$ then the buffer can reach, but never exceed, B_1. We also need conditions to help us piece the segments together.

Assumption 13.54. The solutions to the variational problems are continuous, and possess continuous first derivatives.

The first condition in Assumption 13.54 is obvious, since the only paths in contention to be minimizers are absolutely continuous. The second condition is far from obvious. We give it the name "The Principle of Smooth Fit" and discuss, in Principle 13.63 below, why the principle should be expected to hold for single-queue problems, and in particular for one of our bit dropping models. General conditions for this to hold are given in Theorem C.18. We do not provide a proof for our case; we simply assume that any optimal path satisfies both of these conditions. We see in Lemma 13.55 that Assumption 13.54, together with appropriate boundary conditions, suffice to uniquely specify the optimal path for the problems we consider.

We now define the various segments of solutions that will be pieced together to form solutions of the variational problem (13.101)–(13.102). For $B > B_1$ define

$$B_3 \overset{\triangle}{=} \frac{B - B_1}{\rho}.$$

Next define

$$r_1(t) \overset{\triangle}{=} q + (C - q)e^{(\lambda + \mu)t}, \qquad -\infty < t \leq 0; \qquad (13.103)$$

$r_1(t)$ is the cheapest path from zero to C, and is the beginning of any optimal path for achieving a positive buffer. Define $r_2(t)$ as the extension of $r_1(t)$ for positive time, and T_2 as the time when $r_2(t) = C/\rho$:

$$r_2(t) \overset{\triangle}{=} q + (C - q)e^{(\lambda + \mu)t}, \qquad t \geq 0 \qquad (13.104)$$

$$T_2 \overset{\triangle}{=} \frac{1}{\lambda + \mu} \log \frac{C/\rho - q}{C - q}. \qquad (13.105)$$

Given an $x \in (0, 1)$ define $r_4(t)$ as the minimal cost hyperbolic cosine from zero to T_4 that satisfies

$$r_4(0) = C/\rho \qquad \int_0^{T_4} (\rho r_4(t) - C) \, dt = B - B_1$$

$$r_4(T_4) = x \qquad \text{or} \quad \int_0^{T_4} (r_4(t) - C/\rho) \, dt = B_3 \qquad (13.106)$$

Here "minimal cost" means cost while bit trimming is in effect. That is,

$$r_4(t) = x \cosh(yt + z) + w,$$

where x, y, z, and w are defined by (13.36)–(13.37) with C replaced by C/ρ and with K determined by the last equation in (13.106) with $B = B_3$. Given $T_1 \in [0, T_2]$ and x define $r_3(t)$ as the minimal cost solution (with the original parameter C) of (13.31) with $r_3(T_1) = r_2(T_1)$, $r_3'(T_1) = r_2'(T_1)$, $r_3(T) = x$, $r_3'(T) = r_4'(T)$, and with $T(B_1)$ defined in Lemma 13.55 so that

$$\int_{T_1}^{T} (r_3(t) - C) \, dt = B_1 - \int_0^{T_1} (r_2(t) - C) \, dt. \qquad (13.107)$$

We can find a solution to (13.107) by Lemma 13.55. Note that $r_3(t)$ might be a hyperbolic sine or a hyperbolic cosine. Finally define

$$r_5(t) = z_\infty(t), \quad z_\infty(T_2) = C/\rho. \tag{13.108}$$

This is the final portion of any optimal curve. The time T_2 will be chosen later, depending on the first portion of the optimal curve.

We construct the solutions to the variational problem (13.101)–(13.102) out of segments of solutions of the original AMS model. We piece these segments together using the "Principle of Smooth Fit" 13.63: $r'(t)$ is continuous. The solutions will turn out to have one of the two following forms.

$$
\begin{aligned}
\text{Case I} \qquad r^*(t) &= \begin{cases} r_1(t) & t < 0 \\ r_2(t) & 0 \leq t \leq T \\ r_4(t) & T \leq t \leq T_2 \\ r_5(t) & t \geq T_2, \end{cases} \\[2mm]
\text{Case II} \qquad r^*(t) &= \begin{cases} r_1(t) & t < 0 \\ r_2(t) & 0 \leq t \leq T_1 \\ r_3(t) & T_1 \leq t \leq T \\ r_4(t) & T \leq t \leq T_2 \\ r_5(t) & t \geq T_2. \end{cases}
\end{aligned}
\tag{13.109}
$$

It turns out that the form of the optimal path is determined by the following simple condition:

$$\text{Case I} \qquad S \overset{\triangle}{=} \int_0^{T_2} (r_2(t) - C)\, dt \geq B_1 \tag{13.110}$$

$$\text{Case II} \qquad S \overset{\triangle}{=} \int_0^{T_2} (r_2(t) - C)\, dt < B_1. \tag{13.111}$$

Does there exist a function $r^*(t)$ of a form given by (13.109) with continuous first derivative? Case I naturally has the derivatives of $r_2(T)$ and $r_4(T)$ match at T if $r_2(T) = r_4(T)$ since both are equal to $\left(-\frac{d}{dt} z_\infty(t)\right)$ there. Similarly r_4 and r_5 have matching derivatives. Therefore the question is only about Case II: does there exist a point $x \in (0, 1)$ such that $r_3(T) = x = r_4(T)$ and with $r_3'(T) = r_4'(T)$? Similarly, do $r_2(T_1)$ and $r_3(T_1)$ match smoothly? The answer to this question is affirmative, and is given by the following lemma.

Lemma 13.55. *Let Assumptions 13.53, 13.54 and 13.56 (below) hold and fix $\rho \in (C, 1)$, $B_1 > 0$, and $B_3 > 0$. Given K, L let $r_K(t)$ and $r_L(t)$ solve the differential equations (13.33)–(13.34) with parameters K and L respectively as the Lagrange multipliers, and C is replaced by C/ρ for $r_L(t)$. Then there is a unique $\{K, L\}$ so that*

$$
\begin{aligned}
&1.\ r_K(0) = C, && 4.\ r_L(T') = C/\rho \\[2mm]
&2.\ r_K(T) = r_L(T), && 5.\ \int_0^T \left(r_K(t) - C\right) dt = B_1, \\[2mm]
&3.\ r_K'(T) = r_L'(T), && 6.\ \int_T^{T'} \left(r_L(t) - C/\rho\right) dt = B_3.
\end{aligned}
$$

Remark. The paths $r_K(t)$ and $r_L(t)$ are $r_3(t)$ and $r_4(t)$ respectively. We can imagine $r_L(t)$ as starting at a time $T' > T$ and being defined backwards in time. In the parlance of ordinary differential equations, we shoot $r_4(t)$ backwards from $t = T'$ to $t = T$. Then we view the solution in forward time. The same reasoning shows that r_2 and r_3 match smoothly. The proof of Lemma 13.55 is immediate from the following result. First recall from Corollary 13.21 that for each $x \in [0, 1]$ and $B > 0$ there is a unique T and path $r_{x,B}(t)$ such that $\mathbf{B}r(T) = B$. We let $r'(x, B) \triangleq \frac{d}{dt}r_{x,B}(T)$.

Assumption 13.56. *Given* $B > 0$, $r'(x, B)$ *is monotone increasing in* $x \in (0, 1)$.

Proof (partial). Here is a proof of the assumption for the case $x \in [C, 1)$; but the case $x \in [0, C)$ must, at this point, be described as a conjecture.

Consider paths $r_K(t)$ for K small, so that the paths reach one in finite time. Call this time T_K. We have $T_K < \infty$ for $K < \frac{\mu}{1-C} - \frac{\lambda}{C}$ by the calculation in Exercise 13.10. As K increases the paths are monotone decreasing, and as shown in Figure 13.20 (and proved in Lemma 13.19), the area they make between time zero and time T_K is monotone increasing. Therefore for some smallest value of K, the area between the curve $r_K(t)$ and C between time zero and time T_K is exactly B. Now as we increase K the area continues to increase. Therefore there will be a time $\tau(K) < T_K$ when the area is equal to B. Furthermore it is easy to see that the value of $r_K(\tau(K))$ is decreasing as K increases (again just look at Figure 13.20). Now the value of $r'(r)$ is also strictly decreasing in K for each fixed r. Therefore we have shown, for small enough K, that the lemma holds.

Let us now continue to increase K until $K = K(B)$; that is, $r(\tau(K)) = C$, and $\tau(K) = T$, where T is defined by (13.37). At this point further increases in K will not allow us to find a $\tau(K)$, since the area up to time T will be smaller than B, and the area after time T will be negative. However, there is another solution that bifurcates from the one we have been following. Consider K decreasing from $K(B)$. The area up to time T is larger than B, but $r(t) < C$ for $t > T$. Therefore there will be a solution to the equation $\int_0^{T_2(K)}(r_K(t)-C)\,dt = B$ with $T_2(K) > T$ and $r_K(T_2(K)) < C$. As we continue to decrease K we eventually reach the point when $r_K(T_2(K)) = 0$. At this point it is impossible to decrease K further and still find a solution with $r_K(T_2(K)) > 0$.

It is clear that the value $r_K(T_2(K))$ is decreasing as K decreases, since r_K increases in K, so there is more area before r crosses C, and there is less negative area after it crosses. What is not clear is that $r'_K(T_2(K))$ decreases. We know that it is more negative than the value when $r = C$, but we have not yet been able to prove that it is strictly monotone. We leave that as a conjecture. This finishes our proof and discussion. ∎

Proof of Lemma 13.55. Consider the position $r_K(T)$ as a parameter x. There is a unique pair (K, T) such that items 1 and 5 hold and with $r_K(T) = x$. Furthermore there is a unique pair (L, T') such that items 4 and 6 hold with $r_L(T) = x$. This makes item 2 hold. Now note that Assumption 13.56 implies that r'_K is monotone increasing in x, and r'_L is monotone decreasing in x, that one is positive and the other negative at $x = C$ and the converse holds at $x = 1$; therefore there is a unique x such that 3 also holds.   ∎

Single watermark.

In this section we solve the variational problem (13.101)–(13.102) for the single-watermark model. Fix a number $B > B_1$. (If $B \leq B_1$ then the problem reduces to a problem without a watermark at all.) The functional $\mathbf{B}$ is defined implicitly by (13.97). Let us suppose that there is a time τ such that $b_n(t) < B_1$ for $t \in (-\infty, \tau)$, and $b_n(t) \geq B_1$ for $t \in (\tau, T)$, where T is the first time that $b_n(t) = B$. In each interval $r^*(t)$ will satisfy an Euler equation, so will be equal to a solution as given in §13.1. The solutions must agree at time τ, of course, since $r^*(t)$ is absolutely continuous. Furthermore, the Principle of Smooth Fit says that the derivatives of the solutions must match at time τ also. Now in the interval $(-\infty, \tau)$, $r^*(t)$ is given by a portion of a solution to (13.33). In (τ, ∞), $r^*(t)$ has a different presentation, since we don't know the starting point $r^*(\tau)$ a priori. But using the symmetry of (13.33) with respect to time reversal, we can consider $r^*(t) = s^*(-t)$ for a path s^* that satisfies the initial condition $s^*(0) = C/\rho$ (this is so that the buffer begins to build immediately after time zero when the path follows s^*).

We first consider what transpires when Case I, (13.110) holds. How can the buffer $b_n(t)$ exceed B_1? It must first reach B_1, and also $z_n(t)$ must exceed C/ρ. But if (13.110) holds, then the cheapest way of causing $z_n(t)$ to exceed C/ρ also happens to make $b_n(T_2) = B_1$ (Why?). Therefore we know that the first part of the optimal path is $r_2(t)$, $0 \leq t \leq T_2$. From here the path simply follows a time shift of $r_4(t)$ until b_n reaches B. Therefore we have the following result.

Theorem 13.57. *Let Assumptions 13.53, 13.54, and 13.56 hold. If*

$$\int_0^{T_2} (r_2(t) - C)\, dt \geq B_1$$

then for $B > B_1$

$$r^*(t) = \begin{cases} r_1(t) & t < 0 \\ r_2(t) & 0 \leq t \leq T_2 \\ r_4(t - T_2) & T_2 \leq t \leq T_2 + T_4, \end{cases} \tag{13.112}$$

$$I^*(B) = c_1(\rho) + I_\rho^* \left(\frac{B - B_1}{\rho} \right). \tag{13.113}$$

Now consider Case II, when (13.111) holds. Then the path $r^*(t)$ must follow the solution to (13.31), with the boundary conditions $r(0) = C$, $r'(0) = |z'_\infty|$ where the second equality comes from Principle 13.63, the Principle of Smooth Fit. This forces the additive constant in (13.31) to be zero. It also shows that $T_1 = 0$; that is, there is no $r_2(t)$ in this case. (We usually use the transversality condition to show that this constant must be zero, but we cannot use this condition in this case since our problem does not necessarily fall under its domain of applicability.) The final boundary condition is

$$r(T_3) = r_4(T_3)$$
$$r'(T_3) = r_4'(T_3).$$

Here $r_4(t)$ is the $r_L(t)$ defined in Lemma 13.55. Now Lemma 13.55 shows that there is a unique x such that

$$r_3(T_3) = x = r_4(T_3)$$
$$r_3'(T_3) = r_4'(T_3).$$

This means that we have found the unique solution to the variational problem of achieving a buffer of size $B > B_1$ with minimal cost. We summarize our findings as a theorem.

Theorem 13.58. *Let Assumptions 13.53, 13.54, and 13.56 hold. If*

$$\int_0^{T_2} (r_2(t) - C)\, dt < B_1$$

then, denoting $c(B_1, \rho) = I_0^T(r_3)$, we have for $B > B_1$

$$r^*(t) = \begin{cases} r_1(t) & t < 0 \\ r_3(t) & 0 \le t \le T \\ r_4(t - T) & T \le t \le T + T_4, \end{cases}$$

$$I^*(B) = c_1 + c(B_1, \rho) + I_\rho^*(B_3).$$

Let us now develop some asymptotics for the cost. We consider the case $B_1 \to \infty$, $B - B_1 \to 0$, and also the case $B_1 \to \infty$, $B - B_1 \to \infty$. The first case is probably the most interesting for applications. We would like B_1 to be large so that the control is not in effect very often. That is, if B is the capacity of the buffer, then we would like $B - B_1$ to be small.

When B_1 is large then we know that $r_3(t)$ is nearly an exponential approach to D with exponent $\lambda \frac{1-C}{C} + \mu \frac{C}{1-C}$, followed by a nearly exponential departure from D. Also $r_3(T_3)$ must be near C/ρ since $B - B_1$ is small. The first question is whether or not $D < C/\rho$. If $D < C/\rho$ then $r_3'(T_3) > 0$. But by the monotonicity of r' with respect to K (and the monotonicity of r with respect to K) we know that $r'(t) < |z_\infty'(t)|$ when $r(t) = z_\infty(t) = C/\rho$. Therefore the point x where $r_3(t) = x = r_4(t)$ must be larger than C/ρ. In fact we will find a constant L such that for $B - B_1$ small,

$$B - B_1 \approx L\sqrt{x - C}. \tag{13.114}$$

This enables us to find x, and hence T_3 and $I_0^T(r)$, as functions of B.

Note that at the time τ when $r_3(\tau) = C/\rho$, by (13.59) we have

$$r'(\tau) \approx \left(\frac{C}{\rho} - D \right) \left(\lambda \frac{1 - C}{C} + \mu \frac{C}{1 - C} \right) \triangleq U. \tag{13.115}$$

Now for B_3 small, r_4 is approximately given by (13.45) with C replaced by C/ρ. We wish to see where $r_4' = U$. For each K, $r_4(t)$ is simply a magnified copy of $r_4(t)$ where $K = 1$. So set $K = 1$ and use (13.44) to obtain

$$r' = U \quad \text{at} \quad r = \frac{C}{\rho} + L + M - \sqrt{U^2 + 4LM}. \tag{13.116}$$

This occurs at time t given by (13.45) with $K = 1$

$$\frac{C}{\rho} + L + M + 2\sqrt{LM} \cosh(t - z) = r';$$

substituting (13.116) for r' yields

$$t - z = \cosh^{-1}\sqrt{1 + \frac{U^2}{4LM}}$$

$$t = z + \frac{1}{2} \log\left(\sqrt{1 + \frac{U^2}{4LM}} - \frac{U}{2\sqrt{ML}}\right).$$

Now we know that B scales as $1/K^2$, that I scales as $1/K$, and that $x - C/\rho$ scales as $1/K$ also. This enables us to calculate the optimal path and cost as a function of B.

Now in case $D > C/\rho$ the preceding calculation carries through with one modification. The path r_3 has $r_3'(T) < 0$, and hence $r_4'(T) < 0$. That is, the two paths intersect to the right of the axis of symmetry of r_4, whereas when $D < C/\rho$ they intersected to the left.

If $D = C/\rho$ the calculation changes a bit. The path $r_3(t)$ must rise to a height x where $x = O\left(\sqrt{B_3}\right)$. This incurs a cost of about $x \log \frac{\mu D}{\lambda(1-D)}$. The derivative of r_3 is of order x, so may be regarded as zero with negligible error. There is another contribution to the cost, though: the increase in area as the path r_3 climbs to x is accompanied by a decrease in the time spend at D; this is easily seen to be of order x also. Finally, the path r_4 meets r_3 very nearly at the axis of symmetry of r_4, so the area under r_4 is half the area under the symmetric hyperbolic cosine. This suffices for determining the rate function for B_1 large, B_3 small, and $D = C/\rho$. So:

Theorem 13.59. *Let Assumptions 13.53, 13.54, and 13.56 hold for the single watermark model. Then*

$$\lim_{B_1 \uparrow \infty, B_3 \downarrow 0} \frac{I^*(B) - c_1 - c_2 B_1 - c_3 \sqrt{B_3} - c_4}{B_3} = 0.$$

The path $r^*(t)$ is given by $r_1(t)$ for $t < 0$, by $r_3(t)$ for $0 \le t \le T$, by $r_4(t)$ for $T \le t \le T_4$, and by $z_\infty(t)$ for $t > T_4$. The constant c_1, defined in (13.15), is the cost of r_1, the cheapest way of getting from $q = \lambda/(\lambda + \mu)$ to C.

If B_1 is large and $B - B_1$ is large, then r_3 is a hyperbolic sine that very nearly decomposes into two exponentials. The reason for this is that r_4 approaches $D_2 >$

D for a long period of time, where D_2 is the minimal cost/area point for the model with bit dropping in effect:

$$D_2 \stackrel{\triangle}{=} \frac{\mu C^2}{\mu C^2 + \lambda(\rho - C)^2} > D. \tag{13.117}$$

Therefore, the path r_3 must be rising in order to meet r_4 smoothly. Also r_4 is a hyperbolic cosine that very nearly decomposes into two exponentials. It is easy to find the point x where r_3 and r_4 meet smoothly:

$$r_3' \approx \left(\lambda \frac{1-C}{C} + \mu \frac{C}{1-C}\right)(r_3 - D) = A(r_3 - D)$$

$$r_4' \approx \left(\lambda \frac{\rho - C}{C} + \mu \frac{C}{\rho - C}\right)(D_2 - r_4) = F(D_2 - r_4), \tag{13.118}$$

where (13.118) defines A and F. Setting $r_3 = r_4 = x$ we find

$$x = \frac{AD + FD_2}{A + B} = \frac{\mu C \frac{1}{1-C} + \frac{1}{\rho - C}}{\frac{\lambda(1-C)^2 + \mu C^2}{C(1-C)} + \frac{\lambda(\rho - C)^2 + \mu C^2}{C(\rho - C)}}.$$

This determines both r^* and I^*. Following the arguments of §13.4, we obtain

Theorem 13.60. *Let Assumptions 13.53, 13.54, and 13.56 hold for the single watermark model. Then*

$$\lim_{B_1 \uparrow \infty, \, B_3 \downarrow \infty} \left(I^*(B) - c_5 + B_1 c_3 + \frac{B - B_1}{\rho} c_3(\rho) \right) = 0,$$

where c_3 is given by (13.56), $c_3(\rho)$ is given by (13.56) with C replaced by C/ρ, and c_5 is calculated below.

Following the argument of §13.4, we obtain

$$c_5 = c_1 + \int_{r=C}^{x} \theta(r_3, r_3') \, dr + \int_{r=x}^{D_2} \theta(r_4, r_4') \, dr + \int_{r=D_2}^{C/\rho} \theta(r_4, r_4') \, dr \tag{13.119}$$

where we use

$$\theta(r, r') = \log \frac{r' + \sqrt{r'^2 + 4\lambda\mu r(1 - r)}}{2\lambda(1 - r)}$$

and (13.118) for the first two integrals in (13.119), and $r_4' = -F(D_2 - x)$ for the third integral.

Exercise 13.61. There is another, more geometric, way of calculating c_5. Compare the path $r(t) = D$, $0 \le t \le T_2$, and the path $r(t) = r_3(t)$, $0 \le t \le T_1$. We choose T_2 so that $T_2(D - C) = B_1$. Then our asymptotic cost is based on the fact that $r_3(t) \approx D$ except near the endpoints. But we can calculate the cost of the endpoints directly, using (13.118) or (13.58), and we can also calculate the difference in areas between the two paths. As in (13.58), let

$$s(t) = D - (D - C) \exp\left(-t\left(\lambda\frac{1-C}{C} + \mu\frac{C}{1-C}\right)\right)$$

$$u(t) = D + (x - D) \exp\left(t\left(\lambda\frac{1-C}{C} + \mu\frac{C}{1-C}\right)\right),$$

which is the symmetrically rising path that reaches x at time zero. Then

$$\int_0^{T_1} \ell(r_3, r_3') \, dt \approx T_1 \ell(D, 0) + \int_0^\infty (\ell(s, s') - \ell(D, 0)) \, dt$$

$$+ \int_{-\infty}^0 (\ell(u, u') - \ell(D, 0)) \, dt, \qquad (13.120)$$

$$\int_0^\infty (s(t) - D) \, dt + \int_{-\infty}^0 (u(t) - D) \, dt \approx (T_2 - T_1)D. \qquad (13.121)$$

You can calculate the limiting difference in cost $\int_0^{T_1} \ell(r_3, r_3') \, dt - T_2 D$ using (13.120) and (13.121).   ♠

Full trim.

We now show how to solve the variational problem (13.101)–(13.102) for the full trim model. We analyze this system the same way we did the previous one. We first consider whether we are in Case I or Case II. If we are in Case II then the analysis goes through very much as in the previous case. We find $T_1 = 0$, so that there is no $r_2(t)$ in the path again. We find $r_3(t)$, $r_4(t)$, and x exactly as before, but now

$$\int_{T_3}^T \rho(r_3(t) - C) \, dt = B - \rho B_1 \qquad (13.122)$$

since as soon as the buffer reaches B_1 it is trimmed to size ρB_1. That is, the cost of exceeding a level $B \approx B_1$ is higher than the previous case by exactly $I_\rho((1 - \rho)B_1)$.

Now suppose that we are in Case I. This case is considerably more complex than before. The problem is that $b_n(t)$ might go back and forth several times between zero and B_1 while $r_2(t)$ is rising from C to C/ρ. Let T_1 denote the last time $b_n(t) = 0$ for $0 \le t \le T_2$, assuming that $r(t) = r_2(t)$. We might have $T_1 = 0$, and we also might have $T_1 = T_2$. Clearly between $t = 0$ and $t = T_1$ the buffer $b_n(t)$ has no net increase. Therefore $r_2(t)$ is the optimal path during this interval. Now on $[T_1, T_2]$ the buffer $b_n(t)$ experiences a net increase. Therefore the optimal path must be of the form $r_3(t)$. On the interval $[T_1, T]$, $r_3(t)$ defined by (13.33), where the initial position $r_3(T_1)$ replaces C so that the Principle of Smooth Fit obtains at $t = T_1$.

We finish calculating $r_3(t)$ in the same way as the previous case: find the unique $x \in [C, D_2]$ with

$$r_3(T) = x = r_4(T)$$
$$r_3'(T) = r_4'(T)$$

$$\int_{T_1}^{T} (r_3(t) - C)\, dt = B_1$$

$$\int_{T_1}^{T} (\rho r_4(t) - C)\, dt = B - \rho B_1.$$

In principle this solves the problem. We can find T_1 analytically, but we need to perform a numerical search for x, and hence for r^* and $I^*(B)$.

We can calculate the asymptotics of $I^*(B)$ for B_1 small. As $B_1 \to 0$, the time T_1 approaches T_2 (in an oscillatory fashion, but it approaches). This means that the connection $r_3(t)$ between r_2 and r_4 takes less and less time. Eventually, of course, the system simply approaches one where the activity is always equal to ρ; for such a system, the optimal path is clearly r_1 followed by r_4, and the cost of achieving any prescribed buffer level is simple to calculate—it is the same as in the original AMS model, with C replaced by C/ρ.

Dual watermarks.

We now examine the variational problem (13.101)–(13.102) for the dual water-mark model. As in the previous models, we first determine whether we are in Case I or Case II. Suppose Case II obtains. Then the optimal path is clearly $r_3(t)$ followed by $r_4(t)$, with a match at a time T at a position x where r_3 and r_4 join smoothly. This time and position are calculated exactly as in the previous models. We can easily compute asymptotics of the system for B_1 and $B - B_1$ both large, or for B_1 large and $B - B_1$ small.

In Case I the situation is much as in the single watermark model. The path $b_n(t)$ may oscillate between B_2 and B_1 while $r(t)$ climbs from C to C/ρ. We let T_1 denote the last time $b_n(t) = B_2$. Then as before we obtain $r^* = r_2$ on the interval $[0, T_1]$. Then on the interval $[T_1, T]$ we have that $r^*(t)$ follows a hyperbolic path $r_3(t)$ defined by (13.33), where the initial position $r_3(T_1)$ replaces C so that the Principle of Smooth Fit obtains at $t = T_1$. We then find the unique place and time so that r and r_4 match smoothly, each making the appropriate buffer size. The details are almost exactly the same as in the previous cases.

We see that as $B_2 \uparrow B_1$, the two watermark case approaches the single water-mark case, with $b_n(t)$ remaining at B_1 during the time $r(t)$ climbs from C to C/ρ. The time T_1 approaches T_2, and so the time when the intermediate path r_3 obtains shrinks to zero, as does its contribution to the buffer and to the cost. The details are quite similar to the previous cases, so we simply state some results, whose proofs follow identically to the proofs of the analogous results for the other two bit dropping models.

Theorem 13.62. *Let Assumptions 13.53, 13.54, and 13.56 hold. For $B > B_1$, the cost function I has the following form.*

$$\text{Case I} \qquad I^* = c_1 + I_0^{T_1}(r_2) + I_{T_1}^{T}(r_3) + I_T^{T_2}(r_4).$$

$$\text{Case II} \qquad I^* = c_1 + I_0^{T}(r_3) + I_T^{T_2}(r_4).$$

The constant c_1 is defined in (13.15).

The paths in Theorem 13.62 are defined by the following equations.

Case I: $\qquad$ T_1 is the last time $\mathbf{B}(r_2)(t) = B_2$

$$r_3(T_1) = r_2(T_1)$$
$$r_3'(T_1) = r_2'(T_1)$$

$$\int_{T_1}^{T} (r_3(t) - C)\, dt = B_1 - B_2$$

$$\int_{T}^{T_2} (r_4(t) - C)\, dt = B_3$$

$$r_3(T) = r_4(T)$$
$$r_3'(T) = r_4'(T),$$

Case II: the same as Case I, with $T_1 = 0$. This means $r_3(0) = C$.

The Principle of Smooth Fit.

We now explain why we expect solutions to variational problems arising in queues to have continuous first derivatives when the path is away from any boundaries. We have already seen how useful this can be in solving these variational problems. Note that extremal paths in general might have arbitrarily bad behavior (see, e.g., Young [Yo]). The principle depends heavily on the type of question we ask—we must be asking about a single queue to be sure it holds. The principle must be checked in any particular case, so we just provide a skeleton of a proof. We do not define "boundaries," but typically we mean a place where the jump rates $\lambda_i(x)$ are discontinuous in x. For our bit dropping models, then, there are no boundaries, even thought the equation for the buffer has discontinuities. We already saw in Chapter 11 that an optimal path can have discontinuous derivatives at a boundary (in that case at the point $x = 0$, where x is the queue occupancy). In Chapters 15 and 16 we give more applications where some optimal paths have discontinuous first derivatives at boundaries.

Principle 13.63: *The Principle of Smooth Fit for Queuing Problems. An extremal path for a time-homogeneous single-queue variational problem must have a continuous first derivative whenever it is away from boundaries.*

Proof skeleton. This proposition is based on a scaling property of the rate function $\ell(x, y)$ and of the queue size. Suppose that there is a discontinuity in $r'(t)$ at

$t = 0$, where $r(t)$ is the optimal path for a single-queue variational problem. Then for a small number δ we consider a new path $r_1(t)$ that is equal to $r(t)$ except on the interval $(-\delta, \delta)$ where it is a linear interpolation.

First we claim that there are constants K and δ_0 such that for any $\delta < \delta_0$,

$$I_0^T(r_1) < I_0^T(r) - K\delta. \tag{13.123}$$

This is because $\ell(r(t), r'(t))$ is nearly equal to $\ell(r(0), r'(t))$ for $t \in (-\delta, \delta)$, and we know that constant-coefficient processes have minimal cost for straight line paths. In detail, Lemma 5.13 states that $\ell(x, y)$ is strictly convex in y, uniformly over x in any finite neighborhood. Furthermore, Exercise 5.34 shows that $\ell(x, y)$ is jointly continuous in x and y. Now assume that ℓ is twice continuously differentiable in both arguments, and that $r'(t)$ and $r''(t)$ are continuous on $(-\delta, 0)$ and on $(0, \delta)$ with limits as $t \to 0$ denoted r'_- and r'_+. Then

$$\int_{-\delta}^0 \ell(r(t), r'(t))\, dt = \int_{-\delta}^0 \ell(r(0), r'_-)\, dt + \int_{-\delta}^0 \left(\ell(r(t), r'(t)) - \ell(r(0), r'_-) \right) dt$$

$$= \int_{-\delta}^0 \ell(r(0), r'_-)\, dt + O(\delta^2)$$

with a similar estimate holding for the integral from zero to δ. Therefore

$$\int_{-\delta}^{\delta} \ell(r(t), r'(t))\, dt = \delta \left(\ell(r(0), r'_-) + \ell(r(0), r'_+) \right) + O(\delta^2)$$

$$> 2\delta \ell \left(r(0), \frac{r'_- + r'_+}{2} \right) + K\delta$$

by the uniform convexity of $\ell(x, y)$ with respect to y. But it is easy to see that the last expression is within $O(\delta^2)$ of $I_{-\delta}^{\delta}(r_1)$.

Now we claim that typically the difference in queue size between that caused by the path $r(t)$ and that caused by the path $r_1(t)$ is of order δ^2. This is because typically the queue is a function of the path that is either an integral or something very much like an integral, and the two paths r_1 and r_2 differ by order δ over a time interval of length δ.

Since the queue is changed by order δ^2 when we change the path from r to r_1, the implicit function theorem indicates that as long as the queue changes with a change in r, we can alter the path r_1 over a time of order δ^2 so that the queue caused by r_1 will be exactly the same as the queue caused by the path r. This might be done, for example, by keeping the endpoints of the path r_1 on $(-\delta, \delta)$ fixed, but lengthening or shrinking the interval by a small amount. This alteration will make a change in $I_0^T(r_1)$ of order δ^2. Alternatively, if $\partial b / \partial t = 0$ at $t = 0$ (where b represents the queue size), then the difference in queue size between the two paths r and r_1 is very small [smaller than $O(\delta^2)$], so changing the path r_1 at another time by simply stretching time by less than $O(\delta^2)$ will make the two queues equal with negligible change in cost. This method of changing the queue is valid only if the buffer is not locally constrained, but is only constrained at the end time T or some other time far from zero.

Therefore we have seen that if r is an optimal path for a variational problem associated with a single queue, then it cannot have a discontinuity in its derivative, or else we could find a path with lower cost that still satisfies the queueing constraint. (The reason we restrict this to a single queue is that we use the implicit function theorem, which in this context is easier to verify for one-dimensional problems than higher dimensional problems.) ∎

We now specialize the Principle of Smooth Fit 13.63 for the single watermark model. The interested reader can supply a complete proof, or consider the other models using similar reasoning. We wish to show that a minimal cost path $r(t)$ that causes $b(t) \geq B > B_1$ has a continuous first derivative. Obviously any path $r(t)$ that is extremal must satisfy the Euler equation when $b(t) \neq B_1$. The only point is to show that the first derivative must be smooth when $b(t) = B_1$. Let τ be the first time that $b(t) = B_1$. There are two cases: $r(\tau) < C/\rho$, and $r(\tau) > C/\rho$ [these cases exactly correspond to Case I and Case II with $r(t) = r_2(t)$]. In the first case the buffer holds at B_1 while $r(t)$ increases. In the second case $\frac{d}{dt}b(\tau) > 0$. In the second case the reasoning outlined in the "proof" of the Principle of Smooth Fit holds. In the first case it is clear that the variational problem does not notice the behavior of $b(t)$, since the path $r(t)$ simply has to reach C/ρ in order to make the buffer increase farther than B_1. Therefore the solution to the variational problem, in either case, has a continuous first derivative.

3. Throttling.

Data packets have different requirements than voice packets. Generally speaking, data packets are insensitive to delays, but very sensitive to errors. This means that bit dropping is not a good control scheme for data. However, since data can be buffered at the source with no ill effect, we can try "throttling" the sources to reduce network congestion. Throttling can mean different things, but all of them involve slowing a source's transmission. There are many schemes for throttling, including "leaky bucket," rate feedback such as Jacobson [Jac] or [JRC], Mitra et al.'s proposal [MRS], etc. We look at a simplified model based on some heuristics.

The idea behind leaky bucket, at least according to its inventor John Turner, is to "enforce the mean." That is, if every source transmits at its mean rate q then obviously there can be no buildup in the queue. One way of enforcing the mean is to have sources' transmissions monitored, and as network congestion increases, hold the sources to their mean level of transmission over shorter and shorter time scales. This makes it more unlikely that an overflow will occur. We model this heuristic as follows. We let the rate at which individual sources turn off be a function of $z_n(t)$; specifically, we make

$$\mu(z) = \mu m(z), \tag{13.124}$$

where $m(z)$ is an increasing function of z. Then the larger the offered traffic rate is, the more quickly individual sources turn off. Now in order to model *the cause* of any surge, we make λ depend on z the same way:

$$\lambda(z) = \lambda m(z). \tag{13.125}$$

This both models the propensity of a throttled source to turn on more quickly than usual, because it still has something to send, and it also models the "enforced mean" heuristic well by not changing the average amount of traffic that a source attempts to send. The model is certainly arguable—but it is simple to analyze, and may be a step towards analyzing more detailed and well-founded models.

The analysis of this model of throttling is very simple, given our results of §13.1. Here are some of the results.

1. The frequency of non-empty buffer is unchanged, since

$$I^* = \int_q^C \log \frac{\mu m(r) r}{\lambda m(r)(1 - r)} \, dr = \int_p^C \log \frac{\mu r}{\lambda(1 - r)} \, dr. \tag{13.126}$$

2. For small values of B,

$$\mathbb{P}(b_n(t) \geq B) \approx \exp\left(-nc_1 - nc_2 \sqrt{B} \sqrt{m(C)}\right). \tag{13.127}$$

That is, even though the probability of the buffer being non-empty doesn't change (asymptotically), the probability that the buffer exceeds any level B decreases by a considerable amount.

3. Large buffers again occur by having the process $z_n(t)$ hold at a constant value δ. The value δ minimizes the quantity

$$\frac{\left(\sqrt{\mu m(\delta)\delta} - \sqrt{\lambda m(\delta)(1 - \delta)}\right)^2}{\delta - C}, \tag{13.128}$$

assuming that there is a unique minimum. The value of $I^*(B)$ is then about $c_5 B$, where c_5 is the minimum in the expression above.

We can make a similar heuristic calculation for a caricature of a bit-dropping model. We model the fact that packets are shorter by multiplying μ by $m(z)$, leaving λ fixed. This makes each busy period for each source smaller by an amount equal on average to m, which is in some sense equivalent to removing a fraction of each packet. We assume that $m(q) = 1$. Now we easily find the following.

1. The frequency of non-empty buffer is changed. The new rate is given by

$$I^* = \int_q^C \log \frac{\mu m(r) r}{\lambda(1 - r)} \, dr = c_1 + \int_q^C \log m(r) \, dr. \tag{13.129}$$

2. For small values of B we have

$$\mathbb{P}(b(t) > B) \approx \exp\left(-nI^* - n\sqrt{B} C_7\right), \tag{13.130}$$

where C_7 has the same form as c_2 but with $\mu m(C)$ replacing μ everywhere. It is easy to see that $C_7 > c_2$. That is, the frequency of non-empty buffer decreases, and the buffer is smaller during busy periods as well.

Our previous model of bit dropping had $m(b)$ instead of $m(z)$, and had m a step function from one to ρ.

Exercise 13.64. Compare the results of this model to the previous one for large B, where you might expect both models to be accurate and to agree. ♠

13.8. Multiple Classes

The simple source model has obvious deficiencies, especially for modeling multiple classes of sources. One potential benefit of packet traffic is its universality: all communication, whether of voice, data, video, multimedia or types not yet developed, can be reduced to moving packets of bits. Different types of sources can be expected to have different characteristics, including different requirements or sensitivities to delay and errors.

The simplest system with multiple classes of sources has each source given by a simple on/off process, where the different classes of sources can have different values of λ_i, μ_i, and a_i, the activity. This falls within our general model of sources if we assign a dimension to each source, as follows. Let f_j be the fraction of sources that are in class j, $1 \le j \le k$, and k is the total number of classes of sources; that is,

$$f_j = \frac{1}{n} \cdot \text{(the number of class } j \text{ sources)} \tag{13.131}$$

where as usual n is the total number of sources. The state vector $\vec{z}_n(t)$ is

$$\vec{z}_n(t) = \left(z_n^1(t), \ldots, z_n^k(t) \right)$$

$$z_n^j(t) = \frac{1}{n} \cdot \text{(the number of active class } j \text{ sources at time t) .}$$

The traffic generated at time t is given by $\langle \vec{z}_n(t), \vec{a} \rangle$, where recall that $\vec{a}$ is the vector of activities. The scaled buffer content $b_n(t)$ satisfies

$$\frac{d}{dt} b_n(t) = \begin{cases} \langle \vec{z}_n(t), \vec{a} \rangle - C & \text{if } b_n(t) > 0 \text{ or } \langle \vec{z}_n(t), \vec{a} \rangle > C; \\ 0 & \text{otherwise.} \end{cases}$$

The equilibrium point of the system is

$$\vec{q} = \left(f_1 \frac{\lambda_1}{\lambda_1 + \mu_1}, \ldots, f_k \frac{\lambda_k}{\lambda_k + \mu_k} \right). \tag{13.132}$$

We assume that

$$\langle \vec{a}, \vec{q} \rangle < C < \langle \vec{a}, \vec{f} \rangle \tag{13.133}$$

so that the statistics of the buffer are nontrivial. The path $\vec{z}_\infty(t)$ is given by

$$\frac{d}{dt} \vec{z}_\infty(t) = \sum_{i=1}^{k} \left(\lambda_i (f_i - z_i) \vec{e}_i - \mu_i z_i \vec{e}_i \right). \tag{13.134}$$

The rate function for the system is defined by

$$I_0^T(\vec{r}) = \int_0^T \ell(\vec{r}(t), \vec{r}'(t))\, dt$$

$$\ell(\vec{x}, \vec{y}) = \sup_{\vec{\theta} \in \mathbb{R}^k} \left(\langle \vec{\theta}, \vec{y} \rangle - \sum_{i=1}^k \left[\lambda_i (f_i - x_i) \left(e^{\theta_i} - 1 \right) + \mu_i x_i \left(e^{-\theta_i} - 1 \right) \right] \right).$$

It is easy to see that $\ell(\vec{x}, \vec{y}) = \sum_{i=1}^k \ell_i(x_i, y_i)$, where

$$\ell_i(x_i, y_i) = y_i \log \frac{y_i + \sqrt{y_i^2 - 4\lambda_i(f_i - x_i)\mu_i x_i}}{2\lambda_i(f_i - x_i)} + \lambda_i(f_i - x_i) \qquad (13.135)$$
$$+ \mu_i x_i - \sqrt{y_i^2 - 4\lambda_i(f_i - x_i)\mu_i x_i}\,.$$

The reason that the ℓ-function splits into a sum of independent pieces is that the process is a sum of independent pieces: each dimension is (statistically) independent of the others. This enables us to piece together the solution of various problems from a number of simple parts.

This split enables us to prove the large deviations principle for $\vec{z}_n(t)$ from the corresponding large deviations principle for the components of $\vec{z}_n(t)$. The lower bound follows directly from the lower bound for the components. Using the simple inequality for $\vec{y} \in \mathbb{R}^k$

$$|\vec{y}| \le \sqrt{k} \max |y_i|$$

we have for any T and $\varepsilon > 0$

$$\mathbb{P}_{\vec{x}}\left(\sup_{0 \le t \le T} |\vec{z}_n(t) - \vec{r}(t)| < \varepsilon \right) \ge \prod_i \mathbb{P}_{x_i}\left(\sup_{0 \le t \le T} |z_{n,i}(t) - r_i(t)| < \frac{\varepsilon}{\sqrt{k}} \right)$$

$$\ge e^{\left(-n \sum_i \int_0^T \ell_i(r_i, r_i')\, dt + O(n\varepsilon) + o(n) \right)}$$

$$= e^{-n I_0^T(\vec{r}) + O(n\varepsilon) + o(n)}.$$

The upper bound is almost as easy. Recall the definition

$$\Phi_{\vec{x}}(K) = \left\{ \vec{r} : \vec{r}(0) = \vec{x},\ I_0^T(\vec{r}) \le K \right\}.$$

Define the closed set

$$S_{\vec{x}}(\varepsilon, K) \stackrel{\triangle}{=} \left\{ \vec{r}(t) : d(\vec{r}, \Phi_{\vec{x}}(K)) \ge \varepsilon \right\}.$$

Then define $\vec{y}_n(t)$ as in Chapter 5 as the piecewise linear interpolation of $\vec{z}_n(t)$. Then for each i and $K_i \ge 0$, the large deviations upper bound for the components $z_{n,i}$ of $\vec{z}_n$ gives

$$\limsup_{n \to \infty} \frac{1}{n} \log \mathbb{P}\left(z_{n,i} \in S_x(\varepsilon, K_i) \right) \le -K_i. \qquad (13.136)$$

Divide K into K/ε equal components. Then $d(\vec{y}, \Phi_{\vec{x}}(K)) > \varepsilon$ implies that there is a set of K_i with $\sum_i K_i = K - k\varepsilon$ and such that

$$d(y_i, \Phi_i(K_i)) > \varepsilon/k. \tag{13.137}$$

There are only a finite number of possible K_i to choose from. Recall Lemma 5.57, which states that $\vec{y}_n$ and $\vec{z}_n$ are close with rate function tending to ∞ as $n \to \infty$. Hence (13.136) and (13.137) imply that

$$\limsup_{n\to\infty} \frac{1}{n} \log \mathbb{P}_{\vec{x}}\left(\vec{y}_n \in S_{\vec{x}}(\varepsilon/2, K)\right) \leq -(K - \varepsilon). \tag{13.138}$$

That is, (5.70) holds with slightly different constants. This is the key equation used in proving Theorem 5.64, the large deviations upper bound. That is, the upper bound holds for our process $\vec{z}_n(t)$.

The heuristic derivation of the large deviations upper bound also provides a compact reason why the upper bound should hold for a process $\vec{z}_n(t)$ if it holds for the components $\vec{z}_{n,i}(t)$. Each component should satisfy

$$\mathbb{E}\left(e^{nI_0^T(y_{n,i})}\right) = O(1).$$

Therefore

$$\mathbb{E}\left(e^{nI_0^T(\vec{y}_n)}\right) = \prod \mathbb{E}\left(e^{nI_0^T(y_{n,i})}\right) = O(1)$$

also, which is the essential step in proving the upper bound. This is very nearly the reasoning we use in our proof. See also the note following Equation (13.142) below.

Level crossing.

We begin with the question of how often the buffer becomes non-empty; equivalently, we try to find the cheapest path from $\vec{q}$ to the hyperplane $\langle \vec{x}, \vec{a} \rangle = C$. The variational problem to be solved is

$$I^* = \inf_G I_0^T(\vec{r})$$
$$G = \{\vec{r}(t), T : r(0) = \vec{q}, \langle \vec{r}(T), \vec{a} \rangle = C\}.$$

Either by observing the form of ℓ as the sum of ℓ_i, or by taking special variations in each component separately, we see that the Euler equation for each class is the same as before except that $\lambda(1 - x)$ is replaced by $\lambda_i(f_i - x_i)$. Or even more simply, if the point $r_i(T)$ is given, then we know that the cheapest way to get there is for $r_i(t)$ to follow the level crossing solution given before as the reversed path from $r_i(T)$ to q_i. In any case, we find that

$$r_i^*(t) = q_i + (r_i^*(T) - q_i)\exp((\lambda_i + \mu_i)t),$$

and the associated cost is given by

$$I^* = \sum_{i=1}^{k} \int_{q_i}^{r_i(T)} \log \frac{\mu_i x}{\lambda_i(f_i - x)} \, dx.$$

By making the usual time shift we may take $T = 0$ so that the path crosses the hyperplane at time zero. The only question is to find the place on the hyperplane where the crossing takes place.

Let us minimize I^* using a Lagrange multiplier L on the constraint

$$\sum_{i=1}^{k} a_i r_i(0) = C. \tag{13.139}$$

We obtain the system of equations

$$\log \frac{\mu_i r_i(0)}{\lambda_i(f_i - r_i(0))} = La_i; \qquad i = i, \dots, k.$$

Defining $S \overset{\triangle}{=} \exp(L)$, we obtain

$$\frac{\mu_i r_i(0)}{\lambda_i(f_i - r_i(0))} = S^{a_i}$$

so that

$$r_i(0) = f_i \frac{\lambda_i S^{a_i}}{\mu_i + \lambda_i S^{a_i}}. \tag{13.140}$$

We find S by the constraint (13.139):

$$\sum_{i=1}^{k} a_i f_i \frac{\lambda_i S^{a_i}}{\mu_i + \lambda_i S^{a_i}} = C. \tag{13.141}$$

This equation cannot be solved analytically, in general, but a moment's inspection shows that it is trivial to solve numerically because of the following properties:

1. The left-hand side is monotonically increasing in S.

2. It is equal to $\langle \vec{a}, \vec{q} \rangle$ at $S = 0$, and is equal to $\langle \vec{a}, \vec{f} \rangle$ at $S = \infty$.

To summarize, the level crossing problem is solved by recognizing that it may be reduced to a number of one-dimensional level crossing problems. The only thing that ties the problems together is the position on the hyperplane where the level is crossed, and this is reduced to a one-dimensional root of a monotone function. The final answer is

$$I^* = \sum_{i=1}^{k} \int_{q_i}^{r_i(0)} \log \frac{\mu_i x}{\lambda_i(f_i - x)} \, dx$$

$$= \sum_{i=1}^{k} \left(r_i(0) \log \frac{r_i(0)}{q_i} + (f_i - r_i(0)) \log \frac{f_i - r_i(0)}{f_i - q_i} \right), \tag{13.142}$$

where $r_i(0)$ is given by (13.140) with S the solution of (13.141).

It should not be too surprising that the level crossing problem reduces to a one dimensional problem. Consider the equivalent problem in the setting of i.i.d. random
variables. Suppose we have a collection of nf_1 random variables with distribution
F_1, nf_2 random variables with distribution F_2, up to nf_K random variables with
distribution F_K. The question is now, What is the probability that the sum of all
these random variables exceeds a level $nC > n \sum_i f_i \mu_i$, where we are writing
$\mu_i = \int x \, dF_i(x)$? There are at least two ways to approach this problem. Using
Chernoff's Theorem, we could estimate the probability that the type i random variables achieve a mean of at least r_i, and then compute the cheapest set of levels r_i
satisfying $\sum_i f_i r_i \geq C$, where "cheap" means the sum of the rates $\sum_i f_i \ell_i(r_i)$.
This is equivalent to the problem of finding where the path $\vec{r}(t)$ crosses the hyperplane $\langle \vec{r}, \vec{a} \rangle = C$. However, there is another point of view. We could simply
examine the proof of Chernoff's Theorem for independent random variables, and
come up with a single θ so that

$$\sum_i \frac{\int f_i x_i e^{\theta x_i} \, dx_i}{\int e^{\theta x_i} \, dx_i} = C.$$

That is, instead of taking a different θ in each rate function ℓ_i, the minimum cost
will occur when all the various θ_i are the same. This is the same idea that allowed
us to prove that the various presentations of the rate function for the finite levels
process are the same in Chapter 8.

Buffer statistics.

We may calculate the statistics of the buffer in a manner very similar to the level
crossing problem. We reduce the problem to a number of one-dimensional previously solved problems, then attempt to fit them together. We make use of the
Principle of Smooth Fit 13.63 in the calculation of various asymptotics.

We now write the variational problem and its solution, insofar as we are able.
Given a level B, we introduce a Lagrange multiplier L and examine stationary
points of the functional

$$\int_{T_1}^0 \ell(\vec{r}(t), \vec{r}'(t)) \, dt + \int_0^T \left(\ell(\vec{r}(t), \vec{r}'(t)) - L(\langle \vec{r}(t), \vec{a} \rangle - C) \right) \, dt$$

$$\vec{r}(T_1) = \vec{q}, \quad \langle \vec{r}(0), \vec{a} \rangle = C, \quad \int_0^T (\langle \vec{r}(t), \vec{a} \rangle - C) \, dt = B.$$

Replacing C with $\langle \vec{r}(0), \vec{a} \rangle$ in the second integral, we are led to examine the stationary points of the functional

$$\sum_{i=1}^k \int_0^T \left(\ell_i(r_i(t), r_i'(t)) - La_i(r_i(t) - r_i(0)) \right) \, dt.$$

Again the variational problem has split into k independent problems. We have the solution in the form

$$r_i(t) = x_i \cosh(y_i t - z_i) + w_i.$$

The only question here is to find $r_i(0)$ and $r_i(T)$. Once we do, we are able to calculate all the parameters of $r_i(t)$ since then we have reduced the problem to independent one-dimensional problems. (There is, of course, the problem of calculating L, but we were unable to solve this except numerically in the one-dimensional case as well.) To find $\vec{r}(T)$, we use the Principle of Smooth Fit 13.63. After T the path $\vec{r}^*$ follows $z_\infty(t)$. Also, from $T_1 (= -\infty)$ to zero we have $\vec{r}^*(t) = \vec{z}_\infty(-t)$. Therefore we must have $r'_j(0) = -r'_j(T)$. But this can happen only if $r_j(0) = r_j(T)$. That is, the minimizing path must exit the region $\langle \vec{r}, \vec{a} \rangle \geq C$ at the same point it came in. [This is not to say that different buffer sizes B might not have different places where they hit the hyperplane, only that for a given B, $\vec{r}(0) = \vec{r}(T)$.] Furthermore, given a value for $\vec{r}(0)$, we have $\int_{T_1}^0 \ell(\vec{r}, \vec{r}') \, dt$ is given by (13.142). This is as far as we are able to get with the solution of the general variational problem; we are not able to give a closed form solution for $\vec{r}(0)$ and must leave it for a numerical search. However, we can again find the asymptotics for large and small B, to which we now proceed.

Small buffers.

The small B asymptotics are computed in very much the same way as the simple source model. The only new item is the place where the optimal path crosses the hyperplane. We begin with this calculation.

Consider the point $\vec{y} \triangleq \vec{r}(0)$ defined by the level crossing problem (13.140). Since r is an extremal, the rate function I of paths that reach any point $\vec{x}$ in a small neighborhood of $\vec{y}$ on the hyperplane will be close to I^* to within $O(|\vec{x} - \vec{y}|^2)$. Furthermore, for any $\varepsilon > 0$ small enough there is a $\delta > 0$ such that all paths ending outside a δ-neighborhood of $\vec{y}$ will have a cost $I > I^* + \varepsilon$. These considerations lead us to the following conclusions. For the B-buffer path $\vec{r}_B^*(t)$ with $\langle \vec{r}_B^*(0), \vec{a} \rangle = C$, $\int_0^{T(B)} \langle \vec{r}_B^*(t) - \vec{r}_B^*(0), \vec{a} \rangle \, dt = B$:

1. $\lim_{B \to 0} \vec{r}_B^*(0) = \vec{y}$.

2. $|\vec{r}_B^*(0) - \vec{y}|^2 \leq O(B^{1/4})$, since $\int_0^{T(B)} \ell \leq O(\sqrt{B})$ for an optimal path.

Hence we may approximate $\vec{r}_B^*(0)$ by $\vec{y}$. Furthermore, we may approximate $\ell(\vec{r}, \vec{r}')$ by the constant-coefficient ℓ-function as in the simple source case; that is, we replace each $\mu_i x_i$ by the constant $M_i \triangleq \mu_i y_i$, and each $\lambda_i (f_i - x_i)$ by the constant $L_i \triangleq \lambda_i (f_i - y_i)$. We find that the minimal cost path $\vec{r}(t)$ satisfies

$$\ell_i(\vec{r}, \vec{r}') - r'_i \frac{\partial}{\partial r'_i} \ell_i(\vec{r}, \vec{r}') + K (r_i a_i - C) = \text{constant} \qquad \text{for each } i.$$

We may write this explicitly as

$$\sqrt{(r_i')^2 + 4L_i M_i} = L_i + M_i - K\,(r_i a_i - C) + c_i.$$

Following the arguments leading to (13.45), we obtain

$$r_i(t) = r_i(0) + \frac{L_i + M_i + c_i}{K a_i} - \frac{2\sqrt{L_i M_i}}{K a_i}\cosh(K a_i t - z_i). \qquad (13.143)$$

The transversality condition gives us $\sum_{i=1}^{k} c_i = 0$, and the Principle of Smooth Fit 13.63 give us $r_i(0) = r_i(T)$ for every i. [Why? The paths leading to and from $\vec{y}$ are time reversals of each other, so the only way $\vec{r}(t)$ could match both is to come back to the same place it started from.] Therefore we can calculate c_i and z_i as follows: $K a_i T = 2 z_i$, since $r_i(0) = r_i(T)$. Furthermore,

$$\cosh(z_i) = L_i + M_i + c_i \qquad (13.144)$$

since $\vec{r}\,'(0) = -\vec{r}\,'(T) = \vec{z}_\infty'$. So we obtain

$$KT = \frac{2 z_i}{a_i}, \qquad (13.145)$$

making the constant KT the only unknown. Let us define $\kappa \overset{\triangle}{=} KT$. Then we have that each z_i is monotone increasing from zero to ∞ as κ does the same. If we choose κ small enough, then, since $L_i < M_i$ for each i, we have $c_i < 0$ from (13.144) and the arithmetic-geometric mean inequality. Furthermore, as κ becomes large, eventually all the c_i will be positive. So there is indeed a unique value of κ that makes $\sum_{i=1}^{k} c_i = 0$, and this enables us to calculate T, z_i, and $\vec{r}(t)$ via (13.145), (13.144), and (13.143).

Once we have found κ (numerically), we proceed as follows. Find $B(K)$ by

$$B = \sum_{i=1}^{k} \int_0^T a_i (r_i(t) - r_i(0))\,dt.$$

Since $T(K) = \kappa/K$, we obtain $B = O(1/K^2)$, so that, for small B,

$$I^*(B) \approx c_1 + C_2 \sqrt{B}. \qquad (13.146)$$

Exercise 13.65. Show that the error term in $I^*(B)$ due to the approximation of $\vec{z}_n(t)$ by a constant-coefficient process is $O(B)$. You may wish to show that the difference in jump rates is $O(\sqrt{B})$ by scaling space so that the constant-coefficient path does not depend on B. In this scaling the original process has linear rates with slopes $O(\sqrt{B})$. Also find out how far apart the initial points are in this scaling.   ♠

Exercise 13.66. Give explicit expressions for the constants in (13.146).   ♠

Large buffers.

The large B asymptotics are again calculated in a very similar fashion to the calculations for the simple source model. The way a large buffer is achieved is for the path $\vec{r}^*(t)$ to be nearly constant over a long period of time, and our main object is to calculate this constant. The cost per unit time for holding at a point $\vec{x}$ is

$$\text{cost/unit time} = \ell(\vec{x}, \vec{0}) = \sum_{i=1}^{k} \left(\sqrt{\mu_i x_i} - \sqrt{\lambda_i (f_i - x_i)} \right)^2 \tag{13.147}$$

and the buffer fill rate per unit time while holding at $\vec{x}$ is

$$\text{buffer/unit time} = \langle \vec{x}, \vec{a} \rangle - C. \tag{13.148}$$

We maximize the ratio cost/unit buffer by differentiating with respect to x_i for each i. We obtain the equation

$$\frac{\text{cost}}{\text{unit buffer}} = \frac{1}{a_i} \left(\mu_i - \lambda_i - \sqrt{\mu_i \lambda_i} \frac{f_i - 2x_i}{\sqrt{x_i (f_i - x_i)}} \right) \tag{13.149}$$

for each i. This may be solved iteratively by many methods. One is to start the iteration at $\vec{x} = \vec{f}$, calculate the cost per unit buffer [by dividing (13.147) over (13.148)] for this value of $\vec{x}$, then update each x_i by equation (13.149). In any case, we can numerically find the value of $\vec{D}$, the $\vec{x}$ that minimizes cost per unit buffer, and hence of the linear term in the cost function $I^*(B) \approx$ cost/unit buffer $\cdot B$.

To find a more accurate approximation to $I^*(B)$ we need to find the most probable path from $\vec{q}$ to $\vec{D}$. We again let $\vec{y}$ represent the point where the critical path intersects the hyperplane. We have the cost of the critical path from $\vec{q}$ to $\vec{y}$ is given by (13.142). The Euler equations for the remainder of the path are

$$\ell_i(r_i, r_i') - r_i' \frac{\partial \ell_i}{\partial r_i'} - L(a_i r_i - a_i y_i) = S_i \qquad i = 1, \ldots, k, \tag{13.150}$$

where L is a Lagrange multiplier and the S_i are constants. The transversality condition is $\sum_{i=1}^{k} S_i = 0$, and the value of L is the cost/unit buffer when holding at $\vec{D}$. Now to find S_i consider Equation (13.150) at the point $\vec{r}(t) = \vec{D}$, where $\vec{r}'(t) = \vec{0}$:

$$\ell_i(D_i, 0) - L(a_i D_i - a_i y_i) = S_i.$$

We are then able to write $I^*(B)$ in terms of the only remaining unknown, $\vec{y}$, as follows:

$$I^*(B) \approx c_1(\vec{y}) + \sum_{i=1}^{k} \left(\int_{y_i}^{D_i} \frac{\partial \ell_i}{\partial r_i'} (r_i, r_i') \, dr_i + \int_{D_i}^{y_i} \frac{\partial \ell_i}{\partial r_i'} (r_i, r_i') \, dr_i \right) + LB, \tag{13.151}$$

where we use the positive solution of (13.150) for r_i' in terms of r_i in the first integral of (13.151), and the negative solution for the second. The minimum of this cost over $\vec{y} \in \{\langle \vec{y}, \vec{a} \rangle = C\}$ gives the asymptotic expression for $I^*(B)$ for B large. This completes our discussion of the solution of the variational problem for the multiple-class model.

We wish to point out that flow control schemes for the multiple class model have exactly the same efficacy as they did in the simple source case. That is, cutoff flow control helps only when it keeps the system from reaching the level $\vec{D}$, and then it has a calculable effect; rate-changing schemes have the same sort of effect on level crossing an buffer sizes, etc.

13.9. End Notes

As the chapter's title implies, the class of models we investigated was proposed and largely solved by Anick, Mitra, and Sondhi in their seminal work [AMS]. The approach presented in this chapter was initiated by us in [We2]. The present exposition largely follows that work, with many corrections and amplifications. The subject has been studied extensively of late by many investigators through a variety of methods. See, for example, [Mo1, EM1, EM2, EMS, KuM, BoD].

We glossed over the point that there are several time scales in a good model of traffic. We examined only the time scale of packet generation and transmission. We explicitly modeled the time scale of bits as fluid, taking a limit that might better be left unevaluated. We also assumed that each source was connected to the system forever, and was always in the same sort of activity mode. A better model would either have sources enter and leave the system, or would have collections of states with extremely long holding times within the collection; some could represent being quiescent, or disconnected, while others could represent various activity levels. This leads to a very important subject called access control. When a customer wants a virtual circuit, should the network provide it or not? Two different approaches have recently appeared to analyze this question using large deviations. One, called equivalent bandwidth, uses a simple scaling property of the solution of the multiple-source model to reduce access control for virtual circuits to a circuit switching problem. It was initiated in [Hu1] for the zero-buffer case, and by [GH] and [GAN] for the large buffer case, and has also been investigated by [CW, KWC, EM3, KeF, Wh2], and many others. The other uses the asymptotic solution for large B to make a procedure based on measurements and heuristics. This was created by [CKW]. David Tse [Tse] has also attacked the problem using large deviations in his Ph.D. dissertation, including the important case of multiple time scales.

The models of packet traffic we considered are the beginning of many directions of current research, and the interested student could probably make a contribution just by following some of the directions alluded to above or those we will now indicate. There are other sorts of models that can be analyzed without undue difficulty. Whenever the underlying source model is reversible then the level crossing problem, at least, can be solved since the most probable path to any given point is the time reversal of $\vec{z}_\infty$ from that point. The small- and large-B asymptotics should then be solvable, if not more.

Chapter 14

Aloha

Shouting down a pipe is a good method of getting your message across, as long as you're the only one shouting. This is the main advantage and disadvantage of a *multiple-access channel*: everybody has to listen to only one place (their hole in the pipe), but if more than one person attempts to transmit at the same time, the message is garbled. The problem of communication is reduced to either coordination among the transmitters (so that messages do not collide), recovery procedures for lost messages, or some combination of the two.

"Aloha" is the name of Abramson's invention [Ab], the first and simplest protocol for coordination and error recovery. He assumed that transmitters could determine when a message was garbled by colliding with another message. Here's the protocol in a nutshell: if your transmission is garbled, then wait a random amount of time and try again. That's it. If everyone is choosing statistically independent random amounts of waiting time, then after colliding with someone, you're quite likely not to collide with him at your next try.

We begin with a nonstandard model of Aloha, but one that fits easily into the framework of the theory we developed in Chapter 8. We will compare this model to two more standard ones in §14.3 and §14.6.

There are three novelties in this chapter. One, our model of "instantaneous detection" Aloha, is new. Second, our analysis is new—both the heuristic derivation of the capacity of the system, and our large deviations analysis via the finite levels model. Third, our analysis of Gaver and Fayolle's model in §14.6 is both new and is the first rigorous analysis of the paradox that model offers.

14.1. The I.D. Model and Heuristics

Suppose that packets arrive in a Poisson λ stream, and have unit length. Suppose further that collisions are detected instantly. This means that when a packet arrives at the channel (either a fresh arrival or a retransmission) and the channel is already occupied, then the arriving packet and the occupying packet both instantly abandon the channel and join the retry queue. We call this the i.d. model (i.d. = instantaneous detection).

We approximate this system by a finite levels model. The distribution of packet lengths is not Markovian—it is a point mass at one. We approximate this distribution by a convolution (sum of independent random variables) of K exponential distributions with mean $1/K$ each. We encode this into a finite levels model as follows. We let (x, i) be the state of the system, where $x \geq 0$ is the number of packets in the retry queue (not including the transmitting packet if any). The level

$0 \leq i \leq K$ encodes the "service stage" of the transmitting packet. An empty channel corresponds to $i = 0$. When a packet first arrives at the channel we set $i = 1$. After the first exponential clock goes off we increment i to $i = 2$. We continue to increment i until either i reaches K, or another packet arrives at the channel. If another packet arrives before the first leaves then we set $i = 0$ (the channel is again empty) as both packets go to the retry queue. If i reaches K, then when the last exponential clock goes off the packet leaves the system, so that x decreases by one and i is set to zero. Each packet in the retry queue retransmits (independently) with rate ν. The transition structure is sketched in the upper diagram of Figure 14.1.

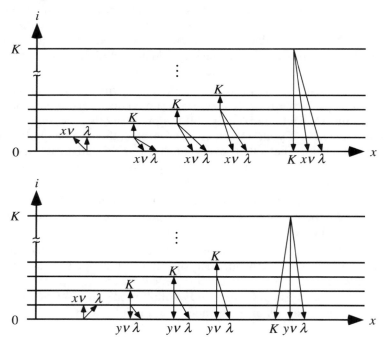

Figure 14.1. Transitions of the i.d. model. $y = x - 1$ is used for brevity for the lower figure (described below).

There are two natural presentations of the i.d. model. One includes the packet in service (if any) in x, the other does not. Our definition of the system above does not count the packet in service, only those in the retry queue. The upper figure corresponds to that definition. It has jump directions $\vec{e}_1$ (corresponding to a new arrival with rate λ) with $\vec{e}_1(0) = 0$, and $\vec{e}_1(m) = 2$ for $1 \leq m \leq K$. The lower figure (which includes the transmitting packet in the state x) has $\vec{e}_1(m) = 1$ for all m. Schematically, the two presentations differ only in that the bottom level is shifted with respect to the upper levels by one unit. The jump rates have a corresponding shift: the jump rate $\lambda_2(m)$, corresponding to retries, is equal to $x\nu$ in one model and is equal to $(x - 1)\nu$ in the other. We would like to be able to ignore the subtraction of one from the rate (that is, have the rate equal to $x\nu$ in both models) in order to have simpler formulas and models. It is easy to show that asymptotically as $\nu \to 0$, the formulas do not see the difference. Both the path $z_\infty(t)$ and the large deviations rate function $\ell(x, y)$ are indifferent to this change. This is a general scaling property of finite levels large deviations models, as you can prove for yourself.

For each K this finite levels model is an approximation to the true i.d. model. We show in §14.5 that this approximation makes sense as $K \to \infty$. Until then we will be cavalier about allowing K to approach ∞, deferring all justification.

The first item of business, as usual, is to define and analyze $z_\infty(t)$. We need to be able to calculate the drift of the finite levels model as $K \to \infty$. We perform the scaling required by Chapter 8. We let $z_n(t)$ have jump sizes $1/n = \nu$ and scale time by n so that the jump rates are $n \cdot \{\lambda, K, z_n\}$. We may now calculate $z_\infty(t)$ by evaluating the local drift of the (x, i) process. Let $\pi(m; z)$ be the local equilibrium probability $\pi(m; z) = \mathbb{P}(i = m \mid x\nu = z)$ of level $i = m$, given that $z_\infty = z$. Here are the equations satisfied by $\pi(m)$.

(A) $\quad \pi(m + 1; z) \cdot (z + \lambda + K) = \pi(m; z) \cdot K$

$\qquad \pi(0; z) \cdot (z + \lambda) = (1 - \pi(0; z))(z + \lambda) + \pi(K; z) \cdot K$

(B) $\quad \pi(1; z) \cdot (z + \lambda + K) = \pi(0; z) \cdot (z + \lambda)$

(C) $\displaystyle \sum_{m=0}^{K} \pi(m; z) = 1.$

Now fix $1 \le m \le K$ and define

$$\alpha \stackrel{\Delta}{=} \frac{1}{\rho} \stackrel{\Delta}{=} \frac{K}{z + \lambda + K} \tag{14.1}$$

by (A): $\quad \pi(m; z) = \alpha^{m-1}\pi(1; z); \quad$ or $\pi(m; z) = \rho^{K-m}\pi(K; z)$

by (B): $\quad \pi(1; z) = (1 - \alpha)\pi(0; z); \quad$ or $\pi(0; z) = \rho^{K-1}\pi(K; z)\dfrac{\rho}{\rho - 1},$

by (C): $\displaystyle \sum_{0}^{K} \pi(m; z) = \pi(K; z)\left[\frac{\rho^K - 1}{\rho - 1} + \frac{\rho^K}{\rho - 1}\right] = \pi(K; z)\frac{2\rho^K - 1}{\rho - 1} = 1.$

That is,

$$\pi(K; z) = \frac{\rho - 1}{2\rho^K - 1}, \tag{14.2}$$

where $\rho = 1 + \frac{z+\lambda}{K}$ by Equation (14.1). Now by (8.6),

$$\frac{d}{dt}z_\infty(t) = \lambda - K\pi(K; z_\infty(t))$$

$$= \lambda - \frac{z_\infty + \lambda}{2\rho^K(z_\infty) - 1}. \tag{14.3}$$

We are interested in the drift as $K \to \infty$. We have

$$\lim_{K \to \infty} \left(1 + \frac{x}{K}\right)^K = e^x,$$

$$\lim_{K \to \infty} \frac{d}{dt} z_\infty(t) = \lambda - \frac{\lambda + z_\infty(t)}{2e^{(\lambda + z_\infty(t))} - 1} . \tag{14.4}$$

The drift of $z_\infty(t)$ (the value of $\frac{d}{dt} z_\infty(t)$ as given in (14.4)) is plotted in Figure 14.2.

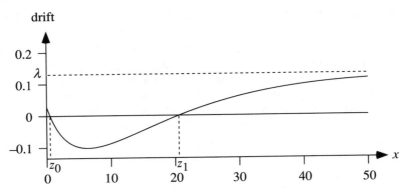

Figure 14.2. The drift of i.d. Aloha, $\lambda = 0.13$.

When is the minimum value of drift negative? When is

$$\min_{0 \le z < \infty} \left(\lambda - \frac{z + \lambda}{2e^{z+\lambda} - 1} \right) < 0 \ ?$$

We maximize the function $\frac{u}{2e^u - 1}$, finding the maximum of $\frac{u}{2e^u - 1}$ numerically to be 0.23196 at $u = 0.76804 = 1 - \max \frac{u}{2e^u - 1}$. Hence the largest value of λ for which $\min \left(\lambda - \frac{\lambda + z}{2e^{\lambda + x} - 1} \right) < 0$ is $\lambda = 0.23196\ldots$. At this value we need

$$x = u - \lambda = 0.53608\ldots .$$

When $\lambda \le 0.23196\ldots$ there are two roots of the drift, labeled z_0 and z_1 in Figure 14.2. The smaller root, z_0, is "stable," while z_1 is "unstable." This means that

$$\lim_{t \to \infty} z_\infty(t) = z_0 \text{ whenever } z_\infty(0) \in [0, z_1),$$

$$\lim_{t \to \infty} z_\infty(t) = \infty \text{ whenever } z_\infty(0) > z_1.$$

The behavior of $z_\nu(t)$ ($z_n(t)$ in our standard notation, where recall $1/n = \nu$) over any finite interval of time is therefore most likely determined by Kurtz's Theorem as $1/n = \nu \to 0$: if $z_\nu(0) < z_1$ then most likely $z_\nu(t)$ will approach z_0. The set $[0, z_1)$ is sometimes called the "basin of attraction" of z_0. The long time behavior of $z_\nu(t)$ is basically the study of the time to exit the basin of attraction, and it should be intuitively clear that this is an exit problem almost exactly of the form studied in Chapter 6.

An alternate derivation.

If you have studied Aloha protocols before you might not have found the preceding derivation intuitive, although we hope you'll agree that it is straightforward from the theory we developed in Chapter 8. We now present an alternative derivation of the drift of i.d. Aloha that is heuristic, but is along the lines of more familiar derivations for other models. Perhaps the complexity and length of this derivation will convince you to go the large deviations way, which is shorter and rigorous! (If you are already converted, you may want to skip directly to §14.2.) We will use similar reasoning in deriving the drift of some other models later, so neophytes to the field might find the arguments here easier to read if they first examine the arguments of §14.2.

Let us assume that there is a Poisson (λ) arrival process, and that retries have rate ν, with the retry queue having a current occupancy $x > 0$. Recall that transmissions take one unit of time. We need to calculate the probability that a packet (new arrival, or from the retry queue) attempting transmission suffers no collision. Suppose that a packet attempts transmission at time zero. Then it may collide immediately, or it may collide during the next unit of time, or it may be successful. We will see below (14.11) that the probability that it does not collide immediately is

$$\frac{1}{2 - \exp(-(\lambda + x\nu))}.$$

The probability that it does not collide during the next unit of time, given that it did not collide on first attempt, is $\exp(-(\lambda + x\nu))$. So, the net transmission rate is the rate of attempting transmission times the probability of success

$$(\lambda + x\nu)\frac{\exp(-(\lambda + x\nu))}{2 - \exp(-(\lambda + x\nu))}.$$

This yields a drift

$$\text{drift} = \lambda - (\lambda + x\nu)\frac{\exp(-(\lambda + x\nu))}{2 - \exp(-(\lambda + x\nu))}, \tag{14.5}$$

which agrees with our previous calculation (14.4).

Here is how to calculate the probability that the channel is free when a packet arrives [the missing ingredient in obtaining (14.5)]. We consider a packet that arrives at time zero. Let p denote the probability that the channel is free at this time; this is the probability that the packet does not immediately collide. Now in steady state, this is also the probability that the channel is free at time (-1). The probability that the channel is free at time zero is equal to the probability that the channel is free at time (-1) times the probability that there are an even number of arrivals in $(-1, 0)$, plus the probability that the channel is not free at time (-1) times the probability that the arriving packets in $(-1, 0)$ cancel each other out. This last probability can be broken down into the probability that there are an odd number of arrivals in $(-1, 0)$ and the first arrival collides with the packet that existed at time (-1), plus the probability that there are an even number of arrivals and the first arrival takes place after the packet that existed at time (-1) already departed.

We calculate these last probabilities by considering the probability that the packet that existed at time (-1) arrived at time $(-1 - t)$ and calculating the chances that it would collide with the first of j subsequent arrivals, for every j. To find the probability that the packet arrived at time $(-1 - t)$, consider that there must have been no arrivals in $(-1 - t, -1)$, which has probability $\exp(-(\lambda + x\nu)t)$. That is, for each $t \in (0, 1)$, the probability that the packet arrived in $(-1-t, -1-t+dt)$ is proportional to $\exp(-\gamma t)\,dt$ where $\gamma \overset{\triangle}{=} \lambda + x\nu$. Normalizing this probability to one we see that the probability density function is

$$\frac{\gamma \exp(-\gamma t)}{1 - \exp(-\gamma)}. \tag{14.6}$$

Now the probability that j subsequent arrivals all occur after the packet that existed at time (-1) has departed is the probability that these packets arrived in time $(-t, 0)$, which is clearly t^j. That is,

$$\mathbb{P}\,(\text{all } j \text{ arrivals miss the original packet}) = \int_0^1 \frac{\gamma \exp(-\gamma t)}{1 - \exp(-\gamma)} t^j \, dt$$

$$= \frac{1}{1 - e^{-\gamma}} \frac{j!}{\gamma^j} \left[1 - e^{-\gamma} \sum_{i=0}^j \frac{\gamma^i}{i!} \right].$$

The term within brackets may be "simplified" as follows:

$$1 - e^{-\gamma} \sum_{i=0}^j \frac{\gamma^i}{i!} = e^{-\gamma} \sum_{i=j+1}^\infty \frac{\gamma^i}{i!}. \tag{14.7}$$

Therefore, the probability that the channel is free at time zero, given that it was occupied at time (-1), is

$$\sum_{j \text{ even}} \frac{j!}{(1 - e^{-\gamma})\gamma^j} \frac{\gamma^j}{j!} e^{-\gamma} \left[1 - e^{-\gamma} \sum_{i=j+1}^\infty \frac{\gamma^i}{i!} \right]$$

$$+ \sum_{j \text{ odd}} e^{-\gamma} \frac{\gamma^j}{j!} \left[1 - \frac{j!}{(1 - e^{-\gamma})\gamma^j} \frac{\gamma^j}{j!} e^{-\gamma} \sum_{i=j+1}^\infty \frac{\gamma^i}{i!} \right]. \tag{14.8}$$

Now interchange the sums on i and j, and note that

$$\sum_{i>j\geq 0 \text{ even}} 1 = \left\lfloor \frac{1+i}{2} \right\rfloor$$

$$\sum_{i>j\geq 1 \text{ odd}} 1 = \left\lfloor \frac{i}{2} \right\rfloor$$

$$\sum_{j\geq 1 \text{ odd}} e^{-\gamma} \frac{\gamma^j}{j!} = e^{-\gamma} \sinh \gamma.$$

We obtain that (14.8) above is equal to

$$\sinh \gamma \left(e^{-\gamma} + \frac{e^{-2\gamma}}{1 - e^{-\gamma}} \right) = \frac{1 - e^{-2\gamma}}{2(1 - e^{-\gamma})}. \tag{14.9}$$

Therefore,

$$
\begin{aligned}
p &\overset{\triangle}{=} \mathbb{P}(\text{the channel is free at zero}) \\
&= \mathbb{P}(\text{the channel is free at } (-1)) \\
&= p \sum_{j \geq 0 \text{ even}} e^{-\gamma} \frac{\gamma^j}{j!} + (1 - p) \frac{1 - e^{-2\gamma}}{2(1 - e^{-\gamma})}.
\end{aligned}
\tag{14.10}
$$

Now the sum in (14.10) is equal to

$$\sum_{j \geq 0 \text{ even}} e^{-\gamma} \frac{\gamma^j}{j!} = e^{-\gamma} \cosh \gamma = \frac{1}{2} \left(1 + e^{-2\gamma} \right).$$

Solving (14.10) for p we obtain

$$p = \frac{1}{2 - e^{-\gamma}}. \tag{14.11}$$

This finishes our derivation of (14.5). The probability p that a packet does not collide on arrival has the properties

$$p \to 1 \text{ as } \gamma \to 0 \quad \text{and} \quad p \to \frac{1}{2} \text{ as } \gamma \to \infty, \tag{14.12}$$

which are to be expected.

Exercise 14.3. Give arguments that are not based on calculations why (14.12) must hold. ♠

Not only is the heuristic justification we just developed less straightforward than the analysis based on a finite levels model, but it is harder to make rigorous [for example, we shall see later that there is no steady-state distribution for the process $x(t)$]. In fact, when we were developing this model, we first derived the drift using the finite levels model, and only later used this more standard argument.

14.2. Related Models

We now describe two other models of Aloha. These models are standard, and our description and analysis of them contains no novelties. They are both outside the scope of the theory developed in this book. We found several compelling reasons for including them, though. The first is that they are standard, and we wished to compare the theoretical performance of the i.d. model to them. The second is to show people who are not familiar with Aloha exactly how our model differs from standard ones. The third reason is that it is easier to analyze slotted Aloha than it is to analyze i.d. Aloha; therefore we include the analysis to show how properties of Aloha systems might be expected to respond to changes in various parameters.

The first standard model is called continuous-time Aloha. In this model packets arrive according to a Poisson process with rate λ, and have unit length. Whenever a packet collides with another, by simultaneously occupying the channel, the packet is deemed to have been garbled, and when it is finished transmitting (one unit of time after it arrived) it joins the retry queue. Unlike the i.d. model, in continuous-time Aloha packets hold the channel for a unit of time, whether or not they are involved in collisions. Packets may overlap for only a small fraction of their existence and still interfere destructively with each other. It is difficult to give a compact Markovian description of this process, since we need to remember the arrival times for all packets in the channel. This is why we cannot analyze this model using our theory.

The second standard model is called slotted Aloha. It operates in discrete time. We suppose that time is divided into integer lengths, and that packets that arrive in the middle of an interval wait until the next integer time, and then attempt to seize the channel. If exactly one packet seizes the channel then it is transmitted successfully. If more than one packet seizes the channel then all are garbled, and at the end of the slot are sent back to the waiting room. Packets wait a geometrically distributed number of slots before attempting to retransmit, with the mean number of slots given by $1/\nu$.

We now give heuristic analyses of the drifts of both continuous-time Aloha and slotted Aloha along the lines of the analysis of the previous section. We start with continuous-time Aloha. Recall that each packet takes a unit of time to transmit. A transmission that begins at time T will be successful only if there are no other arrivals to the channel, either from fresh transmissions or retransmissions, during the interval $(T - 1, T + 1)$. Let us suppose that there are x packets in the retry queue, each attempting to retransmit with rate ν. Then the probability that there are no arrivals to the channel in a time interval of length 2 is

$$\mathbb{P}(\text{success}) = \exp(-2(\lambda + x\nu)).$$

Hence, for each x, the rate of successful transmission is the arrival rate (new and retries) to the channel times the success probability, $(\lambda + x\nu)\exp(-2(\lambda + x\nu))$. Therefore the net drift of $x(t)$ (namely $\frac{d}{dt}\mathbb{E}_x x(t)$) is

$$\text{drift} = \text{rate in} - \text{rate out} = \lambda - (\lambda + x\nu)\exp(-2(\lambda + x\nu)). \quad (14.13)$$

In order for the drift to be negative we must have

$$\lambda < (\lambda + xv) \exp(-2(\lambda + xv)). \qquad (14.14)$$

The maximum of the right-hand side of (14.14) occurs when $\lambda + xv = 1/2$, and there it achieves a value $1/(2e)$. Therefore we may solve (14.14) with equality for every $\lambda < 1/(2e)$. Just as in the i.d. case discussed in §14.1 there are two solutions, $z_0 < z_1$, whenever the minimal value of the drift is negative [i.e., whenever $\lambda < 1/(2e)$]. The drift of $x(t)$ is negative for $z_0 < x < z_1$. Therefore z_0 is a stable equilibrium point of the system, and z_1 is an unstable equilibrium point. The point $x = z_0/v$ tells us how many packets we expect to be backlogged, since $z_v(t)$ tends to stay near the point z_0.

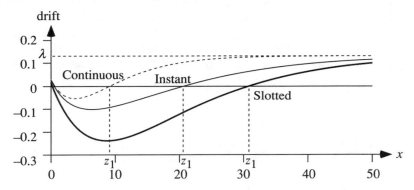

Figure 14.4. Drifts of the various Aloha models for $\lambda = 0.13$. The roots z_0 are not labeled because they are nearly indistinguishable on this scale.

Exercise 14.5. Find an approximation for z_0 as a function of λ for small values of λ. What can you say about z_1 for small values of λ? Hint: you might wish to use Newton's method for finding the root of the drift near $\lambda = 0$, starting from the known value of the drift at $x = 0$. For large values of x ($z_1 \to \infty$ as $\lambda \to 0$) you might wish to use the approximation $\lambda \approx \exp(-2z_1)$ as the first step in a repeated substitution into the expression $x = \frac{1}{2} \frac{\log x}{\lambda}$. ♠

Slotted Aloha has a similar heuristic calculation, with slightly different results. Suppose that the exogenous arrival process in each slot (slot = discrete unit of time) is Poisson with parameter λ, and that the retries occur at geometrically distributed times with mean $1/v$. Suppose further that there are x packets in the retry queue. Then, as $v \to 0$ and $xv \to z$, the number of arrivals at the channel at each slot has approximately a Poisson $(\lambda + xv)$ distribution. Therefore the net departure rate is $(\lambda + xv) \exp(-(\lambda + xv))$ (this is the frequency of slots containing exactly one packet), so the drift of $x(t)$ is

$$\text{drift} = \lambda - (\lambda + xv) \exp(-(\lambda + xv)). \qquad (14.15)$$

There will be a region where the drift is negative for every $\lambda < 1/e$.

Except for the factor of 2 in the exponent, the analysis and commentary for this case is exactly the same as for the previous one. We see that slotted Aloha is much more stable then continuous-time Aloha; it has twice the ultimate capacity ($1/e$ as opposed to $1/2e$), and a stronger "restoring force" to the point z_0 for a given exogenous arrival rate λ.

We can easily see that slotted Aloha where the slot length is equal to the packet length will have less contention than continuous-time Aloha. Imagine a set of arrival times marked on the real line, with all retransmissions also marked. Now round each packet's starting time to the nearest integer multiple of a packet length. The "rounded" system is slotted Aloha. The main point to note is that any packets that collide in the slotted scheme would have collided in the unslotted scheme, since by definition their starting points are within one unit of each other. However, some collisions may be eliminated, as in Figure 14.6. (Actually, of course, packet arrival times are not rounded to the nearest integer, but are truncated. However, this is equivalent to rounding to the nearest [integer + 1/2], so our conclusion holds.)

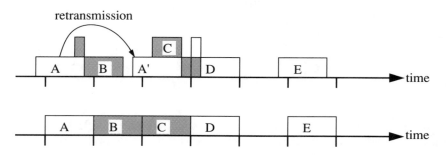

Figure 14.6. Continuous-time Aloha has more collisions than slotted Aloha.

Furthermore, by reducing collisions, some retransmissions will not occur, further reducing the number of collisions. Hence slotted Aloha has, pathwise, greater carrying capacity than continuous-time Aloha. The only thing wrong with this discussion is that the slots must, in practice, be a bit larger than the packets, to allow for transmission delays and other sources of jitter. This reduces the capacity by a fixed amount. Also, variable-length packets can be supported more easily under continuous-time Aloha, since a slot large enough for long packets might be wasteful for short ones.

14.3. Basic Analysis

We now compare the three models in terms of their most likely performance. It is easy to see that the models have similar behavior: the drift is positive for $x = 0$, achieves at most one local minimum, and approaches λ as $x \to \infty$. The drifts are also monotone decreasing to their minima, and are monotone increasing thereafter. Therefore there are at most two values of x that make the drift equal to zero. These points, denoted z_0 and z_1 when they exist, are the only possible stable points of $z_\infty(t)$.

Our main result in comparing the three models is the following

Proposition 14.7. *For any $\lambda > 0$ and any $x \geq 0$,*

$$drift\ (slotted) \ < \ drift\ (i.d.) \ < \ drift\ (continuous).$$

Exercise 14.8. Prove Proposition 14.7. [It's just algebra with Equations (14.4), (14.13), and (14.15).] ♠

Corollary 14.9.

$$z_0\ (slotted) \ < \ z_0\ (i.d.) \ < \ z_0\ (continuous)$$

$$z_1\ (slotted) \ > \ z_1\ (i.d.) \ > \ z_1\ (continuous)$$

The evolution of $z_\infty(t)$ under all three models is clear. If $0 \leq z_\infty(0) < z_1$ then $z_\infty(t)$ approaches z_0 in a monotone fashion. If $z_\infty(0) > z_1$ then $z_\infty(t) \to \infty$ as $t \to \infty$, and $\frac{d}{dt}z_\infty(t) \to \lambda$ as $t \to \infty$. Therefore if $z_n(0) < z_1$ we know from Kurtz's Theorem that $z_n(t)$ most likely approaches z_0 as t increases, at least for a while. Eventually, though, just by random fluctuations $z_n(t)$ will exceed z_1. After that we expect $z_n(t)$ to continue to increase forever. Our analysis of Chapter 6 does not apply to this process without modification, though, since the escape from the region $z < z_1$ occurs at a point where drift = 0. This contradicts Assumption 6.4, and therefore invalidates some of the analysis.

We nevertheless expect that the behavior of Aloha is the following. For a long time, on the exponential order of $e^{I^*/\nu}$, the process $z_n(t)$ will stay near z_0. Then $z_n(t)$ will make an excursion above $z_1(t)$ and will most likely increase forevermore. Our job as large deviationists is to calculate I^* for the various models, and see how it depends on parameters such as λ.

14.4. Large Deviations of Aloha

Our large deviations results on the i.d. model are incomplete, in that we have not yet proved every property we'd like. Nevertheless we include our partial results, in the hope that the reader will find some insight, and will forgive us for not finishing them.

We begin with a derivation of a rigorous upper bound on the rate function related to the length of time the system remains in the region $z_\nu(t) \le z_1$, namely

$$
\begin{aligned}
I^* &= \inf_{(r,T) \in G} I_0^T(r) \\
&= \inf_{(r,T) \in G} \int_0^T \ell(r(t), r'(t)) \, dt \\
&= \int_{z_0}^{z_1} \inf_{y > 0} \frac{\ell(x, y)}{y} \, dx
\end{aligned}
\tag{14.16}
$$

$$
G \triangleq \left\{ (r, T) \; : \; T > 0, \; r \in AC[0, T], \; r(0) = z_0, \; r(T) = z_1 \right\}.
$$

Exercise 14.10. Establish (14.16). Hints: show monotone paths suffice so that zero is never reached. Now establish the hypotheses and follow the path of §C.3.◆

Recall from Theorem 8.19 that the local rate function satisfies

$$
\ell(x, y) = \inf_{\vec\mu \in B} \inf_{\vec v \in \mathcal{S}(\vec\mu)} \sum_{m=1}^{D} \mu_m \left(\sum_{i=1}^{k(m)} v_i(m) \log \frac{v_i(m)}{\lambda_i(m)} - v_i(m) + \lambda_i(m) \right)
\tag{14.17}
$$

$$
B \triangleq \left\{ \vec\mu \in \mathbb{R}^D : \mu_m \ge 0, \; \sum_{m=1}^{D} \mu_m = 1 \right\}
$$

$$
\mathcal{S}(\mu) \triangleq \left\{ \vec v : v_i(m) \ge 0, \; \sum_{m=1}^{D} \mu_m \sum_{i=1}^{k(m)} v_i(m) \vec e_i(m) = \left(\vec y, \vec 0^D \right) \right\}.
$$

(We can in principle calculate such a rate function I_K^* for each fixed K, and then let $I^* = \lim_{K \to \infty} I_K^*$ if this limit exists.) We may obtain an upper bound on I^* by simply defining a change of measure v with its associated invariant measure μ in such a way that the resulting drift y is positive. We actually need to define a sequence of changes of measure, one for each K, and calculate the limit (if it exists) as $K \to \infty$ of the resulting rate. This will give us a value

$$
I_\infty = \lim_{K \to \infty} I_K
$$

which then provides the bound

$$
I^* = \lim_{K \to \infty} I_K^* \le I_\infty.
\tag{14.18}
$$

We propose the following change of measure ν. Each jump rate K remains unchanged. The two jump rates out of level zero, $\lambda(0)$ and $x(0)$, change to $\tilde{\lambda}(0)$ and $\tilde{x}(0)$. The remaining jump rates $\lambda(m)$ and $x(m)$, $1 \leq m \leq K$, change to $\tilde{\lambda}(1)$ and $\tilde{x}(1)$, for all m. That is, we have four parameters $\left[\tilde{\lambda}(0), \tilde{x}(0), \tilde{\lambda}(1), \tilde{x}(1)\right]$ that define our change of measure. For each $x \in (z_0, z_1)$ we choose the four that minimize the expression

$$\frac{\tilde{\ell}(x, y)}{y} \tag{14.19}$$

over $y > 0$. This minimization is performed numerically. Specifically we perform the following steps.

1. Given $\left[\tilde{\lambda}(0), \tilde{x}(0), \tilde{\lambda}(1), \tilde{x}(1)\right]$, compute the invariant distribution $\pi(m)$ for $0 \leq m \leq K$. This is given in (14.1) and (14.2).

2. Compute the drift y. This is given in (14.3), and the limit as $K \to \infty$ is given in (14.4).

3. Compute $\tilde{\ell}(x, y)$, the bound on the cost, given by expression (14.17) with the change of measure and invariant measure given in item 1.

4. Minimize (14.19) numerically over the set of positive parameters $\left[\tilde{\lambda}(0), \tilde{x}(0), \tilde{\lambda}(1), \tilde{x}(1)\right]$. This minimization can be done even more easily by writing

$$\tilde{\lambda}(0) = \lambda e^{\theta_1} \qquad \tilde{x}(0) = x e^{\theta_1 - \theta_0}$$
$$\tilde{\lambda}(1) = \lambda e^{\theta_2 + 2\theta_0} \qquad \tilde{x}(0) = x e^{\theta_2 + \theta_0}$$

and minimizing over the triple $(\theta_0, \theta_1, \theta_2)$. This reduces the problem from four dimensions to three, and also has the advantage that y is monotone increasing in θ_0, so it is simple to find values of the parameters with $y > 0$.

5. Numerically integrate (14.19) from z_0 to z_1 to obtain a bound on I^*.

We present our numerical findings in Figure 14.11. This table also contains a numerical evaluation of the rate for slotted Aloha (this calculation is outlined in §14.7). We see that i.d. Aloha has a smaller rate I^* than slotted Aloha for every value of λ. This can be taken as a numerical proof that not only does slotted Aloha have a stronger stabilizing drift, but its deviations take much longer to occur than those of the i.d. model.

Instant detection		Slotted Models						
		Poisson		Birth-death		Finite population		
λ	I^*	λ	I^*	λ	I^*	n	b	I^*
.05	4.27	.05	6.09	.05	7.03	10	.003	10.7
.10	1.68	.10	3.23	.10	4.06	6	.04	1.36
.15	0.615	.15	1.86	.15	2.60	5	.07	0.413
.20	0.121	.20	1.06	.20	1.70	10	.03	0.389
.23	0.00168	.25	0.539	.25	1.09	6	.07	0.000224
		.30	0.209	.30	0.662			
		.35	0.0256	.35	0.360			
				.40	0.158			
				.45	0.0394			

Figure 14.11. Values of I^* for various Aloha models and parameters. The "Poisson" slotted model is the one we have discussed so far; the others will be developed in §14.7.

14.5. Justification

There are two items that need justification, and several that need more analysis. One glaring deficiency is our blithe reduction of the estimation of the stability of i.d. Aloha to the calculation of $\lim_{K \to \infty} I_K^*$. This is an interchange of limits that needs to be justified: we have estimated

$$\lim_{K \to \infty} \lim_{\nu \to 0} \nu \log \mathbb{E}(\tau), \qquad (14.20)$$

when what we want is

$$I^* = \lim_{\nu \to 0} \lim_{K \to \infty} \nu \log \mathbb{E}(\tau), \qquad (14.21)$$

where τ is the time when $z_{K,\nu}(t)$ escapes the region $z_\nu \leq z_1$. Furthermore, we have assumed that the Freidlin-Wentzell theory applies, so that the calculation of $I^* = \lim_{\nu \to 0} \lim_{K \to \infty} \nu \log \mathbb{E}(\tau)$ reduces to solving a variational problem. The problem here is that the drift of $z_\infty(t)$ is equal to zero at $z = z_1$; that is, Assumption 6.4 of Chapter 6 is not satisfied (and neither is Assumption 6.5). This means that, in order to be completely rigorous, we would have to prove anew some aspects of the Freidlin-Wentzell theory.

There are a few other points worth mentioning. We claimed to have only an upper bound on I^*, yet we conjecture that this bound is sharp, in a strong sense: we believe that the optimal change of measure approaches the four-parameter family $[\tilde{\lambda}(0), \tilde{x}(0), \tilde{\lambda}(1), \tilde{x}(1)]$ as $K \to \infty$, at least in a certain sense to be defined below. Furthermore we did not show that $\inf_{\lambda > 0}(I^*(slotted) - I^*(i.d.)) > 0$, though we also believe that this is true. Finally, we did not calculate I^* for the continuous-time model. This last calculation is one we do not know how to do—we believe that this is a good topic for an interested investigator.

Most of the justifications we present in this section are incomplete. This section, therefore, is simply a sketch of ideas and partial results that we hope will someday be completed by us, or perhaps by you the reader.

We begin with a sketch of a proof of the justification of interchanging the limits $v \to 0$ and $K \to \infty$. The proof is based on specific properties of the i.d. model. There is *no* general result saying that these types of limits will interchange for other finite levels models that approach a limit as the number of levels approaches infinity.

Definition 14.12. *A linear Poisson process (also called a Yule process) is a birth process $y(t)$, $t \geq 0$ whose jump rate at $y(t) = x$ equals x.*

Lemma 14.13. *Let $y(t)$ be a linear Poisson process with $y(0) = 1$. Then*

$$\mathbb{P}_1(y(t) > n) = \left(1 - e^{-t}\right)^n .$$

Proof. Consider a collection $\{w_1, \ldots, w_n\}$ of mean one exponential i.i.d. random variables. Define the order statistics of these random variables as $\{t_1, \ldots, t_n\}$. That is,

$$t_1 = \min\{w_1, \ldots, w_n\}$$
$$t_2 = \text{second smallest } w_i \in \{w_1, \ldots, w_n\}$$
$$\vdots \qquad\qquad (14.22)$$
$$t_n = \max\{w_1, \ldots, w_n\} .$$

Now the first n jumps of the process $y(t)$ can be constructed from the variables $\{w_1, \ldots, w_n\}$ as follows. The first jump t_1 occurs after a time that is exponentially distributed with parameter n. The second jump occurs at a time t_2, which is beyond t_1 by an amount $t_2 - t_1$, which is itself distributed exponentially with parameter $n - 1$. This continues, and we see that $t_i - t_{i-1}$ form a collection of independent exponentially distributed random variables with parameter $n - i + 1$, $1 \leq i \leq n$. (See [TK] for a proof of this equivalence.) Therefore we can construct y from the w_i as follows. The first jump of y can be placed at time $t_n - t_{n-1}$; the second can be placed at time $t_n - t_{n-2} = (t_n - t_{n-1}) + (t_{n-1} - t_{n-2})$; and the nth can be placed at time t_n. We therefore see that

$$\mathbb{P}(y(t) > n) = \mathbb{P}(t > t_n) = \mathbb{P}\left(\max_{1 \leq i \leq n} w_i < t\right) = \left(1 - e^{-t}\right)^n . \qquad \blacksquare$$

(There are several ways of obtaining this result; we gave the one we found cutest. If you don't like this method, you can derive the result using generating functions, as Larry Shepp pointed out to us.)

This shows that at every time t the process $y(t)$ has a geometric distribution with mean $\exp(t)$. This means that $y(t)$ satisfies the conditions of Chernoff's Theorem, with a rate function ℓ_t. Furthermore, you can easily show that if $\zeta_m(t)$ represents a linear Poisson process with $\zeta_m(0) = m$, then we can represent

$$\zeta_m(t) = y_1(t) + \cdots + y_m(t),$$

where the $y_i(t)$ are independent linear Poisson processes satisfying $y_i(0) = 1$. Therefore we obtain the following

Corollary 14.14. *For any* $n > me^t$,

$$\mathbb{P}_m \left(\zeta_m(t) > n \right) \leq \exp(-m\ell_t(n/m)). \tag{14.23}$$

Proof. This is Chernoff's Theorem applied to the random variables y_i. ∎

Proposition 14.15. *For every* $\lambda > 0, C > 0, \delta > 0$, *and* $z < z_1$, *there exists a* $K_0 > 0$ *such that for any* $K > K_0$ *and* $\nu > \nu_0$ *we have*

$$\mathbb{P}_z \left(\max_{0 \leq t \leq \delta} |z_\nu(t) - z_{K,\nu}(t)| > \varepsilon \right) < e^{-C/\nu}.$$

Proof. We now show how to bound the growth in the difference between $z_\nu(t)$ and $z_{K,\nu}(t)$. We will do this by constructing random processes that stochastically bound the growth in this difference. We will look at the unscaled processes, and then scale time and space. Construct the processes $z_\nu(t)$ and $z_{K,\nu}(t)$ on the same space so that the following occur:

1. All exogenous arrivals (associated with the arrival rate λ) occur at the same times in both processes.

2. All retries in both processes occur at the same times, *except* that when $z_\nu \neq z_{K,\nu}$, there can be retries of one and not the other.

Remark. Note that the main difference between the processes $z_\nu(t)$ and $z_{K,\nu}(t)$ is that $z_\nu(t)$ has packets of unit length, while $z_{K,\nu}(t)$ has packets with random lengths.

Let us consider a "chain" defined in terms of $z_\nu(t)$. A chain is a set of two or more arrivals (either exogenous or retries) that have at most one unit of time between neighboring arrivals. The odd-numbered arrivals (first, third, etc.) in a chain see an empty channel, the even-numbered ones cause collisions (freeing the channel). All members of a chain are unsuccessful, except the last member of a chain that has an odd number of participants.

Suppose that we now construct the $z_{K,\nu}$ process with $z_\nu(0) = z_{K,\nu}(0)$, and suppose that K is large so that the packets in the $z_{K,\nu}(t)$ process are likely to be very near unit length. The two processes $z_{K,\nu}(t)$ and $z_\nu(t)$ will be nearly identical; the chains of one will be the chains of the other, until a mistake first occurs. A mistake can occur in one of several ways. A chain might be broken by a service time being short, so that a success occurs in the middle of the chain. Or the last member of a chain (or a singleton) might stretch long enough to collide with the next chain or singleton. Furthermore, while one channel is occupied and the other is not (when, say, a singleton stretches or shrinks), there might be a difference in the arrival processes; we classify this as a rebound. The rate of rebounds at time t is $|z_\nu(t) - z_{K,\nu}(t)|$.

Let us now bound the growth of the difference between $z_\nu(t)$ and $z_{K,\nu}(t)$. Let us consider the effect of a single mistake that occurs in a chain. The length of even-numbered members of a chain is obviously irrelevant, since they spend zero time in the channel (they collide immediately upon arrival). Odd-numbered members can grow to arbitrary lengths (except for the last member of a chain, which we consider separately below). However, if an odd-numbered member of a chain shrinks to the point where it is successful, then the chain is broken, and the processes differ by one (in the backlog, meaning $|z_\nu(t) - z_{K,\nu}(t)| = \nu$). The latter part of the chain remains a chain, but the even members become odd and vice-versa. This leads to two effects. The first effect is that the end of a chain changes from even to odd and vice-versa. Therefore, if the original chain was odd (so that the end member was successful), it becomes even and the end member is unsuccessful. If no other difference between the chains occurs, then the two processes end up equal at the end of the chain, the successful member simply having changed time. However, if the original chain was even, then it becomes odd, so that at the end of the chain the processes differ by two. We can therefore bound the effect of each mistake as an increment of two in the difference between processes.

The second effect of a mistake in a chain is that the periods of time the channel is occupied after the mistake are switched with the periods of time the channel is free. This leads to an extra difference in the rate at which retries occur; again, the difference in the processes is bounded by two for each mistake that is made.

Mistakes made by singletons and by the ends of odd chains can cause new chains to form, for chains to join, or for chains to pick up a singleton. This can happen when the service time for the singleton or end member of an odd chain is so long that it collides with the next arrival. This again can lead to an increment of at most two in the difference between processes, and again we can count this difference as occurring at the time when the mistake is made.

After a mistake is made, rebounds occur according to a Poisson process with rate equal to the difference in the processes (remember that the processes are scaled by ν; the rebounds occur at rate ν times the difference in the actual backlogs, which is the difference between z_ν and $z_{K,\nu}$). Therefore, each mistake can lead to a sequence of rebounds that is stochastically bounded by a linear Poisson process y with arrival rate $2\nu y$. (Equivalently, we could analyze a linear Poisson process that has pairs of arrivals instead of single arrivals, but the process we described is obviously stochastically equivalent.)

To finish the analysis we need an estimate on the rate of occurrence of mistakes. The probability that a packet causes a mistake is equal to the probability that it is smaller than the time until the arrival of the next packet if that time is less than one, or is larger than that time if the time is greater than one. Since the interarrival times are exponentially distributed with parameter $\lambda + z_{K,\nu}(t)$, and since the packets in the $z_{K,\nu}$ process are approximately normally distributed with standard deviation $1/\sqrt{K}$, you can see that there is a constant D independent of $z_{K,\nu}(t)$ such that the number of mistakes that occur in a time interval of length δ is bounded above by a Poisson process with rate $\delta D/\sqrt{K}$. Therefore, for the scaled processes $z_\nu(t)$

and $z_{K,\nu}(t)$, the number of mistakes in a time interval of length δ is bounded by a Poisson random variable with parameter $\frac{\delta D}{\nu\sqrt{K}}$.

Each mistake leads to a growth in the difference of at most two. The process of rebounds due to a particular mistake is then bounded by a linear Poisson process $y(t)$ with rate $2\nu y(t)$; equivalently, the number of rebounds up to time t is bounded by a linear Poisson process at time $2\nu t$.   ∎

This lemma is the key to seeing why the limits $\nu \to 0$ and $K \to \infty$ interchange. We can couple the processes $z_\nu(t)$, the true i.d. process, and $z_{K,\nu}(t)$, its approximating process with K stages, so that they are very nearly equal over any finite time interval of length δ, where we are free to choose any convenient value for δ. Then we choose a $C > I^* + 1$ and apply the lemma. We see that the probability that the two processes differ on an interval of length δ is smaller than $\exp\left((-I^* + 1)/2\right)$. Then consider the processes over sequences of times of length δ, where we choose δ larger than the time it takes r^* to go from $z_0 + \varepsilon$ to $z_1 - \varepsilon$, and larger than the time when the cheapest path that stays in the interval from $z_0 + \varepsilon$ to $z_1 - \varepsilon$ has an I function larger than $I^* + 1$. We can restart the $z_{K,\nu}(t)$ process to be equal to the $z_\nu(t)$ process at the end of each interval. Then we see that the deviations of each process occur at the same exponential rate, and within ε of each other.

We now sketch a proof that the Freidlin-Wentzell theory applies to i.d. Aloha. That is, we show that the escape time for $z_\nu(t)$ to leave the stable region $x < z_1$ is well approximated by $\exp(I^*/\nu)$. The problem is that the process $z_\nu(t)$ might spend an inordinate amount of time in a neighborhood of the point z_1. This would enable the process to have cycle times that are extremely long, without deciding whether or not it leaves the stable interval. This is the problem of "characteristic boundary," which has been extensively studied principally by Marty Day [D2], [D3], although Ross Pinsky and Tom Kurtz have also looked at the problem (and perhaps others as well). The problem is not so difficult in our case, though. Since we have a one-dimensional process (at least, the basic underlying process is essentially one-dimensional) we can obtain simple bounds on the time the process spends in a neighborhood of any point simply by comparing it to a balanced random walk without drift. This will enable us to conclude that the probability that the process spends an exponentially long time in a neighborhood of z_1 without leaving the neighborhood is small, uniformly in starting points in the neighborhood.

Actually, we can conclude more than we do. It is not hard to obtain estimates that show that the process has a positive probability of *never* returning to the stable interval once it has exited. We do not prove this here, even though it is not so difficult. See Rosencrantz and Towsley [RT] or Drmota and Schmidt [DrS] for proofs.

Once we have proved that the process does not hang around near z_1 for extremely long periods of time (perhaps exactly along the lines of the escape from q, the attracting point), we can use the Freidlin-Wentzell theory to show that escapes take about $\exp(I^*/v)$ to occur. This concludes our heuristic discussion of the Freidlin-Wentzell theory as applied to Aloha.

Our last topic is a sketch of why we think our four-parameter change of measure $\left[\tilde{\lambda}(0), \tilde{x}(0), \tilde{\lambda}(1), \tilde{x}(1) \right]$ is asymptotically optimal as $K \to \infty$. Consider an optimal change of measure for the K level approximating process $z_{K,v}(t)$ to make a large deviation from z_0 to z_1. For each $x \in (z_0, z_1)$ we have to find the cheapest change of measure v minimizing $\ell(x, y)/y$, where y is the drift associated with v.

First we should note that this change of measure is definitely not optimal for any finite K. This is because we know that an optimal change of measure is given by a $\vec{\theta}$ such that

$$\lambda_i(m) \to \lambda_i(m)e^{\langle \vec{\theta}, \vec{e}_i(m) \rangle}. \tag{14.24}$$

But it is easy to see that the proposed change of measure leads to the jump rate K at level K changing, with all other jump rates K not changing, and a simple calculation then shows that this is *not* an optimal thing to do for any fixed y.

The optimality we are referring to, then, does not hold uniformly over all levels, but is approached in the following manner. We can achieve a measure that has nearly the same cost and drift as the proposed change of measure with a $\vec{\theta}$ that is constant for indices $i \in (\sqrt{K}, K - \sqrt{K})$. This is achieved in a similar fashion to the way polarized light can be rotated through a series of polarizing filters with arbitrarily small loss of amplitude. The cost of changing θ_i from one value to another can be made arbitrarily small in $O(\sqrt{K})$ space. (The formal calculation showing this is not difficult, but since we do not have a complete proof that the four-parameter family is optimal, we don't want to take the space to derive this result.) That is, there might be a small boundary layer near the levels one and K that do not have the same jump rates as the interior, but the effect of this boundary layer on the cost and on the drift is asymptotically negligible for calculation, and the problem reduces to the four-dimensional one we solved earlier.

14.6. A Paradox—Resolved

Gaver and Fayolle [GF] developed a model of Aloha for analyzing burst arrivals, or equivalently long messages. The idea is that messages may come in batches of more than one packet. In fact, messages might consist of very long series of packets. Therefore, the batch is approximated as a continuous spread of packets.

Their model can be described as a finite levels model with two levels. When a batch arrives it tries to seize the channel. If the channel is free (level zero) then the batch will begin to drain through the channel continuously (the system being in level one), until it either finishes or suffers a collision with another batch. When two batches collide they both immediately join the retry queue and the system goes to level zero. In contrast with other models, though, all the transmission prior to the collision is regarded as successful. This is because the prior transmission actually consists of many "small" packets, and only one packet was involved in the collision. If packet lengths are small compared to batch sizes, this would seem to be a reasonable approximation.

Arrivals of batches are, as before, Poisson with rate λ and batch size is exponentially distributed. Therefore the system is Markovian, and can be described in terms of two parameters: x, the number of backlogged batches, and the level m, which is either zero for an empty channel, or one for an occupied channel. The finite levels model is given by five jump rates and directions:

Jump direction	Jump rate	Interpretation
$\vec{e}_1(0) = (1, 1)$	$\lambda_1(0) = \lambda$	arrival of new batch
$\vec{e}_2(0) = (1, 0)$	$\lambda_2(0) = x$	retransmission
$\vec{e}_1(1) = (-1, 1)$	$\lambda_1(1) = \lambda$	arrival of new batch
$\vec{e}_2(1) = (-1, 0)$	$\lambda_2(1) = x$	retransmission
$\vec{e}_3(1) = (-1, -1)$	$\lambda_3(1) = 1$	successful batch leaves.

Let us calculate the drift $\dfrac{d}{dt} z_\infty(t)$ of this process.

$$
\begin{aligned}
\frac{d}{dt} z_\infty(t) &= \pi_0(z_\infty)\lambda_1(0) + (1 - \pi_0(z_\infty))\,(\lambda_1(1) - \lambda_3(1)) \\
&= \frac{\lambda_1(1) + \lambda_2(1) + \lambda_3(1)}{R(z_\infty)}\lambda_1(0) + \frac{\lambda_1(0) + \lambda_2(0)}{R(z_\infty)}\,(\lambda_1(1) - \lambda_3(1)) \\
&= \lambda - \frac{\lambda + z_\infty}{2(\lambda + z_\infty) + 1}
\end{aligned}
$$

$$
R(x) \overset{\triangle}{=} \lambda_1(0) + \lambda_2(0) + \lambda_1(1) + \lambda_2(1) + \lambda_3(1) = 2(\lambda + x) + 1.
$$

It is clear that this drift is monotonically decreasing as z_∞ increases from zero to ∞, it is equal to $\lambda(2\lambda)/(2\lambda + 1)$ at $z_\infty = 0$, and approaches $\lambda - 1/2$ as z_∞ approaches ∞. Therefore for any $\lambda < 1/2$ there will be a unique stable point z_0 where the drift is zero. It is easy to find this point is given by

$$
z_0 = \frac{2\lambda^2}{1 - 2\lambda}. \tag{14.25}
$$

This leads us to conclude that the process has a unique invariant measure (you can see this by Dai [Da] or Chen [Cn] directly), and that the tail of this measure can be estimated by the Freidlin-Wentzell theory. For example, for $\lambda = 1/4$, $z_0 = 1/4$, and as $\nu \to 0$ we expect the invariant measure to concentrate near this point. We could estimate the steady-state probability that $z_\nu(t) > 1$ by solving a variational problem. Gaver and Fayolle calculated the invariant measure explicitly in [GF].

Exercise 14.16. Write down the variational problem that you would use to estimate the steady-state probability that $z_\nu(t) > 1$ when $\lambda = 1/4$. Solve the problem by recalling that the solution can be represented as

$$
I^* = \int_{1/4}^{1} \inf_{y>0} \frac{\ell(x, y)}{y}\, dx. \qquad \spadesuit
$$

Why is this model different from all other Aloha models? How can it be stable when the others are not? Since the stability is equivalent to examining the behavior of the system when z_ν becomes large (the stability results from the drift being negative as $z_\nu \to \infty$), let us examine the behavior of the system when z_ν is large. When z_ν is large, the most likely behavior is for the level to switch between zero and one with rate about νx. Very rarely [with probability $= O\left(1/(x\nu)\right)$] a transition corresponds to the entrance of a new customer, or the departure of an old one. How is it that a customer manages to depart? His service time must be smaller than $O\left(1/(z_\nu)\right)$—otherwise he gets bumped back to the waiting room. More exactly, the system switches between levels zero and one with rate very nearly z_ν. Each time the system is at level one there is a chance that a customer departs. This chance is approximately equal to the probability that an exponential random variable with mean one is smaller than an exponential random variable with mean $1/z_\nu$. There are approximately z_ν transitions between levels in each unit time interval, so there are about $1/2z_\nu$ chances for a departure in each unit time interval. Therefore we see that the rate of departures from the system is approximately $1/2$ as $z_\nu \to \infty$. The reason that there are departures from the system is that very likely there will be a short batch that attempts to depart, since there are many attempts in a unit time interval. Equivalently, all the work in the system is useful work, and the system is busy about half the time when z_ν is large, so the departure rate is about half the rate it would be if the system were continually busy.

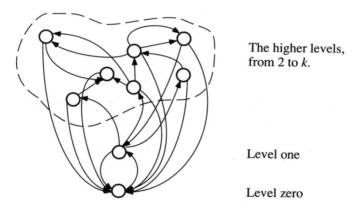

The higher levels, from 2 to k.

Level one

Level zero

Figure 14.17. The structure of a Gaver-Fayolle Aloha model that has a minimum time for successful transmission.

Now let us look at a slightly different system. Suppose that there is some minimum amount of time that the channel must be occupied so that a transmission may be regarded as successful. In reality, at least the header of a packet must be transmitted without error in order that any bits might be received successfully. Or, we could say that a bit takes a small but finite amount of time to transmit. We can model this by the addition of a level to our finite levels model. We can take a single level above zero that has no possibility of successful transmission, that feeds into a group of levels that encode the rest of the distribution of transmission time. The original Gaver-Fayolle model only has one state in this "group," so our analysis below is more general than an analysis of the original model. Note that the i.d. model is of this type.

We claim that every model in this class has the property that as $z_\nu \to \infty$, the drift tends to λ. This means that any such system can only be quasistable, not absolutely stable. Here is how to prove the claim. The rate of transition from level zero to level one is $\lambda + z_\nu$. The rate of transition from level one to the higher levels is a constant, say μ. The rate of transition from the higher levels to level zero is at least $\lambda + z_\nu$, and at most a constant $\xi + \lambda + z_\nu$. Therefore we see that as $z_\nu \to \infty$, the steady state probability among levels $\pi_z(i)$ satisfies

$$\pi_z(0)(\lambda + z) = \sum_{m=2}^{k} \pi_z(m)(\lambda + z + \xi(m)), \tag{14.26}$$

where $\mu(m) \in [0, \xi]$. Furthermore we have

$$\pi_z(0)(\lambda + z) + \sum_{m=2}^{k} \pi_z(m)\eta(m) = \pi_z(1)(\lambda + z + \mu). \tag{14.27}$$

The net flow into the states $\{2, \ldots, k\}$ is $\mu\pi_z(1)$, and there is a rate out of each state of at least $\lambda + z$. Therefore

$$\sum_{m=2}^{k} \pi_z(m)(\lambda + z) \le \mu\pi_z(1),$$

$$\sum_{m=2}^{k} \pi_z(m) \le \frac{\mu}{\lambda + z}\pi_z(1). \tag{14.28}$$

Now looking at the flow into and out of the state one, we see that for some C (representing the maximal rate from any state in $\{2, \ldots, k\}$ to 1)

$$\pi_z(1)(\lambda + z + \mu) \le \pi_z(0) + \sum_{m=2}^{k} \pi_z(m)C$$

$$\pi_z(1) \le \pi_z(0)\frac{\lambda + z}{\lambda + z + \mu - \frac{C\mu}{\lambda+z}} \le \frac{1}{2} \tag{14.29}$$

for z large enough. Then (14.28) and (14.29) show that the drift is at least

$$\lambda - \xi \sum_{m=2}^{k} \pi_z(m) \ge \lambda - \xi\frac{\mu}{\lambda + z}\frac{1}{2}, \tag{14.30}$$

which approaches λ as $z \to \infty$. This shows that the system cannot be stable if there is a lower bound on the amount of time a packet needs an uninterrupted channel in order to count the time as successful.

We could presumably carry this analysis further, to see in more detail how these systems differ from the plain Gaver-Fayolle model and to find the large deviations rate function that corresponds to instability. Lack of time and space prevents us from including that analysis. However, we hope that despite this lack the reader appreciates how a subtle modeling inaccuracy can change a system from having unstable (quasistable) behavior to one that has stable behavior.

14.7. Slotted Aloha Models

This section contains a cursory analysis of the large deviations of some simple slotted Aloha models. We do not provide the requisite theory for this analysis, since these are discrete-time models. (For the relevant theory, see Azencott and Ruget [AR], Wentzell [Wen], Deuschel and Stroock [DeS], Dembo and Zeitouni [DZ], or Dupuis and Ellis [DE2]. For justification of large deviations applied to slotted Aloha, see Jelenkovic and Weiss [JW].) We deemed them worthy of inclusion since they are the most common Aloha models, and their analysis is simpler than the analysis of i.d. Aloha, so more information can be extracted from them (see, e.g., Exercise 14.24).

Here are some standard Markov models of slotted Aloha. We let $x(t)$ be the number of packets waiting to be retransmitted at slot $t = 0, 1, 2, \ldots$. We suppose that:

1. The number of fresh packets $n(t)$ arriving at slot t is an independent random variable with distribution

$$\mathbb{P}(n(t) = k) = p_k(x(t)) .$$

   That is, the distribution may depend on the value of $x(t)$, but is otherwise independent of the past. Previously we had $n(t)$ a Poisson process, and later in this section we will specialize to three distributions for $n(t)$, including Poisson.

2. The number of retransmitted packets arriving at slot t is binomial $(x(t), v)$, where v is a fixed number.

3. If the channel contains exactly one packet during a slot, the packet leaves the system. Otherwise all fresh arrivals and retransmissions fail, and join the waiting packets.

With these assumptions, $x(t)$ is a Markov chain on the positive integers whose transition probabilities $q_k(x) = \mathbb{P}\left(x(t+1) - x(t) = k \mid x(t) = x\right)$ are given by

$$q_k(x) = \begin{cases} p_k(x) & \text{if } k \geq 2 \\ p_1(x)\left(1 - (1-v)^x\right) & \text{if } k = 1 \\ p_1(x)(1-v)^x + p_0(x)\left(1 - xv(1-v)^{x-1}\right) & \text{if } k = 0 \\ p_0(x)xv(1-v)^{x-1} & \text{if } k = -1 \end{cases}$$

if $x \geq 1$, while for $x = 0$,

$$q_k(0) = \begin{cases} p_k(0) & \text{if } k \geq 2 \\ 0 & \text{if } k = 1 \\ p_0(0) + p_1(0) & \text{if } k = 0. \end{cases}$$

The mean change in size of $x(t)$ is the drift, given by

$$\text{drift } (x) = \sum_{k=-1}^{\infty} k q_k(x). \tag{14.31}$$

We analyze $x(t)$ for three specific arrival modes $p_k(x)$: a *birth-death model*, a *Poisson model*, and a *Bernoulli, finite population model*.

The three models.

The birth-death model is included for the ease with which we can calculate its statistics, and the limpid formulae that describe its behavior. The Poisson model is the standard infinite population model. The finite population model is probably the most realistic.

1. **Birth-death model**: We suppose that $p_0 = 1 - \lambda$ and $p_1 = \lambda$; that is, packets arrive singly with probability λ at each slot. Furthermore, when $x(t) = 0$ we suppose that with probability $q > 0$, an arriving packet does not transmit but instead gets queued. [This is to prevent $x(t) = 0$ (waiting room empty) from being an absorbing state.] Hence, $x(t+1)$ is either $x(t), x(t) - 1$ or $x(t) + 1$ when $x(t) > 0$, and is either zero or one when $x(t) = 0$.

2. **Poisson model**: We assume that the probability of k new packets arriving at a given time slot is Poisson distributed:

$$p_k = e^{-\lambda} \frac{\lambda^k}{k!}$$

for some $\lambda > 0$. This is the model we analyzed briefly in §14.2 and §14.3.

3. **Bernoulli, finite population model**: Suppose that there are N stations, each of which attempts to transmit its current packet (if it has one) before generating a new one. At each time slot t while a station is without a packet it generates a new one independently with probability λ. Thus, the arrival probabilities p_k depend on the state x as

$$p_k(x) = \binom{N - x}{k} \lambda^k (1 - \lambda)^{N-x-k}$$

if $0 \leq k \leq N - x$, and $p_k(x) = 0$ otherwise.

Scaling and theory.

Recall that the parameter ν is the probability that a particular packet that is waiting to retransmit will attempt to retransmit at a given slot t. Now define

$$z_\nu(t) = \nu x(t/\nu) . \tag{14.32}$$

As $\nu \to 0$, $z_\nu(t)$ tends to a continuous-time deterministic process we call $z_\infty(t)$, much as $z_n(t) \to z_\infty(t)$ is our standard setup. Suppose that $p_k(x)$ does not depend on x, as in the Bernoulli and Poisson models above. Then if $\nu \to 0$ as $z = \nu x$ is held fixed,

$$q_k \approx \begin{cases} p_k & k \geq 2 \\ p_1(1 - e^{-z}) & k = 1 \\ p_1 e^{-z} + p_0(1 - ze^{-z}) & k = 0 \\ p_0 z e^{-z} & k = -1. \end{cases} \tag{14.33}$$

Hence

$$
\begin{aligned}
d(t) &\overset{\triangle}{=} \frac{d}{dt} \lim_{\nu \to 0} \mathbb{E}\,(z_\nu(t)) \\
&\approx \frac{\mathbb{E}\,(z_\nu\,(t + \nu)) - z_\nu(t)}{\nu} \\
&= \sum_{k=0}^{\infty} k p_k - p_1 e^{-z} - p_0 z e^{-z} \\
&\overset{\triangle}{=} \rho - p_1 e^{-z} - p_0 z e^{-z} ,
\end{aligned}
\tag{14.34}
$$

where ρ is the average packet arrival rate $\sum_{k=0}^{\infty} k p_k$. If the $p_k(x)$ do depend on x, then we assume that as $\nu \to 0$ and $\nu x \to z$, $\bar{p}_k(z) = \lim p_k(x)$ exists. We define $q_k(z)$ through $\bar{p}_k$ in (14.33) instead of using p_k.

We assume that a large deviations principle holds for the processes defined by (14.32). We furthermore assume that the Freidlin-Wentzell theory applies. The rate function $\ell(x, y)$ for these processes is defined by

$$\ell(x, y) \overset{\triangle}{=} \sup_\theta \left(\theta y - \log \sum_k e^{\theta k} q_k(x) \right). \tag{14.35}$$

(The references cited above give this definition; it should be plausible from Chernoff's Theorem.)

Analysis

The first step in analyzing the behavior of any such system is to examine the behavior of $z_\infty(t)$. Since $\frac{d}{dt} z_\infty(t) = $ drift $(z_\infty(t))$, we need to examine where the drift is positive and where it is negative, to find the stable and unstable regions.

Near $z = 0$, drift $\approx \rho - p_1 \geq 0$, and drift $= 0$ at $z = 0$ if and only if $p_k = 0$ for $k \geq 2$. That is, an empty system tends to grow.

Now assume that $p_k(x)$ does not depend on x, as in the Poisson and birth-death models. Then if $p_1 < p_0$, we obtain from (14.34) that

$$d(z) = \rho - p_1 e^{-z} - p_0 z e^{-z} \qquad (14.36)$$

has a unique minimum at $z = 1 - p_1/p_0$, and $d(z) = \rho - p_0 \exp(p_1 - p_0/p_0)$ there. As z increases beyond $1 - p_1/p_0$, we get $d(z) \to \rho$.

If $p_1 > p_0$ then $d(z) > 0$ for all $z > 0$, and the process $z_\infty(t) \to \infty$ as $t \to \infty$. This means that the backlog would tend to increase without bound, a poor sort of performance.

We suppose that $p_k(z)$ has the property that $d(0) > 0$ and that there exist z_0, z_1 such that

$$d(z_0) = 0$$
$$d(z) < 0 \quad \text{for } z \in (z_0, z_1)$$
$$d(z) > 0 \quad \text{for } z < z_0 \quad \text{and for some interval} \quad z > z_1.$$

Assuming that the system starts with $z_\nu(0) < z_1$, Kurtz's Theorem shows that $z_\nu(t) \to z_0$ as t increases. Then the Freidlin-Wentzell theory comes into play, showing that $z_\nu(t)$ will remain near z_0 for time τ, where $\mathbb{E}(\tau) \sim e^{(I^*/\nu)}$, and I^* solves the variational problem (14.16).

While $z \approx z_0$, we can easily estimate the performance of the system. Newly arriving packets arrive alone with probability

$$\frac{p_1}{\sum_{k=1}^{\infty} k p_k} = \frac{p_1}{\rho} < 1.$$

The probability that a slot has no retransmission attempts is approximately e^{-z_0}. Hence the probability that an arriving packet experiences no collision is about $p_1 \cdot e^{-z_0}/\rho$. If a packet experiences a collision, it undergoes a geometrically distributed number of retries, each retry being geometrically $(1/\nu)$ spaced from the one before. The probability of succeeding on retry is $p_0 e^{-z}$; hence the distribution of delay experienced has been determined.

Exercise 14.18. Make the preceding discussion rigorous. Hint: as $\nu \to 0$, $\mathbb{P}(\sup_{0 \le t \le T} |z_\nu(t) - z_\infty(t)| \ge \varepsilon) \le \exp(-\frac{f(\varepsilon)}{\nu} + o(1/\nu))$. This can be used to bound the distribution of the number of packets that attempt to retransmit at any slot in $0 \le$ slot number $\le T/\nu$. Prove

$$\left| \mathbb{P}(\text{ no. of slots } \ge K) - \mathbb{P}'_{geo}(\text{ no. } \ge K) \right| \le \varepsilon. \qquad \spadesuit$$

Definition 14.19. *If the function $d(z)$ has at least two roots z_0 and z_1, (with $0 \le z_0 < z_1 <$ all other roots), then we call the interval $[0, z_1)$ the stable region and z_0 the stable point.*

Let us see how this definition applies to our examples. For the birth-death model,

$$d(z) = \lambda(1 - e^{-z}) - (1 - \lambda)z e^{-z}. \qquad (14.37)$$

Here $\rho = \lambda = p_1$ and $p_0 = 1 - \lambda$, so the minimum of $d(z)$ is

$$\lambda - (1 - \lambda) \exp\left(-\frac{1 - 2\lambda}{1 - \lambda}\right) = \lambda - (1 - \lambda) \exp\left(-1 + \frac{\lambda}{1 - \lambda}\right), \qquad (14.38)$$

which is negative for $\lambda < \frac{1}{2}$. Thus, there is a stable region if $\lambda < \frac{1}{2}$ with $z_0 = 0$; z_1 must be evaluated numerically.

For the Poisson model, $p_0 = e^{-\lambda}$ and $p_1 = \lambda e^{-\lambda}$, so

$$d(z) = \lambda - \lambda e^{-\lambda} e^{-z} - e^{-\lambda} z e^{-z} \qquad (14.39)$$

and the minimum of $d(z)$ is

$$\lambda - \exp(-\lambda) \exp\left(-\frac{\exp(-\lambda) - \lambda \exp(-\lambda)}{\exp(-\lambda)}\right) = \lambda - e^{-\lambda} e^{-(1-\lambda)}$$

$$= \lambda - e^{-1}, \qquad (14.40)$$

which is negative for $\lambda < e^{-1}$. Here, the equation $d(z) = 0$ must be solved numerically for z_0 and z_1.

To ensure $\bar{p}_k(z)$ exists for our finite population Bernoulli model, we scale the parameters N (total number of stations) and λ (probability of generating a new packet) with ν. Specifically, we suppose that there are numbers n and b with

$$N\nu = n$$
$$\frac{\lambda}{\nu} = b. \qquad (14.41)$$

Then we find

$$\bar{p}_k(z) = e^{-b(n-z)} \frac{(b(n - z))^k}{k!};$$

that is, the process is Poisson with (state dependent) rate $b(n - z)$. By (14.34)

$$d(z) = b(n - z)\left(1 - e^{-b(n-z)} e^{-z}\right) - e^{-b(n-z)} z e^{-z}. \qquad (14.42)$$

It is difficult to give closed form regions for this case. See Figures 14.25–14.28 for examples of parameters with stable regions, and Exercise 14.29 for an idea of asymptotic behavior.

Large time behavior of slotted aloha.

When there are at least two roots $0 \le z_0 < z_1$ of the drift function $d(z)$, we know (by a simple extension to Freidlin-Wentzell theory to the case where the drift at the boundary of G, a one-dimensional region, is zero) that the system $z_\nu(t)$ will take about $e^{c/\nu}$ time to leave the basin of attraction of z_0, where $c = I^*$ satisfies (14.16), with ℓ given by (14.35). Since this is a one-dimensional variational problem, we may use the solution as developed in §11.5. To recapitulate,

$$I^* = \int_{z_0}^{z_1} \theta^*(z)\, dz \qquad (14.43)$$

where $\theta^* > 0$ solves

$$\log M(\theta^*, z) = 0 \tag{14.44}$$

$$M(\theta, z) = \sum_{k=-1}^{\infty} e^{\theta k} q_k(z) = \sum_{k=-1}^{\infty} \left(e^{\theta k} - 1\right) q_k(z) + 1. \tag{14.45}$$

We now explore the consequences of this analysis for our three models.

Birth-death model.

For this model, there are two roots, z_0 and z_1, of the drift function when $\lambda < \frac{1}{2}$.

$$M(\theta, z) = \left(e^{\theta} - 1\right) \lambda \left(1 - e^{-z}\right) + \left(e^{-\theta} - 1\right)(1 - \lambda) z e^{-z} + 1. \tag{14.46}$$

Solving (14.44), we obtain

$$\theta^*(z) = \log \left[\frac{(1 - \lambda) z e^{-z}}{\lambda(1 - e^{-z})} \right],$$

$$I^* = \int_0^{z_1} \theta^*(z) \, dz,$$

which can be evaluated numerically once we find z_1. The results are given in Figure 14.11, presented in §14.4.

Exercise 14.20. Show $z_1 \in \left(1 - \frac{\lambda}{1-\lambda}, \frac{1}{\lambda}\right)$. Can you obtain better bounds? An asymptotic expression as $\lambda \to 0$? ♠

Exercise 14.21. Show $I^* \to 0$ as $\lambda \uparrow \frac{1}{2}$. Show $I^* \to \infty$ as $\lambda \downarrow 0$. ♠

Poisson model.

We saw in (14.15) that there are two roots, $z_0 < z_1$, of the drift function when $\lambda < 1/e$. Now by (14.45),

$$M(\theta, z) = \exp\left(\lambda\left(e^{\theta} - 1\right)\right) - \left(e^{\theta} - 1\right) e^{-z} \lambda e^{-\lambda} + \left(e^{-\theta} - 1\right) z e^{-z} e^{-\lambda}.$$

We can solve $M(\theta^*, z) = 1$, $\theta^* > 0$ for $\theta^*(z)$ numerically, and can also find z_0 and z_1 numerically. This gives a numerical procedure for evaluating I^*. The results are given in Figure 14.11, presented in §14.4.

Exercise 14.22. Show $0 < z_0 < 1 - \lambda < z_1 < 1/\lambda$. Hint: where is the drift minimized? ♠

Exercise 14.23. Show that $I^* \to 0$ as $\lambda \uparrow 1/e$, and $I^* \to \infty$ as $\lambda \to 0$. At what rate does this occur? (See Exercise 14.24.) ♠

Exercise 14.24. Show that, as $\lambda \downarrow 0$,

$$I^* \approx \log \left(\frac{1}{\lambda} \right) + \left(\log \frac{1}{\lambda} \right) \log \left(\log \frac{1}{\lambda} \right) + o \left(\log \frac{1}{\lambda} \right).$$

Hint: show first that $z_0 \approx 0$, $z_1 \approx \log \left(\frac{1}{\lambda} \log \frac{1}{\lambda} \right)$. Then if $g(z) = \lambda e^{\theta^*(z)}$, we obtain $e^g - ge^{-z} = 1 + ze^{-z}$, so $g \approx ze^{-z}$ for large t. (See Drmota and Schmidt [DrS].) ♠

Finite population model.

It is difficult to describe the regime for which this model has a quasistability region. We can easily find z_0, z_1, and I^* numerically for any specified values of b and n by using the form

$$M(\theta, z) = \exp \left(b(n - z)(e^\theta - 1) \right) - (e^\theta - 1)e^{-z}b(n - z)e^{-b(n-z)}$$
$$+ (e^{-\theta} - 1)ze^{-z}e^{-b(n-z)}.$$

[This form is just that of a Poisson source, with arrival rate $\lambda(z) = b(n - z)$. Why?] The numerical results are given in Figure 14.11, presented in §14.4.

The drift functions for these parameters are illustrated in Figures 14.25—14.28. We have drawn the root $z_2 > z_1$ in these figures; z_2 is a second stable point for this model. This gives rise to the following behavior of the finite population model. For a long time the process $z_\nu(t)$ remains near z_0. Then it makes a switch, and remains for a long time near z_2. It eventually changes back to z_0, and the cycle repeats. You can formulate some large deviations questions: How long will the process remain in the basin of attraction of z_2, How long will it remain near z_0, and then since the invariant measure of the process degenerates (asymptotically as $\nu \to 0$) to two point masses, What are the respective masses at the two points z_0 and z_2? These questions are of purely theoretical interest, since any real system that wanders to z_2 would most likely be forcefully reset to operate near z_0.

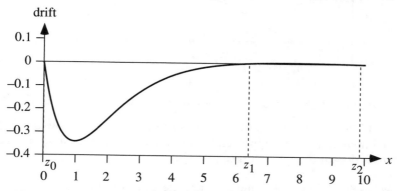

Figure 14.25. Finite population model drift, $n = 10$, $b = 0.003$.

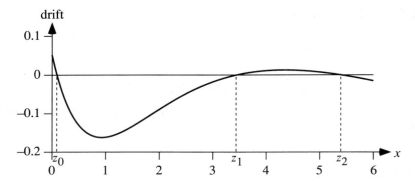

Figure 14.26. Finite population model drift, $n = 6$, $b = 0.04$.

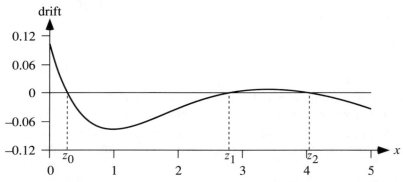

Figure 14.27. Finite population model drift, $n = 5$, $b = 0.07$.

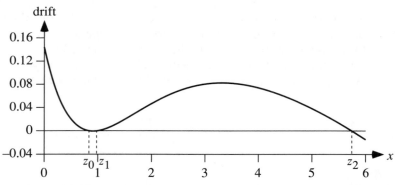

Figure 14.28. Finite population model drift, $n = 6$, $b = 0.07$.

Exercise 14.29. As $n \rightarrow \infty$ and $b \rightarrow 0$ so that $nb \rightarrow \lambda$, how quickly does the finite population model approach the infinite population (Poisson) model? In what senses does this approach take place? ♠

Exercise 14.30. Write a program to verify the entries in Figure 14.11. ♠

14.8. End Notes

The analysis of slotted Aloha models in this chapter is based on an AT&T internal report by Albert Greenberg and A. Weiss. This analysis is largely subsumed by the previous work of Cottrell, Fort, and Malgouyers [CFM]. Furthermore, many other people have investigated Aloha via large deviations; see, e.g., Maier [Mai] and his references.

There is a great deal more work that can be done in these multiple-access channels. David Aldous [Al] has shown, based on a large deviations analysis of a ball-and-bucket model, that the popular Ethernet protocol is unstable when subjected to Poisson arrivals. Goodman et al. [GGM] showed that a finite-population Ethernet model is stable for low enough arrival rates. Many adaptive schemes (usually called "controlled Aloha") have been proposed, and most work pretty well. For a good overview of the field, see Rom and Sidi [RS]; also, the IEEE Trans. Info. Theory volume 31 (1985) has many articles on the subject. Even though it is not hard to think of improvements to the Aloha protocol, it survives because of its simplicity and adaptability.

Chapter 15

Priority Queues

Priority queues are used extensively in modeling the performance of computers and computer algorithms. They are also observed in real life sometimes: at supermarkets or at toll plazas we might join the shortest queue, and at these places extra service people might be put on the longest queue. In this chapter we analyze three simple priority queues: one with two types of customers, "low" and "high" priority; a system of two queues with a single server who serves the longer queue; and a system of two queues where arrivals join the shorter queue. These basic models are, like the $M/M/1$ queue, building blocks for more complicated and realistic models.

"Join the shortest queue" is an optimal policy for many models in terms of lowering the waiting times of arriving customers; see [Web] (although see also [Wh1] for some counterexamples). Priority queuing models shared resources, such as memory. "Serve the longest queue" is an attempt at fairness, or to lower the variance of waiting times. Although conformance to strictly proper English usage would have us call these models "Serve the long**er** queue" and "Join the short**er** queue" (since we analyze only two-queue systems in detail), custom and a hope of generalizing these models will sometimes lead us to lapse.

This chapter can be viewed as a collection of analyses of three related systems consisting of two queues each. It can also be viewed as a complement of Chapter 16, which is another two-queue system. In all these systems we consider questions of "asymptotic correlation" or "induced rare events." For example we consider, as n becomes large,

$$\mathbb{E}_{ss}\left(x_2(t) \mid x_1(t) = n\right), \tag{15.1}$$

where $\mathbb{E}_{ss}$ refers to steady-state expectation, and x_i is the number of customers in queue i, $i = 1, 2$. This type of question has a compelling interpretation. Considering $x_i(t)$ as a "subsystem" of the queuing system, we are asking "If a subsystem does something rare, what can we infer about the rest of the system?"

The mathematical difficulties for all these systems are similar. One is the presence of boundaries, either the usual ones at $x_i = 0$, or ones that arise in the service or arrival policies, such as "serve the longer queue" which has a discontinuity in service rates along the line $x_1 = x_2$. Another difficulty is the possibility of a non-compact state space; the x_i might not be bounded. The last difficulty is the "singular" nature of our conditioning. That is, conditioning on $x_1 = n$ is conditioning on a "thin" set, that furthermore is not compact.

The techniques we use to overcome the mathematical difficulties are the following. The boundaries are generally handled by the flat boundary theory of Chapter 8. There are two additional special arguments we also need. For some processes we have to do a transformation to turn a discontinuity into a standard flat boundary. Also there are the point discontinuities that arise at the intersections of flat boundaries (typically the point $\vec{0}$); for these, we only have to show that the probability of staying near the point is high, and that the point isn't absorbing.

Noncompactness is only a problem when we want to study steady state via the Freidlin-Wentzell theory. For this we simply have to prove positive recurrence in order to use Theorem 6.89. Our processes have constant coefficients within large regions, so these estimates are easy.

The argument for conditioning on a small set is pretty simple for processes on a lattice, since there really isn't a singularity: points carry all the mass, so it isn't very hard to justify conditioning on some set of them, even asymptotically. This might be more of an issue for other types of processes.

While our main results have the same form as (15.1), we obtain much more information from our approach. For example, we also obtain sample path information of how the overflows are likely to occur. To help keep track within this chapter we provide the models with the nicknames PP (Preemptive Priority queue), SL (Serve the Longest queue), and JS (Join the Shortest queue).

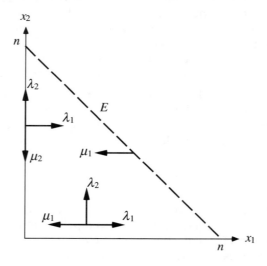

Figure 15.1. Transition diagram for priority queue.

15.1. Preemptive Priority Queue

The simplest model of a priority queue has a single server for a single queue. There are two types of customers arriving at the queue. We call them *high-priority customers*, or Type 1 customers, and *low-priority customers*, or Type 2 customers. (Equivalently, there are two queues, one for each type of customer, and a single server who shuttles between the queues.) Each type arrives by an independent Poisson process of rate λ_1 or λ_2 respectively. Their service times are independent exponentially distributed random variables with rates μ_1 and μ_2 respectively. The server always serves high-priority customers whenever one is in the system; it only serves low-priority customers when no high-priority customers are present. That is, the server will interrupt the service of a low-priority customer in order to serve an arriving high-priority customer. This makes the process $(x_1(t), x_2(t)) =$ (no. Type 1, no. Type 2) customers in the system at time t a Markov process. We are mainly interested in the finite version of this system, where the total capacity of the queue is n (and arrivals to a full queue are rejected). The transition rates and directions for the finite system are given in the following table and picture.

Definition 15.2. *The finite process $\vec{x}(t)$ is defined by the following rates and directions:*

Jump direction	Jump rate	Interpretation
$\vec{e}_1 = (1, 0)$	$\lambda_1(\vec{x}) = \begin{cases} \lambda_1 & x_1 + x_2 < n \\ 0 & x_1 + x_2 = n \end{cases}$	*High priority arrival*
$\vec{e}_2 = (0, 1)$	$\lambda_2(\vec{x}) = \begin{cases} \lambda_2 & x_1 + x_2 < n \\ 0 & x_1 + x_2 = n \end{cases}$	*Low priority arrival*
$\vec{e}_3 = (-1, 0)$	$\lambda_3(\vec{x}) = \begin{cases} \mu_1 & x_1 > 0 \\ 0 & x_1 = 0 \end{cases}$	*High priority service*
$\vec{e}_4 = (0, -1)$	$\lambda_4(\vec{x}) = \begin{cases} \mu_2 & x_1 = 0, \ x_2 > 0 \\ 0 & \text{otherwise} \end{cases}$	*Low priority service.*

Suppose that the queue has capacity n; that is, there is room for only n customers in the queue (including the one in service). There are several questions we might ask about the system: What is the stability condition for the infinite system (that is, if the queue size is infinite)? And, for the infinite and the finite system, what is the probability (in steady state) that the queue is full? What is the most likely way it becomes full? How long does it take to fill?

We make the usual scaling and then try to answer these questions. Let

$$\vec{z}_n(t) \overset{\triangle}{=} \frac{1}{n}\vec{x}(nt); \tag{15.2}$$

this conforms to our standard scaling (5.1). This process satisfies a large deviations principle away from the corners of the triangle according to Chapter 8. The flat boundaries in this case are $x_1 = 0$ and $x_1 + x_2 = 1$ (note that $x_2 = 0$ is not a boundary!). The infinite process extends to $\langle \vec{z}_n, (1, 1) \rangle \geq 1$, and its transitions and rates are obtained by simply ignoring the boundary at one. Therefore we may use the Freidlin-Wentzell theory to analyze the way the process $\vec{z}_n(t)$ (finite or not) escapes from the region $G \stackrel{\triangle}{=} \{\vec{z} : \langle \vec{z}, (1, 1) \rangle < 1\}$ (remember that point singularities such as the corners of the triangle do not affect the Freidlin-Wentzell theory, as long as the corners are not absorbing points; see, e.g., Corollary 6.63 and Corollary 6.65). Denote the boundary arising from the finite capacity by

$$E \stackrel{\triangle}{=} \{\vec{x} : \langle \vec{x}, (1, 1) \rangle = 1\}. \tag{15.3}$$

15.2. Most Probable Behavior—PP

We begin our analysis with an investigation of the fluid limit process $\vec{z}_\infty$. We are interested only in the case where the process is stable; therefore we may pose at the outset $\lambda_1 < \mu_1$, so that the high-priority process (which is just an $M/M/1$ queue) is stable. Let us establish, using (8.11), that that

$$\frac{d}{dt}\vec{z}_\infty(t) = \begin{cases} (\lambda_1 - \mu_1, \lambda_2) & \begin{array}{l}\text{if } z_{\infty,1} > 0, \text{ and either} \\ z_{\infty,1} + z_{\infty,2} < 1 \text{ or } \lambda_2 < \mu_1 - \lambda_1 \end{array} \\[2ex] \left(\frac{-\mu_1\lambda_2}{\lambda_1+\lambda_2}, \frac{\mu_1\lambda_2}{\lambda_1+\lambda_2}\right) & \begin{array}{l}\text{if } z_{\infty,1} > 0 \text{ with } \vec{z}_\infty \in E \\ \text{and } \lambda_2 \geq \mu_1 - \lambda_1 \end{array} \\[2ex] \left(0, \lambda_2 - \left(1 - \frac{\lambda_1}{\mu_1}\right)\mu_2\right) & \text{if } z_{\infty,1} = 0 \text{ and } z_{\infty,2} > 0 \\[2ex] \vec{0} & \text{at } \vec{0}. \end{cases} \tag{15.4}$$

The first line of (15.4) comes from (5.7)—it is the drift in the interior. The second line comes from (8.11), considering E as the flat boundary as in Figure 15.3. The probabilistic interpretation of $\vec{z}_\infty(t)$ was established in Exercise 8.13.

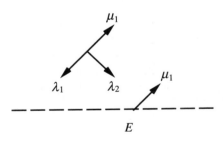

Figure 15.3. The process regarded with E as the flat boundary.

From (8.9)–(8.11) we see that, if $\lambda_2 < \mu_1 - \lambda_1$ then the drift is away from E, so that (5.7) applies. If $\lambda_1 + \lambda_2 > \mu_1$ then the drift is towards E, and the derivative of $\vec{z}_\infty$ in the new coordinates is equal to

$$(\pi\mu_1 + (1 - \pi)(\mu_1 + \lambda_2 - \lambda_1), 0)/\sqrt{2}, \tag{15.5}$$

where π solves

$$(\pi\mu_1 + (1 - \pi)(\mu_1 - \lambda_1 - \lambda_2))/\sqrt{2} = 0. \tag{15.6}$$

This gives the drift along the boundary E; changing coordinates back to $\vec{x}$ gives the second line of (15.4). Similarly, along the line $x_1 = 0$ we obtain the third line of (15.4) from

$$(0, \pi\lambda_2 + (1 - \pi)(\lambda_2 - \mu_2)), \tag{15.7}$$

where, since $\lambda_1 < \mu_1$, π solves

$$\pi\lambda_1 + (1 - \pi)(\lambda_1 - \mu_1) = 0. \tag{15.8}$$

Equation (15.4) has two or three domains of solution, one inside the region $x_1 > 0$, possibly one on the boundary E, the last on the boundary $x_1 = 0$. Clearly $\vec{z}_\infty(t)$ reaches the line $x_1 = 0$ in finite time. From there it drifts with constant rate with velocity $\left(0, \lambda_2 - \left(1 - \frac{\lambda_1}{\mu_1}\right)\mu_2\right)$. So we see that the process $\vec{z}_\infty(t)$ approaches (actually reaches) $\vec{0}$ if and only if

$$\lambda_2 - \left(1 - \frac{\lambda_1}{\mu_1}\right)\mu_2 < 0. \tag{15.9}$$

Thus we have arrived at the stability condition: $\lambda_1 < \mu_1$ and (15.9). (Actually, (15.9) implies $\lambda_1 < \mu_1$ since rates are positive.) Equation (15.9) has a simple interpretation: the arrival rate of low priority customers is λ_2, and the service rate is μ_2 during any time the server is not busy with high priority customers. The server is busy with high priority customers λ_1/μ_1 fraction of the time, so the long term service rate available to low priority customers is at most $(1 - \lambda_1/\mu_1)\mu_2$. Note that $\langle\frac{d}{dt}\vec{z}_\infty(t), \vec{n}\rangle \geq 0$ if $\lambda_2 \geq \mu_1 - \lambda_1$, so the Freidlin-Wentzell boundary condition (Assumption 6.4) might not hold. This is not a problem, as we see in §15.5.

15.3. The Variational Problem—PP

Our first problem is to find the frequency of buffer overflow. We do this, as usual, by writing a rate function and solving an associated variational problem. The process $\vec{z}_\infty(t) \to \vec{0}$ as $t \to \infty$ whenever the stability condition (15.9) holds. Therefore the Freidlin-Wentzell theory suggests that we can estimate $\mathbb{P}_{ss}(\vec{z}_n(t) \in E)$ by

$$\mathbb{P}_{ss}(\vec{z}_n(t) \in E) = e^{-nI^* + o(n)}, \tag{15.10}$$

where

$$I^* \stackrel{\triangle}{=} \inf_{(\vec{r}, T) \in G} I_0^T(\vec{r})$$
$$G \stackrel{\triangle}{=} \left\{(\vec{r}, T) : \vec{r}(0) = \vec{0}, \vec{r}(T) \in E\right\}. \tag{15.11}$$

The rate function $I_0^T(\vec{r})$ corresponds to $\vec{z}_n(t)$ and is defined via the local rate function $\ell(\vec{x}, \vec{y})$ below. We proceed formally with the calculations, mainly the solution of the variational problem (15.11), leaving some justifications and a proof of the connection between the variational problem and the probabilistic estimate to §15.5. The results of this section are summarized in Lemmas 15.20 and 15.21 below.

If $x_1 > 0$ and $\vec{x} \notin E$, then the process is not near a boundary, and (5.2)–(5.4) give

$$\ell(\vec{x}, \vec{y}) \stackrel{\triangle}{=} \sup_{\vec{\theta} \in \mathbb{R}^2} \left(\langle \vec{\theta}, \vec{y} \rangle - \sum_{i=1}^{2} \lambda_i \left(e^{\theta_i} - 1 \right) - \mu_1 \left(e^{-\theta_1} - 1 \right) \right) \tag{15.12}$$
$$= \ell_1(y_1) + \ell_2(y_2),$$

where ℓ_1 corresponds to an $M/M/1$ queue with rates λ_1, μ_1 and ℓ_2 corresponds to a Poisson process with rate λ_2:

$$\ell_1(y) = y \log \frac{y + \sqrt{y^2 + 4\lambda_1\mu_1}}{2\lambda_1} + \lambda_1 + \mu_1 - \sqrt{y^2 + 4\lambda_1\mu_1},$$
$$\ell_2(y) = y \log \frac{y}{\lambda_2} + \lambda_2 - y. \tag{15.13}$$

The same holds for $x_1 = 0$, $y_1 > 0$.

It will be convenient to perform the calculations with an alternate version of the local rate function, as described in Theorem 5.26 (the notation is chosen to facilitate the treatment of the boundaries). In the interior (when $x_1 > 0$ or $y_1 > 0$)

$$\ell(\vec{x}, \vec{y}) = \inf_{\{\lambda_1(1), \lambda_2(1), \mu_1(1)\} \in K_y} h(\lambda_1(1), \lambda_1) + h(\lambda_2(1), \lambda_2) + h(\mu_1(1), \mu_1), \tag{15.14}$$

where $\lambda_1(1), \lambda_2(1), \mu_1(1)$ are the appropriate twisted rates, and

$$K_y \stackrel{\triangle}{=} \{\lambda_1(1), \lambda_2(1), \mu_1(1) \geq 0 : (\lambda_1(1) - \mu_1(1), \lambda_2(1)) = \vec{y}\}$$
$$h(a, b) \stackrel{\triangle}{=} a \log \frac{a}{b} + b - a.$$

Compare this with the form of $\ell_2(y)$ in (15.13).

Exercise 15.4. Prove (15.12), by showing that when $x_1 > 0$, the maximization over $\vec{\theta}$ splits into two independent maximizations over θ_1 and θ_2. Check that (15.14) is indeed equivalent to (15.12). ♠

Let us first calculate the optimal path from $\vec{0}$ to E, among all paths that remain in the interior (except, of course, their endpoints). We shall then establish that an optimal path consists of at most two segments, each a straight line, and calculate the cost of traveling along the boundary.

Extremal interior path.

For paths that remain in the interior (i.e., where $x_1 > 0$ and $\vec{x} \notin E$), ℓ is given by (15.12). The strict convexity of ℓ and Lemma 5.16 guarantee that the lowest cost interior path from $\vec{0}$ to the line E must be a straight line. So, suppose our path is of the form

$$\vec{r}(t) = (y_1, y_2) \cdot t \quad \text{for} \quad 0 \le t \le \frac{1}{y_1 + y_2}$$

(so that indeed it ends on E). By (15.14) the minimal cost of a straight line path is

$$\min_{y_1 \ge 0, y_2 \ge 0} \frac{\ell(\vec{y})}{y_1 + y_2} \ge \min_{\{\lambda_i(1), \mu_1(1) \ge 0\}} \frac{h(\lambda_1(1), \lambda_1) + h(\lambda_2(1), \lambda_2) + h(\mu_1(1), \mu_1)}{\lambda_1(1) - \mu_1(1) + \lambda_2(1)},$$
(15.15)

where the inequality results from the fact that we have removed the restriction $\lambda_1(1) \ge \mu_1(1)$, which is necessary to have a path from $\vec{0}$ to E. However, if the minimizer of the right of (15.15) satisfies this constraint, then we have equality.

Since $\partial h(a, b)/\partial a \to -\infty$ as $a \to 0$, there must be a minimum where the parameters $\lambda_1(1)$, $\lambda_2(1)$, and $\mu_1(1)$ are strictly positive. Therefore, we are justified in setting partial derivatives of (15.15) with respect to $\lambda_1(1)$, $\lambda_2(1)$, and $\mu_1(1)$ to zero to find the extremal point. From this we obtain

$$\frac{\lambda_1}{\lambda_1(1)} = \frac{\lambda_2}{\lambda_2(1)} = \frac{\mu_1(1)}{\mu_1}.$$
(15.16)

Since there is only one point where the derivatives vanish, it must be the global minimum.

We could have derived the relation $\lambda_1/\lambda_1(1) = \mu_1(1)/\mu_1$ by using the relationship between the coefficients appearing in the various definitions of the rate function. Indeed, if $\ell(\vec{x}, \vec{y})$ is given by a particular $\vec{\theta}^*$ in (15.12), and by a particular set of $\lambda_1(1)$, $\lambda_2(1)$, and $\mu_1(1)$ in (15.14), then Exercise 5.29 gives

$$\lambda_1(1) = \lambda_1 e^{\langle \vec{\theta}^*, \vec{e}_1 \rangle}, \quad \lambda_2(1) = \lambda_2 e^{\langle \vec{\theta}^*, \vec{e}_2 \rangle}, \quad \mu_1(1) = \mu_1 e^{\langle \vec{\theta}^*, \vec{e}_3 \rangle}.$$
(15.17)

The relation now follows directly from Definition 15.2.

Relation (15.17) allows us to do partial calculations with each form of the rate function, and then to piece the calculations together. Let us define $C \overset{\triangle}{=} \lambda_1(1)/\lambda_1$. Our minimization of the right of (15.15) becomes, after some algebra

$$\min_{C > 0} \frac{(\lambda_1 + \lambda_2)(C \log C + 1 - C) + \mu_1 \left(\frac{1}{C} \log \frac{1}{C} + 1 - \frac{1}{C} \right)}{(\lambda_1 + \lambda_2)C - \mu_1/C}.$$
(15.18)

Define $\lambda \overset{\triangle}{=} \lambda_1 + \lambda_2$ and $\mu \overset{\triangle}{=} \mu_1$, and compare the derivation for (15.15) to the variational problem (11.8) of the $M/M/1$ queue, starting at zero and exiting at one, and its solution in Theorem 11.15. The minimization (15.18) is the same one encountered in the $M/M/1$ queue and we conclude that the optimal change of measure is where the twisted arrival rate is μ and the twisted service rate is λ, or

$$C = \frac{\mu}{\lambda} = \frac{\mu_1}{\lambda_1 + \lambda_2}. \tag{15.19}$$

This corresponds to a path $\vec{r}^*(t)$ that travels in direction

$$
\begin{aligned}
\vec{r}^{*\prime}(t) &= \lambda_1 C \vec{e}_1 + \lambda_2 C \vec{e}_2 + \frac{\mu_1}{C} \\
&= \left(\frac{\lambda_1 \mu_1}{\lambda_1 + \lambda_2} - (\lambda_1 + \lambda_2), \frac{\lambda_2 \mu_1}{\lambda_1 + \lambda_2} \vec{e}_3 \right),
\end{aligned} \tag{15.20}
$$

and by (11.23), the associated cost is $\log C$. Perhaps the best way to interpret this equation is using the result of §7.4. There we showed that any path $\vec{r}(t)$ is equivalent to a change of measure $\lambda_i \to \lambda_i e^{\langle \vec{\theta}, \vec{e}_i \rangle}$; in the present case,

$$\vec{\theta} = (\log C, \log C)$$

since $\lambda_i \to C \lambda_i$. The path defined in (15.20) clearly goes from $\vec{0}$ to E in time $T^* = (\mu_1 - \lambda_1 - \lambda_2)^{-1}$. The total cost of this path $I_0^{T^*}(\vec{r}^*)$ is

$$I^* = I_0^{T^*}(\vec{r}^*) = \log \frac{\mu_1}{\lambda_1 + \lambda_2}.$$

The only question is, Is this new path truly in the interior of the state space? That is, under this twist, is $r_1^*(t) \geq 0$? From (15.20), $r_1^*(t) \geq 0$ for $t > 0$ if and only if

$$\lambda_2 < \lambda_c \overset{\triangle}{=} \sqrt{\mu_1 \lambda_1} - \lambda_1. \tag{15.21}$$

Exercise 15.5. Derive (15.19) directly from the minimization (15.18), and verify the calculation leading to (15.21). ♠

We have now proved that the condition for the interior extremal path to be strictly away from the boundary is $\lambda_2 < \lambda_c$. In case of equality clearly it is better to move along the boundary. In Exercise 15.6 below we show that if $\lambda_2 > \lambda_c$ then it is also optimal to move along the boundary. Note that the condition does not depend on the value of μ_2; this is reasonable since μ_2 doesn't operate in the interior. Also note that λ_c has the right dimensions (the same as λ_1 and μ_1, inverse time).

Exercise 15.6. Show that if $\lambda_2 > \lambda_c$ then the optimal straight line path is along the boundary $x_1 = 0$. Hint: consider the last expression in (15.15) but add the constraint $\lambda_1(1) \geq \mu_1(1)$. Under the above condition there is no local minimum in the interior of the parameter set $(0 < \lambda_i(1), \lambda_1(1) > \mu_1(1) > 0)$. Use growth estimates on $h(a, b)$ near $a = 0$ and as $a \to \infty$ and continuity to conclude that there must be a minimum on the boundary $\lambda_1(1) = \mu_1(1)$. ♠

This concludes the computation of the optimal interior path. If $x_1 = 0$ and $y_1 = 0$ the local rate function changes. In that case, by (8.12)

$$\ell(\vec{x}, \vec{y}) = \sup_{\theta \in \mathbb{R}^2} \left(y_2 \theta_2 - \left(\sum_{i=1}^{2} \lambda_i \left(e^{\theta_i} - 1 \right) + \max_{j=1,2} \left\{ \mu_j \left(e^{-\theta_j} - 1 \right) \right\} \right) \right).$$
(15.22)

We will not be concerned with paths along the boundary E, so we don't give the local rate function for those paths, although it is easy to do so. Again, the alternative form of the rate function will be useful. From Theorem 8.19.ii

$$\ell((0, x_2), (0, y_2)) = \inf \left((1 - \pi_0) \left[h(\lambda_1(1), \lambda_1) + h(\lambda_2(1), \lambda_2) + h(\mu_1(1), \mu_1) \right] \right.$$

$$\left. + \pi_0 \left[h(\lambda_1(0), \lambda_1) + h(\lambda_2(0), \lambda_2) + h(\mu_2(0), \mu_2) \right] \right), \quad (15.23)$$

where $\lambda_i(m)$ and $\mu_i(m)$ are the twisted rates, and the infimum is over the set

$$\left\{ 0 \leq \pi_0 \leq 1, \ \lambda_i(m), \mu_i(m) \geq 0 : \begin{array}{l} \pi_0 \lambda_1(0) + (1 - \pi_0) (\lambda_1(1) - \mu_1(1)) = 0, \\ \pi_0 (\lambda_2(0) - \mu_2(0)) + (1 - \pi_0)\lambda_2(1) = y_2 \end{array} \right\}.$$

You should have the following interpretation in mind for these quantities. π_0 is the fraction of time $\vec{x}(t)$ spends on the boundary when following the path $(0, y_2) \cdot t$. The jump rates λ_i and μ_i change to $\lambda_i(1)$ and $\mu_i(1)$ in the interior $(x_1 > 0)$, and to $\lambda_i(0)$ and $\mu_i(0)$ on the boundary.

Exercise 15.7. Check that the local rate function has the form (15.22), and that the form given in Equation (15.23) is indeed equivalent. ♠

Exercise 15.8. Show that if $\ell(\vec{x}, \vec{y})$ is given by a particular $\vec{\theta}^*$ in (15.22), and if it is also given by a particular set of $\lambda_i(m)$, $\mu_i(m)$ in (15.23), then

$$\lambda_i(m) = \lambda_i e^{\langle \vec{\theta}^*, \vec{e}_i \rangle} \quad \text{and} \quad \mu_i(m) = \mu_i e^{\langle \vec{\theta}^*, \vec{e}_{i+2} \rangle}. \quad (15.24)$$

Hint: see Chapter 8. ♠

Relation (15.24) allows us to do partial calculations with each form of the rate function, and then to piece the calculations together. Furthermore, it tells us that the critical path has $\lambda_i(0) = \lambda_i(1)$ since their corresponding $\vec{e}_i$ and λ_i are the same.

As for the interior paths, the local rate function (15.22)–(15.23) does not depend on $\vec{x}$ (as long as we stay on the boundary). Therefore, paths on the boundary are constant speed (straight lines). We can now establish the following facts:

(i) The critical path is composed of at most two line segments, one along the boundary $x_1 = 0$, one in the interior.

(ii) Travel along the boundary always has lower cost than travel parallel to the boundary but just in the interior.

(iii) There is a critical path which is composed of exactly one line segment: in particular, the critical path, if unique, is composed of exactly one line segment. That is, in point (i), uniqueness excludes the case of more than one segment.

Exercise 15.9. Establish (i)–(iii). Hints: soft arguments suffice. Use the principle of optimality; see also Proposition 16.10. Show that an optimal path cannot move from the boundary to the interior and back, to obtain (i). Item (ii) follows from Definitions (15.12) and (15.22). Finally, use the independence of ℓ from $\vec{x}$ in each region and hence the linearity of the accrued cost to obtain (iii).   ♠

Extremal boundary path.

Remark 15.10.   When solving variational problems that arise in multidimensional systems, we need to be careful since the rate function $\ell(\vec{x}, \vec{y})$ is not necessarily strictly convex on boundaries (see Exercise 8.21). This means that minimal cost paths need not be straight lines for constant coefficient processes. The convexity of the rate function implies that straight lines are optimal; the point is that there might be other optimal paths, too. We examine several multidimensional constant coefficient processes with flat boundaries in Chapters 15 and 16. We find that all the minimal paths we calculate are composed of straight line segments. This is true not because we *assume* that the portions of the paths that are on the boundary must be linear, but because we *calculate* that they are. (The portions of optimal paths in the interior of the state space are linear because the rate function is strictly convex there by Exercise 5.27; see also Lemma 5.16.) In higher dimensions it might be more difficult to show that the optimal path is unique and is hence linear; we analyze only two-dimensional systems, so the boundaries are one-dimensional, and the calculation is straightforward. Indeed, for one-dimensional boundaries, the only possible ambiguity is the rate of travel, not the direction of travel, so simple algebra suffices to find that the minimum is unique.

Since there is an optimal path consisting of a single straight-line segment, let us now calculate the cost of traveling along the boundary $x_1 = 0$. We use the form (15.23) of the rate function. We are trying to minimize $\ell((0, x_2), (0, y_2))/y_2$ since we are interested in the cheapest straight line path in direction $(0, y_2)$ that goes from $\vec{0}$ to $(0, 1)$. Using the argument leading from (15.15) to (15.16), we can equate derivatives with respect to the parameters $\lambda_i(m)$ and $\mu_i(m)$ to zero. After some algebra (Exercise 15.11) we find that there are constants C and ρ such that

$$\lambda_2(1) = \lambda_2(0) \stackrel{\triangle}{=} \rho\lambda_2$$
$$\mu_2(0) = \mu_2/\rho$$
$$\lambda_1(1) = \lambda_1(0) \stackrel{\triangle}{=} C\lambda_1 \qquad\qquad (15.25)$$
$$\mu_1(1) = \mu_1/C$$
$$\pi_0 = 1 - C^2\lambda_1/\mu_1.$$

As before, we could obtain the first and third equalities as an immediate consequence of Exercise 15.8; but as we need all these, we shall do it the hard way.

Exercise 15.11. Verify (15.25). Hints: write π_0 as an explicit function of the parameters. Show that

$$(1 - \pi_0)\frac{\partial \pi_0}{\partial \lambda_1(0)} = \pi_0 \frac{\partial \pi_0}{\partial \lambda_1(1)} = -\pi_0 \frac{\partial \pi_0}{\partial \mu_1(1)}.$$

Now equate derivatives to zero. ♠

From the definition of y_2 in terms of the twisted rates we see that

$$y_2 = \rho \lambda_2 - \pi_0 \mu_2 / \rho. \tag{15.26}$$

We can now write the minimum cost in terms of ρ and π_0 alone and reduce the problem of finding the minimum cost path along the boundary $x_1 = 0$ to minimizing, over $0 \leq \pi_0 \leq 1$, $\rho > 0$, the function

$$h_b(\pi_0, \rho) \overset{\triangle}{=} \frac{\lambda_2 h(\rho, 1) + \pi_0 \mu_2 h\left(\frac{1}{\rho}, 1\right) + U}{\rho \lambda_2 - \pi_0 \mu_2 / \rho}, \tag{15.27}$$

where we define

$$U \overset{\triangle}{=} \left(\sqrt{(1 - \pi_0)\mu_1} - \sqrt{\lambda_1}\right)^2. \tag{15.28}$$

Note that U represents the cost of staying next to the boundary.

Exercise 15.12. Derive (15.27). Hint: use (15.25) and show that

$$(1 - \pi_0)\left[h(\lambda_1, \lambda_1(1)) + h(\mu_1, \mu_1(1))\right] + \pi_0 h(\lambda_1, \lambda_1(0)) = U. \tag{15.29}$$

Now write C as a function of π_0. ♠

Note first that a minimal value of h_b can come only when $\pi_0 \in [0, 1 - \lambda_1/\mu_1)$; this is because U is increasing for larger π_0 (note that $\sqrt{\mu_1(1 - \pi_0)} - \sqrt{\lambda_1} = 0$ only when $\pi_0 = 1 - \lambda_1/\mu_1$), so that h_b is increasing in that range. The physical intuition is that there is no benefit to putting more effort into the balance in the x_1 direction, since the larger π_0 is, the harder it is for $x_2(t)$ to increase (the low-priority service rate is $\mu_2 \pi_0$). For any given value of $\pi_0 \in [0, 1 - \lambda_1/\mu_1)$, the value of ρ that minimizes h_b occurs where

$$\frac{\partial h_b}{\partial \rho} = 0 \text{ at } \rho = \frac{\lambda_2 + \pi_0 \mu_2 + U + \sqrt{(\lambda_2 + \pi_0 \mu_2 + U)^2 - 4\lambda_2 \mu_2 \pi_0}}{2\lambda_2}. \tag{15.30}$$

Exercise 15.13. Derive (15.30). Hint: show that $h_b \to \infty$ as $\rho \to \infty$ or as $\rho \downarrow \sqrt{\pi_0 \mu_2 / \lambda_2}$ (the point where $y_2 = 0$). Therefore there is a minimum of h_b where $\partial h_b / \partial \rho = 0$. Factor the resulting numerator into two quadratics, one of which has no roots, so the roots of the other contains the correct value of ρ. Use $y_2 > 0$ in (15.26) to choose the correct root. You will find that

$$y_2 = \sqrt{(\lambda_2 + \pi_0\mu_2 + U)^2 - 4\lambda_2\mu_2\pi_0},$$

so that we can write

$$\rho = \frac{y_2 + \sqrt{y_2^2 + 4\lambda_2\pi_0\mu_2}}{2\lambda_2}.$$

You will recognize this $\rho = \exp(\theta^*)$, where θ^* is the optimal twist for an $M/M/1$ queue with arrival rate λ_2 and service rate $\pi_0\mu_2$ traveling with velocity y_2. ◆

By Exercise 15.13, the minimum value of h_b occurs, in the (ρ, π_0) space, for ρ bounded above and bounded away from zero. In order to see whether the minimum occurs at $\pi_0 = 0$ or at $\pi_0 > 0$, we look for a minimum of h_b, where $\rho(\pi_0)$ is given in (15.30). We have already argued that there could not be a minimum to the right of $1 - \lambda_1/\mu_1$, and since U and its derivative vanish there, this point cannot be a minimum. By continuity of h_b in π_0, there must be a minimum in $[0, 1 - \lambda_1/\mu_1)$. Define

$$\lambda_D \overset{\triangle}{=} \lambda_c \left(1 + \frac{\mu_2 - \mu_1}{\sqrt{\mu_1\lambda_1}}\right). \tag{15.31}$$

Lemma 15.14. *If $\lambda_2 > \lambda_D$ then the optimal value of h_b occurs at $\pi_0 > 0$. If $0 \le \lambda_2 \le \lambda_D$ then the optimal value of h_b occurs at $\pi_0 = 0$.*

Proof. The computations below establish the following.

(i) For each fixed ρ, the function $h_b(\pi_0, \rho)$ has only one local minimum for π_0 in the interval $[0, 1 - \lambda_1/\mu_1)$ that is hence the global minimum.

(ii) When $\rho(\pi_0)$ is given by (15.30), $\partial h_b/\partial\pi_0 = 0$ at $\pi_0 = 0$ only if $\lambda_2 = \lambda_D$.

(iii) If $\lambda_D \ge \lambda_2 = 0$ the minimum is at $\pi_0 = 0$, while if $\lambda_2 = \lambda_M \overset{\triangle}{=} \mu_2 (1 - \lambda_1/\mu_1)$ then the minimum is at $\pi_0 > 0$.

By (ii)–(iii), if $\lambda_D > \lambda_2 = 0$ then $\partial h_b/\partial\pi_0 > 0$ at $\pi_0 = 0$, and by continuity in λ_2 and (ii) we conclude that the derivative is never negative for λ in $[0, \lambda_D]$. From (i) we now conclude that in this case the optimal value of π_0 is zero. Similarly, since by Exercise 15.17 $\lambda_M \ne \lambda_D$ we have from (i)–(iii) that $\partial h_b(0, \rho(0))/\partial\pi_0 < 0$ for $\lambda = \lambda_M$, and by continuity and (ii) it must stay negative for $\lambda > \lambda_D$. By (i) now we have that the optimal value of π_0 is strictly positive for $\lambda_2 > \lambda_D$. The result of the lemma now follows from (i) and the continuity in π_0, so that it remains to check (i)–(iii).

To establish (i), compute from (15.27) $\partial h_b/\partial\pi_0 = N/D$ by the usual rule for differentiating quotients. Since B is the square of the denominator y_2 in (15.27), we have $B > 0$. You might want to check the computations leading to (15.32); we used the computer algebra tool Theorist©, and will not write out the complicated expressions. Repeating the procedure for N, denote $\partial N / \partial \pi_0 = C/D_2$. Again D_2 is positive, and

$$C = \frac{\sqrt{\mu_1 \lambda_1} \left(\lambda_2 \rho^2 - \mu_2 \pi_0 \right) \rho}{2\sqrt{1 - \pi_0}}. \tag{15.32}$$

But we know that any value of ρ that might give us a minimum must have $y_2 > 0$ in (15.26). Therefore $C > 0$, and so $\partial A / \partial \pi_0 > 0$, so that A is an increasing function of π_0. This means that there is at most one root of $\partial h_b / \partial \pi_0$ in the interval $[0, 1 - \lambda_1/\mu_1)$, and if $\partial h_b / \partial \pi_0 > 0$ at $\pi_0 = 0$, then $\pi_0 = 0$ must be the unique minimum. (Why? Think about it. ....) One of the two must hold, since the minimum cannot be at $1 - \lambda_1/\mu_1$. This establishes (i).

To establish (ii), note from Exercise 15.13 that

$$\rho(0) = \frac{y_2}{\lambda_2} = \frac{\left(\sqrt{\mu_1} - \sqrt{\lambda_1} \right)^2}{\lambda_2} + 1. \tag{15.33}$$

(You can check that this gives $\rho = \sqrt{\mu_1/\lambda_1}$ when $\lambda_2 = \lambda_c$. This is the same as the twist for the interior path at $\lambda_2 = \lambda_c$; when $\pi_0 = 0$ the interior and boundary costs are hence the same, as they should be.) Now at this value of ρ we have $\partial h_b(0, \rho(0)) / \partial \pi_0 = 0$ when $\lambda_2 = \lambda_D$.

Finally, in Exercise 15.18 below we show that $\pi_0 = 0$ is optimal for $\lambda_2 \approx 0$ provided $\lambda_D \geq 0$, and in Exercise 15.19 below we show that $\pi_0 > 0$ is optimal for $\lambda_2 \approx \lambda_M$, which implies (iii). ∎

Exercise 15.15. Show that λ_D of (15.31) satisfies

$$\lambda_D = \frac{\left(\mu_2 - \mu_1 + \sqrt{\mu_1 \lambda_1} \right) U}{\mu_1 - \sqrt{\mu_1 \lambda_1}}.$$

(This may be the form in which you find λ_D, depending on how you do your calculations, in Exercise 15.16 below.) ♠

Exercise 15.16. Check that equations (15.31) and (15.33) are correct. ♠

Exercise 15.17. Prove that $\lambda_D < \mu_2\left(1 - \frac{\lambda_1}{\mu_1}\right) \overset{\triangle}{=} \lambda_M$, so that, for any given values of λ_1, μ_1, and μ_2, there always exists an interval of λ_2 where the cheapest path spends real time on the boundary. ♠

Exercise 15.18. If $\lambda_D \geq 0$ and $\lambda_2 \approx 0$ then the minimum of h_b is obtained at $\pi_0 = 0$. Hint: show the derivative in (15.27) at $\pi_0 = 0$ is positive. Note that when $\pi_0 = 0$, $\lambda_2 \cdot \rho \to U$ as $\lambda_2 \to 0$. ♠

Exercise 15.19. If $\lambda_2 \approx \lambda_M$ then the minimum of h_b is obtained at $\pi_0 > 0$. Hint: take $\pi_0 = 1 - \lambda_1/\mu_1$ and $\lambda_2 = \mu_2\pi_0(1-x)$ with x small and positive. Use $\rho \approx 1 + x$ and $h(\rho, 1) \approx x^2/2$ to show $h_b(\pi_0, \rho) \approx 0$. ♠

Unfortunately we do not have an explicit expression for the critical value of π_0 when $\lambda_2 > \lambda_D$. The critical value holds where $\partial h_b/\partial \pi_0 = 0$. This can be simplified to

$$\mu_2\lambda_2\frac{(1-\rho)^2}{\rho} + \frac{\mu_2}{\rho}U - \left(\rho\lambda_2 - \frac{\pi_0\mu_2}{\rho}\right)\sqrt{\frac{U\mu_1}{1-\pi_0}} = 0. \qquad (15.34)$$

There is at least one root of this expression in $(0, 1-\lambda_1/\mu_1)$ when $\lambda_2 > \lambda_D$, since $\partial h_b/\partial \pi_0 > 0$ at $\pi_0 = 0$, and $\partial h_b/\partial \pi_0 < 0$ at $\pi_0 = 1 - \lambda_1/\mu_1$. These statements follow from the proof of Lemma 15.14.

We state our conclusions thus far in the form of lemmas. The first is established in Exercise 15.6 and the preceding discussion, and the second follows mainly from Lemma 15.14.

Lemma 15.20. *Consider the path from $\vec{0}$ to E with minimal cost [the cost is given by (15.15)]. If $\lambda_2 < \lambda_c$, then (15.20) gives the unique interior minimizing path, which satisfies $y_1 > 0$. If $\lambda_2 \geq \lambda_c$ then there is a path along the boundary that is cheaper than any interior path.*

Lemma 15.21. *Consider the problem of the path along the boundary $x_1 = 0$, from $\vec{0}$ to $(0, 1)$, with minimal cost (15.27). There exists a unique path with minimal cost. If $\lambda_2 > \lambda_D$, then this minimum is attained at $\pi_0 > 0$, and its speed y_2 is given in (15.26). If $\lambda_2 \leq \lambda_D$, the unique minimum is attained at $\pi_0 = 0$, and its speed is $y_2 = \lambda_2\rho$, with ρ given in (15.33).*

Comparing interior and boundary costs.

Now in order to find the cheapest path from $\vec{0}$ to E we use the following easily demonstrated facts:

1. The critical path is composed of at most two line segments, one along the boundary $x_1 = 0$, one in the interior.

2. Travel along the boundary always has lower cost than travel parallel to the boundary but just in the interior.

3. The critical path, if unique, is composed of exactly one line segment; that is, in point 1, uniqueness excludes the case of more than one segment.

Exercise 15.22. Show that these three points are true. Hint: the first comes from the jump rates being constants and from Lemma 5.16. The second is always true for flat boundary processes. The third comes from the first and Lemma 5.16. ♠

There is only one more calculation to do before we can say that we have found the minimal cost path for travel from $\vec{0}$ to E. We need to see when the path in the interior has lower cost than the path along the boundary. Unfortunately, in general we cannot give a closed form expression for the answer to this question. We have to content ourselves with an existential statement (and an efficient way of calculating the break point). It turns out that there will be a region of λ_2 where the minimum in the interior is higher than the minimum on the boundary if and only if $\mu_2 < \mu_1$. The reason for this is pretty clear: in the interior, the cheapest path is formed by changing λ_1 and λ_2 by the same amount. So if $\mu_2 < \mu_1$ then it will be cheaper to travel by changing μ_2 than by changing μ_1 for a path sufficiently near the boundary (changing μ_1 a little less will push the path to the boundary, but then changing μ_2 so that the path will continue to E at the same rate as before should cost less).

When $\mu_2 < \mu_1$, we clearly see from (15.31) that $\lambda_D < \lambda_c$. That is, there is a region of λ_2 [namely (λ_D, λ_c)] where the cheapest path along the boundary spends positive time on the boundary ($\pi_0 > 0$), but the cheapest path in the interior is strictly away from the boundary. Now at the point $\lambda_2 = \lambda_c$, the interior path travels along the boundary, but the unique cheapest path along the boundary has $\pi_0 > 0$. This means that the boundary path is cheaper than the interior path at $\lambda_2 = \lambda_c$, since the boundary path is unique, and is of minimal cost over a set of paths that includes the cheapest interior path. By the continuity of all relevant functions with respect to all parameters, the boundary path is cheaper than the interior path for all λ_2 that are sufficiently near λ_c.

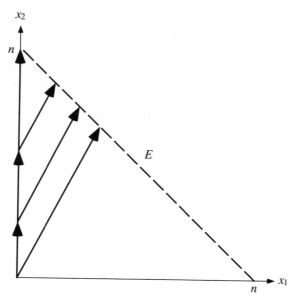

Figure 15.23. All the pictured paths have the same cost at $\lambda_2 = \lambda_B$.

Similarly, at $\lambda_2 = \lambda_D$, the boundary path spends no time on the boundary, hence has the same cost as an interior path along the boundary. But for $\lambda_2 < \lambda_c$, the interior path along the boundary is more expensive than the cheapest path in the interior. Therefore we have demonstrated that the interior path is cheapest for all λ_2 sufficiently near λ_D. Hence by continuity there are points $\lambda_D < \lambda_{B_1} \leq \lambda_{B_2} < \lambda_c$ so that the interior and boundary paths have the same cost whenever

$\lambda_2 = \lambda_{B_i}$, and moreover, for $\lambda_2 < \lambda_{B_1}$ the interior path is the unique cheapest path, and for $\lambda_2 > \lambda_{B_2}$ the boundary path is the unique cheapest path. At $\lambda_2 = \lambda_{B_i}$ there is no uniqueness. Indeed, any path that travels along the boundary for a while, then goes in a direction parallel to the cheapest interior path, will have exactly the same cost. We also have the following

Conjecture 15.24. $\lambda_{B_1} = \lambda_{B_2}$.

We don't know how to calculate λ_{B_i} except by comparing the costs for the interior and boundary paths numerically and finding when they are equal.

This completes our calculations for the simple preemptive priority model. We collect the results on the extremal paths in the form of theorems. The implications to questions concerning probabilities of events are discussed and proved in §15.5.

Recall from (15.21), (15.31) and the proof of (iii), Lemma 15.14 that

$$\lambda_c \overset{\triangle}{=} \lambda_1 \left(\sqrt{\frac{\mu_1}{\lambda_1}} - 1 \right).$$

$$\lambda_D \overset{\triangle}{=} \lambda_c \left(1 + \frac{\mu_2 - \mu_1}{\sqrt{\mu_1 \lambda_1}} \right).$$

$$\lambda_M \overset{\triangle}{=} \mu_2 \left(1 - \frac{\lambda_1}{\mu_1} \right).$$

In the four theorems below we consider the variational problem (15.11), where ℓ is defined through (15.12) and (15.22). Recall that a path $\vec{r}$ is called an interior path if $r_1(t) > 0$ for $t > 0$, and a boundary path if $r_1(t) = 0$ for $t > 0$. We shall refer to the following standard changes of measure.

Definition 15.25. *Define a change of measure by*

$$\lambda_1 \to \lambda_1 C \qquad\qquad \lambda_2 \to \lambda_2 \rho$$
$$\mu_1 \to \mu_1 / C \qquad\qquad \mu_2 \to \mu_2 / \rho,$$

where the constants C and ρ are defined within the theorems below.

Theorem 15.26. *Define the path $\vec{r}\,^*(t), 0 \le t \le T^*$ and the time T^* by*

$$\vec{r}\,^*(t) = (\lambda_1 C - \mu_1 / C, \lambda_2 \rho) \cdot t \tag{15.35}$$

$$T^* = \frac{1}{\lambda_1 C - \mu_1 / C + \lambda_2 \rho} = \frac{1}{\mu_1 - \lambda_1 - \lambda_2}, \tag{15.36}$$

where the change of measure is given by

$$C = \rho = \frac{\mu_1}{\lambda_1 + \lambda_2}. \tag{15.37}$$

Then $\vec{r}\,^*$ is in the interior and is the unique optimal path for the variational problem, with optimal change of measure given by C and ρ, for the following values of the parameters:

(i) $\mu_1 \leq \mu_2$ and $\lambda_2 \in [0, \lambda_c)$,

(ii) $\mu_1 > \mu_2$ and $\lambda_2 \in [0, \lambda_{B_1})$, where $0 < \lambda_{B_1} < \lambda_c$ and λ_{B_1} is the smallest value of λ_2 for which $\vec{r}^*$ has the same cost as the path of Theorem 15.27.

Theorem 15.27. Define the path $\vec{r}^*(t)$, $0 \leq t \leq T^*$ and the time T^* by $\vec{r}^*(t) = (0, y_2) \cdot t$ where $y_2 = \rho\lambda_2 - \pi_0\mu_2/\rho$ and $T^* = 1/y_2$ with the change of measure given by

$$C = \sqrt{\frac{(1 - \pi_0)\mu_1}{\lambda_1}}, \tag{15.38}$$

$$\rho = \frac{\lambda_2 + \pi_0\mu_2 + U + \sqrt{(\lambda_2 + \pi_0\mu_2 + U)^2 - 4\lambda_2\mu_2\pi_0}}{2\lambda_2} \tag{15.39}$$

$$U = \left(\sqrt{(1 - \pi_0)\mu_1} - \sqrt{\lambda_1}\right)^2 \tag{15.40}$$

and $\pi_0 \in (0, 1 - \lambda_1/\mu_1)$ solves

$$\mu_2\lambda_2\frac{(1 - \rho)^2}{\rho} + \frac{\mu_2}{\rho}U - \left(\rho\lambda_2 - \frac{\pi_0\mu_2}{\rho}\right)\sqrt{\frac{U\mu_1}{1 - \pi_0}} = 0.$$

Then $\vec{r}^*$ is on the boundary, $\pi_0 > 0$, and $\vec{r}^*$ is the unique optimal path for the variational problem, with optimal change of measure given by C and ρ, for the following values of the parameters.

(i) $\mu_1 < \mu_2$ and $\lambda_2 \in [\lambda_D, \lambda_M)$, where in this case $\lambda_c < \lambda_D \leq \lambda_M$,

(ii) $\mu_1 = \mu_2$ and $\lambda_2 \in (\lambda_c, \lambda_M)$,

(iii) $\mu_1 > \mu_2$ and $\lambda_2 \in (\lambda_{B_2}, \lambda_M)$, where $0 < \lambda_{B_2} < \lambda_c$ and λ_{B_2} is the largest value of λ_2 for which $\vec{r}^*$ has the same cost as the path of Theorem 15.26.

Theorem 15.28. Define the path $\vec{r}^*(t)$, $0 \leq t \leq T^*$ and the time T^* by $\vec{r}^*(t) = (0, y_2) \cdot t$ where $y_2 = \rho\lambda_2$ and $T^* = 1/y_2$ with the change of measure given by

$$C = \sqrt{\frac{\mu_1}{\lambda_1}} \quad \text{and} \quad \rho = \frac{\left(\sqrt{\mu_1} - \sqrt{\lambda_1}\right)^2}{\lambda_2} + 1. \tag{15.41}$$

Then $\vec{r}^*$ is on the boundary, $\pi_0 = 0$, and $\vec{r}^*$ is the unique optimal path for the variational problem, with optimal change of measure given by C and ρ whenever $\mu_1 \leq \mu_2$ and $\lambda_2 \in [\lambda_c, \lambda_D]$. The interval $[\lambda_c, \lambda_D]$ is not empty, and is a single point if and only if $\mu_1 = \mu_2$. In this last case the definitions of the paths and optimal rates in Theorems 15.26–15.28 agree.

Theorem 15.29. Define λ_{B_1} and λ_{B_2} as in Theorem 15.26(ii) and 15.27(iii), respectively. If $\mu_1 > \mu_2$ then for $\lambda_2 = \lambda_{B_i}$ the critical path is not unique. Any path $\vec{r}^*$ that follows the path $\vec{r}^*(t)$ of Theorem 15.27 for $0 \leq t < T^*$ and then proceeds parallel to the path of Theorem 15.26 is optimal. The value of I^* can be calculated from that of either the boundary path or the interior path.

15.4. Probabilistic Questions—PP

Now that we know the lowest cost path from $\vec{0}$ to E, we expect that several statements about the probabilistic behavior of the preemptive priority system should hold. We justify these statements in §15.5; for now, you should look at them as natural consequences of large deviations theory and Freidlin-Wentzell theory, even if you suspect that we haven't shown that either of these theories apply to the model.

Let τ denote the first time $\vec{z}_n(t) \in E$.

Theorem 15.30. *For the preemptive priority model, with I^*, r^*, and T^* defined in §15.3, for each $\varepsilon > 0$, uniformly over $\vec{x}$: $\langle \vec{x}, (1, 1) \rangle < 1 - \varepsilon$,*

$$\lim_{n \to \infty} \mathbb{P}_{\vec{x}} \left(I^* - \varepsilon < \frac{\log \tau}{n} < I^* + \varepsilon \right) = 1.$$

Theorem 15.31. *For the preemptive priority model, with I^* defined in §15.3,*

$$\lim_{n \to \infty} \frac{1}{n} \log \mathbb{P}_{ss} \left(\vec{z}_n(t) \in E \right) = -I^*. \tag{15.42}$$

Theorem 15.32. *For the preemptive priority model, with I^*, r^*, and T^* defined in §15.3, for each $T > 0$ we have*

$$\lim_{n \to \infty} \mathbb{P}_{ss} \left(\sup_{0 < s < T} |\vec{z}_n(t - s) - \vec{r}^*(T^* - s)| < \varepsilon \,\Big|\, \vec{z}_n(t) \in E \right) = 1. \tag{15.43}$$

[Recall that $\vec{r}^*(t) = \vec{0}$ for $t < 0$.]

As a simple consequence of Theorem 15.31, we can state that

$$\lim_{n \to \infty} \mathbb{P}_{ss} \left(|\vec{z}_n(t) - \vec{r}^*(T^*)| < \varepsilon \,\Big|\, \vec{z}_n(t) \in E \right) = 1.$$

These three theorems contain nearly all the probabilistic insight we have about the preemptive priority model. You might wish to formulate another theorem using the results of §7.4 to the meaning of the twisted rates.

A simple extension.

Now that we have analyzed the simple priority queue, we can immediately answer some questions about the equivalent two-queue model. Instead of asking about the distribution of $\vec{x}$ conditioned on $x_1 + x_2 = 1$, we can investigate the distribution of x_1 given that $x_2 = n$ and conversely the distribution of x_2 given that $x_1 = n$. These will all follow without any calculation or justification, simply from the analysis and justification of the first part of this chapter.

The new model has one of two forms. Either $x_1(t) \leq n$ and $x_2(t) \leq n$ (instead of a boundary at E we have boundaries at $z_{n,1} = 1$ and $z_{n,2} = 1$) or the queues have no upper boundaries. The results will be the same for both systems. We first describe the results and arguments for the bounded system.

Consider the distribution of x_1 given that $x_2 = n$. We know that the larger we make x_1, the larger the cost; this comes from (15.12) and (15.13), where we know $\ell_1(y)$ is increasing in y for $y > 0$ since $\ell_1(y)$ is convex and $\ell_1(y) = 0$ only at $y = \lambda_1 - \mu_1 < 0$. We can therefore say that

$$\mathbb{E}_{ss}\left(\left.\frac{x_1}{n}\right| x_2 = n\right) = 0. \tag{15.44}$$

Furthermore, we know that the sample path for $\vec{x}(t)$ to take is

$$\vec{r}^{\,*}(t) = (0, y_2) \cdot t \quad \text{for} \quad 0 \le t \le T^*,$$

where

$$T^* = \frac{1}{y_2} \quad \text{and} \quad y_2 = \rho \lambda_2 - \pi_0 \mu_2 / \rho$$

and ρ is given in Theorem 15.27 or 15.28 depending on whether $\lambda_2 > \lambda_D$ or $\lambda_2 < \lambda_D$ respectively.

The question of the distribution of x_2 given that $x_1 = n$ is almost as easy to answer. The process x_1 is simply an $M/M/1$ queue, so its upcrossings to the level n were completely analyzed in Chapter 11. During the excursion of x_1, there is no advantage to be had by queue 2 behaving any differently from normal, but there would be a cost for it to do so [that is, by (15.12) and (15.13), we make $\ell_2(y) = 0$ only at $y = \lambda_2$]. Therefore,

$$\mathbb{E}_{ss}\left(\left.\frac{x_2}{n}\right| x_1 = n\right) = \lambda_2 T^*, \tag{15.45}$$

where

$$T^* = \frac{1}{\mu_1 - \lambda_1}.$$

Exercise 15.33. Show that (15.44) and (15.45) are correct, by filling in the details of the calculations that we outlined. ♠

Exercise 15.34. What other conclusions can you draw about these queues without any further calculation? (Consider $\mathbb{P}\left(\left.\left|\frac{x_2}{n} - \lambda_2 T^*\right| < \varepsilon \right| x_1 = n\right)$, consider what Level $1\frac{1}{2}$ (§7.4) has to say about these queues, and consider replacing the conditioning $x_i = n$ by $x_i \ge n$.) ♠

When the state space is not bounded, we can still analyze the asymptotic correlations between $x_i = n$ and x_{3-i}. That is, we consider the simple priority model without a bound on the number of customers allowed in the system. The answers don't change at all from the ones we have already calculated. Not only that, but the conditioning on $x_i = n$ can be changed to $x_i \geq n$ and the theorems again hold without any changes. The only new element in the justification of the calculations is the estimate on the positive recurrence of the process $\vec{x}(t)$.

15.5. Justification—PP

In this section we justify our calculations by showing that the Freidlin-Wentzell theory applies to our process. We first need to prove the large deviations principle for processes in two dimensions with corners. This is not difficult based on the theory of Chapter 8, as we now demonstrate. Let us suppose that the only corner is at $\vec{q}$. We suppose that there are jump directions $\vec{e}_i(m)$ and associated rates $\lambda_i(m)$ for $m = 1, 2, 3, 4$, where $m = 1$ is the corner point $\vec{q}$, $m = 2$ is one boundary, $m = 3$ is another boundary, and $m = 4$ is the interior of the space. We suppose that the log $\lambda_i(m)(\vec{x})$ are bounded and Lipschitz continuous for $m = 2, 3, 4$. We define the local rate function $\ell(\vec{x}, \vec{y})$ by either of the following two formulae at $\vec{x} = \vec{q}$:

(i) $\displaystyle \sup_{\vec{\theta} \in \mathbb{R}^2} \left(\langle \vec{\theta}, \vec{y} \rangle - \max_{1 \leq m \leq 4} \sum_{i=1}^{k(m)} \lambda_i(m) \left(e^{\langle \vec{\theta}, \vec{e}_i(m) \rangle} - 1 \right) \right)$

(ii) $\displaystyle \inf_{\vec{\mu} \in B} \inf_{\vec{v} \in \mathcal{S}(\vec{\mu})} \sum_{m=1}^{D} \mu_m \left(\sum_{i=1}^{k(m)} v_i(m) \log \frac{v_i(m)}{\lambda_i(m)} - v_i(m) + \lambda_i(m) \right)$,

where

$$ B \triangleq \left\{ \vec{\mu} \in \mathbb{R}^D \; : \; \mu_m \geq 0, \; \sum_{m=1}^{D} \mu_m = 1 \right\}, $$

and

$$ \mathcal{S}(\mu) \triangleq \left\{ \vec{v} \; : \; v_i(m) \geq 0, \; \sum_{m=1}^{D} \mu_m \sum_{i=1}^{k(m)} v_i(m) \vec{e}_i(m) = \left(\vec{y}, \vec{0}^D \right) \right\}. $$

When $\vec{x} \neq \vec{q}$ the rate function is given by that for the flat boundary process, which we know has $\ell_i(\vec{x}, \vec{y}) = \ell_{ii}(\vec{x}, \vec{y})$. We show below that $\ell_i(\vec{q}, \vec{0}) = \ell_{ii}(\vec{q}, \vec{0})$. Therefore we can define

$$ I_0^T(\vec{r}) = \int_0^T \ell(\vec{r}(t), \vec{r}\,'(t)) \, dt \tag{15.46} $$

unambiguously, since the only ambiguity is at $\vec{r} = \vec{q}, \vec{r}\,' \neq \vec{0}$, which is a set of measure zero in t.

Theorem 15.35. *Suppose that we have a jump Markov process on $\mathbb{R}^2$ with two flat boundaries that meet at a point $\vec{q}$. Suppose that Assumptions 8.8 (irreducibility) and 8.9 (Lipschitz continuity) hold (except that we allow the point $\vec{q}$ to be a point of discontinuity for the rates on the boundary), that the conclusion of Kurtz's*

Theorem (Theorem 5.3) holds, and that the point $\vec{q}$ is attracting for $\vec{z}_\infty(t)$. Then the large deviations principle holds with good rate function $I_0^T(\vec{r})$.

Remark. The assumption that the process has two flat boundaries that meet at a point $\vec{q}$ means that we assume that the dynamics of the process do not allow for jumps beyond either boundary (Assumption 8.8 for directions normal to each boundary guarantees this, for example).

Proof. The lower bound holds almost trivially. We simply have to find a lower bound on the probability that a path stays near $\vec{q}$ for long periods of time. By the assumption that $\vec{q}$ is attracting, we can see that the lower bound rate function *(ii)* is equal to zero [i.e., $\ell(\vec{q}, \vec{0}) = 0$ where ℓ is defined by *(ii)*], since we do not need to change measure for the process to stay near $\vec{q}$.

The proof of the upper bound is unchanged, since it did not rely on any smoothness of the boundary. Therefore we can say that the upper bound holds with rate function given by form *(i)*. The only thing we need to check is the equivalence of $\ell_i(\vec{q}, \vec{0})$ and $\ell_{ii}(\vec{q}, \vec{0})$ in this new setting. By examining the probability of a neighborhood of the path $\vec{r}(t) = \vec{q}$ we can infer that $\ell_i(\vec{q}, \vec{0}) = 0$, since

$$0 \le \lim_{n \to \infty} \frac{1}{n} \log \mathbb{P}(\vec{z}_n \in N_\varepsilon(\vec{r})) \le -\ell_i(\vec{q}, \vec{0}). \tag{15.47}$$

Nevertheless, here is a direct proof that $\ell_i(\vec{q}, \vec{0}) = 0$. We claim that there is no direction $\vec{\theta} \in \mathbb{R}^2$ such that

$$\sum_i \lambda_i(m)\langle \vec{\theta}, \vec{e}_i(m)\rangle < 0 \quad \text{for} \quad 1 \le m \le 4. \tag{15.48}$$

This follows from the assumption that $\vec{z}_\infty(t) \to \vec{q}$ starting from any point near $\vec{q}$. Combine this with the observation that

$$\sum_i \lambda_i(m)\left(e^{\langle \alpha\vec{\theta}, \vec{e}_i(m)\rangle} - 1\right) \text{ is convex in } \alpha$$

and that

$$\frac{d}{d\alpha}\sum_i \lambda_i(m)\left(e^{\langle \alpha\vec{\theta}, \vec{e}_i(m)\rangle} - 1\right) = \sum_i \lambda_i(m)\langle \vec{\theta}, \vec{e}_i(m)\rangle \tag{15.49}$$

when $\alpha = 0$ easily gives the result $\ell_i(\vec{q}, \vec{0}) = 0$ (just look where a maximum might occur). ∎

Exercise 15.36. Prove that there is no direction $\vec{\theta} \in \mathbb{R}^2$ such that

$$\sum_i \lambda_i(m)\langle \vec{\theta}, \vec{e}_i(m)\rangle < 0 \text{ for } 1 \le m \le 4.$$
♠

We have justified the large deviations principle for processes with a corner only at $\vec{q}$. The preemptive priority process has corners where the line E intersects the coordinate axes, too. We will not show that the large deviations principle holds for the preemptive priority process for paths that touch these corners. We simply make the following observations:

1. Any path that leads to these corners also touches E, since the corners are contained in E.

2. Neither corner is absorbing.

3. Kurtz's Theorem holds for the process starting at either corner (see Exercise 15.37).

Therefore upcrossings to E don't depend on the large deviations principle holding for this process at those points. Furthermore, items 2 and 3 show that steady-state calculations don't depend on the principle, either, since with high probability the process leaves neighborhoods of those points quickly.

In order to show that the Freidlin-Wentzell theory applies to the preemptive priority model, we would like to use Theorem 6.77. That theorem requires us to have a large deviations principle (which we just established, at least enough for that theorem's need). It also requires Kurtz's Theorem for our process, and that $\vec{0}$ is a globally attracting point. This is simple enough that we leave it as an exercise.

Exercise 15.37. Show that $\vec{z}_\infty(t) \to \vec{0}$ from any starting point, and show further that Kurtz's Theorem holds for the PP process. Hint: the only places you need to examine are the corners. Near each corner each coordinate direction $x_i(t)$ grows more slowly than it does on a boundary away from the corner. ♠

We show positive recurrence of $\vec{z}_n(t)$ by referring to the general recurrence lemma for two-dimensional systems, Lemma 16.14. [The idea is that $\vec{z}_n(t)$ must follow $\vec{z}_\infty$ down to $\vec{0}$ relatively quickly, and with overwhelming probability, from any starting point. This is very nearly positive recurrence.]

We used an assumption in proving Theorem 6.77 that is simply not true for this process, or indeed for the other processes we consider in this chapter. Namely, we assumed that the positive cone spanned by the $\vec{e}_i$ is all of $\mathbb{R}^2$. Therefore we need to extend the reasoning leading to Theorem 6.77 in order to be able to apply the Freidlin-Wentzell theory. The result we need is contained in the following lemma.

Lemma 15.38. *There is a constant C such that the following holds. Given $T > 0$ and any path $\vec{r}(t) \in \mathbb{R}^2$, $0 \le t \le T$ with $I_0^T(\vec{r}) < \infty$, for any $\varepsilon > 0$, there is an n_0 such that if $n > n_0$ then for any $\vec{x}$ with $|\vec{x} - \vec{r}(0)| \le 1/n$,*

$$\mathbb{P}_{\vec{x}}\left(\left|\vec{z}_n(T) - \vec{r}(T)\right| \le \frac{C}{n}\right) \ge e^{-n\left(I_0^T(\vec{r})+\varepsilon\right)}.$$

Note that we can take $C = 2$ for the preemptive priority model. We stated the lemma in greater generality so that it may be used for other models.

Proof. The problem is seeing that the endpoint can be reached to within C/n, since we have already proved the large deviations lower bound for this process. The reason we cannot simply use the argument leading to Theorem 11.4 is that the positive cone spanned by the $\vec{e}_i$ in the interior of the state space is not all of $\mathbb{R}^2$—there are no low-priority services. Theorem 11.4 applies to the first coordinate of $\vec{z}_n(t)$ (since that is just an $M/M/1$ queue). For the second coordinate, consider the final portion of the path $\vec{r}(t)$ from the last time it leaves the axis $r_1(t) = 0$ [or time zero if there is no time when $r_1(t) = 0$] until time T; this is the portion of time during which $r_2(t)$ cannot decrease (since $I_0^T(\vec{r}) < \infty$). We simply have to ensure that $z_{n,2}(t) \leq r_2(T)$ during this time, and also ensure that $|z_{n,2}(T) - r_2(T)| < 2/n$. We can accomplish this using the same idea as in the proof of the lower bound, Lemmas 8.81 and 8.82: if the path $\vec{r}(t)$ has small slope in the direction $\vec{u}_2$ (the second coordinate direction), then the cost function is nearly that of a slope zero (constant) function, which is the cost of having no jumps at all in the low-priority queue. Therefore we can (with almost no cost) keep the low-priority queue at the level $r_2(T)$ if it ever reaches that level, since the slope of the curve $\vec{r}(t)$ would have to be small by the large deviations lower bound from the time $\vec{z}_n(t)$ reaches that level. Furthermore, if by time $T - \varepsilon$ the process has not reached the level, then with cost $O(\varepsilon)$ we can force the process $\vec{z}_n(t)$ to that level. ∎

We claim that the arguments we have given thus far have established Theorem 15.30. There is at least one more item that needs to be cleared up before we are finished with the proofs of Theorems 15.31 and 15.32. We need to check that the boundary condition is satisfied for Theorem 6.92; namely, the assumption that "...$\vec{r}$* intersects D at exactly one point even when extended beyond time zero by the path $\vec{z}_\infty$ [this is equivalent to the assumption that $V(\vec{x}) = I^*$ for $\vec{x} \in D$ for a unique $\vec{x} = \vec{r}*(0)$]." By (15.4), the drift $\frac{d}{dt}\vec{z}_\infty$ is directed toward E when

$$\lambda_1 + \lambda_2 \geq \mu_1. \tag{15.50}$$

So the question becomes, Is it possible that (15.50) holds when the minimal cost path $\vec{r}*$ intersects E at a point other than $(0, 1)$? (Then the cheapest path would travel along E, and so all the points along this path would have the same value of V, and hence asymptotically the same probability. This would mean that we could not find a unique cheapest point on the boundary.) The cheapest path is in the interior only when $\lambda_2 < \lambda_c$ (and sometimes not even then, if $\lambda_D < \lambda_c$). But if $\lambda_2 < \lambda_c$ then

$$\lambda_1 + \lambda_2 < \lambda_1 + \lambda_c = \sqrt{\lambda_1 \mu_1} < \mu_1, \tag{15.51}$$

so the drift in the interior is away from the boundary E. Therefore, the cheapest path from $\vec{0}$ to E always hits E at a unique point.

The final point to clear up is the fact that E is not a smooth open set, as was assumed for Theorem 6.92. This is, in fact, irrelevant for our process, since the process can only jump one step at a time, so cannot tell if the set has an interior or not. Nevertheless, here is a formal argument why we can condition on a "thin" set such as E, or (for this and other processes) on sets such as $x_1 = 1$.

Theorem 15.39. *Let $\vec{z}_n(t)$ be a positive recurrent process on a subset G of the integer lattice $\mathcal{Z}^d/n$, satisfying a large deviations principle, and let E be a subset* *of G. Suppose that for each $\vec{y} \in E$,*

$$\mathbb{P}_{ss}(\vec{y}) = e^{-nV(\vec{y})+o(n)} \qquad (15.52)$$

[recall $V(y) \overset{\triangle}{=} \inf\{I_0^T(\vec{r}) : \vec{r}(0) = \vec{q},\ \vec{r}(T) = \vec{y},\ T > 0\}$, where we assume that there exists a unique point $\vec{q} = \lim_t \vec{z}_\infty(t)$ for every starting point $\vec{z}_\infty(0)$]. Suppose further that

$$\mathbb{P}_{ss}(E) = e^{-n \inf_{\vec{x} \in E} V(\vec{x})+o(n)} \qquad (15.53)$$

and suppose that there is a unique $\vec{y}^* \in E$ with

$$V(\vec{y}) = \inf_{\vec{x} \in E} V(\vec{x}). \qquad (15.54)$$

Then for every $\varepsilon > 0$,

$$\lim_{n \to \infty} \mathbb{P}_{ss}\left(|\vec{z}_n(t) - \vec{y}^*| < \varepsilon \;\middle|\; \vec{z}_n(t) \in E \right) = 1. \qquad (15.55)$$

Furthermore, if there exist constants n_0 and K_0 and a function $f(K) \to \infty$ as $K \to \infty$, such that

$$\mathbb{P}_{ss}(|\vec{z}_n(t)| \geq K) \leq e^{-nf(K)}, \qquad (15.56)$$

then

$$\lim_{n \to \infty} \mathbb{E}_{ss}\left(\vec{z}_n(t) \;\middle|\; \vec{z}_n(t) \in E \right) = \vec{y}^*. \qquad (15.57)$$

Exercise 15.42. Prove Theorem 15.39.   ♠

The preemptive priority queue satisfies the hypotheses of Theorem 15.39 [even (15.56), since the process is bounded!]. This completes the justification of Theorems 15.31 and 15.32 for the preemptive priority queue and the set E.

Next we examine the justification of our results for the extended model, where there are two queues that are either bounded by n or are unbounded. For the unbounded queues we need exactly one more lemma: we need to see that (15.56) holds in order to use Theorem 15.39.

Lemma 15.41. *There exists a constant $C > 0$ and an n_0 such that for any $n > n_0$ and any $K \geq 1$,*

$$\mathbb{P}_{ss}\left(\{x_1 \geq nK\} \bigcup \{x_2 \geq nK\} \right) \leq e^{-nKC}. \qquad (15.58)$$

Exercise 15.42. Prove Lemma 15.41. Hint: Kurtz's Theorem applies to the process.   ♠

15.6. Serve the Longest Queue

Suppose that there are two lines in which people can wait for a service, such as tellers' lines in a bank. The lines are such that no one can see the lines before entering the system, or change lines after joining the system. As a result, arrival rates to the queues are equal. There is only one server who serves the longer queue unless queue sizes are equal, in which case the server makes an arbitrary decision (flips a coin). The server interrupts service if the line being served becomes shorter than the other.

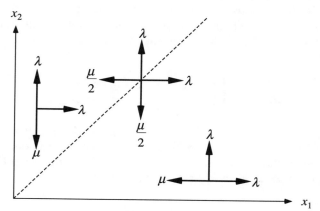

Figure 15.43. Serve the longer queue.

We formalize this system as follows:

Definition 15.44. *The process $\vec{x}(t)$ is defined by the following rates and directions:*

Jump direction	Jump rate		Interpretation
$\vec{e}_1 = (1, 0)$	$\lambda_1(\vec{x}) = \lambda$		*Arrival to queue 1*
$\vec{e}_2 = (0, 1)$	$\lambda_2(\vec{x}) = \lambda$		*Arrival to queue 2*

$$\vec{e}_3 = (-1, 0) \quad \mu_1(\vec{x}) = \begin{cases} \mu & \text{if } x_1 > x_2 \\ \mu/2 & \text{if } x_1 = x_2 > 0 \\ 0 & \text{otherwise} \end{cases} \quad \textit{Departure from queue 1}$$

$$\vec{e}_4 = (0, -1) \quad \mu_2(\vec{x}) = \begin{cases} \mu & \text{if } x_1 < x_2 \\ \mu/2 & \text{if } x_1 = x_2 > 0 \\ 0 & \text{otherwise} \end{cases} \quad \textit{Departure from queue 2}$$

Clearly, the total number in the system $(x_1 + x_2)(t)$ is an $M/M/1$ queue with arrival rate 2λ and service rate μ. Therefore we impose the stability condition $\mu > 2\lambda$.

The first thing you'll probably notice about this system is that it doesn't fall under our theory of flat boundary processes: the "boundary" at $x_1 = x_2$ can be penetrated. We'd ask you to reserve judgment for now. Pretend that a large deviations principle exists for this process, and continue with the program as outlined in the Introduction to the Applications. The justification will appear shortly.

15.7. Most Probable Behavior—SL

It is easy to find $\vec{z}_\infty(t)$, even though the process $\vec{z}_n(t)$ does not fall under the domain of applicability of our theorems to date (since the "boundary" at $x_1 = x_2$ is nonstandard). In the regions $x_1 > x_2$ and $x_2 > x_1$, the process is constant coefficient, and clearly Kurtz's Theorem applies. To justify our calculation on the line $x_1 = x_2$, see Exercise 15.51 below. For now, we calculate $\vec{z}_\infty(t)$ formally. In the region where $x_1 \neq x_2$,

$$\frac{d}{dt}\vec{z}_\infty = \begin{cases} (\lambda - \mu, \lambda) & \text{if } \vec{z}_{\infty,1} > \vec{z}_{\infty,2} \\ (\lambda, \lambda - \mu) & \text{if } \vec{z}_{\infty,1} < \vec{z}_{\infty,2}. \end{cases} \tag{15.59}$$

Since $\mu > 2\lambda$, we have $\langle \frac{d}{dt}\vec{z}_\infty, \vec{n} \rangle < 0$ where $\vec{n} = (1, 1)$. Now what happens when $x_1 = x_2$? If $\vec{z}_\infty$ exists at all it must stay on the line $x_1 = x_2$ once it gets there, since it approaches the line whenever it is not on it. But since the total number in the system $x_1(t) + x_2(t)$ is an $M/M/1$ queue, we have

$$\left\langle \frac{d}{dt}\vec{z}_\infty, \vec{n} \right\rangle = 2\lambda - \mu. \tag{15.60}$$

We have then completely specified $\vec{z}_\infty(t)$ once we insist that $\frac{d}{dt}\vec{z}_\infty = \vec{0}$ when $\vec{z}_\infty = \vec{0}$. Therefore we have found that, with

$$T_s \overset{\triangle}{=} \frac{|x_1 - x_2|}{\mu} \tag{15.61}$$

as the switching time, we have

$$\vec{z}_\infty(t) = \begin{cases} \vec{x} + (\lambda - \mu, \lambda)t, & \text{for } 0 \leq t \leq T_s \quad \text{if } x_1 > x_2 \\ \vec{x} + (\lambda, \lambda - \mu)t, & \text{for } 0 \leq t \leq T_s \quad \text{if } x_2 > x_1 \\ \vec{z}_\infty(T_s) + (\lambda - \mu/2, \lambda - \mu/2)t, & \text{for } T_s \leq t \leq T_s + \frac{z_\infty^1(T_s)}{\mu/2 - \lambda}, \end{cases}$$

where we take $T_s = 0$ and $\vec{z}_\infty(T_s) = \vec{x}$ if $x_1 = x_2$. In Exercise 15.51 below, you will prove that this $\vec{z}_\infty$ satisfies the conclusion of Kurtz's Theorem (note that we have only made a heuristic justification of this theorem along the line $x_1 = x_2$).

15.8. The Main Result—SL

We wish to examine how large $x_2(t)$ is likely to be when $x_1(t)$ is large. Our main result for this system is the following.

Theorem 15.45. *Define*

$$T^* \triangleq \begin{cases} (\mu/2 - \lambda)^{-1} & 4\lambda > \mu \\ (\mu - \lambda)^{-1} & 4\lambda < \mu \end{cases}$$

$$\vec{r}^*(t) \triangleq \begin{cases} \vec{0} & \text{for } t < 0 \\ \left(\frac{\mu}{2} - \lambda, \frac{\mu}{2} - \lambda\right) t & \text{for } 0 \le t \le T^*, \text{ if } 4\lambda > \mu \\ (\mu - \lambda, \lambda) t & \text{for } 0 \le t \le T^*, \text{ if } 4\lambda < \mu \end{cases}$$

Then if $4\lambda \ne \mu$, *for each T and $\varepsilon > 0$,*

$$\lim_{n \to \infty} \mathbb{P}_{ss} \left(\sup_{-T \le t \le 0} |\vec{z}_n(t) - \vec{r}^*(t - T^*)| < \varepsilon \;\middle|\; z_{n,1}(0) = 1 \right) = 1.$$

$$\lim_{n \to \infty} \mathbb{E}_{ss} \left(z_{n,2} \;\middle|\; z_{n,1} = n \right) = r_2^*(T^*).$$

In order to prove this result, we should first prove that the large deviations principle holds for our process. Then we should examine

$$I^* \triangleq \inf \left\{ \int_0^T \ell(\vec{r}, \vec{r}') \, dt : \vec{r}(0) = \vec{0}, \; r_1(T) = 1 \right\}.$$

Since the process is constant coefficient, except for the discontinuity at $x_1 = x_2$, we know that all extremals will be straight lines, or will consist of a straight segment along $x_1 = x_2$ with another segment interior to $x_1 > x_2$ or $x_2 > x_1$. (You can easily see that there are no other possibilities for minimal paths made of straight-line segments: if a path leaves the point $\vec{0}$ along a straight-line segment away from $x_1 = x_2$, then it cannot turn around and hit this boundary at a later time. Furthermore, the "boundaries" at $x_1 = 0$ and $x_2 = 0$ are *not* discontinuities in terms of the jump rates for this model—see Figure 15.43.) Then we should consider the endpoint, $r^*(T) = (1, x_2^*)$, and we would obtain Theorem 15.45.

Before we carry out this program, let's pause to think of a trick. Suppose that $x_1(t)$ approaches n in such a way that $x_2(t) < x_1(t)$ during the approach. Then $x_1(t)$ is an $M/M/1$ queue, and $x_2(t)$ is a Poisson process. We know that $M/M/1$ queues grow by having the arrival and service rates interchange (see Theorem 11.18). That is,

$$x_1(t) \approx (\mu - \lambda)t \quad 0 \le t \le \frac{n}{\mu - \lambda} \tag{15.62}$$

and $x_2(t) \approx \lambda t$. Is this story consistent? That is, is $x_1(t) > x_2(t)$ during this interval? By the stability assumption, $\mu > 2\lambda$, so $\mu - \lambda > \lambda$; therefore consistency holds. The probability of $x_1(t)$ reaching n is about $\exp(-n \log(\mu/\lambda))$ (see §11.2).

So we obtain that x_1 reaches n, and then x_2 is about $\lambda \cdot t = \lambda \frac{n}{\mu - \lambda}$, with probability $(\lambda/\mu)^n$ (that is, the rate function associated with this scenario is $\log \frac{\mu}{\lambda}$).

But what if $x_1(t)$ is not always longer than $x_2(t)$ during the excursion? What if the two queues get long together? In that case we do not know who gets served at any particular time, but we know that the system as a whole needs to have about $2n$ customers for $x_1(t) = n$. Now the system as a whole is an $M/M/1$ queue with arrival rate 2λ and service rate μ. Therefore, if this is the most likely scenario, we would have

$$\mathbb{P}_{ss}(x_1 = n) \approx e^{-2n \log \frac{\mu}{2\lambda}} = \left(\frac{2\lambda}{\mu}\right)^{2n} = \left(4\frac{\lambda}{\mu}\right)^n \left(\frac{\lambda}{\mu}\right)^n. \tag{15.63}$$

Which of these two modes is more probable? We have to compare (15.63) with $(\lambda/\mu)^n$. We see that if $4\lambda/\mu > 1$ then it is more likely that $x_2 \approx x_1$; if $4\lambda/\mu < 1$ then it is more like that $x_1 > x_2$. When $4\lambda/\mu = 1$ we have multiple paths $r^*(t)$ with the same rate function, and our methods do not provide estimates on the distribution of x_2 given $x_1 = n$.

15.9. Justification—SL

The preceding discussion should convince all but the most skeptical investigator. For that person, we now carry out the exact calculations and proofs. We begin with a proof of the large deviations principle for our process. We then prove that the Freidlin-Wentzell theory applies (this is a triviality; once we have the elements in place, we argue as for Theorem 6.92). We end with a justification of our solution of the variational problem. Remark 15.10 is relevant to our discussion; you might wish to review it before continuing.

We obtain the large deviations principle for $\vec{z}_n(t)$ by a change of coordinates. Let

$$u_1(t) \stackrel{\triangle}{=} x_1(t) + x_2(t) \tag{15.64}$$

$$u_2(t) \stackrel{\triangle}{=} |x_1(t) - x_2(t)|. \tag{15.65}$$

Then $\vec{u}(t)$ is a Markov process with transition rates depicted in Figure 15.46 below. Clearly, $\vec{u}(t)$ is a Markov process with a flat boundary and smooth (actually constant) transition rates for $u_1 > 0$ (there is a singularity at $u_1 = u_2 = 0$, but we take care of that problem separately). We use $\vec{u}_n(t)$ for the scaled process derived from $\vec{u}(t)$ in the same manner that $\vec{z}_n(t)$ is derived from $\vec{x}(t)$.

Let E denote the boundary $x_1 = x_2$. The process $\vec{u}(t)$ is obtained from $\vec{x}(t)$ by a continuous mapping. Define a pointwise map R and a mapping of functions $\mathbf{R}$

$$R(\vec{x}) = (u_1, u_2) \stackrel{\triangle}{=} (x_1 + x_2, |x_1 - x_2|)$$

$$\mathbf{R}(\vec{r})(t) = (u_1(t), u_2(t)) \stackrel{\triangle}{=} R(\vec{r}(t)). \tag{15.66}$$

In the notation of Chapter 8, u_2 was denoted u_0 and is the direction perpendicular to the boundary. Note that the sample paths of $\vec{z}_n$ cannot jump across E: that is, if

$\vec{z}_n(t)$ is on one side and $\vec{z}_n(s)$ is on anther side of the boundary, with $s > t$, then necessarily $\vec{z}_n(\tau) \in E$ for some $t < \tau < s$. So, define

$$F_s \triangleq \left\{ \vec{r} \in D^d[0, T] : \vec{r} \text{ does not jump across } E \right\}.$$

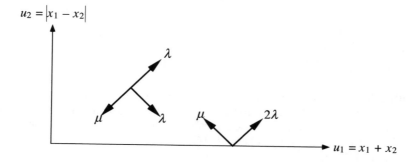

Figure 15.46. Transitions in the variable $\vec{u}$.

Exercise 15.47. $\mathbf{R}$ is a continuous mapping from $D^d[0, T]$ to $D^d[0, T]$ and from $C^d[0, T]$ to $C^d[0, T]$. ♠

Exercise 15.48. The set F_s is closed in $D^d[0, T]$. If $F \subset F_s$ is a closed set in $D^d[0, T]$ then $\mathbf{R}(F) = \{\mathbf{R}(\vec{r}) : \vec{r} \in F\}$ is closed in $D^d[0, T]$. Hint: take a converging subsequence $\{\vec{u}_n = \mathbf{R}(\vec{r}_n), \ n \geq 1\}$, with $\vec{r}_n$ in F. Use Theorem A.57 to show that the set

$$\left\{ \vec{r} : \mathbf{R}(\vec{r}) = \vec{u}_n \quad \text{for some} \quad n \geq 1 \right\}$$

is precompact, hence $\{\vec{r}_n, \ n \geq 1\}$ is precompact. Now take a converging subsequence and use continuity of $\mathbf{R}$. ♠

Since $\mathbf{R}$ is continuous, we could deduce a large deviations principle for $\vec{u}_n(t)$ from a large deviations principle for $\vec{z}_n(t)$ by the contraction principle, Theorem 2.13. We propose to do exactly the opposite: derive the large deviations principle for $\vec{z}_n(t)$ from the large deviations principle for $\vec{u}_n(t)$. The reason we can do this is that the mapping from $\vec{x}(t)$ to $\vec{u}(t)$ is nearly invertible on F_s: the only ambiguity in the inverse map is when a path touches the boundary $u_2 = 0$ (corresponding to $x_1 = x_2$). But then the idea is to count all the paths $\vec{x}(t)$ corresponding to a particular path $\vec{u}(t)$ as equally likely, and since there aren't too many such paths, we just take the rate function for $\vec{z}_n(t)$ as the corresponding rate function for $\vec{u}_n(t)$. Now for the statement and proof.

Define the local rate function $\ell_u(\vec{x}, \vec{y})$ for the process $\vec{u}_n(t)$ as in Chapter 8:

$$\ell_u(\vec{u}, \vec{v}) \triangleq \begin{cases} \sup_{\vec{\theta}} \left(\langle \vec{\theta}, \vec{v} \rangle - g_u(\vec{\theta}) \right) & u_2 > 0 \text{ or } v_2 > 0 \\ \sup_{\vec{\theta}} \left(\langle \vec{\theta}, \vec{v} \rangle - \max \left\{ g_u(\vec{\theta}), g_{ub}(\vec{\theta}) \right\} \right) & \begin{array}{l} u_2 = v_2 = 0, \text{ and} \\ u_1 > 0 \text{ or } v_1 > 0 \end{array} \\ 0 & \vec{u} = \vec{v} = \vec{0} \\ \infty & \text{otherwise,} \end{cases}$$

$$g_u(\vec{\theta}) \triangleq \lambda \left(e^{\theta_1+\theta_2} + e^{\theta_1-\theta_2} - 2 \right) + \mu \left(e^{-\theta_1-\theta_2} - 1 \right)$$

$$g_{ub}(\vec{\theta}) \triangleq 2\lambda \left(e^{\theta_1+\theta_2} - 1 \right) + \mu \left(e^{\theta_2-\theta_1} - 1 \right).$$

We define the local rate function for the process $\vec{z}_n(t)$ in the usual way, via (5.2)–(5.4), for all points not on the boundary; that is, either $x_1 \neq x_2$ or $y_1 \neq y_2$. On the boundary $x_1 = x_2$, $y_1 = y_2$, we use the rate implied by the function for the $\vec{u}_n(t)$ process. That is, $\ell((x, x), (y, y)) = \ell_u((2x, 0), (2y, 0))$; an explicit expression is given in (15.78) below.

Exercise 15.49. Obtain ℓ from ℓ_u using a transformation of coordinates. Conclude that if a continuous path $\vec{r}$ crosses the boundary $x_1 = x_2$ at most a countable number of times, then $I_0^T(\vec{r}) = {}_u I_0^T(\mathbf{R}(\vec{r}))$ (the rate function $I_0^T(\vec{r})$ is defined in the usual way, see Theorem 15.50 below). Hint: R becomes linear if restricted to the correct set. Use Exercise 5.11. For the second part justify ignoring points where $\mathbf{R}(\vec{r})$ is not differentiable, and use symmetry. ♠

Theorem 15.50. *The process $\vec{z}_n(t)$ satisfies a large deviations principle with good rate function*

$$I_0^T(\vec{r}) \triangleq \begin{cases} \int_0^T \ell(\vec{r}(t), \vec{r}'(t)) \, dt & \vec{r} \text{ absolutely continuous,} \\ \infty & \text{otherwise.} \end{cases}$$

Proof. To establish the lower bound, we have to prove that any ε neighborhood of a function $\vec{r}(t)$ has probability at least $\exp\left[-n I_0^T(\vec{r}) + o(n) \right]$. So fix a continuous function $\vec{r}$ and $\varepsilon > 0$ and, to fix the ideas, assume that $r_2(0) > r_1(0) + 2\varepsilon$ (exactly the same arguments apply if the roles of r_1 and r_2 are reversed, or if r_1 and r_2 are equal). Recall that E denotes the boundary $\{\vec{x} : x_1 = x_2 \geq 0\}$ and that $d(\vec{x}, E)$ is the (smallest Euclidean) distance between a point and the set E. Define a set of times in a way reminiscent of a Freidlin-Wentzell construction:

$$\tau_{-1} \triangleq 0 \tag{15.67}$$

$$\tau_{2j} \triangleq \inf \left\{ t \geq \tau_{2j-1} : r_1(t) = r_2(t) \right\} \tag{15.68}$$

$$\tau_{2j+1} \triangleq \inf \left\{ t \geq \tau_{2j} : d(\vec{r}(t), E) \geq \varepsilon/2 \right\} \tag{15.69}$$

$$s_{-1} \triangleq 0 \tag{15.70}$$

$$s_{2j} \triangleq \sup \left\{ t \leq \tau_{2j} : d(\vec{r}(t), E) \geq \varepsilon/2 \right\}. \tag{15.71}$$

Then $s_{2j} < \tau_{2j} < \tau_{2j+1}$ since $\vec{r}$ is continuous, τ_{2j} are the times $\vec{r}$ hits the boundary after being at least $\varepsilon/2$ away, and τ_{2j+1} is the first time after hitting a boundary that $\vec{r}$ escapes an $\varepsilon/2$ neighborhood of the boundary, and s_{2j} are the last times before τ_{2j} that $\vec{r}$ is an $\varepsilon/2$ distance away from the boundary.

Let us show that, since $I_0^T(\vec{r}) < \infty$, there are only a finite number of τ_j. By Lemma 5.17, $\ell(\vec{x}, \vec{y})/|\vec{y}| \to \infty$ uniformly as $|\vec{y}| \to \infty$, and hence there is a K so that if $|\vec{y}| > K$ then $\ell(\vec{x}, \vec{y}) > I_0^T(\vec{r})$. Since the cheapest paths in the interior are straight lines, in an interval of length T there can be at most KT/ε cycles from τ_{2j} to τ_{2j+1} and back with a total cost of up to $I_0^T(\vec{r})$. By repeating the argument with $\varepsilon = 2^{-k}$ we conclude that there are at most a countable number of points where $\vec{r}$ crosses E [that is, points $\vec{r}(t)$ where either $\vec{r}$ is on one side for all small times before t, or on one side for all small times after t]. Therefore, Exercise 15.49 implies that $I_0^T(\vec{r}) = {}_u I_0^T(\mathbf{R}(\vec{r}))$.

We are trying to establish a lower bound for $\mathbb{P}_{\vec{x}}(A_n)$, where $\vec{x} = \vec{r}(0)$ and

$$A_n \triangleq \left\{ \sup_{0 \le t \le T} |\vec{z}_n(t) - \vec{r}(t)| < \varepsilon \right\}. \tag{15.72}$$

By definition of s_j and by continuity of $\vec{r}$, there is a constant $\varepsilon/4 > \delta > 0$ so that

$$\delta \le \min_j \inf_{\tau_{2j-1} \le t \le s_{2j}} d(\vec{r}(t), E). \tag{15.73}$$

Consider the set B_n^r of paths $\vec{z}_n$ satisfying the following conditions: for $j = 0$,

(i) During the interval $\tau_{2j-1} \le t \le s_{2j}$, we have $|\vec{z}_n(t) - \vec{r}(t)| < \delta$.

(ii) During the interval $s_{2j} \le t \le \tau_{2j+1}$, we have $|\vec{z}_n(t) - \vec{r}(t)| < \varepsilon$.

(iii) At some point during the interval $s_{2j} \le t \le \tau_{2j+1}$, $\vec{z}_n$ hits the boundary E.

(iv) $|z_n(\tau_{2j+1}) - \vec{r}(\tau_{2j+1})| < \delta$, and

(v) after τ_{2j+1} the process $\vec{z}_n$ stays within ε of $\vec{r}$.

Clearly $B_n^r \subset A_n$ [the stipulation (iii) is needed so that, even if $\vec{r}$ exits the $\varepsilon/2$ neighborhood of E to the same side it entered from, $\vec{z}_n$ is guaranteed to hit the boundary]. Let now $\vec{v}$ be the path that agrees with $\vec{r}$ until τ_{2j} and, between τ_{2j} and T, is the reflection of $\vec{r}$ across E. That is, $\vec{v}$ leaves E in a direction opposite to $\vec{r}$ and continues as its mirror image. Since our process is symmetric with respect to such reflection and since $\vec{z}_n$ was required to hit E, we have $\mathbb{P}_{\vec{x}}(B_n^r) = \mathbb{P}_{\vec{x}}(B_n^v)$, where B_n^v is defined through (i)–(v) and differs from B_n^r only in that $\vec{v}$ replaces $\vec{r}$ in (iv)–(v). Therefore

$$\mathbb{P}_{\vec{x}}(A_n) \ge \frac{1}{2} \mathbb{P}_{\vec{x}} \left(B_n^r \cup B_n^v \right). \tag{15.74}$$

Repeating exactly the same argument for $j = 1, \ldots, 1 + KT/\varepsilon$ we introduce a factor of $1/2$ each time we invoke symmetry, and obtain a chain of inequalities. The final set of paths we consider allows the process $\vec{z}_n$ to follow, on each interval $[\tau_{2j}, \tau_{2j+1}]$, either $\vec{r}$ or its mirror image. But, since $\vec{z}_n \in F_s$, this is equivalent to requiring that $\mathbf{R}(\vec{z}_n)$ follows $\mathbf{R}(\vec{r})$! Or, more precisely (and reducing the set some more),

$$
\mathbb{P}_{\vec{x}}(A_n) \geq 2^{-KT/\varepsilon} \mathbb{P}_{\vec{x}} \left(\begin{array}{l} \sup_{0 \leq t \leq T} |R(\vec{z}_n(t)) - R(\vec{r}(t))| < \delta, \\ \max_j \inf_{s_{2j} \leq t \leq \tau_{2j+1}} d(\vec{z}_n(t), E) = 0 \end{array} \right)
$$

$$
= 2^{-KT/\varepsilon} \mathbb{P}_{R(\vec{x})} \left(\begin{array}{l} \sup_{0 \leq t \leq T} |R(\vec{z}_n(t)) - R(\vec{r}(t))| < \delta, \\ \max_j \inf_{s_{2j} \leq t \leq \tau_{2j+1}} d(R(\vec{z}_n(t)), E) = 0 \end{array} \right)
$$

where the change in the conditioning is justified by symmetry. From Lemma 6.36 applied to the process $\mathbf{R}(\vec{z}_n)$, we obtain a lower bound for the last probability so that

$$
\mathbb{P}_{\vec{x}}(A_n) \geq 2^{-KT/\varepsilon} \mathbb{P}_{R(\vec{x})} \left(\sup_{0 \leq t \leq T} |R(\vec{z}_n(t)) - R(\vec{r}(t))| < \eta \right) e^{-n\delta C}
$$

for some positive η and C. However, for this probability we have a large deviations principle and, since δ is arbitrary, we conclude that

$$
\liminf_{n \to \infty} \frac{1}{n} \log \mathbb{P}_{\vec{x}} \left(\sup_{0 \leq t \leq T} |\vec{z}_n(t) - \vec{r}(t)| < \varepsilon \right) \geq -I_0^T(\vec{r}). \tag{15.75}
$$

The upper bound is easier to prove. Recall from Exercises 15.47–15.48 that $\mathbf{R}$ is continuous and F_s is closed. For any closed set F,

$$
\mathbb{P}_{\vec{x}}(\vec{z}_n \in F) = \mathbb{P}_{\vec{x}}(\vec{z}_n \in F \cap F_s)
$$
$$
\leq \mathbb{P}_{\vec{x}}(\mathbf{R}(\vec{z}_n) \in \mathbf{R}(F \cap F_s)).
$$

However, by symmetry, for any set A we have

$$
\mathbb{P}_{\vec{x}}(\mathbf{R}(\vec{z}_n) \in A) = \mathbb{P}_{R(\vec{x})}(\mathbf{R}(\vec{z}_n) \in A). \tag{15.76}
$$

But $\mathbf{R}(\vec{z}_n) = \vec{u}_n$ satisfies a large deviations principle with rate function $_u I_0^T$, and we can apply the upper bound since, by Exercise 15.48, $\mathbf{R}(F \cap F_s)$ is closed. The upper bound now follows from Exercise 15.49.

Finally, we show that I_0^T is a good rate function. Since $_u I_0^T$ is lower semicontinuous and $\mathbf{R}$ is continuous, I_0^T is lower semicontinuous, and it is clearly nonnegative. The level sets of I_0^T are compact since

$$
\left\{ \vec{r} : I_0^T(\vec{r}) \leq \alpha \right\} = \left\{ \vec{r} : _u I_0^T(\mathbf{R}(\vec{r})) \leq \alpha \right\} \cap F_s.
$$

The first set on the right is compact by Theorem 8.16, and the second is closed by Exercise 15.48. ∎

Exercise 15.51. Prove that Kurtz's Theorem holds for $\vec{z}_n(t)$. Hint: use the mapping R as in the previous proof—what does Kurtz's Theorem for $\vec{u}_n$ imply about $\vec{z}_n$? ♠

Exercise 15.52. Show that $\vec{0}$ is a removable discontinuity in the sense of Freidlin and Wentzell, by showing that

$$\mathbb{P}_{\vec{0}}\left(\sup_{0\le t\le T} |\vec{z}_n(t)| > \varepsilon\right) \to 0$$

as $n \to \infty$, and that

$$\mathbb{P}_{\vec{0}}\left(\sup_{0\le t\le T} |\vec{z}_n(t)| > \varepsilon\right) > e^{-nf(\varepsilon)}$$

where $f(\varepsilon) \to 0$ as $\varepsilon \to 0$. ♠

This essentially proves the applicability of Freidlin-Wentzell theory for the calculation of steady-state probabilities, and for the calculation of the way in which unlikely events occur.

We can calculate the probability of traveling at rate a along the boundary via ℓ_u which, in this case, takes the form

$$\ell_u((u,0),(a,0)) \overset{\triangle}{=} \sup_{\vec{\theta}\in\mathbb{R}^2}\left(\theta_1 a - \max\left\{\left[2\lambda(e^{\theta_1+\theta_2}-1)+\mu(e^{\theta_2-\theta_1}-1)\right],\right.\right.$$
$$\left.\left.\left[\lambda(e^{\theta_1+\theta_2}-1)+\lambda(e^{\theta_1-\theta_2}-1)+\mu(e^{-\theta_1-\theta_2}-1)\right]\right\}\right). \quad (15.77)$$

This can be evaluated explicitly: in Exercise 15.53 below we obtain

$$\ell_u((u,0),(a,0)) = \sup_{\theta_1}\left(\theta_1 a - 2\lambda(e^{\theta_1}-1) - \mu(e^{-\theta_1}-1)\right)$$

$$= a\log\frac{a+\sqrt{a^2+8\lambda\mu}}{4\lambda} + 2\lambda + \mu - \sqrt{a^2+8\lambda\mu}, \quad (15.78)$$

where the last equality follows since the cost function has been reduced to that of an $M/M/1$ queue with arrival rate 2λ and service rate μ.

Exercise 15.53. Derive (15.78). Hint: for any θ_1 we minimize the "maximum" term when $\theta_2 = 0$, since the two terms in $\{\cdot\}$ in (15.77) are equal at $\theta_2 = 0$, the first is increasing in θ_2 while the second is decreasing for $\theta_2 < 0$. ♠

We have thus reduced the problem of finding the cheapest path that travels along the boundary to that of finding the cheapest way for an $M/M/1$ queue to fill. This is, of course, when the arrival and service rates interchange. So we have

$$\vec{u}_b^*(t) = (\mu - 2\lambda, 0)t, \qquad \vec{r}_b^*(t) = \big((\mu - 2\lambda)/2, (\mu - 2\lambda)/2\big)t. \quad (15.79)$$

The problem of finding when $x_1 = 1$ is now the problem of finding when $u_1^*(t) = 2$; this is clearly when

$$t = T_b^* \triangleq \frac{2}{\mu - 2\lambda}.$$

The cost of this path is

$$I_b^* = 2 \log \frac{\mu}{2\lambda}. \tag{15.80}$$

Exercise 15.54. Perform the calculations that give (15.80). ♠

We must also concern ourselves with paths that travel strictly in the interior of the state space. In the region $x_1 > x_2$, we now show that the cheapest way to get to the line $x_1 = 1$ is for the arrival and service rates for x_1 to interchange, leaving the rates for x_2 unchanged. To prove this, compare the preemptive priority queue of Figure 15.1 to our system, depicted in Figure 15.43. Since the jumps and rates agree for $x_1 > x_2$, the calculation leading to (15.12) apply, and we obtain the following form of the rate function in this region:

$$\ell(\vec{x}, \vec{y}) = \ell_1(y_1) + \ell_2(y_2) \tag{15.81}$$

where ℓ_1 corresponds to an $M/M/1$ queue with arrival rate λ and service rate μ, and ℓ_2 corresponds to a Poisson process with rate λ. The cheapest way of getting from $\vec{0}$ to the line $x_1 = 1$ involves minimizing $\ell_1(y_1)/y_1$; hence we may freely minimize over y_2, and we know that $\ell_2(y_2) = 0$ only when $y_2 = \lambda_2$. Furthermore, minimizing $\ell_2(y_2)/y_2$ is the same as minimizing the cost of an $M/M/1$ queue to fill; by Theorem 11.15 this has cost

$$I_1^* = \log \frac{\mu}{\lambda}.$$

Furthermore, in the optimal mode of travel we have (by, e.g., Theorem 11.18)

$$\vec{r}_1^*(t) = (\mu - \lambda, \lambda)t, \quad 0 \le t \le T_1^* \triangleq \frac{1}{\mu - \lambda}. \tag{15.82}$$

We must check that this path actually lies in the interior of the state space; that is, we have to see that $r_2^*(t) < r_1^*(t)$. Since $\mu > 2\lambda$, we have $\mu - \lambda > \lambda$, so $\vec{r}^*$ does indeed lie in the region $x_1 > x_2$.

There is also a cheapest path in the region $x_2 > x_1$.

Exercise 15.55. Show that there is another mode for x_1 to reach one with $x_2 > x_1$ throughout. Show that the cost of this path is high compared to the other two modes. Hint: write the cost function as in (15.81), where ℓ_1 corresponds to a Poisson process, and ℓ_2 corresponds to an $M/M/1$ queue. Since the path must cross $x_2 = 1$ its cost, by symmetry, must be higher. ♠

We have demonstrated that there are at least two critical paths from the point $\vec{0}$ to the line $x_1 = 1$: one where $r_1^*(t) = r_2^*(t)$, and one where $r_1^*(t) > r_2^*(t)$.

These paths have costs $I_0^T (\vec{r}\,^*)$ given by $2\log(mu/2\lambda)$ and $\log(\mu/\lambda)$ respectively. Therefore, for $4\lambda > \mu$, the first path is cheaper (more likely), whereas for $4\lambda < \mu$ the second path is cheaper. When $4\lambda = \mu$ both paths have the same cost.

Now let's make sure that we have considered all possible critical paths. Since the process is constant coefficient in the regions $x_1 > x_2$ and $x_1 < x_2$, we know that the critical paths must consist of straight-line segments. All possibilities are sketched in Figure 15.56.

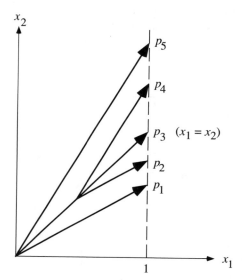

Figure 15.56. The possible critical paths.

The point p_1 represents the path $\vec{r}_1^*$ that goes from $(0, 0)$ to $x_1 = 1$ with cost $\log(\mu/\lambda)$ in the interior defined in (15.82). The point p_3 represents the minimum cost path $\vec{r}_b^*$ that travels along the boundary, defined in (15.79). Its cost is given in (15.80) as $2\log(\mu/\lambda)$. The point p_5 represents the path you calculated in Exercise 15.55, where $x_2(t) > x_1(t)$.

Since the paths are straight lines and the process is constant-coefficient in each region, the cost incurred per unit time along each critical path is $\ell(r, r')$, a constant. Therefore there is a constant cost per unit distance traveled toward $x_1 = 1$. Now if the costs of p_1 and p_3 are the same (that is, if $4\lambda = \mu$), then any path (such as $\vec{0}$—p_2 in Figure 15.56) that follows p_3 for a while and then travels parallel to p_1 thereafter will have the same cost. Similar considerations apply to constructing p_4 from p_3 and p_5, although p_4 and p_5 are never minimal paths, by Exercise 15.55. Again, the case where p_2 is a minimal path is exactly the critical case $4\lambda = \mu$ derived above. This finishes the proof of Theorem 15.45.

Leo Flatto [Fl] has derived the steady state of this system; indeed, his work motivated this study. He found that when $4\lambda = \mu$, the distribution of x_2 conditioned on $x_1 = 1$ is asymptotically (as $n \to \infty$) *uniform* on $x_2 \in \left(\frac{\lambda}{\mu-\lambda}, 1\right)$. Our interpretation is that all paths p_2 have exactly the same probability, so all their endpoints should be equally likely. But of course we cannot prove his result using our methods.

15.10. Join the Shortest Queue

Suppose that a Poisson (λ) stream of customers arrives to a system of two queues. The queues have statistically identical independent exponential (μ) servers. Arriving customers join the shorter queue, flip a 50/50 coin if the queues are the same length, and cannot switch lines once they join. The two-dimensional process $(x_1(t), x_2(t))$ is Markov with the following transition diagram.

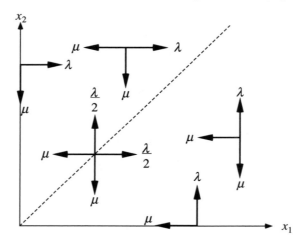

Figure 15.57. Transition diagram for join the shorter queue.

Definition 15.58. *The process $\vec{x}(t)$ is defined by the following rates and directions:*

Jump direction	Jump rate	Interpretation
$\vec{e}_1 = (1, 0)$	$\lambda_1(\vec{x}) = \begin{cases} \lambda & x_1 < x_2 \\ \lambda/2 & x_1 = x_2 \\ 0 & x_1 > x_2 \end{cases}$	*Arrivals to queue one*
$\vec{e}_2 = (0, 1)$	$\lambda_2(\vec{x}) = \begin{cases} \lambda_2 & x_1 > x_2 \\ \lambda/2 & x_1 = x_2 \\ 0 & x_1 < x_2 \end{cases}$	*Arrivals to queue 2*
$\vec{e}_3 = (-1, 0)$	$\lambda_3(\vec{x}) = \begin{cases} \mu & x_1 > 0 \\ 0 & x_1 = 0 \end{cases}$	*Departure from queue one*
$\vec{e}_4 = (0, -1)$	$\lambda_4(\vec{x}) = \begin{cases} \mu & x_2 > 0 \\ 0 & x_2 = 0 \end{cases}$	*Departure from queue 2.*

Our standard scaling (5.1) is obtained as

$$\vec{z}_n(t) \triangleq \frac{\vec{x}(nt)}{n}.$$

Transformation and most probable path.

Although $\vec{z}_n(t)$ does not fall under the domain of flat boundary theory, just as in §15.9 a simple transformation of the state space will enable us to calculate (and prove) limit laws for $\vec{z}_\infty(t)$ and the large deviations principle for $\vec{z}_n(t)$. We use the transformations R and $\mathbf{R}$ defined in (15.66), §15.9, and let $\vec{u} = \mathbf{R}\vec{x}$. In the notation of Chapter 8, u_2 was denoted u_0 and is the direction perpendicular to the boundary.

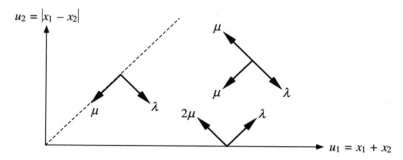

Figure 15.59. The transformed state space.

Then it is a simple matter to calculate the equation of $\vec{u}_\infty$, the fluid limit for the process $\vec{u}_n(t)$ (as in §15.9, $\vec{u}_n$ is the scaled process derived from $\vec{u}$):

$$\frac{d}{dt}\vec{u}_\infty(t) = \begin{cases} (-\mu, -\mu) & \text{if } u_1 = u_2 > 0 \text{ and } \lambda < \mu \\ (\lambda - 2\mu, -\lambda) & \text{if } u_1 = u_2 > 0 \text{ and } \lambda \geq \mu \\ (\lambda - 2\mu, -\lambda) & \text{if } u_1 > u_2 > 0 \\ (\lambda - 2\mu, 0) & \text{if } u_1 > 0 \text{ and } u_2 = 0 \\ \vec{0} & \text{if } \vec{u} = \vec{0}. \end{cases} \tag{15.83}$$

Here is how we derive (15.83): the computations are left to Exercise 15.60 below. In the interior of the state space, where $u_1 > 0$ and $u_2 > u_0$, the third line of (15.83) is just (5.7), which clearly follows from Theorem 5.3. Depending on $\vec{u}_\infty(0)$ and whether or not $\lambda > \mu$, this path may lead to the line $u_1 = 0$ or to the line $u_1 = u_1$ (or it may lead directly to $\vec{0}$, which we analyze below). The line $u_1 = u_2$ will be reached by $\vec{u}_\infty$ only if $\lambda < \mu$. Once $\vec{u}_\infty$ hits either of the boundaries, we calculate its subsequent evolution by Exercise 8.13. This leads to the first and fourth lines of (15.83). Looking at the solutions of (15.83) shows that $\vec{u}_\infty \to \vec{0}$ if and only if $\lambda < 2\mu$. We could have guessed this stability condition—the total number of customers in the system is pretty much the same as an $M/M/1$ queue with arrival rate λ and service rate 2μ. The only difference is when one of the queues is empty: then the rate at which the system empties may only be μ.

Exercise 15.60. Prove that (15.83) is the most likely behavior away from $\vec{0}$. Hint: this is a local result, so the theory of Chapter 8 applies away from $\vec{0}$. ♠

Exercise 15.61. Derive $\vec{z}_\infty(t)$ from $\vec{u}_\infty(t)$. ♠

Upcrossings and steady state.

Now that we know the likely behavior of $\vec{z}_n(t)$, let's examine its unlikely behavior. Specifically, we examine the distribution of $x_2(t)$ conditioned on $x_1(t) = n$. By Theorem 6.89 we expect $\vec{z}_n(t)$ to be at the point $\vec{x}^*$ that minimizes $V(\vec{x})$ over $\{\vec{x} : x_1 = 1\}$. Recall that $V(\vec{x})$ is defined as

$$V(\vec{x}) \overset{\triangle}{=} \inf_{\vec{r} \in G} I_0^T(\vec{r}),$$

where G is the set of paths that start at $\vec{r}(0) = \vec{0}$ and end at a point $\vec{r}(T)$ with $r_1(t) = 1$. But notice that any path from $\vec{0}$ to $x_1 = 1$ must pass through the point $(1, 1)$, since upcrossings to $x_1(t) = n$ can only occur at $x_2(t) = n-1$ or $x_2(t) = n$. Therefore we are led to guess the following

Theorem 15.62.

$$\lim_{n \to \infty} \mathbb{E}_{ss}\left(z_{n,2} \mid z_{n,1} = 1\right) = 1$$

and for all $\varepsilon > 0$,

$$\lim_{n \to \infty} \mathbb{P}_{ss}\left(\left|z_{n,2} - 1\right| < \varepsilon \mid z_{n,1} = 1\right) = 1.$$

Note: This would follow, except we have not proven a large deviations principle for processes with discontinuities that are not flat boundaries, and have not checked that the conditions (specifically, the recurrence condition) of Theorem 6.89 hold. However, note that we did not even need to *pose* a variational problem, let alone *solve* one, to conjecture this theorem.

Proof 1. Theorem 15.35 shows that $\vec{u}_n(t)$ satisfies a large deviations principle. Now we need to show that it is positive recurrent. This follows from Lemma 16.14; therefore, Theorem 15.62 will be proved once we can show that the minimum cost point along the line $x_1 = 1$ is the point $(1, 1)$. But this is easy: all paths that cross up must go through this point, and $\vec{z}_\infty$ immediately heads back to $\vec{0}$ from $(1, 1)$. This means that every other point on the line has strictly greater cost function V. (You should check that the statements we are making about the process $\vec{u}_n(t)$ also hold for $\vec{z}_n(t)$—checking this is almost trivial from the definition of $\vec{u}$.) There is the usual argument to be made about conditioning on the line $x_1 = 1$, namely, the line is not compact and is not the closure of its interior. See Lemma 15.38 (or Lemma 16.13 for a more general method that doesn't apply to this process since the $\vec{e}_i$ don't span $\mathbb{R}^2$) for the way to get around this minor difficulty.   ∎

Proof 2. (Sketch) Since we do not need the full power of a large deviations principle to ascertain where the upcrossing occurs, we have some hope of making a proof that does not rely on the large deviations principle. This proof generalizes to higher dimensions and to more processes than the previous proof. Examine the proof of Theorem (6.89). All that is used is

1. Upcrossings are rare. (We know this since the total queue length is like a stable $M/M/1$ queue. We can also provide a supermartingale that shows positive recurrence; this, combined with the next item, means that upcrossings must be rare.)

2. The process has bounded speed (the λ_i are bounded, not necessarily continuous).

3. We know where upcrossings occur most likely.

4. The process is ergodic (positive recurrent). This is straightforward. As mentioned in point 1, there is a simple supermartingale that shows this: the process $r(t)$, the Euclidean distance between $\vec{0}$ and $\vec{x}(t)$, is a supermartingale if the stability condition holds and if $|\vec{x}|$ is not too small (this is an easy calculation). ∎

Exercise 15.63. Show that, for the problem of finding when $x_1 \geq n$, the optimal path is

$$\vec{r}^{\,*}(t) = \left(\mu - \frac{\lambda}{2}, \mu - \frac{\lambda}{2} \right) t \quad \text{for} \quad 0 \leq t \leq T^* = \frac{1}{\mu - \lambda/2}.$$

[This is pretty obvious from the transition diagram for $\vec{u}(t)$: we see that $u_1(t)$ looks like an $M/M/1$ queue with arrival rate λ and service rate 2μ, so that most likely upcrossings are made by the arrival and service rates switching:

$$u_1(t) = (x_1 + x_2)(t) = (2\mu - \lambda) \cdot t.$$

Also, during this excursion,

$$u_0(t) = |x_1 - x_2| \approx 0.$$

This gives the stated $\vec{r}^{\,*}(t)$.] The exercise is to write the cost function for $\vec{u}(t)$, and solve the appropriate variational problem. ♠

Extensions:

1. The queues may have different service rates. Nothing changes so long as we have $\lambda < \mu_1 + \mu_2$. That is, the conclusion of Theorem 15.62 holds.

2. There can be any number of queues. Each queue has its own service rate μ_i. Customers always join the shortest queue, choosing uniformly at random among all shortest queues at the time of their arrival. We assume the stability condition $\lambda < \sum_i \mu_i$.

Proof 1 breaks down for these extensions, but proof 2 extends verbatim. The reasons that proof 1 breaks down are manifold. We don't have a large deviations principle for processes with discontinuities along the boundary such as an edge. Furthermore, the mapping from $\vec{x}$ to $\vec{u}$ no longer gives a nice mapping of the rates. That is, consider the jump rates for $\vec{u}$ when $\mu_1 \neq \mu_2$ and $x_1 > x_2$; then consider the jump rates when $x_1 < x_2$. Since this mapping is no longer one-to-one, we cannot use a large deviations principle for $\vec{u}_n$ to establish one for $\vec{z}_n$.

However, there is nothing at all wrong with proof 2. We obtain for each $\varepsilon > 0$,

$$\lim_{n\to\infty} \mathbb{P}_{ss}\left(\left|\frac{1}{n}(x_2,\ldots,x_k) - (1,\ldots,1)\right| < \varepsilon \,\middle|\, x_1 = n\right) = 1.$$

15.11. End Notes

Costas Courcoubetis introduced us to the simple priority queue, and asked the questions that led to the analysis of this chapter. Leo Flatto introduced us to the other models, and obtained the exact steady state distributions of (x_1, x_2) for both of them. He used the complex variables methods of Malyshev, so his techniques would not seem to generalize to more than two queues. Of course, his answers are more accurate than ours since they are exact, not asymptotic. He also obtained all the asymptotics we did (and more!), except for the sample path properties.

Leo pointed out an interesting correspondence between the *finite* JSQ model, and the *finite* SLQ model. Suppose that there are only n spaces in each queue in either of the models. Then consider the number of empty slots in each queue. The join the shorter queue model's spaces are pathwise the same as the serve the longer queue model's customers, and vice-versa. How can you reconcile this fact with our calculations showing that the upcrossings (and downcrossings) of the two models are entirely different?

The derivation of a large deviations principle from a principle for a continuous transformation of the process of interest is called "inverse contraction:" cf. [DZ, Theorem 4.2.4]. However, the usual hypotheses do not hold in our case, since our transformations are not invertible (not even on their range).

Dupuis and Ellis [DE2] prove large deviations principles for processes with many types of discontinuities. Their method does not quite apply to our examples, though, since they essentially require that the cone spanned by the $\vec{e}_i$ is $\mathbb{R}^d$ in each region where the process has smooth jump rates, which clearly does not hold for JSQ or SLQ.

Chapter 16

The Flatto-Hahn-Wright model

Suppose that there are two bathrooms[1], one for men and one for women. Each has room for exactly one client at a time. People arrive at the bathrooms either as individuals, or sometimes a (heterosexual) couple causes a simultaneous arrival at each bathroom. The times that people spend in each facility are independent and exponentially distributed, but the different sexes may have different averages. We assume that each type of arrival stream forms an independent Poisson process, and that the two lines (which are $M/M/1$ queues when viewed in isolation) have steady-state distributions; that is, the total arrival rate at each bathroom is less than its associated service rate. Now for the question: If we observe that a large number (n) of men are waiting for their convenience, what can we infer about the number of women waiting?

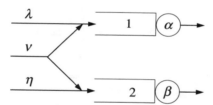

Figure 16.1. The basic Flatto-Hahn-Wright model.

While we have given the formulation of the problem that most appeals to us, there is a more sober and general question that is equivalent to the one we have just asked. Namely, when a subsystem goes into an unlikely state, what can we infer about the rest of the system? The large deviations point of view is particularly well suited to answering this question. We simply look at the cheapest twisted measure (associated with a cheapest path r^*; see §7.4) that would cause the subsystem to go into the unlikely state, and see how long that twist would have to hold for it to do so (that is, what is the time T^* associated with r^*?). Then we see what the effect of holding that twist for that long would be on the rest of the system. We approach this general problem through a specific example partly because we don't know the most general formulation of the question that is amenable to our analysis, and partly because we feel that most of the ideas are contained in our simple example.

This model was posed by Wright [Wr] as a generalization of an earlier model of Flatto and Hahn [FH]. The original motivation was to study asymptotic correlations among parallel processors. That is, if jobs arrive at a machine that consists

---

[1] American usage. British "W.C." or lavatory; "toilet" elsewhere.

of two processors, with some jobs requiring attention from both processors, what can you say about the backlog at one processor given that you know the other processor has a large backlog?

Let us now define the model's parameters—jump rates and directions.

Definition 16.2. *The Markov process $\vec{x}(t)$ is defined through the following jump directions and rates:*

Jump direction	Jump rate	Interpretation
$\vec{e}_1 = (1, 0)$	$\lambda_1 = \lambda$	single man arrives
$\vec{e}_2 = (0, 1)$	$\lambda_2 = \eta$	single woman arrives
$\vec{e}_3 = (1, 1)$	$\lambda_3 = \nu$	couple arrives
$\vec{e}_4 = (-1, 0)$	$\lambda_4(\vec{x}) = \begin{cases} \alpha & x_1 > 0 \\ 0 & x_1 = 0 \end{cases}$	man leaves
$\vec{e}_5 = (0, -1)$	$\lambda_5(\vec{x}) = \begin{cases} \beta & x_2 > 0 \\ 0 & x_2 = 0 \end{cases}$	woman leaves.

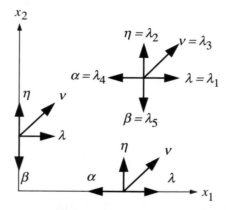

Figure 16.3. Transition diagram for the Flatto-Hahn-Wright model.

We are interested in the steady-state distribution of $(x_1(t), x_2(t))$, conditioned on either $x_1(t) \geq n$ or $x_1(t) = n$. We will show that both problems reduce to evaluating the variational problem

$$\min_{\vec{r}, T \in G} I_0^T(\vec{r})$$

$$G \triangleq \left\{ \vec{r}, \; T > 0 \; : \; \vec{r}(0) = \vec{0}, \; r_1(T) = 1 \right\}, \tag{16.1}$$

where the rate function I_0^T is defined as follows.

$$I_0^T(\vec{r}) = \int_0^T \ell(\vec{r}, \vec{r}') \, dt$$

$$\ell(\vec{x}, \vec{y}) = \sup_{\vec{\theta}} \left(\langle \vec{\theta}, \vec{y} \rangle - g(\vec{x}, \vec{\theta}) \right) \tag{16.2}$$

provided $(\vec{x}, \vec{y})$ correspond to a path that stays in the positive quadrant (including the boundary), that is both components of $\vec{x}$ are nonnegative, and if $\vec{x}_i = 0$ then $\vec{y}_i \geq 0$. We set $\ell(\vec{x}, \vec{y}) = \infty$ otherwise. As in Theorem 8.19,

$$g(\vec{x}, \vec{\theta}) = \lim_{\delta \to 0} \max_{|\vec{z} - \vec{x}| < \delta} \sum_{i=1}^{5} \lambda_i(\vec{z}) \left(e^{\langle \vec{\theta}, \vec{e}_i \rangle} - 1 \right),$$

which takes the usual form (5.4) whenever $x_i > 0$, $i = 1, 2$.

It should be intuitive, or at least believable, that the way a long line forms for the men's room is that the arrival rate of men must be above average for a while, and simultaneously the men must be taking an unusually long time to do their business. Now if more men than usual are arriving, then more women than usual are probably arriving too, since we expect a certain fraction of the men to arrive with partners. Thus the line for the women's room should be longer than usual when the line for the men's room is. This argument can be made quite precise, even more than we will do here; for details, see [SW].

We approach the steady-state distribution of the process $\vec{x}$ using the ideas of Freidlin and Wentzell, as presented in Chapter 6. We need to extend the analysis a little bit since the process $\vec{x}(t)$ is not on a compact state space. Let $\mathbb{P}_{ss}$ (respectively $\mathbb{E}_{ss}$) denote the steady-state distribution (expectation) for the process. Define

$$C \overset{\triangle}{=} \max \left\{ 0, \ \frac{\eta + \frac{\alpha}{\nu + \lambda} \nu - \beta}{\alpha - \lambda - \nu} \right\}. \tag{16.3}$$

Our main result is the following theorem.

Theorem 16.4. *Assume the stability condition*

$$\begin{aligned} \lambda + \nu &< \alpha \\ \eta + \nu &< \beta. \end{aligned} \tag{16.4}$$

For all $\varepsilon > 0$,

$$\lim_{n \to \infty} \mathbb{P}_{ss} \left(\frac{x_2(t)}{n} \in (C - \varepsilon, C + \varepsilon) \ \middle| \ x_1(t) = n \right) = 1, \tag{16.5}$$

$$\lim_{n \to \infty} \mathbb{E}_{ss} \left(\frac{x_2(t)}{n} \ \middle| \ x_1(t) = n \right) = C. \tag{16.6}$$

The conclusions hold if the conditioning "$x_1 = n$" is replaced by "$x_1 \geq n$."

To prove this result we need to prove that the Freidlin-Wentzell calculations are justified, and we need to perform the calculations to obtain C. We shall start from Theorem 6.89, which we need to extend to sets D that are not smooth bounded and open, such as the sets $\{x_1 = 1\}$ and $\{x_1 \geq 1\}$.

The heuristic reasoning behind the Freidlin-Wentzell approach is that steady state probabilities may be approximated by considering upcrossings, as in the proof of Theorem 6.89. So consider the process $\vec{x}(t)$ and the way it goes from its most probable location near $\vec{0}$ (under the stability condition; see §16.1) to the region $x_1 \geq n$. Since $\vec{x}$ is a constant coefficient process in the interior ($x_1 > 0$ and $x_2 > 0$), and is also constant coefficient on each boundary, we expect that the cheapest path—for the variational problem that results in an upcrossing $x_1 \geq n$— consists of straight-line segments (Lemma 5.16). We prove in §16.3 that, in fact, the cheapest path is a simple straight line, and that extra segments can only add to the cost. This should not be surprising: the only reason we might have extra segments is that there might be a lower cost for traveling along a boundary in some direction or other. But if a boundary path is cheaper than an interior path, there will be no reason to switch to the interior, and if an interior path is cheaper than a boundary path, then there is no reason to go along a boundary. Therefore we shall simply calculate the "cost" of following straight-line paths from $\vec{0}$ to $x_1 = n$, and see where x_2 is at the endpoint for the cheapest such line. Then Theorem 6.89 would indicate that (16.5) of Theorem 16.4 holds, and (16.6) follows with a few additional calculations.

As usual, before we prove that the calculation is justified, we perform it. It follows by inspection that the standard scaling given by (5.1) is equivalent to

$$\vec{z}_n(t) \stackrel{\triangle}{=} \frac{1}{n}\vec{x}(nt). \tag{16.7}$$

We start with an examination of the most likely behavior $\vec{z}_\infty(t)$ of the system, then proceed to the large deviations calculations, and conclude this chapter with the justifications.

16.1. Most Probable Behavior

It is easy to see what $\vec{z}_\infty(t)$ must be without any calculation at all. Simply note that $x_i(t)$ is an $M/M/1$ process when viewed in isolation, $i = 1, 2$ (it's only when they are taken together that their correlation makes the system more complicated). But we know what the most likely path for an $M/M/1$ queue is. We conclude that the components of $\vec{z}_\infty(t)$ must be the corresponding most likely paths for the processes $x_1(t)$ and $x_2(t)$. The components of $\vec{z}_\infty$ obviously determine the entire process $z_\infty(t)$. We have found $\vec{z}_\infty(t)$ without calculation:

$$\begin{aligned}
z_{\infty,1}(t) &= \max\left(0,\ z_{\infty,1}(0) + (\lambda + \nu - \alpha)t\right) \\
z_{\infty,2}(t) &= \max\left(0,\ z_{\infty,2}(0) + (\eta + \nu - \beta)t\right).
\end{aligned} \tag{16.8}$$

From this explicit form for $\vec{z}_\infty(t)$ we see that $\vec{z}_\infty(t) \to \vec{0}$ as $t \to \infty$ if and only if the stability condition (16.4) holds. We see that the correlation between $x_1(t)$ and $x_2(t)$ is indeed irrelevant as far as the stability of the system is concerned: the most likely behavior depends only on the *total arrival rates*. Lemma 16.14 shows that $\vec{z}_n(t)$ is positive recurrent if and only if (16.4) holds—we hope that our discussion has made this result entirely believable.

Exercise 16.5. Derive (16.8) directly from the definition (8.11) of $\vec{z}_\infty$.   ♠

16.2. Formal Large Deviations Calculations

We are interested in solving (16.2). The previous argument suggests that we consider straight-line paths $\vec{r}(t)$ from $\vec{0}$ to the line $r_1 = 1$ as depicted in Figure 16.6, and look for the cheapest one. Properly we should also look at other straight-line segments and piece together the cheapest path out of them; however, as proved in §16.3, the minimal path is a simple straight line. Remark 15.10 is relevant here; you might wish to review it now. So, let us try to solve the variational problem (16.1) using straight-line paths. When $\vec{r}(t)$ is a simple straight line and the process is constant coefficient, $I_0^T(\vec{r}) = T\ell(\vec{x}, \vec{r}')$ is independent of $\vec{x}$: we only need to distinguish between the case where the path moves along the boundary $x_2 = 0$, and the case where is in the interior. Thus we need to minimize $T\ell(\vec{x}, \vec{y})$ over all $\vec{y}$ and $T > 0$ so that $y_1 T = 1$. We start by considering paths that have strictly positive slope, so that they travel in the interior. As we show in (16.19) below, for such paths to be optimal it is necessary that

$$\eta + \frac{\alpha}{\lambda + \nu} \nu - \beta \geq 0. \tag{16.9}$$

In fact, to have a strictly positive slope we must have a strict inequality in (16.9). The intuitive interpretation of this condition is given below Lemma 16.7, and we shall assume it holds until the final proof of Theorem 16.4.

We will find it convenient to do our calculations using an alternate form of the local rate function. When $x_2 > 0$ (or $y_2 > 0$), Theorem 5.26 shows [with the notation of (5.20)–(5.22)] that

$$\ell(\vec{x}, \vec{y}) = \inf_{\vec{\mu} \in K_y} \sum_{i=1}^{5} h(\mu_i, \lambda_i), \tag{16.10}$$

where

$$h(a, b) \overset{\triangle}{=} a \log \frac{a}{b} + b - a \tag{16.11}$$

$$K_y \overset{\triangle}{=} \left\{ \vec{\mu} : \mu_i \geq 0, \sum_{i=1}^{5} \mu_i \vec{e}_i = \vec{y} \right\}. \tag{16.12}$$

We will ease our calculations by using the following relationship between the coefficients appearing in the various definitions of the rate function. If $\ell(\vec{x}, \vec{y})$ is given by a particular $\vec{\theta}^*$ in (16.2), and if it is also given by a particular $\vec{\mu}^*$ in (16.10), then Exercise 5.29 gives for interior paths

$$\mu_i^* = \lambda_i e^{\langle \vec{\theta}^*, \vec{e}_i \rangle}. \tag{16.13}$$

This enables us to do partial calculations with each form of the rate function, and then to piece the calculations together.

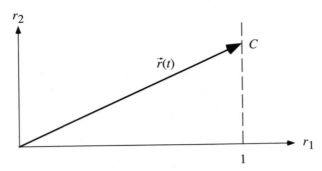

Figure 16.6. A test path $\vec{r}(t)$.

Let us see how the cost varies for different straight-line paths. Such paths have the form $\vec{r}(t) = \vec{b} \cdot t$ for $0 \leq t \leq 1/b_1$; their velocity is $\vec{b}$, and $T = 1/b_1$. In Figure 16.6 the point $(1, C)$ is at $(1, b_2/b_1)$. It is clear that there is no constraint on b_2 other than it be positive; let us ignore even this restriction. As we shall see, the resulting optimal straight-line path turns out to have a positive slope, so that ignoring this condition at this stage would cause no harm. It does, however, simplify the calculation considerably: for we can now minimize the cost freely over b_2 for every b_1. Equivalently, we can change the rates μ_2 and μ_5 arbitrarily in the minimization in (16.10), and the value of $y_2 = b_2$ is free to be any value. Since $h(a, b) = 0$ if and only if $a = b$, it is clear that the optimal twisted rates μ_2 and μ_5 are equal to λ_2 and λ_5 respectively, and that these components contribute nothing to the cost. Now using (16.13) we conclude that $\theta_2^* = 0$; that is, the twisted rates have the form

$$\mu_i = \lambda_i e^{\langle (\theta,0),\vec{e}_i \rangle} \tag{16.14}$$

for some $\theta = \theta_1^*$. This means that we can replace ℓ in the definition (16.2), or in the minimization of $T\ell(\vec{x}, \vec{y})$ by

$$\tilde{\ell}(y_1) \overset{\triangle}{=} \sup_\theta \left(\theta y_1 - \left(\lambda(e^\theta - 1) + \nu(e^\theta - 1) + \alpha(e^{-\theta} - 1) \right) \right). \tag{16.15}$$

But this is exactly the local rate function for the $M/M/1$ process (in the interior), with arrival rate $\lambda + \nu$, and service rate α! Thus we have reduced the problem of calculating $\inf_{\vec{r} \in G} I_0^T(\vec{r})$, restricted to interior paths, to a one-dimensional minimization

$$\inf_{\vec{r} \in G} I_0^T(\vec{r}) = \inf_{b_1} \frac{\tilde{\ell}(b_1)}{b_1}, \tag{16.16}$$

which, in fact, is a problem of the cost for an $M/M/1$ process to go from zero to one. But we have already treated this problem: this is exactly (11.8), which is solved (in greater generality) in Theorem 11.15. Reading off the solution (11.22) and the computation of θ^* we have

$$b_1 = \alpha - \lambda - \nu, \tag{16.17}$$

$$\theta^* = \theta_1^* = \log \frac{\alpha}{\nu + \lambda}. \tag{16.18}$$

By (16.13) and the definition of K_y this value of θ corresponds to

$$b_2 = \eta + v e^{\theta^*} - \beta$$

$$= \eta + \frac{\alpha}{v + \lambda} v - \beta, \tag{16.19}$$

which by our assumption (16.9) is positive. We have proved the following.

Lemma 16.7. *Under assumptions (16.4) and (16.9), the cheapest interior path for the variational problem (16.1) is $\vec{r}(t) = \vec{b} \cdot t$, with $\vec{b}$ given in (16.17) and (16.19), and with $b_2/b_1 = C$.*

If $b_2 > 0$ then we have found a minimal cost path in the interior. Note that the arguments remain valid if $b_2 = 0$: in this case the minimal cost path is certainly on the boundary. However, it is not clear that the cost we have calculated is correct, since it corresponds to spending no time on the boundary—just skimming it. Even if $b_2 > 0$ we don't necessarily know that the lowest cost path in the interior is cheaper than the lowest cost path on the boundary. As we shall show, and not too surprisingly, when $b_2 > 0$ the interior path is indeed the minimal cost path.

Equation (16.18) has a simple interpretation based on our understanding of the $M/M/1$ queue. When queue 1 fills, it does so by its arrival and service rates interchanging for a while (see Theorem 11.18). This means that during this interval, the total arrival rate is α while the service rate is $\lambda + v$. Now the only thing to see is how this change of jump rates affects the second queue. Here is where the change of measure interpretation makes things simple. The jump rates in the x_1 directions are changed by a factor of $\phi \overset{\triangle}{=} \frac{\alpha}{\lambda+v}$, since the jump rates λ_i change to $\lambda_i \exp\langle \vec{\theta}^*, \vec{e}_i \rangle$, and $\vec{\theta}^* = (\theta_1^*, 0)$. That is, the arrival rates change from λ and v to $\phi\lambda$ and ϕv, and the service rate changes from α to α/ϕ. This means that the joint arrival rate for couples is now ϕv. So queue 2 sees a net arrival rate of $\eta + \phi v$, and has a service rate β during this interval. The length of the interval is $T \overset{\triangle}{=} n(\alpha - \lambda - v)^{-1}$, since this is the amount of time it takes this $M/M/1$ queue to fill [see the discussion near (11.23)]. Thus we expect queue 2 to be at about level $T(\eta + \phi v - \beta)$ when $\eta + \phi v - \beta > 0$; otherwise we expect it to be $o(n)$.

There is another reason why the rate v should change to ϕv. We can regard the process of arrivals to queue 1 as being a single Poisson stream of intensity $v + \lambda$, not the merge of two independent streams. We could then have a separate, independent Bernoulli process that selects a fraction $\frac{v}{v+\lambda}$ of this stream to cause arrivals at queue 2. Hence when the original stream has its intensity changes from $v + \lambda$ to α, queue 2 now sees a joint arrival rate of $\alpha \frac{v}{v+\lambda}$. That is, the property of Poisson splitting and merging is encoded in our twisted measure.

We now sketch out the calculation of the minimal cost path on the boundary. In

Chapter 8 we show that the local rate function takes the form

$$\ell(\vec{x}, \vec{y}) = \inf_{\pi_0 \in [0,1]} \inf_{\vec{\mu} \in \mathcal{S}} \left((1 - \pi_0) \sum_{i=1}^{5} h(\mu_i(1), \lambda_i) + \pi_0 \sum_{i=1}^{4} h(\mu_i(0), \lambda_i) \right)$$

(16.20)

where $\mu_i(0)$ are the twisted rates on the boundary and $\mu_i(1)$ are the twisted rates in the interior[2]. In our case,

$$\mathcal{S} = \left\{ \mu_i(m) \geq 0, \; : \; (1 - \pi_0) \sum_{i=1}^{5} \mu_i(1)\vec{e}_i + \pi_0 \sum_{i=1}^{4} \mu_i(0)\vec{e}_i = \vec{y} \right\}.$$

In Chapter 8 [equation (8.21) and its discussion] we establish that (16.14) holds as well when the process is on the boundary, that is when $x_2 = y_2 = 0$. Since for our model all rates and jump directions in the interior are the same as on the boundary—with the exception of λ_5, $\vec{e}_5$—we conclude from (16.14) that the twisted rates agree as well: $\mu_i(0) = \mu_i(1)$ for $i = 1, 2, 3$ and 4. The definition of $\mathcal{S}$ now shows that the only π_0 corresponding to a given $\vec{\mu}$ when $y_2 = 0$ must satisfy

$$0 = \pi_0(\mu_2 + \mu_3) + (1 - \pi_0)(\mu_2 + \mu_3 - \mu_5)$$

$$\pi_0 = 1 - \frac{\mu_2 + \mu_3}{\mu_5}$$

whenever $\mu_5 > \mu_2 + \mu_3$. If the optimal twisted rates satisfy $\mu_5 \leq \mu_2 + \mu_3$ then $\pi_0 = 0$ and $b_2 \geq 0$, so that we are back to the previous case where the optimal path is in the interior (and possibly "skims" the boundary). You should have the following interpretation in mind for these quantities. π_0 is the fraction of time $\vec{x}(t)$ spends on the boundary when following the path $\vec{y} \cdot t$. The jump rates λ_i change to μ_i in the interior and, by the previous argument, change in the same way on the boundary. We show in Proposition 16.10 that, if (16.9) holds, then the fact that the rates change in the same way implies that the interior path is optimal, even in the case $b_2 = 0$. Again, the "twisted process" behaves like an $M/M/1$ queue so that our previous reasoning applies. This is explored in Exercise 16.8 below.

The form of the rate function we have given ignores the boundary at $x_1 = 0$ and also the discontinuity at $\vec{0}$. We did this for simplification; we prove in §16.3 that for the questions we ask, there is no need to calculate the rate for paths that travel along the boundary at $x_1 = 0$, and in any case the reader can easily determine the rate function on that boundary since the system is perfectly symmetrical with respect to a reordering of the indices.

Exercise 16.8. Derive Equation (16.18) by solving the right-hand side of (16.16). How does the calculation change when the path is along a boundary? That is, show that the change of measure we have found holds then, too. You simply have to check that (16.14) holds in this case, too.   ♠

---

[2] In the notation of Theorem 8.19(ii), the rates were ν and π was called μ. Sorry!

Of course, we obtain more from our approach than estimates of the mean of x_2 given x_1; we get the usual sample path properties as well. Define T^* as the time it takes $\vec{r}^*(t)$ to go from $\vec{0}$ to the line $r_1 = 1$:

$$T^* \overset{\triangle}{=} \frac{1}{\alpha - \lambda - \nu}. \tag{16.21}$$

Using the conventions of Chapter 6 and shifting time, $\vec{r}^*(-T^*) = \vec{0}$ and $\vec{r}^*(0) \overset{\triangle}{=} (1, C)$. As in (6.7) we define $\vec{r}^*(t) = \vec{0}$ for $t < -T^*$. The same arguments we use in the proof of Theorem 16.4 (based on the techniques of Chapter 6) would yield

Theorem 16.9. *Under the stability condition (16.4), for each $\varepsilon > 0$ and $T > 0$,*

$$\lim_{n \to \infty} \mathbb{P}_{ss} \left(\sup_{-T \leq t \leq 0} \left| \vec{z}_n(t) - \vec{r}^*(t) \right| < \varepsilon \;\middle|\; \vec{z}_{n,1}(0) = 1 \right) = 1.$$

This sample path property is, of course, nothing but Theorem 6.15 applied to our process. Other theorems in Chapter 6 apply as well; the reader is invited to prove Theorem 16.9 and to see which of the results there have interesting statements to make about the process at hand.

16.3. Justification of the Calculation

As usual we begin our justification with a proof of correctness of our solution to the variational problem. Afterwards we examine the relevance of the variational problem to the probabilistic problem.

Proposition 16.10. *Under Assumptions (16.4) and (16.9), the unique cheapest path for the variational problem (16.1) is $\vec{r}(t) = \vec{b} \cdot t$, with $\vec{b}$ given in (16.17) and (16.19).*

Proof. First we use a "soft" argument to show that the optimal path must be a single straight line. Let $\vec{r}$ be an optimal path with $\vec{r}(0) = \vec{0}$ and $\vec{r}_1(T) = 1$. Consider any piece $\vec{r}(t)$, $0 \leq t_1 \leq t \leq t_2 \leq T$ that lies in the interior. Since $\vec{z}_n(t)$ is a constant coefficient process, its local rate function in the interior is strictly convex and independent of $\vec{x}$. By Lemma 5.16 (see also Appendix C) all optimal paths from $\vec{r}(t_1)$ to $\vec{r}(t_2)$ must be straight lines.

Now consider a piece of a minimal path lying entirely on one boundary. The rate function along the boundary is convex by Corollary 8.18, but not necessarily strictly convex (see Exercise 8.21). However, direct calculation (not included here) shows that ℓ is indeed strictly convex, proving uniqueness of straight-line paths as minimizers.

Consider now the case $r_2(T) > 0$ and assume that $\vec{r}$ consists of at least two line segments—from t_1 to t_2 and from t_2 to T. As we shall see below, the optimal path is unique. But then the segment $[t_1, t_2]$ cannot be on the boundary $x_2 = 0$:

since the process is constant coefficient (in each region), if it is cheaper to follow a positive slope from $\vec{r}(t_2)$ to $\vec{r}(T)$, it would be cheaper to follow a positive slope from $\vec{r}(t_1)$ to $(r_1(t_2), x_2)$! Since the process is constant coefficients, the cost from $(r_1(t_2), x_2)$ to the line $x_1 = 1$ is the same as from $\vec{r}(t_2)$. Exactly the same reasoning shows that if the segment $[t_1, t_2]$ of the optimal path is on the boundary $x_2 = 0$ then $r_2(t)$ must be zero for all $t \geq t_1$. Thus we conclude that any optimal path either makes a straight line from a point $(0, r_2(t_1))$ to the line $x_1 = 1$, or goes down to $x_2 = 0$ and then proceeds in a straight line. But, as our calculation (16.19) shows, the optimal slope to move to the right must be positive. Thus the last segment must be a straight line from $x_1 = 0$ to $x_1 = 1$.

The remaining question is, could it be optimal to go from $\vec{0}$ to a point $(0, x_1)$ and the proceed to $x_1 = 1$? Travel from $\vec{0}$ to a point $(0, r_2(t_1))$ could be useful only if it led to a path $\vec{r}(t)$, $t_1 \leq t \leq T$ with lower cost than the path we already found, since it costs a strictly positive amount to travel to any point which is not on $\vec{z}_\infty$. But note that the path we computed in Lemma 16.7 was optimal among all endpoints $r_2(T)$! And moreover, the cost of a parallel path starting at any $(0, x_2)$ with $x_2 > 0$ is identical, since the process is constant coefficient. Therefore, the cost of going from a point $(0, x_2)$ with $x_2 > 0$ to $x_1 = 1$ cannot be lower then the cost of going from $(0, 0)$ to the line $x_1 = 1$, and hence the optimal path does not travel to a point with $x_1 = 0$.

It remains to compare the cost of going along the boundary $x_2 = 0$ to that of going through the interior in order to establish uniqueness. In fact, the argument below establishes that the optimal path follows a fixed slope, and is thus a single straight line.

Consider the rate function in the form (16.20) it takes on the boundary, and fix b_1. Since the function $h(a, b)$ is positive,

$$\ell(\vec{x}, \vec{y}) \geq \inf_{\pi \in [0,1]} \inf_{\vec{\mu} \in \mathcal{S}_1} \left((1 - \pi) \sum_{i=1,3,4} h(\mu_i, \lambda_i) + \pi \sum_{i=1,3,4} h(\mu_i, \lambda_i) \right)$$

$$\geq \inf_{\vec{\mu} \in \mathcal{S}_1} \sum_{i=1,3,4} h(\mu_i, \lambda_i)$$

where $\mathcal{S}_1$ now places a restriction only on the first coordinate:

$$\sum_{i=1,3,4} \mu_i \langle \vec{e}_i, (1,0) \rangle = b_1. \tag{16.22}$$

But this is exactly the form of $\tilde{\ell}$, and the unique minimum for the level crossing is that obtained for (16.16). If the minimum were obtained for a path on the boundary via some other b_1, our calculation shows we could decrease $\tilde{\ell}$ with the same b_1, a contradiction. We conclude that the unique optimal path is a straight line, with speed $\vec{b}$.   ∎

Now that we know the correct solution to the variational problem, we can invoke the Freidlin-Wentzell theory as proved in Chapters 6 and 15. We use Theorems 15.35 and 15.39, and the analogues of Lemmas 15.38 and 15.41 for our process. We need to check that the conditions of these theorems are satisfied, of course; we do this in a series of lemmas.

Exercise 16.11. Assume (16.4), show that for any T and $\varepsilon > 0$,

$$\lim_{n\to\infty} \mathbb{P}_{\vec{0}}\left(\sup_{0\le t\le T} |\vec{z}_n(t)| < \varepsilon\right) = 1. \tag{16.23}$$

Also show that there is a function $f(\varepsilon) > 0$ with $f(\varepsilon) \to 0$ as $\varepsilon \to 0$ such that

$$\liminf_{n\to\infty} \frac{1}{n} \log \mathbb{P}_{\vec{0}}\left(|\vec{z}_n(T)| \ge \varepsilon\right) \ge -f(\varepsilon). \tag{16.24}$$

(This latter equation is a statement that Lemma 15.41 holds for the FHW model.)
Hint: Each component of $\vec{z}_n(t)$ is an $M/M/1$ queue. ♠

The large deviations principle holds for the FHW model because the hypotheses of Theorem 15.35 are satisfied (see Exercise 16.11, which shows that $\vec{q} = \vec{0}$). It is also easy to see that Kurtz's Theorem holds for the FHW model.

The FHW process is now even easier to analyze than the processes of Chapter 15, since the positive cone spanned by the $\vec{e}_i$ is $\mathbb{R}^2$. Therefore Theorem 6.77 holds for the FHW process.

Furthermore, Theorem 6.92 holds since the FHW process satisfies all its hypotheses, too. With just a bit more arguing we can establish Theorem 6.92 for the FHW model and the set $D = \{x_1 = 1\}$. This is done as in the proof of Theorem 15.39, where again the arguments are even simpler for the FHW model since the $\vec{e}_i$ span $\mathbb{R}^2$. The justifications follow. Therefore we claim that Theorem 16.4 is proved, and even more that Theorem 16.9 holds. We also obtain the following. Let I^* denote the cost for the FHW process to reach the line $x_1 = 1$; by the calculations of §16.2, this is the same as the cost for an $M/M/1$ queue with arrival rate $\lambda + \nu$ and service rate α to reach the level. Therefore

$$I^* = \log \frac{\alpha}{\lambda + \nu}. \tag{16.25}$$

Theorem 16.12. *Let τ denote the first time that the FHW process $\vec{z}_n(t)$ reaches one. Uniformly for $\vec{x}$ in any compact set $G : G \cap \{x_1 \ge 1\} = \emptyset$, for any $\varepsilon > 0$ we have*

$$\lim_{n\to\infty} \mathbb{P}_{\vec{x}}\left(\frac{\log \tau}{n} \in (I^* - \varepsilon, I^* + \varepsilon)\right) = 1 \tag{16.26}$$

and furthermore

$$\lim_{n\to\infty} \frac{1}{n} \log \mathbb{E}_{\vec{x}}(\tau_n(G)) = I^*. \tag{16.27}$$

This is just a restatement of Theorem 6.77.

We can establish that the FHW model satisfies the hypotheses of Theorem 15.39 using a general argument that holds for any process whose positive cone is all of $\mathbb{R}^2$. The idea is that any two nearby points must have nearly the same steady-state probability, since a sample path that visits one point is fairly likely to visit the other.

Lemma 16.13. *Under assumption (16.4), there exists a $\delta > 0$ so that for every $\varepsilon > 0$ there is an n_0 such that the following holds. For every $\vec{x}$ and $\vec{y}$ in $\mathbb{R}_+^2$ with $n\vec{x}$ and $n\vec{y}$ in $\mathbb{Z}^2$ such that $|\vec{x} - \vec{y}| < \delta \cdot \varepsilon$ and for every $n > n_0$,*

$$e^{-n\varepsilon} \leq \frac{\mathbb{P}_{ss}(n\vec{x})}{\mathbb{P}_{ss}(n\vec{y})} \leq e^{n\varepsilon}. \tag{16.28}$$

Proof. This is obvious from representation (6.34) of the steady state and the following observations. For any ε',

$$\frac{\mathbb{P}_{ss}(n\vec{x})}{\mathbb{P}_{ss}(n\vec{y})} = \frac{\mathbb{P}_{ss}(n\vec{x}(t) = \vec{x})}{\mathbb{P}_{ss}(n\vec{y}(t - \varepsilon') = \vec{y})} \tag{16.29}$$

$$\geq \mathbb{P}\left(n\vec{x}(t) = \vec{x} \mid n\vec{y}(t - \varepsilon') = \vec{y}\right)$$

by the Markov property. But for this conditional probability we can easily derive a lower bound (as in the proof of Theorem 11.4 or Exercise 11.20), by considering the probability that the process makes exactly the necessary series of jumps to go from $n\vec{y}$ to $n\vec{x}$ in the prescribed time, and makes no other jumps. With the proper choice of ε' the lower bound follows. Note that we can derive a "universal" bound that depends only on δ. A symmetric argument interchanging the role of $\vec{x}$ and $\vec{y}$ establishes the upper bound.   ∎

Lemma 16.14. *Under Assumption (16.4), the process $\vec{x}(t)$ is positive recurrent.*

Proof. The recurrence follows from more general results for two-dimensional systems; see e.g., [C2] or the original work of Malyshev [Mal]. A modern, intuitive approach is given in [Cn] and in [Da].   ∎

16.4. End Notes

Wright analyzed the FHW model extensively in [Wr], extending the pioneering work of Flatto and Hahn [FH], [Fl]. He obtained an explicit expression for the entire steady-state distribution of $\vec{x}$, as well as the asymptotic formulas we derived here, and even had more accurate expressions in case $C = 0$. His results were based on the complex variables methods of Malyshev, and do not seem to extend to more than two queues, though J. W. Cohen has recently reported some progress in this area. We have also obtained some more detailed asymptotics using time reversal and large deviations in [SW]. That paper also gave some other connections between time reversal and large deviations. Our early work on this problem benefitted from discussions with Richard Ellis and Paul Dupuis.

The techniques we developed here should be applicable to more general queuing systems. For example, it is not hard to find the asymptotics for systems of three or more coupled queues, assuming that the large deviations principle holds (see [SW]). This assumption was our main difficulty in carrying the method forward. However, the new technique in [DE2] seems to provide the requisite theory.

Appendix A

Analysis and Probability

This appendix collects the background information needed to read this book. It is mostly telegraphic, although some intuition is given where, in our judgment, it is called for.

A.1. Topology, Metric Spaces, and Functions

The material in this section is covered is most books on analysis or functional analysis, with the exception of the space $D^d[0, T]$. Dunford and Schwartz [DuS] and Royden [Roy] are the main references; Rockafellar [Roc] is our source for convex analysis, and Billingsley, Kurtz, and Ethier and Kurtz [Bil, Ku1, EK] cover the function spaces of interest.

Let $\mathcal{X}$ be a space, that is, a collection of "points." If A and B are subsets of $\mathcal{X}$, we denote by $A \cap B$ their intersection, that is, the collection of points that belong to both, and by $A \cup B$ their union, that is the collection of points that belong to at least one of these sets. The *complement* A^c of A is the collection of points in $\mathcal{X}$ that are not in A. The set difference $A \setminus B \stackrel{\triangle}{=} \{x : x \in A, \ x \notin B\}$ is sometimes denoted also as $A - B$, and $\emptyset$ is the empty set.

Definition A.1. *A collection τ of subsets of $\mathcal{X}$ is called a* topology *if*

(i) $\emptyset \in \tau$ and $\mathcal{X} \in \tau$,

(ii) Any finite intersection of members of τ belongs to τ,

(iii) Any union of members of τ belongs to τ.

The pair $(\mathcal{X}, \tau)$ is called a topological space.

We sometimes refer to $\mathcal{X}$ as a topological space, if τ is clear from the context. A set is called *open* if and only if it belongs to τ. The complement of an open set is called *closed*. A *neighborhood* of a point x is any open set containing x.

Lemma A.2. *A set A in a topological space $(\mathcal{X}, \tau)$ is open if and only if for every point x in A there is a neighborhood containing x that is contained in A.*

Definition A.3. *Let A be a set in a topological space $(\mathcal{X}, \tau)$. The interior A^o of A is the union of all open sets contained in A. The closure $\overline{A}$ of A is the intersection of all closed sets containing A. The boundary ∂A of A consists of the points in its closure that are not in the interior: $\partial A \stackrel{\triangle}{=} \overline{A} \setminus A^o$. An open cover of a set A is a* collection of open sets whose union contains A, that is, every point in A belongs to at least one of the open sets.

Definition A.4. *A set A in a topological space $(\mathcal{X}, \tau)$ is compact if every open cover of A contains a finite subcollection that is also a cover. It is precompact if its closure is compact.*

Definition A.5. *Let $(\mathcal{X}_1, \tau_1)$ and $(\mathcal{X}_2, \tau_2)$ be topological spaces. The product space $\mathcal{X}_1 \times \mathcal{X}_2$ is the collection of all pairs $(x_1 \in \mathcal{X}_1, x_2 \in \mathcal{X}_2)$. The product topology τ is the smallest topology that includes all sets of the form $\{(x_1, x_2) : x_1 \in A_1, x_2 \in A_2)\}$ for all $A_1 \in \tau_1$ and $A_2 \in \tau_2$.*

Definition A.6. *A metric d on $\mathcal{X}$ is a real-valued function $d(x, y) : \mathcal{X} \times \mathcal{X} \to \mathbb{R}$ with*

  (i) $d(x, y) \geq 0$ for all x, y in $\mathcal{X}$,

  (ii) $d(x, y) = 0$ if and only if $x = y$,

  (iii) $d(x, y) = d(y, x)$, and

  (iv) $d(x, y) \leq d(x, z) + d(z, y)$ for all x, y and z in $\mathcal{X}$.

The requirement (iv) above is the (in)famous *triangle inequality*.

Definition A.7. *Let d be a metric on $\mathcal{X}$. A metric space $(\mathcal{X}, d)$ is a topological space where the topology τ is the smallest one that contains all sets of the form $\{y : d(x, y) < \varepsilon\}$, for all x and ε. These sets are called open balls, with center x and radius ε. Equivalently, every set in τ is the union of finite intersections of open balls. In this case we say τ is induced by d.*

In this setting the set $\{y : d(x, y) < \varepsilon\}$ is called an open ball with radius ε and center x. A set A is called *bounded* if for some x_0 we have $\sup_{x \in A} d(x, x_0) < \infty$ (by the triangle inequality, any x_0 will do). Note that it is possible that there are different topologies τ that make $(\mathcal{X}, \tau)$ also into a topological space. The topology induced by d is termed the Borel topology.

Example A.8. The space $(\mathbb{R}, d)$ of real numbers, where d is the usual (Euclidean) distance, is a metric space. Open balls are just open intervals. In the Euclidean space $\mathbb{R}^d$ (the dimension d is not related to the metric d, sorry),

$$d_1(\vec{x}, \vec{y}) \overset{\triangle}{=} \sum_{i=1}^{d} |x_i - y_i| \quad \text{and} \quad d_2(\vec{x}, \vec{y}) \overset{\triangle}{=} \left(\sum_{i=1}^{d} |x_i - y_i|^2 \right)^{1/2}$$

are both metrics, and both induce the same topology (although open balls are different!). We use the standard distance $|\vec{x} - \vec{y}| \overset{\triangle}{=} d_2(\vec{x}, \vec{y})$, and denote the scalar product by $\langle \vec{x}, \vec{y} \rangle \overset{\triangle}{=} \sum_{i=1}^{d} x_i \cdot y_i$. A set in $\mathbb{R}^d$ is compact if and only if it is closed and bounded (the Heine-Borel theorem).

Theorem A.9. *Let $(\mathcal{X}, \tau)$ be a topological space and A a set contained in $\mathcal{X}$. Then (A, τ_A) is a topological space, where the induced topology τ_A is defined by*

$$B_A \in \tau_A \text{ if and only if } B_A = A \cap B \text{ for some } B \in \tau.$$

If $(\mathcal{X}, d)$ is metric with induced (also called—relative) topology τ, then (A, d) is metric with induced topology τ_A. A set $B_A \subset A$ is compact in the relative topology if and only if it is compact in $\mathcal{X}$. However, a closed set in (A, τ_A) may not be closed in $(\mathcal{X}, \tau)$! In a metric space, open balls in the relative topology take the form $\{y \in A : d(x, y) < \varepsilon\}$.

Definition A.10. *A sequence $x_1, x_2, \ldots$ in a metric space $(\mathcal{X}, d)$ is Cauchy if*

$$\lim_{n, m \to \infty} d(x_n, x_m) = 0.$$

The sequence "converges to a limit" x (written $x_n \to x$ or $\lim_{n \to \infty} x_n = x$) means

$$\lim_{n \to \infty} d(x, x_n) = 0,$$

or, equivalently, means that any neighborhood of x contains all points x_n, $n \geq n_0$ (where n_0 depends on the neighborhood). If x_n and possibly x depend on a parameter $\alpha \in A$, then $x_n(\alpha) \to x(\alpha)$ uniformly in $\alpha \in A$ means

$$\lim_{n \to \infty} \sup_{\alpha \in A} d(x(\alpha), x_n(\alpha)) = 0.$$

The index n need not be an integer.

Definition A.11. *A topology τ_1 is called stronger (or finer) than a topology τ_2 if every set in τ_2 is also a member of τ_1.*

In this case the topological space $(\mathcal{X}, \tau_1)$ has more open (or closed) sets than $(\mathcal{X}, \tau_2)$. The topology τ_2 is called weaker, or coarser. Stronger (or weaker) metrics are defined through the induced topology.

Definition A.12. *Two metrics d_1 and d_2 on $\mathcal{X}$ are equivalent if they induce the same topology.*

Lemma A.13. *If $x_1, x_2, \ldots$ converge to a limit x in $(\mathcal{X}, d_1)$ and d_2 is a weaker metric, then $x_1, x_2, \ldots$ converge to x in $(\mathcal{X}, d_2)$. Conversely, if every sequence that converges under d_1 also converges under d_2, then d_2 is a weaker metric.*

Thus d_1 and d_2 are equivalent if and only if they have the same converging sequences. Note that equivalent metrics may have different Cauchy sequences! The metrics in Example A.8 are equivalent. The speed of convergence can be described through the following notation.

Definition A.14. *Let f and g be real functions.*
   Say $g(\varepsilon) = O(f(\varepsilon))$ if $|g(\varepsilon)|/|f(\varepsilon)|$ is bounded for $|\varepsilon|$ small.

Say $g(\varepsilon) = o(f(\varepsilon))$ if $|g(\varepsilon)|/|f(\varepsilon)| \to 0$ as $|\varepsilon| \to 0$.

The most common usage is when $f(\varepsilon) = \varepsilon$. With some abuse of notation, we

Say $g(n) = O(f(n))$ if $|e(n)|/|f(n)|$ is bounded for n large, and

Say $g(n) = o(f(n))$ if $|g(n)|/|f(n)| \to 0$ as $n \to \infty$.

In $\mathbb{R}$, there are special types of limits that we need.

Definition A.15. *For a real sequence $x_1, x_2, \ldots$ define*

$$\liminf_{n \to \infty} x_n \overset{\triangle}{=} \lim_{n \to \infty} \inf_{j \geq n} x_j \quad and \quad \limsup_{n \to \infty} x_n \overset{\triangle}{=} \lim_{n \to \infty} \sup_{j \geq n} x_j.$$

For a real sequence we write $x_n \to \infty$ if $\liminf_{n \to \infty} x_n = \infty$.

Definition A.16. *A set A in a metric space $(\mathcal{X}, d)$ is sequentially compact if every sequence of points from A contains a converging subsequence.*

Theorem A.17. *A set A in a metric space $(\mathcal{X}, d)$ is compact if and only if it is closed and sequentially compact. The set is open if and only if it contains an open ball around each point. The set is closed if and only if whenever a sequence of points in A converges, the limit is also in A.*

Definition A.18. *A set B is dense in a set A if, for every $x \in A$ and every $\varepsilon > 0$, there is a $y \in B$ with $d(x, y) < \varepsilon$. A topological space is separable if it contains a countable dense subset. A metric space $(\mathcal{X}, d)$ is complete if every Cauchy sequence converges. A complete, separable metric space (sometimes abbreviated as CSMS) is also called a Polish space.*

Theorem A.19. *Let C be closed and K be compact in a complete metric space $(\mathcal{X}, d)$. If $C \cap K = \emptyset$ then there exists some $\varepsilon > 0$ so that $d(x, y) > \varepsilon$ for all $x \in C$ and all $y \in K$.*

Definition A.20. *$\mathcal{X}$ is a vector space if*

(i) For all x in $\mathcal{X}$ and a real, $a \cdot x$ is defined and belongs to $\mathcal{X}$.

(ii) For all x and y in $\mathcal{X}$, $x + y$ is defined and belongs to $\mathcal{X}$.

Definition A.21. *A set A in a vector space is convex if whenever the points x and y are in A, then $\alpha x + (1 - \alpha)y$ is in A for any α in $[0, 1]$.*

Definition A.22. *$(\mathcal{X}, \tau)$ is a topological vector space if it is both a topological space, a vector space, and in addition the functions $f(a, x) \overset{\triangle}{=} a \cdot x$ and $f(x, y) \overset{\triangle}{=} x + y$ are continuous (as functions of two arguments—Definition A.25).*

Example A.23. *The spaces $(\mathbb{R}^d, d_i)$, $i = 1, 2$ of Example A.8 are Polish spaces, and also topological vector spaces.*

Functions.

Let $(\mathcal{X}_1, d_1)$ and $(\mathcal{X}_2, d_2)$ be two metric spaces and f a function $f : \mathcal{X}_1 \to \mathcal{X}_2$.

Definition A.24. *The image of a set $A \subset \mathcal{X}_1$ under f is*

$$f(A) \overset{\triangle}{=} \{y : y = f(x), \text{ for some } x \in A\}.$$

Similarly, the inverse image of a set $A \subset \mathcal{X}_2$ under f is

$$f^{-1}(A) \overset{\triangle}{=} \{x : f(x) \in A\} \subset \mathcal{X}_1.$$

The graph of f is the set in $\mathcal{X}_1 \times \mathcal{X}_2$

$$\text{graph } f \overset{\triangle}{=} \{(x_1, x_2) \in \mathcal{X}_1 \times \mathcal{X}_2 : x_2 = f(x_1)\}.$$

Definition A.25. *The function f is called continuous if $f^{-1}(A) \subset \mathcal{X}_1$ is open whenever the set A is open in $\mathcal{X}_2$. The function f is Lipschitz continuous if for all x, y,*

$$d_2(f(x), f(y)) \leq L \cdot d_1(x, y)$$

for some positive L, called the Lipschitz constant.

Definition A.26. *The local modulus of continuity of a function $f : \mathcal{X} \to \mathbb{R}^d$ at a point x is the function of δ*

$$C_f(x, \delta) \overset{\triangle}{=} \sup\{|f(x) - f(y)| : x, y \in \mathcal{X}, \ d(x, y) \leq \delta\}.$$

The modulus of continuity is the function

$$\rho_f(\delta) \overset{\triangle}{=} \sup_{x \in \mathcal{X}} C_f(x, \delta).$$

A family of functions $\{f(x; \beta), \ \beta \in B\}$ is called continuous in x uniformly in β (over B) if

$$\sup_{\beta \in B} C_{f(\cdot; \beta)}(x, \delta) \to 0 \quad \text{as} \quad \delta \to 0.$$

Theorem A.27. *A function $f : \mathcal{X}_1 \to \mathcal{X}_2$ is continuous if and only if*

$$\lim_{n \to \infty} x_n = x \text{ in } \mathcal{X}_1 \quad \text{implies} \quad \lim_{n \to \infty} f(x_n) = f(x) \text{ in } \mathcal{X}_2.$$

The image of a compact set under a continuous function is compact. A function $f : \mathcal{X} \to \mathbb{R}^d$ is continuous at x if and only if its local modulus of continuity $C(x, \delta) \to 0$ as $\delta \to 0$, and if it is continuous on a compact set K then it is uniformly continuous on K, that is $C(x, \delta) \to 0$ uniformly in $x \in K$.

Definition A.28. *A function $f : \mathcal{X} \to \mathbb{R}$ is lower semicontinuous if the set $f^{-1}((-\infty, a])$ is closed, for every real a. It is upper semicontinuous if the set $f^{-1}([a, \infty))$ is closed.*

Exercise A.29. Show that f is continuous if and only if $f^{-1}(A)$ is closed whenever A is closed, and that f is lower semicontinuous if and only if

$$\lim_{n \to \infty} x_n = x \quad \text{implies} \quad \liminf_{n \to \infty} f(x_n) \geq f(x). \qquad \spadesuit$$

Exercise A.30. Establish the following. If f is lower semicontinuous than $(-f)$ is upper semicontinuous. A function that is both lower semicontinuous and upper semicontinuous is continuous. The supremum of a family of lower semicontinuous functions is lower semicontinuous. The composition $f(g(x))$ of continuous functions is continuous. Hint: use the definitions, not the theorems.   ♠

Theorem A.31. *A lower semicontinuous function attains its minimum on every compact set.*

Definition A.52. *A function of a real variable is right continuous if*

$$\lim_{x \downarrow y} f(x) = f(y),$$

that is, $f(x)$ converges to $f(y)$ as x decreases to y. The definition of left continuous is analogous.

Definition A.33. *A set G in $\mathbb{R}^d$ has smooth boundary if there is a function f from $\mathbb{R}^d$ to $\mathbb{R}$ that is twice continuously differentiable, so that $f(\vec{x}) = 0$ implies $|\nabla f(\vec{x})| \neq 0$, and*

$$\vec{x} \in \partial G \quad \text{if and only if} \quad f(\vec{x}) = 0.$$

Lemma A.34. *Let f be a real-valued function of the variables $x \in A$ and $y \in B$, and $y_1, y_2, \ldots$ an arbitrary sequence in B. Then*

(i)
$$\liminf_n \inf_{x \in A} f(x, y_n) \le \inf_{x \in A} \liminf_n f(x, y_n)$$
$$\limsup_n \inf_{x \in A} f(x, y_n) \le \inf_{x \in A} \limsup_n f(x, y_n)$$
$$\sup_{y \in B} \inf_{x \in A} f(x, y) \le \inf_{x \in A} \sup_{y \in B} f(x, y).$$

(ii) *If, in addition, the following two equalities hold:*

$$\inf_{x \in A} \sup_{y \in B} f(x, y) = \sup_{y \in B} \inf_{x \in A} f(x, y) = \lim_n \inf_{x \in A} f(x, y_n),$$

then $\lim_n \inf_{x \in A} f(x, y_n) = \inf_{x \in A} \liminf_n f(x, y_n) = \inf_{x \in A} \limsup_n f(x, y_n).$

Proof. Denote $f_i(y) \stackrel{\Delta}{=} \inf_{x \in A} f(x, y)$. Then $f_i(y) \le f(x, y)$ for all x, and so

$$\liminf_n f_i(y_n) \le \liminf_n f(x, y)$$

for all x, and hence

$$\liminf_n f_i(y_n) \le \inf_{x \in A} \liminf_n f(x, y),$$

which is exactly (i). The proof with lim sup or sup is identical.

From (i) we obtain

$$\liminf_n \inf_{x \in A} f(x, y_n) \le \inf_{x \in A} \liminf_n f(x, y_n) \le \inf_{x \in A} \sup_{y \in B} f(x, y)$$

since $\liminf_n f(x, y_n) \le \sup_{y \in B} f(x, y)$ for all x. Due to our assumption there is equality throughout, and (ii) is established. ∎

Here are some miscellaneous formulas we need.

Definition A.35. *Hyperbolic sine and cosine are defined by*

$$\sinh x = \frac{1}{2}\left(e^x - e^{-x}\right) \quad \text{and} \quad \cosh x = \frac{1}{2}\left(e^x + e^{-x}\right).$$

Stirling's formula approximating the factorial function is

$$n! = \sqrt{2\pi n}(n)^n e^{-n}(1 + O(1/n)). \tag{A.1}$$

Theorem A.36 (Taylor's Theorem). *Let $f : \mathbb{R} \to \mathbb{R}$ have continuous derivatives in a neighborhood of x_0 up to order n. Then in that neighborhood,*

$$f(x) = f(x_0) + f'(x_0)(x - x_0)$$
$$+ \frac{1}{2!}f''(x_0)(x - x_0)^2 + \ldots + \frac{1}{n!}f^{(n)}(x_0)(x - x_0)^n + R_n,$$

where $f^{(k)}$ is the kth order derivative of f and

$$R_n = \frac{1}{(n-1)!}\int_{x_0}^x (x - t)^n \left(f^{(n)}(t) - f^{(n)}(x_0)\right) dt.$$

A similar expansion is valid for a function of d variables $f : \mathbb{R}^d \to \mathbb{R}$, in which case $R_n/(|x - x_0|^n) \to 0$ as $x \to x_0$.

Convexity.

The standard reference on convexity in $\mathbb{R}^d$ is Rockafellar [Roc].

Definition A.37 [Roc]. *Let C be a subset of $\mathbb{R}^d$ and f a real-valued function on C.*

(i) *C is called affine if $\lambda x + (1 - \lambda)y$ is in C for all real λ whenever x, y are in C (p. 3). It is called convex if the same holds but for $0 \le \lambda \le 1$ (p. 10).*

(ii) *The convex hull of C is the smallest convex set containing C. The affine hull Aff C of C is the smallest affine set containing C (p. 6).*

(iii) *The relative interior ri C of C is the interior of C in the relative topology of Aff C, and the relative boundary of C is the set $\overline{C} \setminus$ ri C.*

(iv) *The epigraph epi $f \triangleq \{(x, \alpha) : x \in C, \alpha \in \mathbb{R}, \alpha \ge f(x)\} \subset \mathbb{R}^{d+1}$ (p. 23).*

(v) *f is called convex if epi f is a convex set in $\mathbb{R}^{d+1}$ (p. 23). A function from a convex set C to $(-\infty, \infty]$ is convex if and only if (Thm. 4.1 p. 25)*

$$f((1 - \lambda)x + \lambda y) \le (1 - \lambda)f(x) + \lambda f(y), \qquad 0 < \lambda < 1.$$

(vi) *The effective domain of a convex function* f, *denoted* dom f, *is the convex set* $\{x : f(x) < \infty\}$ (p. 23).

(vii) *A convex function* f *is proper if* $f(x) > -\infty$ *for all* x, *and* $f(x) < \infty$ *for some* x (p. 24).

(viii) *The closure* cl f *of a convex function* f *is the function whose epigraph is the closure in* $\mathbb{R}^{d+1}$ *of the epigraph of* f, *unless* $f(x) = -\infty$ *for some* x. *In the latter case set* cl $f \equiv \infty$. (p. 52).

(ix) f *is closed if* $f =$ cl f. *A proper convex function is closed if and only if it is lower semicontinuous. The only improper closed convex functions are* $f(x) \equiv \infty$ *or* $(-\infty)$ (p. 52).

(x) *The conjugate of* f *is* $f^*(x^*) \stackrel{\triangle}{=} \sup_x (\langle x, x^* \rangle - f(x))$ (p. 104).

(xi) *A function* f *is* concave *if the function* $(-f)$ *is convex*.

Most of the results on convex functions can be translated to statements about concave functions.

Theorem A.38 [Roc, p. 104 and Thm. 12.2]. *The function* f^* *is convex and closed. If* f *is convex then* f^* *is proper if and only if* f *is and moreover,* (cl $f)^* = f^*$ *and* $f^{**} =$ cl f, *so that for closed convex* f *we have* $f^{**} = f$.

Definition A.39 [Roc p. 215]. *For a convex function* f, *say* x^* *is a* subgradient *at* x *if* $f(z) \geq f(x) + \langle x^*, z - x \rangle$ *for all* z. *That is, the affine function on the right is a supporting hyperplane of* epi f *at* $(x, f(x))$. *The* subdifferential $\partial f(x)$ *is the set of all subgradients of* f *at* x. *The domain* dom ∂f *is the set* $\{x : \partial f(x) \neq \emptyset\}$.

Theorem A.40 [Roc Thm. 23.4 p. 217]. *If* f *is convex and proper then* $\partial f(x)$ *is empty for* $x \notin$ dom f, *and is non-empty for* $x \in$ ri (dom f). *Finally,* $\partial f(x)$ *is a non-empty bounded set if and only if* $x \in$ int (dom f).

Definition A.41 [Roc p. 251]. *A proper convex function is* essentially smooth *if*

(i) $C =$ int (dom f) *is non-empty,*

(ii) f *is differentiable on* C,

(iii) *If* $x_1, x_2, \ldots$ *converges to* $x \in \partial C$ *then* $\lim_{i \to \infty} |\nabla f(x_i)| \to \infty$.

In particular if $C = \mathbb{R}^d$ *then essentially smooth means differentiable.*

Definition A.42 [Roc p. 253]. *A function* f *is* strictly convex *on a convex set* C *if* $f(\lambda x + (1 - \lambda)y) < \lambda f(x) + (1 - \lambda)f(y)$ *for all* $0 < \lambda < 1$ *and all* $x \neq y$. *It is* essentially strictly convex *if it is strictly convex on every convex subset of* dom ∂f.

We shall call f strictly convex, without mentioning a convex set, if it is strictly convex on dom f.

Theorem A.43 [Roc Thm. 26.3 p. 253]. *A closed proper convex function is essentially strictly convex if and only if its conjugate is essentially smooth.*

Theorem A.44 (Min—Max Principle) [Roc, Cor. 37.3.2 p. 393]. *Let* $K(\vec{x}, \vec{y})$ *be a real-valued function, continuous in* $(\vec{x}, \vec{y}) \in \mathbb{R}^A \times \mathbb{R}^B$, *convex in* $\vec{x}$ *for each* $\vec{y}$, *and concave in* $\vec{y}$ *for each* $\vec{x}$. *Let two closed convex sets* $U \in \mathbb{R}^A$ *and* $V \in \mathbb{R}^B$ *be given, at least one of which is bounded. Then*

$$\inf_{\vec{x} \in U} \sup_{\vec{y} \in V} K(\vec{x}, \vec{y}) = \sup_{\vec{y} \in V} \inf_{\vec{x} \in U} K(\vec{x}, \vec{y}).$$

Theorem A.45 [Roc Thm. 10.2 p. 84]. *Let* C *be a polyhedron (intersection of finitely many closed half-spaces) or a polytope (the convex hull of a finite number of points). Let* f *be a proper convex, lower semicontinuous function that is finite at every point of* C. *Then* f *is continuous as a function from* C *to* $\mathbb{R}$.

The definitions extend to sets and functions on more general spaces.

Definition A.46. *A real-valued function* f *on a vector space is convex if for any two points* x *and* y *and any real* $0 < \alpha < 1$,

$$f(\alpha x + (1-\alpha)y) \leq \alpha f(x) + (1-\alpha)f(y).$$

It is called strictly convex near z *if (it is defined on a topological vector space and) in some neighborhood of* z *the inequality is strict* $(<)$ *for all* $x \neq y$.

In our applications convex functions might take the value $+\infty$, but not $(-\infty)$! The results of Theorems A.47 and A.48 are immediate from the definitions.

Theorem A.47. *Let* $\{f_\alpha\}$ *be a collection of convex functions on a metric vector space. Then*

(i) *The function* f *defined by* $f(x) \overset{\triangle}{=} \sup_\alpha f_\alpha(x)$ *is convex.*
(ii) *If for some function* f *the convergence* $f_\alpha(x) \to f(x)$ *holds at each point* x *as, say,* $\alpha \to 0$ *then* f *is convex.*

Theorem A.48. *Let* C *be a convex set in, and* f *a convex function on a metric vector space. Then*

(i) *Any point* x *in* ri C *is a convex combination* $\alpha u + (1-\alpha)v$ *of points* u, v *in the relative boundary of* C.
(ii) *If* f *is strictly convex in* C, *then* f *has at most one minimum in* C.
(iii) *If* C *is a closed and bounded set, then* f *achieves its maximum on the relative boundary of* C. *If* f *is strictly convex then it achieves its maximum only on the relative boundary of* C, *that is* f *cannot have a maximum in the relative interior.*

Spaces of functions: $C^d[0, T]$.

Let $C^d[0, T]$ denote the collection of all continuous functions of a parameter $t \in [0, T]$, with values in $\mathbb{R}^d$. We denote by $AC^d[0, T]$ the subset of absolutely continuous functions (Definition A.83).

Theorem A.49. *The space* $\left(C^d[0, T], d_c\right)$ *is a Polish vector space, where*

$$d_c(\vec{x}, \vec{y}) \stackrel{\Delta}{=} \sup_{0 \le t \le T} |\vec{x} - \vec{y}|.$$

Exercise A.50. $\left(AC^d[0, T], d_c\right)$ *is a separable metric vector space, but is not complete. Hint: take a continuous function that is not absolutely continuous and make a sequence of approximations.*   ♠

We shall suppress the subscript c or, altogether the notation of the metric whenever this "sup norm" is used and no ambiguity arises. This space is standard and is discussed in any text on functional analysis, and many texts on "real analysis," e.g. Royden [Roy].

Theorem A.51 Arzelà-Ascoli. *The set A has compact closure in* $\left(C^d[0, T], d_c\right)$ *if and only if*

 (i) *The initial points are bounded:* $\sup\{|\vec{x}(0)| : \vec{x} \in A\} < \infty$, *and*

 (ii) *The functions in A are equicontinuous, that is, for every t and ε there exists a δ so that, whenever* $|t - s| < \delta$ *we have* $|\vec{x}(t) - \vec{x}(s)| < \varepsilon$, *for all* $\vec{x}$ *in A.*

The space $D^d[0, T]$.

The functions that arise in our applications are either piecewise constant, or are continuous. The sup-norm distance of Theorem A.49 is not appropriate for these applications: a small change in the *time of a jump* means that the new process is far from the original in the sup norm. To allow small fluctuations in the time of jumps, we need the Skorohod topology. The idea is simple: the functions $\vec{x}$ and $\vec{y}$ are close if we can make them close in the sup norm, by making small perturbations to the time-parameter of, say $\vec{y}$. For details and proofs concerning this space, see e.g., [Bil, EK].

Definition A.52. *A function* $\vec{x}$ *of a parameter* $t \in [0, T]$ *with values in a metric space* $(\mathcal{X}, d)$ *is right continuous at t if* $\lim_{s \downarrow t} \vec{x}(s) = \vec{x}(t)$. *The function is said to have left limits at t if* $\lim_{s \uparrow t} \vec{x}(s)$ *exists (although it may not equal* $\vec{x}(t)$). *It is called right continuous (with left limits) if it has this property at every* $t \in [0, T]$.

The space $D^d[0, T]$ contains all functions of a parameter $t \in [0, T]$ with values in $\mathbb{R}^d$, which are right continuous with left limits. Functions in this space have the following simple property.

Theorem A.53. *Every function in* $D^d[0, T]$ *has at most a countable number of points where it is not continuous.*

Let Λ be a collection of strictly increasing (real) functions λ on $[0, T]$, such that $\lambda(0) = 0$ and $\lambda(T) = T$, and such that

$$\gamma(\lambda) \stackrel{\Delta}{=} \sup_{0 \le s \le t \le T} \left| \log \frac{\lambda(s) - \lambda(t)}{s - t} \right| < \infty. \tag{A.2}$$

The standard metric on $D^d[0, T]$ is

$$d_d(\vec{x}, \vec{y}) \overset{\Delta}{=} \inf_{\lambda \in \Lambda} \left\{ \max\left(\gamma(\lambda), \sup_{0 \le t \le T} |x(t) - y(\lambda(t))| \right) \right\}. \qquad (A.3)$$

Below we suppress the subscript d whenever no ambiguity arises.

Definition A.54. *The Skorohod topology is the topology induced on $D^d[0, T]$ by d_d.*

Theorem A.55. $\left(D^d[0, T], d_d\right)$ *is a complete, separable metric space.*

Definition A.56. *The modulus ρ'_f of a function f in $D^d[0, T]$. Let*

$$w_f[t_i, t_{i+1}] \overset{\Delta}{=} \sup\left\{ |f(t) - f(s)| : t_i \le t < s < t_{i+1} \right\}.$$

Given δ consider all finite partitions $\{t_i, \ i = 1, \dots, N\}$ of $[0, T]$ so that

$$0 = t_1 < t_2 < \dots < t_N = T \quad \text{and} \quad t_{i+1} - t_i > \delta.$$

Then $\rho'_f(\delta)$ is the infimum over all such partitions:

$$\rho'_f(\delta) \overset{\Delta}{=} \inf_{\{t_i\}} \max_{1 \le i < N} w_f[t_i, t_{i+1}].$$

Theorem A.57. *A set A has compact closure in $\left(D^d[0, T], d_d\right)$ if and only if*

(i) *The values are bounded:* $\sup\{|\vec{x}(t)| : \vec{x} \in A, \ 0 \le t \le T\} < \infty$, *and*

(ii) $\lim_{\delta \to 0} \sup_{f \in A} \rho'_f(\delta) = 0.$

The second assumption means that the functions in A are equicontinuous at points of continuity, and all have a finite number of large jumps.

Theorem A.58. *A sequence $\vec{x}_1, \vec{x}_2, \dots$ converges to $\vec{x}$ in $\left(D^d[0, T], d\right)$ if and only if there is a sequence $\lambda_1, \lambda_2, \dots$ in Λ so that one of the following equivalent conditions holds:*

$$\gamma(\lambda_n) \to 0 \quad \text{and} \quad \sup_{0 \le t \le T} |x(t) - y(\lambda_n(t))| \to 0 \qquad (A.4a)$$

$$\sup_{0 \le t \le T} |\lambda_n(t) - t| \to 0 \quad \text{and} \quad \sup_{0 \le t \le T} |x(t) - y(\lambda_n(t))| \to 0. \qquad (A.4b)$$

Corollary A.59. $\left(D^d[0, T], d_c\right)$ *is a metric space, and its topology is stronger than that of d_d. Consequently, every open (closed) set in $(D^d[0, T], d_d)$ is also open (respectively closed) in $(D^d[0, T], d_c)$ (the converse is of course false).*

Corollary A.60. *If C is compact in $\left(C^d[0, T], d_c\right)$, then it is compact in $\left(D^d[0, T], d_c\right)$ and in $\left(D^d[0, T], d_d\right)$.*

Proof. The identity maps $f(x) = x$ from $\left(C^d[0, T], d_c\right)$ to $\left(D^d[0, T], d_c\right)$ (or d_d) is continuous, so it maps compact sets into a compact sets (Theorem A.27).∎

Exercise A.61. $(D^d[0, T], d_c)$ is complete, but not separable. Hint: consider the collection $\{f(t) = \mathbf{1}\,[t \geq \alpha]\}$ with $0 \leq \alpha \leq T$. ♠

Lemma A.62. On $C^d[0, T]$, the metrics d_d and d_c are equivalent.

Exercise A.63. Let $\vec{r}$ be a continuous function. Show that in $(D^d[0, T], d_d)$;

$$\left\{\vec{x} \in D^d[0, T] : \sup_{0 \leq t \leq T} |\vec{x}(t) - \vec{r}(t)| < \varepsilon\right\} \quad \text{is an open set, and}$$

$$\left\{\vec{x} \in D^d[0, T] : \sup_{0 \leq t \leq T} |\vec{x}(t) - \vec{r}(t)| \leq \varepsilon\right\} \quad \text{is a closed set.}$$

Hint: use Theorem A.17 and (A.4b).   ♠

Extension to $[0, \infty)$.

Let d^n be a metric on the function space $C^d[0, T]$ or $D^d[0, T]$, with $T = n$.

Theorem A.64. *The space $C^d[0, \infty)$ of continuous functions on $[0, \infty)$ is a Polish space under the metric*

$$d(\vec{x}, \vec{y}) \triangleq \sum_{n=1}^{\infty} \frac{1}{2^n} \frac{d_c^n(\vec{x}_n, \vec{y}_n)}{1 + d_c^n(\vec{x}_n, \vec{y}_n)}, \tag{A.5}$$

where $\vec{x}_n$ is the function on $[0, n]$ defined by $\vec{x}_n(t) = \vec{x}(t)$ for $0 \leq t \leq n$. Similarly, $D^d[0, \infty)$ is a complete metric space under the metric defined through (A.5), with either d_c^n or d_d^n. It is separable (thus Polish) only if d_d^n is used.

The topology induced by this metric is called the product topology.

A.2. Ordinary Differential Equations

The results of this section are quoted from Hale [Hal].

Definition A.65. *Let D be an open set in $\mathbb{R}^{d+1}$. A function $\vec{x}(t) = \vec{x}(t_0, \vec{x}_0, t)$ is called a solution of the ordinary differential equation (ODE) with vector field f and initial condition $\vec{x}_0$ at t_0 on an interval I if for $t \in I$, $\vec{x}(t)$ is continuously differentiable, $(t, \vec{x}(t)) \in D$ and*

$$\frac{d\vec{x}(t)}{dt} = f(t, \vec{x}(t)), \quad \text{for all} \quad t \in I.$$

Theorem A.66 Existence (Peano) [Hal, Thm. 1.1, 2.1]. *If f is continuous in D then for each $(t_0, \vec{x}_0) \in D$ there is at least one solution of the ODE with initial* condition $\vec{x}_0$ at t_0. *There is a maximal interval of existence, and near the endpoints of the interval the solution approaches the boundary of D.*

Theorem A.67 Uniqueness [Hal, Thm. 3.1]. *If f is continuous in D and locally Lipschitz in $\vec{x}$ then for each initial condition there is a unique solution $\vec{x}(t_0, \vec{x}_0, t)$.*

Theorem A.68 Continuous dependence [Hal, Thm. 3.2]. *Let λ belong to some closed set, and suppose $f = f(t, \vec{x}, \lambda)$ is continuous in $(t, \vec{x}, \lambda)$ and Lipschitz in $\vec{x}$, where the Lipschitz constant does not depend on λ. Then for each $(t_0, \vec{x}, \lambda)$ there exists a unique solution $\vec{x}(t_0, \vec{x}_0, \lambda, t)$ with initial conditions $(t_0, \vec{x}_0)$ of the ODE with parameter λ. Moreover, the solution is continuous in $(t_0, \vec{x}_0, \lambda)$.*

Remark. The theorems above apply to higher order equations through the following standard construction. The ODE

$$\frac{d^n \vec{x}(t)}{dt^n} = f\left(t, \frac{d^{n-1}\vec{x}(t)}{dt^{n-1}}, \ldots, \frac{d\vec{x}(t)}{dt}, \vec{x}(t)\right)$$

of order n can be represented, with an appropriate definition of g, as

$$\frac{d\vec{y}(t)}{dt} = g(t, \vec{y}(t))$$

$$\vec{y}(t) \triangleq \left(\frac{d^{n-1}\vec{x}(t)}{dt^{n-1}}, \ldots, \frac{d\vec{x}(t)}{dt}, \vec{x}(t)\right).$$

A.3. Probability and Integration

In this section we review the main results (and notation) of the modern theory of probability. The foundations for the modern approach are the theories of measure and integration, which we also need. For more information consult any modern book on probability theory, e.g. [Br, Chu]. Books on real analysis and integration, e.g. [Roy], will also be helpful. For more advanced subjects see Billingsley, Kurtz, Ethier and Kurtz, and Stroock and Varadhan [Bil, Ku1, EK, SV].

Measure, probability, and integration.

To read this section you need to know the basics of set theory, metric spaces, and functions. The relevant material is reviewed in §A.1.

Definition A.69. *Let Ω be a set, and let $\mathcal{F}$ be a collection of subsets of Ω. The collection $\mathcal{F}$ is called a σ-algebra or a σ-field if it contains the empty set, and is closed under set complements, countable unions and countable intersections. Sets in $\mathcal{F}$ are then called measurable sets. In a probabilistic context they are called events. The pair $(\Omega, \mathcal{F})$ is called a measurable space.*

Definition A.70. *A real-valued function μ whose arguments are sets in $\mathcal{F}$ is called a measure if it is non-negative and countably additive. The last requirement means* *that if A_1, A_2, ... are disjoint sets in $\mathcal{F}$ then*

$$\mu\left(\bigcup_{i=1}^{\infty} A_i\right) = \sum_{i=1}^{\infty} \mu(A_i).$$

The triple $(\Omega, \mathcal{F}, \mu)$ is then called a measure space. A measure μ is σ-finite if there are sets A_1, A_2, ... in $\mathcal{F}$ so that

$$\bigcup_{i=1}^{\infty} A_i = \Omega \quad and \quad \mu(A_i) < \infty.$$

If in addition $\mu(\Omega) = 1$ then μ is called a probability measure, and $(\Omega, \mathcal{F}, \mu)$ is called a probability space. For a set A in $\mathcal{F}$, we say A holds μ-a.s. ("almost surely") if $\mu(A^c) = 0$. If μ is a probability measure then the equivalent term "w.p. 1" (with probability one) is used.

Throughout the book we consider only σ-finite measures.

Exercise A.71. Let A_n be events on a probability space $(\Omega, \mathcal{F}, \mathbb{P})$. Show that for all a and b, $\lim_{n\to\infty} \frac{1}{n} \log\left(e^{na} + e^{nb}\right) = \max\{a, b\}$, and for all positive a and ε,

$$\limsup_{n\to\infty} \frac{1}{n} \log \mathbb{P}(A_n) \leq -a \quad \text{implies} \quad \liminf_{n\to\infty} \frac{1}{n} \log \mathbb{P}(A_n^c) \geq -\varepsilon. \quad \spadesuit$$

Definition A.72. *Let $(\Omega_1, \mathcal{F}_1, \mu_1)$ and $(\Omega_2, \mathcal{F}_2, \mu_2)$ be measure spaces. The product space $\Omega_1 \times \Omega_2$ is the collection of all pairs $\left(\{\omega_1, \omega_2\} : \omega_1 \in \Omega_1, \omega_2 \in \Omega_2\right)$. The product σ-field $\mathcal{F}$ is the smallest σ-field that includes all sets of the form $\left\{\{\omega_1, \omega_2\} : \omega_1 \in A_1, \omega_2 \in A_2\right\}$ for all sets $A_i \in \mathcal{F}_i$. The product measure μ is the (unique) measure on the product space for which $\mu(A_1 \times A_2) = \mu_1(A_1) \cdot \mu_2(A_2)$ for all sets $A_i \in \mathcal{F}_i$.*

Definition A.73. *Let $(\mathcal{X}, d)$ be a Polish space. The smallest σ-field of subsets of $\mathcal{X}$ that contains all open balls is called the Borel σ-field and is denoted $\mathcal{B}(\mathcal{X})$. Sets in $\mathcal{B}(\mathcal{X})$ are called Borel sets.*

Example A.74. There exists a unique measure m on $(\mathbb{R}, \mathcal{B}(\mathbb{R}))$ with the property that $m([a, b]) = b - a$ for all $b > a$. It is called Lebesgue measure. A set A has a Lebesgue measure zero if and only if for any $\varepsilon > 0$ it can be covered by a countable number of intervals, the sum of whose lengths is less than ε.

In fact, the Lebesgue measure is defined on a collection $\mathcal{F}$ of Lebesgue measurable sets that (strictly) contains $\mathcal{B}(\mathbb{R})$. We shall not discuss the resulting fine points: see any book an real analysis or modern integration theory.

Definition A.75. *We say a property holds almost everywhere (abbreviated as a.e.) if it holds outside a set of Lebesgue measure zero. A function $f : \mathbb{R} \to \mathbb{R}^d$ is called essentially bounded if there exists a bound B so that $|f| \leq B$ a.e.*

Throughout the book we hold a probability space $(\Omega, \mathcal{F}, \mathbb{P})$ fixed, where $\mathbb{P}$ is a probability measure. But note that the measures we consider below are not necessarily probability measures, and in particular we might have $\mu(\Omega) = \infty$.

Definition A.76. *A random variable x is a real-valued function on Ω which is Borel measurable, that is $x^{-1}(B) \in \mathcal{F}$ for any Borel set B.*

Definition A.77. *If x is a function on Ω with values in a Polish space $(\mathcal{X}, d)$ and x is measurable, that is, $x^{-1}(B) \in \mathcal{F}$ for every Borel set B in $\mathcal{X}$, then we call x an abstract valued random variable.*

Note that no probability is required in order to define random variables.

Expectations of random variables are special cases of integrals. For details on integration see, e.g., [Roy].

Definition A.78. *A real function x on $(\Omega, \mathcal{F}, \mu)$ is called Lebesgue integrable if it is measurable and*

$$\limsup_{k \to \infty} \sum_{i=1}^{k} \frac{i}{\sqrt{k}} \mu \left(\omega \in \Omega : \frac{i}{\sqrt{k}} \leq |x(\omega)| < \frac{i+1}{\sqrt{k}} \right) < \infty.$$

If x is integrable and non-negative, its Lebesgue integral can be defined by

$$\int_{\Omega} x(\omega) \, d\mu(\omega) \overset{\triangle}{=} \lim_{k \to \infty} \sum_{i=1}^{k} \frac{i}{\sqrt{k}} \mu \left(\omega \in \Omega : \frac{i}{\sqrt{k}} \leq x(\omega) < \frac{i+1}{\sqrt{k}} \right).$$

If x is integrable its integral is defined by

$$\int_{\Omega} x(\omega) \, d\mu(\omega) \overset{\triangle}{=} \int_{\Omega} \left(\max\{x(\omega), 0\} \right) d\mu(\omega) - \int_{\Omega} \left(\max\{-x(\omega), 0\} \right) d\mu(\omega).$$

If μ is a probability measure then this integral is called the expectation of x and is denoted as $\mathbb{E}x$.

This definition of an integral is more explicit than the standard one, but the idea is the same. We approximate x by a function taking a finite number of values. Since x is not necessarily bounded, the number of mesh points has to grow faster than the inverse of the mesh size. Here are some properties of this integral (which of course carry over to expectations).

Theorem A.79. *Let x and y be real, integrable functions.*

(i) *Linearity. For any scalars a and b,*

$$\int_{\Omega} \left(a \cdot x(\omega) + b \cdot y(\omega) \right) d\mu(\omega) = a \int_{\Omega} x(\omega) \, d\mu(\omega) + b \int_{\Omega} y(\omega) \, d\mu(\omega).$$

(ii) *If x is nonnegative then $x(\omega) = 0$ μ-a.s. if and only if $\displaystyle\int_{\Omega} x(\omega) \, d\mu(\omega) = 0$.*

For real-valued functions of a real variable, this integral with respect to the Lebesgue measure is called Lebesgue integral. For "nice enough" functions (in particular, piecewise continuous functions that vanish outside bounded sets) it agrees with the Reimann integral.

Definition A.80. *The space* $L_p\left[(\Omega, \mathcal{F}, \mu), \mathbb{R}\right]$ *(abbreviated* L_p*), with* $1 \le p < \infty$ *is the collection of real-valued functions on the measure space* $(\Omega, \mathcal{F}, \mu)$ *satisfying* $\|x\|_p < \infty$ *where*

$$\|x\|_p \triangleq \left(\int_\Omega |x(\omega)|^p \, d\mu(\omega)\right)^{1/p}.$$

If addition of functions and multiplication are defined in the obvious way (pointwise) then this space is a Polish vector space, with $d(x, y) \triangleq \|x - y\|_p$.

Definition A.81. *Let* x *be a random variable on a probability space* $(\Omega, \mathcal{F}, \mu)$. *Its distribution function* F_x *is defined by* $F_x(t) \triangleq \mathbb{P}(x \le t)$. *Its distribution is a probability measure on* $(\mathbb{R}, \mathcal{B}(\mathbb{R}))$ *defined through* $F_x(B) \triangleq \mathbb{P}(x \in B)$ *for all Borel sets* B. *We say* x *has a density (or* F_x *has a density) if there exists a (measurable) function* $f_x(y)$ *so that for all Borel sets* B,

$$\mathbb{P}(x \in B) = \int_\mathbb{R} \mathbf{1}[B](y) f_x(y) \, dy.$$

If $\vec{x}$ *is an abstract random variable with values in* $(\mathcal{X}, d)$ *then its distribution is a probability measure on* $(\mathcal{X}, \mathcal{B}(\mathcal{X}))$, *defined in the same way.*

Definition A.82. *The empirical distribution of the sample mean of a sequence of random variables* $x_1, x_2, \ldots$ *is*

$$F_n(y) := \frac{1}{n} \sum_{i=1}^n \mathbf{1}[x_i \le y]. \tag{A.6}$$

Let $\vec{x}_1, \vec{x}_2, \ldots$ *be sequence of random variables with values in some metric space* $\mathcal{X}$. *The corresponding empirical measure* μ_n *of a measurable set* A *is then*

$$\mu_n(A) := \frac{1}{n} \sum_{i=1}^n \mathbf{1}[x_i \in A]. \tag{A.7}$$

In terms of the Dirac delta function: $\mu_n := \frac{1}{n} \sum_{i=1}^n \delta_{x_i}$.

We need some special properties for real-valued functions.

Definition A.83. *A real-valued function* f *on* $\mathbb{R}$ *is called absolutely continuous if for any* $\delta > 0$ *there exists an* $\varepsilon > 0$ *so that the following holds. If* $\{[s_i, t_i], i = 1 \ldots I\}$ *is a finite collection of nonoverlapping intervals, then*

$$\sum_{i=1}^I (t_i - s_i) < \varepsilon \quad \text{implies} \quad \sum_{i=1}^I |f(t_i) - f(s_i)| < \delta.$$

A function on $\mathbb{R}$ with values in $\mathbb{R}^d$ is called absolutely continuous if each of its components is absolutely continuous.

Definition A.84. A function f on $\mathbb{R}$ with values in $\mathbb{R}^d$ has a derivative α at a point t if

$$\lim_{\varepsilon \to 0} \frac{f(t + \varepsilon) - f(t)}{\varepsilon} = \alpha,$$

and in this case we denote the value α by $f'(t)$ or by $df(t)/dt$. If the derivative exists at each t, we call the function differentiable. It is differentiable a.e. if the set of points where it is not differentiable has (Lebesgue) measure zero. In this case its derivative f' is only defined Lebesgue-a.e.

Theorem A.85. A function f is absolutely continuous if and only if for all $t > 0$,

$$f(t) = \int_0^t \frac{df(u)}{du}\, du + f(0).$$

In particular, an absolutely continuous function is differentiable Lebesgue-a.e.

Definition A.86. Let $f_1, f_2, \ldots$ be a sequence of measurable functions on the space $(\Omega, \mathcal{F}, \mu)$ with values in $(\mathcal{X}, d)$. We say $f_n \to f$ a.s. (almost surely) if $\mu\big(\omega : f_n(\omega) \not\to f(\omega)\big) = 0$.

Definition A.87. Let $\vec{x}_1, \vec{x}_2, \ldots$ be a sequence of random variables with values in $(\mathcal{X}, d)$. We say $\vec{x}_n \to \vec{x}$ w.p. 1 if $\mathbb{P}\big(\omega : \vec{x}_n(\omega) \to \vec{x}(\omega)\big) = 1$.
We say $\vec{x}_n \to \vec{x}$ in probability if $\mathbb{P}\big(\omega : d(\vec{x}_n(\omega), \vec{x}(\omega)) > \varepsilon\big) \to 0$ for every $\varepsilon > 0$. When x_i are real-valued, we say $x_n \to x$ in distribution if $F_{x_n}(t) \to F_x(t)$ at every point that F_x is continuous. In general, we say $\vec{x}_n \to \vec{x}$ in distribution and denote this as $\vec{x}_n \Rightarrow \vec{x}$ if, for any real-valued, bounded continuous function f, we have $\mathbb{E} f(\vec{x}_n) \to \mathbb{E} f(\vec{x})$. Since this convergence depends only on the distributions, we can say that the distributions converge weakly and, with a slight abuse of notation, that $\vec{x}_1, \vec{x}_2, \ldots$ converges weakly.

Theorem A.88. Under the conditions of Definition A.87, the following implications hold. Convergence w.p. 1 implies convergence in probability which, in turn, implies weak convergence. If the limit $\vec{x}$ is not random then convergence in probability is equivalent to weak convergence. For real valued random variables, convergence in the space L_1 implies convergence in probability.

Definition A.89. Let $\{\vec{x}_\alpha\}$ be a collection of random variables with values in a Polish space $(\mathcal{X}, d)$ and corresponding distributions P_α. The set $\{P_\alpha\}$ of distributions is called tight if, for each $\varepsilon > 0$, there is a compact set K_ε so that

$$P_\alpha(K_\varepsilon) > 1 - \varepsilon \quad \text{for all} \quad \alpha.$$

With a slight abuse of notation, we say in this case that $\{\vec{x}_\alpha\}$ is tight.

Theorem A.90. Let $(\mathcal{X}, d)$ be a Polish space. The collection of probability measures on $(\mathcal{X}, d)$ can be made into a Polish space $(\mathcal{M}(\mathcal{X}), d_w)$, so that a set is com-

pact if and only if it is tight and closed, and convergence in d_w agrees with weak convergence.

The following convergence theorems are used heavily in the text. They can all be found in [Roy].

Theorem A.91. *Let f and $f_1, f_2, \ldots$ be real-valued measurable functions and assume $f_n \to f$ μ-a.s.*

(i) Lebesgue dominated convergence. *If there exists an integrable function g so that μ-a.s. $|f_n| \le |g|$ for each n, then*

$$\lim_{n\to\infty} \int_\Omega |f_n(\omega) - f(\omega)| \, d\mu(\omega) = 0,$$

that is, $f_n \to f$ in the space L_1. In particular, $\lim_{n\to\infty} \int_\Omega f_n \, d\mu = \int_\Omega f \, d\mu$.

(ii) Monotone convergence theorem. *If the functions f_n are non-negative and increasing (in n), then*

$$\lim_{n\to\infty} \int_\Omega f_n(\omega) \, d\mu(\omega) = \int_\Omega f(\omega) \, d\mu(\omega). \tag{A.8}$$

It is possible that one side is infinite—in which case so is the other.

(iii) Bounded convergence theorem. *If $\mu(\Omega) < \infty$ and the $\{f_n\}$ are uniformly bounded then (A.8) holds.*

Exercise A.92. Suppose that $\mu(\Omega) < \infty$, and that f is a positive function on Ω. Assume that for some $\theta = \theta_0 > 0$,

$$\int_\Omega e^{\theta f(\omega)} \, d\mu < \infty.$$

Then for all $0 \le \theta < \theta_0$, the integral is finite and

$$\frac{d}{d\theta} \int_\Omega e^{\theta f(\omega)} \, d\mu = \int_\Omega f(\omega) e^{\theta f(\omega)} \, d\mu.$$

Hint: use dominated convergence and the definition of derivative. ♠

Theorem A.93 (Fatou). *Let f and $f_1, f_2, \ldots$ be real-valued non-negative measurable functions. Then*

$$\int_\Omega \left(\liminf_{n\to\infty} f_n(\omega) \right) d\mu(\omega) \le \liminf_{n\to\infty} \int_\Omega f_n(\omega) \, d\mu(\omega).$$

Definition A.94. *A collection* $\{f_\alpha\}$ *of real integrable functions is called uniformly integrable if*

$$\sup_\alpha \int_\Omega |f_\alpha(\omega)| \mathbf{1}\,[|f_\alpha(\omega)| > N]\,d\mu(\omega) \to 0 \quad \text{as} \quad N \to \infty.$$

The point is that this convergence is uniform over all functions in this collection. As a consequence, the integral converges to zero if we replace the indicator $\mathbf{1}\,[|f_n(\omega)| > N]$ with $\mathbf{1}\,[A_N]$, where $\{A_n\}$ are any measurable sets with $\lim_{n\to\infty} \mu(A_N) = 0$.

Exercise A.95. Let $\{f_n\}$ be uniformly integrable and suppose μ is finite, that is $\mu(\Omega) < \infty$. Show that, if $f_n \to f$ and f is integrable, then

$$\int_\Omega f_n(\omega)\,d\mu \to \int_\Omega f(\omega)\,d\mu.$$

Hint: first approximate by truncating f above N and below $(-N)$, then use uniform integrability to show the error is small as $N \to \infty$.   ♠

Exercise A.96. Let g be a real function so that $t = o(g(t))$ as $t \to \infty$. Then

$$\sup_\alpha \int_\Omega g\big(|f_\alpha(\omega)|\big)d\mu(\omega) < \infty \quad \text{implies} \quad \{f_\alpha\} \text{ is uniformly integrable.}$$ ♠

Theorem A.97 (Fubini, Tonelli). *Let* $f(\omega_1, \omega_2)$ *be an integrable function on* $\Omega_1 \times \Omega_2$. *Then with the notation of Definition A.72*

$$\int_{\Omega_1\times\Omega_2} f(\omega_1, \omega_2)\,d\mu(\omega_1, \omega_2) = \int_{\Omega_1}\left(\int_{\Omega_2} f(\omega_1, \omega_2)\,d\mu_2(\omega_2)\right)d\mu_1(\omega_1).$$

Remark. This is just part of what is usually called "Fubini's Theorem." In addition, this theorem applies to product Lebesgue spaces, and in general to spaces where the notion of completeness is required. See, e.g., [Roy §12.4].

Theorem A.98. *Let* f *and* g *be measurable, positive functions on the measure space* $(\Omega, \mathcal{F}, \mu)$. *Let* $p > 1$ *and define* $1/q = 1 - 1/p$. *Then Holder's inequality states:*

$$\int_\Omega f(\omega)g(\omega)\,d\mu \le \left(\int_\Omega f^p(\omega)\,d\mu\right)^{1/p}\cdot\left(\int_\Omega g^q(\omega)\,d\mu\right)^{1/q}$$

and Minkowski's inequality is:

$$\left(\int_\Omega (f(\omega) + g(\omega))^p\,d\mu\right)^{1/p} \le \left(\int_\Omega f^p(\omega)\,d\mu\right)^{1/p} + \left(\int_\Omega g^p(\omega)\,d\mu\right)^{1/p}$$

If $p = 2$ (so that $q = 2$ also), the first is the Schwarz inequality.

Probabilistic notions.

In the rest of this section we consider only probability measures.

Definition A.99. *The collection of sets*

$$\sigma(x) \triangleq \left\{ x^{-1}(B) : B \in \mathcal{B}(\mathbb{R}) \right\}$$

is called the σ-field generated by the random variable x.

Definition A.100. *Events A and B are called statistically independent or simply independent if $\mathbb{P}(A \cap B) = \mathbb{P}(A) \cdot \mathbb{P}(B)$. The σ-fields $\mathcal{G}_1$ and $\mathcal{G}_2$ are independent if every pair of events $A \in \mathcal{G}_1$ and $B \in \mathcal{G}_2$ is independent. Random variables $\vec{x}$ and $\vec{y}$ are called statistically independent if $\sigma(\vec{x})$ and $\sigma(\vec{y})$ are independent.*

Definition A.101 Bayes' Rule. *The conditional probability of the event C given the event A, where we assume $\mathbb{P}(A) > 0$, is defined to be*

$$\mathbb{P}(C \mid A) \triangleq \frac{\mathbb{P}(C \cap A)}{\mathbb{P}(A)}.$$

Let $\mathcal{G}$ be a sub σ-field of $\mathcal{F}$, denoted $\mathcal{G} \subset \mathcal{F}$: this means that $\mathcal{G}$ is a σ-field, and every set in $\mathcal{G}$ is also in $\mathcal{F}$.

Definition A.102. *Let x be an integrable random variable. The conditional expectation of x given a σ-field $\mathcal{G} \subset \mathcal{F}$, denoted $\mathbb{E}(x \mid \mathcal{G})$, is any random variable measurable with respect to $\mathcal{G}$, so that*

$$\int_A \mathbb{E}(x \mid \mathcal{G}) \, d\mathbb{P}(\omega) = \int_A x \, d\mathbb{P}(\omega) \quad \text{for all} \quad A \text{ in } \mathcal{G}.$$

If $\sigma(y) = \mathcal{G}$ then we denote also $\mathbb{E}(x \mid \mathcal{G}) = \mathbb{E}(x \mid y)$.

A conditional expectation is unique, but only up to changes on sets of probability zero.

Definition A.103. *The conditional probability of an event C given a σ-field $\mathcal{G} \subset \mathcal{F}$ is defined through $\mathbb{P}(C \mid \mathcal{G}) \triangleq \mathbb{E}(\mathbf{1}[C] \mid \mathcal{G})$.*

Exercise A.104. Let A and C be events. Reconcile Definitions A.101 and A.103 by showing that

$$\mathbb{P}(C \mid A)\mathbf{1}[A] = \mathbb{P}(C \mid \mathbf{1}[A]) \quad \text{w.p. 1}.$$

Show that w.p. 1, $\mathbb{E}(x \mid \mathbf{1}[\Omega]) = \mathbb{E}(x)$ for any integrable random variable x. ♠

Definition A.105. *Let $\mathcal{G}$ be a sub σ-field. $P(\cdot \mid \mathcal{G})$ is called a regular conditional probability given $\mathcal{G}$ if*

(i) $P(A \mid \mathcal{G}) = \mathbb{P}(A \mid \mathcal{G})$ $\mathbb{P}$-a.s., for each $A \in \mathcal{F}$.

(ii) For each ω, $P(\cdot \mid \mathcal{G})$ is a probability measure on $(\Omega, \mathcal{F})$.

Definition A.106. Let y be a random variable with range in $(\mathcal{X}, \mathcal{B}(\mathcal{X}))$ and $\mathcal{G}$ a sub σ-field of $\mathcal{F}$. $P_y(\cdot \mid \mathcal{G})$ is called a regular conditional distribution of y given $\mathcal{G}$ if

(i) $P_y(A \mid \mathcal{G}) = \mathbb{P}(y \in A \mid \mathcal{G})$ $\mathbb{P}$-a.s., for each $A \in \mathcal{B}(\mathcal{X})$.

(ii) For each ω, $P_y(\cdot \mid \mathcal{G})$ is a probability on $(\mathcal{X}, \mathcal{B}(\mathcal{X}))$.

Theorem A.107. Let $\vec{y}$ be a random variable with range in $\mathbb{R}^d$ and $\mathcal{G}$ a sub σ-field of $\mathcal{F}$. Then there exists a regular conditional distribution $P_{\vec{y}}(\cdot \mid \mathcal{G})$ of $\vec{y}$ given $\mathcal{G}$. Let f be a Borel measurable function from $\mathbb{R}^d$ to $\mathbb{R}^r$ so that $\mathbb{E}|f(\vec{y})| < \infty$. Then

$$\mathbb{E}(f(\vec{y}) \mid \mathcal{G}) = \int_{\mathbb{R}^d} f(\vec{x}) P_{\vec{y}}(d\vec{x} \mid \mathcal{G}) \quad \mathbb{P} - a.s.$$

Theorem A.108. Let x be a random variable with range in $(\mathcal{X}, \mathcal{B}(\mathcal{X}))$ and let y be a random variable with range in $(\mathcal{X}_1, \mathcal{B}(\mathcal{X}_1))$. There exists a Borel function ϕ from $(\mathcal{X}, \mathcal{B}(\mathcal{X}))$ to $(\mathcal{X}_1, \mathcal{B}(\mathcal{X}_1))$ so that

$$\mathbb{E}(y \mid x) = \phi(x) \quad \mathbb{P}\text{-a.s.}$$

Define $\mathbb{E}(y \mid x = \alpha) \stackrel{\triangle}{=} \phi(\alpha)$. Then

$$\int_B \mathbb{E}(y \mid x = \alpha) \, dP_x(\alpha) = \int_\Omega \mathbf{1}[x \in B] \, y(\omega) \, d\mathbb{P}(\omega) \quad \text{for all} \quad B \in \mathcal{B}(\mathcal{X}).$$

Thus conditional expectation is a form of an integral. A common pitfall is that regular conditional probability may not exist, since the number of null sets in the definition may be so large that they accumulate to an event with non-zero probability. This does not happen with distributions on $\mathbb{R}^d$: [Br §4.3].

Theorem A.109. Let x and y be real, integrable random variables and $\mathcal{G} \subset \mathcal{F}$ a σ-field. Then

(i) Linearity. For any scalars a and b we have ($\mathbb{P}$-a.s.)

$$\mathbb{E}(a \cdot x + b \cdot y \mid \mathcal{G}) = a\mathbb{E}(x \mid \mathcal{G}) + b\mathbb{E}(y \mid \mathcal{G}).$$

(ii) If x is nonnegative then $\mathbb{E}(x \mid \mathcal{G}) \geq 0$ w.p. 1.

Theorem A.110 (Smoothing). Let x be a real, integrable random variable, and let $\mathcal{G}_1$, $\mathcal{G}_2$ be σ-fields with $\mathcal{G}_1 \subset \mathcal{G}_2 \subset \mathcal{F}$. Then

$$\mathbb{E}\big(\mathbb{E}(x \mid \mathcal{G}_1) \mid \mathcal{G}_2\big) = \mathbb{E}\big(\mathbb{E}(x \mid \mathcal{G}_2) \mid \mathcal{G}_1\big) = \mathbb{E}(x \mid \mathcal{G}_1) \quad \text{w.p. 1.}$$

In particular, by Exercise A.104, $\mathbb{E}\big(\mathbb{E}\,(x \mid \{\Omega, \emptyset\})\big) = \mathbb{E}\,(x)$.

Theorem A.111. *Let x and $x_1, x_2, \ldots$ be nonnegative random variables and assume $x_n \uparrow x$ a.s., that is x_n is increasing in n and converges to x. If $\mathbb{E}x < \infty$ then for any sub σ-field $\mathcal{G}$,*

$$\lim_{n \to \infty} \mathbb{E}(x_n \mid \mathcal{G}) \to \mathbb{E}(x \mid \mathcal{G}) \quad \mathbb{P}\text{-a.s.}$$

Theorem A.112. *Let $x_1, x_2, \ldots$ be a sequence of i.i.d. (independent identically distributed) random variables. Then the strong law of large numbers states that*

$$\frac{x_1 + \cdots + x_n}{n} \to \mathbb{E}x_1 \quad \text{w.p.1 provided} \quad \mathbb{E}|x_1| < \infty.$$

The central limit theorem states that if $\mathbb{E}x_1^2 < \infty$ then the distribution of

$$\frac{1}{\sqrt{n}} \sum_{i=1}^{n} \left(\frac{x_i - \mathbb{E}x_i}{\sqrt{\mathbb{E}x_1^2}} \right)$$

converges to the distribution of a standard normal random variable.

The following theorem is sometimes called Chebycheff's Inequality, in particular when $f(x) = x^2$ and x is positive.

Theorem A.113 (Markov inequality). *Let f be a non-negative, monotone non-decreasing function and x a random variable. Then for any number a such that $f(a) > 0$,*

$$\mathbb{P}\,(x \geq a) \leq \frac{\mathbb{E}f(x)}{f(a)}. \tag{A.9}$$

Theorem A.114 (Jensen's Inequality). *Let f be a convex function and x a random variable. Then, provided the right-hand side is well-defined,*

$$f(\mathbb{E}x) \leq \mathbb{E}f(x). \tag{A.10}$$

If f is strictly convex and x is non-degenerate (i.e., not a constant) then the inequality is strict. Moreover, for any set A,

$$f\left(\frac{1}{\mathbb{P}(A)} \mathbb{E}\big(\mathbf{1}\,[A] \cdot x\big) \right) \leq \frac{1}{\mathbb{P}(A)} \mathbb{E}\big(\mathbf{1}\,[A] \cdot f(x)\big). \tag{A.11}$$

Let $\mathcal{G}$ be any sub σ-field. Then, provided the right-hand side is well-defined,

$$f\big(\mathbb{E}(x \mid \mathcal{G})\big) \leq \mathbb{E}\big(f(x) \mid \mathcal{G}\big) \quad \mathbb{P}\text{-a.s.}$$

Lemma A.115 (Union bound). *Let $\{A_1, A_2, \ldots\}$ be events. Then*

$$\mathbb{P}\left(\bigcup_{i=1}^{\infty} A_i \right) \leq \sum_{i=1}^{\infty} \mathbb{P}(A_i). \tag{A.12}$$

Lemma A.116 (Borel-Cantelli Lemma). *Let* $\{A_1, A_2, \ldots\}$ *be events. Let* $B \overset{\triangle}{=} \{\omega : \omega \in A_n$ *for infinitely many* $n\}$. *Then*

$$\sum_{i=1}^{\infty} \mathbb{P}(A_i) < \infty \quad implies \quad \mathbb{P}(B) = 0.$$

If in addition the events are independent, then

$$\sum_{i=1}^{\infty} \mathbb{P}(A_i) = \infty \quad implies \quad \mathbb{P}(B) = 1.$$

A.4. Radon-Nikodym Derivatives

Let $\mathbb{P}_0$ and $\mathbb{P}_1$ be probability measures on $\{\Omega, \mathcal{F}\}$.

Theorem A.117. *There exist random variable* $a(\omega)$ *and event* $N \in \mathcal{F}$ *so that*
(i) $\mathbb{P}_0(N) = 0$,
(ii) *For any event* $A \in \mathcal{F}$,

$$\mathbb{P}_1(A) = \int_A a(\omega)\, d\mathbb{P}_0(\omega) + \mathbb{P}_1(A \cap N)$$

which can also be written as

$$\mathbb{P}_1(A) = \mathbb{E}_0\big(a(\omega)\mathbf{1}\,[A]\big) + \mathbb{P}_1(A \cap N).$$

(iii) *If in addition* x *is any random variable so that* $\mathbb{E}_1|x| < \infty$, *then*

$$\mathbb{E}_1 x \overset{\triangle}{=} \int_{\Omega} x(\omega)\, d\mathbb{P}_1(\omega) = \int_{\Omega \setminus N} a(\omega)x(\omega)\, d\mathbb{P}_0(\omega) + \int_N x(\omega)\, d\mathbb{P}_1(\omega).$$

The random variable $a(\omega)$ is called the Radon-Nikodym derivative of $\mathbb{P}_1$ with respect to $\mathbb{P}_0$, and the standard notation is

$$a \overset{\triangle}{=} \frac{d\mathbb{P}_1}{d\mathbb{P}_0}.$$

Example A.118. Let z be any positive random variable with finite mean $\bar{z}$, and $\mathbb{P}_0$ any probability measure. Define the probability measure $\mathbb{P}_1$ through

$$\mathbb{P}_1(A) \overset{\triangle}{=} \bar{z}^{-1}\mathbb{E}_0\big(z\mathbf{1}\,[A]\big).$$

Then $N = \emptyset$ and $a = z/\bar{z}$.

Exercise JFF A.119. If $\mathbb{P}_0$ is the distribution of z, then the measure $\mathbb{P}_1$ defines in Example A.118 is called the renewal distribution associated with $\mathbb{P}_0$. Show that when $\mathbb{P}_0$ is exponential, i.e., $\mathbb{P}_0(z > t) = e^{-\lambda t}$ then $\mathbb{P}_1 = \mathbb{P}_0$. Compute $\mathbb{P}_1$ in the case $\mathbb{P}_0$ is uniform on $[0, 1]$. Show that if $\mathbb{P}_0$ has a density and $\mathbb{P}_1 = \mathbb{P}_0$ then $\mathbb{P}_0$ is exponential. ♠

Example A.120. Consider the discrete case; $\Omega = \{1, 2, \ldots, d\}$. Fix the probabilities $\mathbb{P}_0(i) \overset{\triangle}{=} p_i$, $i = 1, 2, \ldots, d$ and $\mathbb{P}_1(i) \overset{\triangle}{=} q_i$, $i = 1, 2 \ldots, d$ with $p_d = 0$ and $p_i \neq 0$ otherwise, and with $q_1 = 0$, and $q_i \neq 0$ otherwise. Then from Theorem A.117(i) we obtain $N \subset \{d\}$. The choice $A = \{d\}$ in Theorem A.117(ii) now gives $N = \{d\}$. For $1 \leq i \leq d$ let $x = \mathbf{1}[\omega = i]$. Setting $i = 1$ we get $0 = a(1)p_1$, so that $a(1) = 0$. Finally, with $1 < i < d$ we have $\mathbb{E}_1 x = q_i = a(i)p_i$ so that the Radon-Nikodym derivative is

$$a(i) = \frac{q_i}{p_i}, \quad 1 \leq i \leq d - 1.$$

As might be expected, the value of a is undefined (and indeed is of no consequence) on N.

Example A.121. Let the distribution functions F_0 and F_1 have densities f_i. Assume that the densities have the same support; i.e., they both vanish at the same places. Then for every bounded function g we have by definition

$$\int g(x)\,dF_1(x) \overset{\triangle}{=} \int g(x)f_1(x)\,dx$$

$$= \int \left(\frac{f_1(x)}{f_0(x)}\right) g(x) f_0(x)\,dx$$

$$= \int \left(\frac{f_1(x)}{f_0(x)}\right) g(x)\,dF_0(x)$$

so that the Radon-Nikodym derivative is

$$a(x) = \frac{f_1(x)}{f_0(x)}.$$

That is, dF_1/dF_0 in the Radon-Nikodym sense is equal to dF_1/dF_0 in the usual sense of calculus, which is

$$\frac{dF_1/dx}{dF_0/dx}.$$

A.5. Stochastic Processes

Most of the material in this section can be found in Breiman [Br], and all can be found in the more advanced (and more detailed) book by Liptser and Shiryayev [LiS]. We shall consider only processes on $[0, \infty)$ that are either right-continuous (w.p. 1) (Definition A.52), or are discrete-time processes, and take values is $\mathbb{R}^d$: only these arise in this book. This will also obviate a (rather large) number of measurability issues that arise otherwise. Note that the assumption of right continuity cannot be dispensed with, but will not be restated below.

Definition A.122. We say that $\vec{x} = \{\vec{x}(t), \ 0 \le t < \infty\}$ is a stochastic process on $(\Omega, \mathcal{F}, \mathbb{P})$ with values in $\mathbb{R}^d$ if $\vec{x}(t)$ is a random variable for each t. It is called a measurable stochastic process if it is (jointly) measurable as a function of the arguments (ω, t).

Definition A.123. A filtration $\{\mathcal{F}_t, \ t \ge 0\}$ on $(\Omega, \mathcal{F})$ is a sequence of increasing σ-fields (that is, $\mathcal{F}_t \subset \mathcal{F}_s$ if $s > t$), all of which are sub σ-fields of $\mathcal{F}$. The filtration generated by the process $x(t)$ is $\mathcal{F}_t^x \stackrel{\Delta}{=} \sigma\{x(s), \ s \le t\}$. A filtration $\{\mathcal{F}_t\}$ is called right-continuous if

$$\mathcal{F}_t = \mathcal{F}_{t+} \stackrel{\Delta}{=} \bigcap_{\varepsilon > 0} \mathcal{F}_{t+\varepsilon}. \tag{A.13}$$

Definition A.124. A random variable τ is a stopping time for the process $x(t)$ if for every $t \ge 0$, $\{\tau \le t\} \in \mathcal{F}_t^x$. We allow $\tau = \infty$ with positive probability. The stopped σ-field, denoted $\mathcal{F}_\tau$, is the sub σ-field of $\mathcal{F}$

$$\mathcal{F}_\tau \stackrel{\Delta}{=} \big\{A : A \in \mathcal{F}, \ A \cap \{\tau \le t\} \in \mathcal{F}_t \quad \text{for all} \quad t \ge 0\big\}.$$

Theorem A.125. If τ_1 and τ_2 are both stopping times then so is $\tau = \min\{\tau_1, \tau_2\}$. The degenerate random variable $\tau = c$, a constant, is a stopping time. If x is a measurable process then $x\big(\min\{t, \tau\}\big)$ is a measurable process and, at a fixed t, is measurable with respect to $\mathcal{F}_{\min\{t,\tau\}}$. If in addition $\tau < \infty$ a.s. then $x(\tau)$ is a random variable.

Definition A.126. The pair $(x(t), \mathcal{F}_t)$, where x is a stochastic process and $\mathcal{F}_t$ is a filtration, is called a martingale if, for each t, $x(t)$ is a random variable on $(\Omega, \mathcal{F}_t)$ (i.e., it is measurable with respect to $\mathcal{F}_t$), is integrable and

$$\mathbb{E}(x(t) \mid \mathcal{F}_s) = x(s) \quad \mathbb{P}\text{-a.s.} \quad \text{for all} \quad s < t.$$

$\mathcal{F}_t$ is often deleted from the notation. If $x(t)$ is real-valued and

$$\mathbb{E}(x(t) \mid \mathcal{F}_s) \ge x(s) \quad \mathbb{P}\text{-a.s.} \quad \text{for all} \quad s < t$$

then $x(t)$ is called a submartingale. If the inequality is reversed ($\le$) then $x(t)$ is called a supermartingale.

Exercise A.127. Establish the following. If $(x(t), \mathcal{F}_t)$ is a martingale then so is $(x(t), \sigma\{x(s), \ s \le t\})$, but the converse may not be true. If both x and $(-x)$ are submartingales, then x is a martingale. ♠

Theorem A.128 (Martingale convergence theorem). *Let* $(x(t), \mathcal{F}_t)$ *be a right-continuous real-valued submartingale. If*

$$\sup_{0 \le t} \mathbb{E}\big[\max\{0, x(t)\}\big] < \infty$$

then there exists a random variable x_∞ *so that* $x(t) \to x_\infty$ *a.s. as* $t \to \infty$. *Assume in addition that* $\{x(t),\ t \ge 0\}$ *is uniformly integrable. Then* $\mathbb{E}|x(t) - x_\infty| \to 0$ *as* $t \to \infty$, *and if* x *is in fact a martingale then* $x(t) = \mathbb{E}(x_\infty \mid \mathcal{F}_t)$ *a.s. for all* t.

Theorem A.129 (Optional sampling) [LiS pp. 60–61, EK p. 61]. *Let* $(x(t), \mathcal{F}_t)$ *be a submartingale and let* τ_1 *and* τ_2 *be stopping times with* $\tau_1 \le \tau_2$ *a.s. Then for each finite* t,

$$\mathbb{E}\big[x(\min\{t, \tau_2\}) \mid \mathcal{F}_{\min\{t,\tau_1\}}\big] \ge x(\min\{t, \tau_1\}) \quad \text{a.s.}$$

If $\tau_2 < \infty$ *a.s.,* $\mathbb{E}|x(\tau_2)| < \infty$, *and*

$$\lim_{t \to \infty} \mathbb{E}\big[x(t)\mathbf{1}\,[\tau_2 > t]\big] = 0, \tag{A.14}$$

then

$$\mathbb{E}\big[x(\tau_2) \mid \mathcal{F}_{\tau_1}\big] \ge x(\tau_1) \quad \text{a.s.}$$

If x *is uniformly integrable then so is* $x(\min\{t, \tau_1\})$.

Exercise A.130. *Let* $(x(t), \mathcal{F}_t)$ *be a martingale and* τ *a stopping time. If* (A.14) *holds then* $\big(x(\min\{t, \tau\}), \mathcal{F}_{\min\{t,\tau\}}\big)$ *is a martingale, so that* $\mathbb{E}x(\tau) = \mathbb{E}x(0)$. ♠

Theorem A.131 (Martingale inequality) [Br §14.3]. *If* $x(t)$ *is a right-continuous real-valued martingale on* $[0, T]$, *then*

$$\mathbb{P}\left(\sup_{0 \le t \le T} |x(t)| \ge a\right) \le \frac{\mathbb{E}|x(T)|}{a}. \tag{A.15}$$

Definition A.132. *A pure jump process is a right-continuous, piecewise constant process.*

Definition A.133. *A point process can be described in the following equivalent ways.*

(i) *A sequence of time points* $\{t_1, t_2, \ldots\}$ *so that each* t_i *is a random variable and* $t_{i+1} \ge t_i$ *a.s. These points represent the times the events occurred.*

(ii) *The associated counting process: a pure jump, integer-valued, nondecreasing process* $N(t)$, *so that*

$$N(t) \stackrel{\triangle}{=} \{\text{number of points } t_i \text{ so that } t_i \le t\}.$$

Definition A.134. *A stochastic process with values in* $\mathbb{R}^d$ *is called a Markov process if for any Borel set* B *in* $\mathbb{R}^d$ *and any* $t \ge s \ge 0$,

$$\mathbb{P}\left(x(t) \in B \mid \mathcal{F}_s^x\right) = \mathbb{P}\left(x(t) \in B \mid x(s)\right) \quad \text{a.s.} \tag{A.16}$$

More generally, let μ be a given (initial) distribution and $\{\Omega, \mathcal{F}, \mathbb{P}_\mu\}$ a probability space. The process $\{x_t, \mathcal{F}_t, t \geq 0\}$ is called a Markov process with initial distribution μ if (A.16) holds with $\mathcal{F}_t$ replacing $\mathcal{F}_t^x$, and in addition $\mathbb{P}(x(0) \in B) = \mu(B)$.

In naming $\{x_t, \mathcal{F}_t, t \geq 0\}$ a process we imply that x_t is adapted, that is, for each t, the random variable x_t is measurable with respect to $\mathcal{F}_t$; however, $\mathcal{F}_t$ could be strictly larger than $\mathcal{F}_t^x$. We need to treat simultaneously all possible starting positions for the Markov process, especially in order to define below the strong Markov property. For that, we need the notion of universal measurability: for our purposes it suffices to note that any Borel measurable function is also universally measurable. Moreover, when the state spaces of interest are discrete and countable (as in most cases of interest to us), the whole question of measurability can be ignored. Nonetheless, let us state the precise definitions.

Definition A.135. *A Markov family is an adapted process $\{x_t, \mathcal{F}_t, t \geq 0\}$ together with a family of probabilities $\mathbb{P}_x$, with $x \in \mathbb{R}^d$, so that (A.16) holds, and in addition, for all $x \in \mathbb{R}^d$, Borel sets B of $\mathbb{R}^d$ and $F \in \mathcal{F}$, and all $t \geq s \geq 0$,*

$$\mathbb{P}_x(F) \text{ is a universally measurable function of } x, \tag{A.17}$$
$$\mathbb{P}_x(x(0) = x) = 1, \tag{A.18}$$
$$\mathbb{P}_x\left(x(t) \in B \mid \mathcal{F}_s\right) = \mathbb{P}_x\left(x(t) \in B \mid x(s)\right) \quad \mathbb{P}_x\text{-a.s.}$$
$$\mathbb{P}_x\left(x(t) \in B \mid x(s) = y\right) = \mathbb{P}_y\left(x(t - s) \in B\right) \quad \text{a.s.,}$$

that is, the last equality holds outside a set $Y \subset \mathbb{R}^d$ so that $\mathbb{P}_x(x(s) \in Y) = 0$.

Definition A.136. *Let $\{x_t, \mathcal{F}_t, t \geq 0\}$ be a right-continuous process with a right-continuous filtration (Definition A.123). It is called a strong Markov process with initial distribution μ if, for any Borel set B of $\mathbb{R}^d$, $\mathbb{P}(x(0) \in B) = \mu(B)$ and, for every stopping time τ (of $\mathcal{F}_t$),*

$$\mathbb{P}\left(x(t + \tau) \in B \mid \mathcal{F}_\tau\right) = \mathbb{P}\left(x(t + \tau) \in B \mid x(\tau)\right) \quad \mathbb{P}\text{-a.s.}$$

Such a process together with a family of probabilities $\mathbb{P}_x$ is a strong Markov family if, in addition, (A.17) and (A.18) hold, and

$$\mathbb{P}_x\left(x(t + \tau) \in B \mid x(\tau) = y\right) = \mathbb{P}_y\left(x(\tau) \in B\right) \quad \text{a.s.,}$$

that is, the last equality holds outside a set $Y \subset \mathbb{R}^d$ so that $\mathbb{P}_x(x(\tau) \in Y) = 0$.

We have avoided quite a few technicalities by assuming that all sample paths of the process are right continuous, and that the filtration is right continuous.

Theorem A.137. *Let $\{x_t, \mathcal{F}_t, t \geq 0\}$ satisfy the right continuity condition of Definition A.136. Then it is a strong Markov family if and only if for all stopping* *times τ, all $t \geq 0$, all x and Borel sets B of $\mathbb{R}^d$, one of the following holds:*

$$\mathbb{P}_x \left(x(t + \tau) \in B \mid \mathcal{F}_\tau \right) = \mathbb{P}_{x(\tau)}\left(x(t) \in B \right), \qquad (A.19)$$

$$\mathbb{E}_x \left(f(x(t + \tau)) \mid \mathcal{F}_\tau \right) = \mathbb{E}_{x(\tau)} f(x(t)) \qquad (A.20)$$

for all bounded continuous functions f.

In fact, we can avoid most of this pain by noting the following.

Theorem A.138 [Br Prop. 15.25 p. 328]. *A pure jump Markov process is strong Markov.*

Thus the processes we investigate in Chapters 4–8 as well as in most of the applications are all strong Markov processes. In fact, our processes are even simpler: they are continuous-time countable-state Markov processes.

Definition A.139. *A jump Markov process is called countable or finite according as the number of possible values it takes is countable or finite.*

Countable or finite jump Markov processes are defined through (4.4), §4.2. The *incidence matrix $\mathcal{I}$ of the process is then the matrix whose entry $\mathcal{I}_{ij}$ equals one when $\lambda_{ij} > 0$ and is zero otherwise.*

Definition A.140. *A countable or finite jump Markov process $x(t)$ is called irreducible if, for some state y,*

$$\mathbb{P}_x \left(\inf \{ t > 0 : x(t) = y \} < \infty \right) = 1$$

for all x. It is called ergodic if for each pair of states x, y,

$$\mathbb{E}_x \inf \{ t > 0 : x(t) = y \} < \infty.$$

Note that irreducibility can be determined from the incidence matrix. For processes in discrete time or when the state space is not countable, the corresponding definition is much more involved.

Definition A.141. *For a jump Markov process on $\{1, \ldots, D\}$ with rates λ_{mj} for jump from m to j, define $\vec{\pi}^\lambda = \left(\pi_1^\lambda, \ldots, \pi_D^\lambda \right)$ as its invariant probability measure (provided it exists). That is, π_m^λ satisfies*

$$\pi_m^\lambda \sum_{j \neq m} \lambda_{mj} = \sum_{j \neq m} \pi_j^\lambda \lambda_{jm}, \quad \pi_m^\lambda \geq 0, \quad \sum_{m=1}^{D} \pi_m^\lambda = 1.$$

That is, under this distribution on the states, the rate of transitions out of each state equals the rate of transitions into that state: these are "balance equations."

Theorem A.142. *If $x(t)$ is an ergodic countable (or finite) jump Markov process then it has a unique invariant probability distribution. If $x(t)$ is a finite (not necessarily ergodic) jump Markov process then it has at least one invariant probability distribution (but it may not be unique).*

Appendix B

Discrete-Space Markov Processes

by Robert J. Vanderbei

In this section we introduce the basic concepts of continuous-time Markov processes with discrete state space E. We assume that the reader has some familiarity with Markov processes and that this section serves mostly as a review. A good reference for this material is Dynkin [Dy].

B.1. Generators and Transition Semigroups

In the discrete-space setting, Markov processes are especially easy to grasp. Basically, if the process is at some state $x \in E$, then it will sit there until an exponential clock ticks, at which time it will jump to a new state y. The exponential clock that controls jumps from x to y ticks at rate $a(x, y) \geq 0, x \neq y$. To make an entire array out of $a(\cdot, \cdot)$, we define $a(x, x)$ so that

$$\sum_y a(x, y) = 0. \qquad (B.1)$$

That is, $a(x, x) = -\sum_{y \neq x} a(x, y)$. We let L denote the array that these rates make: $L = [a(x, y)]$. To reduce technicalities, we assume throughout that each row of L has only a finite number of non-zero entries. This array is called the *generator* of the process. We will deal with the generator as a linear operator acting on functions defined on the state space. So, if $f(x)$ is a function on E, then $Lf(x)$ is the function given by

$$Lf(x) = \sum_y a(x, y) f(y) = \sum_y a(x, y)(f(y) - f(x)), \qquad (B.2)$$

where the second formula follows from (B.1).

The transition semigroup.

For discrete-space continuous-time Markov processes, the generator gives the most natural and simplest description of the process. However, there is a related construct that is also important. It is called the *transition semigroup*. It consists of a family of linear operators $T_t = [p_t(x, y)]$, $t \geq 0$, that satisfy the following properties:

$$T_t T_s = T_{t+s} \qquad (B.3)$$

499

$$T_t \geq 0 \qquad\qquad\qquad (B.4)$$
$$T_t 1 = 1 \qquad\qquad\qquad (B.5)$$
$$T_0 = I \qquad\qquad\qquad (B.6)$$
$$\lim_{t \searrow 0} T_t = I \qquad\qquad\qquad (B.7)$$

[where the first condition means that $\sum_y p_t(x, y)p_s(y, z) = p_{t+s}(x, z)$]. Equation (B.4) means the operator T_t maps positive functions to positive functions. Intuitively, the quantity $p_t(x, y)$ represents the probability that at time t the process will be at y given that at time zero it was at x.

There exists a one-to-one correspondence between generators and transition semigroups. Namely, given a generator L, the corresponding semigroup is given by

$$T_t = e^{tL} \overset{\triangle}{=} I + tL + \ldots + \frac{(tL)^n}{n!} + \ldots \qquad\qquad (B.8)$$

Conversely, given a transition semigroup T_t, the generator is

$$L = \lim_{t \searrow 0} \frac{T_t - I}{t}$$

Uniformization.

There is one more representation for discrete-space continuous-time Markov processes that works when the jump rates are uniformly bounded: $\max_x |a(x, x)| < \infty$. Let λ be a positive real number that dominates these rates and put

$$Q = \lambda^{-1}L + I.$$

It is easy to see that Q is nonnegative and has row sums equal to one (i.e., it is a stochastic matrix). A simple calculation shows that

$$T_t = e^{tL} = e^{-\lambda t}e^{\lambda t Q} = \sum_{k=0}^{\infty} e^{-\lambda t}\frac{(\lambda t)^k}{k!}Q^k. \qquad\qquad (B.9)$$

This is called the uniformization theorem. The interpretation of the right-hand side is that there is a single exponential clock that ticks at rate λ and every time it ticks, a jump occurs according to the stochastic matrix Q. Hence, at time t, the number of times this clock has ticked is a Poisson random variable with parameter λt and so the chance that there have been k jumps is the Poisson probability that appears in front of Q^k in (B.9). If this result seems surprising (in particular the fact that λ can be chosen arbitrarily large), it might help to bear in mind that the stochastic matrix Q allows jumps from a state x to itself.

B.2. The Markov Process

So far we have talked about analytical objects that characterize a Markov process but we have yet to introduce the process itself. An actual Markov process consists of several objects. First we need an event space Ω. For discrete-space continuous-time Markov processes, the construction of Ω is quite simple. All we need is a space on which we have defined all the exponential clocks that were mentioned above. Next, we need to construct the trajectories $X_t(\omega)$ of the process. X_t represents the position in the state space at time t. Its construction simply follows the description that we gave above in terms of exponential clocks and jumps between states. Corresponding to each possible starting point $x \in E$, there is a probability measure P_x on Ω such that the probability of being at states $x_1, \ldots, x_k$ at the times $t_1 < t_2 < \ldots < t_k$ given that the process starts in state x is given by

$$P_x(X_{t_1} = x_1, \ldots, X_{t_k} = x_k) = p_{t_1}(x, x_1)p_{t_2-t_1}(x_1, x_2) \ldots p_{t_k-t_{k-1}}(x_{k-1}, x_k).$$

Finally, we need to introduce a family of σ-algebras $\mathcal{F}_t$, $t \geq 0$, representing the events that are observable up to time t. One should think of $\mathcal{F}_t$ as the σ-algebra generated by the random variables X_s, $s \leq t$ (however, to be precise, we want the completed right continuous modification of this family, whatever that means). The process X_t is quite simple in that it remains fixed at a point in E for an interval of time and then it jumps to another point in E. It will turn out to be important that we make the convention that at the instant when a jump occurs, the process is actually at the new point, not the old one. That is, we are assuming that the process is right continuous. It turns out that in the general theory of stochastic process, functions of time should be right continuous with left limits. We call such functions RCLL. All the functions of time that we will consider will be RCLL and in fact will generally be piecewise constant or an integral of a piecewise constant function. An important property of the family of σ-algebras $\mathcal{F}_t$ is that it is increasing: $\mathcal{F}_s \subset \mathcal{F}_t$, $s \leq t$. This captures the intuitive notion that as time progresses, the history of the process increases. Such a family of σ-algebras is called a *filtration*.

The Markov property.

Now that we have a Markov process to work with, we can begin writing probabilistic formulae. For example, the transition semigroup has the following interpretation:

$$T_t f(x) = \mathbb{E}_x f(X_t), \qquad (B.10)$$

where $\mathbb{E}_x$ represents expectations calculated using the measure $\mathbb{P}_x$. As another example, the Markov property which was already visible in (B.3) has the following more sophisticated form:

$$\mathbb{E}_x[f(X_{t+u})|\mathcal{F}_t] = T_u f(X_t). \qquad (B.11)$$

Although this form of the Markov property is sufficient for many purposes, it is not the most general. Indeed, even though we have conditioned on the entire past $\mathcal{F}_t$ in (B.11) we have considered only a very specific type of future event, namely the types that involve only a fixed time in the future. In order to formulate a more all-encompassing version of the Markov property, we need a convenient way to represent future events. To this end, it is convenient to assume that the probability space Ω is actually the space of RCLL paths. That is, a point $\omega \in \Omega$ is a function from $[0, \infty)$ into E that is right continuous and has left limits. Hence, the random variable X_t is simply the evaluation map at time t: $X_t(\omega) = \omega_t$. On this space Ω, we introduce a one-parameter family of shift operators:

$$X_t(\theta_s \omega) = X_{t+s}(\omega), \quad s \in [0, \infty). \tag{B.12}$$

This transformation on Ω induces a transformation on random variables according to the following formula:

$$\theta_s Z(\omega) = Z(\theta_s \omega). \tag{B.13}$$

Using (B.10), (B.12), and (B.13), we see that (B.11) can be written as

$$\mathbb{E}_x[\theta_t f(X_u)|\mathcal{F}_t] = \mathbb{E}_{X_t} f(X_u).$$

A more general form of the Markov property says that we can replace $f(X_u)$ in the above formula by a general random variable:

$$\mathbb{E}_x[\theta_t Z|\mathcal{F}_t] = \mathbb{E}_{X_t} Z, \quad Z \in \mathcal{F}. \tag{B.14}$$

By the definition of conditional expectation, (B.14) is equivalent to

$$\mathbb{E}_x Y \theta_t Z = \mathbb{E}_x Y \mathbb{E}_{X_t} Z, \quad Y \in \mathcal{F}_t, \ Z \in \mathcal{F}. \tag{B.15}$$

Stopping times and the strong Markov property.

We will at times need to apply the Markov Property not at fixed times t, but at random times τ. The property is not true at all random times but it is true if the random time does not look into the future. Such random times are called *stopping times* (Definition A.124). All of the obviously non-clairvoyant random times are actually stopping times.

The strong Markov property says that (B.14) and (B.15) are true even when t is replaced by a stopping time:

$$\mathbb{E}_x[\theta_\tau Z|\mathcal{F}_\tau] = \mathbb{E}_{X_\tau} Z, \quad Z \in \mathcal{F}. \tag{B.16}$$

There is a similar form analogous to (B.15).

Semigroup calculus.

Formula (B.8) says that the generator is the derivative at time zero of the semigroup. Derivatives at later times are almost as simple:

$$\frac{d}{ds} T_s f(x) = \lim_{u \searrow 0} \frac{T_{s+u} f(x) - T_s f(x)}{u} = \lim_{u \searrow 0} T_s \frac{T_u - I}{u} f(x) = T_s L f(x).$$

Hence, the fundamental theorem of calculus gives us the following important semigroup identity:

$$T_t f(x) - f(x) = \int_0^t T_s L f(x) \, ds. \tag{B.17}$$

Written probabilistically, (B.17) becomes

$$\mathbb{E}_x f(X_t) - f(x) = \mathbb{E}_x \int_0^t Lf(X_s)\, ds. \qquad (B.18)$$

Fix a function $q(x)$ defined on the state space and define a family of linear operators $\tilde{T}_t$ according to the following formula:

$$\tilde{T}_t f(x) = \mathbb{E}_x f(X_t) \exp\left(-\int_0^t q(X_u)\, du\right).$$

Using the Markov property (B.15), it is easy to see that $\tilde{T}_t$ satisfies (B.3). In fact, it also satisfies (B.4), (B.6), and (B.7). Hence, it makes sense to define a generator $\tilde{L}$ for $\tilde{T}_t$ just like we would for a transition semigroup. A simple calculation using the fact that

$$\exp\left(-\int_0^t q(X_u)du\right) = 1 - \int_0^t q(X_u)\, du + o(t)$$

shows that

$$\tilde{L} f(x) = Lf(x) - q(x) f(x).$$

Now, given any function $g(x)$, if we pick $q(x) = Lg(x)/g(x)$ and $f(x) = g(x)$, then $\tilde{L} f(x) = 0$ and so (B.17) implies that $\tilde{T}_t g(x) = g(x)$. Rewriting this probabilistically, we get

$$g(x) = \mathbb{E}_x g(X_t) \exp\left(-\int_0^t \frac{Lg(X_u)}{g(X_u)} du\right). \qquad (B.19)$$

This formula can be thought of as the multiplicative counterpart to (B.18).

B.3. Birth-Death Processes

Suppose that $E = \mathbb{Z}^d$, the d-dimensional integer lattice, and that there is a finite collection of step directions e_j, $j = 1, \ldots, m$, and that the transition rates are

$$a(x, y) = \begin{cases} \lambda_j(x) & \text{if } y = x + e_j \text{ for some } j; \\ 0 & \text{otherwise.} \end{cases}$$

In this case the generator takes the following simple form:

$$Lf(x) = \sum_j \lambda_j(x) \left(f(x + e_j) - f(x)\right). \qquad (B.20)$$

Eventually, we will only consider Markov processes that are multidimensional birth-death processes.

For these processes, we introduce a little more notation. Let Y_t^j be the number of times up to and including time t that the process X_t steps in direction e_j. The process Y_t^j is called the *counting process* for the steps in the e_j direction. It is easy to recover X_t from knowledge of all the counting processes:

$$X_t = X_0 + \sum_j Y_t^j e_j.$$

Most Markov processes that arise in queueing models are of this type. For example, the queue length process in an $M/M/1$ queue is the process on the non-negative integers that takes steps to the right at rate λ and steps to the left at rate μ as long as $x > 0$. Hence, for this case, we may take $e_1 = 1, e_2 = -1, \lambda_1 = \lambda$, and $\lambda_2 = \mu 1_{x>0}$. Another example from queueing theory is two queues in tandem. In this case the state space is the non-negative quadrant of Z^2, $e_1 = (1, 0)$ represents arrivals to the first queue which occur at rate $\lambda_1(x) = \lambda_1, e_2 = (-1, 1)$ represents transfers from the first queue to the second which occur at rate $\lambda_2(x) = \lambda_2 1_{x_1>0}$, and $e_3 = (0, -1)$ represents service completions at the second queue which occur at rate $\lambda_3(x) = \lambda_3 1_{x_2>0}$.

Random walk.

Now suppose that $E = Z^d$ and that $a(x, y) = \lambda(y - x), y \neq x$, for some function λ. In this case, $Lf(x)$ has the following form:

$$Lf(x) = \sum_z \lambda(z)(f(x + z) - f(x)).$$

This process has a particularly simple description. Indeed, from the uniformization theorem, we see that $X_t - X_0$ is a Poisson sum of independent random increments:

$$X_t = X_0 + \sum_{k=1}^{N_t} \xi_k.$$

Here N_t is a Poisson random variable with parameter $\tilde{\lambda} = \sum_{z \neq 0} \lambda(z)$ and ξ_k is a random increment that takes value z with probability $\lambda(z)/\tilde{\lambda}$.

Specializing even further, suppose that the state space is the integers and that $\lambda(z) = 1/2$ for $z = \pm 1$ and zero for other values of z. Then,

$$Lf(x) = \frac{f(x + 1) - 2f(x) + f(x - 1)}{2}.$$

This process is called a *simple random walk*. For a simple random walk, N_t is Poisson with parameter one and the ξ_k take values ± 1 with probability $1/2$.

B.4. Martingales

Recall that, for a process $\{M_t, t \geq 0\}$ to be a martingale with respect to a filtration $\{\mathcal{F}_t, t \geq 0\}$ and a measure P, each M_t must be integrable against the measure P and measurable with respect to $\mathcal{F}_t$ and the process $\{M_t, t \geq 0\}$ must satisfy the martingale property:

$$\mathbb{E}[M_{t+u}|\mathcal{F}_t] = M_t, \quad t, u \geq 0.$$

Generally when we claim something is a martingale, the integrability property will be left to the reader to check, the measurability property will be completely obvious, and the martingale property will be shown.

In this section, we will introduce several martingales associated with a Markov process. In later sections, we will then apply certain martingale theorems to these martingales to help us prove large deviations results. The results that we will need about martingales are collected in §A.5.

Additive functionals.

An *additive functional* is a real-valued process $M_u^t, 0 \leq u \leq t$, that satisfies the following properties:

$$M_u^t + M_t^v = M_u^v \tag{B.21}$$

$$\theta_s M_u^t = M_{u+s}^{t+s} \tag{B.22}$$

$$\mathbb{E}_x M_0^t = 0 \qquad \text{for all } x \tag{B.23}$$

$$M_0^t \in \mathcal{F}_t. \tag{B.24}$$

The simplest example of an additive functional is

$$M_u^t = f(X_t) - f(X_u) - \int_u^t Lf(X_s)\,ds.$$

In this case, properties (B.21)–(B.24) are all perfectly straightforward except perhaps (B.23) which follows from (B.17).

Theorem B.1. *If M_u^t is an additive functional, then M_0^t is a martingale with respect to every measure P_x.*

Proof. The fact that M_0^t is a martingale follows from the defining properties of an additive functional and the Markov property of the filtration $\mathcal{F}_t$:

$$\begin{aligned}
\mathbb{E}_x[M_0^{t+u}|\mathcal{F}_t] &= \mathbb{E}_x[M_0^t + M_t^{t+u}|\mathcal{F}_t] \\
&= M_0^t + \mathbb{E}_x[M_t^{t+u}|\mathcal{F}_t] \\
&= M_0^t + \mathbb{E}_x[\theta_t M_0^u|\mathcal{F}_t] \\
&= M_0^t + \mathbb{E}_{X_t} M_0^u \\
&= M_0^t.
\end{aligned}$$

This completes the proof.   ■

From this result, we see that

$$f(X_t) - f(X_0) - \int_0^t Lf(X_s)\, ds \qquad (B.25)$$

is a martingale. We will call such martingales the *linear martingales* associated with the Markov process X_t.

The counting martingale.

Theorem B.2. *If X_t is a multidimensional birth-death process with jump rates $\lambda_j(x)$, then for every j*

$$M_t = Y_t^j - \int_0^t \lambda_j(X_s)ds$$

is a martingale. In addition, $M_u^t = Y_t^j - Y_u^j - \int_u^t \lambda_j(X_s)ds$ is an additive functional.

Proof. It is easy to see that M_u^t satisfies all the properties to be an additive functional except perhaps (B.23). Thus the second claim follows from the first.

Let X_t^j denote the jth coordinate of X. Fix j, and define the $d+1$ dimensional process Z_t by

$$Z_t^1 \overset{\triangle}{=} X_t^1$$

$$\vdots$$

$$Z_t^d \overset{\triangle}{=} X_t^d$$

$$Z_t^{d+1} \overset{\triangle}{=} Y_t^j.$$

Then Z_t is a birth-death process. Its jump rates agree with those of X_t, and by (B.2)–(B.1) its generator is given through

$$Lf(z) = \sum_{i \neq j} \lambda_i(x)[f(z + (e_i, 0)) - f(z)] + \lambda_j(x)[f(z + (e_j, 1)) - f(z)]$$

for any real function on $\mathbb{R}^{d+1}$, where x denotes the first d coordinates of z. Now choose $f(z) = z^{(d+1)}$, the $d+1^{\text{st}}$ coordinate of z. Then by the comment following Theorem B.1,

$$M_t \overset{\triangle}{=} f(Z_t) - \int_0^t Lf(Z_s)\, ds = Y_t^j - \int_0^t \lambda_j(X_s)\, ds$$

is a (linear) martingale.   ■

Multiplicative functionals.

A *multiplicative functional* is a real-valued process $M_u^t, 0 \le u \le t$, that satisfies the following properties:

$$M_u^t M_t^v = M_u^v \qquad (B.26)$$
$$\theta_s M_u^t = M_{u+s}^{t+s} \qquad (B.27)$$
$$\mathbb{E}_x M_0^t = 1 \qquad (B.28)$$
$$M_u^t > 0 \qquad (B.29)$$
$$M_0^t \in \mathcal{F}_t. \qquad (B.30)$$

The simplest example of a multiplicative functional is

$$M_u^t = \frac{g(X_t)}{g(X_u)} \exp\left(-\int_u^t \frac{Lg(X_s)}{g(X_s)} ds\right). \qquad (B.31)$$

In this case, properties (B.26)–(B.30) are all trivial except (B.28) which follows from (B.19).

Theorem B.3. *If M_u^t is a multiplicative functional, then M_0^t is a martingale with respect to every measure P_x.*

Proof. The result follows easily from the defining properties of a multiplicative functional and the Markov property:

$$\begin{aligned}
\mathbb{E}_x[M_0^{t+u} | \mathcal{F}_t] &= \mathbb{E}_x[M_0^t M_t^{t+u} | \mathcal{F}_t] \\
&= M_0^t \mathbb{E}_x[M_t^{t+u} | \mathcal{F}_t] \\
&= M_0^t \mathbb{E}_x[\theta_t M_0^u | \mathcal{F}_t] \\
&= M_0^t \mathbb{E}_{X_t} M_0^u \\
&= M_0^t.
\end{aligned}$$

This completes the proof. ∎

From this result, we see that

$$M_0^t = \frac{g(X_t)}{g(X_0)} \exp\left(-\int_0^t \frac{Lg(X_s)}{g(X_s)} ds\right) \qquad (B.32)$$

is a martingale. We will call such martingales the *exponential martingales* associated with the Markov process X_t.

The Dirichlet problem.

As an example of the utility of the linear martingale, we briefly describe its connection with the Dirichlet problem. Let D be a subset of E and consider the following boundary value problem:

$$Lf(x) = 0, \qquad x \in D,$$
$$f(x) = \phi(x), \qquad x \in D^c.$$

We call this the Dirichlet problem. When the space is Euclidean and the operator L is the Laplacian, this boundary value problem is the classical Dirichlet problem (that's what we'd get if the Markov process were Brownian motion — it's also why the generator is denoted by L). Suppose we have a function f that solves this problem. Let's see what the optional sampling theorem A.129 says when applied to the linear martingale, and τ is the first time X_t exits the domain D:

$$f(x) = \mathbb{E}_x f(X_\tau) - \mathbb{E}_x \int_0^\tau Lf(X_s) \, ds$$
$$= \mathbb{E}_x \phi(X_\tau). \tag{B.33}$$

The integral vanishes since, before time τ, $Lf(X_s) = 0$. Also, at time τ, the process is in the complement of D and so f can be replaced by ϕ in the first term (we have used the right continuity of the process here). The right-hand side in (B.33) does not involve the function f. This proves that the solution to the Dirichlet problem is unique. To prove existence, it suffices to check that the function defined by the right-hand side of (B.33) actually is a solution of the problem. This is an easy exercise using the Markov property that we leave to the reader.

If we choose $\phi(x) = 1_y(x)$, then (B.33) reduces to

$$f(x) = \mathbb{P}_x(X_\tau = y).$$

Hence we see that the Dirichlet problem is closely related to finding exit distributions from domains.

A related problem is to find the expected exit time $g(x) = \mathbb{E}_x \tau$. An analysis similar to the one above shows that $g(x)$ is the unique solution of the following inhomogeneous Dirichlet problem:

$$Lg(x) = -1, \qquad x \in D,$$
$$g(x) = 0, \qquad x \in D^c. \tag{B.34}$$

Sometimes these Dirichlet problems can be solved explicitly, but even when they can't, they yield efficient numerical methods.

This kind of probabilistic analysis of boundary value problems is the basis of a branch of mathematics called probabilistic potential theory. We have only given the briefest exposure just to give a quick appreciation for the value of the linear martingale.

The Feynman-Kac formula.

For a moment let's consider the Shroedinger equation:

$$Lf(x) = \psi(x)f(x), \quad x \in D, \tag{B.35}$$
$$f(x) = \phi(x), \quad x \in D^c. \tag{B.36}$$

Again, let τ be the first exit time from D and apply the optional sampling theorem to the exponential martingale (B.32) to get that any solution to (B.35), (B.36) must be given by

$$f(x) = \mathbb{E}_x \phi(X_\tau) \exp\left(-\int_0^\tau \psi(X_s)\,ds\right). \tag{B.37}$$

This proves uniqueness and as before existence is proved by verifying that the formula given by (B.37) is indeed a solution.

There is one special case of (B.37) that deserves mentioning. When the functions ϕ and ψ are actually the constants one and λ, respectively, then (B.37) becomes

$$f(x) = \mathbb{E}_x e^{-\lambda\tau}.$$

Therefore, solving (B.35), (B.36) gives us the Laplace transform of the first exit time from a domain. This result is often useful.

Change of measure.

Multiplicative functionals are useful for changing measure in such a way that the Markov property is preserved under the new measure. Let M_u^t be a multiplicative functional. We fix a finite time horizon T and consider the process X_t only on the interval $0 \le t \le T$. For each $x \in E$, let $\tilde{P}_x$ be a new measure on $(\Omega, \mathcal{F}_T)$ defined by

$$\tilde{\mathbb{E}}_x Z = \mathbb{E}_x M_0^T Z, \quad Z \in \mathcal{F}_T, \tag{B.38}$$

where $\tilde{\mathbb{E}}_x$ denotes expectation calculated using the measure $\tilde{P}_x$. From (B.38), we see that $\tilde{P}_x$ is absolutely continuous with respect to P_x, and the Radon-Nikodym derivative is simply M_0^T (§A.4). Now for $0 \le t \le T$, define a linear operator $\tilde{T}_t$ as follows:

$$\tilde{T}_t f(x) = \tilde{\mathbb{E}}_x f(X_t). \tag{B.39}$$

Proposition B.4. *If M_u^t is a multiplicative functional, then the operators $\tilde{T}_t$, $t \ge 0$, defined by (B.39) form a transition semigroup and*

$$\tilde{T}_t f(x) = \mathbb{E}_x M_0^t f(X_t). \tag{B.40}$$

Proof. We start by proving (B.40). Using the fact that M_0^t is a martingale, we see that

$$\begin{aligned}
\tilde{T}_t f(x) &= \mathbb{E}_x M_0^T f(X_t) \\
&= \mathbb{E}_x \mathbb{E}_x[M_0^T f(X_t)|\mathcal{F}_t] \\
&= \mathbb{E}_x f(X_t)\mathbb{E}_x[M_0^T|\mathcal{F}_t] \\
&= \mathbb{E}_x f(X_t)M_0^t.
\end{aligned}$$

All the properties that have to be satisfied to be a transition semigroup follow trivially from the fact that M_0^t is positive and has mean one except the most important property, $\tilde{T}_t \tilde{T}_s = \tilde{T}_{t+s}$, which we now verify:

$$
\begin{aligned}
\tilde{T}_t \tilde{T}_s f(x) &= \mathbb{E}_x M_0^t \mathbb{E}_{X_t} M_0^s f(X_s) \\
&= \mathbb{E}_x M_0^t \theta_t [M_0^s f(X_s)] \\
&= \mathbb{E}_x M_0^t M_t^{s+t} f(X_{s+t}) \\
&= \mathbb{E}_x M_0^{s+t} f(X_{s+t}) \\
&= \tilde{T}_{s+t} f(x).
\end{aligned}
$$

Here, the first equality follows from (B.40), the second from the Markov property (B.15), the third from the definitions (B.12) and (B.13) of the shift operator, and the fourth from the multiplicative property (B.26) of M_u^t. This completes the proof. ∎

Since starting with a semigroup, say $\tilde{T}_t$, there is, for each $x \in E$, a unique measure on the space of right-continuous paths such that (B.39) holds, it follows from Proposition B.4 that $\tilde{P}_x$ is this unique measure. The next theorem shows that multidimensional birth-death processes are invariant under this change of measure.

Theorem B.5. *If, under the measure P_x, X_t is a multidimensional birth-death process with step directions e_j and step rates $\lambda_j(x)$, then under the measure $\tilde{P}_x$ the process X_t is again a multidimensional birth-death process with the same step directions and with the rates changed to*

$$
\mu_j(x) = \lambda_j(x) \frac{g(x + e_j)}{g(x)}. \tag{B.41}
$$

Proof. To prove this, we calculate the generator of $\tilde{T}_t$. With overwhelming probability, the process X_t (relative to the original measure P_x) will have taken at most one jump in a small amount of time t. Hence, using (B.40) and (B.31), we get

$$
\tilde{T}_t f(x) = \exp\left(-t \sum_j \lambda_j(x)\right) \exp\left(-t \frac{Lg(x)}{g(x)}\right) f(x)
$$
$$
+ \int_0^t \sum_k \exp\left(-s \sum_j \lambda_j(x)\right) \lambda_k(x) \exp\left(-(t-s)\sum_j \lambda_j(x+e_k)\right)
$$
$$
\times \frac{g(x+e_k)}{g(x)} \exp\left(-s \frac{Lg(x)}{g(x)} - (t-s)\frac{Lg(x+e_k)}{g(x+e_k)}\right) f(x+e_k)\, ds + o(t).
$$

In the second summand we have a jump at s in direction k, and no further jumps until t. The first two terms give the probability of this event. From this simple calculation, using (B.20) we see that

$$\tilde{L}f(x) = \lim_{t \searrow 0} \frac{1}{t}(\tilde{T}_t f(x) - f(x))$$

$$= -\left(\sum_j \lambda_j(x) + \frac{Lg(x)}{g(x)}\right) f(x) + \sum_j \lambda_j(x) \frac{g(x + e_j)}{g(x)} f(x + e_j)$$

$$= \sum_j \lambda_j(x) \frac{g(x + e_j)}{g(x)} (f(x + e_j) - f(x)).$$

Hence, the new process is a multidimensional birth-death process with the same jump directions and with jump rates given by (B.41). ∎

It is interesting to note that there are generally not enough degrees of freedom in (B.41) to be able to get all sets of rates $\mu_j(x)$ by appropriate choices of $g(x)$. But it turns out that we can write the Radon-Nikodym derivative M_0^T purely in terms of $\mu_j(x)$ and that, once we do this, we get a derivative that is correct even if there is no function $g(x)$ connecting the old and the new rates. So, start by assuming that μ_j and λ_j are related through (B.41) for some g. Since $g(X_t)$ is piecewise constant, we can write the difference $\log g(X_T) - \log g(X_0)$ as the sum of the changes over the jumps:

$$\log g(X_T) - \log g(X_0) = \int_0^T \sum_j (\log g(X_s) - \log g(X_{s-})) \, dY_s^j$$

$$= \int_0^T \sum_j \log \frac{g(X_{s-} + e_j)}{g(X_{s-})} dY_s^j$$

$$= \int_0^T \sum_j \log \frac{\mu_j(X_{s-})}{\lambda_j(X_{s-})} dY_s^j, \qquad (B.42)$$

where Y_t^j is the counting process that is incremented every time a jump occurs in the e_j direction. Now, using (B.41), we see that

$$\frac{Lg(x)}{g(x)} = \sum_j \lambda_j(x) \left(\frac{g(x + e_j)}{g(x)} - 1\right)$$

$$= \sum_j (\mu_j(x) - \lambda_j(x)). \qquad (B.43)$$

From (B.43) and (B.42), we see that

$$M_0^T = \exp\left(\log g(X_T) - \log g(X_0) - \int_0^T \frac{Lg(X_s)}{g(X_s)}ds\right)$$

$$= \exp\left(\int_0^T \sum_j \log \frac{\mu_j(X_{s-})}{\lambda_j(X_{s-})}dY_s^j - \int_0^T \sum_j (\mu_j(X_s) - \lambda_j(X_s))\,ds\right).$$

$$(B.44)$$

Equation (B.44) defines a change of measure that maps the multidimensional birth-death process with rates $\lambda_j(x)$ into a process of the same type but having rates $\mu_j(x)$.

Note that formula (B.44) for M_0^T involves the function g only indirectly through formula (B.41). This leads us to the question: Does (B.44) define a change of measure that maps the process with rates $\lambda_j(x)$ to the one with rates $\mu_j(x)$, even when (B.41) does not hold? The answer to this question is yes. Indeed, instead of specifying a function g, we start with desired rates $\mu_j(x)$. We formulate this result as a theorem:

Theorem B.6. *Let X_t be a multidimensional birth-death process with step directions e_j and step rates $\lambda_j(x)$ and let P_x denote the corresponding measure on the space of trajectories. Let $\mu_j(x)$ be an arbitrary set of transition rates and let $\tilde{P}_x$ be a new measure defined by (B.38) with M_0^T defined by (B.44). Then, under this new measure, the process X_t is again a multidimensional birth-death process with the same step directions and with the rates changed to $\mu_j(x)$.*

Proof. Define M_u^t by the formula:

$$M_u^t = \exp\left(\int_u^t \sum_j \log \frac{\mu_j(X_{s-})}{\lambda_j(X_{s-})}dY_s^j - \int_u^t \sum_j (\mu_j(X_s) - \lambda_j(X_s))\,ds\right).$$

$$(B.45)$$

We need to show that M_u^t is a multiplicative functional. All of the defining properties are trivial except (B.28) which can be checked by direct calculation. Indeed, using the law of total probability, we get

$$\mathbb{E}_x M_0^t = \sum_n \int \cdots \int_{0 < s_1 < s_2 < \cdots < s_n \le t} \sum_{j_1=1}^{m} \cdots \sum_{j_n=1}^{m}$$

$$\exp\left(-s_1 \sum_j \lambda_j(x_0)\right) \lambda_{j_1}(x_0)\, ds_1$$

$$\times \exp\left(-(s_2 - s_1) \sum_j \lambda_j(x_1)\right) \lambda_{j_2}(x_1)\, ds_2$$

$$\vdots$$

$$\exp\left(-(s_n - s_{n-1}) \sum_j \lambda_j(x_{n-1})\right)$$

$$\times \lambda_{j_n}(x_{n-1})\, ds_n \, \exp\left(-(t - s_n) \sum_j \lambda_j(x_n)\right)$$

$$\exp\left(\sum_{k=1}^{n} \log \frac{\mu_{j_k}(x_{k-1})}{\lambda_{j_k}(x_{k-1})}\right.$$

$$\left. - \sum_{k=1}^{n+1} \sum_j (\mu_j(x_{k-1}) - \lambda_j(x_{k-1}))(s_k - s_{k-1})\right),$$

where, for notational efficiency, we have put $x_0 = x$, $x_i = x_{i-1} + e_{j_i}$, $s_0 = 0$, and $s_{n+1} = t$. This formula may look messy, but it is easy to explain. The first line corresponds to partitioning the sample space into small pieces. One piece is a trajectory that makes n steps up to time t, with the steps occurring at times $s_1, s_2, \ldots, s_n$, and the ith step being in the direction e_{j_i}. The complicated product on the second through fourth lines is the likelihood of this trajectory. It too is easy to explain. The first factor, $\exp(-s_1 \sum_j \lambda_j(x_0))$, represents the probability that nothing happens in the interval $[0, s_1)$, the second factor, $\lambda_{j_1}(x_0)\, ds_1$, represents the probability that a step in direction e_{j_1} occurs exactly at time s_1, etc. Finally, the last line is the integrand evaluated on this specific trajectory. Now, note that the last line cancels with the previous lines in such a way as to change the previous lines into a formula of exactly the same type but with the λs replaced by μs:

$$
\mathbb{E}_x M_0^t = \sum_n \int \cdots \int_{0 < s_1 < s_2 < \cdots < s_n \le t} \sum_{j_1=1}^{m} \cdots \sum_{j_n=1}^{m}
$$

$$
\exp\left(-s_1 \sum_j \mu_j(x_0)\right) \mu_{j_1}(x_0)\, ds_1
$$

$$
\times \exp\left(-(s_2 - s_1) \sum_j \mu_j(x_1)\right) \mu_{j_2}(x_1)\, ds_2
$$

$$
\vdots
$$

$$
\exp\left(-(s_n - s_{n-1}) \sum_j \mu_j(x_{n-1})\right)
$$

$$
\times \mu_{j_n}(x_{n-1})\, ds_n \ \exp\left(-(t - s_n) \sum_j \mu_j(x_n)\right).
$$

But, again by the law of total probability (now using μs), we see that this is exactly equal to one. Even though this calculation is rather tedious, we have included it so that the reader will see that for multidimensional birth-death processes everything can in principle be calculated explicitly. ∎

Appendix C

Calculus of Variations

To be able to compute things using large deviations, we need to be able to get information from the resulting variational problems. This chapter collects the basics and a number of tricks that make it easier to extract information from, and sometimes even solve, the variational problems. The functionals we minimize are integrals: we shall denote the integrands by f, and shall use ℓ only when we specialize to the local rate function (that is, when we need the special properties of this function). Section §C.1 provides a heuristic derivation of the basic tools: Euler's Equation and the transversality condition. In §C.3 we derive a general approach for one-dimensional level crossing problems. The effort of this independent derivation is worthwhile since the classical elementary treatment in calculus of variations, e.g., [Els, Ew] is restricted to paths that are differentiable, except possibly at a finite number of points. However, we do not know a priori that the optimal solution of our problem satisfies such a condition. In general, the technical conditions imposed by general results on the calculus of variations are fairly strong—see §C.4 below. In particular, conditions under which a candidate solution is indeed the global minimum are usually complicated. On the other hand, the one-dimensional level crossing problem is simple enough so that we can get simple, explicit conditions. In §C.4 we state the precise theorems from the calculus of variations we need for more general variational problems; that is, problems that are not one-dimensional as well as problems that are not of the level crossing type.

C.1. Heuristics of the Calculus of Variations

Fix $a < b$ and a function f and define, for any absolutely continuous function r,

$$I(r) \stackrel{\triangle}{=} \int_a^b f\big(r(t), r'(t)\big)\, dt.$$

The "simplest problem" of the calculus of variations is the problem we call "the level crossing problem." It is the problem of minimizing I over all absolutely continuous functions r that satisfy

$$r(a) = \alpha, \quad r(b) = \beta \qquad (C.1)$$

with α, β fixed. We shall often be interested not only in calculating the minimal value, but also in identifying a function r^* that achieves that minimum. Here is a heuristic derivation of the Euler necessary conditions for optimality of a function r.

Let us perturb a candidate optimal function r in the direction of some function x and in the amount δ. To satisfy the boundary conditions (C.1) we must have

$$x(a) = x(b) = 0.$$

Consider now the function F of the real variable δ

$$F(\delta) \triangleq \int_a^b f(r(t) + \delta x(t), r'(t) + \delta x'(t)) \, dt.$$

For r to be a minimizer it is necessary that the derivative of F vanishes, namely

$$\frac{d}{d\delta} \int_a^b f(r(t) + \delta x(t), r'(t) + \delta x'(t)) \, dt \bigg|_{\delta=0} = 0.$$

Differentiating under the integral sign we get

$$\int_a^b x(t) \frac{\partial}{\partial r} f(r(t), r'(t)) + x'(t) \frac{\partial}{\partial r'} f(r(t), r'(t)) \, dt = 0.$$

Integrating the second term by parts,

$$\int_a^b x'(t) \frac{\partial}{\partial r'} f(r(t), r'(t)) \, dt = - \int_a^b x(t) \frac{d}{dt} \frac{\partial}{\partial r'} f(r(t), r'(t)) \, dt$$

since $x(a) = x(b) = 0$. We obtain, for all such x,

$$\int_a^b x(t) \left[\frac{\partial}{\partial r} f(r(t), r'(t)) - \frac{d}{dt} \frac{\partial}{\partial r'} f(r(t), r'(t)) \right] dt = 0.$$

But this must hold for all x, hence the term in square brackets must vanish for (almost) all t, and we obtain the Euler necessary conditions

$$\frac{\partial}{\partial r} f(r(t), r'(t)) - \frac{d}{dt} \frac{\partial}{\partial r'} f(r(t), r'(t)) = 0. \tag{C.2}$$

In the special case that f does not depend on the first variable, we immediately conclude that

$$\frac{d}{dt} \frac{\partial}{\partial r'} f(r(t), r'(t)) = 0$$

or

$$\frac{\partial}{\partial r'} f(r(t), r'(t)) = \text{constant}.$$

From (C.2) we obtain a useful variation, as follows.

$$\frac{d}{dt} \left[f(r(t), r'(t)) - r'(t) \frac{\partial}{\partial r'} f(r(t), r'(t)) \right]$$

$$= r'(t) \frac{\partial}{\partial r} f(r(t), r'(t)) + r''(t) \frac{\partial}{\partial r'} f(r(t), r'(t))$$

$$- r''(t) \frac{\partial}{\partial r'} f(r(t), r'(t)) - r'(t) \frac{d}{dt} \frac{\partial}{\partial r'} f(r(t), r'(t))$$

$$= r'(t) \left[\frac{\partial}{\partial r} f(r(t), r'(t)) - \frac{d}{dt} \frac{\partial}{\partial r'} f(r(t), r'(t)) \right]$$

$$= 0.$$

Therefore, (C.2) implies the DuBois-Reymond Equation:

$$\ell(r(t), r'(t)) - r'(t) \frac{\partial}{\partial r'} f(r(t), r'(t)) = \text{constant}. \qquad (C.3)$$

Next consider the case where b is free, that is, we wish to minimize I over all r with $r(a) = \alpha$ and $r(b) = \beta$, but also over all possible values of b (assume $\alpha \neq \beta$). Then the Euler Equation (C.2) is still a necessary condition, obtained from the optimality property in r at the best value of b. We now derive an additional equation, obtained from the optimality in b around the best r. We need to be careful here that we only consider functions satisfying the boundary conditions! So assume r is defined on $[a, b+1]$ and consider $\delta > 0$; the calculation for $\delta < 0$ is almost identical. By definition $r(b) = \beta$: to satisfy the boundary condition at $b + \delta$ we need to to choose the perturbation function so as to compensate for the fact that, generally, $r(b + \delta) \neq \beta$. So let δx be a function with

$$\delta x(0) = 0, \quad \delta x(b + \delta) = -\delta \cdot r'(b).$$

Then $r(t) + \delta x(t)$ satisfies the boundary conditions

$$r(b + \delta) + \delta x(b + \delta) \approx r(b) + \delta \cdot r'(b) + \delta x(b + \delta) = \beta$$

(at least to first-order approximation), and δx can (and will) be chosen linear in δ. [Note that we cannot simply choose $r(t) = r(b)$, $t > b$ as this would not work for $\delta < 0$.] Taking derivatives and equating to zero now yields

$$\frac{d}{d\delta} \int_a^{b+\delta} f(r(t) + \delta x(t), r'(t) + \delta x'(t)) \, dt \Big|_{\delta=0}$$

$$= f(r(b), r'(b)) + \int_a^b \frac{d}{d\delta} f(r(t) + \delta x(t), r'(t) + \delta x'(t)) \, dt \Big|_{\delta=0}.$$

But by the argument in the case of fixed b we know that the second term is zero for all x, except that in the integration by parts we no longer have $\delta x(b + \delta) = 0$. Thus we obtain

$$\frac{d}{d\delta} \int_a^{b+\delta} f(r(t) + \delta x(t), r'(t) + \delta x'(t)) \, dt \Big|_{\delta=0}$$

$$= f(r(b), r'(b)) + x(b) \cdot \frac{\partial}{\partial r'} f(r(b), r'(b))$$

and substituting for the value of $x(b)$ we obtain

$$f(r(b), r'(b)) - r'(b) \cdot \frac{\partial}{\partial r'} f(r(b), r'(b)) = 0. \qquad (C.4)$$

This condition is called the "transversality condition."

C.2. Calculus of Variations and Large Deviations

The reason we want to solve problems in the calculus of variations is that the solutions will answer questions about probabilities. However, solving problems in the calculus of variations is notoriously difficult. There are questions of existence of solutions, and most of the results are characterizations (necessary conditions), as will become clear in §C.4. But don't despair! We are fortunate on several counts. First, some properties of the rate function that we derived will address the existence problem, and many special tricks will come to the rescue.

Our approach to the variational problems will go as follows. First, the questions we are interested in are such that existence will be automatic due to lower-semicontinuity and compactness (see Theorem C.1 below). Then, we will usually derive a solution to the Euler equation. Using a variety of techniques, some from the calculus of variations and some specific to the problem under consideration, we will then show that this particular solution of the Euler equation is the only one that can be optimal. This will nail the validity of our formal calculation. Here is an obvious, if somewhat abstract theorem that formalizes this argument.

Theorem C.1. *Assume I is a good rate function, and consider the variational problem*

$$I^*(F) = \inf\{I(r) : r \in F\},$$

where F is a closed set so that $I^*(F) < \infty$. Suppose there is a set S so that

$$\mathcal{O} \overset{\triangle}{=} \{r : r \in F,\ I(r) = I^*(F)\} \subset S.$$

Then $\mathcal{O}$ is not empty, and if $S \cap F = S \cap \mathcal{O}$ contains exactly one point r^* then r^* solves the variational problem. That is, $r^* \in F$ and $I(r^*) = I^*(F)$.

Remark. The set S defines "necessary conditions" for an optimum, in the sense that any optimal point must be in S. However, S may contain many nonoptimal points, and even points outside F.

Proof. By Exercise 2.3, there exists (at least one) point r^* in $\mathcal{O}$. By the hypothesis, every such point must be in $S \cap F$. But there is only one such point r^*. ∎

The next two results concern the function ℓ, defined in Equations (5.2)–(5.4) [or in Theorem 8.19(i), with the obvious definition for g]. They will complement the Euler equations in making the set S of Theorem C.1 small enough so that the theorem can be applied.

Theorem C.2. *Assume* $\ell(x, y)$ *is differentiable in* y *and that*

$$\ell(x, y) = \theta^* y - g(x, \theta^*). \qquad (C.5)$$

If the optimal point $\theta^* = \theta^*(x, y)$ *is differentiable in* y *and* $g(x, \theta)$ *is differentiable in* θ *near* θ^*, *then*

$$\frac{d}{dy}\ell(x, y) = \theta^*(x, y). \qquad (C.6)$$

Proof. Taking derivatives we obtain

$$\frac{d}{dy}\ell(x, y) = \frac{d}{dy}\left(\theta^* y - g(x, \theta^*)\right)$$

$$= \theta^* + \left[y - \frac{\partial}{\partial\theta}g(x, \theta^*)\right]\frac{\partial}{\partial y}\theta^*(x, y)$$

$$= \theta^*$$

since the middle term in square brackets is zero as θ^* is a maximizer.   ∎

Theorem C.3. *Assume* $\ell(x, y)$ *and* $\theta^*(x, y)$ *satisfy (C.5) and (C.6). Then (C.3) implies [for* $(x, y) = (r(t), r'(t))$] *that* $g(x, \theta^*) = C$, *for some constant* C. *If in addition (C.4) holds then* $C = 0$.

Proof. By (C.3), for some constant C

$$y\frac{d}{dy}\ell(x, y) = \ell(x, y) + C$$

$$= \theta^* y - g(x, \theta^*) + C$$

by definition of θ^*. But by (C.6) of Theorem C.2, the left-hand side equals $\theta^* y$ and the first result follows. If (C.4) holds then by the same argument $C = 0$.   ∎

In Chapters 5 and 8 we established that the rate functions I_0^T we deal with are indeed good rate functions. But our applications often involve "free time" problems. To be able to use the reasoning of Theorem C.1, we need to establish a priori that it suffices to look at finite times. So, consider a set G of pairs $(\vec{r}, T_r)$ so that each function $\vec{r}$ is defined on the interval $[0, T_r]$. Let F_G be the following set of functions on $[0, \infty]$. A function $\vec{s}$ is in F if and only if $\vec{s}(t) = \vec{r}(t)$, $0 \le t \le T_r$ for some $\vec{r}$ in G, and in addition $\vec{s}$ follows $\vec{z}_\infty$ for $T_r \le t < \infty$. Thus functions in F_G are just extensions of functions in G. Finally, let F_G^T be the collection of functions in F_G but with each function restricted to the interval $[0, T]$.

Lemma C.4. *Suppose that* I_0^T *is a good rate function for each* $T < \infty$. *Assume that there exists a* T' *so that the following holds: for each* $\vec{r} \in G$ *there is a* $\tilde{r} \in G$ *so that* $T_{\tilde{r}} \le T'$ *and* $I_0^{T_{\tilde{r}}}(\tilde{r}) \le I_0^{T_r}(\vec{r})$. *Then*

$$\inf\left\{I_0^{T'}(\vec{r}) : \vec{r} \in F_G^{T'}\right\} = \inf\left\{I_0^{T_r}(\vec{r}) : \vec{r} \in G, \ T_r \le T'\right\}$$

$$= \inf\left\{I_0^{T_r}(\vec{r}) : \vec{r} \in G\right\}.$$

If $F_G^{T'}$ is closed then the infimum is attained.

Proof. If $\vec{s} \in F_G$ is an extension of $\vec{r} \in G$ and $T_r \leq T'$ then $I_0^{T'}(\vec{s}) = I_0^{T_r}(\vec{r})$, since ℓ vanishes along z_∞ paths. The first claim now follows from the assumptions. The last claim follows since by assumption, $I_0^{T'}$ is a good rate function. $\blacksquare$

C.3. One-Dimensional Tricks

Most of our variational problems turn out to be one-dimensional. In that case, several tricks are available which simplify the computations considerably—often to the point where analytic solutions are possible. We start with the more general properties. Consider the following level crossing problem. Given α, β,

$$\text{minimize} \quad I_0^T(r) \triangleq \int_0^T f\big(r(t), r'(t)\big)\, dt \tag{C.7}$$

$$\text{subject to} \quad T > 0, \ r(0) = \alpha, \ r(T) = \beta, \ r \text{ absolutely continuous,}$$

where $T = \infty$ is acceptable. We denote the minimal value by I^* or $I^*(\alpha, \beta)$, and call this the value of the problem. We say that r^* solves the variational problem (C.7) if r^* satisfies the boundary conditions at zero and T, and $I_0^T(r^*) = I^*$.

Lemma C.5. *Consider the problem (C.7), and assume that $f(x, y) \geq 0$ for each x, y. If $\alpha < \beta$ then it suffices to consider strictly increasing functions. If $\beta < \alpha$ then it suffices to consider strictly decreasing functions.*

Proof. Pick any absolutely continuous function satisfying $r(0) = \alpha$ and $r(T) = \beta$, and assume that $r(u) < r(s)$ for some $0 \leq s < u \leq T$. Since r is continuous, there is some $u' \in [u, T)$ so that $r(u') = r(s)$. Since $f(x, y) \geq 0$ for each x, y, the function

$$\tilde{r}(t) \triangleq \begin{cases} r(t) & t \leq s, \\ r(t + u' - s) & t \geq s, \end{cases}$$

defined on the interval $[0, S]$ where $S = T - (u' - s)$ satisfies the boundary conditions and $I_0^S(\tilde{r}) \leq I_0^T(r)$. Thus we can modify any given function r until it is strictly increasing, without increasing the corresponding integral. Note that by continuity, the number of required modifications is at most countable, so that the change in the derivative of r at these points will not affect the integral. Now repeat the same construction on intervals where r is constant. The proof when $\beta < \alpha$ is identical. $\blacksquare$

For a strictly increasing absolutely continuous function v, denote

$$t(r) = \{t : v(t) = r\} \quad \text{and} \quad v_t'(r) \triangleq \left[\frac{dv(t)}{dt}\right]_{t=t(r)}. \tag{C.8}$$

Note that this is well defined since v is strictly increasing.

Lemma C.6. *Consider the problem (C.7), where $\alpha < \beta$. Let $f(x, y) \geq 0$ for each x, y and let $v(t)$ be a strictly increasing absolutely continuous function with $v(0) = \alpha$, $v(T) = \beta$. If*

$$\int_\alpha^\beta \frac{f(x, v_t'(x))}{v_t'(x)} \, dx = \int_\alpha^\beta \inf_{y>0} \left\{ \frac{f(x, y)}{y} \right\} dx, \qquad (C.9)$$

then v solves problem (C.7). If $\beta < \alpha$ then the same conclusion holds with "decreasing" replacing "increasing" and with the infimum taken over $y < 0$.

Proof. Let r be any strictly increasing, absolutely continuous function satisfying the boundary conditions $r(0) = \alpha$, $r(T') = \beta$. By Lemma C.5 the restriction to strictly increasing functions does not change the minimum. Since $r_t'(x) \geq 0$ for almost all x, changing variables from t to x we obtain $dx = r_t'(x) \, dt$, so

$$
\begin{aligned}
\int_0^{T'} f(r(t), r'(t)) \, dt &= \int_\alpha^\beta \frac{f(x, r_t'(x))}{r_t'(x)} \, dx \\
&\geq \int_\alpha^\beta \inf_{y>0} \left\{ \frac{f(x, y)}{y} \right\} dx \\
&= \int_\alpha^\beta \frac{f(x, v_t'(x))}{v_t'(x)} \, dx \\
&= \int_0^T f(v(t), v'(t)) \, dt.
\end{aligned}
$$

Thus v minimizes the first expression. The proof when $\beta < \alpha$ is identical. ∎

This immediately ties things nicely to the Euler Equation: for suppose f is differentiable in y. Then y achieves the infimum only if

$$\frac{d}{dy} \frac{f(x, y)}{y} = \frac{1}{y^2} \left(f(x, y) - y \frac{\partial f(x, y)}{\partial y} \right) = 0. \qquad (C.10)$$

Let us now apply the structure of our specific problem. Recall Definitions (5.2)–(5.4)

$$\ell(x, y) \overset{\triangle}{=} \sup_\theta \{\theta y - g(x, \theta)\}$$

$$g(x, \theta) \overset{\triangle}{=} \sum_{i=1}^k \lambda_i(x) \left(e^{\langle \theta, \vec{e}_i \rangle} - 1 \right).$$

The following properties of ℓ are established in §5.2. Let us say that x is a nondegenerate point if there are positive rates for jump both to the right and to the left at x (or, more concisely, $\lambda_i(x) \cdot e_i \cdot \lambda_j(x) \cdot e_j < 0$ for some i, j).

Lemma C.7. *Let x be a nondegenerate point. Then*

(i) *By Exercise 5.27 $\ell(x, \cdot)$ is strictly convex. It is nonnegative, and by Lemma 5.32 it is finite for each y.*

(ii) *From the proof of Lemma 5.17 it follows that $\ell(x, y) \geq C_1|y|\log|y| - C_2|y|$ for some positive constants C_1 and C_2.*

(iii) *Consequently $\ell(x, \cdot)$ is differentiable (almost everywhere), and (C.6) holds (for almost every y) with a unique $\theta^*(x, y)$.*

As an immediate consequence we can verify the usefulness of Euler Equation.

Lemma C.8. *Let x be a nondegenerate point. There exists a $y(x) > 0$ so that*

$$\inf_{y>0} \frac{\ell(x, y)}{y} = \frac{\ell(x, y(x))}{y(x)},$$

and it is a solution of Euler's Equation. Assume in addition that $\ell(x, y)$ is (jointly) continuous in both variables, that all points in $[\alpha, \beta]$ are nondegenerate and that

$$\varepsilon < \lambda_i(x) < \frac{1}{\varepsilon} \text{ for some } \varepsilon > 0, \text{ and all } i, \ \alpha \leq x \leq \beta.$$

Then there exists a strictly increasing absolutely continuous function v satisfying (C.9).

Proof. Assume first that $\ell(x, \cdot)$ is differentiable at every point y. From Lemma C.7(ii) we have

$$\frac{\ell(x, y)}{y} \to \infty \quad \text{as} \quad |y| \to \infty \quad \text{or} \quad |y| \to 0.$$

Thus the minimum cannot be achieved either near zero or for large y. But away from zero, by (i) of the same lemma, $\ell(x, y)/y$ is a continuous function, and so it achieves its minimum at some point $0 < y(x) < \infty$. If $\ell(x, y)$ is differentiable in y at $y(x)$, then (C.10) must hold. Now if $\ell(x, y)$ is not differentiable at $y(x)$ then, since it is convex, it has a right and a left derivative at that point. In this case, $y(x)$ solves the Euler Equation (C.10) if the derivative is replaced with the appropriate value, between the right and the left derivatives.

Under the additional conditions, the function $y(x)$ exists for all x in $[\alpha, \beta]$ (to be precise, we need to show that y can be chosen as a measurable function. This is indeed possible under the stated conditions. The mathematical tools involved are "measurable selection theorems.") Suppose we find a solution to the ordinary differential equation

$$\frac{dv(t)}{dt} = y(v(t)).$$

Then v is strictly increasing since $y(v) > 0$. Since the rates are bounded away from zero and bounded above, it follows from the proof of Lemma 5.17 that $y(x)$ is bounded above and bounded away from zero. Therefore v is absolutely continuous. Moreover, the equation

$$\int_\alpha^r \frac{1}{y(x)} dx = \int_0^t ds = t$$

is well defined for all $\alpha \le r \le \beta$ as y is bounded below, and defines a function $t(r)$ that is strictly monotone increasing. But then the inverse function $v(t)$ exists, and is the strictly monotone function determined by $t[v(t)] = t$. Both functions are differentiable due to the bounds on y, and differentiating

$$1 = \frac{dt}{dt} = \frac{dt(v(t))}{dr}\frac{dv(t)}{dt} = \frac{1}{y(v(t))}\frac{dv(t)}{dt}$$

so that v satisfies the desired relation. Finally, the defining relation gives $v(0) = \alpha$, and since $\dot{y}$ is bounded away from zero, there is some T so that $v(T) = \beta$. ■

In the parlance of convex analysis, if ℓ is not differentiable at $y(x)$ then zero belongs to the set of values that are obtained by replacing the derivative with a subgradient. Such points are isolated and do not affect the analysis. In our applications the functions will be differentiable, so we do not elaborate on this point.

Under the conditions of Lemma C.8 the conditions of Lemma C.6 are satisfied, so that v solves the variational problem (C.7). We are now ready to derive an expression for the value of the variational problem, as well as a method to compute an optimal solution.

Lemma C.9. *Consider problem (C.7) with $\alpha < \beta$ and where ℓ is given through (5.2)–(5.4). Assume that all points $x \in (\alpha, \beta)$ are nondegenerate, and that ℓ is differentiable in y at all points x and $y > 0$. For each x, the equation*

$$g(x, \theta^*) = 0 \qquad (C.11)$$

has at most two solutions. If in addition the conditions of Lemma C.8 hold, then with the appropriate choice of $\theta^*(x)$,

$$I^*(\alpha, \beta) = \int_\alpha^\beta \theta^*(x)\,dx. \qquad (C.12)$$

Proof. Since $g(x, \theta)$ is strictly convex in θ, (C.11) has at most two solutions. Let $\theta^*(x, y)$ be a value of θ that achieves the maximum in the definition (5.2)–(5.4) of ℓ: that such a value exists (and is finite) follows from Lemma C.7. From that lemma it also follows that (C.6) holds, that is

$$\frac{\partial}{\partial y}\ell(x, y) = \theta^*(x, y) \qquad (C.13)$$

for all x, y. This and the definition of ℓ yield

$$\ell(x, y) - y\frac{\partial \ell(x, y)}{\partial y} = -g(x, \theta^*(x, y)) \qquad (C.14)$$

for all x, y. Now under the conditions of Lemma C.8 there exists an extremizing path $v(t)$, and from (C.10), along this path (that is, for all x, y so that, for some t, $v(t) = x$ while $v'(t) = y$),

$$\ell(x, y) - y\frac{\partial \ell(x, y)}{\partial y} = 0 \qquad (C.15)$$

since $y > 0$, so that (C.11) holds along such paths, provided $\theta^*(x, y)$ is indeed chosen to achieve the maximum. However, this equation and the choice of solution define $\theta^*(x, y)$ uniquely for each x, so that in fact the condition that we are along an extremizing path implies that $\theta^*(x, y) = \theta^*(x)$ depends on x alone. The definition of I^* together with (C.13)–(C.14) now yields

$$
\begin{aligned}
I^* &= \int_0^T \ell(v(t), v'(t)\,dt \\
&= \int_0^T v'(t)\frac{\partial}{\partial v'}\ell(v(t), v'(t))\,dt \\
&= \int_0^T \theta^*(v(t))v'(t)\,dt \\
&= \int_\alpha^\beta \theta^*(v)\,dv
\end{aligned}
$$

since v is strictly monotone.   ∎

The last lemma provides us with a convenient method to calculate the solution to the level crossing problem. It is often not hard to solve (C.11). Note that $\theta = 0$ is always a solution, and is usually not the one we seek since it corresponds to following z_∞, and thus the boundary conditions will not be satisfied. Using the fact that the desired $\theta^*(x)$ is the maximizer in the definition of ℓ we can compute $y(x)$. Lemma C.8 then derives a formula for an optimal path $v(t)$. In our applications these calculations can sometimes be carried out analytically. This is the case if $k = 2$, that is, there is only one possible jump in each direction. The generic calculation for this case is carried out in §11.5.

Exercise C.10. Assume $\alpha < \beta$ and that the rate of jump to the left goes (say linearly) to zero at α, so that the point α is degenerate. Extend all the results of this section. Hint: it may be convenient to shift the notation and require $r(T) = \alpha$ and $r(0) = \beta$, where $T < 0$ and $T = -\infty$ is acceptable. Construct solutions to the problems with boundary conditions $\alpha + \varepsilon$ and $\alpha + 2\varepsilon$ and patch the solutions. The linear rate will guarantee that the implicit definition of v using $y(x)$ (Lemma C.8) is valid.   ♠

Now consider the general one-dimensional problem with jumps possible in either direction. This process has

$$
g(\theta) \overset{\triangle}{=} \sum_{i=1}^d \lambda_i \left(e^{\theta e_i} - 1\right),
$$

where and $1 \le d_1 < d$ and $\lambda_i > 0$ for all i, $e_i > 0$ for $i = 1, \ldots, d_1$, $e_i < 0$ for $i = d_1 + 1, \ldots, d$.

Lemma C.11. *For a one-dimensional problem with some jumps in each direction the following hold. The function $\ell(y)$ satisfies*

$$
\ell(y) = y\theta^* - g(\theta^*),
$$

where $\theta^* = \theta^*(y)$ is a strictly increasing, continuously differentiable function of y. Consequently, ℓ is continuously differentiable.

Proof. Under the stated conditions it is easy to see that $y\theta - g(\theta)$ is, for each y, differentiable and bounded above in θ. Taking derivatives to obtain the maximum in θ,

$$y = \sum_{i=1}^{d} \lambda_i e_i \left(e^{\theta^* e_i} \right).$$

Since each term in the sum is a strictly monotone increasing analytic function, y is a strictly increasing, analytic function of θ^*. Its derivative is therefore positive at each θ^*, and the results follow. ∎

Note that an analogue of this result is available in higher dimensions; see Exercise C.16.

C.4. Results from the Calculus of Variations

In this section we return to the general setup and use the standard terminology of the Calculus of Variations (the main reference and the source of notation in this section is Cesari [Ce]).

We consider the Lagrange problem:

Minimize $\qquad \displaystyle\int_{t_1}^{t_2} f\left(\vec{r}(t), \vec{r}\,'(t) \right) dt,$

subject to $\qquad \vec{r} \in AC, \ (t, \vec{r}(t)) \in A, \ \left(t_1, \ \vec{r}(t_1), t_2, \vec{r}(t_2) \right) \in B.$

We assume throughout that A and B are closed, and A is the closure of its interior.

Theorem C.12 (Principle of Optimality) [Ce, 2.1.i p. 27]. *Consider the Lagrange problem. Assume $\vec{r}$ is optimal and fix any $a < a' < b' < b$. Then the function $\{\vec{r}(t) : a' \leq t \leq b'\}$ is optimal for the problem on the time interval $[a', b']$ with boundary conditions $\vec{r}(a'), \vec{r}(b')$.*

When discussing a minimizing $\vec{r}$ we shall assume that the integral is finite at some feasible point (and in particular the feasible set is not empty).

Theorem C.13 (Necessary Conditions) [Ce, 2.2.i p. 30]. *Consider the Lagrange problem. Assume f is continuously differentiable in $(\vec{x}, \vec{y})$ and let $\vec{r}$ be a (strong) local minimum which is in the interior of A and such that $f\left(\vec{r}, \vec{r}\,' \right) \in L_1$ (Definition A.80) and $\vec{r}\,'$ is essentially bounded (Definition A.75). Then the Euler Necessary Conditions hold; that is,*

$$\frac{d}{dt} \frac{\partial}{\partial r_i'} f\left(\vec{r}(t), \vec{r}\,'(t) \right) = \frac{\partial}{\partial r_i} f\left(\vec{r}(t), \vec{r}\,'(t) \right) \quad \text{a.e.}$$

and both sides are (a.e. equal to) AC functions. In addition the DuBois-Reymond Equations hold: for some constant C_0,

$$f\left(\vec{r}(t), \vec{r}'(t)\right) - \sum_{i=1}^{n} r_i'(t)\frac{\partial}{\partial r_i'} f\left(\vec{r}(t), \vec{r}'(t)\right) = C_0.$$

The "a.e." refers to Lebesgue measure. Note that if f is positive (as is the case with ℓ), then the assumption $f \in L_1$ is superfluous.

Lemma C.14 [Ce, Remark 1 p. 44]. *Consider the function ℓ of Chapter 5. Let us replace, in Theorem C.13, the assumption that $\vec{r}'$ is essentially bounded by the following. Assume $\vec{r}'$ is integrable on $[t_1, t_2]$, and in a neighborhood of the graph of $\vec{r}$ (Definition A.24) we have, for all $\vec{y}$,*

$$\ell(\vec{x}, \vec{y}) \leq \varphi(|\vec{y}|), \qquad \left|\frac{\partial}{\partial \vec{x}}\ell(\vec{x}, \vec{y})\right| \leq \varphi(|\vec{y}|), \qquad \left|\frac{\partial}{\partial \vec{y}}\ell(\vec{x}, \vec{y})\right| \leq \varphi(|\vec{y}|)$$

where $\varphi(z) = C_1 + C_2 z \log z \mathbf{1}\,[z > 1]$ and is positive. Then the Euler Necessary Condition still holds.

Exercise C.15. Prove Lemma C.14 in the following more general case. The function ℓ satisfies the conditions of Theorem C.13 (but may not be of the form of Chapter 5), and φ is positive and satisfies the conditions

$$\int_{t_1}^{t_2} \varphi(\vec{r}'(t))\, dt < \infty,$$

$$\frac{\varphi(z)}{z} \to \infty \text{ as } z \to \infty \quad \text{and} \quad \varphi(z + K) \leq B_K \cdot \varphi(z) + C_K$$

for each K and all z. Show that the function φ of Lemma C.14 satisfies these conditions. Hint: follow [Ce, Remark 2 p. 40 and Remark 1 p. 44].

Exercise C.16. Assume that $\lambda_i(\vec{x})$ are bounded, let $\ell(\vec{x}, \vec{y})$ be differentiable in $\vec{y}$ and let $\vec{r}(t)$ be any function on $[a, b]$. Then for each t, the function

$$\nabla_y \ell(\vec{r}(t), \vec{y}) \stackrel{\triangle}{=} \left(\frac{\partial}{\partial y_1}\ell(\vec{r}(t), \vec{y}), \ldots, \frac{\partial}{\partial y_d}\ell(\vec{r}(t), \vec{y})\right)$$

never takes the same value twice as $\vec{y}$ traverses $\mathbb{R}^d$. Moreover,

$$\left|\nabla_y \ell(\vec{r}(t), \vec{y})\right| \to \infty \quad \text{as} \quad |\vec{y}| \to \infty$$

uniformly in $t \in [a, b]$. Hint: use strict convexity and the uniform growth: see Lemma 5.17 and Exercise 5.27.   ♠

Lemma C.17 [Ce, 2.6.i p. 58]. *If f is continuously differentiable in both arguments and satisfies the growth condition of Exercise C.16, and if the AC function $\vec{r}$ satisfies the Euler necessary conditions than $\vec{r}$ is Lipschitz continuous, so that its derivative r' is essentially bounded.*

Theorem C.18 [Ce, 2.6.ii p. 58]. *Assume the conditions of both Exercise C.16 and Lemma C.17 hold. Then r' is a continuous function.*

Appendix D

Large Deviations Techniques

The results in this section are of two types. We provide some methods and approaches that are useful in proving large deviations results, but that will not be used in this book. In addition, we quote without proof some state-of-the-art results.

D.1. The Gärtner-Ellis Theorem

The idea here is to prove the large deviations bounds in $\mathbb{R}^d$ by using properties of moment generating functions. This turns out to be quite useful when the random variables of interest are not i.i.d., but are close to it in some sense; see Example D.10 and Exercise below. This is a powerful technique even in the i.i.d. case in $\mathbb{R}^d$. Since we only intend to illustrate the method, we shall restrict the proof to $\mathbb{R}^1$.

Given a sequence $s_1, s_2, \ldots$ of random variables, define

$$\phi_n(\theta) \stackrel{\triangle}{=} \frac{1}{n} \log \mathbb{E}\big(\exp(\theta s_n)\big). \qquad (D.1)$$

We will estimate the asymptotics of $\{s_n/n\}$. Note that if $x_1, x_2, \ldots$ are i.i.d. and $s_n \stackrel{\triangle}{=} \sum_{i=1}^n x_i$ then

$$\phi_n(\theta) = \frac{1}{n} \log \mathbb{E} \prod_{i=1}^n \exp(\theta x_i) = \frac{1}{n} \log \prod_{i=1}^n \mathbb{E} \exp(\theta x_i) = \log M(\theta).$$

Definition D.1. *The effective domain of a function f is $\mathcal{D}_f \stackrel{\triangle}{=} \{x : f(x) < \infty\}$.*

Lemma D.2. *For any random variable z, the function $f(\theta) \stackrel{\triangle}{=} \log \mathbb{E} \exp(\theta z)$ is convex on $\mathcal{D}_f$, and $\mathcal{D}_f$ is convex.*

Proof. Note that $0 \in \mathcal{D}_f$ and that, for $\theta > 0$,

$$\exp(\theta z^+) \leq 1 + \exp(\theta z) \leq 1 + \exp(\theta z^+), \qquad (D.2)$$

where $z^+ \stackrel{\triangle}{=} \max\{z, 0\}$. Since the right- and left-hand sides of (D.2) are monotone increasing in θ, we conclude that if $0 < \theta^* \in \mathcal{D}_f$ then $[0, \theta^*] \subseteq \mathcal{D}_f$. A similar argument applies to $\theta < 0$. Thus $\mathcal{D}_f$ is convex. Since $g(\theta) \stackrel{\triangle}{=} \mathbb{E} \exp(\theta z) > 0$ for each θ, the Monotone Convergence Theorem A.91(ii) establishes that g and hence

f are continuous on $\mathcal{D}_f$. It therefore suffices to prove convexity on the interior of $\mathcal{D}_f$. Now on the interior of $\mathcal{D}_f$, the function f is differentiable, and we can exchange differentiation and expectation. A computation of the second derivative gives

$$\frac{d^2}{d\theta^2} f(\theta) = \frac{\mathbb{E}z^2 e^{\theta z} \mathbb{E}e^{\theta z} - \left(\mathbb{E}ze^{\theta z}\right)^2}{\left(\mathbb{E}e^{\theta z}\right)^2}.$$

Let $x \overset{\triangle}{=} ze^{z\theta/2}$ and $y \overset{\triangle}{=} e^{z\theta/2}$. The inequality $\mathbb{E}x^2 \mathbb{E}y^2 \geq (\mathbb{E}xy)^2$ now implies that the second derivative is positive, and the convexity follows. ∎

The "near i.i.d." property of the sequence is captured by the following assumption:

Assumption D.3. The limit $\lim_n \phi_n(\theta) \overset{\triangle}{=} \phi(\theta)$ exists pointwise.

We allow the limit or any of the ϕ_n to take the value $+\infty$. As we saw above, in the i.i.d. case, $\phi_n \equiv \phi$.

Lemma D.4. ϕ is convex.

Proof: Any pointwise limit of convex functions is convex: for fix the points θ_1, θ_2 and $0 < \alpha < 1$. Then

$$\phi(\alpha\theta_1 + (1-\alpha)\theta_2) = \lim_n \phi_n(\alpha\theta_1 + (1-\alpha)\theta_2) \leq \lim_n(\alpha\phi_n(\theta_1) + (1-\alpha)\phi_n(\theta_2)).$$

Since both limits exist, the result follows. Note that infinite values are allowed, but negative infinite values are never taken by convex functions. ∎

Remark. While convexity is inherited by pointwise limits, differentiability may not be; see Assumption D.7 below.

Definition D.5. $I(x) \overset{\triangle}{=} \sup_\theta (\theta x - \phi(\theta))$.

Note that this definition agrees with the definition of ℓ for the i.i.d. case.

Theorem D.6. *Upper bound: under Assumption D.3, for* $-\infty < a < b < \infty$, *if* $[a, b] \cap \mathcal{D}_I \neq \emptyset$, *then*

$$\limsup_{n\to\infty} \frac{1}{n} \log \mathbb{P}\left(\frac{S_n}{n} \in [a, b]\right) \leq - \inf_{x\in[a,b]} I(x).$$

Proof. Define $\underline{I} \overset{\triangle}{=} \inf\{I(x) : a \leq x \leq b\}$, which is finite by assumption. Now fix some $\varepsilon > 0$ and note that

$$[a, b] \subset \left\{x : I(x) > \underline{I} - \varepsilon\right\} \overset{\triangle}{=} \left\{x : \sup_\theta(\theta x - \phi(\theta)) > \underline{I} - \varepsilon\right\}$$

$$= \bigcup_\theta \{x : \theta x - \phi(\theta) > \underline{I} - \varepsilon\}.$$

Since the function $\theta x - \phi(\theta)$ is obviously continuous in x, for each θ the set $\{x : \theta x - \phi(\theta) > \underline{I} - \varepsilon\}$ is open. From a standard result in analysis (the Heine-Borel theorem) we can find a **finite** number of points $\theta_1, \theta_2, \ldots, \theta_K$ so that

$$[a, b] \subset \bigcup_{i=1}^{K} \{x : \theta_i x - \phi(\theta_i) > \underline{I} - \varepsilon\} .$$

Now let us compute the probability that s_n/n falls into one such set.

$$\mathbb{P}\left(\frac{s_n}{n}\theta_i - \phi(\theta_i) > \underline{I} - \varepsilon\right) = \mathbb{P}\left(s_n\theta_i > n[\phi(\theta_i) + (\underline{I} - \varepsilon)]\right)$$

$$\leq \exp\left(-n[\phi(\theta_i) + (\underline{I} - \varepsilon)]\right)\mathbb{E}e^{\theta_i s_n}$$

by Chebycheff's inequality. Since the last expectation is, by definition $e^{n\phi_n(\theta_i)}$, we have by a union bound (A.12)

$$\limsup_{n\to\infty} \frac{1}{n} \log \mathbb{P}\left(\frac{s_n}{n} \in [a, b]\right)$$

$$\leq \limsup_{n\to\infty} \frac{1}{n} \log \sum_{i=1}^{K} \mathbb{P}\left(\frac{s_n}{n}\theta_i - \phi(\theta_i) > \underline{I} - \varepsilon\right)$$

$$\leq \limsup_{n\to\infty} \frac{1}{n} \log \sum_{i=1}^{K} \exp\left(-n[\phi(\theta_i) - \phi_n(\theta_i) + (\underline{I} - \varepsilon)]\right).$$

However, since for each i we have $\phi_n(\theta_i) \to \phi(\theta_i)$ and K does not depend on n, we finally have

$$\limsup_{n\to\infty} \frac{1}{n} \log \mathbb{P}\left(\frac{s_n}{n} \in [a, b]\right) \leq \underline{I} - \varepsilon$$

and the result follows since ε was arbitrary. ∎

Remark. This bound is weaker than in the i.i.d. case in the sense that we only allow *bounded* sets! To bridge the gap between bounded and unbounded sets, the general technique of §D.3 can be used.

As in the i.i.d. case, stronger assumptions are necessary for the lower bound.

Assumption D.7. ϕ is differentiable on $\mathcal{D}_\phi$.

Note that if $\mathcal{D}_\phi$ is not open, we have to be careful around the boundary.

Theorem D.8. Lower bound: *assume D.3, D.7 and let* $-\infty < a < b < \infty$. *Assume that for any* $v \in (a, b)$, *there exists* θ_v *such that* $\phi'(\theta_v) = v$. *Then*

$$\liminf_{n\to\infty} \frac{1}{n} \log \mathbb{P}\left(\frac{s_n}{n} \in (a, b)\right) \geq - \inf_{x\in(a,b)} I(x) .$$

The last condition is obviously satisfied if ϕ' is continuous, $\mathcal{D}_\phi = \mathbb{R}$ and if $\limsup_{\theta\to\infty} \phi'(\theta) \geq b$ and $\liminf_{\theta\to-\infty} \phi'(\theta) \leq a$. If $\mathcal{D}_\phi \neq \mathbb{R}$, these conditions should hold as θ approaches the boundary of $\mathcal{D}_\phi$.

We try to follow the proof of Theorem 1.5; however, since the i.i.d. assumption is not in force, the change of measure is different, and we cannot use the central-limit argument. The idea is still to change to a measure under which the event we consider is not rare, and bound the effect of the change of measure.

Note that if $I(x) = \infty$ for all $x \in (a, b)$ then there is nothing to prove. If it is finite, then under the hypotheses it is easy to see that $I(v) = \theta_v \cdot v - \phi(\theta_v)$.

The crux of the proof is the following estimate.

Lemma D.9. *Fix* $v \in (a, b)$. *Under the assumptions of Theorem D.8, for any* $\varepsilon > 0$ *small enough,*

$$\liminf_{n\to\infty} \frac{1}{n} \log \mathbb{P}\left(\frac{s_n}{n} \in (a, b)\right) \geq -I(v) - \varepsilon|\theta_v|.$$

Proof of Lemma D.9. Denote the distribution of s_n by $F_n(x)$. Fix a positive ε such that $\varepsilon < b - v$ and $\varepsilon < v - a$. Then

$$\mathbb{P}\left(\frac{s_n}{n} \in (a, b)\right) \geq \mathbb{P}\big(|s_n - nv| < n\varepsilon\big) = \int \mathbf{1}\,[|x - nv| < n\varepsilon]\, dF_n(x).$$

Define a new distribution $G_n(x)$ by the exponential change of measure

$$G_n(x) = \left[\int e^{\theta_v x}\, dF_n(x)\right]^{-1} e^{\theta_v x} F_n(x) = e^{-n\phi_n(\theta_v)} e^{\theta_v x} F_n(x),$$

where the last equality follows from the definition of ϕ_n. By definition of G_n,

$$\mathbb{P}\left(\frac{s_n}{n} \in (a, b)\right) \geq e^{n\phi_n(\theta_v)} \int \mathbf{1}\,[|x - nv| < n\varepsilon]\, e^{-\theta_v x}\, dG_n(x)$$

$$= e^{n[\phi_n(\theta_v) - \theta_v \cdot v]} \int \mathbf{1}\,[|x - nv| < n\varepsilon]\, e^{-\theta_v(x - nv)}\, dG_n(x)$$

$$\geq e^{n[\phi_n(\theta_v) - \theta_v \cdot v]} e^{-n|\theta_v|\varepsilon} \int \mathbf{1}\,[|x - nv| < n\varepsilon]\, dG_n(x). \quad (D.3)$$

In order to bound the integral in (D.3) from below, we use Chebycheff's inequality for the function $x \to e^{rx}$, with r positive and small;

$$\int \mathbf{1}\,[x \geq nv + n\varepsilon]\, dG_n(x) \leq e^{-nr(v+\varepsilon)} \int e^{rx}\, dG_n(x)$$

$$= e^{-nr(v+\varepsilon)} e^{-n\phi_n(\theta_v)} \int e^{rx} e^{\theta_v x}\, dF_n(x)$$

$$= \exp\{n[\phi_n(\theta_v + r) - \phi_n(\theta_v) - rv - r\varepsilon]\}$$

so that

$$\limsup_{n\to\infty} \frac{1}{n} \log \int \mathbf{1}\,[x \geq nv + n\varepsilon]\, dG_n(x) \leq \phi_n(\theta_v + r) - \phi_n(\theta_v) - rv - r\varepsilon.$$

$$(D.4)$$

But $r^{-1}[\phi(\theta_v + r) - \phi(\theta_v)] \to v$ as $r \to 0$, so that for r small enough, the right-hand side of (D.4) is strictly negative. Thus $\int \mathbf{1}[x \geq nv + n\varepsilon] dG_n(x)$ converges to zero exponentially fast. Using the same argument, we also obtain that $\int \mathbf{1}[x \leq nv - n\varepsilon] dG_n(x) \to 0$. Therefore $\int \mathbf{1}[|x - nv| < n\varepsilon] dG_n(x)$ converges to 1 (in fact exponentially fast), and (D.3) now implies

$$\liminf_{n \to \infty} \frac{1}{n} \log \mathbb{P}\left(\frac{s_n}{n} \in (a, b)\right) \geq -(\theta_v \cdot v - \phi_n(\theta_v)) - |\theta_v|\varepsilon$$

$$= -I(v) - |\theta_v|\varepsilon$$

and Lemma D.9 is established. ∎

Proof of Theorem D.8. Since ε in Lemma D.9 is arbitrary, we have

$$\liminf_{n \to \infty} \frac{1}{n} \log \mathbb{P}\left(\frac{s_n}{n} \in (a, b)\right) \geq -I(v),$$

and since this holds for every $v \in (a, b)$, the theorem follows. ∎

Example D.10: The "small noise" case. Let $\{x_1, x_2, \ldots\}$ be i.i.d. with moment generating function $M(\theta)$ that is finite for all θ. Let $\{y_1, y_2, \ldots\}$ be bounded random variables, with $|y_i| \leq \varepsilon_i$ where the ε_i are non-random. Let $s_n \overset{\triangle}{=} \sum_{i=1}^{n}(x_i + y_i)$ and $\tilde{s}_n \overset{\triangle}{=} x_1 + \cdots + x_n$. Clearly

$$\frac{1}{n} \log \mathbb{E} \exp(\theta \tilde{s}_n) \exp[-\theta(\varepsilon_1 + \cdots + \varepsilon_n)] \leq \phi_n(\theta)$$

$$\leq \frac{1}{n} \log \mathbb{E} \exp(\theta \tilde{s}_n) \exp \theta(\varepsilon_1 + \cdots + \varepsilon_n)$$

$$M(\theta) - \frac{\theta}{n}(\varepsilon_1 + \cdots + \varepsilon_n) \leq \phi_n(\theta) \leq M(\theta) + \frac{\theta}{n}(\varepsilon_1 + \cdots + \varepsilon_n)$$

so that D.3 holds, for example, whenever $\varepsilon_i \to 0$. The other assumptions clearly depend only on the distribution of x_1.

The general Gärtner-Ellis Theorem is difficult even to state. It provides bounds for open and closed sets. So, let the sequence $\vec{s}_1, \vec{s}_2, \ldots$ now take values in $\mathbb{R}^d$. Define $\phi_n(\vec{\theta})$ and $I(\vec{x})$ as in (D.1) and Definition D.5, except that now θx is replaced with the scalar product $\langle \vec{\theta}, \vec{x} \rangle$. Let $\mathcal{F}$ be following the set of points in $\mathbb{R}^d$:

$$\left\{ \vec{y} : \exists \vec{\theta} \in \mathcal{D}_I^o \text{ such that } I(\vec{x}) > \langle \vec{\theta}, \vec{x} - \vec{y} \rangle + I(\vec{y}) \text{ for all } \vec{x} \neq \vec{y} \text{ in } \mathbb{R}^d \right\}.$$

That is, $\vec{y} \in \mathcal{F}$ if there is a plane (with slope $\vec{\theta}$) that is tangent to $I(\vec{x})$ at $\vec{y}$ so that $I(\vec{x})$ is strictly above the plane except at $\vec{y}$. Compare this to the conditions of Theorem D.8.

Theorem D.11 Gärtner-Ellis. *Let $\vec{s}_1, \vec{s}_2, \ldots$ be random variables with values in* $\mathbb{R}^d$ *so that Assumption D.3 holds. Then*

(i) *For every closed set F*

$$\limsup_{n \to \infty} \frac{1}{n} \log \mathbb{P}\left(\frac{\vec{s}_n}{n} \in F\right) \leq -\inf_{\vec{x} \in F} I(\vec{x}).$$

(ii) *For every open set G,*

$$\liminf_{n \to \infty} \frac{1}{n} \log \mathbb{P}\left(\frac{\vec{s}_n}{n} \in G\right) \geq -\inf_{\vec{x} \in G \cap \mathcal{F}} I(\vec{x}).$$

(iii) *Assume that ϕ is differentiable in $\mathcal{D}_\phi^o$, and that $\mathcal{D}_\phi^o$ is non-empty. Assume further that ϕ is steep, that is if $\vec{\theta}$ is on the boundary of $\mathcal{D}_\phi$, then*

$$\vec{\theta}_n \to \vec{\theta} \quad \text{implies} \quad |\nabla \phi(\vec{\theta}_n)| \to \infty.$$

If ϕ is also lower semicontinuous, then the large deviations principle holds with a good rate function I.

Note that, unless strong assumptions are made on ϕ, if is possible that the bounds are not tight:

Exercise D.12. Show that if s_n is an exponential random variable with mean n, then the lower bound in Theorem D.11 is not tight. Hint: the set $\mathcal{F}$ contains only the point 0.   ♠

As an immediate application, one can obtain a large deviations principle for finite-state Markov chains.

Example D.13. Let $y_1, y_2, \ldots$ be an ergodic Markov chain with a finite state space, say $1, 2, \ldots N$, and with transition matrix $\Pi = \{p_{ij}\}$. Let $\vec{x}_1, \vec{x}_2, \ldots$ be defined through $\vec{x}_n = f(y_n)$ for some function f, with values in $\mathbb{R}^d$. Define

$$p_{ij}^{\vec{\theta}} \overset{\triangle}{=} p_{ij} e^{\langle \vec{\theta}, f(j) \rangle}.$$

Let $\Pi_{\vec{\theta}} \overset{\triangle}{=} \{p_{ij}^{\vec{\theta}}\}_{i,j=1}^N$ and let $\rho(\Pi_{\vec{\theta}})$ denote the largest eigenvalue of the matrix $\Pi_{\vec{\theta}}$. Define

$$I(\vec{x}) \overset{\triangle}{=} \sup_{\vec{\theta}} \left(\langle \vec{\theta}, \vec{x} \rangle - \log \rho(\Pi_{\vec{\theta}})\right).$$

Then the sample mean $\vec{z}_n = n^{-1} \sum_{k=1}^n \vec{x}_k$ satisfies a large deviations principle with good, convex rate function I. For a proof of this statement, see [DZ, §3.1]. This can be shown [DZ, §3.2] to be equivalent to the level II result we obtain (using a different method) in §D.4.

D.2. Subadditivity Arguments

The only method we pursue for the proof of lower bounds is a direct change-of-measure approach. Consider, however, that for independent random variables,

$$\mathbb{P}\left(\frac{x_1 + x_2 + \ldots + x_n}{n} \geq a\right) \geq \left(\mathbb{P}\left(\frac{x_1 + x_2 + \ldots + x_{n/2}}{n/2} \geq a\right)\right)^2$$

since the mean of n random variables is certainly greater than a if both the first half sum and the second half sum have that property. This implies that the function

$$f(n) \stackrel{\triangle}{=} -\log\left(\mathbb{P}\left(\frac{x_1 + x_2 + \ldots + x_n}{n} \geq a\right)\right)$$

is subadditive, that is $f(n + m) \leq f(n) + f(m)$. But any subadditive function satisfies

$$\lim_{n \to \infty} \frac{f(n)}{n} = \inf_{n \geq N} \frac{f(n)}{n} < \infty,$$

provided $f(n) < \infty$ for all $n \geq N$. This can be exploited to prove the existence of a lower bound, quite easily and in general circumstances. In fact, the same subadditivity property holds for the probability that the sample means of more general i.i.d. random variables lie in a convex set. This method is applicable also to random variables that are "close to i.i.d." See [DZ, DeS] for more details. Our approach, when applicable, has the advantage of giving more information about the way in which the rare event in question takes place.

D.3. Exponential Approximations

Since large deviations deals with probabilistic behavior on an exponential scale, it should be expected that some "exponential approximations" can be used. This subject is discussed in detail in [DZ]. As before, we consider random variables with values in a complete separable metric space $\mathcal{X}$, with metric d.

The first result concerns the upper bound. From Lemma 2.11 it is clear that it suffices to establish the lower bound for small balls around each point. It would be convenient if we could establish the upper bound by checking (2.1) for compact sets only, and not *for every closed set*. This is true under an "exponential tightness" condition.

Definition D.14. *The sequence of measures P_n is called exponentially tight if, for any finite α there exists a compact set K_α so that*

$$\limsup_{n \to \infty} \frac{1}{n} \log P_n(K_n^c) < -\alpha.$$

Theorem D.15. *Let P_n be exponentially tight and I be a rate function. If the upper bound (2.4u) holds for compact sets, it also holds for closed sets. If the lower bound (2.4l) holds for all open sets, then I is a good rate function.*

The next theorem allows to establish a large deviations principle for a process by establishing the principle for another process, and proving that they are close in an appropriate sense.

Theorem D.16. *Assume* $z_1, z_2, \ldots$ *satisfies the large deviations principle with rate function* I. *If* $\tilde{z}_1, \tilde{z}_2, \ldots$ *satisfies*

$$\lim_{n \to \infty} \frac{1}{n} \log \mathbb{P}\left(d(z_n, \tilde{z}_n) > \delta\right) = -\infty \quad \text{for all} \quad \delta > 0,$$

then $\tilde{z}_1, \tilde{z}_2, \ldots$ *satisfies the large deviations principle with rate function* I.

Since the large deviations principle in its abstract formulation (Definition 2.10) deals with probability measures, it is clear that the two processes need not be defined a priori on the same probability space. It suffices that we can find a probability space so that the marginals agree with the laws of z and $\tilde{z}$ respectively. For details, see [DZ, §4.2].

D.4. Level II

Armed with our understanding of jump Markov processes on a finite state space, we can venture into exploring an advanced topic: Level II large deviations. This is the study of how much time a process spends in various states over a long time interval. Contrast this with our previous question of how a jump process moves, and you see that the question is a bit different: we look at the likelihood of an occupation measure, similar to Sanov's Theorem, as opposed to a sample path drift, which in some way is like Chernoff's Theorem. We need to develop a bit of notation before we can state the result precisely. The reader will note that Level II, at least at this introductory level, is difficult only in forcing the reader to come to grips with a new concept; the mathematics is simple, and will be quite familiar to those who have read this far. In fact, we did many level two calculations implicitly for the large deviations principle of finite levels processes; this will become clear as we proceed.

We state and prove a Level II result only for a very simple jump Markov process $\vec{x}(t)$ that lives on exactly D states, as presented in §8.4 or in §7.1. We let $\vec{x}(t) = \vec{u}_i$, the unit vector in the ith coordinate direction, when the process is in the ith state. We let $\lambda_{i,j}$ denote the jump rate from state i to state j. We begin by recalling Definition A.82 of the empirical distribution $\vec{\pi}_t$ of a jump Markov process $\vec{x}(t) \in \mathbb{R}^D$:

$$\vec{\pi}_t \overset{\Delta}{=} \frac{1}{t} \int_0^t \vec{x}(s)\, ds$$

$$= \frac{1}{t} \int_0^t \left(\mathbf{1}\left[x(s) = 1\right], \ldots, \mathbf{1}\left[x(s) = D\right]\right) ds. \tag{D.5}$$

Note that $\vec{\pi}_t$ is itself a stochastic process that lives on a simplex

$$\mathcal{S} = \left\{ \vec{y} \in \mathbb{R}^D \ : \ \sum_i y_i = 1, \ y_i \geq 0 \right\}. \tag{D.6}$$

We define a rate function $I(\vec{\pi})$ for $\vec{\pi} \in \mathcal{S}$ as follows:

$$I(\vec{\pi}) \stackrel{\Delta}{=} \sup_{\vec{\theta} \in \mathbb{R}^D} \left[-\sum_{i,j} \pi(i)\lambda_{i,j} \left(e^{\langle \vec{\theta}, \vec{e}_{i,j} \rangle} - 1 \right) \right]. \qquad (D.7)$$

This should look familiar: it is form *(iv)* of the rate function (see Lemma 8.20) with jump rates $\pi(i)\lambda_{i,j}$, for travel in direction $\vec{y} = \vec{0}$, with $\vec{e}_{i,j}^d = \vec{0}$. This should be believable; if we travel in such a way as to make the average amount of time spent on each level i equal to $\pi(i)$, then it stands to reason that the long term jump rate in direction (i, j) is $\pi(i)\lambda_{i,j}$. It is not hard to see that I is a rate function: positive, convex, and lower semicontinuous. The $\vec{\pi}$ live on $\mathcal{S}$, a compact state space, so of course the level sets are compact. Also, $I(\vec{\pi})$ is obviously bounded by $\sum_{i,j} \pi(i)\lambda_{i,j}$. (This corresponds to no transitions taking place at all.) Our aim in this section is to prove the following large deviations principle.

Theorem D.17. *Let $\vec{\pi}_t$ be defined by (D.5), where $\vec{x}(t)$ is a process on $\mathbb{R}^D$ specified by $\{\lambda_{i,j}\}$. Then $\vec{\pi}_t$ satisfies a large deviations principle with good rate function $I(\vec{\pi})$. That is, if C and G are respectively closed and open sets in $\mathcal{S}$, then*

$$\limsup_{n \to \infty} \frac{1}{n} \log \mathbb{P}(\vec{\pi}_n \in C) \leq - \inf_{\vec{v} \in C} I(\vec{v})$$

$$\liminf_{n \to \infty} \frac{1}{n} \log \mathbb{P}(\vec{\pi}_n \in G) \geq - \inf_{\vec{v} \in G} I(\vec{v}).$$

The stochastic process in Theorem D.17 is $\vec{\pi}_t$. We are asking what the likelihood is that over a very long time interval, the observed occupation distribution is in an unlikely set. Remember, by Lemma 7.4, the occupation measure is very likely to be near the invariant distribution.

This type of large deviations principle is quite useful in practice. Many interesting questions can be formulated in terms of functionals of the occupation distribution. For example, the process $\vec{x}(t)$ would be confined to a subset G of states over an interval $(0, t)$ if the occupation measure $\vec{\pi}_t(G) = 1$. That is, we could formulate a Freidlin-Wentzell escape problem in terms of the occupation distribution. There are, of course, many other examples of interesting functionals of the path that may be expressed in terms of the occupation measure.

The proof of Theorem D.17 follows the usual procedure. The only novelty is the type of process. The upper bound follows from an exponential Chebycheff estimate, and the lower bound follows from a change of measure. Let's get to it.

Proof of the upper bound. We follow Donsker and Varadhan [DV1]. Let $\vec{\theta} \in \mathbb{R}^D$ be given. Then by Theorem B.3 using (B.31) with $g(\vec{x}) = \exp\langle \vec{\theta}, \vec{x} \rangle$,

$$\exp \left(\langle \vec{\theta}, \vec{x}(t) \rangle - \int_0^t \sum_i \mathbf{1}[x(s) = i] \sum_j \lambda_{i,j} \left(e^{\langle \vec{\theta}, \vec{e}_{i,j} \rangle} - 1 \right) ds \right)$$

is a martingale. We can write this martingale in terms of the empirical distribution $\vec{\pi}_t$ as follows:

$$\exp\left(\langle\vec{\theta},\vec{x}(t)\rangle - t\cdot\sum_i\pi_t(i)\sum_j\lambda_{i,j}\left(e^{\langle\vec{\theta},\vec{e}_{i,j}\rangle}-1\right)\right)$$

is a martingale. Therefore, for any $\vec{\theta}\in\mathbb{R}^D$ and any measurable set $C\subset\mathcal{S}$,

$$\max_{\vec{x}\in\{\vec{u}_1,\dots,\vec{u}_D\}} e^{\langle\vec{\theta},\vec{x}\rangle} \geq e^{\langle\vec{\theta},\vec{x}(0)\rangle}$$

$$= \mathbb{E}\left(e^{\langle\vec{\theta},\vec{x}(t)\rangle}\exp\left(-t\cdot\sum_i\pi_t(i)\sum_j\lambda_{i,j}\left(e^{\langle\vec{\theta},\vec{e}_{i,j}\rangle}-1\right)\right)\right)$$

$$\geq \min_{\vec{x}\in\{\vec{u}_1,\dots,\vec{u}_D\}} e^{\langle\vec{\theta},\vec{x}\rangle}\,\mathbb{P}(\vec{\pi}_t\in C)$$

$$\cdot\exp\left(-t\cdot\sup_{\vec{\pi}\in C}\sum_i\pi(i)\sum_j\lambda_{i,j}\left(e^{\langle\vec{\theta},\vec{e}_{i,j}\rangle}-1\right)\right).$$

Therefore, for each $\vec{\theta}\in\mathbb{R}^D$,

$$\mathbb{P}(\vec{\pi}_t\in C) \leq \frac{\max_{\vec{x}} e^{\langle\vec{\theta},\vec{x}\rangle}}{\min_{\vec{x}} e^{\langle\vec{\theta},\vec{x}\rangle}}\exp\left(t\sup_{\vec{\pi}\in C}\sum_i\pi(i)\sum_j\lambda_{i,j}\left(e^{\langle\vec{\theta},\vec{e}_{i,j}\rangle}-1\right)\right),$$

and so

$$\limsup_{n\to\infty}\frac{1}{n}\log\mathbb{P}(\vec{\pi}_n\in C) \leq \sup_{\vec{\pi}\in C}\sum_i\pi(i)\sum_j\lambda_{i,j}\left(e^{\langle\vec{\theta},\vec{e}_{i,j}\rangle}-1\right).$$

Since this holds for each $\vec{\theta}$, we also have

$$\limsup_{n\to\infty}\frac{1}{n}\log\mathbb{P}(\vec{\pi}_n\in C) \leq \inf_{\vec{\theta}\in\mathbb{R}^D}\sup_{\vec{\pi}\in C}\sum_i\pi(i)\sum_j\lambda_{i,j}\left(e^{\langle\vec{\theta},\vec{e}_{i,j}\rangle}-1\right). \quad (D.8)$$

We will have the upper bound if we can exchange sup and inf in (D.8) for closed sets $C\subset\mathcal{S}$. This interchange follows from a general argument given in [DVa]. However, in our case a much simpler argument applies. Note that since C is closed and is a subset of the compact set $\mathcal{S}$, it is compact. Now the function

$$K\left(\vec{\theta},\pi\right) \triangleq \sum_i\pi(i)\sum_j\lambda_{i,j}\left(e^{\langle\vec{\theta},\vec{e}_{i,j}\rangle}-1\right)$$

is convex in $\vec{\theta}$ and concave (actually linear) in $\vec{\pi}$. Therefore by the Min-Max Principle, Theorem A.44 we can interchange the sup and inf, and we are done. ∎

Proof of the lower bound. As usual, the lower bound is computed locally: given any $\vec{v} \in S$ and any $\varepsilon > 0$, we try to find

$$\mathbb{P}(\vec{\pi}_t \in N_\varepsilon(\vec{v})).$$

We use both the equivalence of the upper and lower bounds (Theorem 8.19), and the approximation result Lemma 8.61 (although we really don't need the full power of either of these). First, given $\vec{v} \in S$, we take any set of rates $\mu_{i,j}$ so that

$$\sum_i v(i) \sum_j \mu_{i,j} \log \frac{\mu_{i,j}}{\lambda_{i,j}} + \lambda_{i,j} - \mu_{i,j} = I(\vec{v})$$

(the $\mu_{i,j}$ will have $\vec{v}$ as an invariant measure). That such a set exists follows from the fact that $x \log x + 1 - x \to \infty$ as $x \to \infty$, so that we need only search for a minimizer in a bounded region of rates $\mu_{i,j}$. Then given an $\varepsilon > 0$, using Lemma 8.61 we find a set of jump rates $\gamma_{i,j}$ and an associated invariant measure $\vec{\eta}$ so that

$$\begin{aligned}
|\eta(i) - v(i)| &< \varepsilon/2 && \text{for all } i \\
|\eta(i)\gamma_{i,j} - v(i)\mu_{i,j}| &< \varepsilon/2 && \text{for all } i, j \\
\gamma_{i,j} &> 0 && \text{for all } i, j.
\end{aligned}$$

We will show that

$$\liminf_{n \to \infty} \frac{1}{n} \log \mathbb{P}\left(\vec{\pi}(n) \in N_{\varepsilon/2}(\vec{\eta})\right) \geq I(\vec{\eta}),$$

which implies the lower bound, since $N_{\varepsilon/2}(\vec{\eta}) \subset N_\varepsilon(\vec{v})$ and, for some C,

$$|I(\vec{\eta}) - I(\vec{v})| < C\varepsilon.$$

The key element of the proof is (surprise!) a change of measure. Letting

$$U(\varepsilon) = \left\{ \omega : \vec{\pi}_t \in N_{\varepsilon/2}(\vec{\eta}) \right\},$$

and letting dP represent probabilities associated with jump rates $\lambda_{i,j}$, and dQ represent probabilities associated with jump rates $\gamma_{i,j}$, we calculate

$$\begin{aligned}
\mathbb{P}\left(\vec{\pi}_t \in N_{\varepsilon/2}(\vec{\eta})\right) &= \int_{U(\varepsilon)} dP(\omega) \\
&= \int_{U(\varepsilon)} \frac{dP}{dQ}(\omega)\, dQ(\omega) \\
&= \int 1[\omega \in U(\varepsilon)] \exp\left(-t \sum_i (\lambda_i - \gamma_i)\right) \prod_{\text{jumps}} \eta_{i,j} \log \frac{\gamma_{i,j}}{\lambda_{i,j}}
\end{aligned}$$

(here we used the convention $\gamma_i = \sum_j \gamma_{i,j}$). It is very easy to show (and is done rigorously in Lemma 7.6) that

$$\lim_{n \to \infty} \frac{1}{n} \mathbb{E}\left(\sum_{\text{jumps}} \gamma_{i,j} \log \frac{\gamma_{i,j}}{\lambda_{i,j}}\right) = \sum_i \eta(i) \sum_j \gamma_{i,j} \log \frac{\gamma_{i,j}}{\lambda_{i,j}}$$

and furthermore Lemma 7.4 shows that, under the Q-process, for every $\varepsilon > 0$ there is a $\delta > 0$ and n_0 such that for $n > n_0$,

$$\mathbb{P}^Q \left(\vec{\pi}_n \notin N_{\varepsilon/2}(\vec{\eta}) \right) < e^{-n\delta}.$$

These equations imply that

$$\liminf_{n \to \infty} \frac{1}{n} \log \mathbb{P} \left(\vec{\pi}_n \in N_{\varepsilon/2}(\vec{\eta}) \right) \geq \sum_i \eta(i) \sum_j \gamma_{i,j} \log \frac{\gamma_{i,j}}{\lambda_{i,j}} + \lambda_{i,j} - \gamma_{i,j}.$$

Now we use the following equality, which is shown as part of the proof of Theorem 8.19:

$$\sup_{\vec{\theta} \in \mathbb{R}^D} -\sum_{i,j} \pi(i) \lambda_{i,j} \left(e^{\langle \vec{\theta}, \vec{e}_{i,j} \rangle} - 1 \right) = \inf_{\vec{\mu} \in S(\vec{\pi})} \sum_{i,j} \pi(i) \mu_{i,j} \log \frac{\mu_{i,j}}{\lambda_{i,j}} + \lambda_{i,j} - \mu_{i,j}$$

where $S(\vec{\pi})$ is the set of all transition rates $\mu_{i,j}$ that have $\vec{\pi}$ as an invariant distribution. Then noting that our construction of $\gamma_{i,j}$ and $\vec{\eta}$ makes $I(\vec{\eta})$ close to $I(\vec{v})$ finishes the proof of the lower bound. ∎

References

AKW Abate, J., M. Kijima, and W. Whitt, "Decompositions of the $M/M/1$ transition function," *Queueing Systems* **9** pp. 323–336, 1991.

AW Abate, J., and W. Whitt, "Transient behavior of the M/G/1 workload process," AT&T Tech. Memo., 1991.

Ab Abramson, N., "Development of the ALOHANET," *IEEE Trans. Info. Theory* **31** pp. 119–123, 1985.

Al Aldous, D., "Ultimate instability of exponential backoff protocol for acknowledgment based transmission control of random access communication channels," *IEEE Trans. Info. Th.* **33** pp. 219–233, 1987.

AC Algoet, P.H., and T.M. Cover, "Asymptotic optimality and asymptotic equipartition properties of log-optimum investment," *Ann. Prob.* **16** pp. 876–898, 1988.

An Anantharam, V., "How large delays build up in a $GI/G/1$ queue," *Queueing Systems* **5** pp. 345–368, 1988.

AHT Anantharam, V., P. Heidelberger, P. Tsoucas, "Analysis of rate events in continuous time Markov chains via time reversal and fluid approximation," IBM Research Report RC 16280 (#71858), 1990.

AMS Anick, D., D. Mitra, M.M. Sondhi, "Stochastic theory of a data-handling system and multiple sources," *Bell Sys. Tech. J.* **61** pp. 1871–1894, 1982.

As "Grande deviations et applications statistiques," *Astérisque* **68**, 1979.

Az Azencott, R., "Grandes déviations et applications," In *Ecole d'Été de Probabilités de Saint-Flour VIII–1978*, Edited by P.L. Hennequin, pp. 1–176, Lecture Notes in Math. 774, Springer-Verlag, Berlin, 1980.

AR Azencott, R., and G. Ruget, "Mélanges d'équations differentielles et grands écarts à la loi des grands nombres," *Z. Wahrsch Verw. Geb.* **38** pp. 1–54, 1977.

BaM Baccelli, F., and A.M. Makowski, "Queueing models for systems with synchronization constraints," *Proc. IEEE* **77** pp. 138–161, 1989.

BES Barles, G., L.C. Evans, P.E. Souganidis, "Wavefront propagation for reaction-diffusion systems of PDE," *Duke Math J.* **61** pp. 835–858, 1990.

BP Barlow, R.E., and F. Proschan, *Statistical Theory of Reliability and Life Testing*, Holt, Reinhart and Winston, New York, 1975.

Be Beneš, "The Covariance function of a simple trunk group, with applications to traffic measurement," *Bell Sys. Tech. J.* pp. 117–148, 1961.

B1 Bezuidenhout, C., "A large deviations principle for small perturbations of random evolution equations," *Ann. Probab.* **15** pp. 646–658, 1987.

B2   Bezuidenhout, C., "Singular perturbations of degenerate diffusions," *Ann. Probab.* **15** pp. 1014–1043, 1987.

Big   Biggins, J.D., "Chernoff's theorem in the branching random walk," *J. Appl. Prob.* **14** pp. 630–636, 1977.

BLS   Biggins, J.D., B. Lubachevsky, A. Shwartz, A. Weiss, "A branching random walk with a barrier," *Ann. Appl. Prob.* **1** pp. 573–581, 1991.

Bil   Billingsley, P., *Convergence of probability measures*, John Wiley & Sons, 1968.

BlD   Blinovskii, V.M. and R.L. Dobrushin, "Process level large deviations for a class of piecewise homogeneous random walks," Preprint, 1994.

BoD   Botvich, D.D. and N.G. Duffield, "Large deviations, the shape of the loss curve, and economies of scale in large multiplexers," Preprint, 1994.

BCG   Bramson, M., J.T. Cox, D. Griffeath, "Occupation time large deviations of the voter model," *Prob. Th. Rel. Fields* **77** pp. 401–413, 1988.

Br   Breiman, L., *Probability*, Second Edition, Classics in Applied Mathematics Vol. 7, SIAM, Philadelphia, 1992.

BV   Brémaud, P., and F.J. Vázquez-Abad, "On the Pathwise Computation of Derivatives with Respect to the Rate of a Point Process: The Phantom RPA Method," Preprint, 1990.

Bry   Bryc, W., "On large deviations for uniformly strong mixing sequences," *Stoch. Proc. Appl.* **41** pp. 191–202, 1992.

BrM   Brydges, D.C., and I.M. Maya, "An Application of Berezin Integration to Large Deviations," preprint, 1989.

Bu   Bucklew, J.A., *Large Deviations Techniques in Decision, Simulation, and Estimation*, Wiley, New-York, 1990.

Ce   Cesari, L., *Optimization—Theory and Applications*, Springer-Verlag, NY, 1983.

Cn   Chen, H., "Fluid approximation and stability of multiclass queueing networks I: Work-conserving disciplines," Preprint, 1994.

CM   Chen, H.,A. Mandelbuam, "Hierarchical modeling of stochastic networks, part I: fluid models," In *Applied Probability in Manufacturing Systems*, Edited by D.D. Yao, to appear.

Ch   Chernoff, H., "A measure of asymptotic efficiency for tests of a hypothesis based on the sum of observations," *Ann. Math. Statist.* **23** pp. 493–507, 1952.

Chu   Chung, K.L., *Markov chains with stationary transition probabilities*, Second Edition, Springer Verlag, 1967.

Ci   Çinlar, E., *Introduction to Stochastic Processes*, Prentice Hall, 1975.

C1   Cohen, J.W., *The Single Server Queue (second edition)*, North Holland, NY, 1982.

C2   Cohen, J.W., *Analysis of random walks*, I.S.O. Press, Amsterdam, 1992.

Co   Cooper, R.C., *Introduction to Queueing Theory, Second Edition*, North Holland, NY, 1981.

CFM Cottrell, M., J.-C. Fort, G. Malgouyres, "Large deviations and rare events in the study of stochastic algorithms," *IEEE Trans. Auto. Control* **AC-28** pp. 907–920, 1983.

CKW Courcoubetis, C., G. Kesidis, A. Ridder, J. Walrand, R. Weber, "Admission control and routing in ATM networks using inferences from measured buffer occupancy," To appear in IEEE Trans. Communications, 1995.

CW Courcoubetis, C. and R. Weber, "Effective Bandwidths for Stationary Sources," To appear in Probability in Engineering and Information Sciences, 1995.

Cr Cramér, H., "Sur un nouveau théorème–limite de la théorie des probabilités," In *Actualités Scientifiques et Industrielles*, pp. 5–23, number 736 in Colloque consacré à la théorie des probabilités, Hermann, Paris, 1938.

Da Dai, J.G., "On Positive Harris Recurrence of Multiclass Queueing Networks: A Unified Approach Via Fluid Limit Models," *Annals of Applied Prob.* , To appear, 1995.

DM Dai, J.G., and S. Meyn, "Stability and convergence of moments for multiclass queueing networks via fluid limit models," Preprint, 1994.

D1 Day, M.V., "Recent progress on the small parameter exit problem," *Stochastics* **20** pp. 121–150, 1987.

D2 Day, M.V., "Large deviations results for the exit problem with characteristic boundary," *J. Math Anal. Appl.* **147** pp. 134–153, 1990.

D3 Day, M.V., "Conditional exits for small noise diffusions with characteristic boundary," *Ann. Probab.* **20** pp. 1385–1419, 1992.

dA1 de Acosta, A., "Upper bounds for large deviations of dependent random vectors," *Z. Wahrsch. verw. Geb.* **69** pp. 551–565, 1985.

dA2 de Acosta, A., "Large deviations for empirical measures of Markov chains," *J. Theoretical Prob.* **3** pp. 395–431, 1990.

DZ Dembo, A., and O. Zeitouni, *Large deviations techniques and applications*, Jones and Bartlett, Boston, 1992.

DeS Deuschel, J.D., and D.W. Stroock, *Large Deviations*, Academic Press, Boston, 1989.

Di Dinwoodie, I.H., "A note on the upper bound for i.i.d. large deviations," *Ann. Probab.* **19** pp. 1732–1736, 1991.

DP1 Dobrushin, R.L. and E.A. Pechersky, "Large deviations for tandem queuing systems," Preprint, 1994.

DP2 Dobrushin, R.L. and E.A. Pechersky, "Large deviations for random processes with independent increments on infinite intervals," Preprint, 1994.

DVa Donsker, M.D., and S.R.S. Varadhan, "Asymptotic evaluation of certain Wiener integrals for large time," In *Functional Integration and is Applications*, Edited by A.M. Arthurs, pp. 20–21, , Oxford U. Press, 1975.

DV1 Donsker, M.D., and S.R.S. Varadhan, "Asymptotic evaluation of certain Markov process expectations for large time I," *Comm. Pure Appl. Math.* **28** pp. 1–47, 1975.

DV2 Donsker, M.D., and S.R.S. Varadhan, "Asymptotic evaluation of certain Markov process expectations for large time II," *Comm. Pure Appl. Math.* **28** pp. 279–301, 1975.

DV3 Donsker, M.D., and S.R.S. Varadhan, "Asymptotic evaluation of certain Markov process expectations for large time III," *Comm. Pure Appl. Math.* **29** pp. 389–461, 1976.

DV4 Donsker, M.D., and S.R.S. Varadhan, "Asymptotic evaluation of certain Markov process expectations for large time IV," *Comm. Pure Appl. Math.* **36** pp. 183–212, 1983.

DV5 Donsker, M.D., and S.R.S. Varadhan, "Large deviations from a hydrodynamic scaling limit," *Comm. Pure Appl. Math.* **42** pp. 243–270, 1989.

DrS Drmota, M., and U. Schmid, "Ultimate Characterization of the Successful Operation of Slotted ALOHA," preprint, 1993.

DuS Dunford, N., and J.T. Schwartz, *Linear Operators, Part I: General Theory*, Interscience Publishers, NY, 1957.

DE1 Dupuis, P., and R.S. Ellis, "Large deviations for Markov processes with discontinuous statistics, II. Random Walks," *Prob. Th. Relat. Fields* **91** pp. 153–194, 1992.

DE2 Dupuis, P., and R.S. Ellis, *A Weak Convergence Approach to the Theory of Large Deviations*, J. Wiley, 1995.

DEW Dupuis, P., R.S. Ellis, A. Weiss, "Large deviations for Markov processes with discontinuous statistics, I: General upper bounds," *Ann. Probab.* **19** pp. 1280–1297, 1991.

DI Dupuis, P. and H. Ishii, "SDEs with oblique reflection on nonsmooth domains," *Ann. Probab.* **21** pp. 554–580, 1993.

DIS Dupuis, P., H. Ishii, H.M. Soner, "A viscosity solution approach to the asymptotic analysis of queueing systems," *Ann. Prob.* **18** pp. 226–255, 1990.

DK1 Dupuis, P., and H.J. Kushner, "Stochastic Approximations Via Large Deviations: Asymptotic Properties," *SIAM J. Control Optim.* **23** pp. 675–696, 1985.

DK2 Dupuis, P., and H.J. Kushner, "Stochastic approximation and large deviations: Upper bounds and w.p.1 convergence," *SIAM J. Control Optim.* **27** pp. 1108–1135, 1989.

Dy Dynkin, E.B., *Markov Processes*, Springer-Verlag, New York, 1963.

Ell Ellis, R.S., *Entropy, Large Deviations and Statistical Mechanics*, Springer-Verlag, New York, 1985.

Els Elsgolc, L.E., *Calculus of Variations*, Addison-Wesley, Reading, MA, 1962.

EM1 Elwalid, A.I., and D. Mitra, "Analysis and design of rate-based congestion control of high speed networks, I: stochastic fluid models, access regulation," *Queueing Systems* **9** pp. 29–64, 1991.

EM2 Elwalid, A.I., and D. Mitra, "Fluid models for the analysis and design of statistical multiplexing with loss priorities on multiple classes of bursty traffic," *IEEE Infocom* pp. 3C.4.1–3C.4.11, 1992.

EM3 Elwalid, A.I., and D. Mitra, "Effective Bandwidth of General Markovian Traffic Sources and Admission Control of High Speed Networks," *IEEE/ACM Trans. Networking* **1** pp. 329–343, 1993.

EMS Elwalid, A.I., D. Mitra, T.E. Stern, "Statistical multiplexing of Markov modulated sources: theory and computational algorithms," In *Teletraffic and datraffic in a period of change, ITC-13*, Edited by Jensen, A., and V.B. Iversen, pp. 495–500, Elsevier Science Publishers, , 1991.

EK Ethier, S.N., and T.G. Kurtz, *Markov Processes: Characterization and Convergence*, Wiley, NY, 1986.

Ew Ewing, G.M., *Calculus of Variations with Applications*, Dover, NY (reprint), 1985.

Fe Fendick, K.W., "An Asymptotically Exact Decomposition of Coupled Brownian Systems," AT&T Tech. Memo., 1992.

Fl Flatto, L., "The longer queue model," *Probability in the Engineering and Informational Sciences* **3** pp. 537–559, 1989.

FH Flatto, L., and S. Hahn, "Two parallel queues created by arrivals with two demands," *SIAM J. Appl. Math.* **44** pp. 1041–1053, 1984.

FSS Fleming, W.H., S.-J. Sheu, H.M. Soner, "A Remark on the Large Deviations of an Ergodic Markov Process," *Stochastics* **22** pp. 187–199, 1987.

FW Freidlin, M.I., and A.D. Wentzell, *Random Perturbations of Dynamical Systems*, Springer-Verlag, New York, 1984.

Gal Galambos, J., *The Asymptotic Theory of Extreme Order Statistics*, Wiley, 1978.

Gar Gärtner, J., "On large deviations from the invariant measure," *Th. Prob. Appl.* **22** pp. 24–39, 1977.

GF Gaver, D., G. Fayolle, "A resource conflict resolution problem formulated in continuous time," Report NPS 55-85-018, Naval Postgraduate School, Monterey, California, 1985.

GLM Gazdzicki, P., I. Lambadaris, R. Mazumdar, "Blocking Probabilities for Large Multirate Erlang Loss Systems," *Adv. Appl. Prob.* **25** pp. 997–1009, 1993.

GM Gelenbe, E. and I. Mitrani, *Analysis and Synthesis of Computer Systems*, Academic Press, London, 1980.

GH Gibbens, R.J., and P.J. Hunt, "Effective bandwidths for the multi-type UAS channel," *QUESTA* **9** pp. 17–28, 1991.

GW Glynn, P. and W. Whitt, "Logarithmic asymptotics for steady-state tail probabilities in a single-server queue," to appear in Studies in Applied Probability, 1995.

GGM Goodman, J., A. Greenberg, N. Madras, P. March, "The stability of binary exponential backoff," *JACM* **35** pp. 579-602, 1988.

Gr   Graham, R.L., "Bounds on multiprocessing timing anomalies," *SIAM J. Applied Math.* **17** pp. 416–429, 1989.

GAN   Guerin, R., H. Amadi, and M. Naghshineh, "Equivalent capacity and its application to bandwidth allocation in high-speed networks," *IEEE JSAC* **9** pp. 968–981, 1991.

Gu   Gunther, N., "Path Integral Methods for Computer Performance Analysis," *Inf. Proc. Letters* **32** pp. 7–13, 1989.

GS   Gunther, N., and J.G. Shaw, "Path Integral Evaluation of ALOHA Network Transients," Preprint, 1989.

Hal   Hale, J.K., *Ordinary Differential Equations*, Pure and Applied Mathematics XXI, Robert E. Krieger Publishing Company, Huntington, New York, 1980.

Har   Harrison, J.M., *Brownian motion and stochastic flow systems*, J. Wiley, New York, 1985.

He1   Heyman, D.P., "An Analysis of the Carrier-Sense Multiple-Access Protocol," *Bell System Technical Journal* **61** pp. 2023–2051, 1982.

He2   Heyman, D.P., "The Effects of Random Message Sizes on the Performance of the CSMA/CD Protocol," *IEEE Transactions on Communications* **COM-34** pp. 547–553, 1986.

Hu1   Hui, J.Y., "Resource allocation for broadband networks," *IEEE JSAC* **6** pp. 1598–1608, 1988.

Hu2   Hui, J.Y., *Switching and Traffic Theory for Integrated Broadband Networks*, Kluwer, Boston, 1990.

HL   Hunt, P.J., and C.N. Laws, "Asymptotically Optimal Loss Network Control," *Math of Oper. Res.* **18** pp. 880–900, 1993.

IL   Ibragimov, I.A., and Yu.V. Linnik, *Independent and Stationary Sequences of Random Variables*, Wolters-Noordhoff Publishing, Groningen, The Netherlands, 1971.

IS   Ignatyuk, I.A. and V. Scherbakov, "Large deviations for random walks with discontinuities," Presentation at Stoch. Proc. and Applications meeting, Paris, 1993.

INN   Iscoe I., P. Ney, E. Nummelin, "Large deviations of uniformly recurrent Markov additive processes," *Adv. Appl. Math.* **6** pp. 373–412, 1985.

Jac   Jacobson, V., "Congestion avoidance and control," *Proc. ACM SIGCOM* pp. 314–329, 1988.

Jag   Jagerman, D.L., "Some properties of the Erlang loss function," *Bell Sys. Tech. J.* **53** pp. 525–551, 1974.

JRC   Jain, R. , K. Ramakrishnan, D. Chiu, "Congestion avoidance in computer networks with a connectionless network layer," Innovations in Internetworking, Artech House, 1988.

JW   Jelenkovic, P. and A. Weiss, "Large deviation analysis of slotted Aloha protocols," preprint, 1995.

KS   Karatzas, I., and S.E. Shreve, *Brownian Motion and Stochastic Calculus*, Second Edition, Springer Verlag, New York, 1991.

Ka Kaufman, J.S., "Blocking in a shared resource environment," *IEEE Trans. Comm.* **9** pp. 5–16, 1991.

KeF Kelly, F.P., "Effective Bandwidths at Multi-class Queues," *QUESTA* **29** pp. 1474–1481, 1981.

KeJ Kelly, J.L. Jr., "A new interpretation of information rate," *Bell Sys. Tech. J.* **35** pp. 917-926, 1956.

KWC Kesidis, G., J. Walrand, C.S. Chang, "Effective Bandwidths for multiclass Markov Fluids and other ATM Sources," *IEEE/ACM Trans. Networking* **1** pp. 424–428, 1993.

Ke Key, P.P., "Optimal Control and Trunk Reservation in Loss Networks," *Probability in the Engineering and Informational Sciences* **4** pp. 203–242, 1990.

Kl Kleinrock, L., *Queueing Systems Volume 1: Theory*, John Wiley, New York, 1975.

Kn Knessl, C., "On the transient behavior of the M/M/m/m loss model," *Stochastic Models* **6** pp. 749–776, 1990.

KMS Knessl, C., B.J. Matkowsky, Z. Schuss, C. Tier, "Two parallel queues with dynamic routing," *IEEE Trans. Comm.* **COM-34** pp. 1170–1175, 1986.

KnM Knessl, C., and J.A. Morrison, "Heavy-traffic analysis of a data-handling system with many sources," *SIAM J. Appl. Math.* **51** pp. 187–213, 1991.

Ko Kosten, L., "Stochastic theory of a multi-entry buffer (1)," *Delft Progr. Report* **1** pp. 10–18, 1974.

KrW Kruskal, C.P., and A. Weiss, "Allocating independent subtasks on parallel processors," *IEEE Trans. Software Eng.* **SE-11** pp. 1001–1015, 1985.

KL Kullback, S., and R.A. Leibler, "On information and sufficiency," *Ann. Math. Stat.* **22** pp. 79–86, 1951.

Ku1 Kurtz, T.G., *Approximation of Population Processes*, CBMS-NSF **36**, SIAM, Philadelphia, 1981.

Ku2 Kurtz, T.G., "Strong approximation theorems for density dependent Markov chains," *Stoch. Proc. Appl.* **6** pp. 223–240, 1978.

KuM Kushner, H.J., and L.F. Martins, "Heavy traffic analysis of a data transmission system, with many independent sources," *SIAM J. Appl. Math.* **53** pp. 1095–1122, 1993.

LR1 Lai, T.L., and H. Robbins, "Maximally dependent random variables," *Proc. Nat. Acad. Sci. USA* **73** pp. 286–288, 1976.

LR2 Lai, T.L., and H. Robbins, "A class of dependent random variables and their maxima," *Z. Wahr. verw. Gebiete* **42** pp. 89–111, 1978.

LN Lehtonen, T., and H. Nyrhinen, "Simulating Level Crossing Probabilities by Importance Sampling," *Advances in Applied Probability* **24** pp. 858–874, 1992.

Li Liptser, R.S., "Large Deviations for Simple Closed Queueing Model," Preprint, 1993.

LP   Liptser, R.S., and A.A. Pukhalskii, "Limit theorems on large deviations for semi-martingales," *Stochastics and Stoch. Rep.* **38** pp. 201–249, 1992.

LiS   Liptser, R.S., and A.N. Shiryayev, *Statistics of Random Processes. I: General Theory, II:Applications*, Applications of Mathematics **5–6**, Springer Verlag, NY, 1978.

LSW   Lubachevsky, B., A. Shwartz, and A. Weiss, "An analysis of rollback-based simulation," *ACM Transactions on Modeling and Computer Simulation* **1** pp. 154–193, 1991.

LyS   Lynch, J., and J. Sethuraman, "Large deviations for processes with independent increments," *Ann. Probab.* **15** pp. 610–627, 1987.

Mc   McKean, H.P., *Stochastic Integrals*, Academic Press, 1969.

Mai   Maier, R.S., "The First Exit Times of Multiaccess Broadcast Channels," Preprint, 1994.

Mal   Malyshev, "Classification of two dimensional positive random walks and almost linear semimartingales," *Soviet Math. Dok.* **13 No. 1** pp. 136–139, 1972.

Mas   Massey, W.A., "Open Networks of Queues: Their Algebraic Structure and Estimating Their Transient Behavior," *Adv. Appl. Prob.* **16** pp. 176–201, 1984.

MN   McCullaugh, P., and Nelder, J.A., *Generalized Linear Models*, Chapman and Hall, NY, 1983.

Mi   Mitra, D., "Stochastic theory of a fluid model of producers and comsumers coupled by a buffer," *Adv. Appl. Prob.* **20** pp. 646–676, 1988.

MGH   Mitra, D., R.J. Gibbens, B.D. Huang, "Analysis and optimal design of aggregated-least-busy-alternative routing on symmetric loss networks with trunk reservation," In *Teletraffic and datatraffic in a period of change, ITC-13*, Edited by Jensen, A., and V.B. Iversen, pp. 477–482, Elsevier Science Publishers, , 1991.

MRS   Mitra, D., K.G. Ramakrishnan, J.B. Seery, A. Weiss, "A unified set of proposals for control and design of high speed data networks," *Queueing Systems* **9** pp. 215–234, 1991.

MW   Mitra, D., and A. Weiss, "The transient behavior of Erlang's traffic model for Large Trunk Groups and Various Traffic Conditions," *Proc.* 12th *ITC* pp. 5.1B4.1–5.1B4.8, 1988.

Mo1   Morrison, J.A., "Asymptotic analysis of a data-handling system with many sources," *SIAM J. Appl. Math.* **49** pp. 617–637, April 1989.

Mo2   Morrison, J.A., "Diffusion approximation for head of the line processor sharing for two parallel queues," *SIAM J. Appl. Math.* **53** pp. 471–490, April 1993.

NKM   Naeh, T., M.M. Klosek, B.J. Matkowsky, Z. Schuss, "A Direct Approach to the Exit Problem," *SIAM J. Appl. Math.* **50** pp. 595–627, 1990.

New   Newell, G., *Applications of Queueing Theory*, Chapman and Hall, London, 1971.

Ne1   Ney, P., "Dominating points and the asymptotics of large deviations for random walk on $\mathbb{R}^d$," *Ann. Probab.* **11** pp. 158–167, 1983.

Ne2   Ney, P., "Convexity and large deviations," *Ann. Probab.* **12** pp. 903–906, 1984.

NN Ney, P., and E. Nummelin, "Markov additive processes II: large deviations," *Ann. Probab.* **15** pp. 593–609, 1987.

OV O'Brien, G.L., and W. Vervaat, "Capacities, Large Deviations and Loglog Laws," Report no. 90-19, 1990.

Ol Olver, F.W.J., *Asymptotics and special functions*, Academic Press, 1974.

Pi Pinsky, R., "The I-function for diffusion processes with boundaries," *Ann. Probab.* **13** pp. 676–692, 1985.

PS Plachky, D., and J. Steinebach, "A theorem about probabilities of large deviations with an application to queuing theory," *Periodica Mathematica Hungarica* **6** pp. 343–345, 1975.

Pu1 Puhalskii, A., "Large deviations of semimartingales via convergence of the predictable characteristics," *Stochastics and Stochastics Reports* **49** pp. 27–85, 1994.

Pu2 Puhalskii, A., "Large deviation analysis of the single server queue," Preprint, 1994.

Pu3 Puhalskii, A., "The method of stochastic exponentials for large deviations," To appear in Stochastic Processes, 1995.

Ree Reeds, J., "Correction terms for multinomial large deviations," In *Asymptotic Theory of Statistical Tests and Estimation*, pp. 287–305, Academic Press, New York, 1980.

Re1 Reiman, M., "A Critically Loaded Multiclass Erlang Loss System," *Queueing Systems* **9** pp. 65–82, 1991.

Re2 Reiman, M., "Optimal trunk reservation for a critically loaded link," In *Teletraffic and datraffic in a period of change, ITC-13*, Edited by Jensen, A., and V.B. Iversen, pp. 247–252, Elsevier Science Publishers, 1991.

Ri Righter, R. "Scheduling," In *Stochastic Orders and Their Applications*, Edited by Shaked, M. and G. Shantikumar, Chapter 13, Academic Press, 1994.

Rob Roberts, J.W., "A service system with heterogeneous user requirements—application to multi-services telecommunications systems," In Performance of Data Communication Systems and Their Applications, North-Holland, 1981.

Roc Rockafellar, R.T., *Convex Analysis*, Princeton University Press, Princeton, 1970.

RS Rom, R. and M. Sidi, *Multiple Access Protocols: Performance and Analysis*, Springer-Verlag, New York, 1990.

RT Rosenkrantz, W., and D. Towsley, "On the instability of the slotted ALOHA multiaccess algorithm," *IEEE Trans. Auto. Control* **AC-28** pp. 994–996, 1983.

Ros1 Ross, S.M., *Applied Probability Models with Optimization Applications*, Holden-Day, San Francisco, 1970.

Ros2 Ross, S.M., *Introduction to Probability Models*, Second Edition, Probability and Mathematical Statistics, Academic Press, 1980.

Ros3 Ross, S.M., *Stochastic Processes*, Wiley Series in Probability and Mathematical Statistics, John Wiley, 1983.

Roy Royden, H.L., *Real Analysis*, Second Edition, Collier McMillan, 1968.

Sa Sanov, I.N. "On the probability of large deviations of random variables," In Russian, 1957. (English translation from *Mat. Sb.* **42**) In *Selected Translations in Mathematical Statistics and Probability* I, pp. 213–244, 1961.

Se Seneta, E., *Non Negative Matrices and Markov Chains*, Second Edition, Springer-Verlag, New York, 1981.

SW Shwartz, A., and A. Weiss, "Induced rare events: analysis via large deviations and time reversal," *Adv. Appl. Prob.* **25** pp. 667–689, 1993.

St Stroock, D.W., *An Introduction to the Theory of Large Deviations*, Springer-Verlag, Berlin, 1984.

SV Stroock, D.W., and S.R.S. Varadhan, *Multidimensional Diffusion Processes*, Springer-Verlag, New York, 1979.

Tse Tse, D.N.C., "Variable-rate Lossy Compression and its Effects on Communication Networks," Ph.D. dissertation, MIT, 1994.

Ta1 Takács, L., *Introduction to the Theory of Queues*, Oxford University Press, New York, 1962.

Ta2 Takács, L., *Combinatorial Methods in the Theory of Stochastic Processes*, Krieger, Huntington, New York, 1990 (reprint).

TK Taylor, H.M. and S. Karlin, *Introduction to Stochastic Modeling*, Academic Press, New York, 1984.

Te Tegeder, R.W., *Large Deviations, Hamiltonian Techniques and Applications in Biology*, Ph.D. Thesis, University of Cambridge, England, 1993.

Ti Tijms, H.C., *Stochastic Modeling and Analysis: a Computational Approach*, John Wiley, New York, 1986.

Ts Tsoucas, P., "Rare Events in Series of Queues," *J. Applied Prob.* **29** pp. 168–175, 1992.

Tu Turner, J. "New directions in communications (or which way to the information age?)," *IEEE Communications Magazine* October 1986.

Va Varadhan, S.R.S., *Large Deviations and Applications*, SIAM, Philadelphia, 1984.

VOW deVeciana, G., C. Olivier, J. Walrand, "Large Deviations for Birth Death Markov Fluids," Preprint, 1992.

VF Ventcel, A.D., and M.I. Freidlin, "On small random perturbations of dynamical systems," *Russian Math. Surveys* **25** pp. 1–55, 1970.

Web Weber, R., "On the Optimal Assignment of Customers to Parallel Servers," *J. Appl. Prob.* **15** pp. 406–413, 1983.

We1 Weiss, A., "The large deviations of a Markov process which models traffic generation," Bell Labs. Tech. Memo., 1983.

We2 Weiss, A., "A new technique for analyzing large traffic systems," *Adv. Appl. Prob.* **18** pp. 506–532, 1986.

WeG Weiss, G., "A tutorial on stochastic scheduling," In *Scheduling theory and its applications*, Edited by Chretienne, P., E.G. Coffman, J.K. Lenstra, and Z. Liu, J. Wiley, New York, 1995

Wen Wentzell, A.D., *Limit Theorems on Large Deviations for Markov Stochastic Processes*, Kluwer, Dordrecht, Holland, 1990.

Wh1 Whitt, W., "Deciding which queue to join: some counterexamples," *Op. Res.* **34** pp. 55–62, 1986.

Wh2 Whitt, W., "Tail probabilities with statistical multiplexing and effective bandwidths in multi-class queues," *Telecommunication Systems* **2** pp. 71–107, 1993.

Wo Wolff, R.W., *Stochastic modeling and the theory of queues*, Prentice Hall, Englewood, NJ, 1989.

Wr Wright, P.E., "Two parallel processors with coupled inputs," *Adv. Appl. Prob.* **24** pp. 986–1007, 1992.

Yo Young, L.C., *Calculus of variations and optimal control*, Academic Press, NY, 1968.

Index

*Note: definitions are found on **bold** pages*